Reality leaves a lot to the imagination.
--John Lennon

Always have a vivid imagination, for you never know when you might need it.
--J.K. Rowling

When you were younger or possibly even very recently, someone may have suggested that you do something you thought was incredible, or impossible, so you responded, "How am I supposed to do that?!" Chances are, the answer you got was, "Use your imagination." In this familiar exchange, the message is, "be resourceful, be creative." In dictionaries, the first definition for *imagination* is usually along the lines of "the act or power of forming a mental image of something not present to the senses or never before wholly perceived in reality" (Merriam-Webster). Most of us imagine things fairly regularly: what it would be like to have a child or get a new car, what we would do if we won the lottery, what we would say if we got to meet a celebrity we admire, what it will be like when we've achieved what has been a long-term goal. Using your sociological imagination is not so different.

What is my sociological imagination?

The mid-twentieth century sociologist C. Wright Mills used the term *sociological imagination* to describe sociological reasoning—what he considered the ability to see the relationship between individual experiences and the larger society. When we use our sociological imagination, we are thinking sociologically, viewing our and others' personal experiences in the social contexts in which they occur. To think sociologically, then, we first have to recognize that we are members of a society, that we are not completely autonomous individuals making choices independent of outside influences.

That means your relationship to society does not look like this:

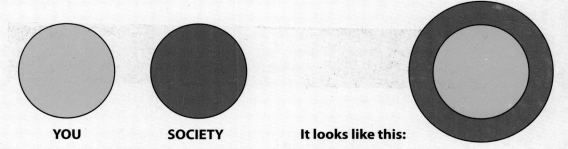

YOU   SOCIETY   It looks like this:

Why do I want to quick-start my sociological imagination?

The sociological imagination is at the center of the course you're taking — and of this textbook. As a result, the earlier you develop an awareness of it and how it can be applied, the greater edge you'll have as you work through the sociological concepts and theories presented in each chapter you're assigned. That edge will speed up your understanding and help you retain what you learn. Better yet, the sooner you start using your sociological imagination, the sooner you'll be able to move beyond established ways of thinking to gain new insights into yourself, and to develop a greater awareness of the connection between your own world and that of other people. You will see that, as another mid-twentieth century sociologist, Peter Berger, put it, "things are not what they seem." As a result, you will develop new ways of approaching problems and making decisions in everyday life.

CHAPTER 6 Deviance and Crime

Quick-start question: *How popular would you consider yourself to have been in high school?*

Additional quick-start questions:

- What is something you do or have that deviates from what your family or neighbors do or have?
- Why do you do this thing or have this possession?
- Do you engage in any type of behavior or own something that might harm or undermine someone else? If yes, how so?
- Under what circumstances do you think you would discard this item or stop this behavior?

CHAPTER 7 Class and Stratification in the United States

Quick-start question: *Are you currently employed? If not, what factors are involved in your being unemployed?*

Additional quick-start questions:

- When you envision your ideal life, what does it look like?
- Do you think it will be possible for you to achieve your ideal life?
- What obstacles do you anticipate might limit your ability to realize your ideal life?

CHAPTER 8 Global Stratification

Quick-start question: *Have you lived in or visited a country other than the one in which you were born?*

Additional quick-start questions:

- Have you spoken to a customer service representative working in a call center in India or another middle-income country?
- Do you think this customer service representative feels she or he has a "good job"? If yes, in what ways?
- What kind of work do you think the representative's parents and grandparents did?
- How do you think the representative's ambitions might be different from those of his or her relatives when they were younger?

CHAPTER 9 Race and Ethnicity

Quick-start question: *Do you think racial/ethnic relations affect sports, and vice versa?*

Additional quick-start questions:

- How many friends or family members do you have whose racial or ethnic background is different from your own?
- What benefits and challenges have you experienced as a result of having friends and family from diverse racial/ethnic backgrounds in your life?
- What factors contribute to people from similar racial/ethnic backgrounds sticking together while seemingly ignoring people from different backgrounds?

CHAPTER 10 Sex and Gender

Quick-start question: *How concerned are you and your closest friends about weight and body image?*

Additional quick-start questions:

- Where would you place yourself on a scale of traditional femininity or masculinity of 1 (barely) to 10 (very)? Why?
- Have you ever tried to move yourself up or down this scale? If yes, why, and how did you do it?
- If you had a child who, at birth, could not be identified as either exclusively female or male, do you think you would want medical practitioners to surgically "assign" your child to one or the other sex? Would you consider raising your child as an "intersexual"? Why or why not?

CHAPTER 11 Families and Intimate Relationships

Quick-start question: *Were you raised in a single- or two-parent or guardian household, or some other arrangement?*

Additional quick-start questions:
- Were the persons you lived with while growing up raised in a household that was similar to or different from the arrangement in which you grew up?
- When you were growing up, how "normal" did your family's living arrangement feel to you?
- To what did you attribute those feelings?

CHAPTER 12 Education and Religion

Quick-start question: *What part, if any, did religion play in your formal education from kindergarten through high school?*

Additional quick-start questions:
- Which, if any, religion (or religions) makes you feel most comfortable when you think of it?
- In what way might you feel uncomfortable about religion, and what do you think might cause this discomfort?
- What connection, if any, do you think exists between the amount of government funding a school receives and the success of students at that school?

CHAPTER 13 Politics and the Economy in Global Perspective

Quick-start question: *What have you learned from the media about politics and the United States' economy? How might this information be different from what you've learned from word of mouth, textbooks, or publications created by and intended for specialists in these fields?*

Additional quick-start questions:
- How, if at all, were you affected by the financial crisis that started in the investment banking industry and dominated the news during the autumn of 2008?
- In what ways did the United States' economy influence your decision to attend college or to choose a specific major?
- What relationships do you think exist between nations' political and economic systems and the wars those nations fight with other countries? What about battles within their own borders?

CHAPTER 14 Health, Health Care, and Disability

Quick-start question: *Do you think smoking should be legally banned in public places?*

Additional quick-start questions:
- Where do you currently receive the health care you need? How is it paid for?
- Have you ever been in a hospital emergency room? How would you describe your experience?
- What are some of the major obstacles faced by persons with a disability?

CHAPTER 15 Population and Urbanization

Quick-start question: *Did either you or your parents emigrate to the United States from another country? How about your grandparents or an even earlier generation?*

Additional quick-start questions:
- Do you live in a mid-size or large city, a suburb, a small city or town, or a rural community?
- What are some of the primary benefits and challenges of residing where you do?
- Do you think that you might want to live in a different kind of community some day? Why or why not?

CHAPTER 16 Collective Behavior, Social Movements, and Social Change

Quick-start question: *Have you ever joined other members of your community in an effort to accomplish some goal or achieve a specific outcome?*

Additional quick-start questions:
- Do you drive a hybrid or other alternative fuel vehicle? Do you know other people who do?
- Do you recycle? If yes, why did you start doing it?
- Are you concerned about the effects of pollution? How so?

Quick-Start Questions by Chapter

INSTRUCTIONS: Try answering the questions below to get yourself thinking sociologically. While responding, the more often you find yourself considering other people's experience, whether those experiences are the same as your own or different, and the more you find yourself considering possible relationships between your and others' experience and the social structures in which our lives occur, the more you're using your sociological imagination.

CHAPTER 1 The Sociological Perspective and Research Process

Quick-start question: *Do you think a personal act like suicide is also a social issue?*

Additional quick-start questions:
- What three values are most important to you?
- Do you know other people who share similar values?
- How are these values reinforced?
- How are these values challenged?

CHAPTER 2 Culture

Quick-start question: *When you think of food, what comes to mind first—a particular dish, meals with family or friends, grocery shopping, gardening or doing farm work, being hungry, or something else?*

Additional quick-start questions:
- Are there activities associated with cultures other than your own that you cannot see yourself doing?
- What is something that you do that you think people from another culture might be unwilling to do?
- Do you believe that cultural differences can lead to misunderstandings between people? Why or why not?

CHAPTER 3 Socialization

Quick-start question: *Who was your primary caregiver when you were a small child—a parent or guardian, a grandparent, staff at a daycare center, a babysitter or nanny, or someone else?*

Additional quick-start questions:
- What childhood experiences did you have that you think had a positive effect on who you are as an adult?
- What childhood experiences did you have that you think had a negative effect on who you are as an adult?
- What circumstances produced each of these experiences?

CHAPTER 4 Social Structure and Interaction in Everyday Life

Quick-start question: *How is where you are currently living similar to (and different from) the place where you primarily lived when you were growing up?*

Additional quick-start questions:
- Where do you usually shop for clothes, and why do you shop there?
- Have you ever been in a clothing store that made you feel uncomfortable?
- What kind of store was it, and what caused the discomfort?

CHAPTER 5 Groups and Organizations

Quick-start question: *Think of a club, organization, or other group you belong to: On a scale of 1 (very little) to 10 (a lot), how much do the other members know about you?*

Additional quick-start questions:
- What is your favorite method of communicating with other people?
- Why do you think this method is your favorite?
- What would cause you to start using a different method of communication?

POLITICAL MAP OF THE WORLD

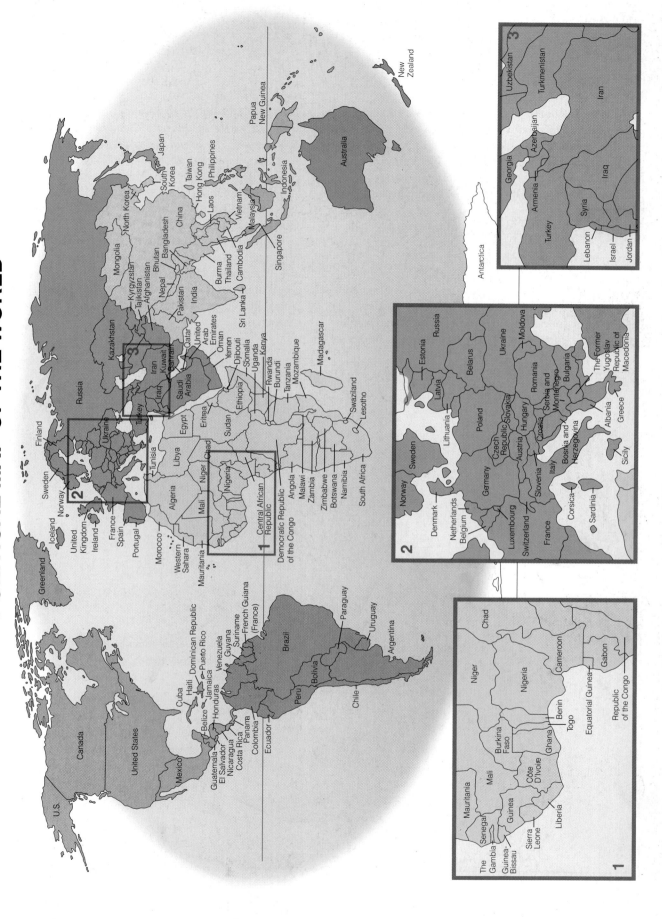

About the Author

Diana Kendall is currently a Professor of Sociology at Baylor University, where she was recognized as an Outstanding Professor for her research. Dr. Kendall has taught a variety of courses, including Introduction to Sociology, Sociological Theory (undergraduate and graduate), Sociology of Medicine, and Race, Class, and Gender. Previously, she enjoyed many years of teaching sociology and serving as chair of the Social and Behavioral Science Division at Austin Community College.

Diana Kendall received her Ph.D. from the University of Texas at Austin, where she was invited to membership in Phi Kappa Phi Honor Society. Her areas of specialization and primary research interests are sociological theory and the sociology of medicine. In addition to *Sociology in Our Times,* she is the author of *The Power of Good Deeds: Privileged Women and the Social Reproduction of the Upper Class* (Rowman & Littlefield, 2002); *Framing Class: Media Representations of Wealth and Poverty in America* (Rowman & Littlefield, 2005); and *Members Only: Elite Clubs and the Process of Exclusion* (Rowman & Littlefield, 2008).

Professor Kendall is actively involved in national and regional sociological associations, including the American Sociological Association, Sociologists for Women in Society, the Society for the Study of Social Problems, and the Southwestern Sociological Association.

edition

7

SOCIOLOGY IN OUR TIMES

The Essentials

DIANA KENDALL

Baylor University

WADSWORTH
CENGAGE Learning™

Australia • Brazil • Japan • Korea • Mexico • Singapore • Spain • United Kingdom • United States

WADSWORTH
CENGAGE Learning

Sociology in Our Times: The Essentials, **Seventh Edition**
Diana Kendall

Senior Acquisitions Editor: Chris Caldeira

Senior Development Editor: Renee Deljon

Assistant Editor: Melanie Cregger

Editorial Assistant: Rachael Krapf

Technology Project Manager: Lauren Keyes

Marketing Manager: Meghan Pease

Marketing Assistant: Jillian Myers

Marketing Communications Manager: Martha Pfeiffer

Project Manager, Editorial Production: Cheri Palmer

Creative Director: Rob Hugel

Art Director: Caryl Gorska

Print Buyer: Karen Hunt

Permissions Editor, Text: Roberta Broyer

Permissions Editor, Images: Leitha Etheridge-Sims

Production Service: Greg Hubit Bookworks

Text Designer: Diane Beasley

Photo Researcher: Laurie Frankenthaler

Copy Editor: Donald Pharr

Illustrator: Graphic World Illustration Studio

Cover Designer: Riezebos Holzbaur Design Group

Cover Image: © Todd Berman (TheArtDontStop.org)

Compositor: Graphic World, Inc.

For product information and technology assistance, contact us at
Cengage Learning Customer & Sales Support
1-800-354-9706.

For permission to use material from this text or product, submit all requests online at **www.cengage.com/permissions**. Further permissions questions can be e-mailed to **permissionrequest@cengage.com**.

Library of Congress Control Number: 2008930456

ISBN-13: 978-0-495-59862-6
ISBN-10: 0-495-59862-3

Wadsworth
10 Davis Drive
Belmont, CA 94002-3098
USA

Cengage Learning is a leading provider of customized learning solutions with office locations around the globe, including Singapore, the United Kingdom, Australia, Mexico, Brazil, and Japan. Locate your local office at **www.cengage.com/international**.

Cengage Learning products are represented in Canada by Nelson Education, Ltd.

To learn more about Wadsworth, visit **www.cengage.com/Wadsworth**
Purchase any of our products at your local college store or at our preferred online store **www.ichapters.com**.

Printed in the United States of America
1 2 3 4 5 6 7 12 11 10 09 08

Brief Contents

Contents

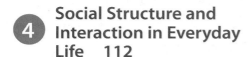

Part 2   Social Groups and Social Control

4   Social Structure and Interaction in Everyday Life   112

5   Groups and Organizations   144

Photo Essay

**How Do We "Do Gender" in
the Twenty-First Century? 346**

Part 4 Social Institutions

11 Families and Intimate Relationships 354

12 Education and Religion 388

Features

Preface

Welcome to the seventh edition of *Sociology in Our Times: The Essentials*! The twenty-first century offers unprecedented challenges and opportunities for each of us as individuals and for our larger society and world. In the United States, we can no longer take for granted the peace and economic prosperity that many—but far from all—people were able to enjoy in previous decades. However, even as some things change, others remain the same, and among the things that have not changed are the significance of education and the profound importance of understanding how and why people act the way they do. It is also important to analyze how societies grapple with issues such as economic hardship and the threat of terrorist attacks and war, and to gain a better understanding of why many of us seek stability in our social institutions—including family, religion, education, government, and media—even if we believe that some of these institutions might benefit from certain changes.

Like previous editions of this widely read text, the seventh edition of *Sociology in Our Times: The Essentials* is up-to-date, forward-looking, and committed to highlighting the presence of sociology in our lives, as well as its relevance to our lives. It does this in at least two ways. First, it helps students connect with sociology by providing a meaningful, concrete context for learning by featuring the stories—the lived experiences—of real individuals and the social issues they face. Therefore, the text presents a diverse array of classical and contemporary theory as well as interesting and relevant research within the framework of engaging personal stories and familiar issues such as child care, the American Dream, and immigration.

Second, this text highlights the relevance of sociology to our lives by showing students how sociology can help them understand the important social questions and circumstances that not only the featured individuals face but that they themselves or other people they know or encounter may also face. By providing students with greater understanding of others' lives, their own lives, and the leading social issues of our time, this text also tries to help students see themselves as *members of their communities* and to show them what can be done in responding to social issues both locally and globally. The result is that students learn how sociology is not only a collection of concepts and theories but also a field that can make a difference in their lives, their communities, and the world at large.

What's New to the Seventh Edition? For Starters, Sociology Works!

The seventh edition builds on the best of previous editions while offering new insights, learning tools, and opportunities to apply the content of each chapter to relevant sociological issues and major concerns of the twenty-first century. As it is my goal to make each edition better than the previous one, I have revised all of the chapters to reflect the latest in sociological theory and research, and have updated examples throughout. Additionally, all statistics, such as data relating to suicide, crime, demographics, health, and poverty, are the latest available at the time of this writing.

The text's new **Sociology Works!** feature shows how sociological theories and research continue to enhance our understanding of contempoary social issues and our interactions in everyday life. Sociology Works! discussions include the following: "Erving Goffman's Impression Management and Facebook" (Chapter 4), "Why *Place* Matters in Global Poverty" (Chapter 8), and "Sociology Sheds Light on the Physician–Patient Relationship" (Chapter 14).

To visually capture some of the significant circumstances and issues of our time, this edition includes four **photo essays,** two of which are new—"Trying to Go It Alone: Adolescent and Teenage Runaways" (Chapter 3) and "How Do We 'Do Gender' in the Twenty-First Century?" (Chapter 11). Expanded for improved usefulness, each essay now has three pages of thought- and conversation-stimulating photos with brief sociological commentary. Each photo essay also has a new companion online **ABC news video** with assignable **Turning to Video** questions to further bring the essays' topics to life.

The text's highly praised **Concept Quick Reviews** (previously called Concept Review Tables) appear in every chapter of the seventh edition and have been redesigned for greater ease of use. Still in table format but more prominently displayed than in past editions, these reading and study aids provide concise overviews of key theories and concepts, including "Social Interaction: The Microlevel Perspective" (Chapter 4), "Theoretical Perspectives on Deviance" (Chapter 6), and "Sociological Perspectives on Health and Medicine" (Chapter 14).

Designed to stimulate both students' critical-thinking ability and sociological imagination, new **Reflect & Analyze** questions are provided at the end of the Sociology Works! feature and photo essays, as well as at the end of the text's Media Framing, Social Policy, and Global Perspectives boxes.

Finally, students will enjoy and benefit from the seventh edition's entirely new—livelier and more functional—**interior design** and its greatly expanded and improved collection of **illustrations and photos,** as well as an innovative new feature at the front of the book, the Quick-Start Guide to Using Your Sociological Imagination, which orients students and poses Quick-Start questions for each chapter so that students are already thinking sociologically even before they begin reading a chapter. *Sociology in Our Times: The Essentials,* Seventh Edition, also offers instructors a significantly **improved Instructor Resource Manual** and **Annotated Instructor's Edition,** which provide more and more targeted resources than ever before. An **electronic version** of the text is also new to this edition. Full descriptions of the student and instructor supplements begin on page xix.

Overview of the Text's Contents with More Details of What's New

Sociology in Our Times: The Essentials, Seventh Edition, contains sixteen carefully written, clearly organized chapters to introduce students to the best of sociological thinking. The length of the text makes full coverage of the book possible in the time typically allocated to the introductory course so that students are not purchasing a book that contains numerous chapters that the instructor may not have the time or desire to cover.

Part 1 establishes the foundation for studying society and social life. **Chapter 1** introduces students to the *sociological perspective and research process.* The chapter sets forth the major theoretical perspectives used by sociologists in analyzing compelling social issues and provides a thorough description of both quantitative and qualitative methods of sociological research. Chapter 1 now opens with an account of Megan Meier's MySpace-related suicide as well as an updated Sociology and Everyday Life quiz. Chapter 1 includes a Media Framing box ("Framing Suicide in the Media: Sociology Versus Sensationalism") that introduces students to the concept of framing and shows them how the media help shape our society and our cultural perceptions of many issues—in this case, suicide. The new Sociology Works! feature appears in this chapter, here highlighting "Durkheim's Sociology of Suicide and Twenty-First Century India." The chapter also has a new You Can Make a Difference box titled "Responding to a Cry for Help."

Chapter 2 spotlights *culture* as either a stabilizing force or a force that can generate discord, conflict, and even violence in societies. Cultural diversity is discussed as a contemporary issue, and unique coverage is given to popular culture and leisure and to divergent perspectives on popular culture. With a new theme of global perspectives on food, the chapter opens with a new lived experience about how food can serve as a powerful cultural symbol and with a new Sociology and Everyday Life quiz. Also new are the chapter's You Can Make a Difference and Media Framing boxes, as well as its Sociology Works! feature ("Schools as Laboratories for Getting Along").

Chapter 3 looks at *socialization* and presents an innovative analysis of gender and racial–ethnic socialization as well as human development and lifespan issues. The chapter opens with actress Drew Barrymore recalling an early childhood experience. A new Sociology in Global Perspective box, "Everybody Else Has a Cell Phone!" focuses on how widespread cell phone use among children, adolescents, and teens influences socialization. The chapter's new Sociology Works! feature, "Good Job!: Mead's Generalized Other and the Issue of Excessive Praise," looks at educational analyst Alfie Kohn's study of the practice of praising children for practically everything they say or do, recalling the earlier sociological insights of George Herbert Mead.

Part 2 examines social groups and social control. **Chapter 4** applies the sociological imagination to *social structure and interaction in everyday life,* opening with a personal account of homelessness, which is returned to throughout to illustrate the dynamic interplay of social structure and human agency in people's daily lives. This chapter has a new You Can Make a Difference box, "Offering a Helping Hand to Homeless People," and its Sociology Works! feature presents a compelling discussion of Erving Goffman's impression management and Facebook. Chapter 4 also includes a new illustration (Figure 4.4) of the social distance rules in three contrasting cultures.

Chapter 5 analyzes *groups and organizations,* and it includes a discussion of innovative forms of social organization and ways in which organizational structures may differentially affect people based on race, class, gender, and age. The chapter opens with college students discussing their use of and attitudes toward the social network site Facebook .com, which sets up the chapter's overall exploration of how groups and organizations are changing as a result of new technological means of communication. The chapter's Sociology and Everyday Life quiz has been updated, and its You Can Make a Difference box now focuses on "Developing Invisible (But Meaningful) Networks on the Web." The Sociology Works! feature, "Ingroups, Outgroups, and 'Members Only' Clubs," highlights how studies on private clubs and exclusive college social organizations show that the sociological concepts of "ingroup" and "outgroup" remain highly relevant today, particularly in regard to the processes of inclusion and exclusion and how such activities affect individuals and groups. The chapter also includes a new Figure 5.5, "Characteristics and Effects of Bureaucracy."

Chapter 6 examines how *deviance and crime* emerge in societies, using diverse theoretical approaches to describe the nature of deviance, crime, and the criminal justice system. The chapter opens with personal commentaries on the underground economy and gang membership, themes that are addressed throughout the chapter and highlighted in the new Sociology Works! feature, "Social Definitions of Deviance: Have You Seen Bigfoot or a UFO Lately?" and the chapter's new You Can Make a Difference box, "Seeing the Writing on the Wall—and Doing Something About It!" The chapter also includes a new Figure 6.6 illustrating the discretionary powers in law enforcement.

Part 3 looks at social differences and social inequality, focusing on issues of class, race/ethnicity, and sex/gender. **Chapter 7** focuses on *class*

and stratification in the United States. The chapter opens with three young men recounting how they achieved what, in the eyes of many people, would be "the American Dream" when they accomplished their educational goals and were named among the forty most influential African Americans by *Essence* magazine. The chapter analyzes the causes and consequences of inequality and poverty, as well as discussing the ideology and accessibility of the American Dream. The chapter's photo essay, "What Keeps the American Dream Alive?" broadens this discussion, while the chapter's new Sociology Works! feature, "Reducing Structural Barriers to Achieving the American Dream," offers another perspective. Chapter 7 includes a new Sociology and Social Policy box that asks the question "Should Our Laws Guarantee People a Living Wage?"

Chapter 8 addresses the issue of *global stratification* and examines differences in wealth and poverty in rich and poor nations around the world. Explanations for these differences are discussed. The chapter opens with Paul Tergat, a world-record holder in the marathon and winner of two silver Olympic medals, describing his early childhood in Kenya, which was marked by poverty and hunger, and the difference meals at school made. The Sociology and Everyday Life quiz has been revised for currency, a new Sociology in Global Perspective box focuses on "Marginal Migration: Moving to a Less-Poor Nation," and the chapter's Sociology Works! feature looks at "Why *Place* Matters in Global Poverty." A new Concept Quick Review summarizes the classification of economies by income.

The focus of **Chapter 9** is *race and ethnicity,* which includes an illustration of the historical relationship (or lack of it) between sports and upward mobility by persons from diverse racial–ethnic groups. A thorough analysis of prejudice, discrimination, theoretical perspectives, and the experiences of diverse racial and ethnic groups is presented, along with global racial and ethnic issues. The contrasting opinions of professional baseball players Milton Bradley and Jeff Kent regarding whether certain remarks by Kent were racially insensitive open the chapter to introduce these issues. This chapter's Sociology Works! feature, titled "Attacking Discrimination to Reduce Prejudice?" presents the relationship between sociological research and court rulings and government initiatives over the past sixty years that have sought to reduce the corrosive effects of racial and ethnic discrimination in the United States.

Chapter 10 examines *sex and gender,* with an emphasis on gender stratification in historical perspective. Linkages between gender socialization and contemporary gender inequality are explored, introduced by new personal accounts of the Miss America pageant and body image. This chapter's Sociology in Everyday Life quiz has been revised, and the subject presented in its Global Perspective box is new to this edition: "The Rise of Islamic Feminism in the Middle East?" The chapter's Sociology Works! feature, "Institutional Discrimination: Women in a Locker-Room Culture," presents compelling sociological research on this topic, and a new photo essay considers the symbolic interactionist perspective of sociologists Candace West and Don H. Zimmerman, who present gender as something that we *do* rather than as a set of masculine or feminine traits that resides within the individual.

Part 4 offers a systematic discussion of social institutions, building students' awareness of the importance of social institutions and showing how a problem in one often has a significant influence on others. *Families and intimate relationships* are explored in **Chapter 11,** which includes both U.S. and global perspectives on family relationships, a view of families throughout the life course, and a discussion of diversity in contemporary U.S. families. A new Sociology in Global Perspective box examines the trend of Western women outsourcing the birth of their children to surrogate mothers in India, and the chapter's new Sociology Works! feature presents research focused on the social factors that influence parenting. A new illustration, Figure 11.2, compares the U.S. 1990–2005 divorce rates for each state.

Education and religion are presented in **Chapter 12,** which highlights important sociological theories pertaining to these pivotal social institutions. Personal experience involving the debate over teaching "intelligent design" in public schools opens the chapter, whose discussion of education and religion in the future has been revised and updated. The chapter also includes a new You Can Make a Difference box that provides suggestions for understanding and tolerating religious and cultural differences, and its Sociology Works! feature highlights research on character building and social norms in regard to bullying. A new Figure 12.2 shows which U.S. school districts have the highest and lowest graduation rates.

Chapter 13 discusses *politics and the economy in global perspective,* highlighting the global context in which contemporary political and economic systems operate and showing the intertwining nature of politics, the economy, and global media outlets. The chapter now opens with personal accounts re-

lated to the phenomenal success of parody news shows, particularly Jon Stewart's *The Daily Show.* Fully updated, the chapter includes a new section on global underground economies, and its Sociology Works! feature looks at how C. Wright Mills's ideas about the media may have been ahead of his time. A new Figure 13.1 illustrates the functional view of government.

Chapter 14 analyzes *health, health care, and disability* in both U.S. and global perspectives. Among the topics included are social epidemiology, lifestyle factors influencing health and illness, health care organization in the United States and other nations, social implications of advanced medical technology, and holistic and alternative medicine. This chapter is unique in that it contains a thorough discussion of sociological perspectives on disability and of social inequalities based on disability. Dr. Atul Gawande's description of the power and the limits of medicine now opens the chapter, and the framing of drug ads is presented in a new Media Framing box. This chapter's Sociology Works! feature sheds light on the physician–patient relationship. This chapter includes two new illustrations: Figure 14.1 ("Smoking Bans Around the World") and Figure 14.2 ("Obesity in the United States").

Part 5 surveys social dynamics and social change. **Chapter 15** examines *population and urbanization,* looking at demography, global population change, and the process and consequences of urbanization. Special attention is given to race- and class-based segregation in urban areas and the crisis in health care in central cities. First-person accounts of undocumented workers open the chapter and reveal some of the reasons that "illegal immigrants" choose to enter this country and why they think that laws should be changed to allow them to stay. Immigration examples are used throughout the chapter, and a new You Can Make a Difference box focuses on ways that links between college students and immigrant children can be created. In addition to the new Sociology Works! feature, which looks at the contributions of Herbert Gans in regard to twenty-first-century urban villagers, the chapter includes a new section on issues of rural communities in the United States. It also includes a new illustration, Figure 15.5, that shows the value of urban space.

Chapter 16 concludes the text with an innovative analysis of *collective behavior, social movements, and social change.* In the opening personal narrative, Rodney Crowell and Jim Lovell explain their concerns about global warming. The example of environmental activism is used throughout the chapter to help students grasp the importance of collective behavior and social movements in producing social change. The Sociology in Global Perspective box now focuses on China's environmental woes and its emergent social activism on behalf of the environment, and this chapter's Sociology Works! feature focuses on fine-tuning theories and data gathering about environmental racism.

Distinctive, Classroom-Tested Features

The following special features are specifically designed to demonstrate the relevance of sociology in our lives, as well as to support students' learning. As the preceding overview of the book's contents shows, these features appear throughout the text—some in every chapter, others in selected chapters.

Unparalleled Coverage of and Attention to Diversity

From its first edition, I have striven to integrate diversity in numerous ways throughout this book. The individuals portrayed and discussed in each chapter accurately mirror the diversity in society itself. As a result, this text speaks to a wide variety of students and captures their interest by taking into account their concerns and perspectives. Moreover, the research used includes the best work of classical and established contemporary sociologists—including many white women and people of color—and it weaves an inclusive treatment of *all* people into the examination of sociology in *all* chapters. Therefore, this text helps students consider the significance of the interlocking nature of individuals' class, race, and gender (and, increasingly, age) in all aspects of social life.

Personal Narratives That Highlight Issues and Serve as Chapter-Length Examples

Authentic first-person vignettes open each chapter to personalize the issue that unifies the chapter's coverage. These lived experiences provide opportunities for students to examine social life beyond their own experiences and for instructors to systematically incorporate into lectures and discussions an array of interesting and relevant topics that help demonstrate to students the value of applying sociology to their everyday lives.

Focus on the Relationship Between Sociology and Everyday Life

Each chapter has a brief quiz that relates the sociological perspective to the pressing social issues presented in the opening vignette. (Answers are provided on a subsequent page.)

Emphasis on the Importance of a Global Perspective

The global implications of all topics are examined throughout each chapter and in the Sociology in Global Perspective boxes, which highlight our interconnected world and reveal how the sociological imagination extends beyond national borders.

Focus on Media Framing

A significant benefit of a sociology course is encouraging critical thinking about such things as how the manner in which the media "package" news and entertainment influences our perception of social issues.

Applying the Sociological Imagination to Social Policy

The Sociology and Social Policy boxes in selected chapters help students understand the connection between sociology and social policy issues in society.

Focus on Making a Difference

Designed to help get students involved in their communities, the You Can Make a Difference boxes look at ways in which students can address, on a personal level, issues raised by the chapter theme.

Census Profiles

This feature highlights current relevant data from the U.S. Census Bureau, providing students with further insight into the United States.

Effective Study Aids

In addition to basic reading and study aids such as chapter outlines, key terms, and running glossary, and the popular online study system CengageNOW™, *Sociology in Our Times: The Essentials* includes the following pedagogical aids to promote students' mastery of the course's content:

- **Concept Quick Review.** These tables categorize and contrast the major theories or perspectives on the specific topics presented in a chapter.

- **Questions for Critical Thinking.** Each chapter concludes with questions for critical thinking to encourage students to reflect on important issues, to develop their own critical-thinking skills, and to highlight how ideas presented in one chapter often build on those developed previously.
- **End-of-Chapter Summaries in Question-and-Answer Format.** Chapter summaries provide a built-in review for students by reexamining material covered in the chapter in an easy-to-read question-and-answer format to review, highlight, and reinforce the most important concepts and issues discussed in each chapter.

Comprehensive Supplements Package

The seventh edition of *Sociology in Our Times: The Essentials* is accompanied by a wide array of supplements developed to create the best teaching and learning experience inside as well as outside the classroom. All of the continuing supplements have been thoroughly revised and updated, and some new supplements have been added. Cengage Learning prepared the following descriptions, and I invite you to start taking full advantage of the teaching and learning tools available to you by reading this overview.

Supplements for Instructors

Annotated Instructor's Edition. The thoroughly revised Annotated Instructor's Edition (AIE) of the text, prepared by D. R. Wilson of Houston Baptist University, offers page-by-page annotations to assist instructors before, during, and after class. The AIE's new annotations are gathered under several diverse headings, including Active Learning, Extra Examples, Global Perspective, Popular Culture, Sociological Imagination, and Applied Sociology. Additionally, annotations flag coverage that fulfills the American Sociological Association (ASA) Task Force's recommendations for teaching sociology.

Instructor's Resource Manual, by D. R. Wilson of Houston Baptist University. Streamline and maximize the effectiveness of your course preparation by using such resources as brief chapter outlines, chapter summaries, extensively detailed chapter lecture outlines, lecture suggestions, video suggestions, and creative lecture and teaching suggestions. This manual also contains student learning objectives, key terms, essay/discussion questions, student activities, InfoTrac® College Edition exercises, discussion exercises, and Internet exercises. Applied ASA recommendations are now a part of this manual, as are group activities and experimental learning activities based on the ASA Task Force recommendations.

Test Bank, **by Richard J. Jenks of Indiana University Southeast.** Simplify testing and assessment by using this printed selection of 100+ multiple-choice questions, 30 true–false questions, 15 short-answer questions, and 10 essay questions for each chapter of the text, all with answer explanations and page references that correspond to the text. All questions are labeled as new, modified, or pickup so that instructors know if the question is new to this edition of the test bank, modified but picked up from the previous edition of the test bank, or picked up straight from the previous edition of the test bank.

PowerLecture with JoinIn™ and ExamView®. On CD or DVD, this one-stop class-preparation tool contains ready-to-use Microsoft® PowerPoint® slides, enabling you to assemble, edit, publish, and present custom lectures with ease. PowerLecture helps you bring together text-specific lecture outlines and art from the text along with videos and your own materials—culminating in powerful, personalized, media-enhanced presentations. The **JoinIn™** content (for use with most "clicker" systems) available within PowerLecture delivers instant classroom assessment and active learning. Take polls and attendance, quiz students, and invite them to actively participate while they learn. Featuring automatic grading, **ExamView®** is also available within PowerLecture, allowing you to create, deliver, and customize tests and study guides (both print and online) in minutes. See assessments onscreen exactly as they will print or display online. Build tests of up to 250 questions using up to 12 question types, and enter an unlimited number of new questions or edit existing questions. PowerLecture also includes the text's *Instructor's Resource Manual* and Test Bank as Word documents.

WebTutor™ on Blackboard® and WebCT®. Jump-start your course with customizable, rich, text-specific content within your Course Management System. Simply load a content cartridge into your course management system to easily blend, add, edit, reorganize, or delete content, all of which is specific to Kendall's *Sociology in Our Times: The Essentials,* Seventh Edition, and includes media resources, quizzing, web links, discussion topics, and interactive games and exercises.

CengageNOW® with Cengage Learning eBook and InfoTrac College Edition. CengageNOW is an online teaching and learning resource that gives you and your students more control in less time and delivers better outcomes—NOW. An online study system, CengageNOW gives students the option of taking a diagnostic pretest for each chapter. The system uses the results of each pretest to create personalized chapter study plans for students. The personalized study plans

- help students save study time by identifying areas on which they should concentrate and gives them one-click access to corresponding pages of the Cengage Learning eBook;
- provide interactive exercises and study tools to help students fully understand chapter concepts; and
- include a posttest for students to take to confirm that they are ready to move on to the next chapter.

Visit **www.cengage.com/tlc** for a product demonstration. To log in to or to set up an instructor eResources account, go to **www.cengage.com/login.**

Classroom Activities for Introductory Sociology Courses. Made up of contributions from instructors teaching introductory sociology around the country, this booklet features classroom activities, student projects, and lecture ideas to help instructors make topics fun and interesting for students. With general teaching tactics as well as topic-focused activities, it's never been easier to find a way to integrate new ideas into your classroom.

Tips for Teaching Sociology, **Third Edition.** Written by veteran instructor Jerry M. Lewis of Kent State University, this booklet contains tips on course goals and syllabi, lecture preparation, exams, class exercises, research projects, course evaluations, and more. It is an invaluable tool for first-time instructors of the introductory course and for veteran instructors in search of new ideas.

Videos. Adopters of *Sociology in Our Times: The Essentials,* Seventh Edition, have numerous video options available for use with the text:

- **Wadsworth's Lecture Launchers for Introductory Sociology.** An exclusive offering jointly created by Wadsworth/Cengage Learning and DALLAS TeleLearning, this video contains a collection of video highlights taken from the *Exploring Society: An Introduction to Sociology* tele-

course (formerly *The Sociological Imagination*). Each 3- to 6-minute-long video segment has been specially chosen to enhance and enliven class lectures and discussions of 20 key topics covered in the introduction to sociology course. Accompanying the video is a brief written description of each clip, along with suggested discussion questions to help instructors incorporate the material into their course. Available on VHS or DVD.

- **Sociology: Core Concepts Video.** Another exclusive offering jointly created by Thomson Wadsworth and DALLAS TeleLearning, this video contains 15- to 20-minute video segments that will enhance student learning of the essential concepts in the introductory course and can be used to initiate class lectures, discussion, and review. The video covers topics such as the sociological imagination, stratification, race and ethnic relations, and social change. Available on VHS or DVD.
- **ABC® videos.** Launch your lectures with exciting video clips from the award-winning news coverage of ABC. Addressing topics covered in a typical course, these videos are divided into short segments—perfect for introducing key concepts in contexts relevant to students' lives.
- **Wadsworth Sociology Video Library.** Bring sociological concepts to life with videos from Wadsworth's Sociology Video Library, which includes thought-provoking offerings from Films for Humanities as well as other excellent educational video sources. This extensive collection illustrates important sociological concepts covered in many sociology courses.
- **AIDS in Africa DVD.** Expand your students' global perspective of HIV/AIDS with this award-winning documentary series, which focuses on controlling HIV/AIDS in southern Africa. Films focus on caregivers in the faith community; how young people share messages of hope through song and dance; the relationship of HIV/AIDS to gender, poverty, stigma, education, and justice; and the story of two HIV-positive women helping others.

Readers. To help you add depth or further broaden the scope of your course, Wadsworth/Cengage Learning offers readers for your consideration. These readers are available at a discount when packaged with *Sociology in Our Times: The Essentials:*

- *Sociological Odyssey: Contemporary Readings in Introductory Sociology,* **Third Edition,** by Patricia A. Adler and Peter Adler

- *Globalization: The Transformation of Social Worlds,* by D. Stanley Eitzen and Maxine Baca Zinn
- *Classic Readings in Sociology,* **Fourth Edition,** by Eve L. Howard
- *Understanding Society: An Introductory Reader,* **Second Edition,** by Margaret L. Andersen, Kim A. Logio, and Howard F. Taylor
- *Sociological Footprints: Introductory Readings in Sociology,* **Eleventh Edition,** by Leonard Cargan and Jeanne H. Ballantine
- **Extension: Wadsworth's Sociology Reader Collection.** Create your own customized reader for your sociology class, drawing from dozens of classic and contemporary articles found on the exclusive Thomson Wadsworth TextChoice database. Using the TextChoice website (**www .TextChoice.com**), you can preview articles, select your content, and add your own original material. TextChoice will then produce your materials as a printed supplementary reader for your class.

Supplements for Students

CengageNOW. Students can sign in, save time, and get the grade they want with CengageNOW, our on-line study system that includes an integrated eBook and a personalized study plans. Access to Cengage-NOW includes access to InfoTrac College Edition, which puts a complete online university library at students' fingertips 24/7, and vMentor™, which gives students access to live online tutoring with an expert in the subject. Details about CengageNOW, including how to order, access, and benefit from this popular system, are provided at the front of the book.

Study Guide with Practice Tests. This student study tool, by Kaythryn Sinast Mueller of Baylor University, contains brief chapter outlines; chapter summaries; student learning objectives; a list of key terms and key people with page references to the text; detailed chapter outlines; Analyzing and Understanding the Boxes; study activities; learning objectives; practice tests consisting of 25–30 multiple-choice questions, 10–15 true–false questions, and 5 essay questions; InfoTrac College Edition readings and exercises; Internet exercises; and student class projects and activities. All multiple-choice and true–false questions include answer explanations and page references to the text.

Introduction to Sociology Group Activities Workbook. This supplement, by Lori Ann Fowler of Tarrant County College, contains both in- and out-of-class group activities (using resources such as Mi-croCase® Online Data exercises from Wadsworth's Online Sociology Resource Center) that you can tear out and turn in to the instructor once they are completed. Also included are ideas for video clips to anchor group discussions, maps, case studies, group quizzes, ethical debates, group questions, group project topics, and ideas for outside readings on which you can base group discussions. Both a workbook for students and a repository of ideas, this is a valuable resource for students and instructors alike!

Student Telecourse Guide. This Telecourse Guide for Kendall's *Sociology in Our Times: The Essentials,* Seventh Edition, is designed to accompany the *Exploring Society: Introduction to Sociology* telecourse produced by DALLAS TeleLearning of the Dallas County Community College District (DCCCD). The Telecourse Guide provides the essential integration of videos and text, providing you with valuable resources designed to direct your daily study in the *Exploring Society* telecourse. Each chapter of the Telecourse Guide contains a lesson that corresponds to each of the 22 video segments in the telecourse. Every lesson includes the following components: Overview, Lesson Assignment, Lesson Goal, Lesson Learning Objectives, Review, Lesson Focus Points, Related Activities, Practice Tests, and Answer Key.

Cengage Learning Audio Study Tools. Students will enjoy the MP3-ready Audio Lecture Overviews for each chapter and a comprehensive audio glossary of key terms for quick study and review. Whether walking to class, doing laundry, or studying at their desk, students now have the freedom to choose when, where, and how they interact with their audio-based educational media.

Wadsworth's Sociology Home Page (www .cengage.com/sociology). Wadsworth's Sociology Home Page provides instructors and students with a wealth of free information and resources, such as Sociology in Action, Census 2000: A Student Guide for Sociology, Research Online, an Sociology Timeline, interactive maps, and a glossary of key sociological terms and concepts in Spanish.

The Kendall Companion Website (www.cengage .com/sociology/kendall). The book's companion site includes chapter-specific resources for instructors and students. For instructors, the site offers a password-protected instructor's manual, Microsoft PowerPoint® presentation slides, and more. For students, there is a multitude of text-specific study aids: tutorial practice quizzes that can be scored and e-mailed to the instructor, the photo essays' Turning to Video videos and online versions of their Reflect & Analyze questions, web links, InfoTrac College Edition exercises, flash cards, MicroCase Online data exercises, crossword puzzles, Virtual Explorations, and much more!

These resources are available to qualified adopters, and ordering options for student supplements are flexible. Please consult your local Cengage Learning sales representative or visit us at **www.cengage.com** for more information, including ISBNs; to receive examination copies of any of these instructor or student resources; or for product demonstrations. Print and ebook versions of this text are available for students to purchase at **www.ichapters.com**.

Acknowledgments

Sociology in Our Times: The Essentials would not have been possible without the insightful critiques of these colleagues, who have reviewed some or all of this book. My profound thanks to each one for engaging in this time-consuming process:

Annette Allen, Troy University
Amy Blackstone, University of Maine
Mark Gaskill, State University of New York at Oneonta
Sara E. Hanna, Oakland Community College
Jane Hellinghausen, Odessa College
Connie Jamison, Volunteer State Community College
Cindy J. Lahar, York County Community College
Chad Litton, University of Mary
Suzanne L. Maughan, University of Nebraska–Kearney
Nix McCray, Community College of Baltimore County
Efren Padilla, California State University–East Bay
K. J. Stone, Los Angeles Southwest College
Dawn Tawwater, Austin Community College
Daniel Van Dussen, Youngstown State University

I'd also like to thank those dedicated instructors who responded to our survey: Cheryl Albers, Buffalo State University; Daniel Boudon, Hofstra University; Robin Brownfield, Rutgers University; Karyn Daniels, Long Beach City College; Danielle Hedegard, University of Arizona; Ursula Hines, Cuyamaca College; Karen Honeycutt, Keene State College; Chris Johnson, University of Louisiana at Monroe; Allison Jones, Ventura College; John C. Kilburn, Jr., Texas A&M International University; Christine Monnier, College of DuPage; and Robert Turley, College of the Canyons.

Previous Reviewers of *Sociology in Our Times: The Essentials*

Jan Abu-Shakrah, Portland Community College; Ted Alleman, Penn State University; Ginna Babcock, University of Idaho; Kay Barches, Skyline College; Sampson Lee Blair, University of Oklahoma; Elizabeth Bodien, Northampton Community College; Barbara Bollmann, Community College of Denver; Gayle Gordon Bouzard, Texas State University, San Marcos; Joni Boye-Bearman, Wayne State College; John Brady, Texas Lutheran University; Valerie Brown, Case Western Reserve University; Deborah Carter, Johnson C. Smith University; Margaret Choka, Pellissippi State Community College; Dennis Cole, Owens Community College; Brooks Cowan, University of Vermont; Russell Curtis, University of Houston; Mary Ann Czarnezki, University of Wisconsin–Milwaukee; Lillian Daughaday, Murray State University; Kevin Delaney, Temple University; Judith Dilorio, Indiana University–Purdue University; Harold Dorton, Southwest Texas State University; Kevin Early, Oakland University; Don Ecklund, Lincoln Land Community College; Cindy Epperson, St. Louis Community College at Meramec; Julian Thomas Euell, Ithaca College; Grant Farr, Portland State University; Joe R. Feagin, University of Florida; Nancy Feather, West Virginia University; Kathryn M. Feltey, University of Akron; Leticia Fernandez, University of Texas at El Paso; William Finlay, University of Georgia; Andrea Fontana, UNLV; Mark Foster, Johnson County Community College; Howard Gabennesch, University of Southern Indiana; Richard Gale, University of Oregon; DeAnn K. Gauthier, University of Southwestern Louisiana; Robert B. Girvan, Clarion University; Tanya Gladney, University of Nebraska–Lincoln; Michael Goslin, Tallahassee Community College; Anna Hall, Delgado Community College; Lee Hamilton, New Mexico State University; Gary D. Hampe, University of Wyoming; Carole Hill, Shelton State Community College; Art Jipson, University of Dayton; Richard Kamei, Glendale Community College; William Kelly, University of Texas, Austin; Bud Khleif, University of New Hampshire; Joachim Kibirige, Missouri Western State College; Michael B. Kleiman, University of South Florida; Janet Koenigamen, College of St. Benedict; Rosalind Kopfstein, Western

Connecticut State University; Amy Krull, Western Kentucky University; Monica Kumba, Buffalo State University; Yechiel Lahavy, Atlantic Community College; Gregory Leavitt, Idaho State University; Amy Leimer, Tennessee Tech University; Abraham Levine, El Camino College; Diane Levy, University of North Carolina, Wilmington; Stephen Lilley, Sacred Heart University; Michael Lovaglia, University of Iowa; Akbar Mahdo, Ohio Wesleyan University; Rebecca Matthews, University of Iowa; M. Cathey Maze, Franklin University; Jane McCandless, State University of West Georgia; Barbara Miller, Pasadena City College; John Molan, Alabama State University; Betty Morrow, Florida International University; Kelly Mosel-Talavera, Texas State University, San Marcos; Tina Mougouris, San Jacinto College; Mani Pande, Kansas State University; Dale Parent, Southeastern Louisiana University; Linda Petroff, Central Community College; Karen Polite, Harrisburg Area Community College–Lancaster Campus; J. Keith Price, West Texas A&M University; Barbara Richardson, Eastern Michigan University; Lois Rood, Tennessee Tech University; Ellen Rosengarten, Sinclair Community College; Phil Rutledge, University of North Carolina, Charlotte; Joseph Scimecca, George Mason University; Marcia Texler Segal, Indiana University Southeast; Dorothy Smith, State University of New York–Plattsburg; Teresa Sobieszczyk, University of Montana; Jackie Stanfield, Northern Colorado University; Mary Texeira, California State University–San Bernardino; Jamie Thompson, University of Central Oklahoma; Gary Tiedeman, Oregon State University; James Unnever, Radford University; Richard Valencia, Fresno City College; Steven Vasser, Mankato State University; Carl Wahlstrom, Genesee Community College; Henry Walker, Cornell University; Susan Waller, University of Central Oklahoma; Judith Ann Warner, Texas A&M International University; N. Ree Wells, Missouri Southern State College; Jerry Williams, Stephen Austin State; and John Zipp, University of Wisconsin–Milwaukee.

I deeply appreciate the energy, creativity, and dedication of the many people responsible for the development and production of *Sociology in Our Times: The Essentials.* I wish to thank Wadsworth/Cengage Learning's Chris Caldeira, Renee Deljon, Melanie Cregger, Lauren Keyes, and Cheri Palmer for their enthusiasm and insights throughout the development of this text. Many other people worked hard on the production of the seventh edition of *Sociology in Our Times: The Essentials,* especially Greg Hubit and Donald Pharr. I am extremely grateful to them. My most profound thanks go to my husband, Terrence Kendall, who made invaluable contributions to *Sociology in Our Times: The Essentials.*

I invite you to send your comments and suggestions about this book to me in care of:

Wadsworth, Cengage Learning
10 Davis Drive
Belmont, CA 94002

SOCIOLOGY
IN OUR TIMES

1

The Sociological Perspective and Research Process

His name was Josh Evans. He was 16 years old. And he was hot.

"Mom! Mom! Mom! Look at him!" Tina Meier recalls her daughter saying.

Josh had contacted Megan Meier through her MySpace page and wanted to be added as a friend.

"Yes, he's cute," Tina Meier told her daughter. "Do you know who he is?"

"No, but look at him! He's hot! Please, please, can I add him?"

Mom said yes. And for six weeks Megan and Josh—under Tina's watchful eye—became acquainted in the virtual world of MySpace.

Josh said he was born in Florida and recently had moved to O'Fallon [Missouri]. He was homeschooled. He played the guitar and drums. . . .

As for 13-year-old Megan . . . [she] loved swimming, boating, fishing, dogs, rap music and boys.

© AP Images/Tom Gannam

Tina Meier, the mother of Megan Meier, displays photographs of her daughter, who committed suicide at the age of thirteen after reacting to fake MySpace postings by the mother of one of her former friends. Studying sociology provides us with new insights on problems such as suicide by making us aware that much more goes on in social life than we initially observe.

But her life had not always been easy, her mother says. She was heavy and for years had tried to lose weight. She had attention deficit disorder and battled depression. . . . But things were going exceptionally well. She had shed 20 pounds, getting down to 175. She was 5 foot $5^{1}/_{2}$ inches tall. . . .

Part of the reason for Megan's rosy outlook was Josh, Tina says. After school Megan would rush to the computer. . . . It did seem odd, Tina says, that Josh never asked for Megan's phone number. And when Megan asked for his, she says, Josh said he didn't have a cell and his mother did not yet have a landline.

And then on Sunday, Oct. 15, 2006, Megan received a puzzling and disturbing message from Josh. Tina recalls that it said, "I don't know if I want to be friends with you anymore because I've heard that you are not very nice to your friends."

Frantic, Megan shot back: "What are you talking about?" (Pokin, 2007)

This and other hostile instant message exchanges set into motion the final, disturbing episode in the life of Megan Meier, as she was suddenly confronted with not only the anger and cynicism of a young man she thought she knew and trusted but also the bullying

Chapter Focus Question

How do sociological theory and research add to our knowledge of human societies and social issues such as suicide?

of other young people gathered on the social networking site MySpace who also sent a barrage of hate-filled messages that called Megan a liar and much worse.

"Mom, they're being Horrible!" Megan said, sobbing into the phone when her mother called. After an hour, Megan ran into her bedroom and hanged herself with a belt.

"She felt there was no way out," Ms. Meier said. (Maag, 2007)

—the parents of Megan Meier recalling the events leading up to her suicide at only thirteen years of age

Clearly, the suicide of Megan deeply touched her parents and friends while raising many issues about the problem of cyber-bullying. In the aftermath of Megan's tragic death, her parents tried to send an instant message to Josh Evans to inform him about the destructive nature of his actions, only to learn that his MySpace account had been deleted. Six weeks after Megan's death, her parents learned that Josh Evans never existed: His fake persona allegedly had been created by a mother whose daughter was once Megan's friend. According to some media reports, this parent created a fake MySpace account for "Josh Evans" so that she could find out what Megan would say about her daughter and other people. Subsequently, other members gathered on MySpace and—not knowing that Josh Evans did not exist—jumped into the fray and began hurling accusations at Megan and bullying her. A local ordinance in Megan's hometown now prohibits any harassment that uses the Internet, text messaging services, or any other electronic medium.

Although we will never know the full story of Megan's life, this tragic occurrence brings us to a larger sociological question: Why does anyone commit suicide? Is suicide purely an individual phenomenon, or is it related to our social interactions and the social environment and society in which we live?

In this chapter, we examine how sociological theories and research can help us understand the seemingly individualistic act of taking one's own life. We will see how sociological theory and research methods might be used to answer complex questions, and we will wrestle with some of the difficulties that sociologists experience as they study human behavior.

Putting Social Life into Perspective

Sociology **is the systematic study of human society and social interaction.** It is a *systematic* study because sociologists apply both theoretical perspectives and research methods (or orderly approaches) to examinations of social behavior. Sociologists study human societies and their social interactions in order to develop theories of how human behavior is shaped by group life and how, in turn, group life is affected by individuals.

Why Study Sociology?

Sociology helps us gain a better understanding of ourselves and our social world. It enables us to see how behavior is largely shaped by the groups to which we belong and the society in which we live. A *society* **is a large social grouping that shares the same geographical territory and is subject to the same political authority and dominant cultural expectations,** such as the United States, Mexico, or Nigeria. Examining the world order helps us understand that each of us is affected by *global interdependence*—a relationship in which the lives of all people

Box 1.1 Sociology and Everyday Life

How Much Do You Know About Suicide?

True	False	
T	F	1. For people thinking of suicide, it is difficult, if not impossible, to see the bright side of life.
T	F	2. People who talk about suicide don't do it.
T	F	3. Once people contemplate or attempt suicide, they must be considered suicidal for the rest of their lives.
T	F	4. In the United States, suicide occurs on the average of one every sixteen minutes.
T	F	5. Accidents and injuries sustained by teenagers and young adults may indicate suicidal inclinations.
T	F	6. Over half of all suicides occur in adult women between the ages of 25 and 65.
T	F	7. Older women have lower rates of both attempted and completed suicide than older men.
T	F	8. Children don't know enough to be able to intentionally kill themselves.
T	F	9. Suicide rates for African Americans are higher than for white Americans.
T	F	10. More teenagers and young adults die from suicide than from cancer, heart disease, AIDS, birth defects, stroke, pneumonia, influenza, and chronic lung disease combined.

Answers on page 6.

are intertwined closely and any one nation's problems are part of a larger global problem. Environmental problems are an example: People throughout the world share the same biosphere—the zone of the Earth's surface and atmosphere that sustains life. When environmental degradation, such as removing natural resources or polluting the air and water, takes place in one region, it may have an adverse effect on people around the globe.

Individuals can make use of sociology on a more personal level. Sociology enables us to move beyond established ways of thinking, thus allowing us to gain new insights into ourselves and to develop a greater awareness of the connection between our own "world" and that of other people. According to the sociologist Peter Berger (1963: 23), sociological inquiry helps us see that "things are not what they seem." Sociology provides new ways of approaching problems and making decisions in everyday life. For this reason, people with a knowledge of sociology are employed in a variety of fields that apply sociological insights to everyday life (see ▶ Figure 1.1).

Sociology promotes understanding and tolerance by enabling each of us to look beyond intuition, common sense, and our personal experiences. Many of us rely on intuition or common sense gained from personal experience to help us understand our daily lives and other people's behavior. *Commonsense knowledge* guides ordinary conduct in everyday life. However, many commonsense notions are actually myths. A *myth* is a popular but false notion that may be used, either intentionally or unintentionally, to perpetuate certain beliefs or "theories" even in the light of conclusive evidence to the contrary. Before reading on, take the quiz in Box 1.1, which lists a number of commonsense notions about suicide.

sociology the systematic study of human society and social interaction.

society a large social grouping that shares the same geographical territory and is subject to the same political authority and dominant cultural expectations.

Box 1.1 Sociology and Everyday Life

Answers to the Sociology Quiz on Suicide

1. True. To people thinking of suicide, an acknowledgment that there is a bright side only confirms and conveys the message that they have failed; otherwise, they could see the bright side of life as well.

2. False. Some people who talk about suicide do kill themselves. Warning signals of possible suicide attempts include talk of suicide, the desire not to exist anymore, and despair.

3. False. Most people think of suicide for only a limited amount of time. When the crisis is over and the problems leading to suicidal thoughts are resolved, people usually cease to think of suicide as an option.

4. True. A suicide occurs on the average of every sixteen minutes in the United States; however, this differs with respect to the sex, race/ethnicity, and age of the individual. For example, men are four times more likely to kill themselves than are women.

5. True. Accidents, injuries, and other types of life-threatening behavior may be signs that a person is on a course of self-destruction. One study concluded that the incidence of suicide was twelve times higher among adolescents and young adults who had been previously hospitalized because of an injury.

6. False. Just the opposite is true: Over half of all suicides occur in adult men aged 25 to 65.

7. True. In the United States, as in other countries, suicide rates are the highest among men over age 70. One theory of why this is true asserts that older women may have a more flexible and diverse coping style than do older men.

8. False. Children do know how to intentionally hurt or kill themselves. They may learn the means and methods from television, movies, or other people. However, the National Center for Health Statistics (the agency responsible for compiling suicide statistics) does not recognize suicides under the age of 10; they are classified as accidents, despite evidence that young children have taken their own lives.

9. False. Suicide rates are much higher among white Americans than African Americans. For example, in 2004 the overall U.S. suicide rate was 11.0 per 100,000 population. For white males, the rate was 19.6; for African American males, it was 9.0. For white females, it was 5.1; for African American females, the rate was 1.8.

10. True. Suicide is a leading cause of death among teenagers and young adults. It is the third leading cause of death among young people between 15 and 24 years of age, following unintentional injuries and homicide.

Sources: Based on American Association of Suicidology, 2006; Leenars, 1991; Levy and Deykin, 1989; and National Center for Injury Prevention and Control, 2006.

By contrast, sociologists strive to use scientific standards, not popular myths or hearsay, in studying society and social interaction. They use systematic research techniques and are accountable to the scientific community for their methods and the presentation of their findings. Whereas some sociologists argue that sociology must be completely value free—free from distorting subjective (personal or emotional) bias—others do not think that total objectivity is an attainable or desirable goal when studying human behavior. However, all sociologists attempt to discover patterns or commonalities in human behavior. When they study suicide, for example, they look for recurring patterns of behavior even though

Health and Human Services	Business	Communication	Academia	Law
Counseling Education Medicine Nursing Social Work	Advertising Labor Relations Management Marketing	Broadcasting Public Relations Journalism	Anthropology Media Studies/ Communication Economics Geography History Information Studies Political Science Psychology Sociology	Law Criminal Justice

▶ **Figure 1.1 Fields That Use Social Science Research**
In many careers, including jobs in academia, business, communication, health and human services, and law, the ability to analyze social science research is an important asset.

Source: Based on Katzer, Cook, and Crouch, 1991.

individual people usually commit the acts and *other individuals* suffer as a result of these actions.

Consequently, sociologists seek out the multiple causes and effects of suicide or other social issues. They analyze the impact of the problem not only from the standpoint of the people directly involved but also from the standpoint of the effects of such behavior on all people.

The Sociological Imagination

Sociologist C. Wright Mills (1959b) described sociological reasoning as the ***sociological imagination**—the ability to see the relationship between individual experiences and the larger society.* This awareness enables us to understand the link between our personal experiences and the social contexts in which they occur. The sociological imagination helps us distinguish between personal troubles and social (or public) issues. *Personal troubles* are private problems that affect individuals and the networks of people with whom they associate regularly. As a result, those problems must be solved by individuals within their immediate social settings. For example, one person being unemployed may be a personal trouble. *Public issues* are problems that affect large numbers of people and often require solutions at the societal level. Widespread unemployment as a result of economic changes such as plant closings is an example of a public issue. The sociological imagi-

nation helps us place seemingly personal troubles, such as losing one's job or feeling like committing suicide, into a larger social context, where we can distinguish whether and how personal troubles may be related to public issues.

Suicide as a Personal Trouble Many of our individual experiences may be largely beyond our own control. They are determined by society as a whole—by its historical development and its organization. In everyday life, we do not define personal experiences in these terms. If a person commits suicide, many people consider it to be the result of his or her own personal problems.

Suicide as a Public Issue We can also use the sociological imagination to look at the problem of suicide as a public issue—a societal problem. Early sociologist Emile Durkheim refused to accept commonsense explanations of suicide. In what is probably the first sociological study to use scientific research methods, he related suicide to the issue of cohesiveness (or lack of cohesiveness) in society instead of viewing suicide as an isolated

sociological imagination C. Wright Mills's term for the ability to see the relationship between individual experiences and the larger society.

act that could be understood only by studying individual personalities or inherited tendencies. In *Suicide* (1964b/1897), Durkheim documented his contention that a high suicide rate was symptomatic of large-scale societal problems.

The Importance of a Global Sociological Imagination

Although existing sociological theory and research provide the foundation for sociological thinking, we must reach beyond past studies that have focused primarily on the United States to develop a more comprehensive *global* approach for the future (see ▶ Map 1.1). In the twenty-first century, we face important challenges in a rapidly changing nation and world. The world's **high-income countries are nations with highly industrialized economies; technologically advanced industrial, administrative, and service occupations; and relatively high levels**

of national and personal income. Some examples are the United States, Canada, Australia, New Zealand, Japan, and the countries of Western Europe.

As compared with other nations of the world, many high-income nations have a high standard of living and a lower death rate due to advances in nutrition and medical technology. However, everyone living in a so-called high-income country does not necessarily have a high income or an outstanding quality of life. In contrast, **middle-income countries are nations with industrializing economies, particularly in urban areas, and moderate levels of national and personal income.** Some examples of middle-income countries are the nations of Eastern Europe and many Latin American countries, where nations such as Brazil and Mexico are industrializing rapidly.

Low-income countries are primarily agrarian nations with little industrialization and low levels of national and personal income. Examples of low-

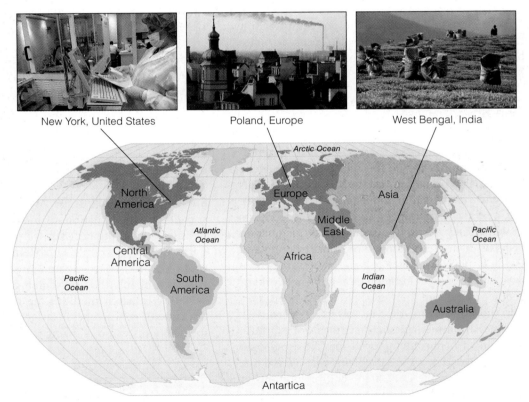

New York, United States   Poland, Europe   West Bengal, India

▶ **Map 1.1** **The World's Economies in the Early Twenty-First Century**
High-income, middle-income, and low-income countries.

income countries are many of the nations of Africa and Asia, particularly India and the People's Republic of China, where people typically work the land and are among the poorest in the world. However, generalizations are difficult to make because there are wide differences in income and standards of living within many nations (see Chapter 8, "Global Stratification"). Throughout this text, we will continue to develop our sociological imaginations by examining social life in the United States and other nations.

Developing a better understanding of diversity and tolerance for people who are different from us is important for our personal, social, and economic well-being. Whatever your race/ethnicity, class, sex, or age, are you able to include in your thinking the perspectives of people who are quite dissimilar in experiences and points of view? Before answering this question, a few definitions are in order. *Race* is a term used by many people to specify groups of people distinguished by physical characteristics such as skin color; in fact, there are no "pure" racial types, and the concept of race is considered by most sociologists to be a social construction that people use to justify existing social inequalities. *Ethnicity* refers to the cultural heritage or identity of a group and is based on factors such as language or country of origin. *Class* is the relative location of a person or group within the larger society, based on wealth, power, prestige, or other valued resources. *Sex* refers to the biological and anatomical differences between females and males. By contrast, *gender* refers to the meanings, beliefs, and practices associated with sex differences, referred to as *femininity* and *masculinity*.

The Development of Sociological Thinking

Throughout history, social philosophers and religious authorities have made countless observations about human behavior. However, the idea of observing how people lived, finding out what they thought, and doing so in a systematic manner that could be verified did not take hold until the nineteenth century and the social upheaval brought about by industrialization and urbanization.

Industrialization **is the process by which societies are transformed from dependence on agriculture and handmade products to an emphasis on manufacturing and related industries.** This process occurred first during the Industrial Revolution in Britain between 1760 and 1850, and was soon repeated throughout Western Europe. By the mid-nineteenth century, industrialization was well under way in the United States. Massive economic, technological, and social changes occurred as machine technology and the factory system shifted the economic base of these nations from agriculture to manufacturing. A new social class of industrialists emerged in textiles, iron smelting, and related industries. Many people who had labored on the land were forced to leave their tightly knit rural communities and sacrifice well-defined social relationships to seek employment as factory workers in the emerging cities, which became the centers of industrial work.

Urbanization **is the process by which an increasing proportion of a population lives in cities rather than in rural areas.** Although cities existed long before the Industrial Revolution, the development of the factory system led to a rapid increase

high-income countries (sometimes referred to as **industrial countries**) nations with highly industrialized economies; technologically advanced industrial, administrative, and service occupations; and relatively high levels of national and personal income.

middle-income countries (sometimes referred to as **developing countries**) nations with industrializing economies, particularly in urban areas, and moderate levels of national and personal income.

low-income countries (sometimes referred to as **underdeveloped countries**) nations with little industrialization and low levels of national and personal income.

industrialization the process by which societies are transformed from dependence on agriculture and handmade products to an emphasis on manufacturing and related industries.

urbanization the process by which an increasing proportion of a population lives in cities rather than in rural areas.

© Hulton Archive/Getty Images

As the Industrial Revolution swept through the United States beginning in the nineteenth century, sights like this became increasingly common. This early automobile assembly line is symbolic of the factory system that shifted the base of the U.S. economy from agriculture to manufacturing. What new technologies are transforming the U.S. economy in the twenty-first century?

in both the number of cities and the size of their populations. People from very diverse backgrounds worked together in the same factory. At the same time, many people shifted from being *producers* to being *consumers*. For example, families living in the cities had to buy food with their wages because they could no longer grow their own crops to consume or to barter for other resources. Similarly, people had to pay rent for their lodging because they could no longer exchange their services for shelter.

These living and working conditions led to the development of new social problems: inadequate housing, crowding, unsanitary conditions, poverty, pollution, and crime. Wages were so low that entire families—including very young children—were forced to work, often under hazardous conditions and with no job security. As these conditions became more visible, a new breed of social thinkers turned its attention to trying to understand why and how society was changing.

Early Thinkers: A Concern with Social Order and Stability

At the same time that urban problems were growing worse, natural scientists had been using reason, or rational thinking, to discover the laws of physics and the movement of the planets. Social thinkers started to believe that by applying the methods developed by the natural sciences, they might discover the laws of human behavior and apply these laws to solve social problems.

Auguste Comte The French philosopher Auguste Comte (1798–1857) coined the term *sociology* from the Latin *socius* ("social, being with others") and the Greek *logos* ("study of") to describe a new science that would engage in the study of society. Even though he never actually conducted sociological

Auguste Comte (1798–1857) (oil on canvas), Etex, Louis Jules (1810–1889)/ Temple de la Religion de l'Humanité, Paris, France/The Bridgeman Art Library International

Auguste Comte

research, Comte is considered by some to be the "founder of sociology." Comte's theory that societies contain *social statics* (forces for social order and stability) and *social dynamics* (forces for conflict and change) continues to be used, although not in these exact terms, in contemporary sociology.

Comte stressed that the methods of the natural sciences should be applied to the objective study of society. He sought to unlock the secrets of society so that intellectuals like him could become the new secular (as contrasted with religious) "high priests" of society (Nisbet, 1979). For Comte, the best policies involved order and authority. He envisioned that a new consensus would emerge on social issues and that the new science of sociology would play a significant part in the reorganization of society (Lenzer, 1998).

Comte's philosophy became known as ***positivism*—a belief that the world can best be understood through scientific inquiry.** He believed that positivism had two dimensions: (1) methodological—the application of scientific knowledge to both physical and social phenomena—and (2) social and political—the use of such knowledge to predict the likely results of different policies so that the best one could be chosen.

Social analysts have praised Comte for his advocacy of sociology and his insights regarding linkages between the social structural elements of society (such as family, religion, and government) and social thinking in specific historical periods. However, a number of contemporary sociologists argue that Comte contributed to an overemphasis on the "natural science model" and focused on the experiences of a privileged few, to the exclusion by class, gender, race, ethnicity, and age of all others.

Harriet Martineau Comte's works were made more accessible for a wide variety of scholars through the efforts of the British sociologist Harriet Martineau (1802–1876). Until recently, Martineau received no recognition in the field of sociology, partly because she was a woman in a male-dominated discipline and society. Not only did she translate and condense Comte's works, but she was also an active sociologist in her own right. Martineau studied the social customs of Britain and the United States, analyzing the consequences of industrialization and capitalism. In *Society in America* (1962/1837), she examined re-

© Spencer Arnold/Getty Images

Harriet Martineau

ligion, politics, child rearing, slavery, and immigration in the United States, paying special attention to social distinctions based on class, race, and gender. Her works explore the status of women, children, and "sufferers" (persons who are considered to be criminal, mentally ill, handicapped, poor, or alcoholic).

Martineau advocated racial and gender equality. She was also committed to creating a science of society that would be grounded in empirical observations and widely accessible to people. She argued that sociologists should be impartial in their assessment of society but that it is entirely appropriate to compare the existing state of society with the principles on which it was founded (Lengermann and Niebrugge-Brantley, 1998). Martineau believed that a better society would emerge if women and men were treated equally, enlightened reform occurred, and cooperation existed among people in all social classes (but led by the middle class).

Herbert Spencer Unlike Comte, who was strongly influenced by the upheavals of the French Revolution, the British social theorist Herbert Spencer

positivism a term describing Auguste Comte's belief that the world can best be understood through scientific inquiry.

Herbert Spencer

(1820–1903) was born in a more peaceful and optimistic period in his country's history. Spencer's major contribution to sociology was an evolutionary perspective on social order and social change. Evolutionary theory is "a theory to explain the mechanisms of organic/social change" (Haines, 1997: 81). According to Spencer's Theory of General Evolution, society, like a biological organism, has various interdependent parts (such as the family, the economy, and the government) that work to ensure the stability and survival of the entire society.

Spencer believed that societies develop through a process of "struggle" (for existence) and "fitness" (for survival), which he referred to as the "survival of the fittest." Because this phrase is often attributed to Charles Darwin, Spencer's view of society is known as **social Darwinism—the belief that those species of animals, including human beings, best adapted to their environment survive and prosper, whereas those poorly adapted die out.** Spencer equated this process of *natural selection* with

progress because only the "fittest" members of society would survive the competition, and the "unfit" would be filtered out of society.

Critics believe that his ideas are flawed because societies are not the same as biological systems; people are able to create and transform the environment in which they live. Moreover, the notion of the survival of the fittest can easily be used to justify class, racial–ethnic, and gender inequalities and to rationalize the lack of action to eliminate harmful practices that contribute to such inequalities.

Emile Durkheim French sociologist Emile Durkheim (1858–1917) stressed that people are the product of their social environment and that behavior cannot be understood fully in terms of *individual* biological and psychological traits. He believed that the limits of human potential are *socially* based, not *biologically* based.

In his work *The Rules of Sociological Method* (1964a/1895), Durkheim set forth one of his most important contributions to sociology: the idea that societies are built on social facts. **Social facts are patterned ways of acting, thinking, and feeling that exist *outside* any one individual but that exert social control over each person.** Durkheim believed that social facts must be explained by other

Emile Durkheim

Sociology *Works!*

Durkheim's Sociology of Suicide and Twenty-First Century India

The bond attaching [people] to life slackens because the bond which attaches [them] to society is itself slack.

—Emile Durkheim, *Suicide* (1964b/1897)

Although this statement described social conditions accompanying the high rates of suicide found in late-nineteenth-century France, Durkheim's words ring true today as we look at contemporary suicide rates for cities such as Bangalore, which some refer to as "India's Suicide City" (Guha, 2004).

At first glance, we might think that the outsourcing of jobs in the technology sector—from high-income nations such as the United States to India—would provide happiness and job satisfaction for individuals in cities such as Bangalore and New Delhi who have gained new opportunities and higher salaries in recent years as a result of outsourcing. News stories have focused on the wealth of opportunities that these outsourced jobs have brought to millions of men and women in India, most of whom are in their twenties and thirties and who now earn larger incomes than do their parents and many of their contemporaries. However, the underlying story of what is really going on in these cities stands in stark contrast: Rapid

urbanization and fast-paced changes in the economy and society are weakening social ties that have been so important to individuals. Social bonds have been weakened or dissolved as people move away from their families and their community. Life in the cities moves at a much faster pace than in the rural areas, and many individuals experience loneliness, sleep disorders, family discord, and major health risks such as heart disease and depression (Mahapatra, 2007). In the words of Ramachandra Guha (2004), a historian residing in India, Durkheim's sociology of suicide remains highly relevant to finding new answers to this challenging problem: "The rash of suicides in city and village is a qualitatively new development in our history. We sense that tragedies are as much social as they are individual. But we know very little of what lies behind them. What we now await, in sum, is an Indian Durkheim."

Reflect & Analyze

How does sociology help us to examine seemingly private acts such as suicide within a larger social context? Why are some people more inclined to commit suicide if they are not part of a strong social fabric?

---

social facts—by reference to the social structure rather than to individual attributes.

Durkheim observed that rapid social change and a more specialized division of labor produce *strains* in society. These strains lead to a breakdown in traditional organization, values, and authority and to a dramatic increase in **anomie—a condition in which social control becomes ineffective as a result of the loss of shared values and of a sense of purpose in society.** According to Durkheim, anomie is most likely to occur during a period of rapid social change. In *Suicide* (1964b/1897), he explored the relationship between anomic social conditions and suicide, a concept that remains important in the twenty-first century (see "Sociology Works!" at the top of this page).

Durkheim's contributions to sociology are so significant that he has been referred to as "*the* crucial

figure in the development of sociology as an academic discipline [and as] one of the deepest roots of the sociological imagination" (Tiryakian, 1978:

---

social Darwinism Herbert Spencer's belief that those species of animals, including human beings, best adapted to their environment survive and prosper, whereas those poorly adapted die out.

social facts Emile Durkheim's term for patterned ways of acting, thinking, and feeling that exist *outside* any one individual but that exert social control over each person.

anomie Emile Durkheim's designation for a condition in which social control becomes ineffective as a result of the loss of shared values and of a sense of purpose in society.

187). He is described as the founding figure of the functionalist theoretical tradition.

Although they acknowledge Durkheim's important contributions, some critics note that his emphasis on societal stability, or the "problem of order"—how society can establish and maintain social stability and cohesiveness—obscured the *subjective meaning* that individuals give to social phenomena such as religion, work, and suicide. From this view, overemphasis on *structure* and the determining power of "society" resulted in a corresponding neglect of *agency* (the beliefs and actions of the actors involved) in much of Durkheim's theorizing (Zeitlin, 1997).

Differing Views on the Status Quo: Stability Versus Change

Together with Karl Marx, Max Weber, and Georg Simmel, Durkheim established the course of modern sociology. We will look first at Marx's and Weber's divergent thoughts about conflict and social change in societies and then at Georg Simmel's microlevel analysis of society.

Karl Marx In sharp contrast to Durkheim's focus on the stability of society, German economist and philosopher Karl Marx (1818–1883) stressed that history is a continuous clash between conflicting ideas and forces. He believed that conflict—especially class conflict—is necessary in order to produce social change and a better society. For Marx, the most important changes are economic. He concluded that the capitalist economic system was responsible for the overwhelming poverty that he observed in London at the beginning of the Industrial Revolution (Marx and Engels, 1967/1848).

In the Marxian framework, *class conflict* is the struggle between the capitalist class and the working class. The capitalist class, or *bourgeoisie,* comprises those who own and control the means of production—the tools, land, factories, and money for investment that form the economic basis of a society. The working class, or *proletariat,* is composed of those who must sell their labor because they have no other means to earn a livelihood. From Marx's viewpoint, the capitalist class controls and exploits the masses of struggling workers by paying less than

Karl Marx

the value of their labor. This exploitation results in workers' *alienation*—a feeling of powerlessness and estrangement from other people and from oneself. Marx predicted that the working class would become aware of its exploitation, overthrow the capitalists, and establish a free and classless society.

Marx is regarded as one of the most profound sociological thinkers; however, his social and economic analyses have also inspired heated debates among generations of social scientists. Central to his view was the belief that society should not just be studied but should also be changed, because the *status quo* (the existing state of society) involved the oppression of most of the population by a small group of wealthy people. Those who believe that sociology should be value free are uncomfortable with Marx's advocacy of what some perceive to be radical social change. Scholars who examine society through the lens of race, gender, and class believe his analysis places too much emphasis on class relations, often to the exclusion of issues regarding race/ethnicity and gender.

Max Weber German social scientist Max Weber (pronounced VAY-ber) (1864–1920) was also concerned about the changes brought about by the Industrial Revolution. Although he disagreed with

© Hulton Archive/Getty Images

Max Weber

Marx's idea that economics is *the* central force in social change, Weber acknowledged that economic interests are important in shaping human action. Even so, he thought that economic systems were heavily influenced by other factors in a society.

Unlike many early analysts who believed that values could not be separated from the research process, Weber emphasized that sociology should be *value free*—research should be conducted in a scientific manner and should exclude the researcher's personal values and economic interests (Turner, Beeghley, and Powers, 2002). However, Weber realized that social behavior cannot be analyzed by the objective criteria that we use to measure such things as temperature or weight. Although he recognized that sociologists cannot be totally value free, Weber stressed that they should employ *verstehen* (German for "understanding" or "insight") to gain the ability to see the world as others see it. In contemporary sociology, Weber's idea is incorporated into the concept of the sociological imagination (discussed earlier in this chapter).

Weber was also concerned that large-scale organizations (bureaucracies) were becoming increasingly oriented toward routine administration and a specialized division of labor, which he believed were destructive to human vitality and freedom. According to Weber, rational bureaucracy, rather than class

struggle, is the most significant factor in determining the social relations between people in industrial societies. From this view, bureaucratic domination can be used to maintain powerful (capitalist) interests in society. As discussed in Chapter 5 ("Groups and Organizations"), Weber's work on bureaucracy has had a far-reaching impact.

Weber made significant contributions to modern sociology by emphasizing the goal of value-free inquiry and the necessity of understanding how others see the world. He also provided important insights on the process of rationalization, bureaucracy, religion, and many other topics. In his writings, Weber was more aware of women's issues than many of the scholars of his day. Perhaps his awareness at least partially resulted from the fact that his wife, Marianne Weber, was an important figure in the women's movement in Germany in the early twentieth century (Roth, 1988).

Georg Simmel At about the same time that Durkheim was developing the field of sociology in France, the German sociologist Georg Simmel (pronounced ZIM-mel) (1858–1918) was theorizing about society as a web of patterned interactions

© The Granger Collection, New York

Georg Simmel

among people. In *The Sociology of Georg Simmel* (1950/1902–1917), he analyzed how social interactions vary depending on the size of the social group. He concluded that interaction patterns differed between a *dyad,* a social group with two members, and a *triad,* a social group with three members. He developed *formal sociology,* an approach that focuses attention on the universal recurring social forms that underlie the varying content of social interaction. Simmel referred to these forms as the "geometry of social life."

Like the other social thinkers of his day, Simmel analyzed the impact of industrialization and urbanization on people's lives. He concluded that class conflict was becoming more pronounced in modern industrial societies. He also linked the increase in individualism, as opposed to concern for the group, to the fact that people now had many cross-cutting "social spheres"—membership in a number of organizations and voluntary associations—rather than having the singular community ties of the past.

© Simon Jarratt/Corbis

According to the sociologist Georg Simmel, society is a web of patterned interactions among people. If we focus on the behavior of individuals only, we may miss the underlying forms that make up the "geometry of social life."

Simmel's contributions to sociology are significant. He wrote more than thirty books and numerous essays on diverse topics, leading some critics to state that his work is fragmentary and piecemeal. However, his thinking has influenced a wide array of sociologists, including the members of the "Chicago School" in the United States.

The Beginnings of Sociology in the United States

From Western Europe, sociology spread in the 1890s to the United States, where it thrived as a result of the intellectual climate and the rapid rate of social change. The first departments of sociology in the United States were located at the University of Chicago and at Atlanta University, then an African American school.

The Chicago School The first department of sociology in the United States was established at the University of Chicago, where the faculty was instrumental in starting the American Sociological Society (now known as the American Sociological Association). Robert E. Park (1864–1944), a member of the Chicago faculty, asserted that urbanization has a disintegrating influence on social life by producing an increase in the crime rate and in racial and class antagonisms that contribute to the segregation and isolation of neighborhoods (Ross, 1991). George Herbert Mead (1863–1931), another member of the faculty at Chicago, founded the symbolic interaction perspective, which is discussed later in this chapter.

Jane Addams Jane Addams (1860–1935) is one of the best-known early women sociologists in the United States because she founded Hull House, one of the most famous settlement houses, in an impoverished area of Chicago. Throughout her career, she was actively engaged in sociological endeavors: She lectured at numerous colleges, was a charter member of the American Sociological Society, and published a number of articles and books. Addams was one of the authors of *Hull-House Maps and Papers,* a ground-breaking book that used a methodological technique employed by sociologists for the next forty

Jane Addams

W. E. B. Du Bois

years (Deegan, 1988). She was also awarded a Nobel Prize for her assistance to the underprivileged.

W. E. B. Du Bois and Atlanta University The second department of sociology in the United States was founded by W. E. B. Du Bois (1868–1963) at Atlanta University. He created a laboratory of sociology, instituted a program of systematic research, founded and conducted regular sociological conferences on research, founded two journals, and established a record of valuable publications. His classic work, *The Philadelphia Negro: A Social Study* (1967/1899), was based on his research into Philadelphia's African American community and stressed the strengths and weaknesses of a community wrestling with overwhelming social problems. Du Bois was one of the first scholars to note that a dual heritage creates conflict for people of color. He called this duality *double-consciousness*—the identity conflict of being both a black and an American. Du Bois pointed out that although people in this country espouse such values as democracy, freedom, and equality, they also accept racism and group discrimination. African Americans are the victims of these conflicting values and the actions that result from them (Benjamin, 1991).

Contemporary Theoretical Perspectives

Given the many and varied ideas and trends that influenced the development of sociology, how do contemporary sociologists view society? Some see it as basically a stable and ongoing entity; others view it in terms of many groups competing for scarce resources; still others describe it based on the everyday, routine interactions among individuals. Each of these views represents a method of examining the same phenomena. Each is based on general ideas about how social life is organized and represents an effort to link specific observations in a meaningful way. Each uses a ***theory*—a set of logically interrelated statements that attempts to describe, explain, and (occasionally) predict social events.** Each theory helps interpret reality in a distinct way by providing a framework in which

theory a set of logically interrelated statements that attempts to describe, explain, and (occasionally) predict social events.

observations may be logically ordered. Sociologists refer to this theoretical framework as a *perspective*—an overall approach to or viewpoint on some subject. Three major theoretical perspectives have been predominant in U.S. sociology: the functionalist, conflict, and symbolic interactionist perspectives. Other perspectives, such as postmodernism, have emerged and gained acceptance among some social thinkers more recently. Before turning to the specifics of these perspectives, we should note that some theorists and theories do not neatly fit into any of these perspectives.

Functionalist Perspectives

Also known as *functionalism* and *structural functionalism,* **functionalist perspectives are based on the assumption that society is a stable, orderly system.** This stable system is characterized by *societal consensus,* whereby the majority of members share a common set of values, beliefs, and behavioral expectations. According to this perspective, a society is composed of interrelated parts, each of which serves a function and (ideally) contributes to the overall stability of the society. Societies develop social structures, or institutions, that persist because they play a part in helping society survive. These institutions include the family, education, government, religion, and the economy. If anything adverse happens to one of these institutions or parts, all other parts are affected, and the system no longer functions properly.

Talcott Parsons and Robert Merton Talcott Parsons (1902–1979), perhaps the most influential contemporary advocate of the functionalist perspective, stressed that all societies must provide for meeting social needs in order to survive. Parsons (1955) suggested, for example, that a division of labor (distinct, specialized functions) between husband and wife is essential for family stability and social order. The husband/father performs the *instrumental tasks,* which involve leadership and decision-making responsibilities in the home and employment outside the home to support the family. The wife/mother is responsible for the *expressive tasks,* including housework, caring for the children, and providing emotional support for the entire family. Parsons

© The Granger Collection, New York

Talcott Parsons

Estate of Robert K. Merton, photo by Sandra Still

Robert Merton

believed that other institutions, including school, church, and government, must function to assist the family and that all institutions must work together to preserve the system over time (Parsons, 1955).

Shopping malls are a reflection of a consumer society. A manifest function of a shopping mall is to sell goods and services to shoppers; however, a latent function may be to provide a communal area in which people can visit friends and eat. For this reason, food courts have proven to be a boon in shopping malls around the globe.

Functionalism was refined further by Robert K. Merton (1910–2003), who distinguished between manifest and latent functions of social institutions. *Manifest functions* **are intended and/or overtly recognized by the participants in a social unit.** In contrast, *latent functions* **are unintended functions that are hidden and remain unacknowledged by participants.** For example, a manifest function of education is the transmission of knowledge and skills from one generation to the next; a latent function is the establishment of social relations and networks. Merton noted that all features of a social system may not be functional at all times; *dysfunctions* are the undesirable consequences of any element of a society. A dysfunction of education in the United States is the perpetuation of gender, racial, and class inequalities. Such dysfunctions may threaten the capacity of a society to adapt and survive (Merton, 1968).

Applying a Functional Perspective to Suicide How might functionalists analyze the problem of suicide among young people? Most functionalists empha-size the importance to a society of shared moral values and strong social bonds. When rapid social change or other disruptive conditions occur, moral values may erode, and people may become more uncertain about how to act and about whether or not their life has meaning.

In sociologist Donna Gaines's (1991) study of the suicide pact of four teenagers in Bergenfield, New Jersey, she concluded that Durkheim's description of both fatalistic and anomic suicide could be applied to the suicides of some teenagers. In regard to fatalistic suicide, people sometimes commit suicide "because they have lost the ability to dream" (Gaines, 1991: 253). According to Gaines, the four teenagers who took their lives in Bergenfield were bound closely together in a shared predicament: They wanted to get out, but they couldn't. As one young man noted, "They were beaten down as far as they could go" because they lacked confidence in themselves and were fearful of the world they faced. In other words, they felt trapped, and to them—without having a sense of meaningful choices—the only way out was to commit suicide (Gaines, 1991: 253). But, according to Gaines, fatalistic suicide does not completely explain the experiences of suicidal young people. Young people may also engage in anomic suicide, where the individual does not feel connected to the society. In fact, as Gaines (1991: 253) notes, "the glue that holds the person to the group isn't strong enough; social bonds are loose, weak, or absent. To be anomic is to feel disengaged, adrift, alienated.

functionalist perspectives the sociological approach that views society as a stable, orderly system.

manifest functions functions that are intended and/or overtly recognized by the participants in a social unit.

latent functions unintended functions that are hidden and remain unacknowledged by participants.

Like you don't fit in anywhere, there is no place for you: in your family, your school, your town—in the social order."

Conflict Perspectives

According to *conflict perspectives,* **groups in society are engaged in a continuous power struggle for control of scarce resources.** Conflict may take the form of politics, litigation, negotiations, or family discussions about financial matters. Simmel, Marx, and Weber contributed significantly to this perspective by focusing on the inevitability of clashes between social groups. Today, advocates of the conflict perspective view social life as a continuous power struggle among competing social groups.

Max Weber and C. Wright Mills As previously discussed, Karl Marx focused on the exploitation and oppression of the proletariat (the workers) by the bourgeoisie (the owners or capitalist class). Max Weber recognized the importance of economic conditions in producing inequality and conflict in society, but he added *power* and *prestige* as other sources of inequality. Weber (1968/1922) defined *power* as the ability of a person within a social relationship to carry out his or her own will despite resistance from others, and *prestige* as a positive or negative social estimation of honor (Weber, 1968/1922).

C. Wright Mills (1916–1962), a key figure in the development of contemporary conflict theory, encouraged sociologists to get involved in social reform. Mills encouraged everyone to look beneath everyday events in order to observe the major resource and power inequalities that exist in society. He believed that the most important decisions in the United States are made largely behind the scenes by the *power elite*—a small clique composed of top corporate, political, and military officials. Mills's power elite theory is discussed in Chapter 13 ("Politics and the Economy in Global Perspective").

The conflict perspective is not one unified theory but rather encompasses several branches. One branch is the neo-Marxist approach, which views struggle between the classes as inevitable and as a prime source of social change. A second branch focuses on racial–ethnic inequalities and the continued exploitation of members of some racial–ethnic

© Hulton Archive/Getty Images

C. Wright Mills

groups. A third branch is the feminist perspective, which focuses on gender issues.

The Feminist Approach A feminist theoretical approach (or "feminism") directs attention to women's experiences and the importance of gender as an element of social structure. This approach is based on the belief that "women and men are equal and should be equally valued as well as have equal rights" (Basow, 1992). According to feminist theorists, we live in a *patriarchy,* a system in which men dominate women and in which things that are considered to be "male" or "masculine" are more highly valued than those considered to be "female" or "feminine." The feminist perspective assumes that gender is socially created, rather than determined by one's biological inheritance, and that change is essential in order for people to achieve their human potential without limits based on gender. Some feminists argue that women's subordination can end only after the patriarchal system becomes obsolete. However,

© AP Images/Denis Farrell

As one of the wealthiest and most-beloved entertainers in the world, Oprah Winfrey is an example of Max Weber's concept of prestige—a positive estimate of honor. Ms. Winfrey has used her success and prestige to do good works for many causes, including starting the Oprah Winfrey Leadership Academy for Girls in South Africa.

feminism is not one single, unified approach; there are several feminist perspectives, which are discussed in Chapter 10 ("Sex and Gender").

Applying Conflict Perspectives to Suicide How might advocates of a conflict approach explain suicide among teenagers and young people?

Social Class Although many other factors may be present, social class pressures may affect rates of suicide among young people when they perceive that they have few educational or employment opportunities in our technologically oriented society and little hope for the future. As a result, young people from low-income or working-class family backgrounds may believe that they are among the most powerless people in society. However, class-based inequality alone cannot explain suicides among young people because teenage suicides also occur among affluent young people (Colt, 1991).

Gender In North America, females are more likely to attempt suicide, whereas males are more likely to actually take their own life. Despite the fact that women's suicidal behavior has traditionally been attributed to problems in their interpersonal relation-

ships, such as loss of a boyfriend, lover, or husband, feminist analysts believe that we must examine social structural pressures that are brought to bear on young women and how these may contribute to their behavior—for example, cultural assumptions about women and what their multiple roles should be in the family, education, and the workplace. Women also experience unequal educational and employment opportunities that may contribute to feelings of powerlessness and alienation. Recent research shows that there are persistent gender gaps in U.S. employment, politics, and other areas of social life that tend to adversely affect women more than men.

Race Racial subordination may be a factor in some suicides. (▶ Figure 1.2 displays U.S. suicides in terms of race and sex.) This fact is most glaringly reflected in the extremely high rate of suicide among Native Americans, who constitute about one percent of the U.S. population. The rate of suicide among young Native American males on government-owned reservations is especially high, and almost 60 percent of Native American suicides involve firearms (National Center for Injury Prevention and Control, 2003). Most research has focused on individualistic reasons why young Native Americans commit suicide; however, analysts using a race-and-ethnic framework focus on the effect of social inequalities and racial discrimination on suicidal behavior.

Symbolic Interactionist Perspectives

The conflict and functionalist perspectives have been criticized for focusing primarily on macrolevel analysis. A *macrolevel analysis* **examines whole societies, large-scale social structures, and social systems** instead of looking at important social

---

conflict perspectives the sociological approach that views groups in society as engaged in a continuous power struggle for control of scarce resources.

macrolevel analysis an approach that examines whole societies, large-scale social structures, and social systems.

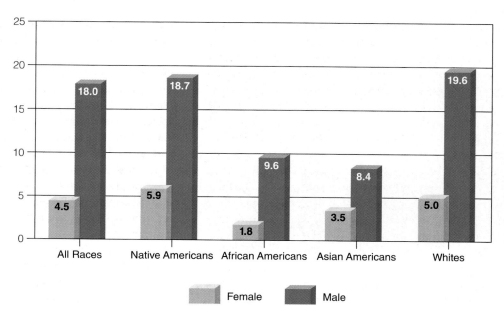

▶ Figure 1.2 Suicide Rates by Race and Sex
Rates are for U.S. suicides and indicate the number of deaths by suicide for every 100,000 people in each category for 2004.

Source: National Center for Health Statistics, 2008.

dynamics in individuals' lives. Our third perspective, symbolic interactionism, fills this void by examining people's day-to-day interactions and their behavior in groups. Thus, symbolic interactionist approaches are based on a *microlevel analysis,* **which focuses on small groups rather than large-scale social structures.**

We can trace the origins of this perspective to the Chicago School, especially George Herbert Mead and the sociologist Herbert Blumer (1900–1986), who is credited with coining the term *symbolic interactionism.* According to **symbolic interactionist** *perspectives,* **society is the sum of the interactions of individuals and groups.** Theorists using this perspective focus on the process of *interaction*—defined as immediate reciprocally oriented communication between two or more people—and the part that symbols play in giving meaning to human communication. A *symbol* is anything that meaningfully represents something else. Examples of symbols include signs, gestures, written language, and shared values. Symbolic interaction occurs when people communicate through the use of symbols—for example, a ring to indicate a couple's engagement.

But symbolic communication occurs in a variety of forms, including facial gestures, posture, tone of voice, and other symbolic gestures (such as a handshake or a clinched fist).

Symbols are instrumental in helping people derive meanings from social interactions. In social encounters, each person's interpretation or definition of a given situation becomes a *subjective reality* from that person's viewpoint. We often assume that what we consider to be "reality" is shared by others; however, this assumption is often incorrect. Subjective reality is acquired and shared through agreed-upon symbols, especially language. If a person shouts "Fire!" in a crowded movie theater, for example, that language produces the same response (attempting to escape) in all of those who hear and understand it. When people in a group do not share the same meaning for a given symbol, however, confusion results: People who do not know the meaning of the word *fire* will not know what the commotion is about. How people *interpret* the messages they receive and the situations they encounter becomes their subjective reality and may strongly influence their behavior.

Applying Symbolic Interactionist Perspectives to Suicide Social analysts applying a symbolic interactionist framework to the study of suicide would focus on a microlevel analysis of the people's face-to-face interactions and the roles they played in society. In our efforts to interact with others, we define the situation according to our own subjective reality. This applies to suicide just as it does to other types of conduct.

From this point of view, a suicide attempt may be a way of moving toward other people—in the form of a cry for help and personal acceptance—rather than a move toward death (Colt, 1991). People may attempt to communicate in such desperate ways because other forms of communication have failed. For example, a thirteen-year-old Illinois girl slashed her wrists shortly before she knew her mother was expected to come into her room; she wanted to show her parents how upset and unhappy she was about their pending divorce. As the girl later said, "I didn't really want to die. I just hoped and prayed that if Mom and Dad knew how upset and unhappy I was, Dad would move back in" (qtd. in Giffin and Felsenthal, 1983: 19).

Postmodern Perspectives

According to *postmodern perspectives,* **existing theories have been unsuccessful in explaining social life in contemporary societies that are characterized by postindustrialization, consumerism, and global communications.** Postmodern social theorists reject the theoretical perspectives we have previously discussed, as well as how those theories were created (Ritzer, 2000b).

Postmodern theories are based on the assumption that the rapid social change that occurs as societies move from modern to postmodern (or postindustrial) conditions has a harmful effect on people. One evident change is a significant decline in the influence of social institutions such as the family, religion, and education on people's lives. Those who live in postmodern societies typically pursue individual freedom and do not want the structural constraints that are imposed by social institutions. However, the collective ties that once bound people together become weakened, placing people at higher levels of risk—"probabilities of physical harm due to given technological or other processes"—than in the past (Beck, 1992: 4). As social institutions grow weaker, people at risk come to believe that social and economic chaos looms before them and that no safety net is provided by families, peers, and the larger community. According to one theorist, there is a relationship between risk and the class structure: "Wealth accumulates at the top, risks at the bottom" (Beck, 1992: 35). Social inequality and class differences increase as people in the lower economic tiers are exposed to increasing levels of personal risk that, in turn, produce depression, fear, and ambivalence.

In sum, postmodern theories tend to emphasize the fragmented nature of contemporary society brought about by constant change, a situation that leads to feelings of ambivalence among individuals because they are uncertain about which course of action they should take and because there is no one present to help them figure this out (Bauman, 1992). Postmodern theory enhances our study of sociology because it opens up broad new avenues of inquiry, even as it challenges other theoretical perspectives and asks pertinent questions about our current belief systems. However, postmodern theory has also been criticized for raising more questions than it attempts to answer and for being so vague and abstract that it is hard to understand and even more difficult to apply (Ritzer and Goodman, 2004).

Applying Postmodern Perspectives to Suicide Although most postmodern social theorists have not addressed suicide as a social issue, some sociologists believe that postmodern theory can help us understand differing rates of suicide, particularly among

---

microlevel analysis sociological theory and research that focus on small groups rather than on large-scale social structures.

symbolic interactionist perspectives the sociological approach that views society as the sum of the interactions of individuals and groups.

postmodern perspectives the sociological approach that attempts to explain social life in modern societies that are characterized by postindustrialization, consumerism, and global communications.

people in certain ethnic categories. Consider, for example, this research question: Why in recent decades has there been a significant increase in suicide rates among young African American males between the ages of fifteen and nineteen? Historically, African Americans at all income levels had lower rates of suicide than other ethnic groups in the United States—a fact that scholars attributed to higher levels of religiosity and group support that protected African Americans from suicidal behavior (Willis, Coombs, Cockerham, and Frison, 2002). However, this pattern changed over the past two decades, particularly in the category of young African American males between the ages of fifteen and nineteen, a statistical group that now has a rate of suicide comparable to that of young white males (Willis et al., 2002).

Although youths across ethnic categories share certain risk factors for suicide, such as social isolation, feelings of marginality, high rates of unemployment, and sometimes depression, young African American males appear to be at greater risk for suicide (based on postmodern theorization) because people in this statistical category are among the most likely to be at risk in other ways in a postmodern society. A major problem for some African American young men is the absence of educational and employment opportunities. This problem is combined with other issues, such as the weakening of the family structure, a breakdown in community values, the diminished importance of the African American church, and the growing influence of the media in revealing the vast inequalities that exist in the United States and in other high-income nations. These problems have been further exacerbated by a dramatic rise in the drug trade and in possession of firearms (Willis et al., 2002).

According to some sociologists, postmodern theory helps explain why there are higher rates of suicide among African American youths than in the past because the transition from a modern to a postmodern society has left them at greater risk than others due to the loss of social support systems that previously would have helped them cope with fear, depression, and suicidal tendencies. In the past, African Americans lived a community-oriented existence; today, they face an "individualistic" society that contributes to personal stress and heightened vulnerability without any place to which to turn (Willis et al., 2002).

Each of the four sociological perspectives we have examined involves different assumptions. Consequently, each leads us to ask different questions and to view the world somewhat differently. Different aspects of reality are the focus of each approach. Whereas functionalism emphasizes social cohesion and order, conflict approaches focus primarily on social tension and change. In contrast, symbolic interactionism primarily examines people's interactions and shared meanings in everyday life. Postmodernism challenges all of the other perspectives and questions current belief systems. The Concept Quick Review for this chapter summarizes all four of these perspectives. Throughout this book, we will be using these perspectives as lenses through which to view our social world.

The Sociological Research Process

Research is the process of systematically collecting information for the purpose of testing an existing theory or generating a new one. What is the relationship between sociological theory and research? The relationship between theory and research has been referred to as a continuous cycle, as shown in ▶ Figure 1.3 (Wallace, 1971).

Not all sociologists conduct research in the same manner. Some researchers primarily engage in quantitative research, whereas others engage in qualitative research. With **quantitative research, the goal is scientific objectivity, and the focus is on data that can be measured numerically.** Quantitative research typically emphasizes complex statistical techniques. Most sociological studies on suicide have used quantitative research. They have compared rates of suicide with almost every conceivable variable, including age, sex, race/ethnicity, education, and even sports participation. For example, researchers in one study examined the effects of church membership, divorce, and migration on suicide rates in the United States at various times between 1933 and 1980. The study concluded that suicide rates were higher where divorce rates were higher, migration higher, and church membership lower (see Breault, 1986). (The box on page 26,

CONCEPT QUICK REVIEW

The Major Theoretical Perspectives

Perspective	Analysis Level	View of Society
Functionalist	Macrolevel	Society is composed of interrelated parts that work together to maintain stability within society. This stability is threatened by dysfunctional acts and institutions.
Conflict	Macrolevel	Society is characterized by social inequality; social life is a struggle for scarce resources. Social arrangements benefit some groups at the expense of others.
Symbolic Interactionist	Microlevel	Society is the sum of the interactions of people and groups. Behavior is learned in interaction with other people; how people define a situation becomes the foundation for how they behave.
Postmodernist	Macrolevel/ Microlevel	Societies characterized by postindustrialization, consumerism, and global communications bring into question existing assumptions about social life and the nature of reality.

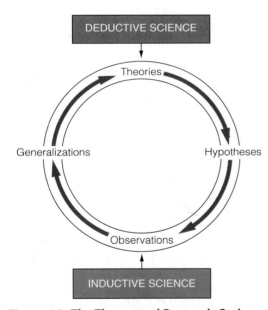

▶ Figure 1.3 **The Theory and Research Cycle**
The theory and research cycle can be compared to a relay race; although all participants do not necessarily start or stop at the same point, they share a common goal—to examine all levels of social life.

Source: Adapted from Walter Wallace, *The Logic of Science in Sociology.* New York: Aldine de Gruyter, 1971.

"Understanding Statistical Data Presentations," explains how to read numerical tables, how to interpret the data and draw conclusions, and how to calculate ratios and rates.)

With *qualitative research,* **interpretive descriptions (words) rather than statistics (numbers) are used to analyze underlying meanings and patterns of social relationships.** An example of qualitative research is a study in which the researchers systematically analyzed the contents of suicide notes to look for recurring themes (such as feelings of despair or failure) in such notes to determine if any patterns could be found that would help in understanding why people kill themselves (Leenaars, 1988).

quantitative research sociological research methods that are based on the goal of scientific objectivity and that focus on data that can be measured numerically.

qualitative research sociological research methods that use interpretive description (words) rather than statistics (numbers) to analyze underlying meanings and patterns of social relationships.

Understanding Statistical Data Presentations

Are men or women more likely to commit suicide? Are suicide rates increasing or decreasing? Such questions can be answered in numerical terms. Sociologists often use statistical tables as a concise way to present data because such tables convey a large amount of information in a relatively small space; ◆ Table 1 gives an example. To understand a table, follow these steps:

1. *Read the title.* The title, "U.S. Suicides, by Sex and Method Used, 1984 and 2004," tells us that the table shows relationships between two variables: sex and method of suicide used. It also indicates that the table contains data for two different time periods: 1984 and 2004.

2. *Check the source and other explanatory notes.* In this case, the source is *National Center for Injury Prevention and Control, 2006.* Checking the source helps determine its reliability and timeliness. The first footnote indicates that the table includes only people who reside in the United States. The next footnote reflects that, due to rounding, the percentages in a column may not total 100.0%. The final two footnotes provide more information about exactly what is included in each category.

3. *Read the headings for each column and each row.* The main column headings in Table 1 are "Method," "Males," and "Females." These latter two column headings are divided into two groups: 1984 and 2004. The columns present information (usually numbers) arranged vertically. The rows present information horizontally. Here, the row headings indicate suicide methods.

4. *Examine and compare the data.* To examine the data, determine what units of measurement have been used. In Table 1, the figures are numerical counts (for example, the total number of reported female suicides by poisoning in 2004 was 2,600) and percentages (for example, in 2004, poisoning accounted for 37.8 percent of all female suicides reported). A *percentage,* or proportion, shows how many of a given item there are in every one hundred. Percentages allow us to compare groups of different sizes.

5. *Draw conclusions.* By looking for patterns, some conclusions can be drawn from Table 1.

   a. *Determining the increase or decrease.* Between 1984 and 2004, reported male suicides by firearms increased from 14,504 to 14,523—an increase of 19—while female suicides by firearms decreased by 382. This represents a *total* increase (for males) and decrease (for females) in suicides by firearms for the two years being compared. The *amount* of increase or decrease can be stated as a percentage: Total male suicides by firearms were about 0.13 percent higher in 2004, calculated by dividing the total increase (19) by the earlier (lower) number. Total female suicides by firearms were about 14.64 percent lower in 2004, calculated by dividing the total decrease (382) by the earlier (higher) number.

   b. *Drawing appropriate conclusions.* The number of female suicides by firearms decreased about 14 percent between 1984 and 2004; the number for poisoning increased by about 8 percent. We might conclude that more women preferred poisoning over firearms as a means of killing themselves in 2004 than in 1984. Does that mean fewer women had access to guns in 2004? That poisoning oneself became more acceptable? Such generalizations do not take into account that we are looking at data from two years and that the difference in statistics for those two years may not really represent a trend.

◆ Table 1 U.S. Suicides, by Sex and Method Used, 1984 and 2004[a]

Method	Males		Females	
	1984	2004	1984	2004
Total	22,689	25,566	6,597	6,873
Firearms (% of total)[b]	14,504 (64.0)	14,523 (56.8)	2,609 (39.5)	2,227 (32.4)
Poisoning[c] (% of total)[b]	3,203 (14.1)	3,200 (12.5)	2,406 (36.5)	2,600 (37.8)
Suffocation[d] (% of total)[b]	3,478 (15.3)	5,980 (23.4)	863 (13.0)	1,356 (19.7)
Other (% of total)[b]	1,504 (6.6)	1,863 (7.3)	719 (10.9)	690 (10.0)

[a]Excludes deaths of nonresidents of the United States.
[b]Due to rounding, the percentages in a column may not add up to 100.0%.
[c]Includes solids, liquids, and gases.
[d]Includes hanging and strangulation.
Source: National Center for Injury Prevention and Control, 2006.

The "Conventional" Research Model

Research models are tailored to the specific problem being investigated and the focus of the researcher. Both quantitative research and qualitative research contribute to our knowledge of society and human social interaction, and involve a series of steps as shown in ▶ Figure 1.4. We will now trace the steps in the "conventional" research model, which focuses on quantitative research. Then we will describe an alternative model that emphasizes qualitative research.

1. *Select and define the research problem.* Sometimes, a specific experience such as knowing someone who committed suicide can trigger your interest in a topic. Other times, you might select topics to fill gaps or challenge misconceptions in existing research or to test a specific theory (Babbie, 2004). Emile Durkheim selected suicide because

he wanted to demonstrate the importance of *society* in situations that might appear to be arbitrary acts by individuals. Suicide was a suitable topic because it was widely believed that suicide was a uniquely individualistic act. However, Durkheim emphasized that *suicide rates* provide better explanations for suicide than do *individual acts* of suicide. He reasoned that if suicide were purely an individual act, then the rate of suicide (the relative number of people who kill themselves each year) should be the same for every group regardless of culture and social structure. Moreover, Durkheim wanted to know why there were different rates of suicide—whether factors such as religion, marital status, sex, and age had an effect on social cohesion.

2. *Review previous research.* Before beginning the research, it is important to analyze what others have written about the topic. You should determine

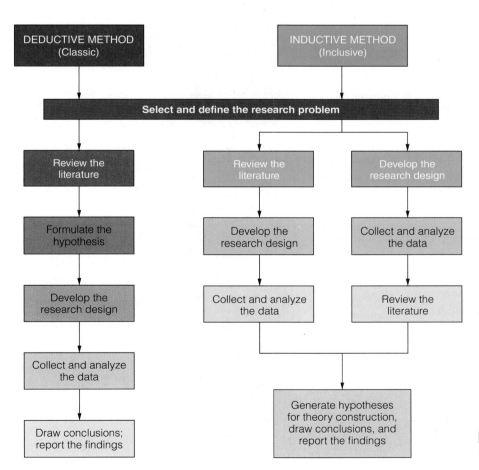

▶ **Figure 1.4 Steps in Sociological Research**

where gaps exist and note mistakes to avoid. When Durkheim began his study, very little sociological literature existed to review; however, he studied the works of several moral philosophers, including Henry Morselli (1975/1881).

3. *Formulate the hypothesis* (if applicable). You may formulate a **hypothesis—a statement of the expected relationship between two or more variables.** A *variable* **is any concept with measurable traits or characteristics that can change or vary from one person, time, situation, or society to another.** The most fundamental relationship in a hypothesis is between a dependent variable and one or more independent variables (see ▶ Figure 1.5). The ***independent variable* is presumed to be the cause of the relationship; the *dependent variable* is assumed to be caused by the independent variable(s)** (Babbie, 2004). Durkheim's hypothesis stated that the rate of suicide varies *inversely* with the degree of social integration. In other words, a low degree of social integration (the independent variable) may "cause" or "be related to" a high rate of suicide (the dependent variable).

Not all social research uses hypotheses. If you plan to conduct an explanatory study (showing

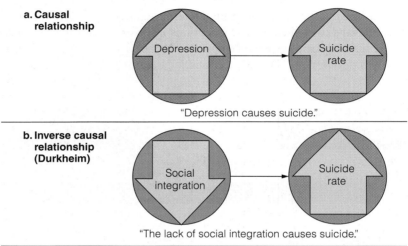

a. Causal relationship

"Depression causes suicide."

b. Inverse causal relationship (Durkheim)

"The lack of social integration causes suicide."

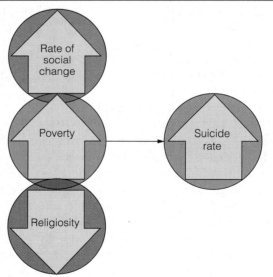

c. Multiple-cause explanation

"Many factors interact to cause suicide."

▶ **Figure 1.5 Hypothesized Relationships Between Variables**

A causal hypothesis connects one or more independent (causal) variables with a dependent (affected) variable. The diagram illustrates three hypotheses about the causes of suicide. To test these hypotheses, social scientists would need to operationalize the variables (define them in measurable terms) and then investigate whether the data support the proposed explanation.

a cause-and-effect relationship), you likely will want to formulate one or more hypotheses to test theories. If you plan to conduct a descriptive study, however, you will be less likely to do so because you may desire only to describe social reality or provide facts.

4. *Develop the research design.* You must determine the unit of analysis to be used in the study. A *unit of analysis* is *what* or *whom* is being studied (Babbie, 2004). In social science research, individuals, social groups (such as families, cities, or geographic regions), organizations (such as clubs, labor unions, or political parties), and social artifacts (such as books, paintings, or weddings) may be units of analysis. Durkheim's unit of analysis was social groups, not individuals, because he believed that the study of individual cases of suicide would not explain the rates of suicide in various European countries.

5. *Collect and analyze the data.* You must decide what population will be observed or questioned and then carefully select a sample. A *sample* is the people who are selected from the population to be studied; the sample should accurately represent the larger population. A *representative sample* is a selection from a larger population that has the essential characteristics of the total population. For example, if you interviewed five students selected haphazardly from your sociology class, they would not be representative of your school's total student body. By contrast, if you selected five students from the total student body by a random sample, they would be closer to being representative (although a random sample of five students would be too small to yield much useful data).

Validity and reliability may be problems in research. *Validity* **is the extent to which a study or research instrument accurately measures what it is supposed to measure.** A recurring issue in studies that analyze the relationship between religious beliefs and suicide is whether "church membership" is an accurate indicator of a person's religious beliefs. In fact, one person may be very religious yet not belong to a specific church, whereas another person may be a member of a church yet not hold very deep religious beliefs. *Reliability* **is the extent to which a study or research instrument yields consistent results when applied to different individuals at one time or to the same individuals over time.** Sociologists have found that different interviewers get different answers from the people being interviewed. For example, how might interviews with college students who have contemplated suicide be influenced by the interviewers themselves?

Once you have collected your data, the data must be analyzed. *Analysis* is the process through which data are organized so that comparisons can be made and conclusions drawn. Sociologists use many techniques to analyze data. After collecting data from vital statistics for approximately 26,000 suicides, Durkheim analyzed his data according to four distinctive categories of suicide. *Egoistic suicide* occurs among people who are isolated from any social group. By contrast, *altruistic suicide* occurs among individuals who are excessively integrated into society (for example, military leaders who kill themselves after defeat in battle). *Anomic suicide* results from a lack of social regulation, whereas *fatalistic suicide* results from excessive regulation and oppressive discipline (for example, slaves).

hypothesis a statement of the expected relationship between two or more variables.

variable in sociological research, any concept with measurable traits or characteristics that can change or vary from one person, time, situation, or society to another.

independent variable in an experiment, the variable assumed to be the cause of the relationship between variables.

dependent variable in an experiment, the variable assumed to be caused by the independent variable(s).

validity in sociological research, the extent to which a study or research instrument accurately measures what it is supposed to measure.

reliability in sociological research, the extent to which a study or research instrument yields consistent results when applied to different individuals at one time or to the same individuals over time.

Box 1.2 Framing Suicide in the Media

Sociology Versus Sensationalism

Front Page: *New York Post,* March 10, 2004:

The daughter of a Silicon Valley executive has become the fourth New York University student to die in a plunge this academic year. [Name of student,] 19, jumped . . . from the roof of her boyfriend's 24-story apartment building Saturday after having a fight with him. (*New York Post,* 2004: 1)

***New York Times,* March 10, 2004:**

The suicide of a New York University student who fell to her death from a rooftop off campus on Saturday chilled students, who learned about it in an e-mail message from university officials on Monday. It was the fourth N.Y.U. student death this year. One N.Y.U. student committed suicide last semester by jumping from a high floor of the Elmer Holmes Bobst Library. A second student also jumped from a high floor in the library, but the city medical examiner's office found that he had been on drugs and ruled his death an accident. They have not yet ruled on the death of a third student, a young woman who fell to her death from the window of a friend's apartment near Washington Square Park. . . . (Arenson, 2004)

Compare these two news accounts of a college student's suicide. One source is a tabloid; the other is a so-called mainstream national newspaper. Tabloids have a newspaper format but are smaller in size. They typically provide readers with a condensed version of the news and contain illustrated, often sensational material that they hope will encourage people to buy that day's paper at the newsstand or vending machine. It is no surprise that in its effort to attract readers, the tabloid placed a large, four-color photo of the student jumping to her death on the front page and suggested a cause of suicide (a fight with her boyfriend) that might entice readers. By contrast, the *New York Times* article begins with a description of how this student's death might affect other students and explains how school officials notified them of the tragedy. No picture was included with the coverage.

As these examples show, the media offer us different vantage points from which to view a given social event based on how they *frame* the information they provide to their audience. The term *media framing* refers to the process by which information and entertainment are packaged by the mass media (newspapers, magazines, radio and television networks and stations, and the Internet) before being presented to an audience. This process includes factors such as the amount of exposure given to a story, where it is placed, the positive or negative tone it conveys, and its accompanying headlines, photographs, or other visual or auditory effects (if any). Through framing, the media emphasize some beliefs and values over others

6. *Draw conclusions and report the findings.* After analyzing the data, your first step in drawing conclusions is to return to your hypothesis or research objective to clarify how the data relate both to the hypothesis and to the larger issues being addressed. At this stage, you note the limitations of the study, such as problems with the sample, the influence of variables over which you had no control, or variables that your study was unable to measure.

Reporting the findings is the final stage. The report generally includes a review of each step taken in the research process in order to make the study available for *replication*—the repetition of the investigation in substantially the same way that it was originally conducted. Social scientists generally present their findings in papers at professional meetings and publish them in technical journals and books. In reporting his findings in *Suicide* (1964b/1897), Durkheim concluded that the suicide rate of a group is a social fact that cannot be explained in terms of the personality traits of individuals (see Box 1.2 for a current example).

We have traced the steps in the "conventional" research process (based on deduction and quantita-

and manipulate salience by directing people's attention to some ideas while ignoring others. As such, a frame constitutes a story line or an unfolding narrative about an issue. These narratives are organizations of experience that bring order to events. Consequently, such narratives wield power because they influence how we make sense of the world (Kendall, 2005).

Thinking sociologically, what problems exist in how the media frame stories about a social issue such as suicide? As discussed in this chapter, the early sociologist Emile Durkheim believed that we should view even seemingly individual actions—such as suicide—from a *sociological perspective* that focuses on the part that social groups and societies play in patterns of behavior rather than focusing on the *individual* attributes of people who commit such acts. If we use a sociological approach to analyzing how the media frame stories about suicide, here are three issues to consider:

- Do television and newspaper reports of suicides simplify the *reasons* for the suicide? Many factors interact in a complex manner to contribute to a person's decision to commit suicide, but one or more factors do not necessarily *cause* a suicide to occur. The media often report a final precipitating situation (such as a fight with a boyfriend or girlfriend, losing one's job, or getting a divorce) that distressed the individual prior to suicide but do not inform audiences that this was not the *only* cause of this suicide.
- Are readers and viewers provided with *repetitive, ongoing, and excessive reporting* on a suicide? Repeated sensationalistic framing of stories about suicide may

contribute to *suicide contagion*. Some people are more at risk because of age, stress, and/or other personal problems. For example, *suicide clusters* (suicides occurring close together in time and location) are most common among people fifteen to twenty-four years of age who may not have known the original suicide victim and only learned of the details from the media.

- Do the media use dramatic photographs (such as of the person committing suicide or friends and family weeping and showing great emotion at the victim's funeral) primarily to sell papers or increase viewer ratings? Photographs and other sensational material are potentially most damaging for people who have long-standing mental health problems or who have limited social networks to provide them with hope, encouragement, and guidance (American Association of Suicidology, 2005).

Reflect & Analyze

In our mass-mediated culture, many sociologists agree that the media do much more than simply *mirror* society: The media help shape our society and our cultural perceptions on many issues, including seemingly individual behavior such as suicide. Are we influenced by how the media frame news and entertainment stories even when we think we are unaffected by such coverage? We will explore this issue in several chapters.

tive research). But what steps might be taken in an alternative approach based on induction and qualitative research?

A Qualitative Research Model

Although the same underlying logic is involved in both quantitative and qualitative sociological research, the *styles* of these two models are very different (King, Keohane, and Verba, 1994). As previously stated, qualitative research is more likely to be used when the research question does not easily lend itself to numbers and statistical methods. As

compared to a quantitative model, a qualitative approach often involves a different type of research question and a smaller number of cases.

How might qualitative research be used to study suicidal behavior? In studying different rates of suicide among women and men, for example, the social psychologist Silvia Canetto (1992) questioned whether existing theories and quantitative research provided an adequate explanation for gender differences in suicidal behavior and decided that she would explore alternate explanations. Analyzing previous research, Canetto learned that most studies linked suicidal behavior in women to problems in their personal relationships, particularly with men. By contrast, most studies of men's suicides focused on their performance and found that men are more likely to be suicidal when their self-esteem and independence are threatened. According to Canetto's analysis, gender differences in suicidal behavior are more closely associated with beliefs about and cultural expectations for men and women rather than purely interpersonal crises.

As in Canetto's study, researchers using a qualitative approach may engage in *problem formulation* to clarify the research question and to formulate questions of concern and interest to people participating in the research (Reinharz, 1992). To create a research design for Canetto's study, we might start with the proposition that studies have attributed women's and men's suicidal behavior to the wrong causes. Next, we might decide to interview individuals who have attempted suicide. Our research design might develop a collaborative approach in which the participants are brought into the research design process, not just treated as passive objects to be studied (Reinharz, 1992).

Although Canetto did not gather data in her study, she reevaluated existing research, concluding that alternate explanations of women's and men's suicidal behavior are justified from existing data.

In a qualitative approach, the next step is collecting and analyzing data to assess the validity of the starting proposition. Data gathering is the foundation of the research. Researchers pursuing a qualitative approach tend to gather data in natural settings, such as where the person lives or works, rather than in a laboratory or other research setting. Data collection and analysis frequently occur concurrently, and the analysis draws heavily on the language of the persons studied, not the researcher.

Research Methods

How do sociologists know which research method to use? Are some approaches best for a particular problem? **Research methods are specific strategies or techniques for systematically conducting research.** We will look at four of these methods:

survey research, analysis of existing statistical data, field research, and experiments.

Survey Research

A *survey* **is a poll in which the researcher gathers facts or attempts to determine the relationships among facts.** Surveys are the most widely used research method in the social sciences because they make it possible to study things that are not directly observable—such as people's attitudes and beliefs—and to describe a population too large to observe directly (Babbie, 2004). Researchers frequently select a representative sample (a small group of respondents) from a larger population (the total group of people) to answer questions about their attitudes, opinions,

© Michael Newman/PhotoEdit

Conducting surveys and polls is an important means of gathering data from respondents. Some surveys take place on street corners; increasingly, however, such surveys are done by telephone, Internet, or other means.

CENSUS PROFILES

How People in the United States Self-Identify as to Race

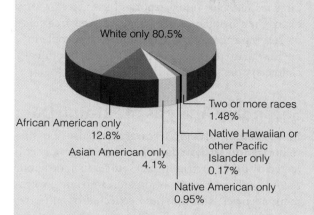

White only 80.5%

Two or more races 1.48%

African American only 12.8%

Native Hawaiian or other Pacific Islander only 0.17%

Asian American only 4.1%

Native American only 0.95%

How can this be? Simply stated, some individuals are counted at least twice, based on the number of racial categories they listed.

Race	Percentage of Total Population
White alone or in combination with one or more other races	81.8
African American alone or in combination with one or more other races	13.3
Asian American alone or in combination with one or more other races	4.6
Native American alone or in combination with one or more other races	1.5
Native Hawaiian or other Pacific Islander alone or in combination with one or more other races	0.3
Total	101.5

Beginning with Census 2000, the U.S. Census Bureau has made it possible for people responding to census questions regarding their race to mark more than one racial category. Although the vast majority of respondents select only one category (see above), the Census Bureau reports that in 2003 approximately 4.3 million people (1.48 percent of the population) in the United States self-identified as being of more than one race. As a result, if you look at the figures as set forth, they total more than 100 percent of the total population.

Source: U.S. Census Bureau, 2007.

or behavior. *Respondents* are people who provide data for analysis through interviews or questionnaires. The Gallup and Harris polls are among the most widely known large-scale surveys; however, government agencies such as the U.S. Census Bureau conduct a variety of surveys as well.

Unlike many polls that use various methods of gaining a representative sample of the larger population, the Census Bureau attempts to gain information from all persons in the United States. The decennial census occurs every 10 years, in the years ending in "0." The purpose of this census is to count the population and housing units of the entire United States. The population count determines how seats in the U.S. House of Representatives are apportioned; however, census figures are also used in formulating public policy and in planning and de-

cision making in the private sector. The Census Bureau attempts to survey the *entire* U.S. population by using two forms—a "short form" of questions asked of *all* respondents, and a "long form" that contains additional questions asked of a *representative sample* of about one in six respondents. Statistics from the Census Bureau provide information that sociologists use in their research. An example is shown in the Census Profiles feature: "How People in the United States Self-Identify as to Race." Note that

research methods specific strategies or techniques for systematically conducting research.

survey a poll in which the researcher gathers facts or attempts to determine the relationships among facts.

because of recent changes in the methods used to collect data by the Census Bureau, information on race from the 2000 census is not directly comparable with data from earlier censuses.

Survey data are collected by using questionnaires and interviews. A *questionnaire* is a printed research instrument containing a series of items to which subjects respond. Items are often in the form of statements with which the respondent is asked to "agree" or "disagree." Questionnaires may be administered by interviewers in face-to-face encounters or by telephone, but the most commonly used technique is the *self-administered questionnaire,* which is either mailed to the respondent's home or administered to groups of respondents gathered at the same place at the same time. For example, the sociologist Kevin E. Early (1992) used survey data collected through questionnaires to test his hypothesis that suicide rates are lower among African Americans than among white Americans due to the influence of the black church. Data from questionnaires filled out by members of six African American churches in Florida supported Early's hypothesis that the church buffers some African Americans against harsh social forces—such as racism—that might otherwise lead to suicide.

Survey data may also be collected by interviews. An **interview** is a data-collection encounter in which an interviewer asks the respondent questions and records the answers. Survey research often uses *structured interviews,* in which the interviewer asks questions from a standardized questionnaire. Structured interviews tend to produce uniform or replicable data that can be elicited time after time by different interviews. For example, in addition to surveying congregation members, Early (1992) conducted interviews with pastors of African American churches to determine the pastors' opinions about the extent to which the African American church reinforces values and beliefs that discourage suicide.

Survey research is useful in describing the characteristics of a large population without having to interview each person in that population. In recent years, computer technology has enhanced researchers' ability to do *multivariate analysis*—research involving more than two independent variables. For example, to assess the influence of religion on suicidal behavior among African Americans, a researcher might look at the effects of age, sex, income level, and other variables all at once to determine which of these independent variables influences suicide the most or least and how influential each variable is relative to the others. However, a weakness of survey research is the use of standardized questions; this approach tends to force respondents into categories in which they may or may not belong. Moreover, survey research relies on self-reported information, and some people may be less than truthful, particularly on emotionally charged issues such as suicide.

Secondary Analysis of Existing Data

In *secondary analysis,* **researchers use existing material and analyze data that were originally collected by others.** Existing data sources include public records, official reports of organizations or government agencies, and *raw data* collected by other researchers. For example, Durkheim used vital statistics (death records) that were originally collected for other purposes to examine the relationship among variables such as age, marital status, and the circumstances surrounding the person's suicide.

Secondary analysis also includes **content analysis**—**the systematic examination of cultural artifacts or various forms of communication to extract thematic data and draw conclusions about social life.** Among the materials studied are written records (such as books, diaries, poems, and graffiti), narratives and visual texts (such as movies, television shows, advertisements, and greeting cards), and material culture (such as music, art, and even garbage). In content analysis, researchers look for regular patterns, such as the frequency of suicide as a topic on television talk shows.

One strength of secondary analysis is that data are readily available and inexpensive. Another is that because the researcher often does not collect the data personally, the chances of bias may be reduced. In addition, the use of existing sources makes it possible to analyze longitudinal data (things that take place over a period of time or at several different points in time) to provide a historical context within which to locate original research. However, secondary analysis has inherent problems. For one thing, the researcher does not always know if the data are incomplete, unauthentic, or inaccurate.

© Addison Geary

Cover used by permission of W. W. Norton & Company, Inc.
Cover photo © Camilo Jose Vergara.

Elijah Anderson (at left in photo) conducted an ethnographic study of two very different Philadelphia neighborhoods that became the basis for his landmark study, *Code of the Street*. What can researchers learn from ethnographic research that might be less apparent if they used other methods to study human behavior?

Field Research

Field research is the study of social life in its natural setting: observing and interviewing people where they live, work, and play. Some kinds of behavior can be best studied by "being there"; a fuller understanding can be developed through observations, face-to-face discussions, and participation in events. Researchers use these methods to generate *qualitative* data: observations that are best described verbally rather than numerically.

Sociologists who are interested in observing social interaction as it occurs may use *participant observation—the process of collecting systematic observations while being part of the activities of the group that the researcher is studying.* Participant observation generates more "inside" information than simply asking questions or observing from the outside. For example, to learn more about how coroners make a ruling of "suicide" in connection with a death and to analyze what (if any) effect such a ruling has on the accuracy of "official" suicide statistics, the sociologist Steve Taylor (1982) engaged in participant observation at a coroner's office over a six-month period. As he followed a number of cases from the initial report of death through the various stages of investigation, Taylor learned that it was important to "be around" so that he could listen to

discussions and ask the coroners questions because intuition and guesswork play a large part in some decisions to rule a death as a suicide.

Another approach to field research is the *ethnography—a detailed study of the life and activities of a group of people by researchers who may live with that group over a period of years* (Feagin, Orum,

---

interview a research method using a data-collection encounter in which an interviewer asks the respondent questions and records the answers.

secondary analysis a research method in which researchers use existing material and analyze data that were originally collected by others.

content analysis the systematic examination of cultural artifacts or various forms of communication to extract thematic data and draw conclusions about social life.

participant observation a research method in which researchers collect data while being part of the activities of the group being studied.

ethnography a detailed study of the life and activities of a group of people by researchers who may live with that group over a period of years.

and Sjoberg, 1991). Unlike participant observation, ethnographic studies usually take place over a longer period of time. For example, the sociologist Elijah Anderson (1990) conducted a study in two areas of a major city—one African American and low-income, the other racially mixed but becoming increasingly middle- to upper-income and white. As Anderson spent numerous hours on the streets, talking and listening to the people, he was able to document changes in residents' everyday lives brought about by increased drug abuse, loss of jobs, decreases in city services despite increases in taxes, and the eventual exodus of middle-income people from the central city.

Do extremely violent video games cause an increase in violent tendencies in their users? Experiments are one way to test this hypothesis.

Experiments

An *experiment* **is a carefully designed situation in which the researcher studies the impact of certain variables on subjects' attitudes or behavior.** Experiments are designed to create "real-life" situations, ideally under controlled circumstances, in which the influence of different variables can be modified and measured. Conventional experiments require that subjects be divided into two groups: an experimental group and a control group. The *experimental group* **contains the subjects who are exposed to an independent variable (the experimental condition) to study its effect on them.** The *control group* **contains the subjects who are not exposed to the independent variable.** For example, the sociologist Arturo Biblarz and colleagues (1991) examined the effects of media violence and depictions of suicide on attitudes toward suicide by showing one group of subjects (an experimental group) a film about suicide, while a second (another experimental group) saw a film about violence, and a third (the control group) saw a film containing neither suicide nor violence. The research found some evidence that people exposed to suicidal acts or violence in the media may be more likely to demonstrate an emotional state favorable to suicidal behavior, particularly if they are already "at risk" for suicide.

Researchers may use experiments when they want to demonstrate that a cause-and-effect relationship exists between variables. In order to show that a change in one variable causes a change in another, three conditions must be satisfied: (1) a correlation between the two variables must be shown to exist (*correlation* **exists when two variables are associated more frequently than could be expected by chance**), (2) the independent variable must have occurred prior to the dependent variable, and (3) any change in the dependent variable must not have been due to an extraneous variable—one outside the stated hypothesis.

The major advantage of an experiment is the researcher's control over the environment and the ability to isolate the experimental variable. Because many experiments require relatively little time and money and can be conducted with limited numbers of subjects, it is possible for researchers to replicate an experiment several times by using different groups of subjects (Babbie, 2004). Perhaps the greatest limitation of experiments is that they are artificial: Social processes that are set up by researchers or that take place in a laboratory setting are often not the same as real-life occurrences.

Ethical Issues in Sociological Research

The study of people ("human subjects") raises vital questions about ethical concerns in sociological research. Researchers are required to obtain written "informed consent" statements from the persons they study—but what constitutes "informed consent"? And how do researchers protect the identity and confidentiality of their sources?

The American Sociological Association (ASA) *Code of Ethics* (1997) sets forth certain basic standards that sociologists must follow in conducting research. Among these standards are the following:

1. Researchers must endeavor to maintain objectivity and integrity in their research by disclosing their research findings in full and including all possible interpretations of the data (even when these interpretations do not support their own viewpoints).
2. Researchers must safeguard the participants' right to privacy and dignity while protecting them from harm.
3. Researchers must protect confidential information provided by participants, even when this information is not considered to be "privileged" (legally protected, as is the case between doctor and patient and between attorney and client) and legal pressure is applied to reveal this information.
4. Researchers must acknowledge research collaboration and assistance they receive from others and disclose all sources of financial support.

Sociologists are obligated to adhere to this code and to protect research participants; however, many ethical issues arise in conducting research. For example, the sociologist William Zellner (1978) wanted to look at fatal single-occupant automobile accidents to determine if some drivers were actually committing suicide. To examine this issue further, he sought to interview the family, friends, and acquaintances of persons killed in single-car crashes to determine if the deaths were possibly intentional. To recruit respondents, Zellner told them that he hoped the research would reduce the number of automobile accidents in the future. He did not mention that he suspected "autocide" might have occurred in the case of their friend or loved one. From his data, Zellner concluded that at least 12 percent of the fatal single-occupant crashes were suicides—and that these crashes sometimes also killed or critically injured other people as well. However, Zellner's research raised important research questions: Was his research unethical? Did he misrepresent the reasons for his study?

In this chapter, we have looked at how theory and research work together to provide us with insights on human behavior. Theory and research are the "lifeblood" of sociology. Theory provides the framework for analysis; research provides opportunities for us to use our sociological imagination to generate new knowledge. Our challenge today is to find new ways of integrating knowledge and action and to include all people in the theory and research process in order to help fill the gaps in our existing knowledge about social life and how it is shaped by gender, race, class, age, and the broader social and cultural context in which everyday life occurs (Cancian, 1992). Each of us can and should find new ways to integrate knowledge and action into our daily lives (see Box 1.3 on the next page).

experiment a research method involving a carefully designed situation in which the researcher studies the impact of certain variables on subjects' attitudes or behavior.

experimental group in an experiment, the group that contains the subjects who are exposed to an independent variable (the experimental condition) to study its effect on them.

control group in an experiment, the group that contains the subjects who are not exposed to the independent variable.

correlation a relationship that exists when two variables are associated more frequently than could be expected by chance.

Box 1.3 You Can Make a Difference

Responding to a Cry for Help

Chad felt that he knew Frank quite well. After all, they had been roommates for two years at State U. As a result, Chad was taken aback when Frank became very withdrawn, sleeping most of the day and mumbling about how unhappy he was. One evening, Chad began to wonder whether he needed to do something because Frank had begun to talk about "ending it all" and saying things like "The world will be better off without me." If you were in Chad's place, would you know the warning signs that you should look for? Do you know what you might do to help someone like Frank?

The American Foundation for Suicide Prevention, a national nonprofit organization dedicated to funding research, education, and treatment programs for depression and suicide prevention, suggests that each of us should be aware of these warning signs of suicide:

- *Talking about death or suicide.* Be alert to such statements as "Everyone would be better off without me." Sometimes, individuals who are thinking about suicide speak as if they are saying good-bye.
- *Making plans.* The person may do such things as giving away valuable items, paying off debts, and otherwise "putting things in order."
- *Showing signs of depression.* Although most depressed people are not suicidal, most suicidal people are depressed. Serious depression tends to be expressed as a loss of pleasure or withdrawal from activities that a person has previously enjoyed. It is especially important to note whether five of the following symptoms are present almost every day for several weeks: change in appetite or weight, change in sleeping patterns, speaking or moving with unusual speed or slowness, loss of interest in usual activities, decrease in sexual

drive, fatigue, feelings of guilt or worthlessness, and indecisiveness or inability to concentrate.

The possibility of suicide must be taken seriously: Most people who commit suicide give some warning to family members or friends. Instead of attempting to argue the person out of suicide or saying "You have so much to live for," let the person know that you care and understand, and that his or her problems can be solved. Urge the person to see a school counselor, a physician, or a mental health professional immediately. If you think the person is in imminent danger of committing suicide, you should take the person to an emergency room or a walk-in clinic at a psychiatric hospital. It is best to remain with the person until help is available.

For more information about suicide prevention, contact the following organizations:

- American Foundation for Suicide Prevention (**http://www.afsp.org**), 120 Wall Street, 22nd Floor, New York, NY 10005. (888)333-2377. AFSP is a leading not-for-profit organization dedicated to understanding and preventing suicide through research and education.

- Suicide Awareness Voices of Education (**http://www.save.org**) is a resource index with links to other valuable resources, such as "Questions Most Frequently Asked on Suicide," "Symptoms of Depression and Danger Signs of Suicide," and "What to Do If Someone You Love Is Suicidal."

- Befrienders Worldwide (**http://www.befrienders.org**) is a website providing information for anyone feeling depressed or suicidal or who is worried about a friend or relative who feels that way. It includes a directory of suicide and crisis helplines.

Chapter Review

● **What is sociology, and how can it help us understand ourselves and others?**

Sociology is the systematic study of human society and social interaction. We study sociology to understand how human behavior is shaped by group life and, in turn, how group life is affected by individuals. Our culture tends to emphasize individualism, and sociology pushes us to consider more complex connections between our personal lives and the larger world.

● **What is the sociological imagination?**

According to C. Wright Mills, the sociological imagination helps us understand how seemingly personal troubles, such as suicide, are actually related to larger social forces. It is the ability to see the relationship between individual experiences and the larger society.

● **What are the major contributions of early sociologists such as Durkheim, Marx, and Weber?**

Durkheim argued that societies are built on social facts, that rapid social change produces strains in society, and that the loss of shared values and purpose can lead to a condition of anomie. Marx stressed that within society there is a continuous clash between the owners of the means of production and the workers, who have no choice but to sell their labor to others. According to Weber, sociology should be value free and people should become more aware of the role that bureaucracies play in daily life.

● **How did Simmel's perspective differ from that of other early sociologists?**

Whereas other sociologists primarily focused on society as a whole, Simmel explored small social groups and argued that society is best seen as a web of patterned interactions among people.

● **What are the major contemporary sociological perspectives?**

Functionalist perspectives assume that society is a stable, orderly system characterized by societal consensus. Conflict perspectives argue that society is a continuous power struggle among competing groups, often based on class, race, ethnicity, or gender. Symbolic interactionist perspectives focus on how people make sense of their everyday social interactions. Postmodern theorists believe that entirely new ways of examining social life are needed and that it is time to move beyond functionalist, conflict, and symbolic interactionist approaches.

● **How does quantitative research differ from qualitative research?**

Quantitative research focuses on data that can be measured numerically (comparing rates of suicide, for example). Qualitative research focuses on interpretive description (words) rather than statistics to analyze underlying meanings and patterns of social relationships.

● **What are the key steps in the conventional research process?**

A conventional research process based on deduction and the quantitative approach has these key steps: (1) selecting and defining the research problem; (2) reviewing previous research; (3) formulating the hypothesis, which involves constructing variables; (4) developing the research design; (5) collecting and analyzing the data; and (6) drawing conclusions and reporting the findings.

● **What steps are often taken by researchers using the qualitative approach?**

A researcher taking the qualitative approach might (1) formulate the problem to be studied instead of creating a hypothesis, (2) collect and analyze the data, and (3) report the results.

● **What are the major types of research methods?**

The main types of research methods are surveys, secondary analysis, field research, and experiments. Surveys are polls used to gather facts about people's attitudes, opinions, or behaviors; a representative sample of respondents provides data through questionnaires or interviews. In secondary analysis, researchers analyze existing data, such as a government census, or cultural artifacts, such as a diary. In field research, sociologists study social life in its natural setting through participant observation, interviews, and ethnography. Through experiments, researchers study the impact of certain variables on their subjects.

www.cengage.com/login

Register for a Student eResource account to maximize your study time online using CengageNOW. First take the system's diagnostic pre-test, and then follow the personalized study plan that is created for you to help you review this chapter. The study plan will

• help you identify areas on which you should concentrate;

• provide interactive exercises to help you master the chapter concepts; and

• provide a post-test to confirm you are ready to move on to the next chapter.

Key Terms

anomie 13
conflict perspectives 20
content analysis 34
control group 36
correlation 36
dependent variable 28
ethnography 35
experiment 36
experimental group 36
functionalist perspectives 18
high-income countries 8
hypothesis 28
independent variable 28
industrialization 9

interview 34
latent functions 19
low-income countries 8
macrolevel analysis 21
manifest functions 19
microlevel analysis 22
middle-income countries 8
participant observation 35
positivism 11
postmodern perspectives 23
qualitative research 25
quantitative research 24
reliability 29
research methods 32

secondary analysis 34
social Darwinism 12
social facts 12
society 4
sociological imagination 7
sociology 4
survey 32
symbolic interactionist perspectives 22
theory 17
urbanization 9
validity 29
variable 28

Questions for Critical Thinking

1. What does C. Wright Mills mean when he says that the sociological imagination helps us "to grasp history and biography and the relations between the two within society"? (Mills, 1959b: 6).

2. As a sociologist, how would you remain objective yet still see the world as others see it? Would you make subjective decisions when trying to understand the perspectives of others?

3. Early social thinkers were concerned about stability in times of rapid change. In our more global world, is stability still a primary goal? Or is constant conflict important for the well-being of all humans? Use the conflict and functionalist perspectives to bolster your analysis.

The Kendall Companion Website

www.cengage.com/sociology/kendall

Supplement your review of this chapter by going to the text's companion website, where you can take tutorial quizzes, use flash cards to master key terms, follow live links to useful websites, and explore the other study and research resources you'll find there, such as a comprehensive interactive sociology timeline, GSS Data, and Census 2000 information, much of it presented visually in maps.

At home, I kept opening the refrigerator and cupboards, wishing for American foods to magically appear. I wanted what the other kids had: Bundt cakes and casseroles, Cheetos and Doritos. . . . The more American foods I ate, the more my desires multiplied, outpacing my interest in Vietnamese food. I had memorized the menu at Dairy Cone, the sugary options in the cereal aisle at Meijer's [grocery], and every inch of the candy display at Gas City: the rows of gum, the rows of chocolate, the rows without chocolate. . . . I knew Reese's peanut butter cups, Twix, Heath Crunch, Nestlé Crunch, Baby Ruth, Bar None, Oh Henry!, Mounds and Almond Joy, Snickers, Mr. Goodbar[,] . . . Milk Duds, [and] Junior Mints. I dreamed of taking it all, plus the freezer full of popsicles and nutty, chocolate-coated ice cream drumsticks. I dreamed of Little Debbie, Dolly Madison,

© Bob Daemmrich/The Image Works

How is the food that we consume linked to our identity and the larger culture of which we are a part? Do people who identify with more than one culture face more complex issues when it comes to food preferences?

Swiss Miss, all the bakeries presided over by prim and proper girls.

—Bich Minh Nguyen (2007: 50–51), an English professor at Purdue University, describing how food served as a powerful cultural symbol in her childhood as a Vietnamese American

Growing up in Oakland . . . I came to dislike Chinese food. That may have been, in part, because I was Chinese and desperately wanted to be American. I *was* American, of course, but being born and raised in Chinatown—in a restaurant my parents operated, in fact—I didn't feel much like the people I saw outside Chinatown, or in books and movies.

It didn't help that for lunch at school, my mother would pack— *Ai ya!*—Chinese food. Barbecued pork sandwiches, not ham and cheese; Chinese pears, not apples. At home—that is, at the New Eastern Café—it was Chinese food night after night. No wonder I would sneak off, on the way to Chinese school, to Hamburger Gus for a helping of thick-cut French fries.

—author Ben Fong-Torres (2007: 11) describing his experiences as a Chinese American who desired to "Americanize" his eating habits

Why are these authors concerned about the food they ate as children? For all of us, the food we consume is linked to our identity and to the larger culture of which we are a part. For people who identify with more than one culture, food and eating patterns may become a very complex issue. To some people, food consumption is nothing more than how we meet a basic biological need; however, many sociologists are interested in food and eating because of their cultural significance in our lives (see Mennell, 1996; Mennell, Murcott, and van Otterloo, 1993).

Chapter Focus Question

What part does culture play in shaping people and the social relations in which they participate?

What is culture? ***Culture* is the knowledge, language, values, customs, and material objects that are passed from person to person and from one generation to the next in a human group or society.** As previously defined, a *society* is a large social grouping that occupies the same geographic territory and is subject to the same political authority and dominant cultural expectations. Whereas a society is composed of people, a culture is composed of ideas, behavior, and material possessions. Society and culture are interdependent; neither could exist without the other.

In this chapter, we examine society and culture, with special attention to how our material culture, including the food we eat, is related to our beliefs, values, and actions. We also analyze culture from functionalist, conflict, symbolic interactionist, and postmodern perspectives. Before reading on, test your knowledge of food and culture by answering the questions in Box 2.1.

Culture and Society in a Changing World

How important is culture in determining how people think and act on a daily basis? Simply stated, culture is essential for our individual survival and our communication with other people. We rely on culture because we are not born with the information we need to survive. We do not know how to take care of ourselves, how to behave, how to dress, what to eat, which gods to worship, or how to make or spend money. We must learn about culture through interaction, observation, and imitation in order to participate as members of the group. Sharing a common culture with others simplifies day-to-day interactions. However, we must also understand other cultures and the world views therein.

Just as culture is essential for individuals, it is also fundamental for the survival of societies. Culture has been described as "the common denominator that makes the actions of individuals intelligible to the group" (Haviland, 1993: 30). Some system of rule making and enforcing necessarily exists in all societies. What would happen, for example, if *all* rules and laws in the United States suddenly disap-

peared? At a basic level, we need rules in order to navigate our bicycles and cars through traffic. At a more abstract level, we need laws to establish and protect our rights.

In order to survive, societies need rules about civility and tolerance. We are not born knowing how to express certain kinds of feelings toward others. When a person shows kindness or hatred toward another individual, some people may say "Well, that's just human nature" when explaining this behavior. Such a statement is built on the assumption that what we do as human beings is determined by *nature* (our biological and genetic makeup) rather than *nurture* (our social environment)—in other words, that our behavior is instinctive. An *instinct* is an unlearned, biologically determined behavior pattern common to all members of a species that predictably occurs whenever certain environmental conditions exist. For example, spiders do not learn to build webs. They build webs because of instincts that are triggered by basic biological needs such as protection and reproduction.

Humans do not have instincts. What we most often think of as instinctive behavior can actually be attributed to reflexes and drives. A *reflex* is an unlearned, biologically determined involuntary response to some physical stimuli (such as a sneeze after breathing some pepper in through the nose or the blinking of an eye when a speck of dust gets in it). *Drives* are unlearned, biologically determined impulses common to all members of a species that satisfy needs such as those for sleep, food, water, or sexual gratification. Reflexes and drives do not determine how people will behave in human societies; even the expression of these biological characteristics is channeled by culture. For example, we may be taught that the "appropriate" way to sneeze (an involuntary response) is to use a tissue or turn our head away from others (a learned response). Similarly, we may learn to sleep on mats or in beds. Most contemporary sociologists agree that culture and social learning, not nature, account for virtually all of our behavior patterns.

Because humans cannot rely on instincts in order to survive, culture is a "tool kit" for survival. According to the sociologist Ann Swidler (1986: 273), culture is a "tool kit of symbols, stories, rituals, and world views, which people may use in varying con-

How Much Do You Know About Global Food and Culture?

True	False	
T	F	1. Cheese is a universal food enjoyed by people of all nations and cultures.
T	F	2. Giving round-shaped foods to the parents of new babies is considered to be lucky in some cultures.
T	F	3. Wedding cakes are made of similar ingredients in all countries, regardless of culture or religion.
T	F	4. Food is an important part of religious observance for many different faiths.
T	F	5. In authentic Chinese cuisine, cooking methods are divided into "yin" and "yang" qualities.
T	F	6. Because of the fast pace of life in the United States, virtually everyone relies on mixes and instant foods at home and fast foods when eating out.
T	F	7. Potatoes are the most popular mainstay in the diet of first- and second-generation immigrants who have arrived in the United States over the past forty years.
T	F	8. According to sociologists, individuals may be offended when a person from another culture does not understand local food preferences or the cultural traditions associated with eating even if the person is obviously an "outsider" or a "tourist."

Answers on page 46.

figurations to solve different kinds of problems." The tools we choose will vary according to our own personality and the situations we face. We are not puppets on a string; we make choices from among the items in our own "tool box."

Material Culture and Nonmaterial Culture

Our cultural tool box is divided into two major parts: material culture and nonmaterial culture (Ogburn, 1966/1922). *Material culture* **consists of the physical or tangible creations that members of a society make, use, and share.** Initially, items of material culture begin as raw materials or resources such as ore, trees, and oil. Through technology, these raw materials are transformed into usable items (ranging from books and computers to guns and tanks). Sociologists define *technology* as the knowledge, techniques, and tools that make it possible for people to transform resources into usable forms, and the knowledge and skills required to use them after they are developed. From this standpoint, technology is

both concrete and abstract. For example, technology includes a pair of scissors and the knowledge and skill necessary to make them from iron, carbon, and chromium (Westrum, 1991). At the most basic level, material culture is important because it is our buffer against the environment. For example, we create shelter to protect ourselves from the weather and to give ourselves privacy. Beyond the survival level, we make, use, and share objects that are interesting and important to us. Why are you wearing the particular clothes you have on today? Perhaps

culture the knowledge, language, values, customs, and material objects that are passed from person to person and from one generation to the next in a human group or society.

material culture a component of culture that consists of the physical or tangible creations (such as clothing, shelter, and art) that members of a society make, use, and share.

Box 2.1 Sociology and Everyday Life

Answers to the Sociology Quiz on Global Food and Culture

1. False. Although cheese is a popular food in many cultures, most of the people living in China find cheese very distasteful and prefer delicacies such as duck's feet.

2. True. Round foods such as pears, grapes, and moon cakes are given to celebrate the birth of babies because the shape of the food is believed to symbolize family unity.

3. False. Although wedding cakes are a tradition in virtually all nations and cultures, the ingredients of the cake—as well as other foods served at the celebration—vary widely at this important family celebration. The traditional wedding cake in Italy is made from biscuits, for example, whereas in Norway the wedding cake is made from bread topped with cream, cheese, and syrup.

4. True. Many faiths, including Christianity, Judaism, Islam, Hinduism, and Buddhism, have dietary rules and rituals that involve food; however, these practices and beliefs vary widely among individuals and communities. For some people, food forms an integral part of religion in their life; for others, food is less relevant.

5. True. Just as foods are divided into yin foods (e.g., bean sprouts, cabbage, and carrots) and yang foods (beef, chicken, eggs, and mushrooms), cooking methods are also referred to as having yin qualities (e.g., boiling, poaching, and steaming) or yang qualities (deep-frying, roasting, and stir-frying). For many Chinese Americans, yin and yang are complementary pairs that should be incorporated into all aspects of social life, including the ingredients and preparation of foods.

6. False. Although more people now rely on fast foods, there is a "slow-food" movement afoot to encourage people to prepare their food from scratch for a healthier lifestyle. Also, some cultural and religious communities—such as the Amish of Ohio, Pennsylvania, and Indiana—encourage families to prepare their food from scratch and to preserve their own fruits, vegetables, and meats. Rural families are more likely to grow their own food or prepare it from scratch than are families residing in urban areas.

7. False. Rice is a popular mainstay in the diets of people from diverse cultural backgrounds who have arrived in the United States over the past four decades. Groups ranging from the Hmong and Vietnamese to Puerto Ricans and Mexican Americans use rice as a central ingredient in their diets. Among some in the younger generations, however, food choices have become increasingly Americanized, and items such as french fries and pizza have become very popular.

8. True. Cultural diversity is a major issue in eating, and people in some cultures, religions, and nations expect that even an "outsider" will have a basic familiarity with, and respect for, their traditions and practices. However, social analysts also suggest that we should not generalize or imply that certain characteristics apply to *all* people in a cultural group or nation.

Sources: Based on Better Health Channel, 2007; Ohio State University, 2007; and PBS, 2005a.

you're communicating something about yourself, such as where you attend school, what kind of music you like, or where you went on vacation.

Nonmaterial culture **consists of the abstract or intangible human creations of society that influence people's behavior.** Language, beliefs, values,

rules of behavior, family patterns, and political systems are examples of nonmaterial culture. Even the gestures that we use in daily conversation are part of the nonmaterial culture in a society. As many international travelers and businesspeople have learned, it is important to know what gestures mean in vari-

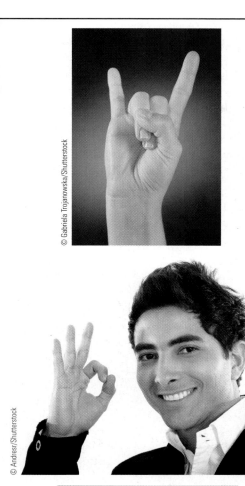

▶ **Figure 2.1 Hand Gestures with Different Meanings**
As international travelers and businesspeople have learned, hand gestures may have very different meanings in different cultures.

ous nations (see ▶ Figure 2.1). Although the "hook 'em Horns" sign—the pinky and index finger raised up and the middle two fingers folded down—is used by fans to express their support for University of Texas at Austin sports teams, for millions of Italians the same gesture means "Your spouse is being unfaithful." In Argentina, rotating one's index finger around the front of the ear means "You have a telephone call," but in the United States it usually suggests that a person is "crazy" (Axtell, 1991). Similarly, making a circle with your thumb and index finger indicates "OK" in the United States, but in Tunisia it means "I'll kill you!" (Samovar and Porter, 1991).

As the example of hand gestures shows, a central component of nonmaterial culture is *beliefs*—**the mental acceptance or conviction that certain things are true or real.** Beliefs may be based on tradition, faith, experience, scientific research, or some combination of these. Faith in a supreme being and trust in another person are examples of beliefs. We may also have a belief in items of material culture. When we travel by airplane, for instance, we believe that it is possible to fly at 33,000 feet and to arrive at our destination even though we know that we could not do this without the airplane itself.

Cultural Universals

Because all humans face the same basic needs (such as for food, clothing, and shelter), we engage in similar activities that contribute to our survival. Anthropologist George Murdock (1945: 124) compiled a list of more than seventy *cultural universals*—**customs and practices that occur across all societies.** His categories included appearance (such

nonmaterial culture a component of culture that consists of the abstract or intangible human creations of society (such as attitudes, beliefs, and values) that influence people's behavior.

beliefs the mental acceptance or conviction that certain things are true or real.

cultural universals customs and practices that occur across all societies.

as bodily adornment and hairstyles), activities (such as sports, dancing, games, joking, and visiting), social institutions (such as family, law, and religion), and customary practices (such as cooking, folklore, gift giving, and hospitality). Whereas these general customs and practices may be present in all cultures, their specific forms vary from one group to another and from one time to another within the same group. For example, although telling jokes may be a universal practice, what is considered to be a joke in one society may be an insult in another.

How do sociologists view cultural universals? In terms of their functions, cultural universals are useful because they ensure the smooth and continual operation of society (Radcliffe-Brown, 1952). A society must meet basic human needs by providing food, shelter, and some degree of safety for its members so that they will survive. Children and other new members (such as immigrants) must be taught the ways of the group. A society must also settle disputes and deal with people's emotions. All the while, the self-interest of individuals must be balanced with the

© Celia Peterson/Getty Images

© Frans Lemmens/Getty Images

© Eddie Gerald/Alamy

Food is a universal type of material culture, but what people eat and how they eat it vary widely, as shown in these cross-cultural examples from the United Arab Emirates (upper left), Holland (upper right), and China (bottom photo). What might be some reasons for the similarities and differences that you see in these photos?

The customs and rituals associated with weddings are one example of nonmaterial culture. What can you infer about beliefs and attitudes concerning marriage in the societies represented by these photographs?

needs of society as a whole. Cultural universals help fulfill these important functions of society.

From another perspective, however, cultural universals are not the result of functional necessity; these practices may have been *imposed* by members of one society on members of another. Similar customs and practices do not necessarily constitute cultural universals. They may be an indication that a conquering nation used its power to enforce certain types of behavior on those who were defeated (Sar-

gent, 1987). Sociologists might ask questions such as "Who determines the dominant cultural patterns?" For example, although religion is a cultural universal, traditional religious practices of indigenous peoples (those who first live in an area) have often been repressed and even stamped out by subsequent settlers or conquerors who hold political and economic power over them. However, many people believe there is cause for optimism in the United States because the democratic ideas of this nation provide more guarantees of religious freedom than do some other nations.

Components of Culture

Even though the specifics of individual cultures vary widely, all cultures have four common nonmaterial cultural components: symbols, language, values, and norms. These components contribute to both harmony and strife in a society.

Symbols

A *symbol* **is anything that meaningfully represents something else.** Culture could not exist without symbols because there would be no shared meanings among people. Symbols can simultaneously produce loyalty and animosity, and love and hate. They help us communicate ideas such as love or patriotism because they express abstract concepts with visible objects. For example, flags can stand for patriotism, nationalism, school spirit, or religious beliefs held by members of a group or society. Symbols can stand for love (a heart on a valentine), peace (a dove), or hate (a Nazi swastika), just as words can be used to convey these meanings. Symbols can also transmit other types of ideas. A siren is a symbol that denotes an emergency situation and sends the message to clear the way immediately. Gestures are also a symbolic form of communication—a movement of the head, body, or hands can express our ideas or feelings to others. For example, in the United States, pointing toward your chest with your thumb or finger is a symbol for "me."

Symbols affect our thoughts about class. For example, how a person is dressed or the kind of car that he or she drives is often at least subconsciously used as a measure of that individual's economic standing or position. With regard to clothing, although many people wear casual clothes on a daily basis, where the clothing was purchased is sometimes used as a symbol of social status. Were the items purchased at Wal-Mart, Old Navy, Abercrombie & Fitch, or Saks Fifth Avenue? What indicators are there on the items of clothing—such as the Nike *swoosh,* some

Would you expect the user of this device to be impoverished or affluent? What do possessions indicate about their owner's social class?

other logo, or a brand name—that indicate something about the status of the product? Automobiles and their logos are also symbols that have cultural meaning beyond the shopping environment in which they originate.

Finally, symbols may be specific to a given culture and have special meaning to individuals who share that culture but not necessarily to other people. Consider, for example, the use of certain foods to celebrate the Chinese New Year: Bamboo shoots and black moss seaweed both represent wealth, peanuts and noodles symbolize a long life, and tangerines represent good luck. What foods in other cultures represent "good luck" or prosperity?

Language

Language **is a set of symbols that expresses ideas and enables people to think and communicate with one another.** Verbal (spoken) language and nonverbal (written or gestured) language help us describe reality. One of our most important human attributes is the ability to use language to share our experiences, feelings, and knowledge with others. Language can create visual images in our head, such as "the kittens look like little cotton balls" (Samovar and Porter, 1991). Language also allows people to distinguish themselves from outsiders and maintain group boundaries and solidarity (Farb, 1973).

Language is not solely a human characteristic. Other animals use sounds, gestures, touch, and smell to communicate with one another, but they use signals with fixed meanings that are limited to the immediate situation (the present) and that cannot encompass past or future situations. For example, chimpanzees can use elements of Standard American Sign Language and manipulate physical objects to make "sentences," but they are not physically endowed with the vocal apparatus needed to form the consonants required for oral language. As a result, nonhuman animals cannot transmit the more complex aspects of culture to their offspring. Humans have a unique ability to manipulate symbols to express abstract concepts and rules and thus to create and transmit culture from one generation to the next.

Language and Social Reality Does language *create* or simply *communicate* reality? Anthropological linguists Edward Sapir and Benjamin Whorf have suggested that language not only expresses our thoughts and perceptions but also influences our perception of reality. According to the *Sapir–Whorf hypothesis,* **language shapes the view of reality of its speakers** (Whorf, 1956; Sapir, 1961). If people are able to think only through language, then language must precede thought. If language actually shapes the reality we perceive and experience, then some aspects of the world are viewed as important and others are virtually neglected because people know the world only in terms of the vocabulary and grammar of their own language.

If language does create reality, are we trapped by our language? Many social scientists agree that the Sapir–Whorf hypothesis overstates the relationship between language and our thoughts and behavior patterns. Although they acknowledge that language has many subtle meanings and that words used by people reflect their central concerns, most sociolo-

The Inuit (Eskimo) language has dozens of words to describe snow. English has only a few. How would you explain this difference?

gists contend that language may *influence* our behavior and interpretation of social reality but does not *determine* it.

Language and Gender What is the relationship between language and gender? What cultural assumptions about women and men does language reflect? Scholars have suggested several ways in which language and gender are intertwined:

- The English language ignores women by using the masculine form to refer to human beings in general (Basow, 1992). For example, the word *man* is used generically in words such as *chairman* and *mankind,* which allegedly include both men and women.
- Use of the pronouns *he* and *she* affects our thinking about gender. Pronouns show the gender of

symbol anything that meaningfully represents something else.

language a set of symbols that expresses ideas and enables people to think and communicate with one another.

Sapir–Whorf hypothesis the proposition that language shapes the view of reality of its speakers.

the person we *expect* to be in a particular occupation. For instance, nurses, secretaries, and schoolteachers are usually referred to as *she,* but doctors, engineers, electricians, and presidents are usually referred to as *he* (Baron, 1986).

- Words have positive connotations when relating to male power, prestige, and leadership; when related to women, they carry negative overtones of weakness, inferiority, and immaturity (Epstein, 1988: 224). ◆ Table 2.1 shows how gender-based language reflects the traditional acceptance of men and women in certain positions, implying that the jobs are different when filled by women rather than men.

- A language-based predisposition to think about women in sexual terms reinforces the notion that women are sexual objects. Women are often described by terms such as *fox, broad, bitch, babe,* or *doll,* which ascribe childlike or even petlike characteristics to them. By contrast, men have performance pressures placed on them by being defined in terms of their sexual prowess, such as *dude, stud,* and *hunk* (Baker, 1993).

Gender in language has been debated and studied extensively in recent years, and some changes have occurred. The preference of many women to be called *Ms.* (rather than *Miss* or *Mrs.* in reference to their marital status) has received a degree of acceptance in public life and the media. Many organizations and publications have established guidelines for the use of nonsexist language and have changed titles such as *chairman* to *chair* or *chairperson.* "Men Working" signs in many areas have been replaced with "People Working." Some occupations have been given "genderless" titles, such as *firefighter* or *flight attendant.* Yet many people resist change, arguing that the English language is being ruined (Epstein, 1988). To develop a more inclusive and equitable society, many scholars suggest that a more inclusive language is needed (see Basow, 1992).

Language, Race, and Ethnicity Language may create and reinforce our perceptions about race and ethnicity by transmitting preconceived ideas about the superiority of one category of people over another. Let's look at a few images conveyed

◆ **Table 2.1 Language and Gender**

Male Term	Female Term	Neutral Term
Teacher	Teacher	Teacher
Chairman	Chairwoman	Chair, chairperson
Congressman	Congresswoman	Representative
Policeman	Policewoman	Police officer
Fireman	Lady fireman	Firefighter
Airline steward	Airline stewardess	Flight attendant
Race car driver	Woman race car driver	Race car driver
Professor	Teacher/female professor	Professor
Doctor	Lady/woman doctor	Doctor
Bachelor	Old maid	Single person
Male prostitute	Prostitute	Prostitute
Welfare recipient	Welfare mother	Welfare recipient
Worker/employee	Working mother	Worker/employee
Janitor/maintenance man	Maid/cleaning lady	Custodial attendant

Sources: Adapted from Korsmeyer, 1981: 122; and Miller and Swift, 1991.

Certain jobs are stereotypically considered to be "men's jobs"; others are "women's jobs." Is your perception of a male flight attendant the same as your perception of a female flight attendant?

- The "voice" of verbs may minimize or incorrectly identify the activities or achievements of people of color. For example, the use of the passive voice in the statement "African Americans *were given* the right to vote" ignores how African Americans *fought* for that right. Active-voice verbs may also inaccurately attribute achievements to people or groups. Some historians argue that cultural bias is shown by the very notion that "Columbus discovered America"—given that America was already inhabited by people who later became known as Native Americans (see Stannard, 1992; Takaki, 1993).

In addition to these concerns about the English language, problems also arise when more than one language is involved. Across the nation, the question of whether or not the United States should have an "official" language continues to arise. Some people believe that there is no need to designate an official language; other people believe that English should be designated as the official language and that the

by words in the English language in regard to race/ethnicity:

- Words may have more than one meaning and create and/or reinforce negative images. Terms such as *blackhearted* (malevolent) and expressions such as *a black mark* (a detrimental fact) and *Chinaman's chance of success* (unlikely to succeed) associate the words *black* or *Chinaman* with negative associations and derogatory imagery. By contrast, expressions such as *that's white of you* and *the good guys wear white hats* reinforce positive associations with the color white.
- Overtly derogatory terms such as *nigger, kike, gook, honkey, chink, spic,* and other racial–ethnic slurs have been "popularized" in movies, music, comic routines, and so on. Such derogatory terms are often used in conjunction with physical threats against persons.
- Words are frequently used to create or reinforce perceptions about a group. For example, Native Americans have been referred to as "savage" and "primitive," and African Americans have been described as "uncivilized," "cannibalistic," and "pagan."

Rapid changes in language and culture in the United States are reflected in this sign at a shopping center. How do functionalist and conflict theorists' views regarding language differ?

use of any other language should be discouraged or negatively sanctioned. Recently, the city council in Farmers Branch—a suburb of Dallas, Texas—adopted a resolution declaring English as the official language of that city. According to the resolution, the use of a common language "removes barriers of misunderstanding and helps to unify the people of Farmers Branch, [the state of Texas,] and the United States and helps to enable the full economic and civic participation of all of its citizens . . ." (City of Farmers Branch, 2006). This resolution was passed at the same time as a local law that banned "illegal immigrants" from renting apartments in Farmers Branch. Are deep-seated social and cultural issues embedded in social policy decisions such as these? Although the United States has always been a nation of immigrants, in recent decades this country has experienced rapid changes in population that have brought about greater diversity in languages and cultures. Recent data gathered by the U.S. Census Bureau (see "Census Profiles: Languages Spoken in U.S. Households") indicate that although more than 80 percent of the people in this country speak only English at home, almost 20 percent speak a language other than English. The largest portion (over 10 percent of the U.S. population) of non-English speakers speak Spanish at home.

Language is an important means of cultural transmission. Through language, children learn about their cultural heritage and develop a sense of personal identity in relationship to their group. For example, Latinos/as in New Mexico and south Texas use *dichos*—proverbs or sayings that are unique to the Spanish language—as a means of expressing themselves and as a reflection of their cultural heritage. Examples of *dichos* include *Anda tu camino sin ayuda de vecino* ("Walk your own road without the help of a neighbor") and *Amor de lejos es para pendejos* ("A long-distance romance is for fools"). *Dichos* are passed from generation to generation as a priceless verbal tradition whereby people can give advice or teach a lesson (Gandara, 1995).

Language is also a source of power and social control; language perpetuates inequalities between people and between groups because words are used (whether or not intentionally) to "keep people in their place." As the linguist Deborah Tannen (1993: B5) has suggested, "The devastating group hatreds that result in so much suffering in our own country and around the world are related in origin to the small intolerances in our everyday conversations—our readiness to attribute good intentions to ourselves and bad intentions to others." Language, then, is a reflection of our feelings and values.

Values

Values **are collective ideas about what is right or wrong, good or bad, and desirable or undesirable in a particular culture** (Williams, 1970). Values do not dictate which behaviors are appropriate and which ones are not, but they provide us with the criteria by which we evaluate people, objects, and events. Values typically come in pairs of positive and negative values, such as being brave or cowardly, hardworking or lazy. Because we use values to justify our behavior, we tend to defend them staunchly.

Core American Values Do we have shared values in the United States? Sociologists disagree about the extent to which all people in this country share a core set of values. Functionalists tend to believe that shared values are essential for the maintenance of a society, and scholars using a functionalist approach have conducted most of the research on core values. Analysts who focus on the importance of core values maintain that the following ten values, identified almost forty years ago by sociologist Robin M. Williams, Jr. (1970), are still very important to people in the United States:

1. *Individualism.* People are responsible for their own success or failure. Individual ability and hard work are the keys to success. Those who do not succeed have only themselves to blame because of their lack of ability, laziness, immorality, or other character defects.
2. *Achievement and success.* Personal achievement results from successful competition with others. Individuals are encouraged to do better than others in school and to work in order to gain wealth, power, and prestige. Material possessions are seen as a sign of personal achievement.
3. *Activity and work.* People who are industrious are praised for their achievement; those perceived as lazy are ridiculed. From the time of the early Puri-

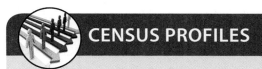

CENSUS PROFILES

Languages Spoken in U.S. Households

Among the categories of information gathered by the U.S. Census Bureau is data on the languages spoken in U.S. households. As shown below, English is the only language spoken at home in more than 80 percent of U.S. households; however, in almost 20 percent of U.S. households, some other language is the primary language spoken at home.

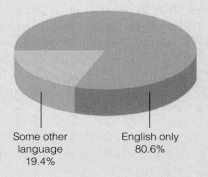

Some other language
19.4%

English only
80.6%

People who speak a language other than English at home are asked not only to indicate which other languages they speak but also how well they speak English. Approximately 44 percent of people who speak a language other than English at home report that they speak English "less than well." The principal languages

other than English that are most frequently spoken at home are shown in the following chart. Do you think that changes in the languages spoken in this country will bring about other significant changes in U.S. culture? Why or why not?

Languages Spoken at Home Other Than English, by Percentage

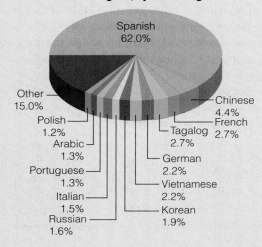

Spanish 62.0%
Other 15.0%
Polish 1.2%
Arabic 1.3%
Portuguese 1.3%
Italian 1.5%
Russian 1.6%
Korean 1.9%
Vietnamese 2.2%
German 2.2%
Tagalog 2.7%
French 2.7%
Chinese 4.4%

Source U.S. Census Bureau 2008.

---

tans, work has been viewed as important. Even during their leisure time, many people "work" in their play. Think, for example, of all the individuals who take exercise classes, run in marathons, garden, repair or restore cars, and so on in their spare time.

4. *Science and technology.* People in the United States have a great deal of faith in science and technology. They expect scientific and technological advances ultimately to control nature, the aging process, and even death.

5. *Progress and material comfort.* The material comforts of life include not only basic necessities (such as adequate shelter, nutrition, and medical care) but also the goods and services that make life easier and more pleasant.

6. *Efficiency and practicality.* People want things to be bigger, better, and faster. As a result, great value is placed on efficiency ("How well does it work?") and practicality ("Is this a realistic thing to do?").

7. *Equality.* Since colonial times, overt class distinctions have been rejected in the United States. However, "equality" has been defined as "equality of *opportunity*"—an assumed equal chance to achieve success—not as "equality of *outcome*."

---

values collective ideas about what is right or wrong, good or bad, and desirable or undesirable in a particular culture.

8. *Morality and humanitarianism.* Aiding others, especially following natural disasters (such as floods or hurricanes), is seen as a value. The notion of helping others was originally a part of religious teachings and tied to the idea of morality. Today, people engage in humanitarian acts without necessarily perceiving that it is the "moral" thing to do.

9. *Freedom and liberty.* Individual freedom is highly valued in the United States. The idea of freedom includes the right to private ownership of property, the ability to engage in private enterprise, freedom of the press, and other freedoms that are considered to be "basic" rights.

10. *Racism and group superiority.* People value their own racial or ethnic group above all others. Such feelings of superiority may lead to discrimination; slavery and segregation laws are classic examples. Many people also believe in the superiority of their country and that "the American way of life" is best.

Do you think that these values are still important today? Are there core values that you believe should be added to this list? Although sociologists have not agreed upon a specific list of emerging core values, various social analysts have suggested that some additional shared values in the United States today include the following:

- Ecological sensitivity, with an increased awareness of global problems such as overpopulation and global warming.
- Emphasis on developing and maintaining relationships through honesty and with openness, fairness, and tolerance of others.
- Spirituality and a need for meaning in life that reaches beyond oneself.

Value Contradictions All societies, including the United States, have value contradictions. *Value contradictions are values that conflict with one another or are mutually exclusive (achieving one makes it difficult, if not impossible, to achieve another).* Core values of morality and humanitarianism may conflict with values of individual achievement and success. For example, humanitarian values reflected in welfare and other government aid programs for people in need come into conflict with values emphasizing hard work and personal achievement.

Ideal Culture Versus Real Culture What is the relationship between values and human behavior? Sociologists stress that a gap always exists between ideal culture and real culture in a society. *Ideal culture* refers to the values and standards of behavior that people in a society profess to hold. *Real culture* refers to the values and standards of behavior that people actually follow. For example, we may claim to be law-abiding (ideal cultural value) but smoke marijuana (real cultural behavior), or we may regularly drive over the speed limit but think of ourselves as "good citizens."

Most of us are not completely honest about how well we adhere to societal values. In a University of Arizona study known as the "Garbage Project," household waste was analyzed to determine the rate of alcohol consumption in Tucson. People were asked about their level of alcohol consumption, and in some areas of the city, they reported very low levels of alcohol use. However, when these people's garbage was analyzed, researchers found that over 80 percent of those households consumed some beer, and more than half discarded eight or more empty beer cans a week (Haviland, 1993). Obviously, this study shows a discrepancy between ideal cultural values and people's actual behavior.

Norms

Values provide ideals or beliefs about behavior but do not state explicitly how we should behave. Norms, on the other hand, do have specific behavioral expectations. *Norms* **are established rules of behavior or standards of conduct.** *Prescriptive norms* state what behavior is appropriate or acceptable. For example, persons making a certain amount of money are expected to file a tax return and pay any taxes they owe. Norms based on custom direct us to open a door for a person carrying a heavy load. By contrast, *proscriptive norms* state what behavior is inappropriate or unacceptable. Laws that prohibit us from driving over the speed limit and "good manners" that preclude you from reading a newspaper during class are examples. Prescriptive and proscriptive norms operate at all levels of society, from our everyday actions to the formulation of laws.

Formal and Informal Norms Not all norms are of equal importance; those that are most crucial are formalized. *Formal norms* are written down and

Crowded conditions exist around the world, yet certain norms prevail in everyday life. Is the behavior of the people in this Osaka, Japan, train station a reflection of formal or informal norms?

Mores Other norms are considered to be highly essential to the stability of society. *Mores* **are strongly held norms with moral and ethical connotations that may not be violated without serious consequences in a particular culture.** Because mores (pronounced MOR-ays) are based on cultural values and are considered to be crucial to the well-being of the group, violators are subject to more severe negative sanctions (such as ridicule, loss of employment, or imprisonment) than are those who fail to adhere to folkways. The strongest mores are referred to as taboos. *Taboos* **are mores so strong that their violation is considered to be extremely offensive and even unmentionable.** Violation of taboos is punishable by the group or even, according to certain belief systems, by a supernatural force. The incest taboo, which prohibits sexual or marital relations between certain categories of kin, is an example of a nearly universal taboo.

Laws *Laws* **are formal, standardized norms that have been enacted by legislatures and are enforced by formal sanctions.** Laws may be either civil or criminal. *Civil law* deals with disputes among persons or groups. Persons who lose civil suits may encounter negative sanctions such as having to

involve specific punishments for violators. Laws are the most common type of formal norms; they have been codified and may be enforced by sanctions. *Sanctions* **are rewards for appropriate behavior or penalties for inappropriate behavior.** Examples of *positive sanctions* include praise, honors, or medals for conformity to specific norms. *Negative sanctions* range from mild disapproval to the death penalty.

Norms considered to be less important are referred to as *informal norms*—unwritten standards of behavior understood by people who share a common identity. When individuals violate informal norms, other people may apply informal sanctions. *Informal sanctions* are not clearly defined and can be applied by any member of a group (such as frowning at someone or making a negative comment or gesture).

Folkways Norms are also classified according to their relative social importance. *Folkways* **are informal norms or everyday customs that may be violated without serious consequences within a particular culture** (Sumner, 1959/1906). They provide rules for conduct but are not considered to be essential to society's survival. In the United States, folkways include using underarm deodorant, brushing our teeth, and wearing appropriate clothing for a specific occasion. Often, folkways are not enforced; when they are enforced, the resulting sanctions tend to be informal and relatively mild.

---

value contradictions values that conflict with one another or are mutually exclusive.

norms established rules of behavior or standards of conduct.

sanctions rewards for appropriate behavior or penalties for inappropriate behavior.

folkways informal norms or everyday customs that may be violated without serious consequences within a particular culture.

mores strongly held norms with moral and ethical connotations that may not be violated without serious consequences in a particular culture.

taboos mores so strong that their violation is considered to be extremely offensive and even unmentionable.

laws formal, standardized norms that have been enacted by legislatures and are enforced by formal sanctions.

pay compensation to the other party or being ordered to stop certain conduct. *Criminal law,* on the other hand, deals with public safety and well-being. When criminal laws are violated, fines and prison sentences are the most likely negative sanctions, although in some states the death penalty is handed down for certain major offenses.

Technology, Cultural Change, and Diversity

Cultures do not generally remain static. There are many forces working toward change and diversity. Some societies and individuals adapt to this change, whereas others suffer culture shock and succumb to ethnocentrism.

Cultural Change

Societies continually experience cultural change at both material and nonmaterial levels. Changes in technology continue to shape the material culture of society. **Technology refers to the knowledge, techniques, and tools that allow people to transform resources into usable forms and the knowledge and skills required to use what is developed.** Although most technological changes are primarily modifications of existing technology, *new technologies* refers to changes that make a significant difference in many people's lives. Examples of new technologies include the introduction of the printing press more than 500 years ago and the advent of computers and electronic communications in the twentieth century. The pace of technological change has increased rapidly in the past 150 years, as contrasted with the 4,000 years prior to that, during which humans advanced from digging sticks and hoes to the plow.

All parts of culture do not change at the same pace. When a change occurs in the material culture of a society, nonmaterial culture must adapt to that change. Frequently, this rate of change is uneven, resulting in a gap between the two. Sociologist William F. Ogburn (1966/1922) referred to this disparity as **cultural lag—a gap between the technical development of a society and its moral and legal insti-** tutions. In other words, cultural lag occurs when material culture changes faster than nonmaterial culture, thus creating a lag between the two cultural components. For example, at the material cultural level, the personal computer and electronic coding have made it possible to create a unique health identifier for each person in the United States. Based on available technology (material culture), it would be possible to create a national data bank that included everyone's individual medical records from birth to death. Using this identifier, health providers and insurance companies could rapidly transfer medical records around the globe, and researchers could access unlimited data on people's diseases, test results, and treatments. However, the availability of this technology does not mean that it will be accepted by people who believe (nonmaterial culture) that such a national data bank constitutes an invasion of privacy and could easily be abused by others. The failure of nonmaterial culture to keep pace with material culture is linked to social conflict and societal problems. As in the above example, such changes are often set in motion by discovery, invention, and diffusion.

Discovery is the process of learning about something previously unknown or unrecognized. Historically, discovery involved unearthing natural elements or existing realities, such as "discovering" fire or the true shape of the Earth. Today, discovery most often results from scientific research. For example, discovery of a polio vaccine virtually eliminated one of the major childhood diseases. A future discovery of a cure for cancer or the common cold could result in longer and more productive lives for many people.

As more discoveries have occurred, people have been able to reconfigure existing material and nonmaterial cultural items through invention. *Invention* is the process of reshaping existing cultural items into a new form. Guns, video games, airplanes, and First Amendment rights are examples of inventions that positively or negatively affect our lives today.

When diverse groups of people come into contact, they begin to adapt one another's discoveries, inventions, and ideas for their own use. *Diffusion* is the transmission of cultural items or social practices from one group or society to another through such means as exploration, war, the media, tourism, and immigration. Today, cultural diffusion moves at a very rapid pace in the global economy.

Cultural Diversity

Cultural diversity refers to the wide range of cultural differences found between and within nations. Cultural diversity between countries may be the result of natural circumstances (such as climate and geography) or social circumstances (such as level of technology and composition of the population). Some nations—such as Sweden—are referred to as *homogeneous societies,* meaning that they include people who share a common culture and who are typically from similar social, religious, political, and economic backgrounds. By contrast, other nations—including the United States—are referred to as *heterogeneous societies,* meaning that they include people who are dissimilar in regard to social characteristics such as religion, income, or race/ethnicity (see ▶ Figure 2.2).

Immigration contributes to cultural diversity in a society. Throughout its history, the United States has been a nation of immigrants. Over the past 175 years, more than 55 million "documented" (legal) immigrants have arrived here; innumerable people have also entered the country as undocumented immigrants. Immigration can cause feelings of frustration and hostility, especially in people who feel threatened by the changes that large numbers of immigrants may produce (Mydans, 1993). Often, people are intolerant of those who are different from themselves. When societal tensions rise, people may look for others on whom they can place blame—or single out persons because they are the "other," the "outsider," the one who does not "belong." Ronald Takaki, an ethnic studies scholar, described his experience of being singled out as an "other":

> I had flown from San Francisco to Norfolk and was riding in a taxi to my hotel to attend a conference on multiculturalism. . . . My driver and I chatted about the weather and the tourists. . . . The rearview mirror reflected a white man in his forties. "How long have you been in this country?" he asked. "All my life," I replied, wincing. "I was born in the United States." With a strong southern drawl, he remarked: "I was wondering because your English is excellent!" Then, as I had many times before, I explained: "My grandfather came here from Japan in the 1880s. My family has been here, in America, for over a hundred years." He glanced at me in the mirror. Somehow I did not look "American" to him; my eyes and complexion looked foreign. (Takaki, 1993: 1)

Have you ever been made to feel like an "outsider"? Each of us receives cultural messages that may make us feel good or bad about ourselves or may give us the perception that we "belong" or "do not belong." Can people overcome such feelings in a culturally diverse society such as the United States? Some analysts believe it is possible to communicate with others despite differences in race, ethnicity, national origin, age, sexual orientation, religion, social class, occupation, leisure pursuits, regionalism, and so on (see Box 2.2). People who differ from the dominant group may also find reassurance and social support in a subculture or a counterculture.

Subcultures A *subculture* **is a category of people who share distinguishing attributes, beliefs, values, and/or norms that set them apart in some significant manner from the dominant culture.** Emerging from the functionalist tradition, this concept has been applied to distinctions ranging from ethnic, religious, regional, and age-based categories

---

technology the knowledge, techniques, and tools that allow people to transform resources into a usable form and the knowledge and skills required to use what is developed.

cultural lag William Ogburn's term for a gap between the technical development of a society (material culture) and its moral and legal institutions (nonmaterial culture).

subculture a group of people who share a distinctive set of cultural beliefs and behaviors that differs in some significant way from that of the larger society.

Religious Affiliation

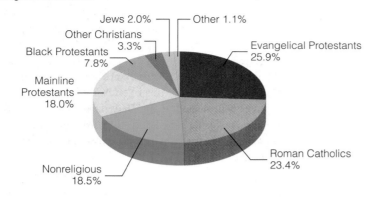

Jews 2.0%
Other 1.1%
Other Christians 3.3%
Black Protestants 7.8%
Evangelical Protestants 25.9%
Mainline Protestants 18.0%
Roman Catholics 23.4%
Nonreligious 18.5%

Household Income[a]

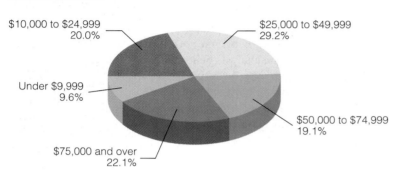

$10,000 to $24,999 20.0%
$25,000 to $49,999 29.2%
Under $9,999 9.6%
$50,000 to $74,999 19.1%
$75,000 and over 22.1%

Race and Ethnic Distribution

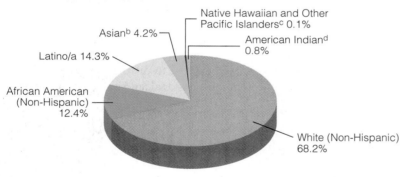

Native Hawaiian and Other Pacific Islanders[c] 0.1%
Asian[b] 4.2%
American Indian[d] 0.8%
Latino/a 14.3%
African American (Non-Hispanic) 12.4%
White (Non-Hispanic) 68.2%

▶ **Figure 2.2 Heterogeneity of U.S. Society**
Throughout history, the United States has been heterogeneous. Today, we represent a wide diversity of social categories, including our religious affiliations, income levels, and racial/ethnic categories.

[a]In Census Bureau terminology, a household consists of people who occupy a housing unit. [b]Includes Chinese, Filipino, Japanese, Asian, Indian, Korean, Vietnamese, and other Asians. [c]Includes Native Hawaiian, Guamanian or Chamorro, Samoan, and other Pacific Islanders. [d]Includes American Indians, Eskimos, and Aleuts.

Source: U.S. Census Bureau, 2008.

to those categories presumed to be "deviant" or marginalized from the larger society. In the broadest use of the concept, thousands of categories of people residing in the United States might be classified as participants in one or more subcultures, including Native Americans, Muslims, Generation Xers, and motorcycle enthusiasts. However, many sociological studies of subcultures have limited the scope of

inquiry to more visible, distinct subcultures such as the Old Order Amish and ethnic enclaves in large urban areas—to see how subcultural participants interact with the dominant U.S. culture.

The Old Order Amish Having arrived in the United States in the early 1700s, members of the Old Order Amish have fought to maintain their

Children (about seven per family) are cherished and seen as an economic asset: They help with the farming and other work. Many of the Old Order Amish speak Pennsylvania Dutch (a dialect of German) as well as English. They dress in traditional clothing, live on farms, and rely on the horse and buggy for transportation.

The Amish are aware that they share distinctive values and look different from other people; these differences provide them with a collective identity and make them feel close to one another (Schaefer and Zellner, 2007). The belief system and group cohesiveness of the Amish remain strong despite the intrusion of corporations and tourists, the vanishing farmlands, and increasing levels of government regulation in their daily lives (Schaefer and Zellner, 2007).

Ethnic Subcultures Some people who have unique shared behaviors linked to a common racial, language, or nationality background identify themselves as members of a specific subculture, whereas others do not. Examples of ethnic subcultures include African Americans, Latinos/Latinas (Hispanic Americans), Asian Americans, and Native Americans. Some analysts include "white ethnics" such

In heterogenous societies such as the United States, people from diverse cultures encourage their children to learn about their heritage. This East Indian mother and daughter in California dance with flower petals.

Although modernization and consumerism have changed the way of life of some subcultures, groups such as the Old Order Amish have preserved some of their historical practices, including traveling by horse-drawn carriage.

distinct identity. Today, over 75 percent of the more than 100,000 Amish live in Pennsylvania, Ohio, and Indiana, where they practice their religious beliefs and remain a relatively closed social network. According to sociologists, this religious community is a subculture because its members share values and norms that differ significantly from those of people who primarily identify with the dominant culture. The Amish have a strong faith in God and reject worldly concerns. Their core values include the joy of work, the primacy of the home, faithfulness, thriftiness, tradition, and humility. The Amish hold a conservative view of the family, believing that women are subordinate to men, birth control is unacceptable, and wives should remain at home.

Box 2.2 You Can Make a Difference

Bonding with Others Through Food and Conversation

[A reader recently said] how meeting for lunch was central to her relationship with her friends. I think that's true for a lot of us. . . . [Food is a] medium through which connections [are] built. My new friends and I started getting together once a month for dinner to celebrate birthdays, with lots of food involved. Most of them were non-native English speakers, so deep discussion of intellectual topics was difficult. Functional English only goes so far, and my smattering of Arabic wasn't up to the challenge, either. So we'd spend hours over dinner in various restaurants, exchanging food off our plates. . . . There's something intimate about sharing food. . . . When we share food, we share ourselves. It bonds us in a way other things don't.

— Terry, in a blog titled "Food as Bonding"
(Dailytroll.com, 2006)

Terry's blog describes how sharing food (an important component of culture) can help us in bonding with and learning more about people from diverse cultures. However, simply meeting with people from other cultural backgrounds or sharing a meal with them is not the same thing as really getting to understand them and helping them to understand us. Daisy Kabagarama, a U.S. college professor who was born in Uganda, suggests in *Breaking the Ice* (1993) that the following techniques can help each of us in communicating across cultures:

- *Get acquainted.* Show genuine interest, have a sense of curiosity and appreciation, feel empathy for others, be nonjudgmental, and demonstrate flexibility.
- *Ask the right questions.* Ask general questions first and specific ones later, making sure that questions are clear and simple and are asked in a relaxed, nonthreatening manner.

- *Consider visual images.* Use compliments carefully; it is easy to misjudge other people based on their physical appearance alone, and appearance norms differ widely across cultures.
- *Deal with stereotypes.* Overcome stereotyping and myths about people from other cultures through sincere self-examination, searching for knowledge, and practicing objectivity.
- *Establish trust and cooperation.* Be available when needed. Give and accept criticism in a positive manner and be spontaneous in interactions with others, but remember that rules regarding spontaneity are different for each culture.

Electronic systems now link people around the world, making it possible for us to communicate with people from diverse racial–ethnic backgrounds and cultures without even leaving home or school. Try these websites for interesting information on multicultural issues and cultural diversity:

- Multicultural Pavilion provides resources on racism, sexism, and classism in the United States, as well as access to multicultural newsgroups, essays, and a large list of multicultural links on the Web:

  http://www.edchange.org/multicultural

- *MultiWorld* is a bilingual (Chinese and English) e-zine that includes information on culture, people, art, and nature, along with sites about nations such as the United States, Canada, Ireland, China, Belgium, and Brazil. You can access *MultiWorld* by going to the following address and clicking the Resources tab:

  http://sunsite.nus.edu

as Irish Americans, Italian Americans, and Polish Americans. Others also include Anglo Americans (Caucasians).

Although people in ethnic subcultures are dispersed throughout the United States, a concen-

tration of members of some ethnic subcultures is visible in many larger communities and cities. For example, Chinatowns, located in cities such as San Francisco, Los Angeles, and New York, are one of the more visible ethnic subcultures in the

United States. By living close to one another and clinging to their original customs and language, first-generation immigrants can survive the abrupt changes they experience in material and nonmaterial cultural patterns. In New York City, for example, Korean Americans and Puerto Rican Americans constitute distinctive subcultures, each with its own food, music, and personal style. In San Antonio, Mexican Americans enjoy different food and music than do Puerto Rican Americans or other groups. Subcultures provide opportunities for expression of distinctive lifestyles, as well as sometimes helping people adapt to abrupt cultural change. Subcultures can also serve as a buffer against the discrimination experienced by many ethnic or religious groups in the United States. However, some people may be forced by economic or social disadvantage to remain in such ethnic enclaves.

Countercultures Some subcultures actively oppose the larger society. A *counterculture* **is a group that strongly rejects dominant societal values and norms and seeks alternative lifestyles** (Yinger, 1960, 1982). Young people are most likely to join countercultural groups, perhaps because younger persons generally have less invested in the existing culture. Examples of countercultures include the beatniks of the 1950s, the flower children of the 1960s, the drug enthusiasts of the 1970s, and contemporary members of nonmainstream religious sects, or cults.

Culture Shock

Culture shock **is the disorientation that people feel when they encounter cultures radically different from their own and believe they cannot depend on their own taken-for-granted assumptions about life.** When people travel to another society, they may not know how to respond to that setting. For example, Napoleon Chagnon (1992) described his initial shock at seeing the Yanomamö (pronounced yah-noh-MAH-mah) tribe of South America on his first trip in 1964.

The Yanomamö (also referred to as the "Yanomami") are a tribe of about 20,000 South American Indians who live in the rain forest. Although Chagnon traveled in a small aluminum motorboat for three days to reach these people, he was not prepared for the sight that met his eyes when he arrived:

> I looked up and gasped to see a dozen burly, naked, sweaty, hideous men staring at us down the shafts of their drawn arrows. Immense wads of green tobacco were stuck between their lower teeth and lips, making them look even more hideous, and strands of dark-green slime dripped from their nostrils—strands so long that they reached down to their pectoral muscles or drizzled down their chins and stuck to their chests and bellies. We arrived as the men were blowing ebene, a hallucinogenic drug, up their noses. . . . I was horrified. What kind of welcome was this for someone who had come to live with these people and learn their way of life—to become friends with them? But when they recognized Barker [a guide], they put their weapons down and returned to their chanting, while keeping a nervous eye on the village entrances. (Chagnon, 1992: 12–14)

The Yanomamö have no written language, system of numbers, or calendar. They lead a nomadic lifestyle, carrying everything they own on their backs. They wear no clothes and paint their bodies; the women insert slender sticks through holes in the lower lip and through the pierced nasal septum. In other words, the Yanomamö—like the members of thousands of other cultures around the world—live in a culture very different from that of the United States.

counterculture a group that strongly rejects dominant societal values and norms and seeks alternative lifestyles.

culture shock the disorientation that people feel when they encounter cultures radically different from their own and believe they cannot depend on their own taken-for-granted assumptions about life.

Even as global travel and the media make us more aware of people around the world, the distinctiveness of the Yanomamö in South America remains apparent. Are people today more or less likely than those in the past to experience culture shock upon encountering diverse groups of people such as these Yanomamö?

Ethnocentrism and Cultural Relativism

When observing people from other cultures, many of us use our own culture as the yardstick by which we judge their behavior. Sociologists refer to this approach as *ethnocentrism*—**the practice of judging all other cultures by one's own culture** (Sumner, 1959/1906). Ethnocentrism is based on the assumption that one's own way of life is superior to all others. For example, most schoolchildren are taught that their own school and country are the best. The school song, the pledge to the flag, and the national anthem are forms of *positive ethnocentrism.* However, *negative ethnocentrism* can also result from constant emphasis on the superiority of one's own group or nation. Negative ethnocentrism is manifested in derogatory stereotypes that ridicule recent immigrants whose customs, dress, eating habits, or religious beliefs are markedly different from those of dominant-group members. Long-term U.S. residents who are members of racial and ethnic minority groups, such as Native Americans, African Americans, and Latinas/os, have also been the target of ethnocentric practices by other groups.

An alternative to ethnocentrism is *cultural relativism*—**the belief that the behaviors and customs of any culture must be viewed and analyzed by the culture's own standards.** For example, the anthropologist Marvin Harris (1974, 1985) uses cultural relativism to explain why cattle, which are viewed as sacred, are not killed and eaten in India, where widespread hunger and malnutrition exist. From an ethnocentric viewpoint, we might conclude that cow worship is the cause of the hunger and poverty in India. However, according to Harris, the Hindu taboo against killing cattle is very important to their economic system. Live cows are more valuable than dead ones because they have more important uses than as a direct source of food. As part of the ecological system, cows consume grasses of little value to humans. Then they produce two valuable resources—oxen (the neutered offspring of cows) to power the plows and manure (for fuel and fertilizer)—as well as milk, floor covering, and leather. As Harris's study reveals, culture must be viewed from the standpoint of those who live in a particular society.

Cultural relativism also has a downside. It may be used to excuse customs and behavior (such as cannibalism) that may violate basic human rights. Cultural relativism is a part of the sociological imagination; researchers must be aware of the customs and norms of the society they are studying and then spell out their background assumptions so that others can spot possible biases in their studies. However, according to some social scientists, issues surrounding ethnocentrism and cultural relativism may become less distinct in the future as people around the globe increasingly share a common popular culture. Others, of course, disagree with this perspective. Let's see what you think.

A Global Popular Culture?

Before taking this course, what was the first thing you thought about when you heard the term *culture*? In everyday life, culture is often used to describe the fine arts, literature, and classical music. When people say that a person is "cultured," they may mean that the individual has a highly developed sense of style or aesthetic appreciation of the "finer" things.

High Culture and Popular Culture

Some sociologists use the concepts of high culture and popular culture to distinguish between different cultural forms. *High culture consists of classical music, opera, ballet, live theater, and other activities usually patronized by elite audiences,* composed primarily of members of the upper-middle and upper classes, who have the time, money, and knowledge assumed to be necessary for its appreciation. In the United States, high culture is often viewed as being international in scope, arriving in this country through the process of diffusion, because many art forms originated in European nations or other countries of the world. By contrast, much of U.S. popular culture is often thought of as "homegrown" in this country. *Popular culture consists of activities, products, and services that are assumed to appeal primarily to members of the middle and working classes.* These include rock concerts, spectator sports, movies, and television soap operas and situation comedies. Although we will distinguish between "high" and "popular" culture in our discussion, it is important to note that

some social analysts believe that the rise of a consumer society in which luxury items have become more widely accessible to the masses has greatly reduced the great divide between activities and possessions associated with wealthy people or a social elite (see Huyssen, 1984; Lash and Urry, 1994).

However, most sociological examinations of high culture and popular culture focus primarily on the link between culture and social class. French sociologist Pierre Bourdieu's (1984) *cultural capital theory* views high culture as a device used by the dominant class to exclude the subordinate classes. According to Bourdieu, people must be trained to appreciate and understand high culture. Individuals learn about high culture in upper-middle- and upper-class families and in elite education systems, especially higher education. Once they acquire this trained capacity, they possess a form of cultural capital. Persons from poor and working-class backgrounds typically do not acquire this cultural capital. Because knowledge and appreciation of high culture are considered a prerequisite for access to the dominant class, its members can use their cultural capital to deny access to subordinate-group members and thus preserve and reproduce the existing class structure.

Forms of Popular Culture

Three prevalent forms of popular culture are fads, fashions, and leisure activities. A *fad* is a temporary but widely copied activity followed enthusiastically

ethnocentrism the practice of judging all other cultures by one's own culture.

cultural relativism the belief that the behaviors and customs of any culture must be viewed and analyzed by the culture's own standards.

high culture classical music, opera, ballet, live theater, and other activities usually patronized by elite audiences.

popular culture the component of culture that consists of activities, products, and services that are assumed to appeal primarily to members of the middle and working classes.

by large numbers of people. Most fads are short-lived novelties. According to the sociologist John Lofland (1993), fads can be divided into four major categories. First, *object fads* are items that people purchase despite the fact that they have little use or intrinsic value. Recent examples include Webkinz, Harry Potter wands, and SpongeBob SquarePants trading cards. Second, *activity fads* include pursuits such as body piercing, "surfing" the Internet, and the "free hugs" campaign, wherein individuals offer hugs to strangers in a public setting as a random act of kindness to make someone feel better. Third are *idea fads,* such as New Age ideologies. Fourth are *personality* fads—for example, those surrounding celebrities such as Paris Hilton, Tiger Woods, Snoop Dogg, and Brad Pitt.

A *fashion* is a currently valued style of behavior, thinking, or appearance that is longer lasting and more widespread than a fad. Examples of fashion are found in many areas, including child rearing, education, arts, clothing, music, and sports. Soccer is an example of a fashion in sports. Until recently, only schoolchildren played soccer in the United States. Now it has become a popular sport, perhaps in part because of immigration from Latin America and other areas of the world where soccer is widely played.

Like soccer, other forms of popular culture move across nations. In fact, popular culture is one of the United States' largest exports to other nations. Of the world's 100 most-attended films in the 1990s, for example, 88 were produced by U.S.-based film companies. Likewise, music, television shows, novels, and street fashions from the United States have become a part of many other cultures. In turn, people in this country continue to be strongly influenced by popular culture from other nations. For example, contemporary music and clothing in the United States reflect African, Caribbean, and Asian cultural influences, among others.

Will the spread of popular culture produce a homogeneous global culture? Critics argue that the world is not developing a global culture; rather, other cultures are becoming westernized. Political and religious leaders in some nations oppose this process, which they view as **cultural imperialism—the extensive infusion of one nation's culture into other nations.** For example, some view the widespread infusion of the English language into countries that speak other languages as a form of cultural imperialism. On the other hand, the concept of cultural imperialism may fail to take into account various cross-cultural influences. For example, cultural diffusion of literature, music, clothing, and food has occurred on a global scale. A global culture, if it comes into existence, will most likely include components from many societies and cultures.

Sociological Analysis of Culture

Sociologists regard culture as a central ingredient in human behavior. Although all sociologists share a similar purpose, they typically see culture through somewhat different lenses as they are guided by different theoretical perspectives in their research. What do these perspectives tell us about culture?

Functionalist Perspectives

As previously discussed, functionalist perspectives are based on the assumption that society is a stable, orderly system with interrelated parts that serve specific functions. Anthropologist Bronislaw Malinowski (1922) suggested that culture helps people meet their *biological needs* (including food and procreation), *instrumental needs* (including law and

© Sonda Dawes / The Image Works

Is this wristband an example of a fad or a fashion?

education), and *integrative needs* (including religion and art). Societies in which people share a common language and core values are more likely to have consensus and harmony.

How might functionalist analysts view popular culture? According to many functionalist theorists, popular culture serves a significant function in society in that it may be the "glue" that holds society together. Regardless of race, class, sex, age, or other characteristics, many people are brought together (at least in spirit) to cheer teams competing in major sporting events such as the Super Bowl or the Olympic Games. Television and the Internet help integrate recent immigrants into the mainstream culture, whereas longer-term residents may become more homogenized as a result of seeing the same images and being exposed to the same beliefs and values (Gerbner et al., 1987).

However, functionalists acknowledge that all societies have dysfunctions that produce a variety of societal problems. When a society contains numerous subcultures, discord results from a lack of consensus about core values. In fact, popular culture may undermine core cultural values rather than reinforce them. For example, movies may glorify crime, rather than hard work, as the quickest way to get ahead. According to some analysts, excessive violence in music videos, movies, and television programs may be harmful

to children and young people. From this perspective, popular culture can be a factor in antisocial behavior as seemingly diverse as hate crimes and fatal shootings in public schools.

A strength of the functionalist perspective on culture is its focus on the needs of society and the fact that stability is essential for society's continued survival. A shortcoming is its overemphasis on harmony and cooperation. This approach also fails to fully account for factors embedded in the structure of society—such as class-based inequalities, racism, and sexism—that may contribute to conflict among people in the United States or to global strife.

Conflict Perspectives

Conflict perspectives are based on the assumption that social life is a continuous struggle in which members of powerful groups seek to control scarce resources. According to this approach, values and norms help create and sustain the privileged position of the powerful in society while excluding others. As early conflict theorist Karl Marx stressed, ideas are *cultural creations* of a society's most powerful members. Thus, it is possible for political, economic, and social leaders to use *ideology*—an integrated system of ideas that is external to, and coercive of, people—to maintain their positions of dominance in a society. As Marx stated,

> The ideas of the ruling class are in every epoch the ruling ideas, i.e., the class which is the ruling material force in society, is at the same time, its ruling intellectual force. The class, which has the means of material production at its disposal, has control at the same time over the means of mental production. . . . The ruling ideas are nothing more than the ideal expression of the dominant material relationships, the dominant material relationships grasped as ideas. (Marx and Engels, 1970/1845–1846: 64)

Many contemporary conflict theorists agree with Marx's assertion that ideas, a nonmaterial component of culture, are used by agents of the ruling class

The decision to hold the 2008 Summer Olympics in China was a controversial one, as demonstrated by the protests that popped up along the Olympic torch relay.

© Guilad Kahn/AFP/Getty Images

cultural imperialism the extensive infusion of one nation's culture into other nations.

to affect the thoughts and actions of members of other classes. The role of the mass media in influencing people's thinking about the foods that they should—or should not—eat is an example of ideological control (see Box 2.3).

How might conflict theorists view popular culture? Some conflict theorists believe that popular culture, which originated with everyday people, has been largely removed from their domain and has become nothing more than a part of the capitalist economy in the United States (Gans, 1974; Cantor, 1980, 1987). From this approach, media conglomerates such as Time Warner, Disney, and Viacom create popular culture, such as films, television shows, and amusement parks, in the same way that they would produce any other product or service. Creating new popular culture also promotes consumption of *commodities*—objects outside ourselves that we purchase to satisfy our human needs or wants (Fjellman, 1992). According to contemporary social analysts, consumption—even of things that we do not necessarily need—has become prevalent at all social levels, and some middle- and lower-income individuals and families now use as their frame of reference the lifestyles of the more affluent in their communities. As a result, many families live on credit in order to purchase the goods and services that they would like to have or that keep them on the competitive edge with their friends, neighbors, and coworkers (Schor, 1999).

Other conflict theorists examine the intertwining relationship among race, gender, and popular culture. According to the sociologist K. Sue Jewell (1993), popular cultural images are often linked to negative stereotypes of people of color, particularly African American women. Jewell believes that cultural images depicting African American women as mammies or domestics—such as those previously used in Aunt Jemima Pancake ads and recent resurrections of films such as *Gone with the Wind*—affect contemporary black women's economic prospects in profound ways (Jewell, 1993).

A strength of the conflict perspective is that it stresses how cultural values and norms may perpetuate social inequalities. It also highlights the inevitability of change and the constant tension between those who want to maintain the status quo and those who desire change. A limitation is its focus on societal discord and the divisiveness of culture.

Symbolic Interactionist Perspectives

Unlike functionalists and conflict theorists, who focus primarily on macrolevel concerns, symbolic interactionists engage in a microlevel analysis that views society as the sum of all people's interactions. From this perspective, people create, maintain, and modify culture as they go about their everyday activities. Symbols make communication with others possible because they provide us with shared meanings.

According to some symbolic interactionists, people continually negotiate their social realities. Values and norms are not independent realities that automatically determine our behavior. Instead, we reinterpret them in each social situation we encounter. However, the classical sociologist Georg Simmel warned that the larger cultural world—including both material culture and nonmaterial culture—eventually takes on a life of its own apart from the actors who daily re-create social life. As a result, individuals may be more controlled by culture than they realize. Simmel (1990/1907) suggested that money is an example of how people may be controlled by their culture. According to Simmel, people initially create money as a means of exchange, but then money acquires a social meaning that extends beyond its purely economic function. Money becomes an end in itself, rather than a means to an end. Today, we are aware of the relative "worth" not only of objects but also of individuals. Many people revere wealthy entrepreneurs and highly paid celebrities, entertainers, and sports figures for the amount of money they make, not for their intrinsic qualities. According to Simmel (1990/1907), money makes it possible for us to *relativize* everything, including our relationships with other people. When social life can be reduced to money, people become cynical, believing that anything—including people, objects, beauty, and truth—can be bought if we can pay the price. Although Simmel acknowledged the positive functions of money, he believed that the social interpretations people give to money often produce individual feelings of cynicism and isolation.

A symbolic interactionist approach highlights how people maintain and change culture through their interactions with others. However, interac-

Box 2.3 Framing Culture in the Media

You Are What You Eat?

The agonizing decision to pick Yale over Harvard didn't come down only to academics for Philip Gant. . . . It also came down to his tummy. And his eco-savvy. . . . When he chose Yale last year, Gant wasn't swayed by its running tab of presidential alumni: President Bush, George H. W. Bush, Bill Clinton, Gerald Ford and William Howard Taft. He was more impressed by Yale's leading-edge dedication to serving "sustainable" food. . . . In addition to wanting sustainable food, students such as Gant want it to be organic: grown without pesticides, herbicides, antibiotics or hormones.

> —*USA Today* reporting that "More University Students Call for Organic, 'Sustainable' Food" (Horovitz, 2006)

You may ask "What does a newspaper article about university cafeterias and organic food have to do with culture?" The answer is simple: Food is very much a part of all cultures. What we eat and how it is grown and prepared are a product of the culture of the society in which we live. Fads and fashions in food may come and go, but we often become aware of them as a result of mass media such as television, magazines, newspapers, and the Internet. There are television networks and magazines devoted solely to the topic of food, and there are stories and articles about food on an almost daily basis in the other forms of mass media.

So why did *USA Today* report on Philip Gant's concerns about campus food? At least in part, the article resulted from a recent emphasis on organic food reporting in the media. Organic food refers to crops that are grown without the use of artificial fertilizers or most pesticides and that are processed without ionizing radiation or food additives, and to meat that is raised without antibiotics or growth hormones.

What most of us know about the "good" and "bad" sides of organic foods comes from the media. According to some media analysts, organic foods contain some nutrients that are not present in commercial foods, and organic foods do not have certain toxins that may be present in commercial foods (Crinnion, 1995). As one journalist stated, organic food methods "honor the fragile complexity of our ecosystem, the health of those who work the land, and

the long-term well-being of customers who enjoy [the] harvest . . ." (Shapin, 2006). However, not all media reports agree on this issue: Some sources note that pesticides *are* used on organic farms (Idaho Association of Soil Conservation Districts, 2004) and that organic foods typically cost the consumer more money.

How stories are framed by journalists and others in the media may influence our thinking about food and how what we eat is related to other cultural beliefs and values. *Media framing* refers to the process by which information and entertainment are packaged by the mass media before being presented to an audience. How a story about food is framed has a major effect on how each of us feels about the subject of that story. When the media report that some type of food—spinach, packaged salads, or some brand of peanut butter—is being recalled by the manufacturer due to health concerns, for example, we may quit buying that particular product for a while.

Thinking specifically about the production of food, the media often use the term "Big Agra" to describe the major corporations around the globe that grow and market much of the food that we eat. These megacorporations own giant cultivated tracts that use procedures intended to maximize the crop yield, harvest that crop (whether plants or animals) at the lowest price possible, and distribute the crop to markets in the United States and other countries. Maximizing crop yield involves the use of chemicals and pesticides—and cheap labor. "Big Agra" obviously has a stake in the battle over how the media frame stories that compare its foods with organic foods; the organic food industry also has a stake in the battle. Accordingly, both sides attempt to influence how stories about their products are framed in the media because that can make a big difference in their respective profits. Whether it is fast food or fresh spinach, they want to have an impact on what you buy, and where.

Reflect & Analyze

What factors affect your perceptions of what is "good" food and "bad" food? Are these factors influenced by advertising and media framing?

tionism does not provide a systematic framework for analyzing how we shape culture and how it, in turn, shapes us. It also does not provide insight into how shared meanings are developed among people, and it does not take into account the many situations in which there is disagreement on meanings. Whereas the functional and conflict approaches tend to overemphasize the macrolevel workings of society, the interactionist viewpoint often fails to take these larger social structures into account.

Postmodernist Perspectives

Postmodernist theorists believe that much of what has been written about culture in the Western world is Eurocentric—that it is based on the uncritical assumption that European culture (including its dispersed versions in countries such as the United States, Australia, and South Africa) is the true, universal culture in which all the world's people ought to believe (Lemert, 1997). By contrast, postmodernists believe that we should speak of *cultures,* rather than *culture.*

However, Jean Baudrillard, one of the best-known French social theorists and key figures in postmod-

ern theory, believes that the world of culture today is based on *simulation,* not reality. According to Baudrillard, social life is much more a spectacle that simulates reality than reality itself. Many U.S. children, upon entering school for the first time, have already watched more hours of television than the total number of hours of classroom instruction they will encounter in their entire school careers (Lemert, 1997). Add to this the number of hours that some children will have spent playing computer games or using the Internet, where they often find that it is more interesting to deal with imaginary heroes and villains than to interact with "real people" in real life. Baudrillard refers to this social creation as *hyperreality*—a situation in which the *simulation* of reality is more real than experiencing the event itself and having any actual connection with what is taking place. For Baudrillard, everyday life has been captured by the signs and symbols generated to represent it, and we ultimately relate to simulations and models as if they were reality.

Baudrillard (1983) uses Disneyland as an example of a simulation—one that conceals the reality that exists outside rather than inside the boundaries of the artificial perimeter. According to Baudrillard, Disney-

People of all ages are spending many hours each week using computers, playing video games, and watching television. How is this behavior different from the ways in which people enjoyed popular culture in previous generations?

like theme parks constitute a form of seduction that substitutes symbolic (seductive) power for real power, particularly the ability to bring about social change. From this perspective, amusement park "guests" may feel like "survivors" after enduring the rapid speed and gravity-defying movements of the roller-coaster rides or see themselves as "winners" after surviving fights with hideous cartoon villains on the "dark rides." In reality, they have been made to *appear* to have power, but they do not actually possess any real power.

Similarly, the anthropologist Stephen M. Fjellman (1992) studied Disney World in Orlando, Florida, and noted that people may forget, at least briefly, that the outside world can be threatening while they stroll Disney World's streets without fear of crime or automobiles. Although this freedom may be temporarily empowering, it may also lull people into accepting a "worldview that presents an idealized United States as heaven. . . . How nice if they could all be like us—with kids, a dog, and General Electric appliances—in a world whose only problems are avoiding Captain Hook, the witch's apple, and Toad Hall weasels" (Fjellman, 1992: 317).

In their examination of culture, postmodernist social theorists thus make us aware of the fact that no single perspective can grasp the complexity and diversity of the social world. There is no one, single, universal culture. They also make us aware that reality may not be what it seems. According to the postmodernist view, no one authority can claim to know social reality, and we should deconstruct—take apart and subject to intense critical scrutiny—existing beliefs and theories about culture in hopes of gaining new insights (Ritzer, 1997).

Although postmodern theories of culture have been criticized on a number of grounds, we will mention only three. One criticism is postmodernism's lack of a clear conceptualization of ideas. Another is the tendency to critique other perspectives as being "grand narratives," whereas postmodernists offer their own varieties of such narratives. Finally, some analysts believe that postmodern analyses of culture lead to profound pessimism about the future.

This chapter's Concept Quick Review summarizes the components of culture as well as how the four major perspectives view it.

CONCEPT QUICK REVIEW

Analysis of Culture

Components of Culture	Symbol	Anything that meaningfully represents something else.
	Language	A set of symbols that expresses ideas and enables people to think and communicate with one another.
	Values	Collective ideas about what is right or wrong, good or bad, and desirable or undesirable in a particular culture.
	Norms	Established rules of behavior or standards of conduct.
Sociological Analysis of Culture	Functionalist Perspectives	Culture helps people meet their biological, instrumental, and expressive needs.
	Conflict Perspectives	Ideas are a cultural creation of society's most powerful members and can be used by the ruling class to affect the thoughts and actions of members of other classes.
	Symbolic Interactionist Perspectives	People create, maintain, and modify culture during their everyday activities; however, cultural creations can take on a life of their own and end up controlling people.
	Postmodern Perspectives	Much of culture today is based on simulation of reality (e.g., what we see on television) rather than reality itself.

Culture in the Future

As we have discussed in this chapter, many changes are occurring in the United States. Increasing cultural diversity can either cause long-simmering racial and ethnic antagonisms to come closer to a boiling point or result in the creation of a truly "rainbow culture" in which diversity is respected and encouraged.

In the future, the issue of cultural diversity will increase in importance, especially in schools. Multicultural education that focuses on the contributions of a wide variety of people from different backgrounds will continue to be an issue of controversy from kindergarten through college. In the Los Angeles school district, for example, students speak more than 114 different languages and dialects. Schools will face the challenge of embracing widespread cultural diversity while conveying a sense of community and national identity to students (see "Sociology Works!").

Technology will continue to have a profound effect on culture. Television and radio, films and videos, and electronic communications will continue to accelerate the flow of information and expand cultural diffusion throughout the world. Global communication devices will move images of people's lives, behavior, and fashions instantaneously among almost all nations. Increasingly, computers and cyberspace will become people's window on the world and, in the process, promote greater integration or fragmentation among nations. Integration occurs when there is a widespread acceptance of ideas and items—such as democracy, rock music, blue jeans, and McDonald's hamburgers—among cultures. By contrast, fragmentation occurs when people in one culture disdain the beliefs and actions of other cultures. As a force for both cul-

© David McNew/Getty Images

In recent years, there has been a significant increase in the number of immigrants who have become U.S. citizens. However, an upsurge in anti-immigrant sentiment has put pressure on the Border Patrol and the U.S. Citizenship and Immigration Service, which are charged with enforcing immigration laws.

© AP Images/John Raoux

Sociology *Works!*

Schools as Laboratories for Getting Along

Sociology makes us aware of the importance of culture in daily life. Research in sociology has also shown the significance of schools and friendship groups in exposing children and young people to cultures that are different from their own. Recent studies have shown that it may be easier for children to set aside their differences and get to know one another than it is for adults to do so. Consider what is happening among some children at International Community School, an innovative Decatur, Georgia, school where some students were born in the United States, but most are refugees from as many as forty war-torn countries: This school has become a "laboratory for getting along," particularly as some of the children have taken the initiative to befriend and help others (St. John, 2007). An excellent example is the friendship that developed between nine-year-old Dante Ramirez and Soung Oo Hlaing, an eleven-year-old Burmese refugee who spoke no English:

> The two boys met on the first day of school this year. Despite the language barrier, Dante managed to invite the newcomer to sit with him at lunch.
>
> "I didn't think he'd make friends at the beginning because he didn't speak that much English," Dante said. "So I thought I should be his friend."
>
> In the next weeks, the boys had a sleepover. They trick-or-treated on Soung's first Halloween. Soung, a gifted artist, gave Dante pointers on how to draw. And Dante helped Soung with his English. "I use simple words that are easy to know and sometimes hand movements," Dante explained. "For 'huge,' I would make my hands bigger. And for 'big,' I would make my hands smaller than for huge." (St. John, 2007: A14)

Over time, as the boys got to know each other better, their mothers also developed a friendship and began to celebrate ethnic holidays together even though they largely relied on gestures (a form of nonverbal communication) to communicate with each other. Only time will tell how successful this "laboratory" will be in helping people from diverse cultures get along, but community efforts such as this are clearly a right start from a sociological perspective.

Sociologists believe that it is important for cross-cultural communications and cooperation to develop among individuals from diverse cultural backgrounds who now share common spaces. If a chance exists for greater understanding and cooperation in the twenty-first century, it may well originate in the small-group interactions of children in settings such as this school.

Reflect & Analyze

What examples can you provide that show how sociology works in regard to culture on your own college campus or in the community where you reside?

tural integration and fragmentation, technology will continue to revolutionize communications, but most of the world's population will not participate in this revolution.

From a sociological perspective, the study of culture helps us not only understand our own "tool kit" of symbols, stories, rituals, and world views but also expand our insights to include those of other people of the world, who also seek strategies for enhancing their own lives. If we understand how culture is used by people, how cultural elements constrain or further certain patterns of action, what aspects of our cultural heritage have enduring effects on our actions, and what specific historical changes undermine the validity of some cultural patterns and give rise to others, we can apply our sociological imagination not only to our own society but to the entire world as well (see Swidler, 1986).

Chapter Review

• What is culture?

Culture is the knowledge, language, values, and customs passed from one generation to the next in a human group or society. Culture can be either material or nonmaterial. Material culture consists of the physical creations of society. Nonmaterial culture is more abstract and reflects the ideas, values, and beliefs of a society.

• What are cultural universals?

Cultural universals are customs and practices that exist in all societies and include activities and institutions such as storytelling, families, and laws. However, specific forms of these universals vary from one cultural group to another.

• What are the four nonmaterial components of culture that are common to all societies?

These components are symbols, language, values, and norms. Symbols express shared meanings; through them, groups communicate cultural ideas and abstract concepts. Language is a set of symbols through which groups communicate. Values are a culture's collective ideas about what is acceptable or not acceptable. Norms are the specific behavioral expectations within a culture.

• What are the main types of norms?

Folkways are norms that express the everyday customs of a group, whereas mores are norms with strong moral and ethical connotations, and are essential to the stability of a culture. Laws are formal, standardized norms that are enforced by formal sanctions.

• What are high culture and popular culture?

High culture consists of classical music, opera, ballet, and other activities usually patronized by elite audiences. Popular culture consists of the activities, products, and services of a culture that appeal primarily to members of the middle and working classes.

• How is cultural diversity reflected in society?

Cultural diversity is reflected through race, ethnicity, age, sexual orientation, religion, occupation, and so forth. A diverse culture also includes subcultures and countercultures. A subculture has distinctive ideas and behaviors that differ from the larger soci-

ety to which it belongs. A counterculture rejects the dominant societal values and norms.

• What are culture shock, ethnocentrism, and cultural relativism?

Culture shock refers to the anxiety that people experience when they encounter cultures radically different from their own. Ethnocentrism is the assumption that one's own culture is superior to others. Cultural relativism views and analyzes another culture in terms of that culture's own values and standards.

• How do the major sociological perspectives view culture?

A functionalist analysis of culture assumes that a common language and shared values help produce consensus and harmony. According to some conflict theorists, culture may be used by certain groups to maintain their privilege and exclude others from society's benefits. Symbolic interactionists suggest that people create, maintain, and modify culture as they go about their everyday activities. Postmodern thinkers believe that there are many cultures within the United States alone. In order to grasp a better understanding of how popular culture may simulate reality rather than be reality, postmodernists believe that we need a new way of conceptualizing culture and society.

www.cengage.com/login

Register for a Student eResource account to maximize your study time online using CengageNOW. First take the system's diagnostic pre-test, and then follow the personalized study plan that is created for you to help you review this chapter. The study plan will

- help you identify areas on which you should concentrate;
- provide interactive exercises to help you master the chapter concepts; and
- provide a post-test to confirm you are ready to move on to the next chapter.

Key Terms

beliefs 47	folkways 57	sanctions 57
counterculture 63	high culture 65	Sapir–Whorf hypothesis 51
cultural imperialism 66	language 50	subculture 59
cultural lag 58	laws 57	symbol 50
cultural relativism 64	material culture 45	taboos 57
cultural universals 47	mores 57	technology 58
culture 44	nonmaterial culture 46	value contradictions 56
culture shock 63	norms 56	values 54
ethnocentrism 64	popular culture 65	

Questions for Critical Thinking

1. Would it be possible today to live in a totally separate culture in the United States? Could you avoid all influences from the mainstream popular culture or from the values and norms of other cultures? How would you be able to avoid any change in your culture?

2. Do fads and fashions reflect and reinforce or challenge and change the values and norms of a society? Consider a wide variety of fads and fash-ions: musical styles; computer and video games and other technologies; literature; and political, social, and religious ideas.

3. You are doing a survey analysis of recent immigrants to the United States to determine the effects of popular culture on their views and behavior. What are some of the questions you would use in your survey?

The Kendall Companion Website

www.cengage.com/sociology/kendall

Supplement your review of this chapter by going to the text's companion website, where you can take tutorial quizzes, use flash cards to master key terms, follow live links to useful websites, and explore the other study and research resources you'll find there, such as a comprehensive interactive sociology timeline, GSS Data, and Census 2000 information, much of it presented visually in maps.

3 Socialization

I don't want to be crippled by things that happened in the past. [My father] was a free bird, you know? He couldn't handle being a father. [When I learned that my father was terminally ill with cancer,] I thought, even if it's seven months, I want to get to know this person. Sometimes I'd just sit there, and he'd say, "Stop staring at me." Then sometimes we'd talk about the past. One day he said, "Perfect. You were made perfect." I just started crying. I was like, "Thank you, God, for letting me have this moment."

—Actor Drew Barrymore explains how she reconciled with her father, from whom she had been estranged for many years, a short time before his death (Lynch and Gold, 2005: 96, 98). As her previous autobiography *Little Girl Lost* described, Barrymore and her father had experienced problems in the past:

© Frank Trapper/CORBIS

I think my mom and dad were boyfriend and girlfriend for a couple of years, but they were apart by the time I was born. . . . The earliest memory I have of my father isn't pleasant. I was three years old. . . . My mom and I were standing in the kitchen, doing the laundry. . . . Suddenly the door

Actor Drew Barrymore's description of childhood maltreatment in her family makes us aware of the importance of early socialization in all our lives. As an adult, she has experienced many happier times.

swung open and there was this man standing there. I yelled, "Daddy!" Even though I didn't know what he looked like, I just automatically knew it was him.

He paused in the doorway, like he was making a dramatic entrance, and I think he said something, but he was so drunk, it was unintelligible. It sounded more like a growl. We stood there, staring at him. I was so excited to see him. . . . I didn't really know what my dad was like, but I learned real fast. In a blur of anger he roared into the room and threw my mom down on the ground. Then he turned on me. I didn't know what was happening. I was still excited to see him, still hearing the echo of my gleeful yell, "Daddy!" when he picked me up and threw me into the wall. Luckily, half of my body landed on a big sack of laundry, and I wasn't hurt. But my dad didn't even look back at me. He turned and grabbed a bottle of tequila, shattered a bunch of glasses all over the floor, and then stormed out of the house. . . . And that was it. That was the first time I remember seeing my dad. (Barrymore, 1994: 185–187)

Chapter Focus Question

What happens when children do not have an environment that supports positive socialization?

For Barrymore and many other people, early interactions with their parents have had a profound influence on their later lives. Clearly, the parent–child relationship is a significant factor in the process of socialization, which is of interest to sociologists. Although most children are nurtured, trusted, and loved by their parents, Barrymore's experience is not an isolated incident: Large numbers of children experience maltreatment at the hands of family members or other caregivers such as babysitters or child-care workers. Child maltreatment includes physical abuse, sexual abuse, physical neglect, and emotional mistreatment of children and young adolescents. Such maltreatment is of interest to sociologists because it has a serious impact on a child's social growth, behavior, and self-image—all of which develop within the process of socialization. By contrast, children who are treated with respect by their parents are more likely to develop a positive self-image and learn healthy conduct because their parents provide appropriate models of behavior.

In this chapter, we examine why socialization is so crucial, and we discuss both sociological and social psychological theories of human development. We look at the dynamics of socialization—how it occurs and what shapes it. Throughout the chapter, we focus on positive and negative aspects of the socialization process. Before reading on, test your knowledge of early socialization and child care by taking the quiz in Box 3.1.

Why Is Socialization Important Around the Globe?

Socialization **is the lifelong process of social interaction through which individuals acquire a self-identity and the physical, mental, and social skills needed for survival in society.** It is the essential link between the individual and society. Socialization enables each of us to develop our human potential and to learn the ways of thinking, talking, and acting that are necessary for social living.

Socialization is essential for the individual's survival and for human development. The many people who met the early material and social needs of each of us were central to our establishing our own identity. During the first three years of our life, we begin to develop both a unique identity and the ability to manipulate things and to walk. We acquire sophisticated cognitive tools for thinking and for analyzing a wide variety of situations, and we learn effective communication skills. In the process, we begin a relatively long socialization process that culminates in our integration into a complex social and cultural system.

Socialization is also essential for the survival and stability of society. Members of a society must be socialized to support and maintain the existing social structure. From a functionalist perspective, individual conformity to existing norms is not taken for granted; rather, basic individual needs and desires must be balanced against the needs of the social structure. The socialization process is most effective when people conform to the norms of society because they believe that doing so is the best course of action. Socialization enables a society to "reproduce" itself by passing on its culture from one generation to the next.

Although the techniques used to teach newcomers the beliefs, values, and rules of behavior are somewhat similar in many nations, the *content* of socialization differs greatly from society to society. How people walk, talk, eat, make love, and wage war are all functions of the culture in which they are raised. At the same time, we are also influenced by our exposure to subcultures of class, race, ethnicity, religion, and gender. In addition, each of us has unique experiences in our family and friendship groupings. The kind of human being that we become depends greatly on the particular society and social groups that surround us at birth and during early childhood. What we believe about ourselves, our society, and the world does not spring full-blown from inside ourselves; rather, we learn these things from our interactions with others.

Human Development: Biology and Society

What does it mean to be "human"? To be human includes being conscious of ourselves as individuals with unique identities, personalities, and relation-

Box 3.1 Sociology and Everyday Life

How Much Do You Know About Early Socialization and Child Care?

True	False	
T	F	1. In the United States, full-day child care often costs as much per year as college tuition at a public college or university.
T	F	2. The cost of child care is a major problem for many U.S. families.
T	F	3. After-school programs have greatly reduced the number of children who are home alone after school.
T	F	4. The average annual salary of a child-care worker is less than the average yearly salaries for funeral attendants or garbage collectors.
T	F	5. All states require teachers in child-care centers to have training in their field and to pass a licensing examination.
T	F	6. In a family in which child abuse occurs, all the children are likely to be victims.
T	F	7. It is against the law to fail to report child abuse.
T	F	8. Some people are "born" child abusers, whereas others learn abusive behavior from their family and friends.

Answers on page 80.

ships with others. As humans, we have ideas, emotions, and values. We have the capacity to think and to make rational decisions. But what is the source of "humanness"? Are we born with these human characteristics, or do we develop them through our interactions with others?

When we are born, we are totally dependent on others for our survival. We cannot turn ourselves over, speak, reason, plan, or do many of the things that are associated with being human. Although we can nurse, wet, and cry, most small mammals can also do those things. As discussed in Chapter 2, we humans differ from nonhuman animals because we lack instincts and must rely on learning for our survival. Human infants have the potential to develop human characteristics if they are exposed to an adequate socialization process.

Every human being is a product of biology, society, and personal experiences—that is, of heredity and environment or, in even more basic terms, "nature" and "nurture." How much of our development can be explained by socialization? How much by our genetic heritage? Sociologists focus on how humans design their own culture and transmit it from generation to generation through socialization. By contrast, sociobiologists assert that nature, in the form of our genetic makeup, is a major factor in shaping human behavior. ***Sociobiology* is the systematic study of how biology affects social behavior** (Wilson, 1975). According to the zoologist Edward O. Wilson, who pioneered sociobiology, genetic inheritance underlies many forms of social behavior such as war and peace, envy of and concern for others, and competition and cooperation. Most sociologists disagree with the notion that biological principles can be used to explain all human behavior. Obviously, however, some aspects of our physical makeup—such as eye color, hair color,

---

socialization the lifelong process of social interaction through which individuals acquire a self-identity and the physical, mental, and social skills needed for survival in society.

sociobiology the systematic study of how biology affects social behavior.

Box 3.1 Sociology and Everyday Life

Answers to the Sociology Quiz on Early Socialization and Child Care

1. True. Full-day child care typically costs between $4,000 and $10,000 per child per year, which is as much or more than tuition at many public colleges and universities.

2. True. Child care outside the home is a major financial burden, particularly for the one out of every three families with young children but with an income of less than $25,000 a year.

3. False. Although after-school programs have slightly reduced the number of children at home alone after school, nearly seven million school-age children are alone each week while their parents work.

4. True. The average salary for a child-care worker is only $15,430 per year, which is less than the yearly salaries for people in many other employment categories.

5. False. Although all states require hairdressers and manicurists to have about 1,500 hours of training at an accredited school, only 11 states require child-care providers to have any early-childhood training prior to taking care of children.

6. False. In some families, one child may be the victim of repeated abuse whereas others are not.

7. True. In the United States, all states have reporting requirements for child maltreatment; however, compliance with these legal mandates has been inconsistent. Some states have requirements that everyone who suspects abuse or neglect must report it. Other states mandate reporting only by certain persons, such as medical personnel and child-care providers.

8. False. No one is "born" to be an abuser. People learn abusive behavior from their family and friends.

Source: Based on Children's Defense Fund, 2002.

height, and weight—are largely determined by our heredity.

How important is social influence ("nurture") in human development? There is hardly a single behavior that is not influenced socially. Except for simple reflexes, most human actions are social, either in their causes or in their consequences. Even solitary actions such as crying or brushing our teeth are ultimately social. We cry because someone has hurt us. We brush our teeth because our parents (or dentist) told us it was important. Social environment probably has a greater effect than heredity on the way we develop and the way we act. However, heredity does provide the basic material from which other people help to mold an individual's human characteristics.

Our biological and emotional needs are related in a complex equation. Children whose needs are met in settings characterized by affection, warmth, and closeness see the world as a safe and comfortable place and see other people as trustworthy and helpful. By contrast, infants and children who receive less-than-adequate care or who are emotionally rejected or abused often view the world as hostile and have feelings of suspicion and fear.

Problems Associated with Social Isolation and Maltreatment

Social environment, then, is a crucial part of an individual's socialization. Even nonhuman primates such as monkeys and chimpanzees need social contact with others of their species in order to develop properly. As we will see, appropriate social contact is even more important for humans.

Isolation and Nonhuman Primates Researchers have attempted to demonstrate the effects of social isolation on nonhuman primates raised without contact with others of their own species. In a series of laboratory experiments, the psychologists Harry and Margaret Harlow (1962, 1977) took infant rhesus monkeys from their mothers and isolated them in separate cages. Each cage contained two nonliving "mother substitutes" made of wire, one with a feeding bottle attached and the other covered with soft terry cloth but without a bottle. The infant monkeys instinctively clung to the cloth "mother" and would not abandon it until hunger drove them to the bottle attached to the wire "mother." As soon as they were full, they went back to the cloth "mother" seeking warmth, affection, and physical comfort.

The Harlows' experiments show the detrimental effects of isolation on nonhuman primates. When the young monkeys were later introduced to other members of their species, they cringed in the corner. Having been deprived of social contact with other monkeys during their first six months of life, they never learned how to relate to other monkeys or to become well-adjusted adults—they were fearful of or hostile toward other monkeys (Harlow and Harlow, 1962, 1977).

Because humans rely more heavily on social learning than do monkeys, the process of socialization is even more important for us.

Isolated Children Of course, sociologists would never place children in isolated circumstances so that they could observe what happened to them. However, some cases have arisen in which parents or other caregivers failed to fulfill their responsibilities, leaving children alone or placing them in isolated circumstances. From analysis of these situations, social scientists have documented cases in which children were deliberately raised in isolation. A look at the lives of two children who suffered such emotional abuse provides important insights into the importance of a positive socialization process and the negative effects of social isolation.

Anna Born in 1932 to an unmarried, mentally impaired woman, Anna was an unwanted child. She was kept in an attic-like room in her grandfather's house. Her mother, who worked on the farm all day and often went out at night, gave Anna just enough care to keep her alive; she received no other care. Sociologist Kingsley Davis (1940) described Anna's condition when she was found in 1938:

> [Anna] had no glimmering of speech, absolutely no ability to walk, no sense of gesture, not the least capacity to feed herself even when the food was put in front of her, and no comprehension of cleanliness. She was so apathetic that it was hard to tell whether or not she could hear. And all of this at the age of nearly six years.

As Harry and Margaret Harlow discovered, humans are not the only primates that need contact with others. Deprived of its mother, this infant monkey found a substitute.

© Martin Rogers/Getty Images

When she was placed in a special school and given the necessary care, Anna slowly learned to walk, talk, and care for herself. Just before her death at the age of ten, Anna reportedly could follow directions, talk in phrases, wash her hands, brush her teeth, and try to help other children (Davis, 1940).

Genie Almost four decades later, Genie was found in 1970 at the age of thirteen. She had been locked in a bedroom alone, alternately strapped down to a child's potty chair or straitjacketed into a sleeping bag, since she was twenty months old. She had been fed baby food and beaten with a wooden paddle when she whimpered. She had not heard the sounds of human speech because no one talked to her and there was no television or radio in her room (Curtiss, 1977; Pines, 1981). Genie was placed in a pe-

diatric hospital, where one of the psychologists described her condition:

> At the time of her admission she was virtually unsocialized. She could not stand erect, salivated continuously, had never been toilet-trained and had no control over her urinary or bowel functions. She was unable to chew solid food and had the weight, height and appearance of a child half her age. (Rigler, 1993: 35)

In addition to her physical condition, Genie showed psychological traits associated with neglect, as described by one of her psychiatrists:

> If you gave [Genie] a toy, she would reach out and touch it, hold it, caress it with her fingertips, as though she didn't trust her eyes. She would rub it against her cheek to feel it. So when I met her and she began to notice me standing beside her bed, I held my hand out and she reached out and took my hand and carefully felt my thumb and fingers individually, and then put my hand against her cheek. She was exactly like a blind child. (Rymer, 1993: 45)

Extensive therapy was used in an attempt to socialize Genie and develop her language abilities (Curtiss, 1977; Pines, 1981). These efforts met with limited success: In the 1990s, Genie was living in a board-and-care home for retarded adults (see Angier, 1993; Rigler, 1993; Rymer, 1993).

Why do we discuss children who have been the victims of maltreatment in a chapter that looks at the socialization process? The answer lies in the fact that such cases are important to our understanding of the socialization process because they show the importance of this process and reflect how detrimental

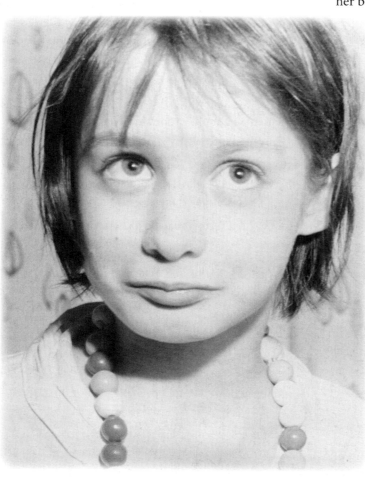

© Bettmann/CORBIS

A victim of extreme child abuse, Genie was isolated from human contact and tortured until she was rescued at the age of thirteen. Subsequent attempts to socialize her were largely unsuccessful.

social isolation and neglect can be to the well-being of people.

Child Maltreatment What do the terms *child maltreatment* and *child abuse* mean to you? When asked what constitutes child maltreatment, many people first think of cases that involve severe physical injuries or sexual abuse. However, neglect is the most frequent form of child maltreatment (Dubowitz et al., 1993). Child neglect occurs when children's basic needs—including emotional warmth and security, adequate shelter, food, health care, education, clothing, and protection—are not met, regardless of cause (Dubowitz et al., 1993: 12). Neglect often involves acts of omission (where parents or caregivers fail to provide adequate physical or emotional care for children) rather than acts of commission (such as physical or sexual abuse). Of course, what constitutes child maltreatment differs from society to society.

Social Psychological Theories of Human Development

Over the past hundred years, a variety of psychological and sociological theories have been developed not only to explain child abuse but also to describe how a positive process of socialization occurs. Let's look first at several social psychological theories that focus primarily on how the individual personality develops.

Freud and the Psychoanalytic Perspective

The basic assumption in Sigmund Freud's (1924) psychoanalytic approach is that human behavior and personality originate from unconscious forces

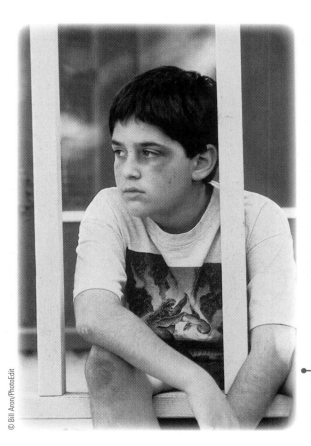

© Bill Aron/PhotoEdit

© Jose Luis Pelaez, Inc./Getty Images

What are the consequences to children of isolation and physical abuse, as contrasted with social interaction and parental affection? Sociologists emphasize that social environment is a crucial part of an individual's socialization.

© Mary Evans//SIGMUND FREUD COPYRIGHTS/The Image Works

Sigmund Freud, founder of the psychoanalytic perspective.

within individuals. Freud (1856–1939), who is known as the founder of psychoanalytic theory, developed his major theories in the Victorian era, when biological explanations of human behavior were prevalent. It was also an era of extreme sexual repression and male dominance when compared to contemporary U.S. standards. Freud's theory was greatly influenced by these cultural factors, as reflected in the importance he assigned to sexual motives in explaining behavior. For example, Freud based his ideas on the belief that people have two basic tendencies: the urge to survive and the urge to procreate.

According to Freud (1924), human development occurs in three states that reflect different levels of the personality, which he referred to as the *id, ego,* and *superego.* The **id is the component of personality that includes all of the individual's basic biological drives and needs that demand immediate gratification.** For Freud, the newborn child's personality is all id, and from birth the child finds that urges for self-gratification—such as wanting to be held, fed, or changed—are not going to be satisfied immediately. However, id remains with people throughout their life in the form of *psychic energy,* the urges and desires that account for behavior. By contrast, the second level of personality—the *ego*—develops as infants discover that their most basic desires are not al-

ways going to be met by others. The **ego is the rational, reality-oriented component of personality that imposes restrictions on the innate pleasure-seeking drives of the id.** The ego channels the desire of the id for immediate gratification into the most advantageous direction for the individual. The third level of personality—the *superego*—is in opposition to both the id and the ego. The **superego, or conscience, consists of the moral and ethical aspects of personality.** It is first expressed as the recognition of parental control and eventually matures as the child learns that parental control is a reflection of the values and moral demands of the larger society. When a person is well adjusted, the ego successfully manages the opposing forces of the id and the superego. ▶ Figure 3.1 illustrates Freud's theory of personality.

Although subject to harsh criticism, Freud's theory made people aware of the importance of early

© AFP/Getty Images

Jean Piaget, a pioneer in the field of cognitive development.

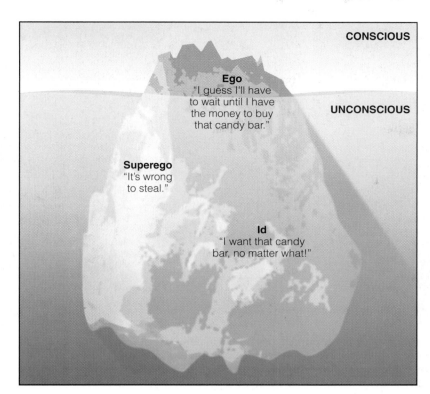

CONSCIOUS

Ego
"I guess I'll have
to wait until I have
the money to buy
that candy bar."

UNCONSCIOUS

Superego
"It's wrong
to steal."

Id
"I want that candy
bar, no matter what!"

▶ **Figure 3.1 Freud's Theory
of Personality**
This illustration shows how Freud
might picture a person's internal
conflict over whether to commit
an antisocial act such as stealing a
candy bar. In addition to dividing
personality into three components,
Freud theorized that our personalities
are largely unconscious—hidden from
our normal awareness. To dramatize
his point, Freud compared conscious
awareness (portions of the ego
and superego) to the visible tip of
an iceberg. Most of personality—
including the id, with its raw desires
and impulses—lies submerged in our
subconscious.

childhood experiences, including abuse and neglect. His theories have also had a profound influence on contemporary mental health practitioners and on other human development theories.

Piaget and Cognitive Development

Jean Piaget (1896–1980), a Swiss psychologist, was a pioneer in the field of cognitive (intellectual) development. Cognitive theorists are interested in how people obtain, process, and use information—that is, in how we think. Cognitive development relates to changes over time in how we think.

Piaget (1954) believed that in each stage of development (from birth through adolescence), children's activities are governed by their perception of the world around them. His four stages of cognitive development are organized around specific tasks that, when mastered, lead to the acquisition of new mental capacities, which then serve as the basis for the next level of development. Piaget emphasized that all children must go through each stage in sequence before moving on to the next one, although some children move through them faster than others.

1. *Sensorimotor stage* (birth to age two). During this period, children understand the world only through sensory contact and immediate action because they cannot engage in symbolic thought or use language. Toward the end of the second year, children comprehend *object permanence*; in other words, they start to realize that objects continue to exist even when the items are out of sight.

---

id Sigmund Freud's term for the component of personality that includes all of the individual's basic biological drives and needs that demand immediate gratification.

ego according to Sigmund Freud, the rational, reality-oriented component of personality that imposes restrictions on the innate pleasure-seeking drives of the id.

superego Sigmund Freud's term for the conscience, consisting of the moral and ethical aspects of personality.

Psychologist Jean Piaget identified four stages of cognitive development, including the preoperative stage, in which children have limited ability to realize that physical objects may change in shape or appearance. Piaget poured liquid from one beaker into a taller, narrower beaker and then asked children about the amounts of liquid in each beaker.

2. *Preoperational stage* (age two to seven). In this stage, children begin to use words as mental symbols and to form mental images. However, they still are limited in their ability to use logic to solve problems or to realize that physical objects may change in shape or appearance while still retaining their physical properties. For example, Piaget showed children two identical beakers filled with the same amount of water. After the children agreed that both beakers held the same amount of water, Piaget poured the water from one beaker into a taller, narrower beaker and then asked them about the amounts of water in each beaker. Those still in the preoperational stage believed that the taller beaker held more water because the water line was higher than in the shorter, wider beaker.

3. *Concrete operational stage* (age seven to eleven). During this stage, children think in terms of tangible objects and actual events. They can draw conclusions about the likely physical consequences of an action without always having to try the action out. Children begin to take the role of others and start to empathize with the viewpoints of others.

4. *Formal operational stage* (age twelve through adolescence). By this stage, adolescents are able to engage in highly abstract thought and understand places, things, and events they have never seen. They can think about the future and evaluate different options or courses of action.

Piaget provided useful insights on the emergence of logical thinking as the result of biological maturation and socialization. However, critics have noted several weaknesses in Piaget's approach to cognitive development. For one thing, the theory says little about individual differences among children, nor does it provide for cultural differences. For another, as the psychologist Carol Gilligan (1982) has noted, Piaget did not take into account how gender affects the process of social development.

Kohlberg and the Stages of Moral Development

Lawrence Kohlberg (1927–1987) elaborated on Piaget's theories of cognitive reasoning by conducting a series of studies in which children, adolescents, and adults were presented with moral dilemmas that took the form of stories. Based on his findings, Kohlberg (1969, 1981) classified moral reasoning into three sequential levels:

1. *Preconventional level* (age seven to ten). Children's perceptions are based on punishment and obedience. Evil behavior is that which is likely to be punished; good conduct is based on obedience and avoidance of unwanted consequences.
2. *Conventional level* (age ten through adulthood). People are most concerned with how they are perceived by their peers and with how one conforms to rules.
3. *Postconventional level* (few adults reach this stage). People view morality in terms of individual rights; "moral conduct" is judged by principles based on human rights that transcend government and laws.

Although Kohlberg presents interesting ideas about the moral judgments of children, some critics have challenged the universality of his stages of moral development. They have also suggested that the elaborate "moral dilemmas" he used are too abstract for children. In one story, for example, a husband contemplates stealing for his critically ill wife medicine that he cannot afford. When questions are made simpler, or when children and adolescents are observed in natural (as opposed to laboratory) settings, they often demonstrate sophisticated levels of moral reasoning (Darley and Shultz, 1990; Lapsley, 1990).

Gilligan's View on Gender and Moral Development

Psychologist Carol Gilligan (b. 1936) is one of the major critics of Kohlberg's theory of moral development. According to Gilligan (1982), Kohlberg's model was developed solely on the basis of research with male respondents, and women and men often have divergent views on morality based on differences in socialization and life experiences. Gilligan believes that men become more concerned with law and order but that women tend to analyze social relationships and the social consequences of behavior. For example, in Kohlberg's story about the man who is thinking about stealing medicine for his wife, Gilligan argues that male respondents are more likely to use *abstract standards* of right and wrong, whereas female respondents are more likely to be concerned about what *consequences* his stealing the drug might have for the man and his family. Does this constitute a "moral deficiency" on the part of either women or men? Not according to Gilligan.

Subsequent research that directly compared women's and men's reasoning about moral dilemmas has supported some of Gilligan's assertions but not others. For example, some other researchers have not found that women are more compassionate than men (Tavris, 1993). Overall, however, Gilligan's argument that people make moral decisions according to both abstract principles of justice and principles of compassion and care is an important contribution to our knowledge about moral reasoning.

Sociological Theories of Human Development

Although social scientists acknowledge the contributions of psychoanalytic and psychologically based explanations of human development, sociologists believe that it is important to bring a sociological perspective to bear on how people develop an awareness of self and learn about the culture in which

How do these teenagers' perceptions of the world differ from their perceptions ten years earlier, according to Piaget?

© PhotoAlto/Alamy

they live. According to a sociological perspective, we cannot form a sense of self or personal identity without intense social contact with others. The self represents the sum total of perceptions and feelings that an individual has of being a distinct, unique person—a sense of who and what one is. When we speak of the "self," we typically use words such as *I, me, my, mine,* and *myself* (Cooley, 1998/1902). This sense of self (also referred to *self-concept*) is not present at birth; it arises in the process of social experience. **Self-concept is the totality of our beliefs and feelings about ourselves.** Four components make up our self-concept: (1) the physical self ("I am tall"), (2) the active self ("I am good at soccer"), (3) the social self ("I am nice to others"), and (4) the psychological self ("I believe in world peace"). Between early and late childhood, a child's focus tends to shift from the physical and active dimensions of self toward the social and psychological aspects. Self-concept is the foundation for communication with others; it continues to develop and change throughout our lives.

Our *self-identity* is our perception about what kind of person we are. As we have seen, socially isolated children do not have typical self-identities because they have had no experience of "humanness." According to symbolic interactionists, we do not know who we are until we see ourselves as we believe that others see us. The perspectives of symbolic interactionists Charles Horton Cooley and George Herbert Mead help us understand how our self-identity is developed through our interactions with others.

Cooley and the Looking-Glass Self

According to the sociologist Charles Horton Cooley (1864–1929), the ***looking-glass self* refers to the way in which a person's sense of self is derived from the perceptions of others.** Our looking-glass self is based on our perception of *how* other people think of us (Cooley, 1998/1902). As ▶ Figure 3.2 shows, the looking-glass self is a self-concept derived from a three-step process:

1. We imagine how our personality and appearance will look to other people.
2. We imagine how other people judge the appearance and personality that we think we present.

▶ **Figure 3.2 How the Looking-Glass Self Works**

Source: Based on Katzer, Cook, and Crouch, 1991.

3. We develop a self-concept. If we think the evaluation of others is favorable, our self-concept is enhanced. If we think the evaluation is unfavorable, our self-concept is diminished. (Cooley, 1998/1902)

The self develops only through contact with others, just as social institutions and societies do not exist independently of the interaction of individuals (Schubert, 1998).

Mead and Role-Taking

George Herbert Mead (1863–1931) extended Cooley's insights by linking the idea of self-concept to *role-taking*—**the process by which a person mentally assumes the role of another person or group in order to understand the world from that person's or group's point of view.** Role-taking often occurs through play and games, as children try out different roles (such as being mommy, daddy, doctor, or teacher) and gain an appreciation of them. First, people come to take the role of the other (role-taking). By taking the roles of others, the individual hopes to ascertain the intention or direction of the acts of others. Then the person begins to construct his or her own roles (role-making) and to anticipate other individuals' responses. Finally, the person plays at her or his particular role (role-playing).

self-concept the totality of our beliefs and feelings about ourselves.

looking-glass self Charles Horton Cooley's term for the way in which a person's sense of self is derived from the perceptions of others.

role-taking the process by which a person mentally assumes the role of another person in order to understand the world from that person's point of view.

According to sociologist George Herbert Mead, the self develops through three stages. In the preparatory stage, children imitate others; in the play stage, children pretend to take the roles of specific people; and in the game stage, children become aware of the "rules of the game" and the expectations of others.

CHAPTER 3 • SOCIALIZATION

Sociology *Works!*

"Good Job!": Mead's Generalized Other and the Issue of Excessive Praise

Hang out at a playground, visit a school, or show up at a child's birthday party, and there's one phrase you can count on hearing repeatedly: "Good job!" Even tiny infants are praised for smacking their hands together ("Good clapping!"). Many of us blurt out these judgments of our children to the point that it has become almost a verbal tic. (Kohn, 2001)

Educational analyst Alfie Kohn describes the common practice of praising children for practically everything they say or do. According to Kohn, excessive praise or unearned compliments may be problematic for children because, rather than bolstering their self-esteem, such praise may increase a child's dependence on adults. As children increasingly rely on constant praise and on significant others to identify what is good or bad about their performance, they may not develop the ability to make meaningful judgments about what they have done. As Kohn suggests (2001), "Sadly, some of these kids will grow into adults who continue to need someone to pat them on the head and tell them whether what they did was OK."

Kohn's ideas remind us of the earlier sociological insights of George Herbert Mead, who described how children learn to take into account the expectations of the larger society and to balance the "I" (the subjective element of the self: the spontaneous and unique traits of each person) with the "me" (the objective element of the self: the internalized attitudes and demands of other members of society and the individual's awareness of those demands). As Mead (1934: 160) stated, "What goes on in the game goes on in the life of the child at all times. He is continually taking the attitudes of those about him, especially the roles of those who in some sense control him and on whom he depends." According to Mead, role-taking is vital to the formation of a mature sense of self as each individual learns to visualize the intentions and expectations of other people and groups. Excessively praising children may make it more difficult for them to develop a positive self-concept and visualize an accurate picture of what is expected of them as they grow into young adulthood.

Does this mean that children should not be praised? Definitely not! It means that we should think about when and how to praise children. What children may need sometimes is not praise, but encouragement. As child development specialist Docia Zavitkovsky has stated,

I sometimes say that praise is fine "when praise is due." We get into the habit of praising when it isn't praise that is appropriate but encouragement. For example, we're always saying to young children: "Oh, what a beautiful picture," even when their pictures aren't necessarily beautiful. So why not really look at each picture? Maybe a child has painted a picture with many wonderful colors. Why don't we comment on that—on the reality of the picture? (qtd. in Scholastic Parent & Child, 2007)

From this perspective, positive feedback can have a very important influence on a child's self-esteem because he or she can learn how to do a "good job" when engaging in a specific activity or accomplishing a task rather than simply being praised for any effort expended. Mead's concept of the *generalized other* makes us aware of the importance of other people's actions in how self-concept develops.

Reflect & Analyze

To what extent are parents responsible for how self-concept develops in their child? Also, when we are dealing with peers, how might we thoughtfully use the phrase "Good job!" without making it into an overworked expression?

According to Mead (1934), in the early months of life, children do not realize that they are separate from others. However, they do begin early on to see a mirrored image of themselves in others. Shortly after birth, infants start to notice the faces of those around them, especially the significant others, whose faces start to have meaning because they are associated with experiences such as feeding and cuddling. ***Significant others*** **are those persons whose care, affection, and approval are especially desired and**

who are most important in the development of the self. Gradually, we distinguish ourselves from our caregivers and begin to perceive ourselves in contrast to them. As we develop language skills and learn to understand symbols, we begin to develop a self-concept. When we can represent ourselves in our minds as objects distinct from everything else, our self has been formed.

Mead (1934) divided the self into the "I" and the "me." The "I" is the subjective element of the self and represents the spontaneous and unique traits of each person. The "me" is the objective element of the self, which is composed of the internalized attitudes and demands of other members of society and the individual's awareness of those demands. Both the "I" and the "me" are needed to form the social self. The unity of the two constitutes the full development of the individual. According to Mead, the "I" develops first, and the "me" takes form during the three stages of self-development:

1. During the *preparatory stage,* up to about age three, interactions lack meaning, and children largely imitate the people around them. At this stage, children are preparing for role-taking.
2. In the *play stage,* from about age three to five, children learn to use language and other symbols, thus enabling them to pretend to take the roles of specific people. At this stage, they begin to see themselves in relation to others, but they do not see role-taking as something they have to do.
3. During the *game stage,* which begins in the early school years, children understand not only their own social position but also the positions of others around them. In contrast to play, games are structured by rules, are often competitive, and involve a number of other "players." At this time, children become concerned about the demands and expectations of others and of the larger society. Mead used the example of a baseball game to describe this stage because children, like baseball players, must take into account the roles of all the other players at the same time. Mead's concept of the ***generalized other*** **refers to the child's awareness of the demands and expectations of the society as a whole or of the child's subculture.**

How useful are symbolic interactionist perspectives in enhancing our understanding of the social-

ization process? Certainly, this approach contributes to our understanding of how the self develops (see "Sociology Works!"). However, this approach has certain limitations. Sociologist Anne Kaspar (1986) suggests that Mead's ideas about the social self may be more applicable to men than to women because women are more likely to experience inherent conflicts between the meanings they derive from their personal experiences and those they take from the culture, particularly in regard to balancing the responsibilities of family life and paid employment. (This chapter's Concept Quick Review summarizes the major theories of human development.)

Recent Symbolic Interactionist Perspectives

The symbolic interactionist approach emphasizes that socialization is a collective process in which children are active and creative agents, not just passive recipients of the socialization process. From this view, childhood is a *socially constructed* category (Adler and Adler, 1998). Children are capable of actively constructing their own shared meanings as they acquire language skills and accumulate interactive experiences (Qvortrup, 1990). According to the "orb web model" of the sociologist William A. Corsaro (1985, 1997), children's cultural knowledge reflects not only the beliefs of the adult world but also the unique interpretations and aspects of the children's own peer culture. Corsaro (1992: 162) states that *peer culture* is "a stable set of activities or routines, artifacts, values, and concerns that children produce and share." This peer culture emerges through interactions as children "borrow" from the adult culture but transform it so that it fits their own situation. In fact, Corsaro (1992) believes that the peer group is the

---

significant others those persons whose care, affection, and approval are especially desired and who are most important in the development of the self.

generalized other George Herbert Mead's term for the child's awareness of the demands and expectations of the society as a whole or of the child's subculture.

CONCEPT QUICK REVIEW

Psychological and Sociological Theories of Human Development

Social Psychological Theories of Human Development	Freud's psychoanalytic perspective	Children first develop the id (drives and needs), then the ego (restrictions on the id), and then the superego (moral and ethical aspects of personality).
	Piaget's cognitive development	Children go through four stages of cognitive (intellectual) development, going from understanding only through sensory contact to engaging in highly abstract thought.
	Kohlberg's stages of moral development	People go through three stages of moral development, from avoidance of unwanted consequences to viewing morality based on human rights.
	Gilligan: gender and moral development	Women go through stages of moral development from personal wants to the greatest good for themselves and others.
Sociological Theories of Human Development	Cooley's looking-glass self	A person's sense of self is derived from his or her perception of how others view him or her.
	Mead's three stages of self-development	In the preparatory stage, children imitate the people around them; in the play stage, children pretend to take the roles of specific people; and in the game stage, children learn the demands and expectations of roles.

most significant arena in which children and young people acquire cultural knowledge.

Agents of Socialization

Agents of socialization **are the persons, groups, or institutions that teach us what we need to know in order to participate in society.** We are exposed to many agents of socialization throughout our lifetime; in turn, we have an influence on those socializing agents and organizations. Here, we look at the most pervasive ones in childhood—the family, the school, peer groups, and the mass media.

The Family

The family is the most important agent of socialization in all societies. From infancy, our families transmit cultural and social values to us. As discussed later

in this book, families vary in size and structure. Some families consist of two parents and their biological children, whereas others consist of a single parent and one or more children. Still other families reflect changing patterns of divorce and remarriage, and an increasing number are made up of same-sex partners and their children. Over time, patterns have changed in some two-parent families so that fathers, rather than mothers, are the primary daytime agents of socialization for their young children.

Theorists using a functionalist perspective emphasize that families serve important functions in society because they are the primary locus for the procreation and socialization of children. Most of us form an emerging sense of self and acquire most of our beliefs and values within the family context. We also learn about the larger dominant culture (including language, attitudes, beliefs, values, and norms) and the primary subcultures to which our parents and other relatives belong.

Families are also the primary source of emotional support. Ideally, people receive love, understanding,

As this celebration attended by three generations of family members illustrates, socialization enables society to "reproduce" itself.

security, acceptance, intimacy, and companionship within families. The role of the family is especially significant because young children have little social experience beyond the family's boundaries; they have no basis for comparing or evaluating how they are treated by their own family.

To a large extent, the family is where we acquire our specific social position in society. From birth, we are a part of the specific racial, ethnic, class, religious, and regional subcultural grouping of our family. Studies show that families socialize their children somewhat differently based on race, ethnicity, and class (Kohn, 1977; Kohn et al., 1990; Harrison et al., 1990). For example, sociologist Melvin Kohn (1977; Kohn et al., 1990) has suggested that social class (as measured by parental occupation) is one of the strongest influences on what and how parents teach their children. On the one hand, working-class parents, who are closely supervised and expected to follow orders at work, typically emphasize to their children the importance of obedience and conformity. On the other hand, parents from the middle and professional classes, who have more freedom and flexibility at work, tend to give their children more freedom to make their own decisions and to be creative. Kohn

concluded that differences in the parents' occupations are a better predictor of child-rearing practices than is social class itself.

Whether or not Kohn's findings are valid today, the issues he examined make us aware that not everyone has the same family experiences. Many factors—including our cultural background, nation of origin, religion, and gender—are important in determining how we are socialized by family members and others who are a part of our daily life.

Conflict theorists stress that socialization contributes to false consciousness—a lack of awareness and a distorted perception of the reality of class as it affects all aspects of social life. As a result, socialization reaffirms and reproduces the class structure in the next generation rather than challenging the conditions that presently exist. For example, children in low-income families may be unintentionally socialized to believe that acquiring an education and aspiring to lofty ambitions are pointless because of existing economic conditions in the family. By contrast, middle- and upper-income families typically instill ideas of monetary and social success in children while encouraging them to think and behave in "socially acceptable" ways.

The School

As the amount of specialized technical and scientific knowledge has expanded rapidly and as the amount of time that children are in educational settings has increased, schools continue to play an enormous role in the socialization of young people. For many people, the formal education process is an undertaking that lasts up to twenty years.

agents of socialization the persons, groups, or institutions that teach us what we need to know in order to participate in society.

Trying to Go It Alone: Runaway Adolescents and Teens

Home. *Family*. The ideal is that these words evoke a sense of well-being—feelings such as love, understanding, acceptance, and security. The reality however is that for many of us, especially adolescents and teens, these words trigger negative feelings—fear, anxiety, dread—and an urgent desire to escape. The most recent statistics available indicate that between 1.6 and 2.8 million youth run away each year in the United States. That means one out of every seven children in this country will run away before the age of 18. Included in these numbers are not just runaway but also "throwaway" children, youth who have been forced out of their homes by parents or guardians, or, because they have turned 18, forced out of a foster care system.

Based on research conducted in 2002, it is estimated that over 70% of runaway and throwaway youth are endangered, either having already been harmed or being potentially at risk of being harmed, most commonly physically or sexually. Youths' substance dependency is the second most common source of being at risk for harm. Consider the following additional statistics:

- Thirty-two percent of runaway and homeless youth have attempted suicide at some point in their lives.
- Approximately 48.2% of female youths living on the street and 33.2% of female youths living in a shelter reported having been pregnant at least once.
- Fifty percent of homeless youth age 16 or older reported having dropped out of school, having been expelled, or having been suspended.
- Seventy-five percent of runaways who are on the street for two or more weeks will become involved in theft, drugs, or pornography, while one out of every three teens on the street will be lured into prostitution within 48 hours of leaving home.

© SW Productions/Getty Images

Conflict, Instability, Crisis, Poor Communication, Abuse

Forty-seven percent of runaway and homeless youth indicate that conflict between them and their parent or guardian is a major problem. Thirty-four percent of runaway youth (girls and boys) report sexual abuse and 43% of runaway youth (girls and boys) report physical abuse before leaving home, abuse that increases youths' risk of being assaulted or raped—or both—on the street. Other problems that runaways report include alcohol and drug use, sexual orientation issues, and mental or physical health issues. National data indicate that the majority—52%—of kids in crisis are aged 15–17. However, the number of very young (under 12) and over 18 is growing at a faster rate than the middle age group.

© Jeff Greenberg/The Image Works

Life on the Street Is Hard

This homeless girl on a sidewalk in the SoHo neighborhood of Manhattan reflects the reality that youth aged 12–17 are at higher risk for homelessness than adults. Available data show that 12% of runaway and homeless youth spend at least one night outside in a park, on a street, under a bridge or overhang, or on a rooftop. Seven percent of youth in runaway and homeless shelters and 14% of youth on the street surveyed in 1995 reported having traded sex for money, food, shelter, or drugs in the previous twelve months. Other means of survival include shelters and soup kitchens, panhandling, and stealing. In addition to malnutrition, HIV and other sexually transmitted diseases, unwanted pregnancies, and drug and alcohol abuse, major depression, conduct disorders, and post-traumatic stress disorder are higher among runaway than non-runaway youth.

What happens to humans' socialization when, as youths, they must fend for themselves to meet their own physical and emotional needs? The images in this essay give you a chance to look at the lives of runaways, from risk factors and precipitating events to means of survival and resources, and the longer-term effects running away or being thrown away has on individuals and their communities. This essay also provides a glimpse of homeless children in global perspective. As you look at these images,

consider the short- and long-term impact persons aged 12 (sometimes younger) to 18 attempting self-sufficiency has on a society.

© Tim Boyle/Getty Images

Society's Safety Nets

Males and females run away in equal numbers, although females are more likely to seek help through shelters and hotlines such as The National Runaway Switchboard (NRS), where callers are 77% female. Established in 1971 in Chicago, Illinois, and offering services provided in part through the U.S. Department of Health and Human Services, the NRS's mission is "to help keep America's runaway and at-risk youth safe and off the streets." In 2007, NRS handled over 176,000 calls. NRS data shows that the organization is serving more youth who are contemplating running away, instead of already having run away, than in the past. Children under 12 are the fastest-growing group of callers, and in 2007, NRS received more crisis calls than it had in the past. Friends and relatives are 51% of crisis callers' means of survival, and 56% of crisis callers have been on the street for one week or less. However, the organization has seen a large increase in the number of callers who have been on the street one to four weeks and two to six months.

Home Free Program

In addition to providing its hotline and online services and a professionally developed runaway prevention curriculum titled "Let's Talk," the NRS partners with several organizations, most notably Greyhound Lines, Inc., to offer its Home Free Program, promoted in this poster by musician Ludacris. More than 10,000 youth have been reunited with their families, free of charge, since the program was started. But Home Free provides more than a bus ticket home, from helping kids and their parents or guardians to develop a plan before the kids' return, to contacting the kids to make sure they got home safely and to see how their return is going, to connecting kids with resources in their community.

Courtesy of the National Runaway Switchboard

Don't Runaway Love, you are not on your own.
Open your eyes and come back with me!

READY TO GO HOME?

GREYHOUND® AND THE NATIONAL RUNAWAY SWITCHBOARD CAN HELP.

If you're a runaway between the ages of 12 and 20 and you need help returning home or someone to talk to, call

1-800-RUNAWAY OR VISIT
WWW.1800RUNAWAY.ORG

© DEAN J. KOEPFLER/Tacoma News Tribune/MCT/Landov

Catching Up

As noted in this essay's introduction, 50% of homeless youth age 16 or older reported having dropped out of school, having been expelled, or having been suspended. Even if school was not a major problem before running away, once adolescents and teens, especially teens, have run away, the disruption to their education may be so significant that they end up dropping out of school. Since 40% of youth in shelters and on the street come from families that receive public assistance or live in publicly supported housing, losing additional ground educationally can be especially devastating. Catching up and moving ahead takes determination—and opportunities such as ones provided by the Skills, Training, Employment, Preparation Services (STEPS) program at the Tacoma, Washington, Goodwill, where teens can study for a high school General Equivalency Development (GED) test.

© RAFIQUR RAHMAN/Reuters/Landov

Global Perspective

These homeless children, or "street kids," sleeping in a traffic island in Dhaka, Bangladesh, in early 2008, are among the 25 million children, adolescents, and teens in developing countries in Asia, Africa, and the Americas who live, work, and sleep on the streets or in shelters such as railway stations. Nongovernmental aid organizations (NGOs) and community members do try to protect and help street kids, and recently governments such as India's have been working with NGOs to address the problems of homeless children. Nonetheless, whether orphaned by war or disease, abandoned, lost, or runaways, the world's street children lack access to adequate health care, nutrition, and hygiene. They are also at serious risk of being recruited or entrapped and then transported for sexual exploitation or forced labor in the underground economy known as human trafficking, whose victims number an estimated 2.5 million at any given time.

Reflect & Analyze

1. Apply the symbolic interactionist perspective to runaways and other street kids. How does being a runaway or a throwaway child affect a child's self-identity and influence her or his ability to thrive, or simply survive, in society?
2. What might other primary agents of socialization such as schools, peer groups, and the mass media do to help reduce the likelihood that children in their communities will run away, and how would these changes help?
3. Did you ever run away, or do you know someone who did? Do you or the person you know still experience the effects of that action? How so, and how if at all do you imagine the experience will affect the course of the rest of your life?

Turning to Video

Watch the ABC video *Girls Behaving Badly: Violent Girls* (running time 8:29), available on the Kendall Companion Website and through Cengage Learning eResources accounts. In 1999, Judge Cindy Lederman of Miami–Dade County Juvenile Court founded a program known as the Girls Advocacy Project, or GAP, which for seven years helped girls in Florida's Miami–Dade County Juvenile Detention Center, filling gaps in both the juvenile justice system and the girls' lives through educational group discussions and other support. As you watch the video, think about the photographs, commentary, and questions that you encountered in this photo essay. After you've watched the video, consider these questions: What risk factors do violent girls have in common with runaway and throwaway youths, and in what ways are running away and acting violently similar—and different?

girls_behaving_badly

File   Edit   View   Window   Help

00:03:24

As the number of one-parent families and families in which both parents work outside the home has increased dramatically, the number of children in day-care and preschool programs has also grown rapidly. Currently, about 60 percent of all U.S. preschool children are in day care, either in private homes or institutional settings, and this percentage continues to climb (Children's Defense Fund, 2002). Generally, studies have found that quality day-care and preschool programs have a positive effect on the overall socialization of children. These programs provide children with the opportunity to have frequent interactions with teachers and to learn how to build their language and literacy skills. High-quality programs also have a positive effect on the academic performance of children, particularly those from low-income families. Today, however, the cost of child-care programs has become a major concern for many families.

Although schools teach specific knowledge and skills, they also have a profound effect on children's self-image, beliefs, and values. As children enter school for the first time, they are evaluated and systematically compared with one another by the teacher. A permanent, official record is kept of each child's personal behavior and academic activities. From a functionalist perspective, schools are responsible for (1) socialization, or teaching students to be productive members of society; (2) transmission of culture; (3) social control and personal development; and (4) the selection, training, and placement of individuals on different rungs in the society (Ballantine, 2001).

In contrast, conflict theorists assert that students have different experiences in the school system depending on their social class, their racial–ethnic background, the neighborhood in which they live, their gender, and other factors. According to the sociologists Samuel Bowles and Herbert Gintis (1976), much of what happens in school amounts to teaching a hidden curriculum in which children learn to be neat, to be on time, to be quiet, to wait their turn, and to remain attentive to their work. Thus, schools do not socialize children for their

© Tony Freeman/PhotoEdit

Students are sent to school to be educated. However, what else will they learn in school beyond the academic curriculum? Sociologists differ in their responses to this question.

own well-being but rather for their later roles in the workforce, where it is important to be punctual and to show deference to supervisors. Students who are destined for leadership or elite positions acquire different skills and knowledge than those who will enter working-class and middle-class occupations (see Cookson and Persell, 1985).

Peer Groups

As soon as we are old enough to have acquaintances outside the home, most of us begin to rely heavily on peer groups as a source of information and approval about social behavior. A *peer group* **is a group of people who are linked by common in-**

terests, equal social position, and (usually) similar age. In early childhood, peer groups are often composed of classmates in day care, preschool, and elementary school. Recent studies have found that preadolescence—the latter part of the elementary school years—is an age period in which children's peer culture has an important effect on how children perceive themselves and how they internalize society's expectations (Adler and Adler, 1998). In adolescence, peer groups are typically made up of people with similar interests and social activities. As adults, we continue to participate in peer groups of people with whom we share common interests and comparable occupations, income, and/or social position.

Peer groups function as agents of socialization by contributing to our sense of "belonging" and our feelings of self-worth. As early as the preschool years, peer groups provide children with an opportunity for successful adaptation to situations such as gaining access to ongoing play, protecting shared activities from intruders, and building solidarity and mutual trust during ongoing activities (Corsaro, 1985; Rizzo and Corsaro, 1995). Unlike families and schools, peer groups provide children and adolescents with some degree of freedom from

parents and other authority figures (Corsaro, 1992). Although peer groups afford children some degree of freedom, they also teach cultural norms such as what constitutes "acceptable" behavior in a specific situation. Peer groups simultaneously reflect the larger culture and serve as a conduit for passing on culture to young people. As a result, the peer group is both a product of culture and one of its major transmitters (Elkin and Handel, 1989).

Is there such a thing as "peer pressure"? Individuals must earn their acceptance with their peers by conforming to a given group's norms, attitudes, speech patterns, and dress codes. When we conform to our peer group's expectations, we are rewarded; if we do not conform, we may be ridiculed or even expelled from the group. Conforming to the demands of peers frequently places children and adolescents at cross-purposes with their parents. For example, young people are frequently under pressure to obtain certain valued material possessions (such as toys, clothing, athletic shoes, or cell phones); they then pass the pressure on to their parents through emotional pleas to purchase the desired items. Peer pressure and the adult tensions that often accompany this kind of pressure are not unique to families in the United States (see Box 3.2).

Mass Media

An agent of socialization that has a profound impact on both children and adults is the *mass media,* composed of large-scale organizations that use print or electronic means (such as radio, television, film, and the Internet) to communicate with large numbers of people. The media function as socializing agents in several ways: (1) they inform us about events; (2) they introduce us to a wide variety of people; (3) they provide an array of viewpoints on current issues; (4) they make us aware of products and services that, if we purchase them, will supposedly help us to be accepted by others; and (5) they entertain us by providing the opportunity to live vicariously (through other people's experiences). Although most of us take for granted that the media play an important part in contemporary socialization, we frequently underestimate the enormous influence this agent of socialization may have on children's attitudes and behavior.

© Michael Newman/PhotoEdit

The pleasure of participating in activities with friends is one of the many attractions of adolescent peer groups. What groups have contributed the most to your sense of belonging and self-worth?

Recent studies have shown that U.S. children, on average, are spending more time each year in front of TV sets, computers, and video games. According to the Annenberg Public Policy Center (University of Pennsylvania) study on media in the home, "The introduction of new media continues to transform the environment in American homes with children. . . . Rather than displacing television as the dominant medium, new technologies have supplemented it, resulting in an aggregate increase in electronic media penetration and use by America's youth" (qtd. in Dart, 1999: A5). It is estimated that U.S. children spend 2.5 hours per day watching television programs and about 2 hours with computers, video games, or a VCR/DVD, which adds up to about 1,642 hours per year. By contrast, U.S. children spend about 1,000 hours per year in school. Considering television watching time alone, by the time that students graduate from high school, they will have spent more time in front of the television set than sitting in the classroom.

Parents, educators, social scientists, and public officials have widely debated the consequences of young people watching that much television. Television has been praised for offering numerous positive experiences to children. Some scholars suggest that television (when used wisely) can enhance children's development by improving their language abilities, concept-formation skills, and reading skills and by encouraging prosocial development (Winn, 1985). However, other studies have shown that children and adolescents who spend a lot of time watching television often have lower grades in school, read fewer books, exercise less, and are overweight (American Academy of Child and Adolescent Psychiatry, 1997).

Of special concern to many people is the issue of television violence. It is estimated that the typical young person who watches 28 hours of television per week will have seen 16,000 simulated murders and 200,000 acts of violence by the time he or she reaches age 18. A report by the American Psychological Association states that about 80 percent of all television programs contain acts of violence and that commercial television for children is 50 to 60 times more violent than prime-time television for adults. For example, some car-

toons average more than 80 violent acts per hour (APA Online, 2000). The violent content of media programming and the marketing and advertising practices of mass media industries that routinely target children under age 17 have come under the scrutiny of government agencies such as the Federal Trade Commission due to concerns raised by parents and social analysts.

In addition to concerns about violence in television programming, motion pictures, and electronic games, television shows have been criticized for projecting negative images of women and people of color. Although the mass media have changed some of the roles that they depict women as playing, some newer characters tend to reinforce existing stereotypes of women as sex objects even when they are in professional roles such as doctors or lawyers. Throughout this text, we will look at additional examples of how the media socialize us in ways that we may or may not realize.

Gender and Racial–Ethnic Socialization

Gender socialization **is the aspect of socialization that contains specific messages and practices concerning the nature of being female or male in a specific group or society.** In some families, gender socialization starts before birth. Parents who learn the sex of the fetus through ultrasound or amniocentesis often purchase color-coded and gender-

peer group a group of people who are linked by common interests, equal social position, and (usually) similar age.

mass media large-scale organizations that use print or electronic means (such as radio, television, film, and the Internet) to communicate with large numbers of people.

gender socialization the aspect of socialization that contains specific messages and practices concerning the nature of being female or male in a specific group or society.

Box 3.2 **Sociology in Global Perspective**

The Youthful Cry Heard Around the World: "Everybody Else Has a Cell Phone! Why Can't I Have One?"

- More than 80 percent of high school students and 25 percent of junior high (middle school) students in Japan have cell phones.
- About 75 percent of all teenagers in Scandinavia have cell phones.
- More than 50 percent of all 7- to 16-year-olds in the United Kingdom have cell phones (Magid, 2004).
- About 6.6 million of the 20 million children between the ages of 8 and 12 years in the United States have a cell phone, and it is estimated that there will be at least 10.5 million preteen cell phone users by 2010 (Foderaro, 2007).

As these figures reflect, children and adolescents in high-income nations around the world are increasingly connected with other people by their own cell phones (referred to as "mobile phones" in some countries), which are not entirely under the control of their parents or other adult supervisors. Increasing numbers of elementary-school children as young as age six view a cell phone as a "must-have techno-toy" and as a status symbol that will impress their friends (Foderaro, 2007).

How do cell phones relate to socialization? Child-oriented cell phone use constitutes a shift from earlier times, when parents or other relatives typically were the most significant agents of socialization in a child's life. When a family had only one "landline" telephone in the household, children's interactions with their peers were most often on a face-to-face basis, and their use of the family phone was limited to speaking with "parentally approved" friends for specified periods of time. As a result, parents typically

exerted more influence over their children because these adults were the first individuals to whom a child turned when he or she had a question or problem. Now, peers may be the first to hear a child's thoughts or concerns, and parents may be left out of the conversational loop.

In the United States and various high-income nations in Western Europe and Asia, many parents have welcomed the use of cell phones by their young children because Mom and Dad view the cell phone as an "electronic security blanket"—a way for parents to keep in touch with their children and protect them from harm. With many parents' busy schedules, the increasing numbers of two-career households, and split-custody arrangements brought about by divorce, the cell phone seems like an ideal way for parents and kids to be constantly in touch with each other (Foderaro, 2007: E2). Dr. Cornelia Brunner of the Center for Children and Technology in New York City suggests that cell phones may serve as "transitional objects" that help young children who are suffering separation anxiety when they are away from their parents. In turn, some parents believe that the Global Positioning System, the satellite-based navigation network feature available on some cell phones, will help them keep up with their children's whereabouts. Others question the use of "tracking" technology because it limits a young person's privacy and can produce undue concern about a child's safety if the technology malfunctions (Stross, 2006).

Technological advances—such as the computer, high-speed Internet access, and the cell phone—have not only changed how we communicate with one another but have

typed clothes, toys, and nursery decorations in anticipation of their daughter's or son's arrival. After birth, parents may respond differently toward male and female infants; they often play more roughly with boys and talk more lovingly to girls. Throughout childhood and adolescence, boys and girls are typically assigned different household chores and given different privileges (such as how late they may stay out at night).

When we look at the relationship between gender socialization and social class, the picture becomes more complex. Although some studies have found less-rigid gender stereotyping in higher-income families (Seegmiller, Suter, and Duviant, 1980; Brooks-Gunn, 1986), others have found more (Bardwell, Cochran, and Walker, 1986). One study found that higher-income families are more likely than low-income families to give "male-oriented"

also necessitated new forms of socialization. Children with cell phones should be socialized to use their phones *wisely* so that they will not become the objects of sexual exploitation or of harassment by bullies. In Japan and Europe (and in the United States, to some extent), parents are particularly concerned about the problem of sexual exploitation because a number of children have "met" another person on an Internet chatroom, exchanged phone numbers with the person online, and, after a series of cell phone conversations, agreed to meet the other individual face-to-face, only to learn the other person is much older and perhaps a sexual predator (Magid, 2004). Bullying by cell phone is another concern of some parents and public officials. In the United Kingdom, a recent survey found that many young people had received at least one bullying or threatening call or text message on their phone. Sometimes the messages were from someone they knew at school, but other times the messages were from a stranger who had somehow obtained their phone number or who had made random calls until someone answered (Magid, 2004).

Socialization for the electronic age is very important around the world because, for many people, a cell phone is like a part of their body. As one analyst stated, if you forget your cell phone, "it's like leaving the house without one of your ears" (qtd. in Kim, 2006). Children and adults alike need to be informed about cell phone safety. Childnet International maintains a website (**http://www.chatdanger.com/mobiles**) to inform young people on how to use their cell phone wisely. The site also informs parents about what they should know when purchasing a mobile phone for their child.

© Stu Smucker/OnAsia/Jupterimages

In our digital age, cell phones make it possible for young people, such as this girl in Vietnam, to be connected with other people around the clock. How have cell phones changed our interactions and created new issues in the socialization of young people?

Reflect & Analyze

What do you think are the positive aspects of cell phone use for children and adults? What are the problems associated with cell phone use? Do we need new social rules to deal with the new culture that cell phones may be creating around the world?

toys (which develop visual spatial and problem-solving skills) to children of both sexes (Serbin et al., 1990). Working-class families tend to adhere to more-rigid gender expectations than do middle-class families (Canter and Ageton, 1984; Brooks-Gunn, 1986).

We are limited in our knowledge about gender socialization practices among racial–ethnic groups because most studies have focused on white, middle-class families. In a study of African American families, the sociologist Janice Hale-Benson (1986) found that children typically are not taught to think of gender strictly in "male–female" terms. Both daughters and sons are socialized toward autonomy, independence, self-confidence, and nurturance of children (Bardwell, Cochran, and Walker, 1986). Sociologist Patricia Hill Collins (1990) has suggested that "othermothers" (women other than a child's biological mother) play an important part in the gender socialization and motivation of African American children, especially girls. Othermothers often serve as gender role models and encourage women to become activists on behalf of their children and community (Collins, 1990). By contrast, studies of Korean American and Latino/a families have found more traditional gender socialization (Min, 1988), although some evidence indicates that this pattern may be changing (Jaramillo and Zapata, 1987).

Schools, peer groups, and the media also contribute to our gender socialization. From kindergarten through college, teachers and peers reward gender-appropriate attitudes and behavior. Sports reinforce traditional gender roles through a rigid division of events into male and female catego-

© David Young-Wolff/PhotoEdit

An important rite of passage for many Latinas is the *quinceañera*—a celebration of their fifteenth birthday and their passage into womanhood. Can you see how this occasion might also be a form of anticipatory socialization?

© AP Images/Mark Humphrey

Do you believe that what this child is learning here will have an influence on her actions in the future? What other childhood experiences might offset early negative racial socialization?

ries. The media are also a powerful source of gender socialization; starting very early in childhood, children's books, television programs, movies, and music provide subtle and not-so-subtle messages about how boys and girls should act (see Chapter 10, "Sex and Gender").

In addition to gender-role socialization, we receive racial socialization throughout our lives. *Racial socialization* **is the aspect of socialization that contains specific messages and practices concerning the nature of one's racial or ethnic status** as it relates to our identity, interpersonal relationships, and location in the social hierarchy. Racial socialization includes direct statements regarding race, modeling behavior (wherein a child imitates the behavior of a parent or other caregiver), and indirect activities such as exposure to an environment that conveys a specific message about a racial or ethnic group ("We are better than they are," for example).

The most important aspects of racial identity and attitudes toward other racial–ethnic groups are passed down in families from generation to generation. As the sociologist Martin Marger (1994: 97) notes, "Fear of, dislike for, and antipathy toward one group or another is learned in much the same way that people learn to eat with a knife or fork rather than with their bare hands or to respect others' privacy in personal matters." These beliefs can be transmitted in subtle and largely unconscious ways; they do not have to be taught directly or intentionally. Scholars have found that ethnic values and attitudes begin to crystallize among children as young as age four (Van Ausdale and Feagin, 2001). By this age, the society's ethnic hierarchy has become apparent to the child. Some minority parents feel that racial socialization is essential because it provides children with the skills and abilities they will need to survive in the larger society.

Socialization Through the Life Course

Why is socialization a lifelong process? Throughout our lives, we continue to learn. Each time we experience a change in status (such as becoming a college student or getting married), we learn a new set of rules, roles, and relationships. Even before we achieve a new status, we often participate in *anticipatory socialization*—**the process by which knowledge and skills are learned for future roles.** Many societies organize social activities according to age and gather data regarding the age composition of the people who live in that society. Some societies have distinct *rites of passage,* based on age or other factors, that publicly dramatize and validate changes in a person's status. In the United States and other industrialized societies, the most common categories of age are childhood, adolescence, and adulthood (often subdivided into young adulthood, middle adulthood, and older adulthood).

Childhood

Some social scientists believe that a child's sense of self is formed at a very early age and that it is difficult to change this self-perception later in life. Symbolic interactionists emphasize that during infancy and early childhood, family support and guidance are crucial to a child's developing self-concept. In some families, children are provided with emotional warmth, feelings of mutual trust, and a sense of security. These families come closer to our ideal cultural belief that childhood should be a time of carefree play, safety, and freedom from economic, political, and sexual responsibilities. However, other families reflect the discrepancy between cultural ideals and reality—children grow up in a setting characterized by fear, danger, and risks that are created by parental neglect, emotional maltreatment, or premature economic and sexual demands (Knudsen, 1992). Abused children often experience low self-esteem, an inability to trust others, feelings of isolationism and powerlessness, and denial of their feelings.

Adolescence

In industrialized societies, the adolescent (or teenage) years represent a buffer between childhood and adulthood. In the United States, no specific rites of passage exist to mark children's move into adulthood; therefore, young people have to pursue their own routes to self-identity and adulthood (Gilmore, 1990). Anticipatory socialization is often associated with adolescence, during which many young people spend much of their time planning or being educated for future roles they hope to occupy. However, other adolescents (such as eleven- and twelve-year-old mothers) may have to plunge into adult responsibilities at this time.

Adolescence is often characterized by emotional and social unrest. In the process of developing their own identities, some young people come into conflict with parents, teachers, and other authority figures who attempt to restrict their freedom. Adolescents may also find themselves caught between the demands of adulthood and their own lack of financial independence and experience in the job market. The experiences of individuals during adolescence vary according to race, class, and gender. Based on their family's economic situation, some

racial socialization the aspect of socialization that contains specific messages and practices concerning the nature of one's racial or ethnic status.

anticipatory socialization the process by which knowledge and skills are learned for future roles.

young people move directly into the adult world of work. However, those from upper-middle-class and upper-class families may extend adolescence into their late twenties or early thirties by attending graduate or professional school and then receiving additional advice and financial support from their parents as they start their own families, careers, or businesses.

Adulthood

One of the major differences between child socialization and adult socialization is the degree of freedom of choice. If young adults are able to support themselves financially, they gain the ability to make more choices about their own lives. In early adulthood (usually until about age forty), people work toward their own goals of creating meaningful relationships with others, finding employment, and seeking personal fulfillment. Of course, young adults continue to be socialized by their parents, teachers, peers, and the media, but they also learn new attitudes and behaviors. For example, when we marry or have children, we learn new roles as partners or parents. Adults often learn about fads and fashions in clothing, music, and language from their children.

Workplace (occupational) socialization is one of the most important types of early adult socializa-

tion. This type of socialization tends to be most intense immediately after a person makes the transition from school to the workplace; however, this process may continue throughout our years of employment. Many people experience continuous workplace socialization as a result of having more than one career in their lifetime.

In middle adulthood—between the ages of forty and sixty-five—people begin to compare their accomplishments with their earlier expectations. This is the point at which people either decide that they have reached their goals or recognize that they have attained as much as they are likely to achieve.

Late adulthood may be divided into three categories: (1) the "young-old" (ages sixty-five to seventy-four), (2) the "old-old" (ages seventy-five to eighty-five), and (3) the "oldest-old" (over age eighty-five). Although these are somewhat arbitrary divisions, the "young-old" are less likely to suffer from disabling illnesses, whereas some of the "old-old" are more likely to suffer such illnesses.

Late Adulthood and Ageism

In older adulthood, some people are quite happy and content; others are not (▶ Figure 3.3). Erik Erikson noted that difficult changes in adult attitudes and behavior occur in the last years of life, when people experience decreased physical ability,

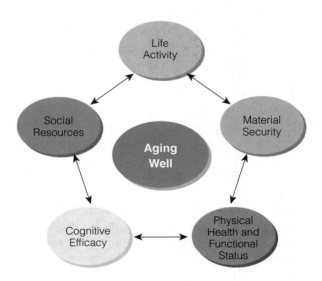

▶ Figure 3.3 **Keys to Aging Well**
Graphic from the Global Aging Initiative–Aging Research Project at Indiana University.
Reprinted by permission of Barbara Hawkins, Indiana University.

lower prestige, and the prospect of death. Older adults in industrialized societies may experience *social devaluation*—wherein a person or group is considered to have less social value than other persons or groups. Social devaluation is especially acute when people are leaving roles that have defined their sense of social identity and provided them with meaningful activity.

Negative images regarding older persons reinforce *ageism*—prejudice and discrimination against people on the basis of age, particularly against older persons. Ageism is reinforced by stereotypes, whereby people have narrow, fixed images of certain groups. Older persons are often stereotyped as thinking and moving slowly; as being bound to themselves and their past, unable to change and grow; as being unable to move forward and often moving backward.

Negative images also contribute to the view that women are "old" ten or fifteen years sooner than men (Bell, 1989). The multibillion-dollar cosmetics industry helps perpetuate the myth that age reduces the "sexual value" of women but increases it for men. Men's sexual value is defined more in terms of personality, intelligence, and earning power than by physical appearance. For women, however, sexual attractiveness is based on youthful appearance. By idealizing this "youthful" image of women and playing up the fear of growing older, sponsors sell thousands of products and services that claim to prevent or fix the "ravages" of aging.

Although not all people act on appearances alone, Patricia Moore, an industrial designer, found that many do. At age twenty-seven, Moore disguised herself as an eighty-five-year-old woman by donning age-appropriate clothing and placing baby oil in her eyes to create the appearance of cataracts. With the help of a makeup artist, Moore supplemented the "aging process" with latex wrinkles, stained teeth, and a gray wig. For three years, "Old Pat Moore" went to various locations, including a grocery store, to see how people responded to her:

> When I did my grocery shopping while in character, I learned quickly that the Old Pat Moore behaved—and was treated—differently from the Young Pat Moore. When I was 85, people were more likely to jockey ahead of me in the

© Sonda Dawes/The Image Works

Throughout life, our self-concept is influenced by our interactions with others.

checkout line. And even more interesting, I found that when it happened, I didn't say anything to the offender, as I certainly would at age 27. It seemed somehow, even to me, that it was okay for them to do this to the Old Pat Moore, since they were undoubtedly busier than I was anyway. And further, they apparently thought it was okay, too! After all, little old ladies have plenty of time, don't they? And then when I did get to the checkout counter, the clerk might start yelling, assuming I was deaf, or becoming immediately testy, assuming I would take a long time to get my money out, or would ask to have the price repeated, or somehow become confused about the transaction. What it all added

social devaluation a situation in which a person or group is considered to have less social value than other individuals or groups.

ageism prejudice and discrimination against people on the basis of age, particularly against older people.

up to was that people feared I would be trouble, so they tried to have as little to do with me as possible. And the amazing thing is that I began almost to believe it myself. . . . I think perhaps the worst thing about aging may be the overwhelming sense that everything around you is letting you know that you are not terribly important any more. (Moore with Conn, 1985: 75–76)

If we apply our sociological imagination to Moore's study, we find that "Old Pat Moore's" experiences reflect what many older persons already know—it is other people's *reactions* to their age, not their age itself, that place them at a disadvantage.

Many older people buffer themselves against ageism by continuing to view themselves as being in middle adulthood long after their actual chronological age would suggest otherwise. Other people begin a process of resocialization to redefine their own identity as mature adults.

Resocialization

Resocialization is the process of learning a new and different set of attitudes, values, and behaviors from those in one's background and previous experience. Resocialization may be voluntary or involuntary. In either case, people undergo changes that are much more rapid and pervasive than the gradual adaptations that socialization usually involves.

Voluntary Resocialization

Resocialization is voluntary when we assume a new status (such as becoming a student, an employee, or a retiree) of our own free will. Sometimes, voluntary resocialization involves medical or psychological treatment or religious conversion, in which case the person's existing attitudes, beliefs, and behaviors must undergo strenuous modification to a new regime and a new way of life. For example, resocialization for adult survivors of emotional/physical child abuse includes extensive therapy in order to form new patterns of thinking and action, somewhat like

Alcoholics Anonymous and its twelve-step program, which has become the basis for many other programs dealing with addictive behavior (Parrish, 1990).

Involuntary Resocialization

Involuntary resocialization occurs against a person's wishes and generally takes place within a *total institution*—**a place where people are isolated from the rest of society for a set period of time and come under the control of the officials who run the institution** (Goffman, 1961a). Military boot camps, jails and prisons, concentration camps, and some mental hospitals are total institutions. Resocialization is a two-step process. First, people are totally stripped of their former selves—or depersonalized—through a degradation ceremony (Goffman, 1961a). For example, inmates entering prison are required to strip, shower, and wear assigned institutional clothing. In the process, they are searched, weighed, fingerprinted, photographed, and given no privacy even in showers and restrooms. Their official identification becomes not a name but a number. In this abrupt break from their former existence, they must leave behind their personal possessions and their family and friends. The depersonalization process continues as they are required to obey rigid rules and to conform to their new environment.

The second step in the resocialization process occurs when the staff at an institution attempt to build a more compliant person. A system of rewards and punishments (such as providing or withholding television or exercise privileges) encourages conformity to institutional norms.

Individuals respond to resocialization in different ways. Some people are rehabilitated; others become angry and hostile toward the system that has taken away their freedom. Although the assumed purpose of involuntary resocialization is to reform persons so that they will conform to societal standards of conduct after their release, the ability of total institutions to modify offenders' behavior in a meaningful manner has been widely questioned. In many prisons, for example, inmates may conform to the norms of the prison or of other inmates but

© Journal Courier/The Image Works

New inmates are taught how to order their meals. Two fingers raised means two portions. There is no talking in line. Inmates must eat all their meal. This "ceremony" suggests how much freedom and dignity an inmate loses when beginning the resocialization process.

have little respect for the norms and the laws of the larger society.

Socialization in the Future

What will socialization be like in the future? The family is likely to remain the institution that most fundamentally shapes and nurtures people's personal values and self-identity. However, parents may increasingly feel overburdened by this responsibility, especially without high-quality, affordable child care. Some analysts have suggested that there may be an increase in known cases of child maltreatment and in the number of children who experience delayed psychosocial development, learning problems, and emotional and behavioral difficulties because of family problems. (See Box

3.3 for suggestions on how to prevent child maltreatment.)

A central value-oriented issue facing parents and teachers as they attempt to socialize children is the growing dominance of television and the Internet, which makes it possible for children to experience many things outside their own homes and schools and to communicate regularly with people around the world. It is very likely that socialization in the future will be vastly different in the world of global instant communication than it has been in the past.

---

resocialization the process of learning a new and different set of attitudes, values, and behaviors from those in one's background and experience.

total institution Erving Goffman's term for a place where people are isolated from the rest of society for a set period of time and come under the control of the officials who run the institution.

Box 3.3 You Can Make a Difference

Helping a Child Reach Adulthood

After Tina—one of your best friends—moves into a large apartment complex near her university, she keeps hearing a baby cry at all hours of the day and night. Although the crying is coming from the apartment next to Tina's, she never sees anyone come or go from it. On several occasions, she knocks on the door, but no one answers. At first Tina tries to ignore the situation, but eventually she can't sleep or study because the baby keeps crying. Tina decides she must take action and asks you, "What do you think I ought to do?" What advice could you give Tina?

Like Tina, many of us do not know if we should get involved in other people's lives. We also do not know how to report child maltreatment. However, social workers and researchers suggest that bystanders must be willing to get involved

© Lisette LeBon/SuperStock

It has been said that it takes a village to raise a child. In contemporary societies, it takes many people pulling together to help a child have a safe and happy childhood and a productive adulthood. Are there ways in which you, like the man in this photo, can help a young person in your community?

Chapter Review

● **What is socialization, and why is it important for human beings?**

Socialization is the lifelong process through which individuals acquire their self-identity and learn the physical, mental, and social skills needed for survival in society. The kind of person we become depends greatly on what we learn during our formative years from our surrounding social groups and social environment.

● **How much of our unique human characteristics comes from heredity and how much from our social environment?**

As individual human beings, we have unique identities, personalities, and relationships with others. Each of us is a product of two forces: (1) heredity, referred to as "nature," and (2) the social environment, referred to as "nurture." Whereas biology dictates our physical makeup, the social envi-

in cases of possible abuse or neglect to save a child from harm by others. They also note the importance of people knowing how to report incidents of maltreatment:

- *Report child maltreatment.* Cases of child maltreatment can be reported to any social service or law enforcement agency.
- *Identify yourself to authorities.* Although most agencies are willing to accept anonymous reports, many staff members prefer to know your name, address, telephone number, and other basic information so that they can determine that you are not a self-interested person such as a hostile relative, ex-spouse, or vindictive neighbor.
- *Follow up with authorities.* Once an agency has validated a report of child maltreatment, the agency's first goal is to stop the neglect or abuse of that child, whose health and safety are paramount concerns. However, intervention also has long-term goals. Sometimes, the situation can be improved simply by teaching the parents different values about child rearing or by pointing them to other agencies and organizations that can provide needed help. Other times, it may be necessary to remove the child from the parents' custody and place the child in a foster home, at least temporarily. Either way, the situation for the child will be better than if he or she had been left in an abusive or neglectful home environment.

So the best advice for Tina—or for anyone else who has reason to believe that child maltreatment is occurring—is to report it to the appropriate authorities. In most telephone directories, the number can be located in the government listings section. Online, use keywords such as "helping children" and "child welfare" to search for sources of information and assistance. Here are some other resources for help:

- Childhelp offers a 24-hour crisis hotline, national information, and referral network for support groups and therapists and for reporting suspected abuse: 15757 North 78th Street, Scottsdale, AZ 85260. (800) 422-4453.
  http://www.childhelp.org

- Child Welfare League of America, a Washington, D.C., association of nearly 800 public and private nonprofit agencies, serves as an advocacy group for children who have experienced maltreatment: 440 First Street, NW, Suite 310, Washington, DC 20001-2085. (202) 638-2952.
  http://www.cwla.org

- The National Center for Missing and Exploited Children provides brochures about child safety and child protection on request.
  http://www.missingkids.com

ronment largely determines how we develop and behave.

• Why is social contact essential for human beings?

Social contact is essential in developing a self, or self-concept, which represents an individual's perceptions and feelings of being a distinct or separate person. Much of what we think about ourselves is gained from our interactions with others and from what we perceive that others think of us.

• What are the main social psychological theories on human development?

According to Sigmund Freud, the self emerges from three interrelated forces: the id, the ego, and the superego. When a person is well adjusted, the three forces act in balance. Jean Piaget identified four cognitive stages of development; each child must go through each stage in sequence before moving on to the next one, although some children move through them faster than others.

● **How do sociologists believe that we develop a self-concept?**

According to Charles Horton Cooley's concept of the looking-glass self, we develop a self-concept as we see ourselves through the perceptions of others. Our initial sense of self is typically based on how families perceive and treat us. George Herbert Mead suggested that we develop a self-concept through role-taking and learning the rules of social interaction. According to Mead, the self is divided into the "I" and the "me." The "I" represents the spontaneous and unique traits of each person. The "me" represents the internalized attitudes and demands of other members of society.

● **What are the primary agents of socialization?**

The agents of socialization include the family, schools, peer groups, and the media. Our families, which transmit cultural and social values to us, are the most important agents of socialization in all societies, serving these functions: (1) procreating and socializing children, (2) providing emotional support, and (3) assigning social position. Schools primarily teach knowledge and skills but also have a profound influence on the self-image, beliefs, and values of children. Peer groups contribute to our sense of belonging and self-worth, and are a key source of information about acceptable behavior. The media function as socializing agents by (1) informing us about world events, (2) introducing us to a wide variety of people, and (3) providing an opportunity to live vicariously through other people's experiences.

● **When does socialization end?**

Socialization is ongoing throughout the life course. We learn knowledge and skills for future roles through anticipatory socialization. Parents are socialized by their own children, and adults learn through workplace socialization. Resocialization is the process of learning new attitudes, values, and behaviors, either voluntarily or involuntarily.

www.cengage.com/login

Register for a Student eResource account to maximize your study time online using CengageNOW. First take the system's diagnostic pre-test, and then follow the personalized study plan that is created for you to help you review this chapter. The study plan will

● help you identify areas on which you should concentrate;
● provide interactive exercises to help you master the chapter concepts; and
● provide a post-test to confirm you are ready to move on to the next chapter.

Key Terms

ageism 105
agents of socialization 92
anticipatory socialization 103
ego 84
gender socialization 99
generalized other 91
id 84

looking-glass self 88
mass media 98
peer group 97
racial socialization 102
resocialization 106
role-taking 89
self-concept 88

significant others 90
social devaluation 105
socialization 78
sociobiology 79
superego 84
total institution 106

Questions for Critical Thinking

1. Consider the concept of the looking-glass self. How do you think others perceive you? Do you think most people perceive you correctly?

2. What are your "I" traits? What are your "me" traits? Which ones are stronger?

3. What are some different ways that you might study the effect of toys on the socialization of children? How could you isolate the toy variable from other variables that influence children's socialization?

4. Is the attempted rehabilitation of criminal offenders—through boot camp programs, for example—a form of socialization or resocialization?

The Kendall Companion Website

www.cengage.com/sociology/kendall

Supplement your review of this chapter by going to the text's companion website, where you can take tutorial quizzes, use flash cards to master key terms, follow live links to useful websites, and explore the other study and research resources you'll find there, such as a comprehensive interactive sociology timeline, GSS Data, and Census 2000 information, much of it presented visually in maps.

Social Structure and Interaction in Everyday Life

I began Dumpster diving [scavenging in a large garbage bin] about a year before I became homeless. . . . The area I frequent is inhabited by many affluent college students. I am not here by chance; the Dumpsters in this area are very rich. Students throw out many good things, including food. In particular they tend to throw everything out when they move at the end of a semester, before and after breaks, and around midterm, when many of them despair of college. So I find it advantageous to keep an eye on the academic calendar. I learned to scavenge gradually, on my own. Since then I have initiated several companions into the trade. I have learned that there is a predictable series of stages a person goes through in learning to scavenge.

At first the new scavenger is filled with disgust and self-loathing. He is ashamed of being seen and may lurk around, trying to duck behind things, or he may dive at night. (In fact, most people

© Bob Collins/The Image Works

All activities in life—including scavenging in garbage bins and living "on the streets"—are social in nature.

instinctively look away from a scavenger. By skulking around, the novice calls attention to himself and arouses suspicion. Diving at night is ineffective and needlessly messy.) . . . That stage passes with experience. The scavenger finds a pair of running shoes that fit and look and smell brand-new. . . . He begins to understand: People throw away perfectly good stuff, a lot of perfectly good stuff.

At this stage, Dumpster shyness begins to dissipate. The diver, after all, has the last laugh. He is finding all manner of good things that are his for the taking. Those who disparage his profession are the fools, not he.

—author Lars Eighner recalls his experiences as a Dumpster diver while living under a shower curtain in a stand of bamboo in a public park. Eighner became homeless when he was evicted from his "shack" after being unemployed for about a year. (Eighner, 1993: 111–119)

Eighner's "diving" activities reflect a specific pattern of social behavior. All activities in life—including scavenging in garbage bins and living "on the streets"—are social in nature. Homeless persons and domiciled persons (those with homes) live in social worlds that have predictable patterns of social interaction. *Social interaction* **is the process by which people act toward or respond to other people and is the foundation for all relationships and groups in society.** In this chapter, we look at the relationship between social structure and social interaction. In the process, homelessness is used as an example of how social problems occur and how they may be perpetuated within social structures and patterns of interaction.

Social structure **is the complex framework of societal institutions (such as the economy, politics, and religion) and the social practices (such as rules and social roles) that make up a society**

Chapter Focus Question

How is homelessness related to the social structure of a society?

and that organize and establish limits on people's behavior. This structure is essential for the survival of society and for the well-being of individuals because it provides a social web of familial support and social relationships that connects each of us to the larger society. Many homeless people have lost this vital linkage. As a result, they often experience a loss of personal dignity and a sense of moral worth because of their "homeless" condition (Snow and Anderson, 1993).

Homeless persons such as Eighner come from all walks of life. They include undocumented workers, parolees, runaway youths and children, Vietnam veterans, and the elderly. They live in cities, suburbs, and rural areas. Contrary to popular myths, most of the homeless are not on the streets by choice or because they were deinstitutionalized by mental hospitals. Not all of the homeless are unemployed. About 22 percent of homeless people hold full- or part-time jobs but earn too little to find an affordable place to live (U.S. Conference of Mayors, 2005). Before reading on, take the quiz on homelessness in Box 4.1.

Social Structure: The Macrolevel Perspective

Social structure provides the framework within which we interact with others. This framework is an orderly, fixed arrangement of parts that together make up the whole group or society (see ▶ Figure 4.1). As defined in Chapter 1, a *society* is a large social grouping that shares the same geographical territory and is subject to the same political authority and dominant cultural expectations. At the macrolevel, the social structure of a society has several essential elements: social institutions, groups, statuses, roles, and norms.

Functional theorists emphasize that social structure is essential because it creates order and predictability in a society (Parsons, 1951). Social structure is also important for our human development. As discussed in Chapter 2, we develop a self-concept as we learn the attitudes, values, and behaviors of the people around us. When these attitudes and values are part of a predictable structure, it is easier to develop that self-concept.

Social structure gives us the ability to interpret the social situations we encounter. For example, we expect our families to care for us, our schools to educate us, and our police to protect us. When our circumstances change dramatically, most of us feel an acute sense of anxiety because we do not know what to expect or what is expected of us. For example, newly homeless individuals may feel disoriented because they do not know how to function in their new setting. The person is likely to ask questions: "How will I survive on the streets?" "Where do I go to get help?" "Should I stay at a shelter?" "Where can I get a job?" Social structure helps people make sense out of their environment even when they find themselves on the streets.

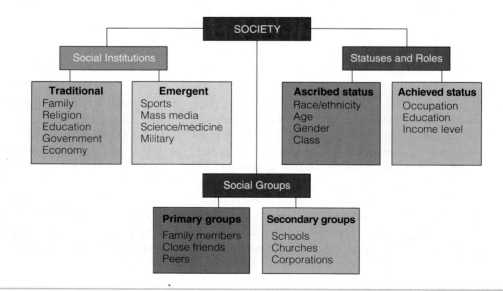

▶ Figure 4.1 Social Structure Framework

Box 4.1 Sociology and Everyday Life

How Much Do You Know About Homeless Persons?

True	False	
T	F	1. Most homeless people choose to be homeless.
T	F	2. Homelessness is largely a self-inflicted condition.
T	F	3. Homeless people do not work.
T	F	4. Most homeless people are mentally ill.
T	F	5. Homeless people typically panhandle (beg for money) so that they can buy alcohol or drugs.
T	F	6. Most homeless people are heavy drug users.
T	F	7. A large number of homeless persons are dangerous.
T	F	8. Homeless persons have existed throughout the history of the United States.
T	F	9. One out of every four homeless persons is a child.
T	F	10. Some homeless people have attended college and graduate school.

Answers on page 116.

However, conflict theorists maintain that there is more to social structure than is readily visible and that we must explore the deeper, underlying structures that determine social relations in a society. For example, Karl Marx suggested that the way economic production is organized is the most important structural aspect of any society. In capitalistic societies, where a few people control the labor of many, the social structure reflects a system of relationships of domination among categories of people (for example, owner–worker and employer–employee).

Social structure creates boundaries that define which persons or groups will be the "insiders" and which will be the "outsiders." *Social marginality* is the state of being part insider and part outsider in the social structure. Sociologist Robert Park (1928) coined this term to refer to persons (such as immigrants) who simultaneously share the life and traditions of two distinct groups. Social marginality results in stigmatization. A *stigma* is any physical or social attribute or sign that so devalues a person's social identity that it disqualifies that person from full social acceptance (Goffman, 1963b). A convicted criminal, wearing a prison uniform, is an example of a person who has been stigmatized; the uniform says that the person has done something wrong and should not be allowed unsupervised outside the prison walls.

Components of Social Structure

The social structure of a society includes its social positions, the relationships among those positions, and the kinds of resources attached to each of the positions. Social structure also includes all the groups that make up society and the relationships among those groups (Smelser, 1988). We begin by examining the social positions that are closest to the individual.

Status

A *status* is a socially defined position in a group or society characterized by certain expectations, rights, and duties. Statuses exist independently of

---

social interaction the process by which people act toward or respond to other people; the foundation for all relationships and groups in society.

social structure the stable pattern of social relationships that exists within a particular group or society.

status a socially defined position in a group or society characterized by certain expectations, rights, and duties.

Box 4.1 Sociology and Everyday Life

Answers to the Sociology Quiz on Homeless Persons

1. **False.** Less than 6 percent of all homeless people are that way by choice.

2. **False.** Most homeless persons did not inflict upon themselves the conditions that produced their homelessness. Some are the victims of child abuse or violence.

3. **False.** Many homeless people are among the working poor. Minimum-wage jobs do not pay enough for an individual to support a family or pay inner-city rent.

4. **False.** Most homeless people are not mentally ill; estimates suggest that about one-fourth of the homeless are emotionally disturbed.

5. **False.** Many homeless persons panhandle to pay for food, a bed at a shelter, or other survival needs.

6. **False.** Most homeless people are not heavy drug users. Estimates suggest that about one-third of the homeless are substance abusers. Many of these are part of the same one-fourth of the homeless who are mentally ill.

7. **False.** Although an encounter with a homeless person occasionally ends in tragedy, most homeless persons are among the least threatening members of society. They are more often the victims of crime, not the perpetrators.

8. **True.** Scholars have found that homelessness has always existed in the United States. However, the number of homeless persons has increased or decreased with fluctuations in the national economy.

9. **True.** Families with children are the fastest growing category of homeless persons in the United States. The number of such families nearly doubled between 1984 and 1989, and continues to do so. Many homeless children are alone. They may be runaways or "throwaways" whose parents do not want them to return home.

10. **True.** Some homeless persons have attended college and graduate school, and many have completed high school.

Sources: Based on Kroloff, 1993; Liebow, 1993; Snow and Anderson, 1993; U.S. Conference of Mayors, 2005; Vissing, 1996; and Waxman and Hinderliter, 1996.

---

the specific people occupying them (Linton, 1936); the statuses of professional athlete, rock musician, professor, college student, and homeless person all exist exclusive of the specific individuals who occupy these social positions. For example, although thousands of new students arrive on college campuses each year to occupy the status of first-year student, the status of college student and the expectations attached to that position have remained relatively unchanged for the past century.

Does the term *status* refer only to high-level positions in society? No, not in a sociological sense. Although many people equate the term *status* with high levels of prestige, sociologists use it to refer to all socially defined positions—high rank and low rank. For example, both the position of director of the Department of Health and Human Services in Washington, D.C., and that of a homeless person who is paid about five dollars a week (plus bed and board) to clean up the dining room at a homeless shelter are social statuses (see Snow and Anderson, 1993).

Take a moment to answer the question "Who am I?" To determine who you are, you must think about your social identity, which is derived from the statuses you occupy and is based on your status set. A ***status set*** comprises all the statuses that a person occupies at a given time. For example, Maria may be a psychologist, a professor, a wife, a mother, a Catholic, a school volunteer, a Texas resident, and a Mexican American. All of these socially defined positions constitute her status set.

Ascribed Status and Achieved Status Statuses are distinguished by the manner in which we acquire them. An *ascribed status* is a social position conferred at birth or received involuntarily later in life, based on attributes over which the individual has little or no control, such as race/ethnicity, age, and gender. For example, Maria is a female born to Mexican American parents; she was assigned these statuses at birth. She is an adult and—if she lives long enough—will someday become an "older adult," which is an ascribed status received involuntarily later in life. An *achieved status* is a social position a person assumes voluntarily as a result of personal choice, merit, or direct effort. Achieved statuses (such as occupation, education, and income) are thought to be gained as a result of personal ability or successful competition. Most occupational positions in modern societies are achieved statuses. For instance, Maria voluntarily assumed the statuses of psychologist, professor, wife, mother, and school volunteer. However, not all achieved statuses are positions most people would want to attain; for example, being a criminal, a drug addict, or a homeless person is a negative achieved status.

Ascribed statuses have a significant influence on the achieved statuses that we occupy. Race/ethnicity, gender, and age affect each person's opportunity to acquire certain achieved statuses. Those who are privileged by their positive ascribed statuses are more likely to achieve the more prestigious positions in a society. Those who are disadvantaged by their ascribed statuses may more easily acquire negative achieved statuses.

Master Status If we occupy many different statuses, how can we determine which is the most important? Sociologist Everett Hughes has stated that societies resolve this ambiguity by determining master statuses. A *master status* is the most important status a person occupies; it dominates all of the individual's other statuses and is the overriding ingredient in determining a person's general social position (Hughes, 1945). Being poor or rich

© Jeff Brass/Getty Images

In the past, a person's status was primarily linked to his or her family background, education, occupation, and other sociological attributes. Today, some sociologists believe that celebrity status has overtaken the more traditional social indicators of status. The rock star Bono, shown here performing at a concert, is an example of celebrity status.

is a master status that influences many other areas of life, including health, education, and life opportunities. Historically, the most common master statuses

status set all the statuses that a person occupies at a given time.

ascribed status a social position conferred at birth or received involuntarily later in life based on attributes over which the individual has little or no control, such as race/ethnicity, age, and gender.

achieved status a social position that a person assumes voluntarily as a result of personal choice, merit, or direct effort.

master status the most important status that a person occupies.

for women have related to positions in the family, such as daughter, wife, and mother. For men, occupation has usually been the most important status, although occupation is increasingly a master status for many women as well. "What do you do?" is one of the first questions most people ask when meeting another. Occupation provides important clues to a person's educational level, income, and family background. An individual's race/ethnicity may also constitute a master status in a society in which dominant-group members single out members of other groups as "inferior" on the basis of real or alleged physical, cultural, or nationality characteristics (see Feagin and Feagin, 2003).

Master statuses confer high or low levels of personal worth and dignity on people. These are not characteristics that we inherently possess; they are derived from the statuses we occupy. For those who have no residence, being a homeless person readily becomes a master status regardless of the person's other attributes. Homelessness is a stigmatized master status that confers disrepute on its occupant because domiciled people often believe that a homeless person has a "character flaw." Sometimes this assumption is supported by how the media frame stories about homeless people (see Box 4.2). The circumstances under which someone becomes homeless determine the extent to which that person is stigmatized. For example, individuals who become homeless as a result of natural disasters (such as a hurricane or a brush fire) are not seen as causing their homelessness or as being a threat to the community. Thus, they are less likely to be stigmatized. However, in cases in which homeless persons are viewed as the cause of their own problems, they are more likely to be stigmatized and marginalized by others. Snow and Anderson (1993: 199) observed the effects of homelessness as a master status:

It was late afternoon, and the homeless were congregated in front of [the Salvation Army shelter] for dinner. A school bus approached that was packed with Anglo junior high school students being bused from an eastside barrio school to their upper-middle and upper-class homes in the city's northwest neighborhoods. As the bus rolled by, a fusillade of coins came flying out the windows, as the students made obscene gestures and shouted, "Get a job." Some of the homeless gestured back, some scrambled for the scattered coins—mostly pennies—others angrily threw the coins at the bus, and a few seemed oblivious to the encounter. For the passing junior high schoolers, the exchange was harmless fun, a way to work off the restless energy built up in school; but for the homeless it was a stark reminder of their stigma-

© Xinhua/Landov

© Rachel Epstein/PhotoEdit

Sociologists believe that being rich or poor may be a master status in the United States. How do the lifestyles of these two men differ based on their master statuses?

tized status and of the extent to which they are the objects of negative attention.

Status Symbols When people are proud of a particular social status that they occupy, they often choose to use visible means to let others know about their position. ***Status symbols are material signs that inform others of a person's specific status.*** For example, just as wearing a wedding ring proclaims that a person is married, owning a Rolls-Royce announces that one has "made it." As we saw in Chapter 2, achievement and success are core U.S. values. For this reason, people who have "made it" tend to want to display symbols to inform others of their accomplishments.

Status symbols for the domiciled and for the homeless may have different meanings. Among affluent persons, a full shopping cart in the grocery store and bags of merchandise from expensive department stores indicate a lofty financial position. By contrast, among the homeless, bulging shopping bags and overloaded grocery carts suggest a completely different status. Carts and bags are essential to street life; there is no other place to keep things, as shown by this description of Darian, a homeless woman in New York City:

> The possessions in her postal cart consist of a whole house full of things, from pots and pans to books, shoes, magazines, toilet articles, personal papers and clothing, most of which she made herself. . . .
>
> Because of its weight and size, Darian cannot get the cart up over the curb. She keeps it in the street near the cars. This means that as she pushes it slowly up and down the street all day long, she is living almost her entire life directly in traffic. She stops off along her route to sit or sleep for awhile and to be both stared at as a spectacle and to stare back. Every aspect of her life including sleeping, eating, and going to the bathroom is constantly in public view. . . . [S]he has no space to call her own and she never has a moment's privacy. Her privacy, her home, is her cart with all its possessions. (Rousseau, 1981: 141)

For homeless women and men, possessions are not status symbols as much as they are a link with the past, a hope for the future, and a potential source of immediate cash. As Snow and Anderson (1993: 147) note, selling personal possessions is not uncommon among most social classes; members of the working and middle classes hold garage sales, and those in the upper classes have estate sales. However, when homeless persons sell their personal possessions, they do so to meet their immediate needs, not because they want to "clean house."

Roles

A role is the dynamic aspect of a status. Whereas we occupy a status, we play a role. A ***role is a set of behavioral expectations associated with a given status.*** For example, a carpenter (employee) hired to remodel a kitchen is not expected to sit down uninvited and join the family (employer) for dinner.

***Role expectation* is a group's or society's definition of the way a specific role ought to be played.** By contrast, ***role performance* is how a person actually plays the role.** Role performance does not always match role expectation. Some statuses have role expectations that are highly specific, such as that of surgeon or college professor. Other statuses, such as friend or significant other, have less-structured expectations. The role expectations tied to the status of student are more specific than those of being a friend. Role expectations are typically based on a range of acceptable behavior rather than on strictly defined standards.

Our roles are relational (or complementary); that is, they are defined in the context of roles performed by others. We can play the role of student because someone else fulfills the role of professor. Conversely, to perform the role of professor, the teacher must have one or more students.

---

status symbol a material sign that informs others of a person's specific status.

role a set of behavioral expectations associated with a given status.

role expectation a group's or society's definition of the way that a specific role *ought* to be played.

role performance how a person *actually* plays a role.

Box 4.2 Framing Homelessness in the Media

Thematic and Episodic Framing

They live—and die—on a traffic island in the middle of a busy downtown street, surviving by panhandling drivers or turning tricks. Everyone in their colony is hooked on drugs or alcohol. They are the harsh face of the homeless in San Francisco.

The traffic island where these homeless people live is a 40-by-75 foot triangle chunk of concrete just west of San Francisco's downtown. . . . The little concrete divider wouldn't get a second glance, or have a name—if not for the colony that lives there in a jumble of shopping carts loaded with everything they own. It's called Homeless Island by the shopkeepers who work near it and the street sweepers who clean it; to the homeless, it is just the Island. The inhabitants live hand-to-mouth, sleep on the cement and abuse booze and drugs, mostly heroin. There are at least 3,000 others like them in San Francisco, social workers say. They are known as the "hard core," the people most visible on the streets, the most difficult to help. . . . (Fagan, 2003)

This news article is an example of typical media framing of stories about homeless people. The full article includes statements about how the homeless of San Francisco use drugs, lack ambition, and present a generally disreputable appearance on the streets. This type of framing of stories about the homeless is not unique. According to the media scholar Eungjun Min (1999: ix), media images typically portray the homeless as "drunk, stoned, crazy, sick, and drug abusers." Such representations of homeless people limit our understanding of the larger issues surrounding the problem of homelessness in the United States.

Most media framing of newspaper articles and television reports about the problem of homelessness can be classified into one of two major categories: *thematic framing* and *episodic framing*. Thematic framing refers to news stories that focus primarily on statistics about the homeless population and recent trends in homelessness. Examples include stories about changes in the U.S. poverty rate and articles about states and cities that have had the largest increases in poverty. Most articles of this type are abstract and impersonal, primarily presenting data and some expert's interpretation of what those data mean. Media representations of this type convey a message to readers that "the poor and homeless are faceless." According to some analysts, thematic framing of poverty is often dehumanizing because it "ignores the human tragedy of poverty—the suffering, indignities, and misery endured by millions of children and adults" (Mantsios, 2003: 101).

By contrast, *episodic framing* presents public issues such as poverty and homelessness as concrete events, showing them to be specific instances that occur more or less in isolation. For example, a news article may focus on the problems of one homeless family, describing how the parents and kids live in a car and eat meals from a soup kitchen. Often, what is not included is the *big picture of homelessness:* How many people throughout the city or nation are living in their cars or in shelters? What larger structural factors (such as reductions in public and private assistance to the poor, or high rates of unemployment in some regions) contribute to or intensify the problem of homelessness in this country?

For many years, the poor have been a topic of interest to journalists and social commentators. Between 1851 and 1995, the *New York Times* alone printed 4,126 articles that had the word *poverty* in the headline. How stories about the poor and homeless are framed in the media has been and remains an important concern for each of us because these reports influence how we view the less fortunate in our society. If we come to see the problem of homelessness as nothing more than isolated statistical data or as marginal situations that affect only a few people, then we are unable to make a balanced assessment of the larger social problems involved.

Reflect & Analyze

How are the poor and homeless represented in the news reports and television entertainment shows you watch? Are the larger social issues surrounding homelessness discussed within the context of these shows? Should they be?

Role ambiguity occurs when the expectations associated with a role are unclear. For example, it is not always clear when the provider–dependent aspect of the parent–child relationship ends. Should it end at age eighteen or twenty-one? When a person is no longer in school? Different people will answer these questions differently depending on their experiences and socialization, as well as on the parents' financial capability and psychological willingness to continue contributing to the welfare of their adult children.

Role Conflict and Role Strain Most people occupy a number of statuses, each of which has numerous role expectations attached. For example, Charles is a student who attends morning classes at the university, and he is an employee at a fast-food restaurant, where he works from 3:00 to 10:00 P.M. He is also Stephanie's boyfriend, and she would like to see him more often. On December 7, Charles has a final exam at 7:00 P.M., when he is supposed to be working. Meanwhile, Stephanie is pressuring him to take her to a movie. To top it off, his mother calls, asking him to fly home because his father is going to have emergency surgery. How can Charles be in all these places at once? Such experiences of role conflict can be overwhelming.

Role conflict **occurs when incompatible role demands are placed on a person by two or more statuses held at the same time.** When role conflict occurs, we may feel pulled in different directions. To deal with this problem, we may prioritize our roles and first complete the one we consider to be most important. Or we may compartmentalize our lives and "insulate" our various roles (Merton, 1968). That is, we may perform the activities linked to one role for part of the day and then engage in the activities associated with another role in some other time period or elsewhere. For example, under routine circumstances, Charles would fulfill his student role for part of the day and his employee role for another part of the day. In his current situation, however, he is unable to compartmentalize his roles.

© Kim Eriksen/zefa/CORBIS

Parents sometimes experience role conflict when they are faced with societal expectations that they will earn a living for their family and that they will also be good parents to their children. Obviously, this father needs to leave for work; however, his son has other needs.

Role conflict may occur as a result of changing statuses and roles in society. Research has found that women who engage in behavior that is gender-typed as "masculine" tend to have higher rates of role conflict than those who engage in traditional "feminine" behavior (Basow, 1992). According to the sociologist Tracey Watson (1987), role conflict can sometimes be attributed not to the roles themselves but to the pressures that people feel when they do not fit into culturally prescribed roles. In her study of women athletes in college sports programs, Watson found role conflict in the traditionally incongruent identities of being a woman and being an athlete. Even though the women athletes

role conflict a situation in which incompatible role demands are placed on a person by two or more statuses held at the same time.

in her study wore makeup and presented a conventional image when they were not on the basketball court, their peers in school still saw them as "female jocks," thus leading to role conflict.

Whereas role conflict occurs between two or more statuses (such as being homeless and being a temporary employee of a social service agency), role strain takes place within one status. ***Role strain occurs when incompatible demands are built into a single status that a person occupies*** (Goode, 1960). For example, many women experience role strain in the labor force because they hold jobs that are "less satisfying and more stressful than men's jobs since they involve less money, less prestige, fewer job openings, more career roadblocks, and so forth" (Basow, 1992: 192). Similarly, married women may experience more role strain than married men because of work overload, marital inequality with their spouse, exclusive parenting responsibilities, unclear expectations, and lack of emotional support.

Recent social changes may have increased role strain for men. In the family, men's traditional po-sition of dominance has eroded as more women have entered the paid labor force and demanded more assistance in child-rearing and homemaking responsibilities. The concepts of role expectation, role performance, role conflict, and role strain are illustrated in ▶ Figure 4.2.

Individuals frequently distance themselves from a role they find extremely stressful or otherwise problematic. *Role distancing* occurs when people consciously foster the impression of a lack of commitment or attachment to a particular role and merely go through the motions of role performance (Goffman, 1961b). People use distancing techniques when they do not want others to take them as the "self" implied in a particular role, especially if they think the role is "beneath them." While Charles is working in the fast-food restaurant, for example, he does not want people to think of him as a "loser in a dead-end job." He wants them to view him as a college student who is working there just to "pick up a few bucks" until he graduates. When customers from the university come in, Charles talks to

▶ Figure 4.2 Role Expectation, Performance, Conflict, and Strain
When playing the role of "student," do you sometimes personally encounter these concepts?

Role Expectation: a group's or society's definition of the way a specific role *ought* to be played.

Role Performance: how a person *actually* plays a role.

Role Conflict: occurs when incompatible demands are placed on a person by two or more statuses held at the same time.

Role Strain: occurs when incompatible demands are built into a single status that a person holds.

Los Angeles Times columnist Steve Lopez (left) met a homeless man, Nathaniel Ayers (above), and learned that he had been a promising musician studying at the Julliard School who had dropped out because of his struggle with mental illness. In his 2008 book titled *The Soloist*, Lopez chronicles the relationship he developed with Ayers and how he eventually helped get him off the street and be treated for his schizophrenia. *The Soloist* is being made into a movie starring Jamie Foxx as Ayers and Robert Downey Jr. as Lopez.

them about what courses they are taking, what they are majoring in, and what professors they have. He does not discuss whether the bacon cheeseburger is better than the chili burger. When Charles is really involved in role distancing, he tells his friends that he "works there but wouldn't eat there."

Role Exit *Role exit* **occurs when people disengage from social roles that have been central to their self-identity** (Ebaugh, 1988). Sociologist Helen Rose Fuchs Ebaugh studied this process by interviewing ex-convicts, ex-nuns, retirees, divorced men and women, and others who had exited voluntarily from significant social roles. According to Ebaugh, role exit occurs in four stages. The first stage is doubt, in which people experience frustration or burnout when they reflect on their existing roles. The second stage involves a search for alternatives; here, people may take a leave of absence from their work or temporarily separate from their marriage partner. The third stage is the turning point at which people realize that they must take some final action, such as quitting their job or getting a divorce. The fourth and final stage involves the creation of a new identity.

Exiting the "homeless" role is often very difficult. The longer a person remains on the streets, the more difficult it becomes to exit this role. Personal resources diminish over time. Possessions are often stolen, lost, sold, or pawned. Work experience and skills become outdated, and physical disabilities that prevent individuals from working are likely to develop.

Groups

Groups are another important component of social structure. To sociologists, a *social group* **consists of two or more people who interact frequently and share a common identity and a feeling of interdependence.** Throughout our lives, most of us participate in groups, from our families and childhood friends, to our college classes, to our work and community organizations, and even to society.

role strain a condition that occurs when incompatible demands are built into a single status that a person occupies.

role exit a situation in which people disengage from social roles that have been central to their self-identity.

social group a group that consists of two or more people who interact frequently and share a common identity and a feeling of interdependence.

Primary and secondary groups are the two basic types of social groups. A *primary group* **is a small, less specialized group in which members engage in face-to-face, emotion-based interactions over an extended period of time.** Primary groups include our family, close friends, and school- or work-related peer groups. By contrast, a *secondary group* **is a larger, more specialized group in which members engage in more impersonal, goal-oriented relationships for a limited period of time.** Schools, churches, and corporations are examples of secondary groups. In secondary groups, people have few, if any, emotional ties to one another. Instead, they come together for some specific, practical purpose, such as getting a degree or a paycheck. Secondary groups are more specialized than primary ones; individuals relate to one another in terms of specific roles (such as professor and student) and more limited activities (such as course-related endeavors). Primary and secondary groups are further discussed in Chapter 5 ("Groups and Organizations").

Social solidarity, or cohesion, refers to a group's ability to maintain itself in the face of obstacles. Social solidarity exists when social bonds, attractions, or other forces hold members of a group in interaction over a period of time (Jary and Jary, 1991). For example, if a local church is destroyed by fire and congregation members still worship together in a makeshift setting, then they have a high degree of social solidarity.

Many of us build social networks that involve our personal friends in primary groups and our acquaintances in secondary groups. A *social network* is a series of social relationships that links an individual to others. Social networks work differently for men and women, for different races/ethnicities, and for members of different social classes. Traditionally, people of color and white women have been excluded from powerful "old-boy" social networks. At the middle- and upper-class levels, individuals tap social networks to find employment, make business deals, and win political elections. However, social networks typically do not work effectively for poor and homeless individuals. Snow and Anderson (1993) found that homeless men have fragile social networks that are plagued with instability. Most of the avenues for exiting the homeless role and acquiring housing are intertwined with the large-scale, secondary groups that sociologists refer to as formal organizations.

A *formal organization* **is a highly structured group formed for the purpose of completing certain tasks or achieving specific goals.** Many of us spend most of our time in formal organizations such as colleges, corporations, or the government. In Chapter 5 ("Groups and Organizations"), we analyze the characteristics of bureaucratic organizations; however, at this point we should note that these organizations are a very important component of social structure in all industrialized societies. We

For many years, capitalism has been dominated by powerful "old boy" social networks. Professional women have increasingly created their own social networks to enhance their business opportunities.

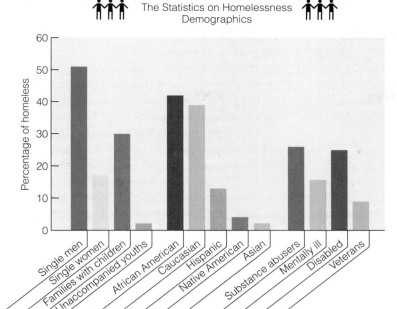

The Statistics on Homelessness
Demographics

▶ **Figure 4.3 Who Are the Homeless?**
Source: National Law Center on Homelessness and Poverty, 2004. Reprinted courtesy of HowStuffWorks.com

expect such organizations to educate us, solve our social problems (such as crime and homelessness), and provide us work opportunities.

Today, formal organizations such as the U.S. Conference of Mayors and the National Law Center on Homelessness and Poverty work with groups around the country to make people aware that homelessness must be viewed within the larger context of poverty and to educate the public on the nature and extent

of homelessness among various categories of people in the United States (see ▶ Figure 4.3 for statistics on homelessness).

Social Institutions

At the macrolevel of all societies, certain basic activities routinely occur—children are born and socialized, goods and services are produced and distributed, order is preserved, and a sense of purpose is maintained (Aberle et al., 1950; Mack and Bradford, 1979). Social institutions are the means by which

As a formal organization, the Salvation Army completes certain tasks and achieves certain goals that otherwise might not be fulfilled in contemporary societies, such as providing meals for the homeless.

primary group Charles Horton Cooley's term for a small, less specialized group in which members engage in face-to-face, emotion-based interactions over an extended period of time.

secondary group a larger, more specialized group in which members engage in more-impersonal, goal-oriented relationships for a limited period of time.

formal organization a highly structured group formed for the purpose of completing certain tasks or achieving specific goals.

these basic needs are met. A *social institution* is a **set of organized beliefs and rules that establishes how a society will attempt to meet its basic social needs.** In the past, these needs have centered around five basic social institutions: the family, religion, education, the economy, and the government or politics. Today, mass media, sports, science and medicine, and the military are also considered to be social institutions.

What is the difference between a group and a social institution? A group is composed of specific, identifiable people; an institution is a standardized way of doing something. The concept of "family" helps to distinguish between the two. When we talk about "your family" or "my family," we are referring to a specific family. When we refer to the family as a social institution, we are talking about ideologies and standardized patterns of behavior that organize family life. For example, the family as a social institution contains certain statuses organized into well-defined relationships, such as husband–wife, parent–child, and brother–sister. Specific families do not always conform to these ideologies and behavior patterns.

Functional theorists emphasize that social institutions exist because they perform five essential tasks:

1. *Replacing members.* Societies and groups must have socially approved ways of replacing members who move away or die.
2. *Teaching new members.* People who are born into a society or move into it must learn the group's values and customs.
3. *Producing, distributing, and consuming goods and services.* All societies must provide and distribute goods and services for their members.
4. *Preserving order.* Every group or society must preserve order within its boundaries and protect itself from attack by outsiders.
5. *Providing and maintaining a sense of purpose.* In order to motivate people to cooperate with one another, a sense of purpose is needed.

Although this list of functional prerequisites is shared by all societies, the institutions in each society perform these tasks in somewhat different ways depending on their specific cultural values and norms.

Conflict theorists agree with functionalists that social institutions are originally organized to meet basic social needs. However, they do not believe that social institutions work for the common good of everyone in society. For example, the homeless lack the power and resources to promote their own interests when they are opposed by dominant social groups. From the conflict perspective, social institutions such as the government maintain the privileges of the wealthy and powerful while contributing to the powerlessness of others (see Domhoff, 2002). For example, U.S. government policies in urban areas have benefited some people but exacerbated the problems of others. Urban renewal and transportation projects have caused the destruction of low-cost housing and put large numbers of people "on the street" (Katz, 1989). Similarly, the shift in governmental policies regarding the mentally ill and welfare recipients has resulted in more people struggling—and often failing—to find affordable housing. Meanwhile, many wealthy and privileged bankers, investors, developers, and builders have benefited at the expense of the low-income casualties of those policies.

Societies: Changes in Social Structure

Changes in social structure have a dramatic impact on individuals, groups, and societies. Social arrangements in contemporary societies have grown more complex with the introduction of new technology, changes in values and norms, and the rapidly shrinking "global village." How do societies maintain some degree of social solidarity in the face of such changes? Sociologists Emile Durkheim and Ferdinand Tönnies developed typologies to explain the processes of stability and change in the social structure of societies. A *typology* is a classification scheme containing two or more mutually exclusive categories that are used to compare different kinds of behavior or types of societies.

Durkheim: Mechanical and Organic Solidarity

Emile Durkheim (1933/1893) was concerned with the question "How do societies manage to hold together?" He asserted that preindustrial societies are

Compare these photos of Reno, Nevada, to see how this city has changed over time. From its origins as a town tied to the mining industry to a city that focuses on tourism, Reno is an example of how change in social structure affects individuals, families, and communities.

held together by strong traditions and by the members' shared moral beliefs and values. As societies industrialized and developed more specialized economic activities, social solidarity came to be rooted in the members' shared dependence on one another. From Durkheim's perspective, social solidarity derives from a society's social structure, which, in turn, is based on the society's division of labor. *Division of labor* **refers to how the various tasks of a society are divided up and performed.** People in diverse societies (or in the same society at different points in time) divide their tasks somewhat differently, based on their own history, physical environment, and level of technological development.

To explain social change, Durkheim categorized societies as having either mechanical or organic solidarity. *Mechanical solidarity* **refers to the social cohesion of preindustrial societies, in which there is minimal division of labor and people feel united by shared values and common social bonds.** Durkheim used the term *mechanical solidarity* because he believed that people in such preindustrial societies feel a more or less automatic sense of belonging. Social interaction is characterized by face-to-face, intimate, primary-group relationships. Everyone is engaged in similar work, and little specialization is found in the division of labor.

Organic solidarity **refers to the social cohesion found in industrial (and perhaps postindustrial)**

societies, in which people perform very specialized tasks and feel united by their mutual dependence. Durkheim chose the term *organic solidarity* because he believed that individuals in industrial societies come to rely on one another in much the same way that the organs of the human body function interdependently. Social interaction is less personal, more status oriented, and more focused on specific goals and objectives. People no longer rely on morality or shared values for social solidarity; instead, they are bound together by practical considerations.

social institution a set of organized beliefs and rules that establishes how a society will attempt to meet its basic social needs.

division of labor how the various tasks of a society are divided up and performed.

mechanical solidarity Emile Durkheim's term for the social cohesion in preindustrial societies, in which there is minimal division of labor and people feel united by shared values and common social bonds.

organic solidarity Emile Durkheim's term for the social cohesion found in industrial societies, in which people perform very specialized tasks and feel united by their mutual dependence.

Tönnies: *Gemeinschaft* and *Gesellschaft*

Sociologist Ferdinand Tönnies (1855–1936) used the terms *Gemeinschaft* and *Gesellschaft* to characterize the degree of social solidarity and social control found in societies. He was especially concerned about what happens to social solidarity in a society when a "loss of community" occurs.

The *Gemeinschaft* (guh-MINE-shoft) is a traditional society in which social relationships are based on personal bonds of friendship and kinship and on intergenerational stability. These relationships are based on ascribed rather than achieved status. In such societies, people have a commitment to the entire group and feel a sense of togetherness. Tönnies (1963/1887) used the German term *Gemeinschaft* because it means "commune" or "community"; social solidarity and social control are maintained by the community. Members have a strong sense of belonging, but they also have very limited privacy.

By contrast, the *Gesellschaft* (guh-ZELL-shoft) is a large, urban society in which social bonds are based on impersonal and specialized relationships, with little long-term commitment to the group or consensus on values. In such societies, most people are "strangers" who perceive that they have very little in common with most other people. Consequently, self-interest dominates, and little consensus exists regarding values. Tönnies (1963/1887) selected the German term *Gesellschaft* because it means "association"; relationships are based on achieved statuses, and interactions among people are both rational and calculated.

Industrial and Postindustrial Societies

Industrial societies are based on technology that mechanizes production. Take a look around you: Most of what you see would not exist if it were not for industrialization. Cars, computers, electric lights, sound systems, cell phones, and virtually every other possession we own is a product of an industrial society.

In industrial societies, a large proportion of the population lives in or near cities. Large corporations and government bureaucracies grow in size and complexity. The nature of social life changes as people come to know one another more as statuses than as individuals. In fact, a person's occupation becomes a key defining characteristic in industrial societies, meaning that those people who are unemployed do not share the same status markers as those who have jobs.

The shift that has taken place toward a postindustrial society in the United States and some other high-income nations has produced new opportunities and problems. A *postindustrial society* is one in which technology supports a service- and information-based economy. Postindustrial societies are characterized by an *information explosion* and an economy in which large numbers of people either provide or apply information or are employed in service jobs (such as fast-food server or health care worker). For example, banking, law, and the travel industry are characteristic forms of employment in postindustrial societies, whereas producing steel or automobiles is representative of employment in industrial societies.

Postindustrial societies produce knowledge that becomes a commodity. This knowledge can be leased or sold to others, or it can be used to generate goods, services, or more knowledge. In the previous types of societies we have examined, machinery or raw materials are crucial to how the economy operates. In postindustrial societies, the economy is based on involvement with people and communications technologies such as the mass media, computers, and the World Wide Web. For example, recent information from the U.S. Census Bureau indicates that more than 60 percent of all U.S. households have at least one computer (see "Census Profiles: Computer and Internet Access in U.S. Households"). Some analysts refer to postindustrial societies as "service economies," based on the assumption that many workers provide services for others. Examples include home health care workers and airline flight attendants. However, most of the new service occupations pay relatively low wages and offer limited opportunities for advancement.

Some sociological theories and research focus on the systematic examination of societies at the macrolevel, including the effects of the postindustrial economy on other social institutions; how-

CENSUS PROFILES

...puter and Internet ...ss in U.S. Households

...U.S. Census Bureau collects extensive data on U.S. ...useholds in addition to the questions it used for ...ensus 2000. For example, Current Population Survey ...data, collected from about 50,000 U.S. households during 2003, show an increase in the percentage of homes with computers and access to the Internet, as the following figure illustrates:

Computers and Internet Access in the Home: 1984 to 2003
(civilian noninstitutional population)

Percentage of households with a computer
Percentage of households with Internet access

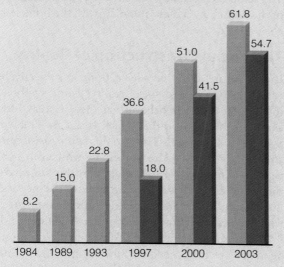

Note: Data on Internet access were not collected before 1997.

Since 1984, the first year in which the Census Bureau collected data on computer ownership and use, there has been more than a 700 percent increase in the percentage of households with computers. However, in Chapter 7 ("Class and Stratification in the United States"), we will see that computer ownership varies widely by income and educational level.

Source: Newburger, 2001; U.S. Census Bureau, 2007.

ever, other approaches primarily use a microlevel perspective in their analysis of social interaction in contemporary societies.

Social Interaction: The Microlevel Perspective

So far in this chapter, we have focused on society and social structure from a macrolevel perspective, seeing how the structure of society affects the statuses we occupy, the roles we play, and the groups and organizations to which we belong. Functionalist and conflict perspectives provide a macrosociological overview because they concentrate on large-scale events and broad social features. By contrast, the symbolic interactionist perspective takes a microsociological approach, asking how social institutions affect our daily lives.

Social Interaction and Meaning

When you are with other people, do you often wonder what they think of you? If so, you are not alone! Because most of us are concerned about the meanings that others ascribe to our behavior, we try to interpret their words and actions so that we can plan how we will react toward them (Blumer, 1969). We know that others have expectations of us. We also have certain expectations about them. For example, if we enter an elevator that has only one other person

Gemeinschaft (guh-MINE-shoft) a traditional society in which social relationships are based on personal bonds of friendship and kinship and on intergenerational stability.

Gesellschaft (guh-ZELL-shoft) a large, urban society in which social bonds are based on impersonal and specialized relationships, with little long-term commitment to the group or consensus on values.

industrial society a society based on technology that mechanizes production.

postindustrial society a society in which technology supports a service- and information-based economy.

in it, we do not expect that individual to confront us and stare into our eyes. As a matter of fact, we would be quite upset if the person did so.

Social interaction within a given society has certain shared meanings across situations. For instance, our reaction would be the same regardless of *which* elevator we rode in *which* building. Sociologist Erving Goffman (1961b) described these shared meanings in his observation about two pedestrians approaching each other on a public sidewalk. He noted that each will tend to look at the other just long enough to acknowledge the other's presence. By the time they are about eight feet away from each other, both individuals will tend to look downward. Goffman referred to this behavior as *civil inattention*—the ways in which an individual shows an awareness that another is present without making this person the object of particular attention. The fact that people engage in civil inattention demonstrates that interaction does have a pattern, or *interaction order,* which regulates the form and processes (but not the content) of social interaction.

Does everyone interpret social interaction rituals in the same way? No. Race/ethnicity, gender, and social class play a part in the meanings we give to our interactions with others, including chance encounters on elevators or the street. Our perceptions about the meaning of a situation vary widely based on the statuses we occupy and our unique personal experiences. For example, sociologist Carol Brooks Gardner (1989) found that women frequently do not perceive street encounters to be "routine" rituals. They fear for their personal safety and try to avoid comments and propositions that are sexual in nature. African Americans may also feel uncomfortable in street encounters. A middle-class African American college student described his experiences walking home at night from a campus job:

> So, even if you wanted to, it's difficult just to live a life where you don't come into conflict with others. . . . Every day that you live as a black person you're reminded how you're perceived in society. You walk the streets at night; white people cross the streets. I've seen white couples and individuals dart in front of cars to not be on the same side of the street. Just the other day, I was walking down the street, and this white female with a child, I saw her pass a young white

> male about 20 yards ahead. When she saw me, she quickly dragged the child and herself across the busy street. . . . [When I pass,] the men tighten their grip on their women. The men people turn around and seem like they're going to take blows from me. . . . So, even you realize [you're black]. Even though you not doing anything wrong; you're just existi. You're just a person. But you're a black perso. perceived in an unblack world. (qtd. in Feagin, 1991: 111–112)

As this passage indicates, social encounters have different meanings for men and women, whites and people of color, and individuals from different social classes. Members of the dominant classes regard the poor, unemployed, and working class as less worthy of attention, frequently subjecting them to subtle yet systematic "attention deprivation" (Derber, 1983). The same can certainly be said about how members of the dominant classes "interact" with the homeless.

The Social Construction of Reality

If we interpret other people's actions so subjectively, can we have a shared social reality? Some symbolic interaction theorists believe that there is very little shared reality beyond that which is socially created. Symbolic interactionists refer to this as the *social construction of reality*—**the process by which our perception of reality is largely shaped by the subjective meaning that we give to an experience** (Berger and Luckmann, 1967). This meaning strongly influences what we "see" and how we respond to situations.

As discussed previously, our perceptions and behavior are influenced by how we initially define situations: We act on reality as we see it. Sociologists describe this process as the *definition of the situation,* meaning that we analyze a social context in which we find ourselves, determine what is in our best interest, and adjust our attitudes and actions accordingly. This process can result in a *self-fulfilling prophecy*—**a false belief or prediction that produces behavior that makes the originally false belief come true** (Merton, 1968). An example would be a person who has been told repeatedly that she or he is not a good student; eventually, this person might come to believe it to be true, stop studying, and receive failing grades.

People can have sharply contrasting perceptions of the same reality.

People may define a given situation in very different ways, a tendency demonstrated by the sociologist Jacqueline Wiseman (1970) in her study of "Pacific City's" skid row. She wanted to know how people who live or work on skid row (a run-down area found in all cities) felt about it. Wiseman found that homeless persons living on skid row evaluated it very differently from the social workers who dealt with them there. On the one hand, many of the social workers "saw" skid row as a smelly, depressing

area filled with men who were "down-and-out," alcoholic, and often physically and mentally ill. On the other hand, the men who lived on skid row did not see it in such a negative light. They experienced some degree of satisfaction with their "bottle clubs [and a] remarkably indomitable and creative spirit"—at least initially (Wiseman, 1970: 18). As this study shows, we define situations from our own frame of reference, based on the statuses that we occupy and the roles that we play.

Dominant-group members with prestigious statuses may have the ability to establish how other people define "reality" (Berger and Luckmann, 1967: 109). Some sociologists have suggested that dominant groups, particularly higher-income white males in powerful economic and political statuses, perpetuate their own world view through ideologies that are frequently seen as "social reality." For example, the sociologist Dorothy E. Smith (1999) points out that the term "Standard North American Family" (meaning a heterosexual two-parent family) is an ideological code promulgated by the dominant group to identify how people's family life *should* be arranged. According to Smith (1999), this code plays a powerful role in determining how people in organizations such as the government and schools believe that a family should be. Likewise, the sociologist Patricia Hill Collins (1998) argues that "reality" may be viewed differently by African American women and other historically oppressed

social construction of reality the process by which our perception of reality is shaped largely by the subjective meaning that we give to an experience.

self-fulfilling prophecy the situation in which a false belief or prediction produces behavior that makes the originally false belief come true

groups when compared to the perspectives of dominant-group members. However, according to Collins (1998), mainstream, dominant-group members sometimes fail to realize how much they could learn about "reality" from "outsiders." As these theorists state, social reality and social structure are often hotly debated issues in contemporary societies.

Ethnomethodology

How do we know how to interact in a given situation? What rules do we follow? Ethnomethodologists are interested in the answers to these questions. *Ethnomethodology* **is the study of the commonsense knowledge that people use to understand the situations in which they find themselves** (Heritage, 1984: 4). Sociologist Harold Garfinkel (1967) initiated this approach and coined the term: *ethno* for "people" or "folk" and *methodology* for "a system of methods." Garfinkel was critical of mainstream sociology for not recognizing the ongoing ways in which people create reality and produce their own world. Consequently, ethnomethodologists examine existing patterns of conventional behavior in order to uncover people's background expectancies—that is, their shared interpretation of objects and events, as well as their resulting actions. According to ethnomethodologists, interaction is based on assumptions of shared expectancies. For example, when you are talking with someone, what expectations do you have that you will take turns? Based on your background expectancies, would you be surprised if the other person talked for an hour and never gave you a chance to speak?

To uncover people's background expectancies, ethnomethodologists frequently break "rules" or act as though they do not understand some basic rule of social life so that they can observe other people's responses. In a series of *breaching experiments,* Garfinkel assigned different activities to his students to see how breaking the unspoken rules of behavior created confusion.

The ethnomethodological approach contributes to our knowledge of social interaction by making us aware of subconscious social realities in our daily lives. However, a number of sociologists regard ethnomethodology as a frivolous approach to studying human behavior because it does not examine the impact of macrolevel social institutions—such as the economy and education—on people's expectancies. Some scholars suggest that ethnomethodologists fail to do what they claim to do: look at how social realities are created. Rather, they take ascribed statuses (such as race, class, gender, and age) as "givens," not as *socially created* realities.

Dramaturgical Analysis

Erving Goffman suggested that day-to-day interactions have much in common with being on stage or in a dramatic production. *Dramaturgical analysis* **is the study of social interaction that compares everyday life to a theatrical presentation.** Members of our "audience" judge our performance and are aware that we may slip and reveal our true character (Goffman, 1959, 1963a). Consequently, most of us attempt to play our role as well as possible and to control the impressions we give to others. *Impression management (presentation of self)* **refers to people's efforts to present themselves to others in ways that are most favorable to their own interests or image.**

For example, suppose that a professor has returned graded exams to your class. Will you discuss the exam and your grade with others in the class? If you are like most people, you probably play your student role differently depending on whom you are talking to and what grade you received on the exam. Your "presentation" may vary depending on the grade earned by the other person (your "audience"). In one study, students who all received high grades ("Ace–Ace encounters") willingly talked with one another about their grades and sometimes engaged in a little bragging about how they had "aced" the test. However, encounters between students who had received high grades and those who had received low or failing grades ("Ace–Bomber encounters") were uncomfortable. The Aces felt as if they had to minimize their own grade. Consequently, they tended to attribute their success to "luck" and were quick to offer the Bombers words of encouragement. On the other hand, the Bombers believed that they had to praise the Aces and hide their own feelings of frustration and disappointment. Students who received low or failing grades ("Bomber–Bomber encounters") were more comfortable when they talked with

Erving Goffman believed that people spend a great amount of time and effort managing the impression that they present. How do political candidates use impression management as they seek to accomplish their goal of being elected to public office?

the Aces attempted to help them save face by asserting that the test was unfair or that it was only a small part of the final grade. Why would the Aces and Bombers both participate in face-saving behavior? In most social interactions, all role players have an interest in keeping the "play" going so that they can maintain their overall definition of the situation in which they perform their roles.

Goffman noted that people consciously participate in *studied nonobservance,* a face-saving technique in which one role player ignores the flaws in another's performance to avoid embarrassment for everyone involved. Most of us remember times when we have failed in our role and know that it is likely to happen again; thus, we may be more forgiving of the role failures of others.

Social interaction, like a theater, has a front stage and a back stage. The *front stage* is the area where a player performs a specific role before an audience. The *back stage* is the area where a player is not required to perform a specific role because it is out of view of a given audience. For example, when the Aces and Bombers were talking with one another at school, they were on the "front stage." When they were in the privacy of their own residences, they were in "back stage" settings—they no longer had to perform the Ace and Bomber roles and could be themselves.

The need for impression management is most intense when role players have widely divergent or devalued statuses. As we have seen

one another because they could share their negative emotions. They often indulged in self-pity and relied on face-saving excuses (such as an illness or an unfair exam) for their poor performances (Albas and Albas, 1988).

In Goffman's terminology, **face-saving behavior refers to the strategies we use to rescue our performance when we experience a potential or actual loss of face.** When the Bombers made excuses for their low scores, they were engaged in face-saving;

ethnomethodology the study of the common-sense knowledge that people use to understand the situations in which they find themselves.

dramaturgical analysis the study of social interaction that compares everyday life to a theatrical presentation.

impression management (presentation of self) Erving Goffman's term for people's efforts to present themselves to others in ways that are most favorable to their own interests or image.

face-saving behavior Erving Goffman's term for the strategies we use to rescue our performance when we experience a potential or actual loss of face.

with the Aces and Bombers, the participants often play different roles under different circumstances and keep their various audiences separated from one another. If one audience becomes aware of other roles that a person plays, the impression being given at that time may be ruined. For example, homeless people may lose jobs or the opportunity to get them when their homelessness becomes known. One woman had worked as a receptionist in a doctor's office for several weeks but was fired when the doctor learned that she was living in a shelter (Liebow, 1993). However, the homeless do not passively accept the roles into which they are cast. For the most part, they attempt—as we all do—to engage in impression management in their everyday life.

The dramaturgical approach helps us think about the roles we play and the audiences who judge our presentation of self. Today, many people are concerned not only about the impressions they make in face-to-face encounters but also in cyberspace (see "Sociology Works!"). However, the dramaturgical approach has been criticized for focusing on appearances and not the underlying substance. This approach may not place enough emphasis on the ways in which our everyday interactions with other people are influenced by occurrences within the larger society. For example, if some members of Congress belittle the homeless as being lazy and unwilling to work, it may become easier for people walking down a street to do likewise. Even so, Goffman's work has been influential in the development of the sociology of emotions, an important area of theory and research.

The Sociology of Emotions

Why do we laugh, cry, or become angry? Are these emotional expressions biological or social in nature? To some extent, emotions are a biologically given sense (like hearing, smell, and touch), but they are also social in origin. We are socialized to feel certain emotions, and we learn how and when to express (or not express) those emotions (Hochschild, 1983).

How do we know which emotions are appropriate for a given role? Sociologist Arlie Hochschild (1983) suggests that we acquire a set of *feeling rules*

that shapes the appropriate emotions for a given role or specific situation. These rules include how, where, when, and with whom an emotion should be expressed. For example, for the role of a mourner at a funeral, feeling rules tell us which emotions are required (sadness and grief, for example), which are acceptable (a sense of relief that the deceased no longer has to suffer), and which are unacceptable (enjoyment of the occasion expressed by laughing out loud) (see Hochschild, 1983: 63–68).

Feeling rules also apply to our occupational roles. For example, the truck driver who handles explosive cargos must be able to suppress fear. Although all jobs place some burden on our feelings, *emotional labor* occurs only in jobs that require personal contact with the public or the production of a state of mind (such as hope, desire, or fear) in others (Hochschild, 1983). With emotional labor, employees must display only certain carefully selected emotions. For example, flight attendants are required to act friendly toward passengers, to be helpful and open to requests, and to maintain an "omnipresent smile" in order to enhance the customers' status. By contrast, bill collectors are encouraged to show anger and make threats to customers, thereby supposedly deflating the customers' status and wearing down their presumed resistance to paying past-due bills. In both jobs, the employees are expected to show feelings that are often not their true ones (Hochschild, 1983).

Social class and race are determinants in managed expression and emotion management. Emotional labor is emphasized in middle- and upper-class families. Because middle- and upper-class parents often work with people, they are more likely to teach their children the importance of emotional labor in their own careers than are working-class parents, who tend to work with things, not people (Hochschild, 1983). Race is also an important factor in emotional labor. People of color spend much of their life engaged in emotional labor because racist attitudes and discrimination make it continually necessary to manage one's feelings.

Emotional labor may produce feelings of estrangement from one's "true" self. C. Wright Mills (1956) suggested that when we "sell our personality" in the course of selling goods or services, we en-

Sociology *Works!*

Erving Goffman's Impression Management and Facebook

Ethan [pseudonym] is in his early 20's and joined [a large software development company] as an entry level consultant six months ago. He joined Facebook in college to keep up with his current friends and used it primarily for getting to know new friends better. He now uses the site to keep in touch with these friends, but his usage has gone from an hour a week to 10 minutes a week. He has over 200 Facebook friends and most of the new employees he met at company orientation are listed as friends. Before starting his job, he purposefully "cleansed" all information about himself on the Internet: from Facebook, his blog, and his personal website. In particular, he removed all photos of himself involving "drinking alcohol." Because of that he is not concerned about strangers, managers, or mentors seeing his information online. (DiMicco and Millen, 2007)

In their recent study regarding how people engage in identity management on Facebook, the social networking website, research scientists Joan Morris DiMicco and David R. Millen (2007) found that Ethan was a good example of how individuals engage in impression management on such websites. When Erving Goffman defined impression management (presentation of self) as people's efforts to present themselves to others in ways that are most favorable to their own interests or image, he was thinking about the face-to-face encounters that each of us has in daily life. Today, many sociologists and social analysts believe that Goffman's ideas may also be applicable to social interactions that take place on the Internet, particularly social networking websites such as Facebook, MySpace, and LinkedIn. According to the journalist Stephanie Rosenbloom (2008: E1), "Now that first impressions are of-

ten made in cyberspace, not face-to-face, people are not only strategizing about how to virtually convey who they are, but also grappling with how to craft an e-version of themselves that appeals to multiple audiences—co-workers, fraternity brothers, Mom and Dad."

Although Facebook originated with—and remains popular among—college students, many people who are beyond their college years, including Ethan, still update their information on this networking website. A number of individuals employed by the software development company where DiMicco and Millen (2007) conducted their study stated that they work to balance the presentation of themselves as professionals versus nonprofessionals on the Web. They engage in a conscious process of determining what to include (or to exclude) from the personal profiles, photos, and blogs they post online.

Like Ethan, many of us believe that managing our personal identity on the Internet is important. We are concerned about what *people we know* might think if they view our personal information, yet we might be even more anxious about the impression we may make on *people we do not know* but who might become acquaintances or even prospective employers in the future. Looking at the dramaturgical perspective and Goffman's ideas of impression management in online communication offers rich new opportunities for application of classical sociological insights to our interactions with others.

Reflect & Analyze

How might you apply *impression management* and *face-saving behavior* to an analysis of your communications with others on the Internet?

gage in a seriously self-alienating process. In other words, the "commercialization" of our feelings may dehumanize our work role performance and create alienation and contempt that spill over into other aspects of our life (Hochschild, 1983; Smith and Kleinman, 1989).

Clearly, the sociology of emotions helps us understand the social context of our feelings and the relationship between the roles we play and the emotions we experience. However, it may overemphasize the cost of emotional labor and the emotional controls that exist outside the individual (Wouters, 1989).

© Tom Prettyman/PhotoEdit

© David Silverman/Getty Images

Are there different gender-based expectations in the United States about the kinds of emotions that men, as compared with women, are supposed to show? What feeling rules shape the emotions of the men in these two roles?

Nonverbal Communication

In a typical stage drama, the players not only speak their lines but also use nonverbal communication to convey information. ***Nonverbal communication is the transfer of information between persons without the use of words.*** It includes not only visual cues (gestures, appearances) but also vocal features (inflection, volume, pitch) and environmental factors (use of space, position) that affect meanings (Wood, 1999). Facial expressions, head movements, body positions, and other gestures carry as much of the total meaning of our communication with others as our spoken words do (Wood, 1999).

Functions of Nonverbal Communication We obtain first impressions of others from various kinds of nonverbal communication, such as the clothing they wear and their body positions. Head and facial movements may provide us with information about other people's emotional states, and others receive similar information from us (Samovar and Porter, 1991). Through our body posture and eye contact, we signal that we do or do not wish to speak to someone. For example, we may look down at the sidewalk or off into the distance when we pass homeless persons who look as if they are going to ask for money.

Nonverbal communication establishes the relationship among people in terms of their responsiveness to and power over one another (Wood, 1999). For example, we show that we are responsive toward or like another person by maintaining eye contact and attentive body posture and perhaps by touching and standing close. We can even express power or control over others through nonverbal communication. Goffman (1956) suggested that *demeanor* (how

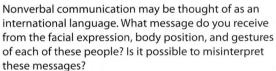
Nonverbal communication may be thought of as an international language. What message do you receive from the facial expression, body position, and gestures of each of these people? Is it possible to misinterpret these messages?

we behave or conduct ourselves) is relative to social power. People in positions of dominance are allowed a wider range of permissible actions than are their subordinates, who are expected to show deference. *Deference* is the symbolic means by which subordinates give a required permissive response to those in power; it confirms the existence of inequality and reaffirms each person's relationship to the other (Rollins, 1985).

Facial Expression, Eye Contact, and Touching
Deference behavior is important in regard to facial expression, eye contact, and touching. This type of nonverbal communication is symbolic of our relationships with others. Who smiles? Who stares? Who makes and sustains eye contact? Who touches whom? All these questions relate to demeanor and

deference; the key issue is the status of the person who is doing the smiling, staring, or touching relative to the status of the recipient (Goffman, 1967).

Facial expressions, especially smiles and eye contact, also reflect gender-based patterns of dominance and subordination in society. Typically, white women have been socialized to smile and frequently do so even when they are not actually happy (Halberstadt and Saitta, 1987). Jobs held predominantly by women (including flight attendant, secretary, elementary schoolteacher, and nurse) are more closely

nonverbal communication the transfer of information between persons without the use of speech.

associated with being pleasant and smiling than are "men's jobs." By contrast, men tend to display less emotion through smiles or other facial expressions and instead seek to show that they are reserved and in control (Wood, 1999).

Women are more likely to sustain eye contact during conversations (but not otherwise) as a means of showing their interest in and involvement with others. By contrast, men are less likely to maintain prolonged eye contact during conversations but are more likely to stare at other people (especially men) in order to challenge them and assert their own status (Pearson, 1985).

Eye contact can be a sign of domination or deference. For example, in a participant observation study of domestic (household) workers and their employers, the sociologist Judith Rollins (1985) found that the domestics were supposed to show deference by averting their eyes when they talked to their employers. Deference also required that they present an "exaggeratedly subservient demeanor" by standing less erect and walking tentatively.

Touching is another form of nonverbal behavior that has many different shades of meaning. Gender and power differences are evident in tactile communication from birth. Studies have shown that touching has variable meanings to parents: Boys are touched more roughly and playfully, whereas girls are handled more gently and protectively (Condry, Condry, and Pogatshnik, 1983). This pattern continues into adulthood, with women touched more frequently than men. Sociologist Nancy Henley (1977) attributed this pattern to power differentials between men and women and to the nature of women's roles as mothers, nurses, teachers, and secretaries. Clearly, touching has a different meaning to women than to men. Women may hug and touch others to indicate affection and emotional support, but men are more likely to touch others to give directions, assert power, and express sexual interest (Wood, 1999). The "meaning" we give to touching is related to its "duration, intensity, frequency, and the body parts touching and being touched" (Wood, 1994: 162).

Personal Space Physical space is an important component of nonverbal communication. Anthropologist Edward Hall (1966) analyzed the physical distance between people speaking to each

other and found the amount of personal space that people prefer varies from one culture to another. *Personal space* **is the immediate area surrounding a person that the person claims as private.** Our personal space is contained within an invisible boundary surrounding our body, much like a snail's shell. When others invade our space, we may retreat, stand our ground, or even lash out, depending on our cultural background (Samovar and Porter, 1991). ▶ Figure 4.4 illustrates differences in social distance rules between two contrasting cultures.

Age, gender, kind of relationship, and social class are important factors in the allocation of personal space. Power differentials between people (including adults and children, men and women, and dominant-group members and people of color) are reflected in personal space and privacy issues. With regard to age, adults generally do not hesitate to enter the personal space of a child (Thorne, Kramarae, and Henley, 1983). Similarly, young children who invade the personal space of an adult tend to elicit a more favorable response than do older uninvited visitors (Dean, Willis, and la Rocco, 1976). The

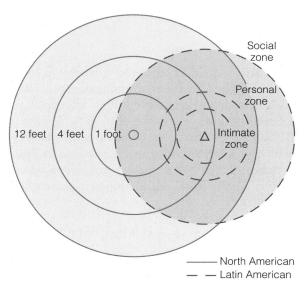

North American
Latin American

▶ **Figure 4.4 North American and Latin American Social Distance Rules**

Source: http://home.snu.edu/~hculbert/space.gif.

CONCEPT QUICK REVIEW

Social Interaction: The Microlevel Perspective

Social interaction and meaning	In a given society, forms of social interaction have shared meanings, although these may vary to some extent based on race/ethnicity, gender, and social class.
Social construction of reality	The process by which our perception of reality is largely shaped by the subjective meaning that we give to an experience.
Ethnomethodology	Studying the commonsense knowledge that people use to understand the situations in which they find themselves makes us aware of subconscious social realities in daily life.
Dramaturgical analysis	The study of social interaction that compares everyday life to a theatrical presentation. This approach includes impression management (people's efforts to present themselves favorably to others).
Sociology of emotions	We are socialized to feel certain emotions, and we learn how and when to express (or not express) them.
Nonverbal communication	The transfer of information between persons without the use of speech, such as by facial expressions, head movements, and gestures.

© David Young-Wolff/PhotoEdit

Have you ever watched how people react to one another in an elevator? How might we explain the lack of eye contact and the general demeanor of the individuals pictured here?

need for personal space appears to increase with age (Baxter, 1970; Aiello and Jones, 1971), although it may begin to decrease at about age forty (Heshka and Nelson, 1972).

In sum, all forms of nonverbal communication are influenced by gender, race, social class, and the personal contexts in which they occur. Although it is difficult to generalize about people's nonverbal behavior, we still need to think about our own non-verbal communication patterns. Recognizing that differences in social interaction exist is important. We should be wary of making value judgments—the differences are simply differences. Learning to understand and respect alternative styles of social interaction enhances our personal effectiveness by increasing the range of options we have for communicating with different people in diverse contexts and for varied reasons (Wood, 1999). (The Concept Quick Review summarizes the microlevel approach to social interaction.)

personal space the immediate area surrounding a person that the person claims as private.

Box 4.3 You Can Make a Difference

Offering a Helping Hand to Homeless People

When you pull up at an intersection and see a person holding a torn piece of cardboard with a handwritten sign on it, how do you react? Many of us shy away from chance encounters such as this because we know, without actually looking, that the sign says something like "Homeless, please help." In an attempt to avoid eye contact with the person on the street corner, we suddenly look with newfound interest at something lying on our car seat, or we check our appearance in the rearview mirror, or we adjust the radio. In fact, we do just about whatever it takes to divert our attention, making eye contact with this person impossible until the traffic light changes and we can be on our way.

Does this scenario sound familiar? Many of us see homeless individuals on street corners and elsewhere as we go about our daily routine. We are uncomfortable in their presence because we don't know what we can do to help them, or even if we should. Frequently, we hear media reports stating that some allegedly homeless people abuse the practice of asking for money on the streets and that many are faking injury or poverty so that they can take advantage of generous individuals. Stereotypes such as this are commonplace when some laypersons, members of the media, and politicians describe the homeless in America. But this is far from the entire picture: Many homeless people are in need of assistance, and many of the homeless are children, persons with disabilities, and people with other problems that make it difficult, if not impossible, for them to earn enough money to pay for housing in many cities.

Do all of these "big picture" problems in our society mean that we have no individual responsibility to help homeless people? We do not necessarily have to hand money over to the person on the street to help individuals who are homeless. There are other, and perhaps even better, ways in which we can provide help to the homeless through our small acts of generosity and kindness.

In some communities, college students lead the way in helping homeless individuals and families. Some programs help homeless children by providing them with clothing, other basic necessities, and even school supplies so that the children will feel comfortable in a classroom setting. Still other college students work in, or run, homeless shelters in their communities. For example, Harvard University students, along with some city officials and church leaders in Cambridge, Massachusetts, created the Harvard Square Homeless Shelter in 1983 to address the housing needs of the area's poorest residents. Although it was hoped that the shelter would be a temporary project that would be rendered unnecessary when society recognized and dealt with its homeless problem, the shelter was still in existence in the 2000s. According to Alina Das, a former volunteer director at the shelter, "I have learned more about humanity and life within these walls than I have learned anywhere else. For students, the shelter is more than a place to stay ... most important, we try to foster a sense of dignity" (qtd. in Powell, 2001).

As organizers of some college groups that seek to help the homeless have suggested, individuals without homes need food, clothing, and shelter, but they also need com-

Changing Social Structure and Interaction in the Future

The social structure in the United States has been changing rapidly in recent decades. Currently, there are more possible statuses for persons to occupy and roles to play than at any other time in history. Although achieved statuses are considered very important, ascribed statuses still have a significant effect on people's options and opportunities.

Ironically, at a time when we have more technological capability, more leisure activities and types of entertainment, and more quantities of material

passion and caring that extend beyond what most bureaucratic organizations can offer. Below are a few ways in which you and others at your school might help homeless individuals and families in your community:

- *Understand who the homeless are* so that you can help dispel the stereotypes often associated with homeless people. Learn what causes homelessness, and remember that each person's story is unique.
- *Buy* Street News *if you live in an urban area where this biweekly newspaper is sold.* Homeless individuals receive

a small amount from every paper they sell, and this money goes into a special savings account earmarked for rent.

- *Give money, clothing, and/or recyclables to organizations that aid the homeless.* In addition to money or clean, usable clothing, recyclable cans and bottles are helpful because they can be turned into small sums of money for living expenses.
- *Volunteer at a shelter, soup kitchen, or battered women's shelter* where you can help staff and other volunteers meet the daily needs of people who are without shelter and food, as well as women and children who need assistance in getting away from abusive relationships with family members.
- *Look for campus organizations that work with the homeless,* or create your own and enlist friends and existing organizations (such as your service organization, sorority, or fraternity) to engage in community service projects that will benefit both the temporarily and permanently homeless.

For additional ways you can help the homeless, check with shelters in your area. You may also want to visit the websites of organizations such as the following:

- Just Give
  http://www.justgive.org
- The Doe Fund
  http://www.doe.org
- U.S. Department of Housing and Urban Development
  http://www.hud.gov/homeless

Courtesy of Gretchen Otto

A unique way that some college students recycle items they no longer want is to conduct a garage sale that benefits a local charity or community organization.

goods available for consumption than ever before, many people experience high levels of stress, fear for their lives because of crime, and face problems such as homelessness. In a society that can send astronauts into space to perform complex scientific experiments, is it impossible to solve some of the problems that plague us here on Earth?

Individuals and groups often show initiative in trying to solve some of our pressing problems, at least on a local level (see an example in Box 4.3). However, the future of this country rests on our collective ability to deal with major social problems at both the macrolevel and the microlevel of society.

Chapter Review

● How does social structure shape our social interactions?

The stable patterns of social relationships within a particular society make up its social structure. Social structure is a macrolevel influence because it shapes and determines the overall patterns in which social interaction occurs. Social structure provides an ordered framework for society and for our interactions with others.

● What are the main components of social structure?

Social structure comprises statuses, roles, groups, and social institutions. A status is a specific position in a group or society and is characterized by certain expectations, rights, and duties. Ascribed statuses, such as gender, class, and race/ethnicity, are acquired at birth or involuntarily later in life. Achieved statuses, such as education and occupation, are assumed voluntarily as a result of personal choice, merit, or direct effort. We occupy a status, but a role is the set of behavioral expectations associated with a given status. A social group consists of two or more people who interact frequently and share a common identity and sense of interdependence. A formal organization is a highly structured group formed to complete certain tasks or achieve specific goals. A social institution is a set of organized beliefs and rules that establishes how a society attempts to meet its basic needs.

● What are the functionalist and conflict perspectives on social institutions?

According to functionalist theorists, social institutions perform several prerequisites of all societies: replace members; teach new members; produce, distribute, and consume goods and services; preserve order; and provide and maintain a sense of purpose. Conflict theorists suggest that social institutions do not work for the common good of all individuals: Institutions may enhance and uphold the power of some groups but exclude others, such as the homeless.

● How do societies maintain stability in times of social change?

According to Emile Durkheim, although changes in social structure may dramatically affect individuals and groups, societies manage to maintain some degree of stability. People in preindustrial societies are united by mechanical solidarity because they have shared values and common social bonds. Industrial societies are characterized by organic solidarity, which refers to the cohesion that results when people perform specialized tasks and are united by mutual dependence.

● How do *Gemeinschaft* and *Gesellschaft* societies differ in social solidarity?

According to Ferdinand Tönnies, the *Gemeinschaft* is a traditional society in which relationships are based on personal bonds of friendship and kinship and on intergenerational stability. The *Gesellschaft* is an urban society in which social bonds are based on impersonal and specialized relationships, with little group commitment or consensus on values.

● What is the dramaturgical perspective?

According to Erving Goffman's dramaturgical analysis, our daily interactions are similar to dramatic productions. Presentation of self refers to efforts to present our own self to others in ways that are most favorable to our interests or self-image.

● Why are feeling rules important?

Feeling rules shape the appropriate emotions for a given role or specific situation. Our emotions are not always private, and specific emotions may be demanded of us on certain occasions.

www.cengage.com/login

Register for a Student eResource account to maximize your study time online using CengageNOW. First take the system's diagnostic pre-test, and then follow the personalized study plan that is created for you to help you review this chapter. The study plan will

- help you identify areas on which you should concentrate;

- provide interactive exercises to help you master the chapter concepts; and
- provide a post-test to confirm you are ready to move on to the next chapter.

Key Terms

achieved status 117
ascribed status 117
division of labor 127
dramaturgical analysis 132
ethnomethodology 132
face-saving behavior 133
formal organization 124
Gemeinschaft 128
Gesellschaft 128
impression management (presentation of self) 132
industrial society 128

master status 117
mechanical solidarity 127
nonverbal communication 136
organic solidarity 127
personal space 138
postindustrial society 128
primary group 124
role 119
role conflict 121
role exit 123
role expectation 119
role performance 119

role strain 122
secondary group 124
self-fulfilling prophecy 130
social construction of reality 130
social group 123
social institution 126
social interaction 113
social structure 113
status 115
status set 116
status symbol 119

Questions for Critical Thinking

1. Think of a person you know well who often irritates you or whose behavior grates on your nerves (it could be a parent, friend, relative, teacher). First, list that person's statuses and roles. Then, analyze the person's possible role expectations, role performance, role conflicts, and role strains. Does anything you find in your analysis help to explain the irritating behavior? How helpful are the concepts of social structure in analyzing individual behavior?

2. You are conducting field research on gender differences in nonverbal communication styles. How are you going to account for variations among age, race, and social class?

3. When communicating with other genders, races, and ages, is it better to express and acknowledge different styles or to develop a common, uniform style?

The Kendall Companion Website

www.cengage.com/sociology/kendall

Supplement your review of this chapter by going to the text's companion website, where you can take tutorial quizzes, use flash cards to master key terms, follow live links to useful websites, and explore the other study and research resources you'll find there, such as a comprehensive interactive sociology timeline, GSS Data, and Census 2000 information, much of it presented visually in maps.

5 Groups and Organizations

O kay, I think I'll share why Facebook works for me and keeps me coming back. I was hesitant to sign up in the first place, I was afraid it would be a lame fad. . . . Since college is fairly dynamic (new classes every quarter), a directory of friends and students remains very dynamic and gives me a reason to come back (to see how friends are doing and what classes they are taking). Also it is cool to look up people you have in class and see what they are interested in. Who knows, it might help you start a conversation sometime (although, it might freak them out if you already know their interests). . . .

—a male college student in Washington state (in a comment posted at Linden, 2005) explaining why he likes Facebook, a highly popular website that tallies 250 million hits every day and ranks ninth on overall traffic on the Internet

Do you think that even 5 percent of people [on Facebook and other networking sites] really connect with people in their network who want them to do the same kewl stuff as them? I doubt it, unless they already knew that person

Have Facebook and other networking websites influenced our social interactions and group participation? Why are face-to-face encounters in groups and organizations still important in everyday life?

© AP Images/L. G. Patterson

IN THIS CHAPTER

- Social Groups
- Group Characteristics and Dynamics
- Formal Organizations in Global Perspective
- Alternative Forms of Organization
- Organizations in the Future

and just didn't know about their hobbies. The truth is none of these sites really connects people. That requires ongoing new information (like web bulletin boards, attending meetings). Or heaven forbid, actual human contact.

—another male college student (in a comment posted at Linden, 2005) claiming that individuals do not really connect with each other online despite the amount of time that they may spend on websites claiming to connect students through their common interests, lifestyles, and attitudes

The problem with facebook.com is people are putting their pictures, cell phone numbers and addresses on the Internet. The Internet is open to anyone. That's just asking for someone to knock on your door. . . . It may seem cool that you know a bunch of people, but it won't be cool if a strange person knows too much about you.

—a junior in a Texas college describing her concerns about facebook.com (Sheppard, 2005)

According to sociologists, we need groups and organizations—just as we need culture and socialization—to live and participate in a society. Historically, the basic premise of groups and organizations was that individuals engage in face-to-face interactions in order to be part of such a group; however, millions of people today communicate with others through the Internet, cell phones, and other forms of information technology that make it possible for them to "talk" with individuals they have never met and who may live thousands of miles away. A variety of networking websites, including Facebook, MySpace, Friendster,

Chapter Focus Question

Why is it important for groups and organizations to enhance communication among participants and improve the flow of information while protecting the privacy of individuals?

and xuqa, now compete with, or in some cases replace, live, person-to-person communications. For many college students, Facebook has become a fun way to get to know other people, to join online groups with similar interests or activities, and to plan "real-life" encounters. Despite the wealth of information and opportunities for new social connections that such websites offer, many of our daily activities require that we participate in social groups and formal organizations where *face-time*—time spent interacting with others on a face-to-face basis, rather than via Internet or cell phone—is necessary.

What do social groups and formal organizations mean to us in an age of rapid telecommunications? What is the relationship between information and social organizations in societies such as ours? How can we balance the information that we provide to other people about us with our own right to privacy and need for security? These questions are of interest to sociologists who seek to apply the sociological imagination to their studies of social groups, bureaucratic organizations, social networking, and virtual communities. Before we take a closer look at groups and organizations, take the quiz in Box 5.1 on personal privacy in groups and formal organizations.

Social Groups

Three strangers are standing at a street corner waiting for a traffic light to change. Do they constitute a group? Five hundred women and men are first-year graduate students at a university. Do they constitute a group? In everyday usage, we use the word *group* to mean any collection of people. According to sociologists, however, the answer to these questions is no; individuals who happen to share a common feature or to be in the same place at the same time do not constitute social groups.

Groups, Aggregates, and Categories

As we saw in Chapter 4, a *social group* is a collection of two or more people who interact frequently with one another, share a sense of belonging, and have

a feeling of interdependence. Several people waiting for a traffic light to change constitute an **aggregate—a collection of people who happen to be in the same place at the same time but share little else in common.** Shoppers in a department store and passengers on an airplane flight are also examples of aggregates. People in aggregates share a common purpose (such as purchasing items or arriving at their destination) but generally do not interact with one another, except perhaps briefly. The first-year graduate students, at least initially, constitute a **category—a number of people who may never have met one another but share a similar characteristic (such as education level, age, race, or gender).** Men and women make up categories, as do Native Americans and Latinos/as, and victims of sexual or racial harassment. Categories are not social groups because the people in them do not usually create a social structure or have anything in common other than a particular trait.

Occasionally, people in aggregates and categories form social groups. For instance, people within the category known as "graduate students" may become an aggregate when they get together for an orientation to graduate school. Some of them may form social groups as they interact with one another in classes and seminars, find that they have mutual interests and concerns, and develop a sense of belonging to the group. Information technology raises new and interesting questions about what constitutes a group. For example, some people question whether we can form a social group on the Internet (see Box 5.2).

Types of Groups

As you will recall from Chapter 4, groups have varying degrees of social solidarity and structure. This structure is flexible in some groups and more rigid in others. Some groups are small and personal; others are large and impersonal. We more closely identify with the members of some groups than we do with others.

Cooley's Primary and Secondary Groups Sociologist Charles H. Cooley (1963/1909) used the term *primary group* to describe a small, less specialized group in which members engage in face-to-face,

Box 5.1 Sociology and Everyday Life

How Much Do You Know About Privacy in Groups and Organizations?

True	False	
T	F	1. A college student's privacy is protected when using a school-owned computer as long as he or she deletes from the computer all e-mails or other documents he or she has worked on and thus prevents anyone else from examining those documents.
T	F	2. Parents of students at all U.S. colleges and universities are entitled to obtain a transcript of their children's college grades, regardless of the student's age.
T	F	3. If you work for a business that monitors phone calls with a pen register, your employer has the right to maintain and examine a list of phone numbers dialed by your extension and how long each call lasted.
T	F	4. Members of a high school football team can be required to submit to periodic, unannounced drug testing.
T	F	5. A company has the right to keep its employees under video surveillance anywhere at the company's place of business—even in the restrooms.
T	F	6. A professor can legally post students' grades in public, using the student's Social Security number as an identifier, as long as the student's name does not appear with the number.
T	F	7. Students at a church youth group meeting who hear one member of the group confess to an illegal act can be required to divulge what that member said.
T	F	8. If you apply for a job in a job setting that has more than 25 employees, your employer can require that you provide a history of your medical background or take a physical examination prior to offering you a job.

Answers on page 148.

emotion-based interactions over an extended period of time. We have primary relationships with other individuals in our primary groups—that is, with our *significant others,* who frequently serve as role models.

In contrast, you will recall, a *secondary group* is a larger, more specialized group in which the members engage in more impersonal, goal-oriented relationships for a limited period of time. The size of a secondary group may vary. Twelve students in a graduate seminar may start out as a secondary group but eventually become a primary group as they get to know one another and communicate on a more personal basis. Formal organizations are secondary groups, but they also contain many primary groups within them. For example, how many primary groups do you think there are within the secondary-group setting of your college?

Sumner's Ingroups and Outgroups All groups set boundaries by distinguishing between insiders who are members and outsiders who are not. Sociologist William Graham Sumner (1959/1906)

aggregate a collection of people who happen to be in the same place at the same time but share little else in common.

category a number of people who may never have met one another but share a similar characteristic, such as education level, age, race, or gender.

Box 5.1 Sociology and Everyday Life

Answers to the Sociology Quiz on Privacy

1. **False.** Deleting an e-mail or other document from a computer does not actually remove it from the computer's memory. Until other files are entered that write over the space where the document was located, experts can retrieve the document that was deleted.

2. **False.** The Family Educational Right to Privacy Act, which allows parents of a student under age eighteen to obtain their child's grades, requires the student's consent once he or she has attained age eighteen; however, that law applies only to institutions that receive federal educational funds.

3. **True.** Telephone numbers dialed from a company's phone extensions can be recorded on a pen register, which allows the employer to see a list of numbers you have dialed and the length of each call, and this information can be used by the employer in evaluating the amount of time you have spent talking with clients—or with other people.

4. **True.** The U.S. Supreme Court has ruled that schools may require students to submit to random drug testing as a condition of participating in extracurricular activities such as sports teams, the school band, the future homemakers' club, the cheerleading squad, and the choir.

5. **False.** An employer may not engage in video surveillance of its employees in situations where they have a reasonable right of privacy. At least in the absence of a sign warning of such surveillance, employees have such a right in company restrooms.

6. **False.** The Federal Educational Rights and Privacy Act states that Social Security numbers are considered to be "personally identifiable information" that may not be released without written consent from the student. Posting grades by Social Security number violates this provision unless the student has consented to the number being disclosed to others.

7. **True.** Although confidential communications made privately to a minister, priest, rabbi, or other religious leader (or to an individual the person reasonably believes to hold such a position) generally cannot be divulged without the consent of the person making the communication, this does not apply when other people are present who are likely to hear the statement.

8. **False.** The Americans with Disabilities Act prohibits employers in job settings with more than 25 employees from asking job applicants about medical information or requiring a physical examination prior to employment.

coined the terms *ingroup* and *outgroup* to describe people's feelings toward members of their own and other groups. An ***ingroup* is a group to which a person belongs and with which the person feels a sense of identity. An *outgroup* is a group to which a person does not belong and toward which the person may feel a sense of competitiveness or hostility.** Distinguishing between our ingroups and our outgroups helps us establish our individual identity and self-worth. Likewise, groups are solidified by ingroup and outgroup distinctions; the presence of

an enemy or a hostile group binds members more closely together (Coser, 1956).

Group boundaries may be formal, with clearly defined criteria for membership. For example, a country club that requires an applicant for membership to be recommended by four current members and to pay a $25,000 initiation fee has clearly set requirements for its members (see "Sociology Works!"). However, group boundaries are not always that formal. For example, friendship groups usually do not have clear guidelines for member-

Box 5.2 Framing "Community" in the Media

"Virtual Communities" on the Internet

Meeting new friends,
Imagining smiles . . .
Across the networks
Spanning the miles. . . .

From all walks of life
We come to the net.
A community of friends
Who have never met.
    —from "Thoughts of Internet Friendships" by
    Jamie Wilkerson (1996)

As this excerpt from a poem posted on the Internet suggests, many people believe that they can make new friends and establish a community online. We are encouraged to establish such friendships by joining chat groups maintained by various Internet service providers. Chat groups are framed as a public service offered as part of the fee a subscriber pays for an Internet connection. To participate, people fill out a profile listing their hobbies and interests so that they can be matched with other participants. Many people hope to become part of the larger Internet community, which has been described as "a body of people looking for similar information, dealing with similar conditions, and abiding by the same general rules" (thewritemarket.com, 2003).

Although chat groups are framed as a new way to make friends, get dates, and establish a cyber community, as you study sociology you might ask whether this form of "community" is actually a true community. Because sociologists define a *social group* as a collection of two or more people who interact frequently with one another, share a sense of belonging, and have a feeling of interdependence, this definition suggests that people must have a sense of place (be in the same place at the same time at least part of the time) in order to establish a true social group or community. However, this definition was developed before the

Internet provided people with the rapid communications that link them with others around the world today. Are we able to form groups and establish communities with people whom we have never actually met?

Some social scientists believe that virtual communities established on the Internet constitute true communities (see Wellman, 2001). However, the sociologists Robyn Bateman Driskell and Larry Lyon examined existing theories and research on this topic and concluded that true communities cannot be established in the digital environment of cyberspace. According to Driskell and Lyon, although the Internet provides us with the opportunity to share interests with others whom we have not met (such as through chat groups) and to communicate with people we already know (such as by e-mail and instant messaging), the original concept of community, which "emphasized local place, common ties, and social interaction that is intimate, holistic, and all-encompassing," is lacking (Driskell and Lyon, 2002: 6). Virtual communities on the Internet do not have geographic and social boundaries, are limited in their scope to specific areas of interest, are psychologically detached from close interpersonal ties, and have only limited concern for their "members" (Driskell and Lyon, 2002). In fact, if we spend many hours in social isolation doing impersonal searches for information, the Internet may *reduce* community rather than enhance it. Even so, it is possible that the Internet will create a "weak community replacement" for people based on a virtual community of specialized ties developed by e-mail correspondence and chat-room discussions (Driskell and Lyon, 2002).

Reflect & Analyze

Do you think that chat groups are accurately framed in descriptions by Internet service providers, or are they overhyped to potential participants?

---

ship; rather, the boundaries tend to be very informal and vaguely defined.

Ingroup and outgroup distinctions may encourage social cohesion among members, but they may also promote classism, racism, sexism, and ageism. Ingroup members typically view themselves positively

ingroup a group to which a person belongs and with which the person feels a sense of identity.

outgroup a group to which a person does not belong and toward which the person may feel a sense of competitiveness or hostility.

Sociology *Works!*

Ingroups, Outgroups, and "Members Only" Clubs

In this country we have a God-given right to associate with whomever we please. And frankly, this includes my right to *not* associate with people I don't want to. If I don't want to be around somebody, why should I have to let them in my club? Let them go start their own club.

—Phil, a white, male attorney who is a member of several prestigious private clubs, explaining why he believes he has the right to establish his own ingroup through private club memberships (qtd. in Kendall, 2008)

A key characteristic of the city clubs and country clubs where Phil is a member is that each organization has formal group boundaries, and people become members "by invitation only." In other words, prospective members must be nominated by current members and be voted into the club: They cannot simply decide to join the organization. For this reason, people who are invited to join typically feel special (like "insiders") because they know that club membership is not available to everyone. Club members such as Phil often develop *consciousness of kind*—a term used by sociologists to describe the awareness that individuals may have when they believe

that they share important commonalities with certain other people. Consciousness of kind is strengthened by membership in clubs ranging from country clubs to college sororities, fraternities, and other by-invitation-only university social clubs. Members of ingroups typically share strong feelings of consciousness of kind and believe that they have little in common with people in the outgroup.

Recent studies on private clubs and exclusive college social organizations show that the sociological concepts of "ingroup" and "outgroup" remain highly relevant today when we conduct research on the processes of inclusion and exclusion to learn more about how such activities affect individuals and groups (see Kendall, 2008). Most of us are aware that our ingroups are very important to us: They provide us with a unique sense of identity, but they also give us the ability to exclude those individuals whom we do not want in our inner circle of friends and acquaintances. The early sociologist Max Weber captured this idea in his description of the *closed relationship*—a setting in which the "participation of certain persons is excluded, limited, or subjected to conditions" (Gerth and Mills, 1946: 139). Exclusive clubs typically have signs posted on gates, fences, or buildings that state "Mem-

and members of outgroups negatively. These feelings of group superiority, or *ethnocentrism,* are somewhat inevitable. However, members of some groups feel more free than others to act on their beliefs. If groups are embedded in larger groups and organizations, the large organization may discourage such beliefs and their consequences (Merton, 1968). Conversely, organizations may covertly foster these ingroup/outgroup distinctions by denying their existence or by failing to take action when misconduct occurs.

Reference Groups Ingroups provide us not only with a source of identity but also with a point of reference. A *reference group* is a group that strongly **influences a person's behavior and social attitudes, regardless of whether that individual is an**

actual member. When we attempt to evaluate our appearance, ideas, or goals, we automatically refer to the standards of some group. Sometimes, we will refer to our membership groups, such as family or friends. Other times, we will rely on groups to which we do not currently belong but that we might wish to join in the future, such as a social club or a profession.

Reference groups help explain why our behavior and attitudes sometimes differ from those of our membership groups. We may accept the values and norms of a group with which we identify rather than one to which we belong. We may also act more like members of a group we want to join than members of groups to which we already belong. In this case, reference groups are a source of anticipatory social-

bers Only." These organizations do not welcome outsiders within their walls, and members are often pledged to loyalty and secrecy about their club's activities. Similarly, many college fraternities and sororities thrive on rituals, secrecy, and the importance of what it means to pledge—to have accepted a bid to join but not having yet been initiated into—the group of one's choice (Robbins, 2004: 342).

Reflect & Analyze

What areas of sociological research or personal interest can you think of that might benefit from applying the ingroup/outgroup concept to your analysis? How might these concepts be applied to other areas of college life beside invitational social organizations?

Sometimes, the distinction between what constitutes an ingroup and an outgroup is subtle. Other times, it is not subtle at all. Would you feel comfortable entering either of these establishments if you were not a member?

ization. For most of us, our reference-group attachments change many times during our life course, especially when we acquire a new status in a formal organization.

Group Characteristics and Dynamics

What purpose do groups serve? Why are individuals willing to relinquish some of their freedom to participate in groups? According to functionalists, people form groups to meet instrumental and expressive needs. *Instrumental,* or task-oriented, needs cannot always be met by one person, so the group works cooperatively to fulfill a specific goal. Groups help members do jobs that are impossible to do alone or that would be very difficult and time-consuming at best. For example, think of how hard it would be to function as a one-person football team or to single-handedly build a skyscraper. In addition to instrumental needs, groups also help people meet their *expressive,* or emotional, needs, especially those involving self-expression and support from family, friends, and peers.

reference group a group that strongly influences a person's behavior and social attitudes, regardless of whether that individual is an actual member.

Although not disputing that groups ideally perform such functions, conflict theorists suggest that groups also involve a series of power relationships whereby the needs of individual members may not be equally served. Symbolic interactionists focus on how the size of a group influences the kind of interaction that takes place among members. To many postmodernists, groups and organizations—like other aspects of postmodern societies—are generally characterized by superficiality and depthlessness in social relationships (Jameson, 1984). One postmodern thinker who focuses on this issue is the literary theorist Fredric Jameson, who believes that people experience a waning of emotion in organizations where fragmentation and superficiality are a way of life (Ritzer, 1997). For example, fast-food restaurant employees and customers interact in extremely superficial ways that are largely scripted: The employees follow scripts in taking and filling customers' orders ("Would you like fries and a drink with that?"), and the customers respond with their own "recipied" action. According to the sociologist George Ritzer (1997: 226), "[C]ustomers are mindlessly following what they consider tried-and-true social recipes, either learned or created by them previously, on how to deal with restaurant employees and, more generally, how to work their way through the system associated with the fast-food restaurant."

We will now look at certain characteristics of groups, such as how size affects group dynamics.

Group Size

The size of a group is one of its most important features. Interactions are more personal and intense in a *small group,* **a collectivity small enough for all members to be acquainted with one another and to interact simultaneously.**

Sociologist Georg Simmel (1950/1902–1917) suggested that small groups have distinctive interaction patterns that do not exist in larger groups. According to Simmel, in a *dyad*—**a group composed of two members**—the active participation of both members is crucial to the group's survival. If one member withdraws from interaction or "quits," the group ceases to exist. Examples of dyads include two people who are best friends, married couples,

According to the sociologist Georg Simmel, interaction patterns change when a third person joins a dyad—a group composed of two members. How might the conversation between these two women change when another person arrives to talk with them?

and domestic partnerships. Dyads provide members with an intense bond and a sense of unity not found in most larger groups.

When a third person is added to a dyad, a *triad,* **a group composed of three members,** is formed. The nature of the relationship and interaction patterns changes with the addition of the third person. In a triad, even if one member ignores another or declines to participate, the group can still function. In addition, two members may unite to create a coalition that can subject the third member to group pressure to conform. A *coalition* is an alliance created in an attempt to reach a shared objective or goal. If two members form a coalition, the other member may be seen as an outsider or intruder.

As the size of a group increases beyond three people, members tend to specialize in different tasks, and everyday communication patterns change. For instance, in groups of more than six or seven people, it becomes increasingly difficult for everyone to take part in the same conversation; therefore, several conversations will probably take place simultaneously. Members are also likely to take sides on issues and form a number of coalitions. In groups of more than ten or twelve people, it becomes virtually impossible for all members to participate in a single conversation unless one person serves as moderator and guides the discussion. As shown in ▶ Figure 5.1, when the size of the group increases, the number of possible social interactions also increases.

Although large groups typically have less social solidarity than small ones, they may have more power. However, the relationship between size and power is more complicated than it might initially seem. The power relationship depends on both a group's *absolute* size and its *relative* size (Simmel, 1950/1902–1917; Merton, 1968). The absolute size is the number of members the group actually has; the relative size is the number of potential members. For example, suppose that three hundred people band together to "march on Washington" and demand enactment of a law on some issue that they feel is important. Although three hundred people is a large number in some contexts, opponents of this group would argue that the low turnout (compared with the number of people in this country)

demonstrates that most people don't believe the issue is important. At the same time, the power of a small group to demand change may be based on a "strength in numbers" factor if the group is seen as speaking on behalf of a large number of other people (who are also voters).

Larger groups typically have more formalized leadership structures. Their leaders are expected to perform a variety of roles, some related to the internal workings of the group and others related to external relationships with other groups.

Group Leadership

What role do leaders play in groups? Leaders are responsible for directing plans and activities so that the group completes its task or fulfills its goals. Primary groups generally have informal leadership. For example, most of us do not elect or appoint leaders

small group a collectivity small enough for all members to be acquainted with one another and to interact simultaneously.

dyad a group composed of two members.

triad a group composed of three members.

Group size: 2
Only one interaction possible

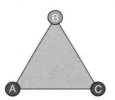

Group size: 3
Three interactions possible

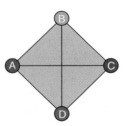

Group size: 4
Six interactions possible

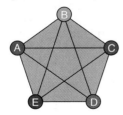

Group size: 5
Ten interactions possible

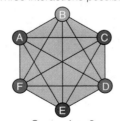

Group size: 6
Fifteen interactions possible

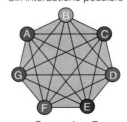

Group size: 7
Twenty-one interactions possible

▶ Figure 5.1 **Growth of Possible Social Interaction Based on Group Size**

in our own families. Various family members may assume a leadership role at various times or act as leaders for specific tasks. In traditional families, the father or eldest male is usually the leader. However, in today's more diverse families, leadership and power are frequently in question, and power relationships may be quite different, as discussed later in this text. By comparison, larger groups typically have more formalized leadership structures. Their leaders are expected to perform a variety of roles, some related to the internal workings of the group and others related to external relationships with other groups. For example, leadership in secondary groups (such as colleges, governmental agencies, and corporations) involves a clearly defined chain of command, with written responsibilities assigned to each position in the organizational structure.

Leadership Functions Both primary and secondary groups have some type of leadership or positions that enable certain people to be leaders, or at least to wield power over others. From a functionalist perspective, if groups exist to meet the instrumental and expressive needs of their members, then leaders are responsible for helping the group meet those needs. *Instrumental leadership* **is goal or task oriented;** this type of leadership is most appropriate when the group's purpose is to complete a task or reach a particular goal. *Expressive leadership* **provides emotional support for members;** this type of leadership is most appropriate when the group is dealing with emotional issues, and harmony, solidarity, and high morale are needed. Both kinds of leadership are needed for groups to work effectively.

Leadership Styles Three major styles of leadership exist in groups: authoritarian, democratic, and laissez-faire. *Authoritarian leaders* **make all major group decisions and assign tasks to members.** These leaders focus on the instrumental tasks of the group and demand compliance from others. In times of crisis, such as a war or natural disaster, authoritarian leaders may be commended for their decisive actions. In other situations, however, they may be criticized for being dictatorial and for fostering intergroup hostility. By contrast, *democratic leaders* **encourage group discussion and decision making through consensus building.** These leaders may be praised for their expressive, supportive behavior toward group members, but they may also be blamed for being indecisive in times of crisis.

Laissez-faire literally means "to leave alone." *Laissez-faire leaders* **are only minimally involved in decision making and encourage group members to make their own decisions.** On the one hand, laissez-faire leaders may be viewed positively by group members because they do not flaunt their power or position. On the other hand, a group that needs active leadership is not likely to find it with this style of leadership, which does not work vigorously to promote group goals.

Studies of kinds of leadership and decision-making styles have certain inherent limitations. They tend to focus on leadership that is imposed externally on a group (such as bosses or political leaders) rather than leadership that arises within a group. Different decision-making styles may be more effective in one setting than another. For example, imagine attending a college class in which the professor asked the students to determine what should be covered in the course, what the course requirements should be, and how students should be graded. It would be a difficult and cumbersome way to start the semester; students might spend the entire term negotiating these matters and never actually learn anything.

Group Conformity

To what extent do groups exert a powerful influence in our lives? Groups have a significant amount of influence on our values, attitudes, and behavior. In order to gain and then retain our membership in groups, most of us are willing to exhibit a high level of conformity to the wishes of other group members. *Conformity* **is the process of maintaining or changing behavior to comply with the norms established by a society, subculture, or other group.** We often experience powerful pressure from other group members to conform. In some situations, this pressure may be almost overwhelming.

In several studies (which would be impossible to conduct today for ethical reasons), researchers found that the pressure to conform may cause group members to say they see something that is contradictory to what they are actually seeing or to do something that they would otherwise be unwilling to do. As we look at two of these studies, ask yourself what you might have done if you had been involved in this research.

Asch's Research Pressure to conform is especially strong in small groups in which members want to fit in with the group. In a series of experiments conducted by Solomon Asch (1955, 1956), the pressure toward group conformity was so great that participants were willing to contradict their own best judgment if the rest of the group disagreed with them.

One of Asch's experiments involved groups of undergraduate men (seven in each group) who were allegedly recruited for a study of visual perception. All the men were seated in chairs. However, the person in the sixth chair did not know that he was the only actual subject; all the others were assisting the researcher. The participants were first shown a large card with a vertical line on it and then a second card with three vertical lines (see ▶ Figure 5.2). Each of the seven participants was asked to indicate which of the three lines on the second card was identical in length to the "standard line" on the first card.

In the first test with each group, all seven men selected the correct matching line. In the second trial, all seven still answered correctly. In the third trial, however, the actual subject became very uncomfort-

able when all the others selected the incorrect line. The subject could not understand what was happening and became even more confused as the others continued to give incorrect responses on eleven out of the next fifteen trials.

Asch (1955) found that about one-third of all subjects chose to conform by giving the same (incorrect) responses as Asch's assistants. In discussing the experiment afterward, most of the subjects who gave incorrect responses indicated that they had known the answers were wrong but decided to go along with the group in order to avoid ridicule or ostracism.

Asch concluded that the size of the group and the degree of social cohesion felt by participants were important influences on the extent to which individuals respond to group pressure. If you had been in the position of the subject, how would you have responded? Would you have continued to give the correct answer, or would you have been swayed by the others?

Milgram's Research How willing are we to do something because someone in a position of authority has told us to do it? How far are we willing to go to follow the demands of that individual? Stanley Milgram (1963, 1974) conducted a series of controversial

▶ Figure 5.2 Asch's Cards
Although Line 2 is clearly the same length as the line in the lower card, Solomon Asch's research assistants tried to influence "actual" participants by deliberately picking Line 1 or Line 3 as the correct match. Many of the participants went along rather than risking the opposition of the "group."

Source: Asch, 1955.

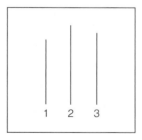
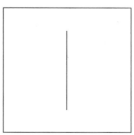

instrumental leadership goal- or task-oriented leadership.

expressive leadership an approach to leadership that provides emotional support for members.

authoritarian leaders people who make all major group decisions and assign tasks to members.

democratic leaders leaders who encourage group discussion and decision making through consensus building.

laissez-faire leaders leaders who are only minimally involved in decision making and who encourage group members to make their own decisions.

conformity the process of maintaining or changing behavior to comply with the norms established by a society, subculture, or other group.

experiments to find answers to these questions about people's obedience to authority. *Obedience* is a form of compliance in which people follow direct orders from someone in a position of authority.

Milgram's subjects were men who had responded to an advertisement for participants in an experiment. When the first (actual) subject arrived, he was told that the study concerned the effects of punishment on learning. After the second subject (an assistant of Milgram's) arrived, the two men were instructed to draw slips of paper from a hat to get their assignments as either the "teacher" or the "learner." Because the drawing was rigged, the actual subject always became the teacher, and the assistant the learner. Next, the learner was strapped into a chair with protruding electrodes that looked something like an electric chair. The teacher was placed in an adjoining room and given a realistic-looking but nonoperative shock generator. The "generator's" control panel showed levels that went from "Slight Shock" (15 volts) on the left, to "Intense Shock" (255 volts) in the middle, to "DANGER: SEVERE SHOCK" (375 volts), and finally "XXX" (450 volts) on the right.

The teacher was instructed to read aloud a pair of words and then repeat the first of the two words.

At that time, the learner was supposed to respond with the second of the two words. If the learner could not provide the second word, the teacher was instructed to press the lever on the shock generator so that the learner would be punished for forgetting the word. Each time the learner gave an incorrect response, the teacher was supposed to increase the shock level by 15 volts. The alleged purpose of the shock was to determine if punishment improves a person's memory.

What was the maximum level of shock that a "teacher" was willing to inflict on a "learner"? The learner had been instructed (in advance) to beat on the wall between him and the teacher as the experiment continued, pretending that he was in intense pain. The teacher was told that the shocks might be "extremely painful" but that they would cause no permanent damage. At about 300 volts, when the learner quit responding at all to questions, the teacher often turned to the experimenter to see what he should do next. When the experimenter indicated that the teacher should give increasingly painful shocks, 65 percent of the teachers administered shocks all the way up to the "XXX" (450-volt) level (see ▶ Figure 5.3). By this point in the process, the teachers were

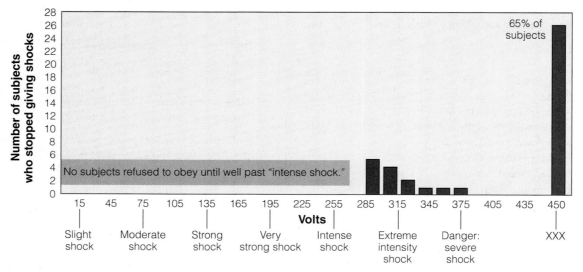

Level of shock (as labeled on Milgram's shock machine)

▶ **Figure 5.3 Results of Milgram's Obedience Experiment**
Even Milgram was surprised by subjects' willingness to administer what they thought were severely painful and even dangerous shocks to a helpless "learner."

Source: Milgram, 1963.

frequently sweating, stuttering, or biting on their lip. According to Milgram, the teachers (who were free to leave whenever they wanted to) continued in the experiment because they were being given directions by a person in a position of authority (a university scientist wearing a white coat).

What can we learn from Milgram's study? The study provides evidence that obedience to authority may be more common than most of us would like to believe. None of the "teachers" challenged the process before they had applied 300 volts. Almost two-thirds went all the way to what could have been a deadly jolt of electricity if the shock generator had been real. For many years, Milgram's findings were found to be consistent in a number of different settings and with variations in the research design (Miller, 1986).

This research once again raises some questions concerning research ethics. As was true of Asch's research, Milgram's subjects were deceived about the nature of the study in which they were asked to participate. Many of them found the experiment extremely stressful. Such conditions cannot be ignored by social scientists because subjects may receive lasting emotional scars from this kind of research. Today, it would be virtually impossible to obtain permission to replicate this experiment in a university setting.

Groupthink

As we have seen, individuals often respond differently in a group context than they might if they were alone. Social psychologist Irving Janis (1972, 1989) examined group decision making among political experts and found that major blunders in U.S. history can be attributed to pressure toward group conformity. To describe this phenomenon, he coined the term *groupthink*—**the process by which members of a cohesive group arrive at a decision that many individual members privately believe is unwise.** Why not speak up at the time? Members usually want to be "team players." They may not want to be the ones who undermine the group's consensus or who challenge the group's leaders. Consequently, members often limit or withhold their opinions and focus on consensus rather than on exploring all of the options and determining the best course of action. ▶ Figure 5.4 summarizes the dynamics and results of groupthink.

The tragic 2003 explosion of the space shuttle *Columbia* while preparing to land has been cited as an example of this process. During takeoff, a chunk of insulated foam fell off the bipod ramp of the external fuel tank, striking and damaging the shuttle's left wing. Although some NASA engineers had previously raised concerns that hardened foam popping off the fuel tank could cause damage to the ceramic tiles protecting the shuttle, and although these concerns were again raised following *Columbia*'s liftoff, these concerns were overruled by NASA officials prior to and during the flight (Glanz and Wong, 2003; Schwartz, 2003). One analyst subsequently described the way that NASA dealt with these concerns as an example of "the ways that smart people working collectively can be dumber than the sum of their brains" (Schwartz and Wald, 2003: WK3).

Formal Organizations in Global Perspective

Over the past century, the number of formal organizations has increased dramatically in the United States and other industrialized nations. Previously, everyday life was centered in small, informal, primary groups, such as the family and the village. With the advent of industrialization and urbanization (as discussed in Chapter 1), people's lives became increasingly dominated by large, formal, secondary organizations. A *formal organization*, you will recall, is a highly structured secondary group formed for the purpose of achieving specific goals in the most efficient manner. Formal organizations (such as corporations, schools, and government agencies) usually keep their basic structure for many years in order to meet their specific goals.

Types of Formal Organizations

We join some organizations voluntarily and others out of necessity. Sociologist Amitai Etzioni (1975) classified formal organizations into three categories—

groupthink the process by which members of a cohesive group arrive at a decision that many individual members privately believe is unwise.

Process of Groupthink

PRIOR CONDITIONS
Isolated, cohesive, homogeneous decision-making group
Lack of impartial leadership
High stress

↓

SYMPTOMS OF GROUPTHINK
Closed-mindedness
Rationalization
Squelching of dissent
"Mindguards"
Feelings of righteousness and invulnerability

↓

DEFECTIVE DECISION MAKING
Incomplete examination of alternatives
Failure to examine risks and contingencies
Incomplete search for information

↓

CONSEQUENCES
Poor decisions

Example: *Columbia* Explosion

NASA had previously orchestrated many successful shuttle missions and was under pressure to complete additional space missions that would fulfill agency goals and keep its budget intact.

↓

Although *Columbia*'s left wing had been damaged on takeoff when a chunk of insulated foam from the external fuel tank struck it, NASA did not regard this as a serious problem because it had occurred on previous launches. Some NASA engineers stated that they did not feel free to raise questions about problems.

↓

The debate among engineers regarding whether the shuttle had been damaged to the extent that the wing might burn off on reentry was not passed on to the shuttle crew or to NASA's top officials in a timely manner because either the engineers harbored doubts about their concerns or were unwilling to believe that the mission was truly imperiled.

↓

The shuttle *Columbia* was destroyed during reentry into the Earth's atmosphere, killing all seven crew members and strewing debris across large portions of the United States.

© AP Images/Chris O'Meara

NASA Kennedy Space Center (NASA-KSC)

© AP Images/Dr. Scott Lieberman

▶ **Figure 5.4 Janis's Description of Groupthink**
In Janis's model, prior conditions such as a highly homogeneous group with committed leadership can lead to potentially disastrous "groupthink," which short-circuits careful and impartial deliberation. Events leading up to the tragic 2003 explosion of the space shuttle Columbia have been cited as an example of this process.

Sources: Broder, 2003; Glanz and Wong, 2003; Schwartz (with Wald), 2003; Schwartz and Broder, 2003; Schwartz and Wald, 2003.

normative, coercive, and utilitarian—based on the nature of membership in each.

Normative Organizations We voluntarily join *normative organizations* when we want to pursue some common interest or gain personal satisfaction or prestige from being a member. Political parties, ecological activist groups, religious organizations, parent–teacher associations, and college sororities and fraternities are examples of normative, or voluntary, associations.

Class, gender, and race are important determinants of a person's participation in a normative association. Class (socioeconomic status based on a person's education, occupation, and income) is the most significant predictor of whether a person will participate in mainstream normative organizations; membership costs may exclude some from joining. Those with higher socioeconomic status are more likely to be not only members but also active participants in these groups. Gender is also an important determinant. Half of the voluntary associations in the United States have all-female memberships; one-fifth are all male. However, all-male organizations usually have higher levels of prestige than do all-female ones (Odendahl, 1990).

Throughout history, people of all racial–ethnic categories have participated in voluntary organizations, but the involvement of women in these groups has largely gone unrecognized. For example, African American women were actively involved in antislavery societies in the nineteenth century and in the civil rights movement in the twentieth century (see Scott, 1990). Other normative organizations focusing on civil rights, self-help, and philanthropic activities in which African American women and men have been involved include the National Association for the Advancement of Colored People (NAACP) and the Urban League. Similarly, Native American women have participated in the American Indian Movement, a group organized to fight problems ranging from police brutality to housing and employment discrimination (Feagin and Feagin, 2003). Mexican American women (as well as men) have held a wide range of leadership positions in La Raza Unida Party and the League of United Latin American Citizens, organizations oriented toward civic activities and protests against injustices (Amott and Matthaei, 1996).

Coercive Organizations People do not voluntarily become members of *coercive organizations*—associations that people are forced to join. Total institutions, such as boot camps, prisons, and some mental hospitals, are examples of coercive organizations. As discussed in Chapter 3, the assumed goal of total institutions is to resocialize people through incarceration. These environments are characterized by restrictive barriers (such as locks, bars, and security guards) that make it impossible for people to leave freely. When people leave without being officially dismissed, their exit is referred to as an "escape."

Utilitarian Organizations We voluntarily join *utilitarian organizations* when they can provide us with a material reward that we seek. To make a living or earn a college degree, we must participate in organizations that can provide us these opportunities. Although we have some choice regarding where we work or attend school, utilitarian organizations are not always completely voluntary. For example, most people must continue to work even if the conditions of their employment are less than ideal. (This chapter's Concept Quick Review summarizes the types of groups, sizes of groups, and types of formal organizations.)

Bureaucracies

The bureaucratic model of organization remains the most universal organizational form in government, business, education, and religion. A **bureaucracy is an organizational model characterized by a hierarchy of authority, a clear division of labor, explicit rules and procedures, and impersonality in personnel matters.**

Sociologist Max Weber (1968/1922) was interested in the historical trend toward bureaucratization that accelerated during the Industrial Revolution. To Weber, bureaucracy was the most "rational" and efficient means of attaining organizational goals because it contributed to coordination and control. According to Weber, **rationality is the process by which traditional methods of social organization, characterized by informality and spontaneity, are gradually replaced by efficiently administered formal rules and procedures.** Bureaucracy can be seen in all aspects of our lives, from small colleges with perhaps a thousand students to multinational

bureaucracy an organizational model characterized by a hierarchy of authority, a clear division of labor, explicit rules and procedures, and impersonality in personnel matters.

rationality the process by which traditional methods of social organization, characterized by informality and spontaneity, are gradually replaced by efficiently administered formal rules and procedures.

CONCEPT QUICK REVIEW

Characteristics of Groups and Organizations

Types of Social Groups	Primary group	Small, less specialized group in which members engage in face-to-face, emotion-based interaction over an extended period of time
	Secondary group	Larger, more specialized group in which members engage in more impersonal, goal-oriented relationships for a limited period of time
	Ingroup	A group to which a person belongs and with which the person feels a sense of identity
	Outgroup	A group to which a person does not belong and toward which the person may feel a sense of competitiveness or hostility
	Reference group	A group that strongly influences a person's behavior and social attitudes, regardless of whether the person is actually a member
Group Size	Dyad	A group composed of two members
	Triad	A group composed of three members
	Formal organization	A highly structured secondary group formed for the purpose of achieving specific goals
Types of Formal Organizations	Normative	Organizations that we join voluntarily to pursue some common interest or gain personal satisfaction or prestige by joining
	Coercive	Associations that people are forced to join (total institutions such as boot camps and prisons are examples)
	Utilitarian	Organizations that we join voluntarily when they can provide us with a material reward that we seek

corporations employing many thousands of workers worldwide.

In his study of bureaucracies, Weber relied on an ideal-type analysis, which he adapted from the field of economics. An *ideal type* **is an abstract model that describes the recurring characteristics of some phenomenon** (such as bureaucracy). To develop this ideal type, Weber abstracted the most characteristic bureaucratic aspects of religious, educational, political, and business organizations. Weber acknowledged that no existing organization would exactly fit his ideal type of bureaucracy (Blau and Meyer, 1987).

Ideal Characteristics of Bureaucracy Weber set forth several ideal-type characteristics of bureaucratic organizations. His model (see ▶ Figure 5.5)

highlights the organizational efficiency and productivity that bureaucracies strive for in these five central elements of the ideal organization:

Division of Labor Bureaucratic organizations are characterized by specialization, and each member has highly specialized tasks to fulfill.

Hierarchy of Authority In a bureaucracy, each lower office is under the control and supervision of a higher one. Those few individuals at the top of the hierarchy have more power and exercise more control than do the many at the lower levels. Those who are lower in the hierarchy report to (and often take orders from) those above them in the organizational pyramid. Persons at the upper levels are responsible

Characteristics

- Division of labor
- Hierarchy of authority
- Rules and regulations
- Qualification-based employment
- Impersonality

Effects

- Inefficiency and rigidity
- Resistance to change
- Perpetuation of race, class, and gender inequalities

▶ **Figure 5.5 Characteristics and Effects of Bureaucracy**
The very characteristics that define Weber's idealized bureaucracy can create or exacerbate the problems that many people associate with this type of organization. Can you apply this model to an organization with which you are familiar?

not only for their own actions but also for those of the individuals they supervise.

Rules and Regulations Rules and regulations establish authority within an organization. These rules are typically standardized and provided to members in a written format. In theory, written rules and regulations offer clear-cut standards for determining satisfactory performance so that each new member does not have to reinvent the rules.

Qualification-Based Employment Bureaucracies require competence and hire staff members and professional employees based on specific qualifications. Individual performance is evaluated against specific standards, and promotions are based on merit as spelled out in personnel policies.

Impersonality Bureaucracies require that everyone must play by the same rules and be treated the same. Personal feelings should not interfere with organizational decisions.

Contemporary Applications of Weber's Theory
How well do Weber's theory of rationality and his ideal-type characteristics of bureaucracy withstand the test of time? Over 100 years later, many organizational theorists still apply Weber's perspective. For example, the sociologist George Ritzer used Weber's theories to examine fast-food restaurants such as McDonald's. According to Ritzer, the process of "McDonaldization" has become a global phenomenon that can be seen in fast-food restaurants and other "speedy" or "jiffy" businesses (such as Sir Speedy Printing and Jiffy Lube). Ritzer (2000a: 433) identifies four dimensions of formal rationality—efficiency, predictability, emphasis on quantity rather than quality, and control through nonhuman

technologies—that are found in today's fast-food restaurants:

> Efficiency means the search for the best means to the end; in the fast-food restaurant, the drive-through window is a good example of heightening the efficiency of obtaining a meal. Predictability means a world of no surprises; the Big Mac in Los Angeles is indistinguishable from the one in New York; similarly, the one we consume tomorrow or next year will be just like the one we eat today. Rational systems tend to emphasize quantity, usually large quantities, rather than quality. The Big Mac is a good example of this emphasis on quantity rather than quality. Instead of the human qualities of a chef, fast-food restaurants rely on nonhuman technologies like unskilled cooks following detailed directions and assembly-line methods applied to the cooking and serving of food. Finally, such a formally rational system brings with it various irrationalities, most notably the demystification and dehumanization of the dining experience.

While still useful today, Weber's ideal type largely failed to take into account the informal side of bureaucracy.

The Informal Side of Bureaucracy When we look at an organizational chart, the official, formal structure of a bureaucracy is readily apparent. In practice, however, a bureaucracy has patterns of activities and interactions that cannot be accounted for

---

ideal type an abstract model that describes the recurring characteristics of some phenomenon (such as bureaucracy).

© Andy Nelson/The Christian Science Monitor/Getty Images

Colleges and universities rely a great deal on the use of standardized tests to assess student applications. How do such tests relate to Weber's model of bureaucracy?

degrees of accuracy) much faster than do official channels of communication, which tend to be slow and unresponsive. The informal structure has also been referred to as *work culture* because it includes the ideology and practices of workers on the job. Workers create this work culture in order to confront, resist, or adapt to the constraints of their jobs, as well as to guide and interpret social relations on the job (Zavella, 1987).

Is the informal side of bureaucracy good or bad? Should it be controlled or encouraged? Two schools of thought have emerged with regard to these questions. One approach emphasizes control of informal groups in order to ensure greater worker productivity. By contrast, the other school of thought asserts that informal groups should be nurtured because such networks may serve as a means of communication and cohesion among individuals. Large organizations would be unable to function without strong informal norms and relations among participants (Blau and Meyer, 1987).

Informal networks thrive in contemporary organizations because e-mail and websites have made it possible for people to communicate throughout the day without ever having to engage in face-to-face interaction. The need to meet at the water fountain or the copy machine in order to exchange information is long gone: Workers now have an opportunity to tell one another—and higher-ups, as well—what they think.

by its organizational chart. These have been referred to as *bureaucracy's other face* (Page, 1946).

The *informal side of a bureaucracy* is composed of those aspects of participants' day-to-day activities and interactions that ignore, bypass, or do not correspond with the official rules and procedures of the bureaucracy. An example is an informal "grapevine" that spreads information (with varying

Problems of Bureaucracies

The characteristics that make up Weber's "rational" model of bureaucracy have a dark side that has frequently given this type of organization a bad name. Three of the major problems of bureaucracies are (1) inefficiency and rigidity, (2) resistance to change, and (3) perpetuation of race, class, and gender inequalities.

© Sean Justice/Getty Images

How do people use the informal "grapevine" to spread information? Is this faster than the organization's official channels of communication? Is it more or less accurate than official channels?

Inefficiency and Rigidity Bureaucracies experience inefficiency and rigidity at both the upper and lower levels of the organization. The self-protective behavior of officials at the top may render the organization inefficient. One type of self-protective behavior is the monopolization of information in order to maintain control over subordinates and outsiders. Information is a valuable commodity in organizations, and those persons in positions of authority guard information because it is a source of power for them—others cannot "second-guess" their decisions without access to relevant (and often "confidential") information (Blau and Meyer, 1987).

When those at the top tend to use their power and authority to monopolize information, they also fail to communicate with workers at the lower levels. As a result, they are often unaware of potential problems facing the organization and of high levels of worker frustration. Bureaucratic regulations are written in far greater detail than is necessary in order to ensure that almost all conceivable situations are covered. *Goal displacement* **occurs when the rules become an end in themselves rather than a means to an end, and organizational survival becomes more important than achievement of goals** (Merton, 1968).

Inefficiency and rigidity occur at the lower levels of the organization as well. Workers often engage in *ritualism;* that is, they become most concerned with "going through the motions" and "following the rules." According to Robert Merton (1968), the term *bureaucratic personality* **describes those workers who are more concerned with following correct procedures than they are with getting the job done correctly.** Such workers are usually able to handle routine situations effectively but are frequently incapable of handling a unique problem or

informal side of a bureaucracy those aspects of participants' day-to-day activities and interactions that ignore, bypass, or do not correspond with the official rules and procedures of the bureaucracy.

goal displacement a process that occurs in organizations when the rules become an end in themselves rather than a means to an end, and organizational survival becomes more important than achievement of goals.

bureaucratic personality a psychological construct that describes those workers who are more concerned with following correct procedures than they are with getting the job done correctly.

an emergency. Thorstein Veblen (1967/1899) used the term *trained incapacity* to characterize situations in which workers have become so highly specialized, or have been given such fragmented jobs to do, that they are unable to come up with creative solutions to problems. Workers who have reached this point also tend to experience bureaucratic alienation—they really do not care what is happening around them.

Resistance to Change Once bureaucratic organizations are created, they tend to resist change. This resistance not only makes bureaucracies virtually impossible to eliminate but also contributes to bureaucratic enlargement. Because of the assumed relationship between size and importance, officials tend to press for larger budgets and more staff and office space. To justify growth, administrators and managers must come up with more tasks for workers to perform.

Resistance to change may also lead to incompetence. Based on organizational policy, bureaucracies tend to promote people from within the organization. As a consequence, a person who performs satisfactorily in one position is promoted to a higher level in the organization. Eventually, people reach a level that is beyond their knowledge, experience, and capabilities.

Perpetuation of Race, Class, and Gender Inequalities Some bureaucracies perpetuate inequalities of race, class, and gender because this form of organizational structure creates a specific type of work or learning environment. This structure was typically created for middle- and upper-middle-class white men, who for many years were the predominant organizational participants.

For people of color, *entry* into dominant white bureaucratic organizations does not equal actual *integration* (Feagin, 1991). Instead, many have experienced an internal conflict between the bureaucratic ideals of equal opportunity and fairness and the prevailing norms of discrimination and hostility that exist in many organizations. Research has found that people of color are more adversely affected than dominant-group members by hierarchical bureaucratic structures and exclusion from informal networks.

Like racial inequality, social class divisions may be perpetuated in bureaucracies (Blau and Meyer, 1987). The theory of a "dual labor market" has been developed to explain how social class distinctions are perpetuated through different types of employment. Middle- and upper-middle-class employees are more likely to have careers characterized by higher wages, more job security, and opportunities for advancement. By contrast, poor and working-

© Blend Images/Alamy

© Mark Richards/PhotoEdit

Corporate employees in different work settings vary widely in their manner and appearance. How does the environment in which we work affect how we dress and act?

According to conflict theorists, members of the capitalist class benefit from the work of laborers such as the people shown here, who are harvesting onions on a farm in the Texas Rio Grande Valley. How do low wages and lack of job security contribute to class-based inequalities in the United States?

class employees work in occupations characterized by low wages, lack of job security, and few opportunities for promotion. The "dual economy" not only reflects but may also perpetuate people's current class position.

Gender inequalities are also perpetuated in bureaucracies. Women in traditionally male organizations may feel more visible and experience greater performance pressure. They may also find it harder to gain credibility in management positions.

Inequality in organizations has many consequences. People who lack opportunities for integration and advancement tend to be pessimistic and to have lower self-esteem. Believing that they have few opportunities, they resign themselves to staying put and surviving at that level. By contrast, those who enjoy full access to organizational opportunities tend to have high aspirations and high self-esteem. They feel loyalty to the organization and typically see their job as a means for mobility and growth.

Bureaucracy and Oligarchy

Why do a small number of leaders at the top make all the important organizational decisions? According to the German political sociologist Robert Michels (1949/1911), all organizations encounter the *iron law of oligarchy*—**the tendency to become a bureaucracy ruled by the few.** His central idea was that those who control bureaucracies not only wield power but also have an interest in retaining their power. Michels found that the hierarchical structures of bureaucracies and oligarchies go hand in hand. On the one hand, power may be concentrated in the hands of a few people because rank-and-file members must inevitably delegate a certain

iron law of oligarchy according to Robert Michels, the tendency of bureaucracies to be ruled by a few people.

amount of decision-making authority to their leaders. Leaders have access to information that other members do not have, and they have "clout," which they may use to protect their own interests. On the other hand, oligarchy may result when individuals have certain outstanding qualities that make it possible for them to manage, if not control, others. The members choose to look to their leaders for direction; the leaders are strongly motivated to maintain the power and privileges that go with their leadership positions.

Are there limits to the iron law of oligarchy? The leaders in most organizations do not have unlimited power. Divergent groups within a large-scale organization often compete for power, and informal networks can be used to "go behind the backs" of leaders. In addition, members routinely challenge, and sometimes they (or the organization's governing board) remove leaders when they are not pleased with their actions.

Alternative Forms of Organization

Many organizations have sought new and innovative ways to organize work more efficiently than the traditional hierarchical model. In the early 1980s, there was a movement in the United States to *humanize bureaucracy*—to establish an organizational environment that develops rather than impedes human resources. More-humane bureaucracies are characterized by (1) less-rigid hierarchical structures and greater sharing of power and responsibility by all participants, (2) encouragement of participants to share their ideas and try new approaches to problem solving, and (3) efforts to reduce the number of people in dead-end jobs, train people in needed skills and competencies, and help people meet outside family responsibilities while still receiving equal treatment inside the organization (Kanter, 1983, 1985, 1993/1977). However, this movement has been overshadowed by globalization and the perceived strengths of systems of organizing work in other nations, such as Japan.

Organizational Structure in Japan

For several decades, the Japanese model of organization has been widely praised for its innovative structure. Let's briefly compare the characteristics of large Japanese corporations with their U.S.-based counterparts.

Long-Term Employment and Company Loyalty

Until recently, many Japanese employees remained with the same company for their entire career, whereas their U.S. counterparts often changed employers every few years. Likewise, Japanese employers in the past had an obligation not to "downsize" by laying off workers or cutting their wages. Although the practice of lifetime employment has been replaced by the concept of long-term employment, workers in Japan often have higher levels of job security than do workers in the United States.

According to advocates, the Japanese system encourages worker loyalty and a high level of productivity. Managers move through various parts of the organization and acquire technical knowledge about the workings of many aspects of the corporation, unlike their U.S. counterparts, who tend to become highly specialized (Sengoku, 1985). Unlike top managers in the United States who have given themselves pay raises and bonuses even when their companies were financially strapped and laying off workers, many Japanese managers have taken pay cuts under similar circumstances.

Japanese management is characterized as being people oriented, taking a long-term view, and having a culture that focuses on *how* work gets done rather than on the result alone. Having respect for employees and emphasizing long-term goals appear to have made Japanese automobile manufacturers more successful than some U.S. companies that are now laying off large numbers of workers. How work is organized, such as by the use of quality circles, may also affect job satisfaction and worker productivity.

Quality Circles
Small work groups made up of about five to fifteen workers who meet regularly with one or two managers to discuss the group's performance and working conditions are known as

© TWPhoto/CORBIS

The Japanese model of organization—including planned group-exercise sessions for employees—has become a part of the workplace in many nations. Would it be a positive change if more workplace settings, such as the one shown here, were viewed as an extension of the family? Why or why not?

quality circles. The purpose of this team approach to management is both to improve product quality and to lower product costs. Workers are motivated to save the corporation money because they, in turn, receive bonuses or higher wages for their efforts. Quality circles have been praised for creating worker satisfaction, helping employees develop their potential, and improving productivity (Ishikawa, 1984). Because quality circles focus on both productivity and worker satisfaction, they (at least ideally) meet the needs of both the corporation and the workers.

Would the Japanese model work in the United States? Although the possibility of implementing the Japanese approach in U.S.-based corporations has been widely discussed, its large-scale accep-

tance is doubtful. Cultural traditions in Japan place greater emphasis on the importance of the group rather than the individual, and workers in the United States are not likely to embrace this idea because it directly conflicts with the values of individualism and personal achievement so strongly held by many in this country (Ouchi, 1981). Many U.S. workers are also unwilling to make a long-term commitment to one corporation for fear that they may be laid off or forced into early retirement.

In recent years, however, more organizations in the United States have developed a participatory management style in hopes of producing greater worker satisfaction and higher rates of productivity and profits (see Florida and Kenney, 1991).

Box 5.3 You Can Make a Difference

Developing Invisible (but Meaningful) Networks on the Web

Sept. 7

Hello! Hello! Is anybody getting this message?? I'm lost in Cyberspace and I don't know if I am getting through to anyone! Please write back if you have received this message!!

Sept. 8

We got your message. Who are you? Where are you? We are 5th graders at Perry Central School. We live in Perry County, Indiana. Where are you from? Can we help? Signed, Darrin, Jenna, Becky (qtd. in Kranning and Ehman, 1999)

As a result of this initial e-mail exchange, fifth-grade students at a rural Indiana elementary school began an extended project ("Mysteries from History") with college students taking a computer education class at Indiana University. Although this collaboration was initially established by a fifth-grade teacher (Antoinette Kranning) and a college professor (Lee Ehman), projects such as this could be established by computer-savvy students in other college classes or by a campus service organization.

Let's look at how "Mysteries from History" works. Students in the computer education class pose historical mysteries for the fifth graders to solve through research and group problem-solving skills. For example, students are told that one mystery person worked in a hospital as a nurse during the Civil War and later became a famous writer, and the students are to guess the identity of this individual. Although the fifth graders' initial choice was Clara Barton, they realized that Louisa May Alcott was the correct choice after their research because only Alcott had gone on to become a famous writer. Other mystery questions were posed about other famous people in history, and the fifth graders became very excited as the project progressed. As one student stated, "I'm really enjoying e-mail. I felt very excited when I got my first message. It felt great. I felt like I was a detective searching for clues to a mystery" (qtd. in Kranning and Ehman, 1999).

Through this collaborative project, both the fifth graders and the college students benefited. The college students learned how to interact as a group with a class of young students via the Internet, and they also picked up valuable teaching tips from the fifth grade teacher, particularly regarding how elementary school students learn about history (Kranning and Ehman, 1999). Likewise, the fifth graders learned how to find information and how to work collaboratively on a project.

Although the nature of groups and organizations has changed with new technologies, and although some people believe that computer networks are not equal to face-to-face interactions, all of us are increasingly involved in computer networks because they allow us to create a range of new social spaces in which we can meet and interact with one another. These social spaces might otherwise not exist: Collaborations between a group of fifth graders and a group of college students would be less likely to occur without technologies such as the Internet. From this perspective, if we are able to create social spaces where we can help others to do things—such as learn history, become savvy Internet users, or feel that they are part of a special group—we can make a difference in their lives by connecting people to people.

What kinds of meaningful social spaces might be created through a class you are taking or a service organization to which you belong? Social spaces on the Internet can be used for a variety of purposes, including discussing topics of mutual interest, learning from one another, working on collaborative projects, entertaining one another, and playing games. Are there ways in which your class or organization (with guidance from a faculty member) could create cross-age or cross-cultural collaborations that would link people together in a shared activity or a discussion that otherwise might not take place?

Organizations in the Future

What is the best organizational structure for the future? Of course, this question is difficult to answer because it requires the ability to predict economic, political, and social conditions. Nevertheless, we can make several observations.

Organizational theorists have suggested a *horizontal* model for corporations in which both hierarchy and functional or departmental boundaries would largely be eliminated. In the horizontal structure, a limited number of senior executives would still fill support roles (such as finance and human resources) while everyone else would work in multidisciplinary teams and perform core processes (such as product development or sales generation). Organizations would have fewer layers between company heads and the staffers responsible for any given process. Performance objectives would be related to the needs of customers; people would be rewarded not just for individual performance but for skills development and team performance. If such organizations become a reality, organizational charts will more closely resemble a pepperoni pizza than the traditional pyramid-shaped stack of boxes connected by lines.

Ultimately, everyone has a stake in seeing that organizations operate in as humane a fashion as possible and that channels for opportunity are widely available to all people regardless of race, gender, or class. Workers and students alike can benefit from organizational environments that make it possible for people to explore their joint interests without fear of losing their privacy or being pitted against one another in a competitive struggle for advantage. (For an example of students working together on a meaningful activity that benefits others, see Box 5.3.)

Chapter Review

● **How do sociologists distinguish among social groups, aggregates, and categories?**

Sociologists define a social group as a collection of two or more people who interact frequently, share a sense of belonging, and depend on one another. People who happen to be in the same place at the same time are considered an aggregate. Those who share a similar characteristic are considered a category. Neither aggregates nor categories are considered social groups.

● **How do sociologists classify groups?**

Sociologists distinguish between primary groups and secondary groups. Primary groups are small and personal, and members engage in emotion-based interactions over an extended period. Secondary groups are larger and more specialized, and members have less personal and more formal, goal-oriented relationships. Sociologists also divide groups into ingroups, outgroups, and reference groups. Ingroups are groups to which we belong and with which we identify. Outgroups are groups we do not belong to or perhaps feel hostile toward. Reference groups are groups that strongly influence people's behavior whether or not they are actually members.

● **What is the significance of group size?**

In small groups, all members know one another and interact simultaneously. In groups with more than three members, communication dynamics change, and members tend to assume specialized tasks.

● **What are the major styles of leadership?**

Leadership may be authoritarian, democratic, or laissez-faire. Authoritarian leaders make major decisions and assign tasks to individual members. Democratic leaders encourage discussion and collaborative decision making. Laissez-faire leaders are minimally involved and encourage members to make their own decisions.

● **What do experiments on conformity show us about the importance of groups?**

Groups may have significant influence on members' values, attitudes, and behaviors. In order to maintain ties with a group, many members are willing to conform to norms established and reinforced by group members.

● **What are the strengths and weaknesses of bureaucracies?**

A bureaucracy is a formal organization characterized by hierarchical authority, division of labor, explicit procedures, and impersonality. According to Max Weber, bureaucracy supplies a rational means of attaining organizational goals because it contributes to coordination and control. A bureaucracy also has an informal structure, which includes the daily activities and interactions that bypass the official rules and procedures. The informal structure may enhance productivity or may be counterproductive to the organization. A bureaucracy may be inefficient, resistant to change, and a vehicle for perpetuating class, race, and gender inequalities.

www.cengage.com/login

Register for a Student eResource account to maximize your study time online using CengageNOW. First take the system's diagnostic pre-test, and then follow the personalized study plan that is created for you to help you review this chapter. The study plan will

● help you identify areas on which you should concentrate;
● provide interactive exercises to help you master the chapter concepts; and
● provide a post-test to confirm you are ready to move on to the next chapter.

Key Terms

aggregate 146

authoritarian leaders 154

bureaucracy 159

bureaucratic personality 163

category 146

conformity 154

democratic leaders 154

dyad 152

expressive leadership 154

goal displacement 163

groupthink 157

ideal type 160

informal side of a bureaucracy 162

ingroup 148

instrumental leadership 154

iron law of oligarchy 165

laissez-faire leaders 154

outgroup 148

rationality 159

reference group 150

small group 152

triad 152

Questions for Critical Thinking

1. Who might be more likely to conform in a bureaucracy, those with power or those wanting more power?
2. Do the insights gained from Milgram's research on obedience outweigh the elements of deception and stress that were forced on its subjects?
3. If you were forming a company based on humane organizational principles, would you base the promotional policies on merit and performance or on affirmative-action goals?

The Kendall Companion Website

www.cengage.com/sociology/kendall

Supplement your review of this chapter by going to the text's companion website, where you can take tutorial quizzes, use flash cards to master key terms, follow live links to useful websites, and explore the other study and research resources you'll find there, such as a comprehensive interactive sociology timeline, GSS Data, and Census 2000 information, much of it presented visually in maps.

6 Deviance and Crime

"**W**hat worries me," said Marlene [a Southside Chicago block-club president], "is that there's about seventy children on my block who use that park—and that's not counting the ones who live on the other side. Can't have them around your boys [gang members]."

"You all are something else," Big Cat [a gang leader for the "Black Kings"] said shaking his head. "I been cooperating with you all for years now, never complaining that I'm losing money. . . . I don't get no respect for that?"

"If you're in our park, we can't be. It's as simple as that," Marlene replied. "I'll give you the nighttime. Maybe I can convince folks that you all need to work at night, but that's going to be tough. But, bottom line, baby, is we can't have you all there during the day." . . .

"Okay," interjected [local pastor] Wilkins. "Now you have to stop for the summer, Big Cat. We're not asking for a two-year thing, or nothing like that. Just when the kids are outside."

Members of the California group known as the Culver City Boyz typify how gang members use items of clothing and gang signs made with their hands to assert their membership in the group and solidarity with one another. Some people might view this conduct as deviant behavior, whereas many gang members view it as an act of conformity.

"I guess I could work it on 59th, but that [business owner] keeps telling us he doesn't want us around, keeps calling the cops. . . ."

"If I get him to leave you alone during the day, and you can hang out in the parking lot on the other side of the store, you'll leave the park for the summer."

"Yeah," Big Cat replied, dejected at the compromise. "Okay, we'll be gone."

—sociologist Sudhir Alladi Venkatesh (2006: 294–295) describing a conversation between three people he interviewed during his research into the underground economy in Chicago and the gangs that constitute part of that economy

Sociologists and criminologists typically define a *gang* as a group of people, usually young, who band together for purposes generally considered to be deviant or criminal by the larger society. Throughout the past century, gang behavior has been of special interest to sociologists (see Puffer, 1912), who generally agree that youth gangs can be found in many settings and among all racial and ethnic categories. The Office of Juvenile Justice and Delinquency Prevention (2002) estimates that there are more than 24,500 gangs with about 772,500 members in the United States.

As unusual as it initially may sound, some important similarities exist between youth gangs and peer cliques, which are typically viewed as conforming to most social norms. At the most basic level, *cliques* are friendship circles, whose members identify one another as mutually connected (Adler and Adler, 1998). However, cliques are much more complex than this definition suggests. According to the sociologists Patricia A. Adler and Peter Adler (1998: 56), cliques "have a hierarchical structure, being dominated by leaders, and are exclusive in nature, so that not all individuals who desire membership are accepted." Moreover, sociologists have found that cliques function as "bodies of power" in schools by "incorporating the most popular individuals, offering the most exciting social lives, and

Chapter Focus Question

What do studies of peer cliques and youth gangs tell us about deviance?

commanding the most interest and attention from classmates" (Adler and Adler, 1998: 56).

Although cliques may have some similarities with gangs, there are also significant differences: Gangs play a large role in the economy of many low-income urban neighborhoods where residents often believe that they must do whatever is necessary to survive. Some activities in the underground economy include the performance of unregulated, unreported, and untaxed work, whereas others involve more widely recognized criminal activities such as the sale of drugs by gang members. According to Venkatesh (2006), one remarkable thing about studying deviance and crime in settings such as "Maquis Park" (a pseudonym for a real Southside Chicago neighborhood) was learning that residents and gang members sometimes forge temporary alliances and engage in self-initiated policing so that neighborhood children may play safely at the park or enjoy other everyday activities without fear of harm. Venkatesh's study reveals people's efforts to survive with the resources they amass in the underground economy, as well as residents' willingness to negotiate with gang members if it will help restore a sense of order to their neighborhood. This unique form of community policing often takes place without the assistance of law enforcement officials.

In this chapter, we look at the relationship among conformity, deviance, and crime; even in times of national crisis and war, "everyday" deviance and crime occur as usual. People do not stop activities that might be viewed by others—or by law enforcement officials—as violating social norms. An example is gang behavior, which is used in this chapter as an example of deviant behavior. For individuals who find a source of identity, self-worth, and a feeling of protection by virtue of gang membership, no radical change occurs in daily life even as events around them may change. Youth gangs have been present in the United States for many years because they meet the perceived needs of members. Some gangs may be thought of as being very similar to youth cliques, whereas other gangs engage in activities that constitute crime. Before reading on, take the quiz on peer cliques, youth gangs, and deviance in Box 6.1.

What Is Deviance?

Deviance **is any behavior, belief, or condition that violates significant social norms in the society or group in which it occurs.** We are most familiar with *behavioral* deviance, based on a person's intentional or inadvertent actions. For example, a person may engage in intentional deviance by drinking too much or robbing a bank, or in inadvertent deviance by losing money in a Las Vegas casino or laughing at a funeral.

Although we usually think of deviance as a type of behavior, people may be regarded as deviant if

© Paramount/courtesy Everett Collection

Although most people think of a high school clique as being far different from a gang, patterns of inclusion and exclusion operate similarly in both groups. In this still from the movie *Mean Girls,* note the three young women on the right-hand side. In what ways are they excluding the young woman (played by Lindsey Lohan) on the left?

How Much Do You Know About Peer Cliques, Youth Gangs, and Deviance?

True	False	
T	F	1. According to some sociologists, deviance may serve a useful purpose in society.
T	F	2. Peer cliques on high school campuses have few similarities to youth gangs.
T	F	3. Most people join gangs to escape from broken homes caused by divorce or the death of a parent.
T	F	4. Juvenile gangs are an urban problem; few rural areas have problems with gangs.
T	F	5. Street crime has a much higher economic cost to society than crimes committed in executive suites or by government officials.
T	F	6. Persons aged 15 to 24 account for more than half of all arrests for property crimes such as burglary, larceny, arson, and vandalism.
T	F	7. Studies have shown that peer cliques have become increasingly important to adolescents over the past two decades.
T	F	8. Gangs are an international problem.

Answers on page 176.

they express a radical or unusual *belief system.* Members of cults (such as Moonies and satanists) or of far-right-wing or far-left-wing political groups may be considered deviant when their religious or political beliefs become known to people with more conventional cultural beliefs. However, individuals who are considered to be "deviant" by one category of people may be seen as conformists by another group. For example, adolescents in some peer cliques and youth gangs may shun mainstream cultural beliefs and values but routinely conform to subcultural codes of dress, attitude (such as defiant individualism), and behavior (Jankowski, 1991). Those who think of themselves as "Goths" may wear black trench coats, paint their fingernails black, and listen to countercultural musicians.

In addition to their behavior and beliefs, individuals may also be regarded as deviant because they possess a specific *condition or characteristic.* A wide range of conditions have been identified as "deviant," including being obese (Degher and Hughes, 1991; Goode, 1996) and having AIDS (Weitz, 2004). For example, research by the sociologist Rose Weitz (2004) has shown that persons with AIDS live with a stigma that affects their relationships with other

people, including family members, friends, lovers, colleagues, and health care workers. Chapter 4 defines a *stigma* as any physical or social attribute or sign that so devalues a person's social identity that it disqualifies the person from full social acceptance (Goffman, 1963b). Based on this definition, the stigmatized person has a "spoiled identity" as a result of being negatively evaluated by others (Goffman, 1963b). To avoid or reduce stigma, many people seek to conceal the characteristic or condition that might lead to stigmatization.

Who Defines Deviance?

Are some behaviors, beliefs, and conditions inherently deviant? In commonsense thinking, deviance is often viewed as inherent in certain kinds of behavior or people. For sociologists, however, deviance is a formal property of social situations and

deviance any behavior, belief, or condition that violates significant social norms in the society or group in which it occurs.

Box 6.1 Sociology and Everyday Life

Answers to the Sociology Quiz on Peer Cliques, Youth Gangs, and Deviance

1. True. From Durkheim to contemporary functionalists, theorists have regarded some degree of deviance as functional for societies.

2. False. Many social scientists believe that there are striking similarities between adolescent cliques and youth gangs, including the demands that are placed on members in each category to conform to group norms pertaining to behavior, appearance, and the people with whom one is allowed to associate.

3. False. Recent studies have found that people join gangs for a variety of reasons, including the desire to gain access to money, recreation, and protection.

4. False. Gangs are frequently thought of as an urban problem because central-city gangs organized around drug dealing have become prominent in recent years; however, gangs are found in rural areas throughout the country as well.

5. False. Although street crime—such as assault and robbery—often has a greater psychological cost, crimes committed by persons in top positions in business (such as accounting and tax fraud) or government (including the Pentagon) have a far greater economic cost, especially for U.S. taxpayers.

6. True. This age group accounts for almost 55 percent of all arrests for property crimes, the most common crimes committed in the United States.

7. True. As more youths grow up in single-parent households or in households where both parents are employed, many adolescents have turned to members of their peer cliques to satisfy their emotional needs and to gain information.

8. True. Gangs are found in nations around the world. In countries such as Japan, youth gangs are often points of entry into adult crime organizations.

Sources: Based on Adler and Adler, 1998, 2003; Inciardi, Horowitz, and Pottieger, 1993; and Jankowski, 1991.

---

social structure. As the sociologist Kai T. Erikson (1964: 11) explains,

> Deviance is not a property inherent in certain forms of behavior; it is a property conferred upon these forms by the audiences which directly or indirectly witness them. The critical variable in the study of deviance, then, is the social audience rather than the individual actor, since it is the audience which eventually determines whether or not any episode of behavior or any class of episodes is labeled deviant.

Based on this statement, we can conclude that deviance is *relative*—that is, an act becomes deviant when it is socially defined as such. Definitions of deviance vary widely from place to place, from time to time, and from group to group (see "Sociology Works!"). Today, for example, some women wear blue jeans and very short hair to college classes; some men wear an earring and long hair. In the past, such looks violated established dress codes in many schools, and administrators probably would have asked these students to change their appearance or leave school.

Deviant behavior also varies in its *degree of seriousness,* ranging from mild transgressions of folkways, to more serious infringements of mores, to quite serious violations of the law. Have you kept a library book past its due date or cut classes? If so, you have violated folkways. Others probably view your infraction as relatively minor; at most, you might have to pay a fine or receive a lower grade.

Sociology Works!

Social Definitions of Deviance:
Have You Seen Bigfoot or a UFO Lately?

My mind's open to anything. After all, they just found another planet. So, who knows? Anything's possible.
—Jim Maier, a resident of Seneca, Illinois, explaining why he thinks it might be possible that a group of observers actually saw Bigfoot (a so-called "wild man" who is allegedly covered in hair, stands about eight feet tall, has a strong odor, and walks on much larger feet than those of a typical human being) near his community, about seventy miles southwest of Chicago (qtd. in Wischnowsky, 2005)

Bigfoot is one of those things that people like to believe in. . . . Regardless of whether there are such things as Bigfoot, people like that thrill of uncertainty, that sense of danger. It's exciting to try and discover the unknown. And it's a lot more fun to have that little bit of doubt when you're sitting out in the woods.
—sociologist Christopher Bader describing why tales of the improbable, such as sightings of Bigfoot (also known as Sasquatch), are exciting and believable to some individuals, whereas others think that people who spend countless hours waiting in a densely wooded area to catch sight of Bigfoot are engaged in deviant behavior (qtd. in Wischnowsky, 2005)

Sociology contributes to our thinking about conformity and deviance by making us aware that the people we are around help us define what we think of as "normal" beliefs and actions. If we are surrounded by individuals who believe that a Bigfoot or UFO (unidentified flying object) sighting is just around the corner, we may think of such beliefs as normal and gain a personal sense of belonging when we go out and wait with these individuals for Bigfoot

or a flying saucer to show up. For this reason, some people join groups such as the Bigfoot Field Researchers Organization (**http://www.bfro.net**) so that they can share their outings, compare field notes on recent sightings, and feel that they are part of an important group or a clique. Among other Bigfoot believers, followers are treated with respect when they record sightings rather than receiving blank stares or comments like "You've got to be kidding?" all the while they are being labeled as "weird" or "deviant" by outsiders who are nonbelievers (Wischnowsky, 2005).

Looking at the seemingly deviant behavior of going out on Bigfoot or UFO sightings from a sociological perspective, researchers such as Christopher Bader place these actions within a larger social context. One context Bader uses for studying people's fascination with Bigfoot sightings and other paranormal occurrences is the sociology of religion. According to Bader, many people who believe in Bigfoot or UFOs "believe without the kinds of evidence that would convince outsiders—it's a matter of faith" (qtd. in Weiss, 2004). This faith may be intensified by use of the Internet, where true believers may easily report their sightings without fear of ridicule or being identified as deviant by outsiders.

Reflect & Analyze

At your college or university, what beliefs and actions of individuals and groups might be classified as conformity by some people but identified as deviance by others? For example, do some students and/or professors believe that certain buildings are haunted and stay away from those areas?

Violations of mores—such as falsifying a college application or cheating on an examination—are viewed as more serious infractions and are punishable by stronger sanctions, such as academic probation or expulsion. Some forms of deviant behavior violate the criminal law, which defines the behaviors that society labels as criminal. A *crime* is a behavior that violates criminal law and is punishable with

fines, jail terms, and/or other negative sanctions. Crimes range from minor offenses (such as traffic violations) to major offenses (such as murder). A

crime behavior that violates criminal law and is punishable with fines, jail terms, and other sanctions.

subcategory, *juvenile delinquency,* **refers to a violation of law or the commission of a status offense by young people.** Note that the legal concept of juvenile delinquency includes not only crimes but also status offenses, which are illegal only when committed by younger people (such as cutting school or running away from home).

What Is Social Control?

Societies not only have norms and laws that govern acceptable behavior; they also have various mechanisms to control people's behavior. ***Social control* refers to the systematic practices that social groups develop in order to encourage conformity to norms, rules, and laws and to discourage deviance.** Social control mechanisms may be either internal or external. Internal social control takes place through the socialization process: Individuals *internalize* societal norms and values that prescribe how people should behave and then follow those norms and values in their everyday lives. By contrast, external social control involves the use of negative sanctions that proscribe certain behaviors and set forth the punishments for rule breakers and nonconformists. In contemporary societies, the criminal justice system, which includes the police,

the courts, and the prisons, is the primary mechanism of external social control.

If most actions deemed deviant do little or no direct harm to society or its members, why is social control so important to groups and societies? Why is the same belief or action punished in one group or society and not in another? These questions pose interesting theoretical concerns and research topics for sociologists and criminologists who examine issues pertaining to law, social control, and the criminal justice system. ***Criminology* is the systematic study of crime and the criminal justice system, including the police, courts, and prisons.**

The primary interest of sociologists and criminologists is not questions of how crime and criminals can best be controlled but rather social control as a social product. Sociologists do not judge certain kinds of behavior or people as being "good" or "bad." Instead, they attempt to determine what types of behavior are defined as deviant, who does the defining, how and why people become deviants, and how society deals with deviants. Although sociologists have developed a number of theories to explain deviance and crime, no one perspective is a comprehensive explanation of all deviance. Each theory provides a different lens through which we can examine aspects of deviant behavior.

© AP Images/Tony Gutierrez

In April 2008, rumors of polygamy and the forced marriages of young girls led to a raid on a compound in West Texas. Many of the people arrested complained that their privacy had been violated; however, other observers pointed out the need for social control and protecting the rights and safety of minors.

Functionalist Perspectives on Deviance

As we have seen in previous chapters, functionalists focus on societal stability and the ways in which various parts of society contribute to the whole. According to functionalists, a certain amount of deviance contributes to the smooth functioning of society.

What Causes Deviance, and Why Is It Functional for Society?

Sociologist Emile Durkheim believed that deviance is rooted in societal factors such as rapid social change and lack of social integration among people. As you will recall, Durkheim attributed the social upheaval he saw at the end of the nineteenth century to the shift from mechanical to organic solidarity, which was brought about by rapid industrialization and urbanization. Although many people continued to follow the dominant morals (norms, values, and laws) as best they could, rapid social change contributed to *anomie*—a social condition in which people experience a sense of futility because social norms are weak, absent, or conflicting. According to Durkheim, as social integration (bonding and community involvement) decreased, deviance and crime increased. However, from his perspective, this was not altogether bad because he believed that deviance has positive social functions in terms of its consequences. For Durkheim (1964a/1895), deviance is a natural and inevitable part of all societies. Likewise, contemporary functionalist theorists suggest that deviance is universal because it serves three important functions:

1. *Deviance clarifies rules.* By punishing deviant behavior, society reaffirms its commitment to the rules and clarifies their meaning.
2. *Deviance unites a group.* When deviant behavior is seen as a threat to group solidarity and people unite in opposition to that behavior, their loyalties to society are reinforced.
3. *Deviance promotes social change.* Deviants may violate norms in order to get them changed. For example, acts of *civil disobedience*—including lunch counter sit-ins and bus boycotts—were used

to protest and eventually correct injustices such as segregated buses and lunch counters in the South.

Functionalists acknowledge that deviance may also be dysfunctional for society. If too many people violate the norms, everyday existence may become unpredictable, chaotic, and even violent. If even a few people commit acts that are so violent that they threaten the survival of a society, then deviant acts move into the realm of the criminal and even the unthinkable. Of course, the example that stands out in everyone's mind is terrorist attacks around the world and the fear that remains constantly present as a result.

Although there is a wide array of contemporary functionalist theories regarding deviance and crime, many of these theories focus on social structure. For this reason, the first theory we will discuss is referred to as a structural functionalist approach. It describes the relationship between the society's economic structure and why people might engage in various forms of deviant behavior.

Strain Theory: Goals and Means to Achieve Them

Modifying Durkheim's (1964a/1895) concept of *anomie*, the sociologist Robert Merton (1938, 1968) developed strain theory. According to **strain theory, people feel strain when they are exposed to cultural goals that they are unable to obtain because they do not have access to culturally approved means of achieving those goals.** The goals may

juvenile delinquency a violation of law or the commission of a status offense by young people.

social control systematic practices developed by social groups to encourage conformity to norms, rules, and laws and to discourage deviance.

criminology the systematic study of crime and the criminal justice system, including the police, courts, and prisons.

strain theory the proposition that people feel strain when they are exposed to cultural goals that they are unable to obtain because they do not have access to culturally approved means of achieving those goals.

© AP Images/John Miller

The sociologist Robert Merton identified five ways in which people adapt to cultural goals and approved ways of achieving them. Consider the two people shown here. Which of Merton's modes of adaptation might best explain their views on social life?

© Paul Hartnett/PYMCA/Jupiterimages

be material possessions and money; the approved means may include an education and jobs. When denied legitimate access to these goals, some people seek access through deviant means.

Merton identified five ways in which people adapt to cultural goals and approved ways of achieving them: conformity, innovation, ritualism, retreatism, and rebellion (see ◆ Table 6.1). According to Merton, *conformity* occurs when people accept culturally approved goals and pursue them through approved means. Persons who want to achieve success through conformity work hard, save their money, and so on. Even people who find that they are blocked from achieving a high level of education or a lucrative career may take a lower-paying job and attend school part time, join the military, or seek alternative (but legal) avenues, such as playing the lottery, to "strike it rich."

Conformity is also crucial for members of middle- and upper-class teen cliques, who often gather in small groups to share activities and confidences. Some youths are members of a variety of cliques, and peer approval is of crucial significance to them—being one of the "in" crowd, not a "loner," is a significant goal for many teenagers. In the aftermath of the recent school shootings, for example, journalists trekked to school campuses to report that athletes ("jocks"), cheerleaders, and other "popular" students enforce the social code at high schools (Adler, 1999; Cohen, 1999).

Merton classified the remaining four types of adaptation as deviance:

- *Innovation* occurs when people accept society's goals but adopt disapproved means for achieving them. Innovations for acquiring material possessions or money cover a wide variety of illegal activities, including theft and drug dealing.
- *Ritualism* occurs when people give up on societal goals but still adhere to the socially approved

◆ **Table 6.1** Merton's Strain Theory of Deviance

Mode of Adaptation	Method of Adaptation	Seeks Culture's Goals	Follows Culture's Approved Ways
Conformity	Accepts culturally approved goals; pursues them through culturally approved means	Yes	Yes
Innovation	Accepts culturally approved goals; adopts disapproved means of achieving them	Yes	No
Ritualism	Abandons society's goals but continues to conform to approved means	No	Yes
Retreatism	Abandons both approved goals and the approved means to achieve them	No	No
Rebellion	Challenges both the approved goals and the approved means to achieve them	No—seeks to replace	No—seeks to replace

means for achieving them. Ritualism is the opposite of innovation; persons who cannot obtain expensive material possessions or wealth may nevertheless seek to maintain the respect of others by being a "hard worker" or "good citizen."

• *Retreatism* occurs when people abandon both the approved goals and the approved means of achieving them. Merton included persons such as skid-row alcoholics and drug addicts in this category; however, not all retreatists are destitute. Some may be middle- or upper-income individuals who see themselves as rejecting the conventional trappings of success or the means necessary to acquire them.

• *Rebellion* occurs when people challenge both the approved goals and the approved means for achieving them and advocate an alternative set of goals or means. To achieve their alternative goals, rebels may use violence (such as vandalism or rioting) or may register their displeasure with society through acts of vandalism or graffiti (as further discussed in Box 6.2, "You Can Make a Difference").

Opportunity Theory: Access to Illegitimate Opportunities

Expanding on Merton's strain theory, sociologists Richard Cloward and Lloyd Ohlin (1960) suggested that for deviance to occur, people must have access to **illegitimate opportunity structures**—circumstances that provide an opportunity for people to acquire through illegitimate activities what they cannot achieve through legitimate channels. For example, gang members may have insufficient legitimate means to achieve conventional goals of status and wealth but have illegitimate opportunity structures—such as theft, drug dealing, or robbery—through which they can achieve these goals. In his study of the "Diamonds," a Chicago street gang whose members are second-generation Puerto Rican youths, sociologist Felix M. Padilla (1993) found that gang membership was linked to the members' belief that they might reach their aspirations by transforming the gang into a business enterprise. Coco, one of the Diamonds, explains the importance of sticking together in the gang's income-generating business organization:

> We are a group, a community, a family—we have to learn to live together. If we separate, we will never have a chance. We need each other even to make sure that we have a spot for selling our supply [of drugs]. You know, there is people around here, like some opposition, that want to take over your *negocio* [business]. And they think that they

illegitimate opportunity structures circumstances that provide an opportunity for people to acquire through illegitimate activities what they cannot achieve through legitimate channels.

Box 6.2 You Can Make a Difference

Seeing the Writing on the Wall—and Doing Something About It!

"Pirus Rule The Streets of Bompton Fools"

What does this graffiti—written on a Compton, California, wall—mean? To figure out the message, it helps to know that Pirus is a collective made up of several Blood gangs (street gangs identified, among other things, by the red color worn by members) in Los Angeles County. Gang graffiti such as "Pirus Rule" is one way the gang claims its supremacy not only over rival gangs but also over the city as a whole, as indicated by the way the letter "C" in Compton was replaced with a "B" to emphasize the gang's Blood/Pirus identity (Alonso, 1998: 17).

Although not all graffiti is done by street gangs, some gang member use graffiti as an illegal form of communication. Although "taggers" primarily use graffiti as (at least in their opinion) an art form, street gangs use graffiti to increase their visibility, mark their territory, threaten rival gangs, and intimidate local residents (Salt Lake City Sheriff's Department, 2007). According to law enforcement officials, taggers are usually less violent than are members of traditional street gangs, and their "art" is usually more "artistic" and less threatening than street-gang graffiti. However, the work of both taggers and street-gang members defaces walls, buses, subways, and other public areas.

Is there anything we can do when we see graffiti to get it removed and to improve the appearance of our community? How might we lessen the opportunities for gang members to use graffiti to communicate with one another and to threaten outsiders? Here are some suggestions from law enforcement officials:

- Do not confront or challenge a person who is tagging a wall or writing graffiti on a public space. Whether they are gang members or not, some taggers are armed, and even if unarmed, they may assault a challenger with spray paint or a physical attack.
- Make sure that owners of private property or public officials are notified about the graffiti because it is im-

portant that the graffiti be painted over immediately. Studies show that if graffiti is left up, it becomes a status symbol, and the area is likely to be hit again and again.

- Look for adopt-a-wall programs or other groups in which volunteers assist in cleaning off or painting over graffiti.
- Find out if your city has a graffiti hotline where you can report graffiti. Many cities have instituted these hotlines so that graffiti can be quickly removed from both public and private property.
- Be aware of graffiti done by children and young people that might indicate that they are thinking about, or have become involved with, gangs. Look for graffiti on or around a residence, such as drawings or "doodles" of gang-related figures, themes of violence, or gang symbols. Also look for the use of substitute letters, such as replacing the "C" in Compton with a "B" (as discussed above) or intentionally misspelling a word (such as when "cigarette" becomes "bigarette") because the removed letter is in a rival gang's name. (Center for Problem-Oriented Policing, 2002; NAGIA, 2007)

Although graffiti may appear to be a small issue when it is examined within the larger context of gang-related activity and urban crimes, this kind of behavior is one telling sign that law enforcement officials use when identifying possible criminal trends. A dramatic increase in the amount of graffiti in a community is often a sign that gang membership is growing and that gang activities are becoming more confrontational toward rival gangs and toward society as a whole.

For additional information on graffiti and gang indicators, conduct searches on these websites:

- Center for Problem-Oriented Policing (**http://www.popcenter.org**)
- National Alliance of Gang Investigators' Associations (**http://www.nagia.org**)

can do this very easy. So we stick together, and that makes other people think twice about trying to take over what is yours. In our case, the opposition has never tried messing with our hood, and that's because they know it's protected real good by us fellas. (qtd. in Padilla, 1993: 104)

Based on their research, Cloward and Ohlin (1960) identified three basic gang types—criminal, conflict, and retreatist—which emerge on the basis of what type of illegitimate opportunity structure is available in a specific area. The *criminal gang* is devoted to theft, extortion, and other illegal means of securing an income. For young men who grow up in a criminal gang, running drug houses and selling drugs on street corners make it possible for them to support themselves and their families as well as purchase material possessions to impress others. By contrast, *conflict gangs* emerge in communities that do not provide either legitimate or illegitimate

opportunities. Members of conflict gangs seek to acquire a "rep" (reputation) by fighting over "turf" (territory) and adopting a value system of toughness, courage, and similar qualities. Unlike criminal and conflict gangs, members of *retreatist gangs* are unable to gain success through legitimate means and are unwilling to do so through illegal ones. As a result, the consumption of drugs is stressed, and addiction is prevalent.

Sociologist Lewis Yablonsky (1997) has updated Cloward and Ohlin's findings on delinquent gangs. According to Yablonsky, today's gangs are more likely to use and sell drugs, and to carry more lethal weapons than gang members did in the past. Today's gangs have become more varied in their activities and are more likely to engage in intraracial conflicts, with "black on black and Chicano on Chicano violence," whereas minority gangs in the past tended to band together to defend their turf from gangs of different racial and ethnic backgrounds (Yablonsky, 1997: 3).

How useful are social structural approaches such as opportunity theory and strain theory in explaining deviant behavior? Although there are weaknesses in these approaches, they focus our attention on one crucial issue: the close association between certain forms of deviance and social class position. According to criminologist Anne Campbell (1984: 267), gangs are a "microcosm of American society, a mirror image in which power, possession, rank, and role . . . are found within a subcultural life of poverty and crime."

© AP Images/Luke Palmisano

Conflict theorists suggest that criminal law is unequally enforced along class lines. Consider this setting, in which low-income defendants are arraigned by a judge who sees them only on a television monitor. Do you think, as a rule, that these defendants will be as well represented by attorneys as a wealthier defendant might be?

Conflict Perspectives on Deviance

Who determines what kinds of behavior are deviant or criminal? Different branches of conflict theory offer somewhat divergent answers to this question. One branch emphasizes power as the central factor in defining deviance and crime: People in positions of power maintain their advantage by using the law to protect their interests. Another branch emphasizes the relationship between deviance and capitalism, whereas a third focuses on feminist perspectives and the confluence of race, class, and gender issues in regard to deviance and crime.

Deviance and Power Relations

Conflict theorists who focus on power relations in society suggest that the lifestyles considered deviant by political and economic elites are often defined as illegal. According to this approach, norms and laws are established for the benefit of those in power and do not reflect any absolute standard of right and wrong (Turk, 1969, 1977). As a result, the activities of poor and lower-income individuals are more likely to be defined as criminal than those of persons from middle- and upper-income backgrounds. Moreover, the criminal justice system is more focused on, and is less forgiving of, deviant and criminal behavior engaged in by people in specific categories. For example, research shows that young, single, urban males are more likely to be perceived as members of the *dangerous classes* and receive stricter sentences in criminal courts (Miethe and Moore, 1987). Power differentials are also evident in how victims of crime are treated. When the victims are wealthy, white, and male, law enforcement officials are more likely to put forth more extensive efforts to apprehend the perpetrator as contrasted with cases in which the victims are poor, black, and female (Smith, Visher, and Davidson, 1984). Recent research generally supports this assertion (Wonders, 1996).

Deviance and Capitalism

A second branch of conflict theory—Marxist/critical theory—views deviance and crime as a function of the capitalist economic system. Although the early economist and social thinker Karl Marx wrote very little about deviance and crime, many of his ideas are found in a critical approach that has emerged from earlier Marxist and radical perspectives on criminology. The critical approach is based on the assumption that the laws and the criminal justice system protect the power and privilege of the capitalist class. As you may recall from Chapter 1, Marx based his critique of capitalism on the inherent conflict that he believed existed between the capitalists (bourgeoisie) and the working class (proletariat). In a capitalist society, social institutions (such as law, politics, and education, which make up the superstructure) legitimize existing class inequalities and maintain the capitalists' superior position in the class structure. According to Marx, capitalism produces haves and have-nots, who engage in different forms of deviance and crime.

According to the sociologist Richard Quinney (2001/1974), people with economic and political power define as criminal any behavior that threatens their own interests. The powerful use law to control those who are without power. For example, drug laws enacted early in the twentieth century were actively enforced in an effort to control immigrant workers, especially the Chinese, who were being exploited by the railroads and other industries (Tracy, 1980). By contrast, antitrust legislation passed at about the same time was seldom enforced against large corporations owned by prominent families such as the Rockefellers, Carnegies, and Mellons. Having antitrust laws on the books merely shored up the government's legitimacy by making it appear responsive to public concerns about big business (Barnett, 1979).

In sum, the Marxist/critical approach argues that criminal law protects the interests of the affluent and powerful. The way that laws are written and enforced benefits the capitalist class by ensuring that

© Nick Koudis/Getty Images

According to Karl Marx, capitalism produces haves and have-nots, and each group engages in different types of crime. Statistically, the man being arrested here is much more likely to be suspected of a financial crime than a violent crime.

individuals at the bottom of the social class structure do not infringe on the property or threaten the safety of those at the top (Reiman, 1998). However, others assert that critical theorists have not shown that powerful economic and political elites actually manipulate law-making and law enforcement for their own benefit. Rather, people of all classes share a consensus about the criminality of certain acts. For example, laws that prohibit murder, rape, and armed robbery protect not only middle- and upper-income people but also low-income people, who are frequently the victims of such violent crimes.

© LEZLIE STERLING/MCT/Landov

After this young prostitute advertised her services on Craigslist, the Sacramento vice squad and the FBI arranged a meeting that led to her arrest. Which of the feminist theories of women's crime best explains this young woman's offense?

Feminist Approaches

Can theories developed to explain male behavior be used to understand female deviance and crime? According to feminist scholars, the answer is no. A new interest in women and deviance developed in 1975 when two books—Freda Adler's *Sisters in Crime* and Rita James Simons's *Women and Crime*—declared that women's crime rates were going to increase significantly as a result of the women's liberation movement. Although this so-called *emancipation theory* of female crime has been refuted by subsequent analysts, Adler's and Simons's works encouraged feminist scholars (both women and men) to examine more closely the relationship among gender, deviance, and crime. More recently, feminist scholars such as Kathleen Daly and Meda Chesney-Lind (1988) have developed theories and conducted research to fill the void in our knowledge about gender and crime. For example, in a study of the female offender, Chesney-Lind (1997) examined the cultural factors in women's lives that may contribute to their involvement in criminal behavior. Although there is no single feminist perspective on deviance and crime, three schools of thought have emerged.

Why do women engage in deviant behavior and commit crimes? According to the *liberal feminist approach,* women's deviance and crime are a rational response to the gender discrimination that women experience in families and the workplace. From this view, lower-income and minority women typically have fewer opportunities not only for education and good jobs but also for "high-end" criminal endeavors. As some feminist theorists have noted, a woman is no more likely to be a big-time drug dealer or an organized crime boss than she is to be a corporate director (Daly and Chesney-Lind, 1988; Simpson, 1989).

By contrast, the *radical feminist approach* views the cause of women's crime as originating in patriarchy (male domination over females). This approach focuses on social forces that shape women's lives and experiences and shows how exploitation may trigger deviant behavior and criminal activities. From this view, arrests and prosecution for crimes such as prostitution reflect our society's sexual double standard whereby it is acceptable for a man to pay for sex but unacceptable for a woman to accept money for such services. Although state laws usually view both the female prostitute and the male customer as violating the law, in most states the woman is far more likely than the man to be arrested, brought to trial, convicted, and sentenced.

The third school of feminist thought, the *Marxist (socialist) feminist approach,* is based on the assumption that women are exploited by both capitalism and patriarchy. Because most females have relatively low-wage jobs (if any) and few economic resources, crimes such as prostitution and shoplifting become a means to earn money or acquire consumer goods. However, instead of freeing women from their problems, prostitution institutionalizes women's dependence on men and results in a form of female sexual slavery (Vito and Holmes, 1994). Lower-income women are further victimized by the fact that they are often the targets of violent acts by lower-class males, who perceive themselves as being powerless in the capitalist economic system.

Some feminist scholars have noted that these approaches to explaining deviance and crime neglect the centrality of race and ethnicity and focus on the problems and perspectives of women who are white, middle and upper income, and heterosexual without taking into account the views of women of color, lesbians, and women with disabilities (Martin and Jurik, 1996).

Approaches Focusing on Race, Class, and Gender

Some studies have focused on the simultaneous effects of race, class, and gender on deviant behavior. In one study, the sociologist Regina Arnold (1990) examined the relationship between women's earlier victimization in their family and their subsequent involvement in the criminal justice system. Arnold interviewed African American women serving criminal sentences and found that adolescent females are often "labeled and processed as deviants—and subsequently as criminals—for refusing to accept or participate in their own victimization." Arnold attributes many of the women's offenses to living in families in which sexual abuse, incest, and other violence left them few choices except to engage in deviance. Economic marginality and racism also contributed to their victimization: "To be young, Black, poor, and female is to be in a high-risk category for victimization and stigmatization on many levels" (Arnold, 1990: 156). According to Arnold, the criminal behavior of the women in her study was linked to class, gender, and racial oppression, which they experienced daily in their families and at school and work.

Symbolic Interactionist Perspectives on Deviance

As we discussed in Chapter 3, symbolic interactionists focus on *social processes,* such as how people develop a self-concept and learn conforming behavior through socialization. According to this approach, deviance is learned in the same way as conformity—through interaction with others. Although there are a number of symbolic interactionist perspectives on deviance, we will examine three major approaches—differential association and differential reinforcement theories, control theory, and labeling theory.

Differential Association Theory and Differential Reinforcement Theory

How do people learn deviant behavior through their interactions with others? According to the sociologist Edwin Sutherland (1939), people learn the necessary techniques and the motives, drives, rationalizations, and attitudes of deviant behavior from people with whom they associate. *Differential association theory* states that people have a greater tendency to deviate from societal norms when they frequently associate with individuals who are more favorable toward deviance than conformity. From this approach, criminal behavior is learned within intimate personal groups such as one's family and peer groups.

Differential association theory contributes to our knowledge of how deviant behavior reflects the individual's learned techniques, values, attitudes, motives, and rationalizations. It calls attention to the fact that criminal activity is more likely to occur when a person has frequent, intense, and long-lasting interactions with others who violate the law. However, it does not explain why many individuals who have been heavily exposed to people who violate the law still engage in conventional behavior most of the time.

Criminologist Ronald Akers (1998) has combined differential association theory with elements of psychological learning theory to create *differential reinforcement theory,* which suggests that both deviant behavior and conventional behavior are

Is this example of graffiti likely to be the work of an isolated artist or of a gang member? In what ways do gangs reinforce such behavior?

© AP Images/Kevork Djansezian

learned through the same social processes. Akers starts with the fact that people learn to evaluate their own behavior through interactions with significant others. If the persons and groups that a particular individual considers most significant in his or her life define deviant behavior as being "right," the individual is more likely to engage in deviant behavior; likewise, if the person's most significant friends and groups define deviant behavior as "wrong," the person is less likely to engage in that behavior. This approach helps explain not only juvenile gang behavior but also how peer cliques on high school campuses have such a powerful influence on people's behavior. For example, when clique members at Glenbrook, a suburban Chicago high school, jealously "guarded" their favorite locations at the school, one student described her response to the powerful pressures to conform as follows:

> As an experiment . . . Lauren Barry, a pink-haired trophy-case kid at Glenbrook, switched identities with a well-dressed girl from "the wall." Barry walked around all day in the girl's expensive jeans and Doc Martens, carrying a shopping bag from Abercrombie & Fitch. "People kept saying, 'Oh, you look so pretty,'" she recalls. "I felt really uncomfortable." It was interesting, but the next day, and ever since, she's been back in her regular clothes. (qtd. in Adler, 1999: 58)

Another such approach to studying deviance is control theory, which suggests that conformity is often associated with a person's bonds to other people.

Control Theory: Social Bonding

According to the sociologist Walter Reckless (1967), society produces pushes and pulls that move people toward criminal behavior; however, some people "insulate" themselves from such pressures by having positive self-esteem and good group cohesion. Reckless suggests that many people do not resort to deviance because of *inner containments*—such as self-control, a sense of responsibility, and resistance to diversions—and *outer containments*—such as supportive family and friends, reasonable social expectations, and supervision by others. Those with the strongest containment mechanisms are able to withstand external pressures that might cause them to participate in deviant behavior.

Extending Reckless's containment theory, sociologist Travis Hirschi's (1969) social control theory

differential association theory the proposition that individuals have a greater tendency to deviate from societal norms when they frequently associate with persons who are more favorable toward deviance than conformity.

is based on the assumption that deviant behavior is minimized when people have strong bonds that bind them to families, schools, peers, churches, and other social institutions. ***Social bond theory* holds that the probability of deviant behavior increases when a person's ties to society are weakened or broken.** According to Hirschi, social bonding consists of (1) *attachment* to other people, (2) *commitment* to conformity, (3) *involvement* in conventional activities, and (4) *belief* in the legitimacy of conventional values and norms. Although Hirschi did not include females in his study, others who have replicated that study with both females and males have found that the theory appears to apply to each (see Naffine, 1987).

What does control theory have to say about delinquency and crime? Control theories suggest that the probability of delinquency increases when a person's social bonds are weak and when peers promote antisocial values and violent behavior. However, some critics assert that Hirschi was mistaken in his assumption that a weakened social bond leads to deviant behavior. The chain of events may be just the opposite: People who routinely engage in deviant behavior may find that their bonds to people who would be positive influences are weakened over time (Agnew, 1985; Siegel, 2007). Or, as labeling theory suggests, people may engage in deviant and criminal behavior because of destructive social interactions and encounters (Siegel, 2007).

Labeling Theory

Labeling theory states that deviance is a socially constructed process in which social control agencies designate certain people as deviants, and

According to control theory, strong bonds—including close family ties—are a factor in explaining why many people do not engage in deviant behavior. Why do some sociologists believe that quality family time is more important in discouraging delinquent behavior than is time spent with other young people?

Primary deviance	→	Secondary deviance	→	Tertiary deviance	▶ Figure 6.1 A Closer Look at
Initial rule breaking		New identity accepted, deviance continues		Individual relabels behavior nondeviant	Labeling Theory

they, in turn, come to accept the label placed upon them and begin to act accordingly. Based on the symbolic interaction theory of Charles H. Cooley and George H. Mead (see Chapter 3), labeling theory focuses on the variety of symbolic labels that people are given in their interactions with others.

How does the process of labeling occur? The act of fixing a person with a negative identity, such as "criminal" or "mentally ill," is directly related to the power and status of those persons who *do* the labeling and those who are *being labeled.* Behavior, then, is not deviant in and of itself; it is defined as such by a social audience (Erikson, 1962). According to the sociologist Howard Becker (1963), *moral entrepreneurs* are often the ones who create the rules about what constitutes deviant or conventional behavior. Becker believes that moral entrepreneurs use their own perspectives on "right" and "wrong" to establish the rules by which they expect other people to live. They also label others as deviant. Often these rules are enforced on persons with less power than the moral entrepreneurs. Becker (1963: 9) concludes that the deviant is "one to whom the label has successfully been applied; deviant behavior is behavior that people so label."

As the definition of labeling theory suggests, several stages may occur in the labeling process (▶ Figure 6.1). *Primary deviance* **refers to the initial act of rule breaking** (Lemert, 1951). However, if individuals accept the negative label that has been applied to them as a result of the primary deviance, they are more likely to continue to participate in the type of behavior that the label was initially meant to control. *Secondary deviance* **occurs when a person who has been labeled a deviant accepts that new identity and continues the deviant behavior.** For example, a person may shoplift an item of clothing from a department store but not be apprehended or labeled as a deviant. The person may subsequently decide to forgo such behavior in the future. However, if the person shoplifts the item, is apprehended, is labeled as a "thief," and subsequently accepts that label, then the person may shoplift items from

stores on numerous occasions. A few people engage in *tertiary deviance,* **which occurs when a person who has been labeled a deviant seeks to normalize the behavior by relabeling it as nondeviant** (Kitsuse, 1980). An example would be drug users who believe that using marijuana or other illegal drugs is no more deviant than drinking alcoholic beverages and therefore should not be stigmatized.

Can labeling theory be applied to high school peer groups and gangs? In a classic study, the sociologist William Chambliss (1973) documented how the labeling process works in some high schools when he studied two groups of adolescent boys: the "Saints" and the "Roughnecks." Members of both groups were constantly involved in acts of truancy, drinking, wild parties, petty theft, and vandalism. Although the Saints committed more offenses than the Roughnecks, the Roughnecks were the ones who were labeled as "troublemakers" and arrested by law enforcement officials. By contrast, the Saints were described as being the "most likely to succeed," and none of the Saints were ever arrested. According to Chambliss (1973), the Roughnecks were more likely to be labeled as deviants because they came

social bond theory the proposition that the probability of deviant behavior increases when a person's ties to society are weakened or broken.

labeling theory the proposition that deviants are those people who have been successfully labeled as such by others.

primary deviance the initial act of rule-breaking.

secondary deviance the process that occurs when a person who has been labeled a deviant accepts that new identity and continues the deviant behavior.

tertiary deviance deviance that occurs when a person who has been labeled a deviant seeks to normalize the behavior by relabeling it as nondeviant.

from lower-income families, did poorly in school, and were generally viewed negatively, whereas the Saints came from "good families," did well in school, and were generally viewed positively. Although both groups engaged in similar behavior, only the Roughnecks were stigmatized by a deviant label.

How successful is labeling theory in explaining deviance and social control? One contribution of labeling theory is that it calls attention to the way in which social control and personal identity are intertwined: Labeling may contribute to the acceptance of deviant roles and self-images. Critics argue that this does not explain what caused the original acts that constituted primary deviance, nor does it provide insight into why some people accept deviant labels and others do not (Cavender, 1995).

Postmodernist Perspectives on Deviance

Departing from other theoretical perspectives on deviance, some postmodern theorists emphasize that the study of deviance reveals how the powerful exert control over the powerless by taking away their free will to think and act as they might choose. From this approach, institutions such as schools, prisons, and mental hospitals use knowledge, norms, and values to categorize people into "deviant" subgroups such as slow learners, convicted felons, or criminally insane individuals, and then to control them through specific patterns of discipline.

An example of this idea is found in social theorist Michel Foucault's *Discipline and Punish* (1979), in which Foucault examines the intertwining nature of power, knowledge, and social control. In this study of prisons from the mid-1800s to the early 1900s, Foucault found that many penal institutions ceased torturing prisoners who disobeyed the rules and began using new surveillance techniques to maintain social control. Although the prisons appeared to be more humane in the post-torture era, Foucault contends that the new means of surveillance impinged more on prisoners and brought greater power to prison officials. To explain, he described the *Panoptican*—a structure that gives prison officials the possibility of complete observation of criminals at all times. Typically, the

Panoptican was a tower located in the center of a circular prison from which guards could see all the cells. Although the prisoners knew they could be observed at any time, they did not actually know when their behavior was being scrutinized. As a result, prison officials were able to use their knowledge as a form of power over the inmates. Eventually, the guards did not even have to be present all the time because prisoners believed that they were under constant scrutiny by officials in the observation post. If we think of this in contemporary times, we can see how cameras, computers, and other devices have made continual surveillance quite easy in virtually all institutions. In such cases, social control and discipline are based on the use of knowledge, power, and technology.

Foucault's view on deviance and social control has influenced other social analysts, including Shoshana Zuboff (1988), who views the computer as a modern Panoptican that gives workplace supervisors virtually unlimited capabilities for surveillance over subordinates. Today, cell phones and the Internet provide new opportunities for

© AP Images/Brett Coomer

Michel Foucault contended that new means of surveillance would make it possible for prison officials to use their knowledge of prisoners' activities as a form of power over the inmates. These guards are able to monitor the activities of many prisoners without ever leaving their station.

CONCEPT QUICK REVIEW

Theoretical Perspectives on Deviance

	Theory	Key Elements
Functionalist Perspectives		
Robert Merton	Strain theory	Deviance occurs when access to the approved means of reaching culturally approved goals is blocked. Innovation, ritualism, retreatism, or rebellion may result.
Richard Cloward/Lloyd Ohlin	Opportunity theory	Lower-class delinquents subscribe to middle-class values but cannot attain them. As a result, they form gangs to gain social status and may achieve their goals through illegitimate means.
Conflict Perspectives		
Karl Marx Richard Quinney	Critical approach	The powerful use law and the criminal justice system to protect their own class interests.
Kathleen Daly Meda Chesney-Lind	Feminist approach	Historically, women have been ignored in research on crime. Liberal feminism views women's deviance as arising from gender discrimination, radical feminism focuses on patriarchy, and socialist feminism emphasizes the effects of capitalism and patriarchy on women's deviance.
Symbolic Interactionist Perspectives		
Edwin Sutherland	Differential association	Deviant behavior is learned in interaction with others. A person becomes delinquent when exposure to law-breaking attitudes is more extensive than exposure to law-abiding attitudes.
Travis Hirschi	Social control/social bonding	Social bonds keep people from becoming criminals. When ties to family, friends, and others become weak, an individual is most likely to engage in criminal behavior.
Howard Becker	Labeling theory	Acts are deviant or criminal because they have been labeled as such. Powerful groups often label less-powerful individuals.
Edwin Lemert	Primary/secondary deviance	Primary deviance is the initial act. Secondary deviance occurs when a person accepts the label of "deviant" and continues to engage in the behavior that initially produced the label
Postmodernist Perspective		
Michel Foucault	Knowledge as power	Power, knowledge, and social control are intertwined. In prisons, for example, new means of surveillance that make prisoners think they are being watched all the time give officials knowledge that inmates do not have. Thus, the officials have a form of power over the inmates.

surveillance by government officials and others who are not visible to the individuals who are being watched.

We have examined functionalist, conflict, interactionist, and postmodernist perspectives on social control, deviance, and crime (see the Concept Quick Review). All of these explanations contribute to our understanding of the causes and consequences of deviant behavior; however, we now turn to the subject of crime itself.

Crime Classifications and Statistics

Crime in the United States can be divided into different categories. We will look first at the legal classifications of crime and then at categories typically used by sociologists and criminologists.

How the Law Classifies Crime

Crimes are divided into felonies and misdemeanors. The distinction between the two is based on the seriousness of the crime. A *felony* is a serious crime such as rape, homicide, or aggravated assault, for which punishment typically ranges from more than a year's imprisonment to death. A *misdemeanor* is a minor crime that is typically punished by less than one year in jail. In either event, a fine may be part of the sanction as well. Actions that constitute felonies and misdemeanors are determined by the legislatures in the various states; thus, their definitions vary from jurisdiction to jurisdiction.

Other Crime Categories

The *Uniform Crime Report* (UCR) is the major source of information on crimes reported in the United States. The UCR has been compiled since 1930 by the Federal Bureau of Investigation based on information filed by law enforcement agencies throughout the country. When we read that the rate of certain types of crimes has increased or decreased when compared with prior years, for example, this information is usually based on UCR data. The UCR focuses on violent crime and property crime (which, prior to 2004, were jointly referred to in that report as "index crimes"), but also contains data on other types of crime (see ▶ Figure 6.2). In 2006 about 14 million arrests were made in the United States for all criminal infractions (excluding traffic violations). Although the UCR gives some indication of crime in the United States,

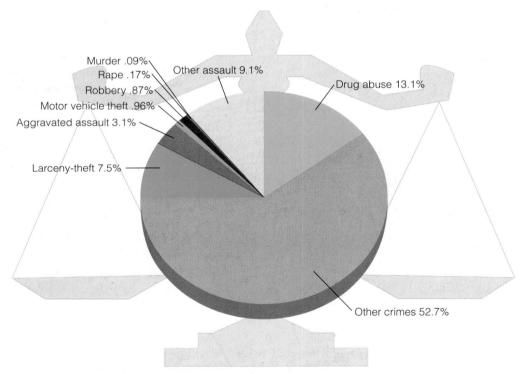

Murder .09%
Rape .17%
Robbery .87%
Motor vehicle theft .96%
Aggravated assault 3.1%
Larceny-theft 7.5%
Other assault 9.1%
Drug abuse 13.1%
Other crimes 52.7%

▶ **Figure 6.2 Distribution of Arrests by Type of Offense, 2006**

Source: FBI, 2007.

the figures do not reflect the actual number and kinds of crimes, as will be discussed later.

Violent Crime *Violent crime* **consists of actions—murder, rape, robbery, and aggravated assault—involving force or the threat of force against others.** Although only 4 percent of all arrests in the United States in 2006 were for violent crimes, this category is probably the most anxiety-provoking of all criminal behavior: Most of us know someone who has been a victim of violent crime, or we have been so ourselves. Victims are often physically injured or even lose their lives; the psychological trauma may last for years after the event (Parker and Anderson-Facile, 2000). Violent crime receives the most sustained attention from law enforcement officials and the media (see Warr, 2000).

Nationwide, there is growing concern over juvenile violence. Beginning in 1988, juvenile violent-crime arrest rates started to rise, a trend that has been linked by some scholars to gang membership (see Inciardi, Horowitz, and Pottieger, 1993; Thornberry et al., 1993). Fear of violence is felt not only by the general public but by gang members themselves, as Charles Campbell commented

> The generation I'm in is going to be lost. Of the circle of friends I grew up in, three are dead, four are in jail, and another is out of school and just does nothing. When he runs out of money he'll sell a couple bags of weed. . . .
>
> I would carry a gun because I am worried about that brother on the fringe. There are some people, there is nothing out there for them. They will blow you away because they have nothing to lose. There are no jobs out there. It's hard to get money to go to school. (qtd. in Lee, 1993: 21)

Property Crime *Property crimes* **include burglary (breaking into private property to commit a serious crime), motor vehicle theft, larceny-theft (theft of property worth $50 or more), and arson.** Some offenses, such as robbery, are both violent crimes and property crimes. In the United States, a property crime occurs, on average, once every three seconds; a violent crime occurs, on average, once every twenty-two seconds (see ▶ Figure 6.3). In most property crimes, the primary motive is to obtain money or some other desired valuable.

Public Order Crime *Public order crimes* involve an illegal action voluntarily engaged in by the participants, such as prostitution, illegal gambling, the private use of illegal drugs, and illegal pornography. Many people assert that such conduct should not be labeled as a crime; these offenses are often referred to as *victimless crimes* because they **involve a willing exchange of illegal goods or services among adults.** However, morals crimes can include children and adolescents as well as adults. Young children and adolescents may unwillingly become child pornography "stars" or prostitutes.

Occupational and Corporate Crime Although the sociologist Edwin Sutherland (1949) developed the theory of white-collar crime about sixty years ago, it was not until the 1980s that the public became fully aware of its nature. *Occupational (white-collar) crime* **comprises illegal activities committed by people in the course of their employment or financial affairs.**

In addition to acting for their own financial benefit, some white-collar offenders become involved in criminal conspiracies designed to improve the market share or profitability of their companies. This is known as *corporate crime*—**illegal acts committed by corporate employees on behalf of the corporation and with its support.** Examples include antitrust violations; tax evasion; misrepresentations in

---

violent crime actions—murder, forcible rape, robbery, and aggravated assault—involving force or the threat of force against others.

property crimes burglary (breaking into private property to commit a serious crime), motor vehicle theft, larceny-theft (theft of property worth $50 or more), and arson.

victimless crimes crimes involving a willing exchange of illegal goods or services among adults.

occupational (white-collar) crime illegal activities committed by people in the course of their employment or financial affairs.

corporate crime illegal acts committed by corporate employees on behalf of the corporation and with its support.

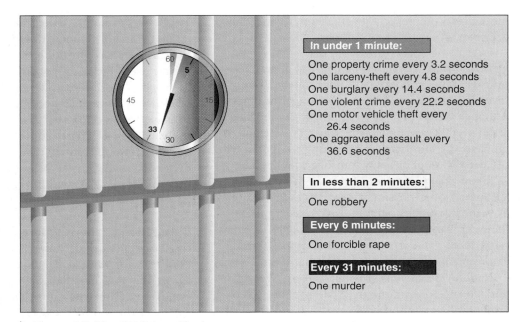

In under 1 minute:

One property crime every 3.2 seconds
One larceny-theft every 4.8 seconds
One burglary every 14.4 seconds
One violent crime every 22.2 seconds
One motor vehicle theft every
  26.4 seconds
One aggravated assault every
  36.6 seconds

In less than 2 minutes:

One robbery

Every 6 minutes:

One forcible rape

Every 31 minutes:

One murder

▶ **Figure 6.3 The FBI Crime Clock**

Source: FBI, 2007.

advertising; infringements on patents, copyrights, and trademarks; price fixing; and financial fraud. These crimes are a result of deliberate decisions made by corporate personnel to enhance resources or profits at the expense of competitors, consumers, and the general public.

Although people who commit occupational and corporate crimes can be arrested, fined, and sent to prison, many people often have not regarded such behavior as "criminal." People who tend to condemn street crime are less sure of how their own (or their friends') financial and corporate behavior should be judged. At most, punishment for such offenses has usually been a fine or a relatively brief prison sentence.

Until recently, public concern and media attention focused primarily on the street crimes disproportionately committed by persons who are poor, powerless, and nonwhite. Today, however, part of our focus has shifted to crimes committed in corporate suites, such as fraud, tax evasion, and insider trading by executives at some large and well-known corporations. Bernard Ebbers, former chief executive officer of communications giant WorldCom, was convicted of fraud and conspiracy in connec-

tion with that company's accounting scandal and was sentenced to 25 years in prison. Dennis Kozlowski, chief executive of Tyco International, and a subordinate were convicted of looting more than $600 million from their company and were sentenced to up to 25 years in prison.

Corporate crimes are often more costly in terms of money and lives lost than street crimes. Thousands of jobs and billions of dollars were lost as a result of corporate crime in the year 2006 alone. Deaths resulting from corporate crimes such as polluting the air and water, manufacturing defective products, and selling unsafe foods and drugs far exceed the number of deaths due to homicides each year. Other costs include the effect on the moral climate of society (Clinard and Yeager, 1980; Simon, 1996). Throughout the United States, the confidence of everyday people in the nation's economy has been shaken badly by the greedy and illegal behavior of corporate insiders.

Organized Crime *Organized crime* **is a business operation that supplies illegal goods and services for profit.** Premeditated, continuous illegal activities of organized crime include drug traffick-

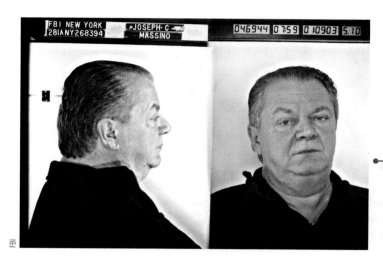

Over the years, there have been many notorious leaders of organized crime syndicates. Shown here is Joseph ("Big Joey") Massino, who was sentenced to prison for racketeering, murder, arson, and other charges. Does organized crime still exist today?

ing, prostitution, loan-sharking, money laundering, and large-scale theft such as truck hijackings (Simon, 1996). No single organization controls all organized crime; rather, many groups operate at all levels of society. Organized crime thrives because there is great demand for illegal goods and services. Criminal organizations initially gain control of illegal activities by combining threats and promises. For example, small-time operators running drug or prostitution rings may be threatened with violence if they compete with organized crime or fail to make required payoffs (Cressey, 1969).

Apart from their illegal enterprises, organized crime groups have infiltrated the world of legitimate business. Known linkages between legitimate businesses and organized crime exist in banking, hotels and motels, real estate, garbage collection, vending machines, construction, delivery and long-distance hauling, garment manufacture, insurance, stocks and bonds, vacation resorts, and funeral homes (National Council on Crime and Delinquency, 1969). In addition, some law enforcement and government officials are corrupted through bribery, campaign contributions, and favors intended to buy them off.

Political Crime The term *political crime* refers to illegal or unethical acts involving the usurpation of power by government officials, or illegal/unethical acts perpetrated against the government by outsiders seeking to make a political statement, undermine the government, or overthrow it. Gov-

ernment officials may use their authority unethically or illegally for the purpose of material gain or political power (Simon, 1996). They may engage in graft (taking advantage of political position to gain money or property) through bribery, kickbacks, or "insider" deals that financially benefit them. In the late 1980s, for example, several top Pentagon officials were found guilty of receiving bribes for passing classified information on to major defense contractors that had garnered many lucrative contracts from the government (Simon, 1996).

Other types of corruption have been costly for taxpayers, including dubious use of public funds and public property, corruption in the regulation of commercial activities (such as food inspection), graft in zoning and land-use decisions, and campaign contributions and other favors to legislators that corrupt the legislative process. Whereas some political crimes are for personal material gain, others (such as illegal wiretapping and political "dirty tricks") are aimed at gaining or maintaining political office or influence.

organized crime a business operation that supplies illegal goods and services for profit.

political crime illegal or unethical acts involving the usurpation of power by government officials, or illegal/unethical acts perpetrated against the government by outsiders seeking to make a political statement, undermine the government, or overthrow it.

Some acts committed by agents of the government against persons and groups believed to be threats to national security are also classified as political crimes. Four types of political deviance have been attributed to some officials: (1) secrecy and deception designed to manipulate public opinion, (2) abuse of power, (3) prosecution of individuals due to their political activities, and (4) official violence, such as police brutality against people of color or the use of citizens as unwilling guinea pigs in scientific research (Simon, 1996).

Political crimes also include illegal or unethical acts perpetrated against the government by outsiders seeking to make a political statement or to undermine or overthrow the government. Examples include treason, acts of political sabotage, and terrorist attacks on public buildings.

Crime Statistics

How useful are crime statistics as a source of information about crime? As mentioned previously, official crime statistics provide important information on crime; however, the data reflect only those crimes that have been reported to the police. Although—except in 2001, when there was a slight rise—the rates have been decreasing slightly during the past few years, the UCR reflects that overall levels of crime have increased by nearly two-thirds over the past twenty-five years. However, this increase may reflect (at least partially) an increase in the number of crimes *reported,* not necessarily a change in the number of crimes *committed.* Why are some crimes not reported? People are more likely to report crime when they believe that something can be done about it (apprehension of the perpetrator or retrieval of their property, for example). About half of all assault and robbery victims do not report the crime because they may be embarrassed or fear reprisal by the perpetrator. Thus, the number of crimes reported to police represents only the proverbial "tip of the iceberg" when compared with all offenses actually committed. Official statistics are problematic in social science research because of these limitations.

The *National Crime Victimization Survey* was developed by the Bureau of Justice Statistics as an alternative means of collecting crime statistics. In this annual survey, the members of 100,000 randomly selected households are interviewed to determine whether they have been the victims of crime, even if the crime was not reported to the police. The most recent victimization survey indicates that 51 percent of all violent crimes and 62 percent of all property crimes are not reported to the police and are thus not reflected in the UCR (U.S. Bureau of Justice Statistics, 2007).

Studies based on anonymous self-reports of criminal behavior also reveal much higher rates of crime than those found in official statistics. For example, self-reports tend to indicate that adolescents of all social classes violate criminal laws. However, official statistics show that those who are arrested and placed in juvenile facilities typically have limited financial resources, have repeatedly committed serious offenses, or both (Steffensmeier and Allan, 2000). Data collected for the Juvenile Court Statistics program also reflect class and racial bias in criminal justice enforcement. Not all children who commit juvenile offenses are apprehended and referred to court. Children from white, affluent families are more likely to have their cases handled outside the juvenile justice system (for example, a youth may be sent to a private school or hospital rather than to a juvenile correctional facility).

Many crimes committed by persons of higher socioeconomic status in the course of business are handled by administrative or quasi-judicial bodies, such as the Securities and Exchange Commission or the Federal Trade Commission, or by civil courts. As a result, many elite crimes are never classified as "crimes," nor are the businesspeople who commit them labeled as "criminals."

Terrorism and Crime

In the twenty-first century, the United States and other nations are confronted with a difficult prospect: how to deal with terrorism. **Terrorism is the calculated, unlawful use of physical force or threats of violence against persons or property in order to intimidate or coerce a government, organization, or individual for the purpose of gaining some political, religious, economic, or social ob-**

Global Perspective") further discusses the issue of terrorism.

Street Crimes and Criminals

Given the limitations of official statistics, is it possible to determine who commits crimes? We have much more information available about conventional (street) crime than elite crime; therefore, statistics concerning street crime do not show who commits all types of crime. Gender, age, class, and race are important factors in official statistics pertaining to street crime.

Gender and Crime Before considering differences in crime rates by males and females, three similarities should be noted. First, the three most common arrest categories for both men and women are driving under the influence of alcohol or drugs (DUI), larceny, and minor or criminal mischief types of offenses. These three categories account for about 40 percent of all male arrests and about 46 percent of all female arrests. Second, liquor law violations (such as underage drinking), simple assault, and disorderly conduct are middle-range offenses for both men and women. Third, the rate of arrests for murder, arson, and embezzlement is relatively low for both men and women (Steffensmeier and Allan, 2000).

The most important gender differences in arrest rates are reflected in the proportionately greater involvement of men in major property crimes (such as robbery and larceny-theft) and violent crime, as shown in ▶ Figure 6.4. In 2006, men accounted for almost 89 percent of robberies and murders and

What effects might a mother's imprisonment have on the future of her infant?

© Sean Cayton/The Image Works

jective. How are sociologists and criminologists to explain world terrorism, which may have its origins in more than one nation and include diverse "cells" of terrorists who operate in a somewhat gang-like manner but are believed to be following directives from leaders elsewhere? In order to deal with the aftermath of terrorist attacks, government officials typically focus on "known enemies" such as Osama bin Laden. The nebulous nature of the "enemy" and the problems faced by any one government trying to identify and apprehend the perpetrators of acts of terrorism have resulted in a global "war on terror." Social scientists who use a rational choice approach suggest that terrorists are rational actors who constantly calculate the gains and losses of participation in violent—and sometimes suicidal—acts against others. Chapter 13 ("Politics and the Economy in

terrorism the calculated unlawful use of physical force or threats of violence against persons or property in order to intimidate or coerce a government, organization, or individual for the purpose of gaining some political, religious, economic, or social objective.

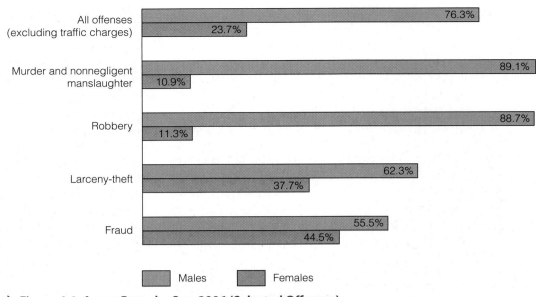

All offenses (excluding traffic charges)
76.3%
23.7%

Murder and nonnegligent manslaughter
89.1%
10.9%

Robbery
88.7%
11.3%

Larceny-theft
62.3%
37.7%

Fraud
55.5%
44.5%

Males   Females

▶ **Figure 6.4 Arrest Rates by Sex, 2006 (Selected Offenses)**

Source: FBI, 2007.

more than 62 percent of all larceny-theft arrests in the United States. Of those types of offenses, males under age 18 accounted for approximately 20 percent of the 2006 arrests. The property crimes for which women are most frequently arrested are nonviolent in nature, including shoplifting, theft of services, passing bad checks, credit card fraud, and employee pilferage. When women are arrested for serious violent and property crimes, they are typically seen as accomplices of the men who planned the crime and instigated its commission (Steffensmeier and Allan, 2000). However, one study found that some women play an active role in planning and carrying out robberies (Sommers and Baskin, 1993).

Age and Crime Of all factors associated with crime, the age of the offender is one of the most significant. Arrest rates for violent crime and property crime are highest for people between the ages of 13 and 25, with the peak being between ages 16 and 17. In 2006, persons under age 25 accounted for more than 45 percent of all arrests for violent crime and almost 55 percent of all arrests for property crime (FBI, 2007). Individuals under age 18 accounted for over 28 percent of all arrests for robbery and 26 percent of all arrests for larceny-theft.

Scholars do not agree on the reasons for this age distribution. In one study, the sociologist Mark Warr (1993) found that peer influences (defined as exposure to delinquent peers, time spent with peers, and loyalty to peers) tend to be more significant in explaining delinquent behavior than age itself.

The median age of those arrested for aggravated assault and homicide is somewhat older, generally in the late twenties. Typically, white-collar criminals are even older because it takes time to acquire both a high-ranking position and the skills needed to commit this particular type of crime.

Rates of arrest remain higher for males than females at every age and for nearly all offenses. This female-to-male ratio remains fairly constant across all age categories. The most significant gender difference in the age curve is for prostitution (a nonviolent crime). In 2006, 57 percent of all women arrested for prostitution were under age 35. For individuals over age 45, many more men than women are arrested for sex-related offenses (including procuring the services of a prostitute). This difference has been attributed to a more stringent enforcement of prostitution statutes when young females are involved (Chesney-Lind, 1997). It has also been suggested that opportunities for prostitution are greater for younger women. This age difference may not have the same impact on males, who continue to purchase sexual services from young females or males (see Steffensmeier and Allan, 2000).

Social Class and Crime Individuals from all social classes commit crimes; they simply commit different kinds of crimes. Persons from lower socioeconomic backgrounds are more likely to be arrested for violent and property crimes. By contrast, persons from the upper part of the class structure generally commit white-collar or elite crimes, although only a very small proportion of these individuals will ever be arrested or convicted of a crime.

What about social class and recent violence by youths? Between 1992 and 2005, there were 582 violent deaths in U.S. schools (U.S. Department of Education, 2007). Most of these deaths were not attributed to lower-income, inner-city youths, as popular stereotypes might suggest. Instead, some of these acts of violence were perpetrated by young people who lived in houses that cost anywhere from $75,000 to $5 million or more.

Similarly, membership in today's youth gangs cannot be identified with just one social class. The U.S. Department of Education estimates that 50 percent of gang members are part of the nation's underclass—the class comprising those persons who are poor, seldom employed, and caught in long-term deprivation—but that 35 percent are working class, 12 percent are middle class, and 3 percent are upper-middle class (Egley, 2000). Today, females are more visible in some previously all-male gangs as well as in female gangs (Egley, 2000).

In any case, official statistics are not an accurate reflection of the relationship between social class and crime. Self-report data from offenders themselves may be used to gain information on family income, years of education, and occupational status; however, such reports rely on respondents to report information accurately and truthfully.

Race and Crime In 2006, whites (including Latinos/as) accounted for almost 70 percent of all arrests, as shown in ▶ Figure 6.5. Compared with African Americans, arrest rates for whites were higher for nonviolent property crimes such as fraud and larceny-theft but were lower for violent crimes such

as robbery. In 2006, whites accounted for about 68 percent of all arrests for property crimes and about 59 percent of arrests for violent crimes. African Americans accounted for 39 percent of arrests for violent crimes and 29 percent of arrests for property crimes (FBI, 2007).

Although official arrest records reveal certain trends, these data tell us very little about the actual dynamics of crime by racial–ethnic category. According to official statistics, African Americans are overrepresented in arrest data. In 2006, African Americans made up about 12 percent of the U.S. population but accounted for 28 percent of all arrests. Latinos/as made up about 13 percent of the U.S. population and accounted for about 13 percent of all arrests. Over two-thirds of their offenses were for nonviolent crimes such as alcohol- and drug-related offenses, and disorderly conduct. In 2006, about 1 percent of all arrests were of Asian Americans or Pacific Islanders, and about 1 percent were of Native Americans (designated in the UCR as "American Indian" or "Alaskan Native"). For the general population, the majority of arrests were for larceny-theft, assaults, vandalism, and alcohol- and drug-related violations (FBI, 2007).

Criminologist Coramae Richey Mann (1993) has argued that arrest statistics are not an accurate reflection of the crimes actually committed in our society. Reporting practices differ in accordance with race and social class. Arrest statistics reflect the UCR's focus on violent and property crimes, especially property crimes, which are committed primarily by low-income people. This emphasis draws attention away from the white-collar and elite crimes committed by middle- and upper-income people (Harris and Shaw, 2000). Police may also demonstrate bias and racism in their decisions regarding whom to question, detain, or arrest under certain circumstances (Mann, 1993).

Another reason that statistics may show a disproportionate number of people of color being arrested is because of the focus of law enforcement on certain types of crime and certain neighborhoods in which crime is considered more prevalent. As discussed previously, many poor, young, central-city males turn to forms of criminal activity due to their belief that no opportunities exist for them to earn a living wage through legitimate employment. Because

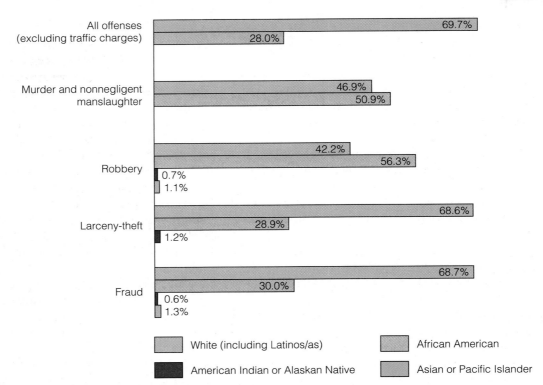

All offenses (excluding traffic charges): 69.7% / 28.0%

Murder and nonnegligent manslaughter: 46.9% / 50.9%

Robbery: 42.2% / 56.3% / 0.7% / 1.1%

Larceny-theft: 68.6% / 28.9% / 1.2%

Fraud: 68.7% / 30.0% / 0.6% / 1.3%

White (including Latinos/as)   African American
American Indian or Alaskan Native   Asian or Pacific Islander

▶ **Figure 6.5 Arrests by Race, 2006 (Selected Offenses)**
Note: Classifications as used in Uniform Crime Report

Source: FBI, 2007.

of the trend of law enforcement efforts to focus on drug-related offenses, arrest rates for young people of color have risen rapidly. These young people are also more likely to live in central-city areas, where there are more police patrols to make arrests.

Finally, we should remember that being arrested does not mean that a person is guilty of the crime with which he or she has been charged. In the United States, individuals accused of crimes are, at least theoretically, "innocent until proven guilty" (Mann, 1993).

Crime Victims

Based on the National Crime Victimization Survey (NCVS), men are more likely to be victimized by crime, although women tend to be more fearful of crime, particularly crimes directed toward them, such as forcible rape (Warr, 2000). Victimization surveys indicate that men are the most frequent vic-

tims of most crimes of violence and theft. Among males who are now 12 years old, an estimated 89 percent will be the victims of a violent crime at least once during their lifetime, as compared with 73 percent of females. The elderly also tend to be more fearful of crime but are the least likely to be victimized. Young men of color between the ages of 12 and 24 have the highest criminal victimization rates.

A study by the Justice Department found that Native Americans are more likely to be victims of violent crimes than are members of any other racial category and that the rate of violent crimes against Native American women was about 50 percent higher than that for African American men (Perry, 2004). During the period covered in the study (from 1992 to 2002), Native Americans were the victims of violent crimes at a rate more than twice the national average. They were also more likely to be the victims of violent crimes committed by members of a race other than their own (Perry, 2004). There has been a shift over the past twenty years in which more Native Americans have moved from reservations to urban areas. In the cities, they do not tend to live in segregated areas, so they come into contact more often with people of other racial and ethnic groups, whereas African Americans and whites are more likely to live in segregated areas of the city and commit violent crimes against other people in their same racial or ethnic category. According to the survey, the average annual rate at which Native Americans were victims of violent crime—101 crimes per 1,000 people, ages 12 or older—is about two-and-a-half times the national average of 41 crimes per 1,000 people who are above the age of 12. By comparison, the average annual rate for whites was 41 crimes per 1,000 people, for African Americans, 50 per 1,000, and for Asian Americans, 22 per 1,000 (Perry, 2004).

The burden of robbery victimization falls more heavily on some categories of people than others. NCVS data indicate that males are robbed at almost twice the rate of females. African Americans are more than twice as likely to be robbed as whites. Young people have a much greater likelihood of being robbed than do middle-age and older persons. Persons from lower-income families are more likely to be robbed than people from higher-income families (U.S. Bureau of Justice Statistics, 2007).

The Criminal Justice System

Of all of the agencies of social control (including families, schools, and churches) in contemporary societies, only the criminal justice system has the power to control crime and punish those who are convicted of criminal conduct. The **criminal justice system refers to the more than 55,000 local, state, and federal agencies that enforce laws, adjudicate crimes, and treat and rehabilitate criminals.** The system includes the police, the courts, and corrections facilities, and it employs more than 2,000,000 people in 17,000 police agencies, nearly 17,000 courts, more than 8,000 prosecutorial agencies, about 6,000 correctional institutions, and more than 3,500 probation and parole departments. More than

$150 billion is spent annually for civil and criminal justice, which amounts to more than $500 for every person living in the United States (Siegel, 2006).

The term *criminal justice system* is somewhat misleading because it implies that law enforcement agencies, courts, and correctional facilities constitute one large, integrated system when, in reality, the criminal justice system is made up of many bureaucracies that have considerable discretion in how decisions are made. *Discretion* refers to the use of personal judgment by police officers, prosecutors, judges, and other criminal justice system officials regarding whether and how to proceed in a given situation (see ▶ Figure 6.6). The police are a prime example of discretionary processes because they have the power to selectively enforce the law and have on many occasions been accused of being too harsh or too lenient on alleged offenders.

The Police

The role of the police in the criminal justice system continues to expand. The police are responsible for crime control and maintenance of order, but local police departments now serve numerous other human service functions, including improving community relations, resolving family disputes, and helping people during emergencies. It should be remembered that not all "police officers" are employed by local police departments; they are employed in more than 25,000 governmental agencies ranging from local jurisdictions to federal levels. However, we will focus primarily on metropolitan police departments because they constitute the vast majority of the law enforcement community.

Metropolitan police departments are made up of a chain of command (similar to the military), with ranks such as officer, sergeant, lieutenant, and captain, and each rank must follow specific rules and procedures. However, individual officers maintain a degree of discretion in the decisions they make as they respond to calls and try to apprehend fleeing or

criminal justice system the more than 55,000 local, state, and federal agencies that enforce laws, adjudicate crimes, and treat and rehabilitate criminals.

Police

- Enforce specific laws
- Investigate specific crimes
- Search people, vicinities, buildings
- Arrest or detain people

Prosecutors

- File charges or petitions for judicial decision
- Seek indictments
- Drop cases
- Reduce charges
- Recommend sentences

Judges or Magistrates

- Set bail or conditions for release
- Accept pleas
- Determine delinquency
- Dismiss charges
- Impose sentences
- Revoke probation

▶ **Figure 6.6 Discretionary Powers in Law Enforcement**

violent offenders. The problem of police discretion is most acute when decisions are made to use force (such as grabbing, pushing, or hitting a suspect) or deadly force (shooting and killing a suspect). Generally, deadly force is allowed only in situations in which a suspect is engaged in a felony, is fleeing the scene of a felony, or is resisting arrest and has endangered someone's life.

Although many police departments have worked to improve their public image in recent years, the practice of *racial profiling*—the use of ethnic or racial background as a means of identifying criminal suspects—remains a highly charged issue. Officers in some police departments have singled out for discriminatory treatment African Americans, Latinos/as, and other people of color, treating them more harshly than white (Euro-American) individuals. However, police department officials typically contend that race is only one factor in determining why individuals are questioned or detained as they go about everyday activities such as driving a car or walking down the street. By contrast, equal-justice advocacy groups argue that differential treatment of minority-group members amounts to a race-based double standard, which they believe exists not only in police work but throughout the criminal justice system (see Cole, 2000).

The belief that differential treatment takes place on the basis of race contributes to a negative image of police among many people of color who believe that they have been hassled by police officers, and this assumption is intensified by the fact that police departments have typically been made up of white male personnel at all levels. In recent years, this sit-

uation has slowly begun to change. Currently, about 22 percent of all *sworn officers*—those who have taken an oath and been given the powers to make arrests and use necessary force in accordance with their duties—are women and minorities (Cole and Smith, 2004). The largest percentage of minority

© AP Images/Phil Sandlin

In 2008, actor Wesley Snipes was convicted on federal income tax evasion charges in Ocala, Florida. Conflict theorists believe that people of color face discrimination in the legal system but that celebrities often receive advantages that others don't. The interplay of these two factors was debated extensively before, during, and after Snipes's trial.

and women police officers are located in cities with a population of 250,000 or more. African Americans make up a larger percentage of the police department in cities with a larger proportion of African American residents (such as Detroit), but Latinos/as constitute a larger percentage in cities such as San Antonio and El Paso, Texas, where Latinos/as make up a larger proportion of the population. Women officers of all races are more likely to be employed in departments in cities of more than 250,000 (where they make up 16 percent of all officers) as compared with smaller communities (cities of less than 50,000), where women officers constitute only 2 to 5 percent of the force (Cole and Smith, 2004). In the past, women were excluded from police departments and other law enforcement careers largely because of stereotypical beliefs that they were not physically and psychologically strong enough to enforce the law. However, studies have indicated that as more females have entered police work, they receive similar evaluations to male officers from their administrators and that fewer complaints are filed against women officers, which some researchers believe is a function of how female officers more effectively control potentially violent encounters (Brandl, Stroshine, and Frank, 2001).

In the future, the image of police departments may change as greater emphasis is placed on *community-oriented policing*—an approach to law enforcement in which officers maintain a presence in the community, walking up and down the streets or riding bicycles, getting to know people, and holding public service meetings at schools, churches, and other neighborhood settings. Community-oriented policing is often limited by budget constraints and the lack of available personnel to conduct this type of "hands-on" community involvement. In many jurisdictions, police officers believe that they have only enough time to keep up with reports of serious crime and life-threatening occurrences and that the level of available personnel and resources does not allow officers to take on a greatly expanded role in the community.

The Courts

Criminal courts determine the guilt or innocence of those persons accused of committing a crime. In theory, justice is determined in an adversarial process in which the prosecutor (an attorney who represents the state) argues that the accused is guilty, and the defense attorney asserts that the accused is innocent. In reality, judges wield a great deal of discretion. Working with prosecutors, they decide whom to release and whom to hold for further hearings, and what sentences to impose on those persons who are convicted.

Prosecuting attorneys also have considerable leeway in deciding which cases to prosecute and when to negotiate a plea bargain with a defense attorney. As cases are sorted through the legal machinery, a steady attrition occurs. At each stage, various officials determine what alternatives will be available for those cases still remaining in the system.

About 90 percent of criminal cases are never tried in court; instead, they are resolved by plea bargaining, a process in which the prosecution negotiates a reduced sentence for the accused in exchange for a guilty plea (Senna and Siegel, 2002). Defendants (especially those who are poor and cannot afford to pay an attorney) may be urged to plead guilty to a lesser crime in return for not being tried for the more serious crime for which they were arrested. Prison sentences given in plea bargains vary widely from one region to another and even from judge to judge within one state. For example, although women typically commit less serious crimes than men, they do not fare as well as men in negotiating or bargaining for sentence reductions (Chesney-Lind, 1997).

Those who advocate the practice of plea bargaining believe that it allows for individualized justice for alleged offenders because judges, prosecutors, and defense attorneys can agree to a plea and to a punishment that best fits the offense and the offender. They also believe that this process helps reduce the backlog of criminal cases in the court system as well as the lengthy process often involved in a criminal trial. However, those who seek to abolish plea bargaining believe that this practice leads to innocent people pleading guilty to crimes they have not committed or pleading guilty to a crime other than the one they actually committed because they are offered a lesser sentence (Cole and Smith, 2004).

More serious crimes, such as murder, felonious assault, and rape, are more likely to proceed to trial than other forms of criminal conduct; however, many of these cases do not reach the trial stage. For example, one study of 75 of the largest counties in the United States found that only 26 percent of murder cases actually went to trial. By contrast, only about 6 percent of all other cases proceeded to trial (Reaves, 2001).

One of the most important activities of the court system is establishing the sentence of the accused after he or she has been found guilty or has pleaded guilty. Typically, sentencing involves the following kinds of sentences or dispositions: fines, probation, alternative or intermediate sanctions (such as house arrest or electronic monitoring), incarceration, and capital punishment (Siegel, 2006).

Punishment and Corrections

Punishment **is any action designed to deprive a person of things of value (including liberty) because of some offense the person is thought to have committed** (Barlow and Kauzlarich, 2002). Historically, punishment has had four major goals:

1. *Retribution* is punishment that a person receives for infringing on the rights of others (Cole and Smith, 2004). Retribution imposes a penalty on the offender and is based on the premise that the punishment should fit the crime: The greater the degree of social harm, the more the offender should be punished. For example, an individual who murders should be punished more severely than one who shoplifts.
2. *General deterrence* seeks to reduce criminal activity by instilling a fear of punishment in the general public. However, we most often focus on *specific deterrence,* which inflicts punishment on specific criminals to discourage them from committing future crimes. Recently, criminologists have debated whether imprisonment has a deterrent effect, given the fact that high rates of those who are released from prison become recidivists (previous offenders who commit new crimes).
3. *Incapacitation* is based on the assumption that offenders who are detained in prison or are executed will be unable to commit additional crimes. This approach is often expressed as "lock 'em' up and throw away the key!" In recent years, more emphasis has been placed on *selective incapacitation,* which means that offenders who repeat cer-

tain kinds of crimes are sentenced to long prison terms (Cole and Smith, 2004).

4. *Rehabilitation* seeks to return offenders to the community as law-abiding citizens by providing therapy or vocational or educational training. Based on this approach, offenders are treated, not punished, so that they will not continue their criminal activity. However, many correctional facilities are seriously understaffed and underfunded in the rehabilitation programs that exist. The job skills (such as agricultural work) that many offenders learn in prison do not transfer to the outside world, nor are offenders given any assistance in finding work that fits their skills once they are released.

Recently, newer approaches have been advocated for dealing with criminal behavior. Key among these is the idea of *restoration,* which is designed to repair the damage done to the victim and the community by an offender's criminal act (Cole and Smith, 2004). This approach is based on the *restorative justice perspective,* which states that the criminal justice system should promote a peaceful and just society; therefore, the system should focus on peacemaking rather than on punishing offenders. Advocates of this approach believe that punishment of offenders actually encourages crime rather than deterring it and are in favor of approaches such as probation with treatment. Opponents of this approach suggest that increased punishment of offenders leads to lower crime rates and that the restorative justice approach amounts to "coddling criminals." However, numerous restorative justice programs are now in operation, and many are associated with community policing programs as they seek to help offenders realize the damage that they have done to their victims and the community and to be reintegrated into society (Senna and Siegel, 2002).

Instead of the term *punishment,* the term *corrections* is often used. Criminologists George F. Cole and Christopher E. Smith (2004: 409) explain corrections as follows:

Corrections refers to the great number of programs, services, facilities, and organizations responsible for the management of people accused or convicted of criminal offenses. In addition to prisons and jails, corrections includes probation, halfway houses, education and work release programs, parole supervision, counseling, and community service. Correctional programs operate in Salvation Army hostels, forest camps, medical clinics, and urban storefronts.

As Cole and Smith (2004) explain, corrections is a major activity in the United States today. Consider the fact that about 6.5 million adults (more than one out of every twenty men and one out of every hundred women) are under some form of correctional control. The rate of African American males under some form of correctional supervision is even greater (one out of every six African American adult men and one out of every three African American men in their twenties). Some analysts believe that these figures are a reflection of centuries of underlying racial, ethnic, and class-based inequalities in the United States as well as sentencing disparities that reflect race-based differences in the criminal justice system. However, others argue that newer practices such as determinate or mandatory sentences may help to reduce such disparities over time. A *determinate sentence* sets the term of imprisonment at a fixed period of time (such as three years) for a specific offense. *Mandatory sentencing guidelines* are established by law and require that a person convicted of a specific offense or series of offenses be given a penalty within a fixed range. Although these practices limit judicial discretion in sentencing, many critics are concerned about the effects of these sentencing approaches. Another area of great discord within and outside the criminal justice system is the issue of the death penalty.

Historically, removal from the group has been considered one of the ultimate forms of punishment. For many years, capital punishment, or the death penalty, has been used in the United States as an appropriate and justifiable response to very serious crimes. About four thousand persons have been executed in the United States since 1930, when the federal government began collecting data on executions.

In 1972 the U.S. Supreme Court ruled (in *Furman v. Georgia*) that *arbitrary* application of the death penalty violates the Eighth Amendment to the Constitution but that the death penalty itself is not unconstitutional. In other words, determining who receives the death penalty and who receives a prison term for similar offenses should not be done on a "lotterylike" basis. To be constitutional, the death penalty must be imposed for reasons other than the race/ethnicity, gender, and social class of the offender.

The ex-slave states are more likely to execute criminals than are other states (see ▶ Figure 6.7). African Americans are eight to ten times more likely to be sentenced to death for homicidal rape than are whites (non-Latinos/as) who have committed the same crime (Marquart, Ekland-Olson, and Sorensen, 1994).

People who have lost relatives and friends as a result of criminal activity often see the death penalty as justified. However, capital punishment raises many doubts for those who fear that innocent individuals may be executed for crimes they did not commit. For still others, the problem of racial discrimination in the sentencing process poses troubling questions.

Deviance and Crime in the United States in the Future

Two pressing questions pertaining to deviance and crime will face us in the future: Is the solution to our "crime problem" more law and order? Is equal justice under the law possible?

Although many people in the United States agree that crime is one of the most important problems in this country, they are divided over what to do about it. Some of the frustration about crime might be based on unfounded fears; studies show that the overall crime rate has been decreasing slightly in recent years.

One thing is clear: The existing criminal justice system cannot solve the "crime problem." If roughly 20 percent of all crimes result in arrest, only half of those lead to a conviction in serious cases, and less than 5 percent of those result in a jail term, the "lock 'em up and throw the key away" approach has little

punishment any action designed to deprive a person of things of value (including liberty) because of some offense the person is thought to have committed.

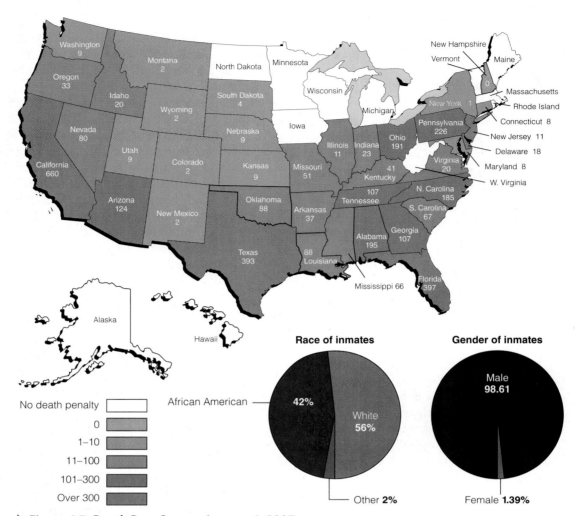

▶ **Figure 6.7 Death Row Census, January 1, 2007**

Death row inmates are heavily concentrated in certain states. African Americans, who make up about 12 percent of the U.S. population, account for approximately 42 percent of inmates on death row. In addition to those persons held by state governments, 53 inmates were held on death row by the federal government or U.S. military.

Sources: Death Penalty Information Center, 2007.

chance of succeeding. Nor does the high rate of recidivism among those who have been incarcerated speak well for the rehabilitative efforts of our existing correctional facilities. Reducing street crime may hinge on finding ways to short-circuit criminal behavior.

One of the greatest challenges is juvenile offenders, who may become the adult criminals of tomorrow. However, instead of military-style boot camps or other stopgap measures, *structural solutions*—such as more and better education and jobs, affordable housing, more equality and less discrimination, and socially productive activities—are needed to reduce street crime. In the past, structural solutions such as these have made it possible for immigrants who initially committed street crimes to leave the streets, get jobs, and lead productive lives. Ultimately, the best approach for reducing delinquency and crime would be prevention: to work with young people *before* they become juvenile offenders to help them establish family relationships, build self-esteem, choose a career, and get an education that will help them pursue that career. Sociologist Elliott Currie (1998) has proposed that an initial goal in working to prevent delinquency and crime is to pinpoint specifically what kinds of preventive programs work and to establish priorities that make prevention possible. Among these priorities are preventing child abuse and neglect, enhancing children's intellectual and social development, providing support and guidance to vulnerable adolescents, and working intensively with juvenile offenders (Currie, 1998).

Is equal justice under the law possible? As long as racism, sexism, classism, and ageism exist in our society, people will see deviant and criminal behavior through a selective lens. To solve the problems addressed in this chapter, we must ask ourselves what we can do to ensure the rights of everyone, including the poor, people of color, and women and men alike. Many of us can counter classism, racism, sexism, and ageism where they occur. Perhaps the only way that the United States can have equal justice under the law (and, perhaps, less crime as a result) in the future is to promote social justice for individuals regardless of their race, class, gender, or age.

The Global Criminal Economy

Consider this scenario:

> Con men operating out of Amsterdam sell bogus U.S. securities by telephone to Germans; the operation is controlled by an Englishman residing in Monaco, with his profits in Panama. Which police force should investigate? In which jurisdiction should a prosecution be mounted? There may even be a question about whether a crime has been committed, although if all the actions had taken place in a single country there would be little doubt. (United Nations Development Programme, 1999: 104)

As this example shows, international criminal activity poses new and interesting questions not only for those who are the victims of such actions but also for governmental agencies mandated to control crime.

Global crime—the networking of powerful criminal organizations and their associates in shared activities around the world—is a relatively new phenomenon (Castells, 1998). However, it is an extremely lucrative endeavor as criminal organizations have increasingly set up their operations on a transnational basis, using the latest communication and transportation technologies (see Box 6.3).

How much money and other resources change hands in the global criminal economy? Although the exact amount of profits and financial flows originating in the global criminal economy is impossible to determine, the 1994 United Nations Conference on Global Organized Crime estimated that about $500 billion (in U.S. currency) per year is accrued in the global trade in drugs alone. Today, profits from all kinds of global criminal activities are estimated to range from $750 billion to more than $1.5 trillion per year (United Nations Development Programme, 1999). Some analysts believe that even these figures underestimate the true nature and extent of the global criminal economy (Castells, 1998). The highest-income-producing activities of global criminal organizations include trafficking in drugs, weapons, and nuclear material; smuggling of things and people (including many migrants); trafficking in women and children for the sex industry; and trafficking in body parts such as corneas and major organs for the medical industry. Undergirding the entire criminal system is money laundering and various complex financial schemes and international trade networks that make it possible for people to use the resources they obtain through illegal activity for the purposes of consumption and investment in the ("legitimate") formal economy.

Can anything be done about global crime? Recent studies have concluded that reducing global crime will require a global response, including the cooperation of law enforcement agencies, prosecutors, and intelligence services across geopolitical boundaries. However, this approach is problematic because countries such as the United States often have difficulty getting the various law enforcement agencies to cooperate within their own nation. Similarly, law enforcement agencies

Box 6.3 **Sociology in Global Perspective**

The Global Reach of Russian Organized Crime

On a spring day when warm sunshine flooded the narrow, potholed streets, I took a taxi to Metropolitan Correctional Center (MCC), an imposing collection of tomblike cinder block towers in lower Manhattan, to interview Monya Elson—one of the most dangerous Russian mobsters the feds ever netted. I passed through several layers of security before I was shepherded by an armed guard up an elevator and deposited in a small, antiseptic cubicle with booming acoustics where lawyers meet their clients. . . . At least half a dozen armed guards stood outside the door, which was closed but had an observation window. Elson, an edgy man with a dark mien, was brought into the room, his hands and feet chained. He is considered a maximum-security risk, and for good reason: a natural-born extortionist and killing machine, Elson is perhaps the most prolific hit man in Russian mob history. (Friedman, 2000: 1)

Does this passage sound like the beginning of a best-selling novel? Although it has all the mystery and intrigue of a good novel, this paragraph is the beginning of a nonfiction book, *Red Mafiya: How the Russian Mob Has Invaded America* by the journalist Robert I. Friedman (2000). Friedman and other analysts have documented how more than thirty Russian crime syndicates operate in the United States, particularly in cities such as New York, Miami, San Francisco, Los Angeles, and Denver. And, as one New York state tax agent stated, "The Russian [mob] didn't come here to enjoy the American dream. They came here to steal it" (Friedman, 2000: Intro).

When and how did the Russian mob first become an international problem? Although Russian organized crime first gained a stronghold in the United States in the 1970s—a period of détente between this country and the Soviet Union—when Moscow allowed many Soviet Jews to emigrate to the United States and is believed to have sent many hard-core Russian criminals with them (*CBS News*, 2000), many experts believe that the breakup of the former Soviet Union in the late 1980s created a weakened

and impoverished Russia that produced an active and vigorous Russian mob that rapidly extended its grip to more than fifty countries worldwide.

Today, in cities ranging from New York and San Francisco to Toronto and Hong Kong, the powerful Russian mafia traffics in prostitution and drugs such as heroin; commits acts of extortion, arson, murder, burglary, and money laundering; and engages in the illegal sale of guns and missiles (Lindberg and Markovic, 2001). It maintains its position by a combination of three attributes that are characteristic of other major international crime groups, as well: (1) *non-ideology*—an absence of political motivations or goals, (2) *hierarchy*—a well-organized structure of specialized criminal cells controlled by a boss, and (3) *limited membership*—membership restrictions based on ethnicity, kinship, race, criminal record, and other factors deemed relevant by the group (Lindberg and Markovic, 2001).

What is the future of Russian organized crime on a global basis? Many analysts believe that organized crime not only remains out of control in Russia but that it is also growing in power and strength in the United States and other nations where it has infiltrated major financial institutions such as banks and brokerage firms. Is it something that we should be concerned about? Probably so. As Monya Elson, the mobster described at the beginning of this box, admits, "I am a criminal. And for you this is bad. . . . I am proud of what I am" (qtd. in Friedman, 2000: 1). Elson purportedly killed more than one hundred people before he was caught and sent to prison.

Reflect & Analyze

Since 2001, most of U.S. attention to international crime has focused on terrorism. Should we turn some of our attention to the threat posed by international organized crime? Which type of crime concerns you more?

in high-income nations such as the United States and Canada are often suspicious of law enforcement agencies in low-income countries, believing that these law enforcement officers are corrupt. Regulation by the international community (for example, through the United Nations) would also be necessary to control global criminal activities such as international money laundering and trafficking in people and controlled substances such

as drugs and weapons. However, development and enforcement of international agreements on activities such as the smuggling of migrants or the trafficking of women and children for the sex industry have been extremely limited thus far. Many analysts acknowledge that economic globalization has provided great opportunities for wealth through global organized crime (Castells, 1998; United Nations Development Programme, 1999).

Chapter Review

• How do sociologists view deviance?

Sociologists are interested in what types of behavior are defined by societies as "deviant," who does that defining, how individuals become deviant, and how those individuals are dealt with by society.

• What are the main functionalist theories for explaining deviance?

Functionalist perspectives on deviance include strain theory and opportunity theory. Strain theory focuses on the idea that when people are denied legitimate access to cultural goals, such as a good job or a nice home, they may engage in illegal behavior to obtain them. Opportunity theory suggests that for deviance to occur, people must have access to illegitimate means to acquire what they want but cannot obtain through legitimate means.

• How do conflict and feminist perspectives explain deviance?

Conflict perspectives on deviance focus on inequalities in society. Marxist conflict theorists link deviance and crime to the capitalist society, which divides people into haves and have-nots, leaving crime as the only source of support for those at the bottom of the economic ladder. Feminist approaches to de-

viance focus on the relationship between gender and deviance.

• How do symbolic interactionists view deviance?

According to symbolic interactionists, deviance is learned through interaction with others. Differential association theory states that individuals have a greater tendency to deviate from societal norms when they frequently associate with persons who tend toward deviance instead of conformity. According to social control theories, everyone is capable of committing crimes, but social bonding (attachments to family and to other social institutions) keeps many from doing so. According to labeling theory, deviant behavior is that which is labeled deviant by those in powerful positions.

• What is the postmodernist view on deviance?

Postmodernist views on deviance focus on how the powerful control others through discipline and surveillance. This control may be maintained through largely invisible forces such as the Panoptican, as described by Michel Foucault, or by newer technologies that place everyone—not just "deviants"—under constant surveillance by authorities who use their knowledge as power over others.

● How do sociologists classify crime?

Sociologists identify six main categories of crime: violent crime (murder, forcible rape, robbery, and aggravated assault), property crime (burglary, motor vehicle theft, larceny-theft, and arson), public order crimes (sometimes referred to as "morals" crimes), occupational (white-collar) crime, organized crime, and political crime.

● What are the main sources of crime statistics?

Official crime statistics are taken from the Uniform Crime Report, which lists crimes reported to the police, and the National Crime Victimization Survey, which interviews households to determine the incidence of crimes, including those not reported to police. Studies show that many more crimes are committed than are officially reported.

● How are age and class related to crime statistics?

Age is the key factor in crime statistics. In 2006, persons under age 25 accounted for more than 45 percent of all arrests for violent crime and almost 55 percent of all arrests for property crime. Persons from lower socioeconomic backgrounds are more likely to be arrested for violent and property crimes; white-collar crime is more likely to occur among the upper socioeconomic classes.

● Who are the most frequent victims of crime?

Young males of color between ages 12 and 24 have the highest criminal victimization rates. The elderly tend to be fearful of crime but are the least likely to be victimized.

● How is discretion used in the criminal justice system?

The criminal justice system, including the police, the courts, and prisons, often has considerable discretion in dealing with offenders. The police use discretion in deciding whether to act on a situation. Prosecutors and judges use discretion in deciding which cases to pursue and how to handle them.

www.cengage.com/login

Register for a Student eResource account to maximize your study time online using CengageNOW. First take the system's diagnostic pre-test, and then follow the personalized study plan that is created for you to help you review this chapter. The study plan will

- help you identify areas on which you should concentrate;
- provide interactive exercises to help you master the chapter concepts; and
- provide a post-test to confirm you are ready to move on to the next chapter.

Key Terms

corporate crime 193

crime 177

criminal justice system 201

criminology 178

deviance 174

differential association theory 186

illegitimate opportunity structures 181

juvenile delinquency 178

labeling theory 188

occupational (white-collar) crime 193

organized crime 194

political crime 195

primary deviance 189

property crimes 193

punishment 204

secondary deviance 189

social bond theory 188

social control 178

strain theory 179

terrorism 196

tertiary deviance 189

victimless crimes 193

violent crime 193

Questions for Critical Thinking

1. Does public toleration of deviance lead to increased crime rates? If people were forced to conform to stricter standards of behavior, would there be less crime in the United States?

2. Should so-called victimless crimes, such as prostitution and recreational drug use, be decriminalized? Do these crimes harm society?

3. As a sociologist armed with a sociological imagination, how would you propose to deal with the problem of crime in the United States? What programs would you suggest enhancing? What programs would you reduce?

The Kendall Companion Website

www.cengage.com/sociology/kendall

Supplement your review of this chapter by going to the text's companion website, where you can take tutorial quizzes, use flash cards to master key terms, follow live links to useful websites, and explore the other study and research resources you'll find there, such as a comprehensive interactive sociology timeline, GSS Data, and Census 2000 information, much of it presented visually in maps

7 Class and Stratification in the United States

We treat them in hospitals every day.

They are young brothers, often drug dealers, gang members, or small-time criminals, who show up shot, stabbed, or beaten after a hustle gone bad. To some of our medical colleagues, they are just nameless thugs, perpetuating crime and death in neighborhoods that have seen far too much of these things. But when we look into their faces, we see ourselves as teenagers, we see our friends, we see what we easily could have become as young adults. And we're reminded of the thin line that separates us—three twenty-nine-year-old doctors (an emergency-room physician, an internist, and a dentist)—from those patients whose lives are filled with danger and desperation.

We grew up in poor, broken homes in New Jersey neighborhoods riddled with crime, drugs, and death, and came of age in the 1980s at the height of a crack epidemic that ravaged

© Anthony Barboza

communities like ours throughout the nation. . . . Two of us landed in juvenile-detention centers before our eighteenth birthdays. But inspired early by caring and imaginative role models, one of us in childhood latched on to a dream of becoming a dentist, steered clear of trouble, and in his senior year of high school persuaded his two best friends to apply to a college program for minority students interested in becoming doctors. We knew we'd never survive if we went after it alone. And so we made a pact: we'd help one another through, no matter what.

—Drs. Sampson Davis, George Jenkins, and Rameck Hunt (2003: 1–2) describing their path from the streets of Newark to being named among the forty most influential African Americans by *Essence* magazine and thus, in the eyes of many people, achieving the American Dream

The remarkable success of Sampson Davis, George Jenkins, and Rameck Hunt as they stuck together and worked diligently to get out of graffiti-covered New Jersey public-housing projects and to ultimately complete their education in medical and dental schools might be described as a contemporary version of the American Dream. What is the American Dream? Simply stated, the American Dream is the belief that if people work hard and play by the rules, they will have a chance to get ahead (see Hochschild, 1995). Moreover, each generation will be able to have a higher standard of living than that of its parents (Danziger and Gottschalk, 1995). The American Dream is based on the assumption that people in the United States have equality of opportunity regardless of their race, creed, color, national origin, gender, or religion.

For middle- and upper-income people, the American Dream typically means that each subsequent generation will be able to acquire more material possessions and wealth than people in the preceding generations. To some people, achieving

Chapter Focus Question

How is the American Dream influenced by social stratification?

the American Dream means having a secure job, owning a home, and getting a good education for their children. To others, it is the promise that anyone may rise from poverty to wealth (from "rags to riches") if he or she works hard enough. In this chapter, we examine systems of social stratification, particularly the U.S. class structure, to see how people's opportunities are affected by their position in that structure. However, before we explore class and stratification, test your knowledge of wealth, poverty, and the American Dream by taking the quiz in Box 7.1.

What Is Social Stratification?

Social stratification **is the hierarchical arrangement of large social groups based on their control over basic resources** (Feagin and Feagin, 2008). Stratification involves patterns of structural inequality that are associated with membership in each of these groups, as well as the ideologies that support inequality. Sociologists examine the social groups that make up the hierarchy in a society and seek to determine how inequalities are structured and persist over time.

Max Weber's term *life chances* **refers to the extent to which individuals have access to important societal resources such as food, clothing, shelter, education, and health care.** According to sociologists, more-affluent people typically have better life chances than the less affluent because they have greater access to quality education, safe neighborhoods, high-quality nutrition and health care, police and private security protection, and an extensive array of other goods and services. In contrast, persons with low- and poverty-level incomes tend to have limited access to these resources. *Resources* are anything valued in a society, ranging from money and property to medical care and education; they are considered to be scarce because of their unequal distribution among social categories. If we think about the valued resources available in the United States, for example, the differences in life chances are readily apparent. As one analyst suggested, "Poverty narrows and closes life chances.

The victims of poverty experience a kind of arteriosclerosis of opportunity. Being poor not only means economic insecurity, it also wreaks havoc on one's mental and physical health" (Ropers, 1991: 25). Our life chances are intertwined with our class, race, gender, and age.

All societies distinguish among people by age. Young children typically have less authority and responsibility than older persons. Older persons, especially those without wealth or power, may find themselves at the bottom of the social hierarchy. Similarly, all societies differentiate between females and males: Women are often treated as subordinate to men. From society to society, people are treated differently as a result of their religion, race/ethnicity, appearance, physical strength, disabilities, or other distinguishing characteristics. All of these differentiations result in inequality. However, systems of stratification are also linked to the specific economic and social structure of a society and to a nation's position in the system of global stratification, which is so significant for understanding social inequality that we will devote the next chapter to this topic (Chapter 8).

Systems of Stratification

Around the globe, one of the most important characteristics of systems of stratification is their degree of flexibility. Sociologists distinguish among such systems based on the extent to which they are open or closed. In an *open system,* the boundaries between levels in the hierarchies are more flexible and may be influenced (positively or negatively) by people's achieved statuses. Open systems are assumed to have some degree of social mobility. *Social mobility* **is the movement of individuals or groups from one level in a stratification system to another** (Rothman, 2001). This movement can be either upward or downward. *Intergenerational mobility* **is the social movement experienced by family members from one generation to the next.** For example, Sarah's father is a carpenter who makes good wages in good economic times but is often unemployed when the construction industry slows to a standstill. Sarah be-

How Much Do You Know About Wealth, Poverty, and the American Dream?

True	False	
T	F	1. People no longer believe in the American Dream.
T	F	2. Individuals over age 65 have the highest rate of poverty.
T	F	3. Men account for two out of three impoverished adults in the United States.
T	F	4. About 5 percent of U.S. residents live in households whose members sometimes do not get enough to eat.
T	F	5. Income is more unevenly distributed than wealth.
T	F	6. People who are poor usually have personal attributes that contribute to their impoverishment.
T	F	7. A number of people living below the official poverty line have full-time jobs.
T	F	8. One in three U.S. children will be poor at some point of their childhood.

Answers on page 216.

comes a neurologist, earning $350,000 a year, and moves from the working class to the upper-middle class. Between her father's generation and her own, Sarah has experienced upward social mobility.

By contrast, ***intragenerational mobility is the social movement of individuals within their own lifetime.*** Consider, for example, RaShandra, who began her career as a high-tech factory worker and through increased experience and taking specialized courses in her field became an entrepreneur, starting her own highly successful "dot.com" business. RaShandra's advancement is an example of upward intragenerational social mobility. However, both intragenerational mobility and intergenerational mobility may be downward as well as upward.

In a *closed system*, the boundaries between levels in the hierarchies of social stratification are rigid, and people's positions are set by ascribed status. Open and closed systems are ideal-type constructs; no actual stratification system is completely open or closed. The systems of stratification that we will examine—slavery, caste, and class—are characterized by different hierarchical structures and varying degrees of mobility. Let's examine these three systems of stratification to determine how people acquire their positions in each and what potential for social movement they have.

Slavery

***Slavery* is an extreme form of stratification in which some people are owned by others.** It is a closed system in which people designated as "slaves"

social stratification the hierarchical arrangement of large social groups based on their control over basic resources.

life chances Max Weber's term for the extent to which individuals have access to important societal resources such as food, clothing, shelter, education, and health care.

social mobility the movement of individuals or groups from one level in a stratification system to another.

intergenerational mobility the social movement (upward or downward) experienced by family members from one generation to the next.

intragenerational mobility the social movement (upward or downward) of individuals within their own lifetime.

slavery an extreme form of stratification in which some people are owned by others.

Box 7.1 Sociology and Everyday Life

Answers to the Sociology Quiz on Wealth, Poverty, and the American Dream

1. False. The American Dream appears to be alive and well. U.S. culture places a strong emphasis on the goal of monetary success, and many people use legal or illegal means to attempt to achieve that goal.

2. False. As a group, children have a higher rate of poverty than the elderly. Government programs such as Social Security have been indexed for inflation, whereas many of the programs for the young have been scaled back or eliminated. However, many elderly individuals still live in poverty.

3. False. Women, not men, account for two out of three impoverished adults in the United States. Reasons include the lack of job opportunities for women, lower pay than men for comparable jobs, lack of affordable day care for children, sexism in the workplace, and a number of other factors.

4. True. It is estimated that about 5 percent of the U.S. population (1 in 20 people) resides in household units where members do not get enough to eat.

5. False. Wealth is more unevenly distributed among the U.S. population than is income. However, both wealth and income are concentrated in very few hands compared with the size of the overall population.

6. False. According to one widely held stereotype, the poor are lazy and do not want to work. Rather than looking at the structural characteristics of society, people cite the alleged personal attributes of the poor as the reason for their plight.

7. True. Many of those who fall below the official poverty line are referred to as the "working poor" because they work full time but earn such low wages that they are still considered to be impoverished.

8. True. According to recent data from the Children's Defense Fund, one in three U.S. children will live in a family that is below the official poverty line at some point in their childhood. For some of these children, poverty will be a persistent problem throughout their childhood and youth.

Sources: Based on Children's Defense Fund, 2001; Gilbert, 2003; and U.S. Census Bureau, 2007.

---

are treated as property and have little or no control over their lives. According to some social analysts, throughout recorded history only five societies have been slave societies—those in which the social and economic impact of slavery was extensive: ancient Greece, the Roman Empire, the United States, the Caribbean, and Brazil (Finley, 1980). Others suggest that slavery also existed in the Americas prior to European settlement, and throughout Africa and Asia (Engerman, 1995).

Those of us living in the United States are most aware of the legacy of slavery in our own country. Beginning in the 1600s, slaves were forcibly imported to the United States as a source of cheap labor. Slavery was defined in law and custom by the 1750s, making it possible for one person to own another person (Healey, 2002). In fact, early U.S. presidents including George Washington, James Madison, and Thomas Jefferson owned slaves. As practiced in the United States, slavery had four primary characteristics: (1) it was for life and was inherited (children of slaves were considered to be slaves); (2) slaves were considered property, not human beings; (3) slaves were denied rights; and (4) coercion was used to keep slaves "in their place" (Noel, 1972). Although most slaves were powerless to bring about change, some were able to challenge slavery—or at least their position in the system—by engaging in activities such as sabotage, intentional carelessness, work slowdowns, or running away from owners and

Social mobility is the movement from one level in a stratification system to another. The background of the photo shows a traditional Indian marketplace; however, the man in the foreground shows signs of upward mobility.

working for the abolition of slavery (Healey, 2002). Despite the fact that slavery in this country officially ended many years ago, sociologists such as Patricia Hill Collins (1990) believe that its legacy is deeply embedded in current patterns of prejudice and discrimination against African Americans.

Slavery is not simply an unfortunate historical legacy. Although legal slavery no longer exists, economist Stanley L. Engerman (1995: 175) believes that the world will not be completely free of slavery as long as there are "debt bondage, child labor, contract labor, and other varieties of coerced work for limited periods of time, with limited opportunities for mobility, and with limited political and economic power."

The Caste System

Like slavery, caste is a closed system of social stratification. A *caste system* **is a system of social inequality in which people's status is per-** **manently determined at birth based on their parents' ascribed characteristics.** Vestiges of caste systems exist in contemporary India and South Africa.

In India, caste is based in part on occupation; thus, families typically perform the same type of work from generation to generation. By contrast, the caste system of South Africa was based on racial classifications and the belief of white South Africans (Afrikaners) that they were morally superior to the black majority. Until the 1990s, the Afrikaners controlled the government, the police, and the military by enforcing *apartheid*—the separation of the races. Blacks were denied full citizenship and restricted to segregated hospitals, schools, residential neighborhoods, and other facilities. Whites held almost all of the desirable jobs; blacks worked as manual laborers and servants.

In a caste system, marriage is endogamous, meaning that people are allowed to marry only within their own group. In India, parents traditionally have selected marriage partners for their children. In South Africa, interracial marriage was illegal until 1985.

Cultural beliefs and values sustain caste systems. Hinduism, the primary religion of India, reinforced the caste system by teaching that people should accept their fate in life and work hard as a moral duty. Caste systems grow weaker as societies industrialize; the values reinforcing the system break down, and people start to focus on the types of skills needed for industrialization.

As we have seen, in closed systems of stratification, group membership is hereditary, and it is almost impossible to move up within the structure. Custom and law frequently perpetuate privilege and ensure that higher-level positions are reserved for the children of the advantaged (Rothman, 2001).

caste system a system of social inequality in which people's status is permanently determined at birth based on their parents' ascribed characteristics.

Systems of stratification include slavery, caste, and class. As shown in these photos, the life chances of people living in each of these systems differ widely.

© Hulton Archive/Getty Images

© Alan Sussman/The Image Works

© John Lund/Drew Kelly/Getty Images

Although some of Karl Marx's ideas have been discredited, his concept of class conflict between the capitalist and working classes continues to be visible in events such as strikes. Mexican, Mexican American, and many other workers in the United States took a day off from work to express their concern about stricter immigration laws in this country.

The Class System

The *class system* is a type of stratification based on the ownership and control of resources and on the type of work people do (Rothman, 2001). At least theoretically, a class system is more open than a caste system because the boundaries between classes are less distinct than the boundaries between castes. In a class system, status comes at least partly through achievement rather than entirely by ascription.

In class systems, people may become members of a class other than that of their parents through both intergenerational and intragenerational mobility, either upward or downward. Horizontal mobility occurs when people experience a gain or loss in position and/or income that does not produce a change in their place in the class structure. For example, a person may get a pay increase and a more prestigious title but still not move from one class to another. By contrast, movement up or down the class structure is *vertical mobility*. Martin, a commercial artist who owns his own firm, is an example of vertical, intergenerational mobility:

> My family came out of a lot of poverty and were eager to escape it. . . . My [mother's parents] worked in a sweatshop. My grandfather to the day he died never earned more than $14 a week. My grandmother worked in knitting mills while she had five children. . . . My father quit school when he was in eighth grade and supported his mother and his two sisters when he was twelve years old. My grandfather died when my father was four and he basically raised his sisters. He got a man's job when he was twelve and took care of the three of them. (qtd. in Newman, 1993: 65)

Martin's situation reflects upward mobility; however, people may also experience downward mobility, caused by any number of reasons, including a

© Robyn Beck/AFP/Getty Images

lack of jobs, low wages and employment instability, marriage to someone with fewer resources and less power than oneself, and changing social conditions (Newman, 1988, 1993).

Classical Perspectives on Social Class

Early sociologists grappled with the definition of class and the criteria for determining people's location in the class structure. Both Karl Marx and Max Weber viewed class as an important determinant of social inequality and social change, and their works have had a profound influence on how we view the U.S. class system today.

---

class system a type of stratification based on the ownership and control of resources and on the type of work that people do.

Karl Marx: Relationship to the Means of Production

According to Karl Marx, class position and the extent of our income and wealth are determined by our work situation, or our relationship to the means of production. As we have previously seen, Marx stated that capitalistic societies consist of two classes—the capitalists and the workers. The ***capitalist class (bourgeoisie)* consists of those who own the means of production**—the land and capital necessary for factories and mines, for example. The ***working class (proletariat)* consists of those who must sell their labor to the owners in order to earn enough money to survive** (see ▶ Figure 7.1).

According to Marx, class relationships involve inequality and exploitation. The workers are exploited as capitalists maximize their profits by paying workers less than the resale value of what they produce but do not own. This exploitation results in workers' ***alienation*—a feeling of powerlessness and estrangement from other people and from oneself.** In Marx's view, alienation develops as workers manufacture goods that embody their creative talents but the goods do not belong to them. Workers are also alienated from the work itself because they are forced to perform it in order to live. Because the workers' activities are not their own, they feel self-estrangement. Moreover, the workers are separated from others in the factory because they individually sell their labor power to the capitalists as a commodity.

In Marx's view, the capitalist class maintains its position at the top of the class structure by control of the society's *superstructure,* which is composed of the government, schools, churches, and other social institutions that produce and disseminate ideas perpetuating the existing system of exploitation. Marx predicted that the exploitation of workers by the capitalist class would ultimately lead to ***class conflict*—the struggle between the capitalist class and the working class.** According to Marx, when the workers realized that capitalists were the source of their oppression, they would overthrow the capitalists and their agents of social control, leading to the end of capitalism. The workers would then take over the government and create a more egalitarian society.

Why has no workers' revolution occurred? According to the sociologist Ralf Dahrendorf (1959), capitalism may have persisted because it has changed significantly since Marx's time. Individual capitalists no longer own and control factories and other means of production; today, ownership and control have largely been separated. For example, contemporary transnational corporations are owned by a multitude of stockholders but run by paid officers and managers. Similarly, many (but by no means all) workers have experienced a rising standard of living, which may have contributed to a feeling of complacency. During the twentieth century, workers pressed for salary increases and improvements in the workplace through their activism and labor union membership. They also gained more legal protection in the form of workers' rights and benefits such as workers' compensation insurance for job-related injuries and disabilities (Dahrendorf, 1959). For these reasons, and because of a myriad of other complex factors, the workers' revolution predicted by Marx never came to pass. However, the failure of his prediction does not mean that his analysis of capitalism and his theoretical contributions to sociology are without validity.

Marx had a number of important insights into capitalist societies. First, he recognized the eco-

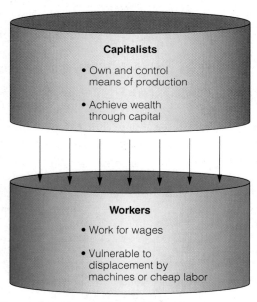

▶ Figure 7.1 **Marx's View of Stratification**

In the twentieth century, labor unions helped to bring about higher wages, shorter workweeks, and improved workplace safety. These factors, among others, short-circuited the workers revolution predicted by Karl Marx.

nomic basis of class systems (Gilbert, 2003). Second, he noted the relationship between people's social location in the class structure and their values, beliefs, and behavior. Finally, he acknowledged that classes may have opposing (rather than complementary) interests. For example, capitalists' best interests are served by a decrease in labor costs and other expenses and a corresponding increase in profits; workers' best interests are served by well-paid jobs, safe working conditions, and job security.

Max Weber: Wealth, Prestige, and Power

Max Weber's analysis of class builds upon earlier theories of capitalism (particularly those by Marx) and of money (particularly those by Georg Simmel, as discussed in Chapter 2). Living in the late nineteenth and early twentieth centuries, Weber was in a unique position to see the transformation that occurred as individual, competitive, entrepreneurial capitalism went through the process of shifting to bureaucratic, industrial, corporate capitalism. As a result, Weber had more opportunity than Marx to see how capitalism changed over time.

Weber agreed with Marx's assertion that economic factors are important in understanding individual and group behavior. However, Weber em-

phasized that no single factor (such as economic divisions between capitalists and workers) was sufficient for defining the location of categories of people within the class structure. According to Weber, the access that people have to important societal resources (such as economic, social, and political power) is crucial in determining people's life chances. To highlight the importance of life chances for categories of people, Weber developed a multidimensional approach to social stratification that reflects the interplay among wealth, prestige, and power. In his analysis of these dimensions of class structure, Weber viewed the concept of "class" as an *ideal type* (that can be used to compare and contrast various societies) rather than as a specific social category of "real" people (Bourdieu, 1984).

Wealth **is the value of all of a person's or family's economic assets, including income, personal property, and income-producing property.** Weber placed categories of people who have a similar level of wealth and income in the same class. For example, he identified a privileged commercial class of *entrepreneurs*—wealthy bankers, ship owners, professionals, and merchants who possess similar financial resources. He also described a class of *rentiers*—wealthy individuals who live off their investments and do not have to work. According to Weber, entrepreneurs and rentiers

capitalist class (or **bourgeoisie**) Karl Marx's term for the class that consists of those who own and control the means of production.

working class (or **proletariat**) those who must sell their labor to the owners in order to earn enough money to survive.

alienation a feeling of powerlessness and estrangement from other people and from oneself.

class conflict Karl Marx's term for the struggle between the capitalist class and the working class.

wealth the value of all of a person's or family's economic assets, including income, personal property, and income-producing property.

have much in common. Both are able to purchase expensive consumer goods, control other people's opportunities to acquire wealth and property, and monopolize costly status privileges (such as education) that provide contacts and skills for their children.

Weber divided those who work for wages into two classes: the middle class and the working class. The middle class consists of white-collar workers, public officials, managers, and professionals. The working class consists of skilled, semiskilled, and unskilled workers.

The second dimension of Weber's system of stratification is *prestige*—**the respect or regard with which a person or status position is regarded by others.** Fame, respect, honor, and esteem are the most common forms of prestige. A person who has a high level of prestige is assumed to receive deferential and respectful treatment from others. Weber suggested that individuals who share a common level of social prestige belong to the same status group regardless of their level of wealth. They tend

© New Line/courtesy Everett Collection

Max Weber believed that people who are at a similar level of social prestige belong to the same status group. In this still from *Wedding Crashers,* U.S. Treasury Secretary William Cleary (played by Christopher Walken) is surrounded by people who are obviously his social peers.

to socialize with one another, marry within their own group of social equals, spend their leisure time together, and safeguard their status by restricting outsiders' opportunities to join their ranks (Beeghley, 2000).

The other dimension of Weber's system is *power*—**the ability of people or groups to achieve their goals despite opposition from others.** The powerful can shape society in accordance with their own interests and direct the actions of others (Tumin, 1953). According to Weber, social power in modern societies is held by bureaucracies; individual power depends on a person's position within the bureaucracy. Weber suggested that the power of modern bureaucracies was so strong that even a workers' revolution (as predicted by Marx) would not lessen social inequality (Hurst, 2007).

Weber stated that wealth, prestige, and power are separate continuums on which people can be ranked from high to low, as shown in ▶ Figure 7.2. Individuals may be high on one dimension while being low on another. For example, people may be very wealthy but have little political power (for example, a recluse who has inherited a large sum of money). They may also have prestige but not wealth (for instance, a college professor who receives teaching excellence awards but lives on a relatively low income). In Weber's multidimensional approach, people are ranked on all three dimensions. Sociologists often use the term *socioeconomic status (SES)* **to refer to a combined measure that attempts to classify individuals, families, or households in terms of factors such as income, occupation, and education to determine class location.**

Weber's analysis of social stratification contributes to our understanding by emphasizing that people behave according to both their economic interests and their values. He also added to Marx's insights by developing a multidimensional explanation of the class structure and by identifying additional classes. Both Marx and Weber emphasized that capitalists and workers are the primary players in a class society, and both noted the importance of class to people's life chances. However, they saw different futures for capitalism and the social system. Marx saw these structures being overthrown; We-

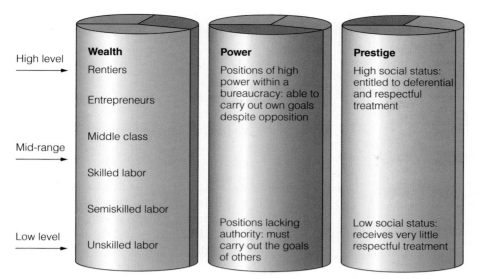

	Wealth	Power	Prestige
High level →	Rentiers	Positions of high power within a bureaucracy: able to carry out own goals despite opposition	High social status: entitled to deferential and respectful treatment
	Entrepreneurs		
	Middle class		
Mid-range →	Skilled labor		
	Semiskilled labor		
Low level →	Unskilled labor	Positions lacking authority: must carry out the goals of others	Low social status: receives very little respectful treatment

▶ **Figure 7.2 Weber's Multidimensional Approach to Social Stratification**
According to Max Weber, wealth, power, and prestige are separate continuums. Individuals may rank high in one dimension and low in another, or they may rank high or low in more than one dimension. Also, individuals may use their high rank in one dimension to achieve a comparable rank in another. How does Weber's model compare with Marx's approach as shown in Figure 7.1?

ber saw the increasing bureaucratization of life even without capitalism.

Contemporary Sociological Models of the U.S. Class Structure

How many social classes exist in the United States? What criteria are used for determining class membership? No broad consensus exists about how to characterize the class structure in this country. In fact, many people deny that class distinctions exist (see Eisler, 1983; Parenti, 1994). Most people like to think of themselves as middle class; it puts them in a comfortable middle position—neither rich nor poor. Sociologists have developed two models of the class structure: One is based on a Weberian approach, the other on a Marxian approach. We will examine both models briefly.

The Weberian Model of the U.S. Class Structure

Expanding on Weber's analysis of class structure, the sociologists Dennis Gilbert (2003) and Joseph A. Kahl developed a widely used model of social classes based on three elements: (1) education, (2) occupation of family head, and (3) family income (see ▶ Figure 7.3).

prestige the respect or regard with which a person or status position is regarded by others.

power according to Max Weber, the ability of people or groups to achieve their goals despite opposition from others.

socioeconomic status (SES) a combined measure that, in order to determine class location, attempts to classify individuals, families, or households in terms of factors such as income, occupation, and education.

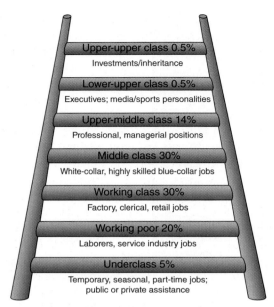

▶ **Figure 7.3 Stratification Based on Education, Occupation, and Income**

The Upper (Capitalist) Class The upper class is the wealthiest and most powerful class in the United States. About 1 percent of the population is included in this class, whose members own substantial income-producing assets and operate on both the national and international levels. According to Gilbert (2003), people in this class have an influence on the economy and society far beyond their numbers.

Some models further divide the upper class into upper-upper ("old money") and lower-upper ("new money") categories (Warner and Lunt, 1941; Coleman and Rainwater, 1978; Kendall, 2002). Members of the upper-upper class come from prominent families, which possess great wealth that they have held for several generations. Family names—such as Rockefeller, Mellon, Du Pont, and Kennedy—are well-known and often held in high esteem. Persons in the upper-upper class tend to have strong feelings of ingroup solidarity. They belong to the same exclusive clubs and support high culture (such as the opera, symphony orchestras, ballet, and art museums). Children are educated in prestigious private schools and Ivy League universities; many acquire strong feelings of privilege from birth, as upper-class author Lewis H. Lapham (1988: 14) states:

Together with my classmates and peers, I was given to understand that it was sufficient accomplishment merely to have been born. Not that anybody ever said precisely that in so many words, but the assumption was plain enough, and I could confirm it by observing the mechanics of the local society. A man might become a drunkard, a concert pianist or an owner of companies, but none of these occupations would have an important bearing on his social rank.

Children of the upper class are socialized to view themselves as different from others; they also learn that they are expected to marry within their own class (Warner and Lunt, 1941; Mills, 1959a; Domhoff, 1983; Kendall, 2002).

Members of the lower-upper class may be extremely wealthy but not have attained as much prestige as members of the upper-upper class. The "new rich" have earned most of their money in their own lifetime as entrepreneurs, presidents of major corporations, sports or entertainment celebrities, or top-level professionals. For some members of the lower-upper class, the American Dream has become a reality. Others still desire the respect of members of the upper-upper class.

The Upper-Middle Class Persons in the upper-middle class are often highly educated professionals who have built careers as physicians, attorneys, stockbrokers, or corporate managers. Others derive their income from family-owned businesses. According to Gilbert (2003), about 14 percent of the U.S. population is in this category. A combination of three factors qualifies people for the upper-middle class: university degrees, authority and independence on the job, and high income. Of all the class categories, the upper-middle class is the one that is most shaped by formal education. Over the past fifty years, Asian Americans, Latinos/as, and African Americans have placed great importance on education as a means of attaining the American Dream. Many people of color have moved into the upper-middle class by acquiring higher levels of education.

The Middle Class In past decades, a high school diploma was necessary to qualify for most middle-class jobs. Today, two-year or four-year college degrees have

Members of the upper-middle class typically have a combination of education and income that allows them to enjoy the "finer things of life."

replaced the high school diploma as an entry-level requirement for employment in many middle-class occupations, including medical technicians, nurses, legal and medical assistants, lower-level managers, semiprofessionals, and nonretail salesworkers. An estimated 30 percent of the U.S. population is in the middle class even though most people in this country think of themselves as middle class. Nowhere is this myth of the vast middle class more prevalent than in television situation comedies, which for decades have focused on idealized notions of the middle class or the debunking of that myth.

Traditionally, most middle-class occupations have been relatively secure and have provided more opportunities for advancement (especially with increasing levels of education and experience) than working-class positions. Recently, however, four factors have eroded the American Dream for this class: (1) escalating housing prices, (2) occupational insecurity, (3) blocked mobility on the job, and (4) the cost-of-living squeeze that has penalized younger workers, even when they have more education and better jobs than their parents (Newman, 1993).

The Working Class An estimated 30 percent of the U.S. population is in the working class. The core of this class is made up of semiskilled machine opera-

tors who work in factories and elsewhere. Members of the working class also include some workers in the service sector, as well as clerks and salespeople whose job responsibilities involve routine, mechanized tasks requiring little skill beyond basic literacy and a brief period of on-the-job training (Gilbert, 2003). Some people in the working class are employed in *pink-collar occupations—***relatively low-paying, nonmanual, semiskilled positions primarily held by women,** such as day-care workers, checkout clerks, cashiers, and waitpersons.

How does life in the working-class family compare with that of individuals in middle-class families? According to sociologists, working-class families not only earn less than middle-class families, but they also have less financial security, particularly because of high rates of layoffs and plant closings in some regions of the country. Few people in the working class have more than a high school diploma, and many have less, which makes job opportunities scarce for them in a "high-tech" society (Gilbert, 2003). Others find themselves in low-paying jobs in the service sector of the economy, particularly fast-food restaurants, a condition that often places them among the working poor.

The Working Poor The working poor account for about 20 percent of the U.S. population. Members of the working-poor class live from just above to just below the poverty line; they typically hold unskilled jobs, seasonal migrant jobs in agriculture, lower-paid factory jobs, and service jobs (such as counter help at restaurants). Employed single mothers often belong to this class; consequently, children are overrepresented in this category. African Americans and other people of color are also overrepresented among the working poor. To cite only one example, in the United States today, there are two white hospital orderlies to every one white physician, whereas there are twenty-five African American orderlies to every one African American physician (Gilbert, 2003). For the working

pink-collar occupations relatively low-paying, nonmanual, semiskilled positions primarily held by women, such as day-care workers, checkout clerks, cashiers, and waitpersons.

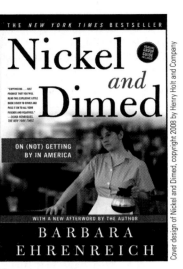

In *Nickel and Dimed: On (Not) Getting by in America*, Barbara Ehrenreich (left) recounts her attempt to replicate the lives of the working poor by working a series of low-paying jobs and trying to survive on her wages.

poor, living from paycheck to paycheck makes it impossible to save money for emergencies such as periodic or seasonal unemployment, which is a constant threat to any economic stability they may have.

Social critic and journalist Barbara Ehrenreich (2001) left her upper-middle-class lifestyle for a period of time to see if it was possible for the working poor to live on the wages that they were being paid as restaurant servers, sales clerks at discount department stores, aides at nursing homes, house cleaners for franchise maid services, or similar jobs. She conducted her research by actually holding those jobs for periods of time and seeing if she could live on the wages that she received. Through her research, Ehrenreich persuasively demonstrated that people who work full time, year-round, for poverty-level wages must develop survival strategies that include such things as help from relatives or constantly moving from one residence to another in order to have a place to live. Like many other researchers, Ehrenreich found that minimum-wage jobs cannot cover the full cost of living, such as rent, food, and the rest of an adult's monthly needs, even without taking into consideration the needs of children or other family members (see also Newman, 1999).

The Underclass According to Gilbert (2003), **people in the *underclass* are poor, seldom employed,** and caught in long-term deprivation that results from low levels of education and income and high rates of unemployment. Some are unable to work because of age or disability; others experience discrimination based on race/ethnicity. Single mothers are overrepresented in this class because of the lack of jobs, lack of affordable child care, and many other impediments to the mother's future and that of her children. People without a "living wage" often must rely on public or private assistance programs for their survival. About 3 to 5 percent of the U.S. population is in this category, and the chances of their children moving out of poverty are about fifty-fifty (Gilbert, 2003).

Studies by various social scientists have found that meaningful employment opportunities are the critical missing link for people on the lowest rungs of the class ladder. According to these analysts, job creation is essential in order for people to have the opportunity to earn a decent wage; have medical

Here, a food-bank worker delivers a package to a single mother and her child. Despite popular stereotypes to the contrary, many such parents would rather have jobs and pay their own way, but the high cost of child care can make this goal impossible to achieve.

coverage; live meaningful, productive lives; and raise their children in a safe environment (see Fine and Weis, 1998; Nelson and Smith, 1999; Newman, 1999; Wilson, 1996). These issues are closely tied to the American Dream we have been discussing in this chapter.

The Marxian Model of the U.S. Class Structure

The earliest Marxian model of class structure identified ownership or nonownership of the means of production as the distinguishing feature of classes. From this perspective, classes are social groups organized around property ownership, and social stratification is created and maintained by one group in order to protect and enhance its own economic interests. Moreover, societies are organized around classes in conflict over scarce resources. Inequality results from the more powerful exploiting the less powerful.

Contemporary Marxian (or conflict) models examine class in terms of people's relationship to others in the production process. For example, conflict theorists attempt to determine the degree of control that workers have over the decision-making process

© Billy Hustace/Getty Images

In which segment of the class structure would sociologists place clerical workers such as those in this office mail room? What are the key elements of that social class?

and the extent to which they are able to plan and implement their own work. They also analyze the type of supervisory authority, if any, that a worker has over other workers. According to this approach, most employees are a part of the working class because they do not control either their own labor or that of others.

Erik Olin Wright (1979, 1985, 1997), one of the leading stratification theorists to examine social class from a Marxian perspective, has concluded that Marx's definition of "workers" does not fit the occupations found in advanced capitalist societies. For example, many top executives, managers, and supervisors who do not own the means of production (and thus would be "workers" in Marx's model) act like capitalists in their zeal to control workers and maximize profits. Likewise, some experts hold positions in which they have control over money and the use of their own time even though they are not owners. Wright views Marx's category of "capitalist" as being too broad as well. For instance, small-business owners might be viewed as capitalists because they own their own tools and have a few people working for them, but they have little in common with large-scale capitalists and do not share the interests of factory workers.

Wright (1979) argues that classes in modern capitalism cannot be defined simply in terms of different levels of wealth, power, and prestige, as in the Weberian model. Consequently, he outlines four criteria for placement in the class structure: (1) ownership of the means of production, (2) purchase of the labor of others (employing others), (3) control of the labor of others (supervising others on the job), and (4) sale of one's own labor (being employed by someone else). Wright (1978) assumes that these criteria can be used to determine the class placement of all workers, regardless of race/ethnicity, in a capitalist society. Let's take a brief look at Wright's (1979, 1985) four classes—(1) the capitalist class, (2) the managerial class, (3) the small-business class, and (4) the working class—so that

underclass those who are poor, seldom employed, and caught in long-term deprivation that results from low levels of education and income and high rates of unemployment.

you can compare them to those found in the Weberian model.

The Capitalist Class According to Wright, this class holds most of the wealth and power in society through ownership of capital—for example, banks, corporations, factories, mines, news and entertainment industries, and agribusiness firms. The "ruling elites," or "ruling class," within the capitalist class hold political power and are often elected or appointed to influential political and regulatory positions (Parenti, 1994).

This class is composed of individuals who have inherited fortunes, own major corporations, or are top corporate executives with extensive stock holdings or control of company investments. Even though many top executives have only limited *legal ownership* of their corporations, they have substantial economic ownership and exert extensive control over investments, distribution of profits, and management of resources. The major sources of income for the capitalist class are profits, interest, and very high salaries. Members of this class make important decisions about the workplace, including which products and services to make available to consumers and how many workers to hire or fire.

Sociologist Erik Olin Wright has focused on the interplay of money and political power. The family of former U.S. President George H. W. and Barbara Bush, shown here, is a current example of Wright's theory.

© ERIC DRAPER/WHITE HOUSE/UPI/Landov

According to *Forbes* magazine's 2007 list of the richest people in the world, Bill Gates (cofounder of Microsoft Corporation, the world's largest microcomputer software company) was the wealthiest capitalist, with a net worth of $56 billion, down slightly from his $63-billion figure in 2000 as a result of charitable gifts and decreases in the stock market (*Forbes*, 2007). Investor Warren Buffet came in second with $52 billion. The number of billion-dollar fortunes in the United States rose from 129 in 1995 to 371 in 2007 (*Forbes*, 2007). Although some of the men who made the *Forbes* list of the wealthiest people have gained their fortunes through entrepreneurship or being CEOs of large corporations, women who made the list have acquired their wealth typically through inheritance, marriage, or both. In 2007, only eleven women were heads of Fortune 500 companies.

The Managerial Class People in the managerial class have substantial control over the means of production and over workers. However, these upper-level managers, supervisors, and professionals typically do not participate in key corporate decisions such as how to invest profits. Lower-level managers may have some control over employment practices, including the hiring and firing of some workers.

Top professionals such as physicians, attorneys, accountants, and engineers may control the structure of their own work; however, they typically do not own the means of production and may not have supervisory authority over more than a few people. Even so, they may influence the organization of work and the treatment of other workers. Members of the capitalist class often depend on these professionals for their specialized knowledge.

The Small-Business Class This class consists of small-business owners and craftspeople who may hire a small number of employees but largely do their own work. Some members own businesses such as "mom and pop" grocery stores, retail clothing stores, and jewelry stores. Others are doctors and lawyers who receive relatively high incomes from selling their own services. Some of these professionals now share attributes with members of the capitalist class because they have formed corporations that hire and control the employees who produce profits for the professionals.

After founding Microsoft and becoming one of the world's richest people, Bill Gates has devoted recent years to the foundation that he began with his wife, Melinda, to combat poverty and disease in Africa.

It is in the small-business class that we find many people's hopes of achieving the American Dream. Recent economic trends, including corporate downsizing, telecommuting, and the movement of jobs to other countries, have encouraged more people to think about starting their own business. As a result, more people today are self-employed or own a small business than at any time in the past (U.S. Department of Labor, 2003). More women of all races and people of color are in the small-business class than was true previously. According to recent statistics, for example, women own 34 percent of all small businesses. However, gaps in revenues persist between businesses owned by women and those owned by men, with women-owned businesses earning on average about 40 percent less than businesses owned by men (U.S. Department of Labor, 2003).

Throughout U.S. history, immigrants and people of color have owned small businesses (Butler, 1991), seeing such enterprises as a way to achieve the American Dream. Over the past decade, the number of businesses owned by subordinate-group members has increased dramatically, but the share of such businesses owned by people of color is still not proportionate to their numbers in the overall population. African Americans make up more than 12 percent of the U.S. population but own less than 4 percent of businesses; Latinos/as make up more than 13 percent of the population yet own slightly more than 4 percent of all businesses. Asian Americans are closest to being proportional in business ownership: They constitute about 3.6 percent of the population and own about 3.5 percent of all U.S. businesses (U.S. Department of Labor, 2003).

The Working Class The working class is made up of a number of subgroups, one of which is blue-collar workers, some of whom are highly skilled and well paid and others of whom are unskilled and poorly paid. Skilled blue-collar workers include electricians, plumbers, and carpenters; unskilled blue-collar workers include janitors and gardeners.

Many immigrants believe that they can achieve the American Dream by starting a small business. San Francisco's Chinatown contains many stores like the ones shown here.

© Michael Newman/PhotoEdit

Skilled laborers and tradespeople sometimes take another approach to small business by offering services such as landscaping, plumbing, carpentry, and, as seen here, mobile dog grooming.

White-collar workers are another subgroup of the working class. Referred to by some as a "new middle class," these workers are actually members of the working class because they do not own the means of production, do not control the work of others, and are relatively powerless in the workplace. Secretaries, other clerical workers, and salesworkers are members of the white-collar faction of the working class. They take orders from others and tend to work under constant supervision. Thus, these workers are at the bottom of the class structure in terms of domination and control in the workplace. The working class contains about half of all employees in the United States.

Although Marxian and Weberian models of the U.S. class structure show differences in people's occupations and access to valued resources, neither fully reflects the nature and extent of inequality in the United States. In the next section, we will take a closer look at the unequal distribution of income and wealth in the United States and the effects of inequality on people's opportunities and life chances.

Inequality in the United States

Throughout human history, people have argued about the distribution of scarce resources in society. Disagreements often center on whether the share we get is a fair reward for our effort and hard work. Recently, social analysts have pointed out that (except during temporary economic downturns) the old maxim "the rich get richer" continues to be valid in the United States. To understand how this happens, we must take a closer look at the distribution of income and wealth in this country.

Distribution of Income and Wealth

Money is essential for acquiring goods and services. People without money cannot purchase food, shelter, clothing, medical care, legal aid, education, and the other things they need or desire. Money—in the form of both income and wealth—is very unevenly distributed in the United States. Median household income varies widely from one state to another, for example (see ▶ Map 7.1). Among prosperous nations, the United States is number one in inequality of income distribution (Rothchild, 1995).

Income Inequality *Income* **is the economic gain derived from wages, salaries, income transfers (governmental aid), and ownership of property** (Beeghley, 2000). Or, to put it another way, "income refers to money, wages, and payments that periodically are received as returns for an occupation or investment" (Kerbo, 2000: 19).

Sociologist Dennis Gilbert (2003) compares the distribution of income to a national pie that has been sliced into portions, ranging in size from stingy to generous, for distribution among seg-

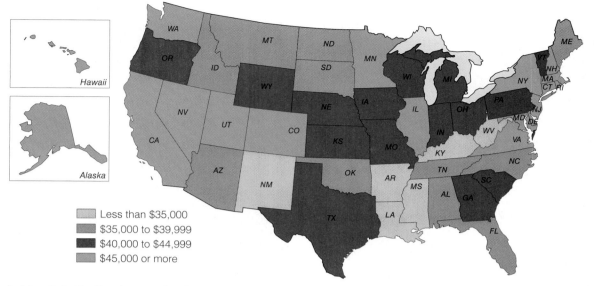

▶ **Map 7.1 Median Income by State**

What factors contribute to the uneven distribution of income in the United States?

Source: U.S. Census Bureau, 2008.

Legend:
- Less than $35,000
- $35,000 to $39,999
- $40,000 to $44,999
- $45,000 or more

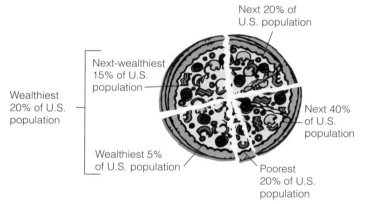

▶ **Figure 7.4 Distribution of Pretax Income in the United States**

Thinking of personal income in the United States (before taxes) as a large pizza helps us to see which segments of the population receive the largest and smallest portions. What part do taxes play in redistributing parts of the pizza?

Source: U.S. Census Bureau, 2008.

ments of the population. As shown in ▶ Figure 7.4, in 2006 the wealthiest 20 percent of households received almost 50 percent of the total income "pie" while the poorest 20 percent of households received less than 4 percent of all income. The top 5 percent *alone* received more than 20 percent of all income—an amount greater than that received by the bottom 40 percent of all households (Bucks, Kennickell, and Moore, 2006).

In the last two decades of the twentieth century and the first few years of the twenty-first century, the gulf between the rich and the poor widened in

the United States. Since the early 1990s, the poor have been more likely to stay poor, and the affluent have been more likely to stay affluent. Between 1994 and 2004, the income of the top one-fifth of U.S. families increased by more than 42 percent; during that same period of time, the income of the bottom one-fifth

income the economic gain derived from wages, salaries, income transfers (governmental aid), and ownership of property.

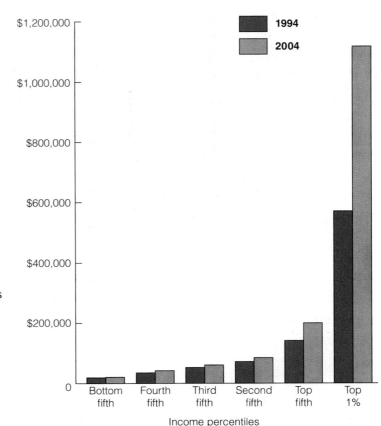

▶ **Figure 7.5 Average After-Tax Family Income in the United States**
As contrasted with Figure 7.4, this chart shows the distribution of after-tax family income in the United States. Notice the dramatic increase in income for the top 1 percent of U.S. families. During the past decade the difference in income between the richest and poorest people in this nation has become even more pronounced.

Source: Tax Policy Center, 2007.

of families increased by only 16 percent (Tax Policy Center, 2007) (see ▶ Figure 7.5).

Income distribution varies by race/ethnicity as well as class. ▶ Figure 7.6 compares median household income by race/ethnicity, showing not only the disparity among groups but also the consistency of that disparity over the last decade. Although households across racial/ethnic categories have experienced some increase in real annual median income, the gap between African American households and white and Asian and Pacific Islanders is particularly striking. In 2006, African American households had the lowest median income, $31,969, which was 61 percent of the median for non-Hispanic white households, $52,423. By contrast, Asian households had the highest median, $64,238, which was about 123 percent of the median for non-Hispanic white households. Median income for Hispanic households was $37,781, which was 72 percent of the median for non-Hispanic white households (DeNavas-Walt, Proctor, and Smith, 2007).

Wealth Inequality Income is only one aspect of wealth. Wealth includes property such as buildings, land, farms, houses, factories, and cars, as well as other assets such as bank accounts, corporate stocks, bonds, and insurance policies. Wealth is computed by subtracting all debt obligations and converting the remaining assets into cash (U.S. Congress, 1986). For most people in the United States, wealth is invested primarily in property that generates no income, such as a home or car. By contrast, the wealth of an elite minority is often in the form of income-producing property.

To see how wealth inequality has increased in recent decades, let's compare two studies. A study by the Joint Economic Committee of Congress divided the population into four categories: (1) the super-rich (0.5 percent of households), who own 35 percent of the nation's wealth, with net assets averaging almost $9 million; (2) the very rich (the next 0.5 percent of households), who own about 7 percent of the nation's wealth, with net assets ranging from

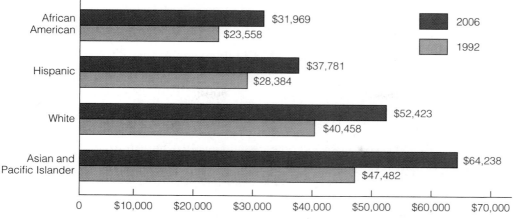

▶ **Figure 7.6** **Median Household Income by Race/Ethnicity in the United States**
Note: Amounts shown in constant dollars.

Sources: DeNavas-Walt, Cleveland, and Webster, 2003; DeNavas-Walt; Proctor, and Lee, 2007.

$1.4 million to $2.5 million; (3) the rich (9 percent of households), who own 30 percent of the wealth, with net assets of a little over $400,000; and (4) everybody else (the bottom 90 percent), who own about 28 percent of the nation's wealth. However, by 1995, another study indicated that the holdings of super-rich households had risen from 35 percent to almost 40 percent of all assets in the nation (stocks, bonds, cash, life insurance policies, paintings, jewelry, and other tangible assets) (Rothchild, 1995).

For the upper class, wealth often comes from interest, dividends, and inheritance (Haseler, 2000). One analysis of the Forbes 400 list of the wealthiest U.S. citizens found that 42 percent of the people on that list had inherited sufficient wealth to put them on the list (Gilbert, 2003). Inheritors are often three or four generations removed from individuals who amassed the original wealth (Odendahl, 1990). After inheriting a fortune, John D. Rockefeller, Jr., stated that "I was born into [wealth] and there was nothing I could do about it. It was there, like air or food or any other element. The only question with wealth is what to do with it" (qtd. in Glastris, 1990: 26).

Disparities in wealth are more pronounced when compared across racial and ethnic categories. According to the Census Bureau, the net worth of the average white household in 2000 was more than ten times that of the average African American household and more than eight times that of the average Latina/o household (Orzechowski and Sepielli, 2001). Married couples have a higher net worth than the unmarried, and households headed by people age 55 and older are wealthier than those headed by younger persons (Orzechowski and Sepielli, 2001).

Consequences of Inequality

Income and wealth are not simply statistics; they are intricately related to the American Dream and our individual life chances. Persons with a high income or substantial wealth have more control over their own lives. They have greater access to goods and services; they can afford better housing, more education, and a wider range of medical services. Persons with less income, especially those living in poverty, must spend their limited resources to acquire the basic necessities of life.

Physical Health, Mental Health, and Nutrition

People who are wealthy and well educated and who have high-paying jobs are much more likely to be healthy than are poor people. As people's economic status increases, so does their health status. The poor have shorter life expectancies and are at greater risk for chronic illnesses such as diabetes, heart disease, and cancer, as well as infectious diseases such as tuberculosis.

What Keeps the American Dream Alive?

Although the American Dream of rags to riches may be an illusive goal for many people, most of us still believe that a person in the United States can get ahead by gaining a good education, through hard work, by marketing a creative idea, by winning the lottery, or by some other means. Whether or not they can ultimately rise to the top economic and social tiers of society, many people still strive to attain their personal—although perhaps scaled down—version of the American Dream.

Some sociological perspectives suggest that vast inequalities between the rich and the poor create such a large divide that upward mobility is virtually impossible for those in the lower economic tiers of society. However, other perspectives are based on the assumption that human capital—in the form of education, hard work, and outstanding achievement—can help a person move up the socioeconomic ladder.

Regardless of which perspective you or I might subscribe to, millions of people in the United States and around the world see this country as the land in which dreams can come true and in which a person can create a better life for his or her family.

As you view the pictures on these two pages, think about the ways in which various people seek out their own American Dream. Doing so helps us gain a better understanding of some of the issues relating to social stratification

Quick and easy ways to attain the American Dream—such as winning a very large lottery drawing—have great appeal to many people in the United States and throughout the world. However, despite the widespread publicity that winners receive, only a very small fraction of those persons who attain the American Dream of wealth do so through lotteries or gambling.

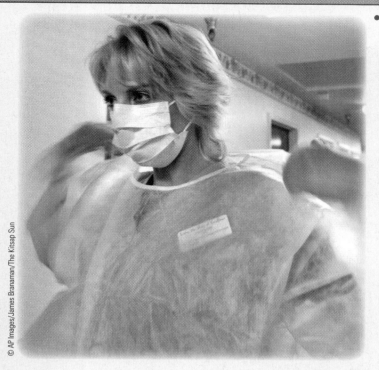

© AP Images/James Branaman/The Kitsap Sun

Long Hours and Hard Work

This single mother with four children seeks the American Dream by working three different jobs as a practical nurse while attending college. For some people, getting ahead requires 24/7 commitment, but she believes that it will be worth the effort to become a registered nurse and earn better pay with fewer hours than she now works.

Small Businesses and the American Dream

We often think of "big-ticket" entrepreneurs such as Bill Gates or Michael Dell as having achieved the American Dream. However, sidewalk vendors who own their own business may believe that they, too, have achieved their dream, especially when they come from nations where no similar dream would have even been possible.

© Frances M. Roberts/Alamy

Social Interaction and Resilience

From homeless person to millionaire stockbroker sounds like the plot line of a movie, which it is: *The Pursuit of Happyness* (2006), starring Will Smith. But it is also the real-life story of Chris Gardner, chief executive of Gardner Rich LLC, a multimillion-dollar Chicago brokerage firm, on whose autobiography *The Pursuit of Happyness* is based. Although Gardner never attended college, he has been highly successful in his financial endeavors and is now hoping to get investors to help him create a billion-dollar investment fund to promote economic opportunities for South Africans. Here you see him on a visit to the soup kitchen at Glide Memorial Church in San Francisco, California, where he used to eat.

© Liz Hafalia/San Francisco Chronicle/CORBIS

© AP Images/Tim Boyd

The Cost of an Education

For individuals who were not born into affluent families, education is important for attaining the American Dream of upward mobility. Here you see a Latina high school student looking over a federal application for student aid during a conference held at the University of New Hampshire. The university is hosting the conference to encourage minority students to consider attending their school. Students often turn to student aid programs as a source of funding in the hope that they can obtain a college education. What will happen to the American Dream for students such as the woman shown here if such funding is reduced or eliminated in the future?

Reflect & Analyze

1. Since the collapse of the housing boom in 2007–2008, the number of undocumented workers in this country has shrunk dramatically. Do you think that the workers have given up their hope of achieving the American Dream, or have they deferred it?
2. Does the U.S. government have the obligation to keep the American Dream alive for people of lower incomes? Are there alternatives to government-funded student aid programs, for example?
3. Do you know anyone whose live was changed as a result of playing a state lottery? Was the change positive or detrimental to the person's life? What conclusions can you draw from this person's experience?

Turning to Video

Watch the ABC video *India Inc: Economic Explosion* (running time 2:38), available on the Kendall Companion Website and through Cengage Learning eResources accounts. Because of India's growing economy (it's the second-fastest growing economy in the world), there are more middle-class Indians with buying power than the entire U.S. population. Many U.S. jobs are being outsourced to India, and India has even begun hiring out-of-work Americans. As you watch this news report, think about the photographs, commentary, and questions that you encountered in this photo essay. After you've watched the video, consider two more questions: How does another country's growth affect the American Dream, and to what degree is it possible that India will replace the United States as the land of opportunity?

Children born into poor families are at much greater risk of dying during their first year of life. Some die from disease, accidents, or violence. Others are unable to survive because they are born with low birth weight, a condition linked to birth defects and increased probability of infant mortality (Rogers, 1986). Low birth weight in infants is attributed, at least in part, to the inadequate nutrition received by many low-income pregnant women. Most of the poor do not receive preventive medical and dental checkups; many do not receive adequate medical care after they experience illness or injury.

Many high-poverty areas lack an adequate supply of doctors and medical facilities. Even in areas where such services are available, the inability to pay often prevents people from seeking medical care when it is needed. Some "charity" clinics and hospitals may provide indigent patients (those who cannot pay) with minimal emergency care but make them feel stigmatized in the process. For many of the working poor, medical insurance is out of the question. Approximately 47 million people in the United States were without health insurance coverage in 2006—an increase of more than 2 million from the preceding year (DeNavas-Walt, Proctor, and Smith, 2007). Many people rely on their employers for health coverage; however, some employers are cutting back on health coverage, particularly for employees' family members. Despite passage of the 1996 Kassenbaum–Kennedy bill by Congress, which makes insurance more readily available for millions of people who change their jobs or lose them, many unemployed workers and their families remain without medical coverage. However, the uninsured are a changing group in the United States—not everyone who becomes uninsured for a month or more remains uninsured throughout a given year. Of all age groups, persons age 18 to 24 are the most likely to be uninsured; Medicare and other benefit programs provide medical care to most persons 65 and over (DeNavas-Walt, Proctor, and Smith, 2007). As shown in Figure ▶ 7.7, a high percentage of poor persons do not have health insurance.

Many lower-paying jobs are often the most dangerous and have the greatest health hazards. Black lung disease, cancer caused by asbestos, and other environmental hazards found in the workplace

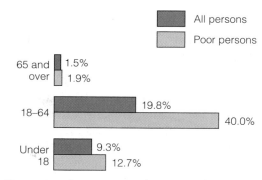

▶ **Figure 7.7 Percentage of U.S. Population Without Health Insurance, 2006**

Sources: Cohen and Martinez, 2007; DeNavas-Walt, Proctor, and Smith, 2007.

are more likely to affect manual laborers and low-income workers, as are job-related accidents.

Although the precise relationship between class and health is not known, analysts suggest that people with higher income and wealth tend to smoke less, exercise more, maintain a healthy body weight, and eat nutritious meals. As a category, more-affluent persons tend to be less depressed and face less psychological stress, conditions that tend to be directly proportional to income, education, and job status (*Mental Medicine*, 1994).

Good health is basic to good life chances; in turn, adequate nutrition is essential for good health. Hunger is related to class position and income inequality. Recent surveys estimate that 13 percent of children under age 12 are hungry or at risk of being hungry. Among the working poor, almost 75 percent of the children are thought to be in this category. After spending 60 percent of their income on housing, low-income families are unable to provide adequate food for their children. Between one-third and one-half of all children living in poverty consume significantly less than the federally recommended guidelines for caloric and nutritional intake (Children's Defense Fund, 2002). Lack of adequate nutrition has been linked to children's problems in school.

Housing As discussed in Chapter 4 ("Social Structure and Interaction in Everyday Life"), homelessness is a major problem in the United States. The lack of affordable housing is a pressing concern for many low-income individuals and families. With the economic prosperity of the 1990s, low-cost housing units in many cities were replaced with expensive condominiums and luxury single-family residences for affluent people. As unemployment rose dramatically beginning in 2001, partly due to terrorism and a faltering economy, housing costs remained high compared to many families' ability to pay for food, shelter, clothing, and other necessities.

Lack of *affordable* housing is one central problem brought about by economic inequality. Another concern is *substandard* housing, which refers to facilities that have inadequate heating, air conditioning, plumbing, electricity, or structural durability. Structural problems—due to faulty construction or lack of adequate maintenance—exacerbate the potential for other problems such as damage from fire, falling objects, or floors and stairways collapsing.

Education Educational opportunities and life chances are directly linked. Some functionalist theorists view education as the "elevator" to social mobility. Improvements in the educational achievement levels (measured in number of years of schooling completed) of the poor, people of color, and white women have been cited as evidence that students' abilities are now more important than their class, race, or gender. From this perspective, inequality in education is declining, and students have an opportunity to achieve upward mobility through achievements at school. Functionalists generally see the education system as flexible, allowing most students

the opportunity to attend college if they apply themselves (Ballantine, 2001).

In contrast, most conflict theorists stress that schools are agencies for reproducing the capitalist class system and perpetuating inequality in society. From this perspective, education perpetuates poverty. Parents with limited income are not able to provide the same educational opportunities for their children as are families with greater financial resources.

Today, great disparities exist in the distribution of educational resources. Because funding for education comes primarily from local property taxes, school districts in wealthy suburban areas generally pay higher teachers' salaries, have newer buildings, and provide state-of-the-art equipment. By contrast, schools in poorer areas have a limited funding base. Students in central-city schools and poverty-stricken rural areas often attend dilapidated schools that lack essential equipment and teaching materials. Author Jonathan Kozol (1991, qtd. in Feagin and Feagin, 1994: 191) documented the effect of a two-tiered system on students:

> Kindergartners are so full of hope, cheerfulness, high expectations. By the time they get into fourth grade, many begin to lose heart. They see the score, understanding they're not getting what others are getting. . . . They see suburban schools on television. . . . They begin to get the point that they are not valued much in our society. By the

© Image Source/Getty Images

Conflict theorists see schools as agents of the capitalist class system that perpetuate social inequality: Upper-class students are educated in well-appointed environments such as the one shown here, whereas children of the poor tend to go to antiquated schools with limited facilities.

time they are in junior high, they understand it. "We have eyes and we can see; we have hearts and we can feel. . . . We know the difference."

Poverty extracts such a toll that many young people will not have the opportunity to finish high school, much less enter college.

Poverty in the United States

When many people think about poverty, they think of people who are unemployed or on welfare. However, many hardworking people with full-time jobs live in poverty. The U.S. Social Security Administration has established an *official poverty line,* **which is based on what is considered to be the minimum amount of money required for living at a subsistence level.** The poverty level is computed by determining the cost of a minimally nutritious diet (a low-cost food budget on which a family could survive nutritionally on a short-term, emergency basis) and multiplying this figure by three to allow for nonfood costs. In 2006, 37 million people lived below the official government poverty level of $20,794 for a family of four, an increase of more than 3 million people in poverty since 2005 (DeNavas-Walt, Proctor, and Smith, 2007). Many of the people below the poverty line hold full-time jobs at low wages (see Box 7.2).

When sociologists define poverty, they distinguish between absolute and relative poverty. *Absolute poverty* **exists when people do not have the means to secure the most basic necessities of life.** This definition comes closest to that used by the federal government. Absolute poverty often has life-threatening consequences, such as when a homeless person freezes to death on a park bench. By comparison, *relative poverty* **exists when people may be able to afford basic necessities but are still unable to maintain an average standard of living.** A family must have income substantially above the official poverty line in order to afford the basic necessities, even when these are purchased at the lowest possible cost. At about 155 percent of the official poverty line, families could live on an economy budget. What is it like to live on the economy budget? John

Schwarz and Thomas Volgy (1992: 43) offer the following distressing description:

> Members of families existing on the economy budget never go out to eat, for it is not included in the food budget; they never go out to a movie, concert, or ball game or indeed to any public or private establishment that charges admission, for there is no entertainment budget; they have no cable television, for the same reason; they never purchase alcohol or cigarettes; never take a vacation or holiday that involves any motel or hotel or, again, any meals out; never hire a babysitter or have any other paid child care; never give an allowance or other spending money to the children; never purchase any lessons or home-learning tools for the children; never buy books or records for the adults or children, or any toys, except in the small amounts available for birthday or Christmas presents ($50 per person over the year); never pay for a haircut; never buy a magazine; have no money for the feeding or veterinary care of any pets; and, never spend any money for preschool for the children, or educational trips for them away from home, or any summer camp or other activity with a fee.

Who Are the Poor?

Poverty in the United States is not randomly distributed, but rather is highly concentrated according to age and race/ethnicity, as indicated in ◆ Table 7.1, as well as gender. See page 241.

---

official poverty line the federal income standard that is based on what is considered to be the minimum amount of money required for living at a subsistence level.

absolute poverty a level of economic deprivation that exists when people do not have the means to secure the most basic necessities of life.

relative poverty a condition that exists when people may be able to afford basic necessities but are still unable to maintain an average standard of living.

Box 7.2 Sociology and Social Policy

Should Our Laws Guarantee People a Living Wage?

One [of the most surprising things I learned] is just how difficult it is, how stressful it is to live check to check. It was an incredible strain on my relationship with [my fiancée] Alex. Suddenly we were exhausted when we were around each other. We had no energy to really give to one another. We were so tired at the end of the day. We ate dinner together, and then we were just done. You know, you see how the quality of your life devoted to relationships can really deteriorate quickly. One thing we talk about on [FX Network's *30 Days*] is that it's no surprise that families that make less than $25,000 a year are twice as likely to get divorced as a family that makes $50,000 a year.

—Morgan Spurlock, the producer and director of *Super Size Me,* discussing what it was like when he and his fiancée tried working for and living on minimum wage for thirty days (qtd. in Campus Progress, 2006)

At some point in our lives, most of us have held a job paying the minimum wage, and we know the limitations of trying to survive on such low earnings. The federal *minimum wage* is the hourly rate that (with certain exceptions) is the lowest amount an employer can legally pay its employees (each state may adopt a higher minimum wage, but not a lower one). In 2007 the minimum wage was set at $5.85 per hour, and the rate will increase to $6.55 per hour in July of 2008 and to $7.25 per hour in July of 2009. Although that represents a substantial increase, a person earning minimum wage and working forty hours every week, fifty-two weeks per year (in other words, no time off, no vacation) would still earn only $15,080 per year—an amount just slightly above the *official poverty line* (and slightly below that line for a person with two children). The low hourly rate paid by many employers to their employees and the high compensation "packages" received by many companies' chief executive officers—who often earn nearly 400 times as much money as their employees (Mintz, 2007)—are one of the major causes of social inequality in this country.

Morgan Spurlock and other social analysts have called our attention to the fact that living at or below the poverty line (the minimum amount of money required for living at a subsistence level) can be difficult and stressful. According to some analysts, we should do away with the idea of a minimum wage (a "poverty wage," as some describe it) and instead focus on a minimum *living wage*—a wage sufficient, based on a forty-hour week, to provide the necessities and comforts essential to an acceptable standard of living in the community in which the individual resides.

Proponents of a living wage assert that it is not only more humane (eliminating the necessity of many low-wage workers to hold more than one job in order to "make ends meet" and reducing the number of families living in poverty) but also that it makes good sense financially: A living wage would lower taxes by reducing the amount of money the government must pay for services provided to poverty-level families such as food stamps, emergency medical treatment, and low-income housing. Critics of the living wage argue that a living wage requirement—or, for that matter, any increase in the federal minimum wage—causes inflation, increases the cost of living for everyone, and increases unemployment because employers are not able to afford this added cost of doing business.

Reflect & Analyze

What do you think? Should your city, or your state, or the federal government require that employers pay their employees, at the very least, a living wage? Why or why not?

Age Today, children are at a much greater risk of living in poverty than are older persons. A generation ago, persons over age 65 were at the greatest risk of being poor; however, government programs such as Social Security and pension plans have been indexed for inflation and thus provide for something closer to an adequate standard of living than do other social welfare programs. Even so, older women are twice as likely to be poor as older men; older African Americans and Latinos/as are much

◆ **Table 7.1 Percentage Distribution of Poverty in the United States**

	All Races[a]	White[b]	African American	Asian American	Hispanic[c]
By Age					
Under 18 years	17.8	10.5	33.6	10.0	28.9
18–24 years	18.1	14.5	28.1	17.9	22.6
25–44	11.2	7.8	20.2	8.1	18.4
45–64	8.8	7.0	16.8	7.9	14.4
65 and above	9.8	7.5	23.9	13.6	18.7
By Education					
No high school diploma	21.8	15.7	34.8	15.8	26.7
4 years of high school	11.9	9.4	22.0	11.3	15.4
Some college (no degree)	8.5	7.0	14.9	11.5	10.6
College degree or more	4.3	3.7	7.1	6.1	7.5

[a]Includes other races/ethnicities not shown separately.
[b]Non-Hispanic white.
[c]Includes Hispanic persons of any race.
Source: U.S. Census Bureau, 2008.

more likely to live below the poverty line than are non-Latino/a whites.

The age category most vulnerable to poverty today is the very young. One out of every three persons below the poverty line is under 18 years of age, and a large number of children hover just above the official poverty line. The precarious position of African American and Latino/a children is even more striking. In 2006, 33.0 percent of all African Americans under age 18 lived in poverty; 26.9 percent of Latino/a children were also poor, as compared with 10.0 percent of non-Latino/a white children (DeNavas-Walt, Proctor, and Smith, 2007).

What do such statistics indicate about the future of our society? Children as a group are poorer now than they were at the beginning of the 1980s, whether they live in one- or two-parent families. The majority live in two-parent families in which one or both parents are employed. However, children in single-parent households headed by women have a much greater likelihood of living in poverty: Approximately 29 percent of white (non-Latino/a) children under age 18 in female-headed households live below the poverty line, as sharply contrasted with about 50 percent of Latina/o and 49 percent

of African American children in the same category (DeNavas-Walt, Proctor, and Mills, 2004). Nor does the future look bright: Many governmental programs established to alleviate childhood poverty and malnutrition have been seriously cut back or eliminated altogether.

Gender About two-thirds of all adults living in poverty are women. In 2006, single-parent families headed by women had a 28-percent poverty rate as compared with a 5-percent rate for two-parent families. Sociologist Diana Pearce (1978) coined a term to describe this problem: The *feminization of poverty* **refers to the trend in which women are disproportionately represented among individuals living in poverty.** According to Pearce (1978), women have a higher risk of being poor because they bear the major economic and emotional burdens of raising children when they are single heads

feminization of poverty the trend in which women are disproportionately represented among individuals living in poverty.

of households but earn between 70 and 80 cents for every dollar a male worker earns. More women than men are unable to obtain regular, full-time, year-round employment, and lack of adequate, affordable day care exacerbates this problem.

Does the feminization of poverty explain poverty in the United States today? Is poverty primarily a women's issue? On the one hand, this thesis highlights a genuine problem—the link between gender and poverty. On the other hand, several major problems exist with this argument. First, women's poverty is not a new phenomenon. Women have always been more vulnerable to poverty (see Katz, 1989). Second, all women are not equally vulnerable to poverty. Many in the upper and upper-middle classes have the financial resources, education, and skills to support themselves regardless of the presence of a man in the household. Third, event-driven poverty does not explain the realities of poverty for many women of color, who instead may experience "reshuffled poverty"—a condition of deprivation that follows them regardless of their marital status or the type of family in which they live. Research by Mary Jo Bane (1986; Bane and Ellwood, 1994) demonstrates that two out of three African American families headed by a woman were poor before the family event that made the woman a single mother. In addition, the poverty risk for a two-parent African American family is more than twice that for a white two-parent family.

Finally, poverty is everyone's problem, not just women's. When women are impoverished, so are their children. Moreover, many of the poor in our society are men, especially the chronically unemployed, older persons, the homeless, persons with disabilities, and men of color who have spent their adult lives without hope of finding work.

Race/Ethnicity According to some stereotypes, most of the poor and virtually all welfare recipients are people of color. However, this stereotype is false; white Americans (non-Latinos/as) account for approximately two-thirds of those below the official poverty line. However, such stereotypes are perpetuated because a disproportionate percentage of the impoverished in the United States are made up of African Americans, Latinos/as, and Native Americans. About 24 percent of African Americans and

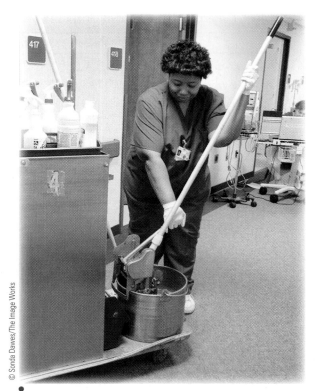

© Sonda Dawes/The Image Works

Many women are among the "working poor," who, although employed full time, have jobs in service occupations that are typically lower paying and less secure than jobs in other sectors of the labor market. Does the nature of women's work contribute to the feminization of poverty in the United States?

21 percent of Latinas/os were among the officially poor in 2006 as compared with about 8 percent of non-Latino/a whites (DeNavas-Walt, Proctor, and Smith, 2007). Native Americans are among the most severely disadvantaged persons in the United States. About one-third live below the poverty line, and some of these individuals live in conditions of extreme poverty.

Economic and Structural Sources of Poverty

Social inequality and poverty have both economic and structural sources. The low wages paid for many jobs is the major cause: Half of all families living in poverty are headed by someone who is employed, and one-third of those family heads work full time.

A person with full-time employment in a minimum-wage job cannot keep a family of four from sinking below the official poverty line.

Structural problems contribute to both unemployment and underemployment. Corporations have been disinvesting in the United States, displacing millions of people from their jobs. Economists refer to this displacement as the *deindustrialization* of America (Bluestone and Harrison, 1982). Even as they have closed their U.S. factories and plants, many corporations have opened new facilities in other countries where "cheap labor" exists because people will, of necessity, work for lower wages. *Job deskilling—a reduction in the proficiency needed to perform a specific job that leads to a corresponding reduction in the wages for that job*—has resulted from the introduction of computers and other technology (Hodson and Parker, 1988). The shift from manufacturing to service occupations has resulted in the loss of higher-paying positions and their replacement with lower-paying and less-secure positions that do not offer the wages, job stability, or advancement potential of the disappearing manufacturing jobs. Many of the new jobs are located in the suburbs, thus making them inaccessible to central-city residents.

The problems of unemployment, underemployment, and poverty-level wages are even greater for people of color and young people in declining central cities. The unemployment rate for African Americans is almost double that of whites (U.S. Bureau of Labor Statistics, 2007b). African Americans have also experienced gender differences in employment that may produce different types of economic vulnerability. African American men who find employment typically earn more than African American women; however, the men's employment is often less secure. In the past decade, African American men have been more likely to lose their jobs because of declining employment in the manufacturing sector (Bane, 1986; Collins, 1990; Bane and Ellwood, 1994).

Solving the Poverty Problem

The United States has attempted to solve the poverty problem in several ways. One of the most enduring is referred to as social welfare. When most people think of "welfare," they think of food stamps

© AP Images/Paul Beaty

This Illinois Link card represents a modern approach to helping people of limited income purchase groceries. Data-encoded cards such as this one were developed to prevent the trading or selling of traditional food stamps.

and programs such as Temporary Assistance for Needy Families (TANF) or the earlier program it replaced, Aid to Families with Dependent Children (AFDC). However, the primary beneficiaries of social welfare programs are not poor. Some analysts estimate that approximately 80 percent of all social welfare benefits are paid to people who do not qualify as "poor." For example, many recipients of Social Security are older people in middle- and upper-income categories.

When older persons, including members of Congress, accept Social Security payments, they are not

job deskilling a reduction in the proficiency needed to perform a specific job that leads to a corresponding reduction in the wages for that job.

Sociology *Works!*

Reducing Structural Barriers to Achieving the American Dream

Our society recognizes a moral obligation to provide a helping hand to those in need, but those in poverty have been getting only the back of the hand. They receive little or no public assistance. Instead, they are scolded and told that they have caused their own misfortunes. This is our "compassion gap"—a deep divide between our moral commitments and how we actually treat those in poverty.

In this statement the sociologists Fred Block, Anna C. Korteweg, and Kerry Woodward (2008: 166) describe the contradiction between our nation's alleged moral commitment to alleviating poverty and how we actually treat people who live in poverty. Children, single mothers with children, and people of color (particularly African Americans and Latinos/as) make up a disproportionate segment of the nation's poorest groups, and individuals in these categories are the persons most disadvantaged by arguments asserting that the poor have no one but themselves to blame for their poverty.

Numerous sociological studies regarding wealth and poverty demonstrate how structural factors contribute to the ability of some individuals to achieve the American Dream whereas others are hampered in achieving that goal by factors that are beyond their control. Yet, according to the Economic Mobility Project, policy makers do little to alleviate poverty in the United States because of the widely held belief that the American Dream should provide everyone with equality of *opportunity* but not necessarily equality of *outcome:* "The belief in America as a land of opportunity may also explain why rising inequality in the United States has yielded so little in terms of responsiveness from policy makers: if the American Dream is alive and well, then there is no need for government intervention to smooth the rough edges of capitalism. Diligence and skill, the argument goes, will yield a fair distribution of rewards" (Pew Charitable Trusts, 2007). However, research continues to reveal that structural factors beyond the control of individuals are important in determining where a person's place will be in the U.S. class system.

Can we keep the American Dream alive for all? According to Block, Korteweg, and Woodward (2008), we must take a number of specific steps to revitalize the American Dream for more people in this country and to reverse the compassion gap. We must make people aware of how far social reality has departed from the ideals of the American Dream. As a nation, we must also take action to deal with the costs of four critical services that have risen much more rapidly than people's wages and the rate of inflation. These four critical services are *health care, higher education, high-quality child care,* and *housing.* As sociologists and other social analysts have suggested, if we as a nation are to claim that we have a commitment to compassion, we must make it our collective responsibility to help remove the structural barriers that currently reduce opportunities, mobility, and a chance for a better way of life for millions of Americans: "True compassion requires that we build a society in which every person has a first chance, a second chance, and, if needed, a third and fourth chance, to achieve the American Dream. We . . . need to use every instrument we have—faith groups, unions, community groups, and most of all government programs—to address the structural problems that reproduce poverty in our affluent society" (Block, Korteweg, and Woodward, 2008: 175).

Reflect & Analyze

Consider the area where you live. Do you know of people there whose situations could be improved if they were given a greater opportunity to take charge of their lives? How could such a change happen?

stigmatized. Similarly, veterans who receive benefits from the Veterans Benefits Administration are not viewed as "slackers," and farmers who profit because of price supports are not considered to be lazy and unwilling to work. Unemployed workers who receive unemployment compensation are viewed with sympathy because of the financial plight of their families. By contrast, poor women and children who receive minimal benefits from welfare programs tend to be stigmatized and sometimes humiliated, even when our nation describes itself as having compassion for the less fortunate (see "Sociology Works!").

Sociological Explanations of Social Inequality in the United States

Obviously, some people are disadvantaged as a result of social inequality. Therefore, is inequality always harmful to society?

Functionalist Perspectives

According to the sociologists Kingsley Davis and Wilbert Moore (1945), inequality is not only inevitable but also necessary for the smooth functioning of society. The Davis–Moore thesis, which has become the definitive functionalist explanation for social inequality, can be summarized as follows:

1. All societies have important tasks that must be accomplished and certain positions that must be filled.
2. Some positions are more important for the survival of society than others.
3. The most important positions must be filled by the most qualified people.
4. The positions that are the most important for society and that require scarce talent, extensive training, or both must be the most highly rewarded.
5. The most highly rewarded positions should be those that are functionally unique (no other position can perform the same function) and on which other positions rely for expertise, direction, or financing.

Davis and Moore use the physician as an example of a functionally unique position. Doctors are very important to society and require extensive training, but individuals would not be motivated to go through years of costly and stressful medical training without incentives to do so. The Davis–Moore thesis assumes that social stratification results in *meritocracy*—**a hierarchy in which all positions are rewarded based on people's ability and credentials.**

Critics have suggested that the Davis–Moore thesis ignores inequalities based on inherited wealth and intergenerational family status (Rossides, 1986). The thesis assumes that economic rewards and prestige are the only effective motivators for people and fails to take into account other intrinsic aspects of work, such as self-fulfillment (Tumin, 1953). It also does not adequately explain how such a reward system guarantees that the most qualified people will gain access to the most highly rewarded positions.

Conflict Perspectives

From a conflict perspective, people with economic and political power are able to shape and distribute the rewards, resources, privileges, and opportunities in society for their own benefit. Conflict theorists do not believe that inequality serves as a motivating force for people; they argue that powerful individuals and groups use ideology to maintain their favored positions at the expense of others. Core values in the United States emphasize the importance of material possessions, hard work, individual initiative to get ahead, and behavior that supports the existing social structure. These same values support the prevailing resource distribution system and contribute to social inequality.

Are wealthy people smarter than others? According to conflict theorists, certain stereotypes suggest that this is the case; however, the wealthy may actually be "smarter" than others only in the sense of having "chosen" to be born to wealthy parents from whom they could inherit assets. Conflict theorists also note that laws and informal social norms support inequality in the United States. For the first half of the twentieth century, both legalized and institutionalized segregation and discrimination reinforced employment discrimination and produced higher levels of economic inequality. Although laws

meritocracy a hierarchy in which all positions are rewarded based on people's ability and credentials.

have been passed to make these overt acts of discrimination illegal, many forms of discrimination still exist in educational and employment opportunities.

Symbolic Interactionist Perspectives

Symbolic interactionists focus on microlevel concerns and usually do not analyze larger structural factors that contribute to inequality and poverty. However, many significant insights on the effects of wealth and poverty on people's lives and social interactions can be derived from applying a symbolic interactionist approach. Using qualitative research methods and influenced by a symbolic interactionist approach, researchers have collected the personal narratives of people across all social classes,

ranging from the wealthiest to the poorest people in the United States.

A few studies provide rare insights into the social interactions between people from vastly divergent class locations. Sociologist Judith Rollins's (1985) study of the relationship between household workers and their employers is one example. Based on in-depth interviews and participant observation, Rollins examined rituals of deference that were often demanded by elite white women of their domestic workers, who were frequently women of color. According to the sociologist Erving Goffman (1967), *deference* is a type of ceremonial activity that functions as a symbolic means whereby appreciation is regularly conveyed to a recipient. In fact, deferential behavior between nonequals (such as employers and employees) confirms the inequality of the relationship and each party's position in

According to a functionalist perspective, people such as these Harvard Law School graduates attain high positions in society because they are the most qualified and they work the hardest. Is our society a meritocracy? How would conflict theorists answer this question?

the relationship relative to the other. Rollins identified three types of linguistic deference between domestic workers and their employers: use of the first names of the workers, contrasted with titles and last names (Mrs. Adams, for example) of the employers; use of the term *girls* to refer to female household workers regardless of their age; and deferential references to employers, such as "Yes, ma'am." Spatial demeanor, including touching and how close one person stands to another, is an additional factor in deference rituals across class lines. Rollins (1985: 232) concludes that

> The employer, in her more powerful position, sets the essential tone of the relationship; and that tone . . . is one that functions to reinforce the inequality of the relationship, to strengthen the employer's belief in the rightness of her advantaged class and racial position, and to provide her with justification for the inegalitarian social system.

Many concepts introduced by the sociologist Erving Goffman (1959, 1967) could be used as springboards for examining microlevel relationships between inequality and people's everyday interactions. What could you learn about class-based inequality in the United States by using a symbolic interactionist approach to examine a setting with which you are familiar?

The Concept Quick Review summarizes the three major perspectives on social inequality in the United States.

U.S. Stratification in the Future

Will social inequality in the United States increase, decrease, or remain the same in the future? Many social scientists believe that existing trends point to an increase. First, the purchasing power of the dollar has stagnated or declined since the early 1970s. As families started to lose ground financially, more family members (especially women) entered the labor force in an attempt to support themselves and their families (Gilbert, 2003). Economist and former Secretary of Labor Robert Reich (1993) has noted that in recent years the employed have been traveling on two escalators—one going up and the other going down. The gap between the earnings of workers and the income of managers and top executives has widened.

Second, wealth continues to become more concentrated at the top of the U.S. class structure. As the rich have grown richer, more people have found themselves among the ranks of the poor. Third, federal tax laws in recent years have benefited corporations and wealthy families at the expense of middle- and lower-income families. Finally, structural sources of upward mobility are shrinking, whereas the rate of downward mobility has increased.

Are we sabotaging our future if we do not work constructively to eliminate poverty? It has been said that a chain is no stronger than its weakest link. If we apply this idea to the problem of poverty, then it is to our advantage to see that those who cannot find

CONCEPT QUICK REVIEW

Sociological Explanations of Social Inequality in the United States	
Functionalist perspectives	Some degree of social inequality is necessary for the smooth functioning of society (in order to fill the most important positions) and thus is inevitable.
Conflict perspectives	Powerful individuals and groups use ideology to maintain their favored positions in society at the expense of others, and wealth is not necessary in order to motivate people.
Symbolic interactionist perspectives	The beliefs and actions of people reflect their class location in society.

Box 7.3 You Can Make a Difference

Feeding the Hungry

The great fear among us all is that we are going to have to feed even more people. . . . It's not enough to just hand food out anymore.

—Robert Egger, director of the nonprofit Central Kitchen in Washington, D.C. (qtd. in Clines, 1996)

Egger is one of the people responsible for an innovative chef's training program that feeds hope as well as hunger. At the Central Kitchen, located in the nation's capital, staff and guest chefs annually train around 48 homeless persons in three-month-long kitchen-arts courses. While the trainees are learning about food preparation, which will help them get starting jobs in the restaurant industry, they are also helping feed about 3,000 homeless persons each day. Much of the food is prepared using donated goods such as turkeys that people have received as gifts at office parties and given to the kitchen, and leftover food from grocery stores including 7-Eleven stores, restaurants such as Pizza Hut, hotel food services, and college cafeterias. Central Kitchen got its start using leftovers from President George H. W. Bush's inaugural banquet in the late 1980s (Clines, 1996). Recently, donated food has gotten a boost from the Good Samaritan law passed in 1996, which exempts nonprofit organizations and *gleaners*—volunteers who collect what is left in the field after harvesting—from liability for problems with food that they contribute in good faith (see Burros, 1996).

Can you think of ways that leftover food could be recovered from places where you eat so the food could be redistributed to persons in need? Have you thought about suggesting that members of an organization to which you belong might donate their time to help the Salvation Army, Red Cross, or other voluntary organization to collect, prepare, and serve food to others? If you would like to know more,

- "A Citizens Guide to Food Recovery" is available to help individuals participate in food recovery. Call 800-GLEAN-IT, or call the Salvation Army in your community. On the Internet:
  http://www.salvationarmy.org

- World Hunger Year has projects such as Reinvesting in America that try to end hunger:
  http://www.worldhungeryear.org

- Contact the American Red Cross:
  http://www.redcross.org

© Joe Sohm/The Image Works

Many community volunteers try to make a difference during holiday seasons by providing food for people who otherwise might have none. Serving dinner for the homeless at a Los Angeles mission is an example. However, some programs empower homeless persons by teaching them kitchen arts so that they can prepare food for themselves and for others.

work or do not have a job that provides a living wage receive adequate training and employment. Innovative programs can combine job training with producing something useful to meet the immediate needs of people living in poverty. Children of today—the adults of tomorrow—need nutrition, education, health care, and safety as they grow up (see Box 7.3).

Some social analysts believe that the United States will become a better nation if it attempts to regain the American Dream by attacking poverty. According to the sociologist Michael Harrington (1985: 13), if we join in solidarity with the poor, we will "rediscover our own best selves . . . we will regain the vision of America."

Chapter Review

● What is social stratification, and how does it affect our daily life?

Social stratification is the hierarchical arrangement of large social groups based on their control over basic resources. People are treated differently based on where they are positioned within the social hierarchies of class, race, gender, and age.

● What are the major systems of stratification?

Stratification systems include slavery, caste, and class. Slavery, an extreme form of stratification in which people are owned by others, is a closed system. The caste system is also a closed one in which people's status is determined at birth based on their parents' position in society. The class system, which exists in the United States, is a type of stratification based on ownership of resources and on the type of work that people do.

● How did classical sociologists such as Karl Marx and Max Weber view social class?

Karl Marx and Max Weber acknowledged social class as a key determinant of social inequality and social change. For Marx, people's relationship to the means of production determines their class position. Weber developed a multidimensional concept of stratification that focuses on the interplay of wealth, prestige, and power.

● What are some of the consequences of inequality in the United States?

The stratification of society into different social groups results in wide discrepancies in income and wealth and in variable access to available goods and services. People with high income or wealth have greater opportunity to control their own lives. People with less income have fewer life chances and must spend their limited resources to acquire basic necessities.

● How do sociologists view poverty?

Sociologists distinguish between absolute poverty and relative poverty. Absolute poverty exists when people do not have the means to secure the basic necessities of life. Relative poverty exists when people may be able to afford basic necessities but still are unable to maintain an average standard of living.

● Who are the poor?

Age, gender, and race tend to be factors in poverty. Children have a greater risk of being poor than do the elderly, and women have a higher rate of poverty than do men. Although whites account for approximately two-thirds of those below the poverty line, people of color account for a disproportionate share of the impoverished in the United States.

● What is the functionalist view on class?

Functionalist perspectives view classes as broad groupings of people who share similar levels of privilege on the basis of their roles in the occupational structure. According to the Davis–Moore thesis, stratification exists in all societies, and some inequality is not only inevitable but also necessary for the ongoing functioning of society. The positions that are most important within society and that require the most talent and training must be highly rewarded.

• What is the conflict view on class?

Conflict perspectives on class are based on the assumption that social stratification is created and maintained by one group (typically the capitalist class) in order to enhance and protect its own economic interests. Conflict theorists measure class according to people's relationships with others in the production process.

• What is the symbolic interactionist view on class?

Unlike functionalist and conflict perspectives that focus on macrolevel inequalities in societies, symbolic interactionist views focus on microlevel inequalities such as how class location may positively or negatively influence one's identity and everyday social interactions. Symbolic interactionists use terms such as "social cohesion" and "deference" to explain how class binds some individuals together while categorically separating out others.

www.cengage.com/login

Register for a Student eResource account to maximize your study time online using CengageNOW. First take the system's diagnostic pre-test, and then follow the personalized study plan that is created for you to help you review this chapter. The study plan will

- help you identify areas on which you should concentrate;
- provide interactive exercises to help you master the chapter concepts; and
- provide a post-test to confirm you are ready to move on to the next chapter.

Key Terms

absolute poverty 239

alienation 220

capitalist class (bourgeoisie) 220

caste system 217

class conflict 220

class system 219

feminization of poverty 241

income 230

intergenerational mobility 214

intragenerational mobility 215

job deskilling 243

life chances 214

meritocracy 245

official poverty line 239

pink-collar occupations 225

power 222

prestige 222

relative poverty 239

slavery 215

social mobility 214

social stratification 214

socioeconomic status (SES) 220

underclass 226

wealth 221

working class (proletariat) 220

Questions for Critical Thinking

1. Based on the Weberian and Marxian models of class structure, what is the class location of each of your ten closest friends or acquaintances? What is their location in relationship to yours? To one another's? What does their location tell you about friendship and social class?

2. Should employment be based on meritocracy, need, or affirmative action policies?

3. What might happen in the United States if the gap between rich and poor continues to widen?

The Kendall Companion Website

www.cengage.com/sociology/kendall

Supplement your review of this chapter by going to the text's companion website, where you can take tutorial quizzes, use flash cards to master key terms, follow live links to useful websites, and explore the other study and research resources you'll find there, such as a comprehensive interactive sociology timeline, GSS Data, and Census 2000 information, much of it presented visually in maps.

Marathon running has taken me a long way from my roots in the small town of Baringo in Kenya's Rift Valley. I grew up knowing what it was like to be poor and hungry. Whenever I come to London or other cities in the developed world to compete in marathons, I enter a different universe where choice, opulence and opportunity characterize people's lives.

It has been fascinating to follow the debate in Britain about school meals. I have listened to the arguments about whether children should be allowed to eat Turkey Twizzlers, or beefburgers and chips. I wish it could be the same the world over. While nutrition is a serious matter for any child, for me and my classmates [in Kenya] it was never really a case of what we might choose to eat, but rather whether we would eat at all.

Most kids in Baringo had to help their families earn a living. Education was out of the question or, at best, something only one child in the family could pursue.

© AP Images/Herbert Knosowski

The success story of marathon world-record-holder Paul Tergat, who grew up poor and hungry, calls our attention to issues of global stratification and inequality.

For the lucky ones like me, who could go to school, the three-mile trek each morning on an empty stomach made it difficult, and sometimes impossible, to concentrate on lessons.

When I was eight, that changed. The United Nations began distributing food at the schools in the area and a heavy burden was lifted from our shoulders. My friends and I no longer worried about being hungry in class. We ate a simple meal each day and could stay focused during lessons. . . . I often ask myself: without the benefit of school meals, would I have become a literate, healthy, successful long-distance runner?

—Paul Tergat (2005), a world marathon record holder and winner of two silver Olympic medals, describing his early childhood, marked by poverty and hunger in Kenya

Marathoner Paul Tergat speaks for millions of people around the world who have experienced poverty and hunger. In his role as Ambassador Against Hunger for the World Food Program, Tergat encourages others to get involved in campaigns against hunger, illiteracy, pollution, homelessness, and other problems that limit people's life chances and opportunities. He also highlights the fact that although students in high-income nations have many food choices, some of which may be bad for them, students in low-income nations have very little food and extremely limited choices in life without intervention from the outside (Hattori, 2006).

Regardless of where people live in the world, social and economic inequalities are pressing daily concerns. Poverty and inequality know no political boundaries or national borders. In this chapter, we examine global stratification and inequality, and discuss perspectives that have been developed to explain the nature and extent of this problem. Before reading on, test your knowledge of global wealth and poverty (see Box 8.1).

Chapter Focus Question

How do global stratification and economic inequality affect the life chances of people around the world?

Wealth and Poverty in Global Perspective

What do we mean by global stratification? ***Global stratification* refers to the unequal distribution of wealth, power, and prestige on a global basis, resulting in people having vastly different lifestyles and life chances both within and among the nations of the world.** Just as the United States is divided into classes, the world is divided into unequal segments characterized by extreme differences in wealth and poverty. For example, the income gap between the richest and the poorest 20 percent of the world population continues to widen (see ▶ Figure 8.1). However, when we compare social and economic inequality within other nations,

we find gaps that are more pronounced than they are in the United States.

As previously defined, *high-income countries* are nations characterized by highly industrialized economies; technologically advanced industrial, administrative, and service occupations; and relatively high levels of national and per capita (per person) income. In contrast, *middle-income countries* are nations with industrializing economies, particularly in urban areas, and moderate levels of national and personal income. *Low-income countries* are primarily agrarian nations with little industrialization and low levels of national and personal income. Within some nations, the poorest one-fifth of the population has an income that is only a slight fraction of the overall average per capita income for that country. For example, in Brazil, Bolivia, and Honduras,

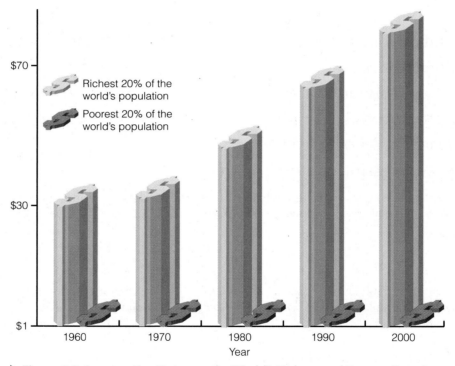

▶ **Figure 8.1 Income Gap Between the World's Richest and Poorest People**
The income gap between the richest and poorest people in the world continued to grow between 1960 and 2000. As this figure shows, in 1960 the highest-income 20 percent of the world's population received $30 for each dollar received by the lowest-income 20 percent. By 2000, the disparity had increased: $74 to $1.

Sources: International Monetary Fund, 1992; United Nations Development Programme, 2003.

Box 8.1 Sociology and Everyday Life

How Much Do You Know About Global Wealth and Poverty?

True	False	
T	F	1. The world's ten richest people are U.S. citizens.
T	F	2. The assets of the 200 richest people are more than the combined income of over 40 percent of the world's population.
T	F	3. More than one billion people worldwide live below the international poverty line, earning less than $1 each day.
T	F	4. Although poverty is a problem in most areas of the world, relatively few people die of causes arising from poverty.
T	F	5. In low-income countries, the problem of poverty is unequally shared between men and women.
T	F	6. The majority of people with incomes below the poverty line live in urban areas of the world.
T	F	7. Most analysts agree that the World Bank was created to serve the poor of the world and their borrowing governments.
T	F	8. Poor people in low-income countries meet most of their energy needs by burning wood, dung, and agricultural wastes, which increases health hazards and environmental degradation.

Answers on page 256.

less than 3 percent of total national income accrues to the poorest one-fifth of the population (World Bank, 2005).

Just as the differences between the richest and poorest people in the world have increased, the gap in global income differences between rich and poor countries has continued to widen over the past fifty years. In 1960, the wealthiest 20 percent of the world population had more than thirty times the income of the poorest 20 percent. By 2000, the wealthiest 20 percent of the world population had almost seventy-five times the income of the poorest 20 percent (United Nations Development Programme, 2003). Income disparities *within* countries were even more pronounced.

Many people have sought to address the issue of world poverty and to determine ways in which resources can be used to meet the urgent challenge of poverty. However, not much progress has been made on this front (Lummis, 1992) despite a great deal of talk and billions of dollars in "foreign aid" flowing from high-income to low-income nations.

The idea of "development" has become the primary means used in attempts to reduce social and economic inequalities and alleviate the worst effects of poverty in the less industrialized nations of the world. Often, the nations that have not been able to reduce or eliminate poverty are chastised for not making the necessary social and economic reforms to make change possible (Myrdal, 1970). Or, as another social analyst has suggested,

> The *problem* of inequality lies not in poverty, but in excess. "The problem of the world's poor," defined more accurately, turns out to be "the problem of the world's rich." This means that the solution to the problem is not a massive change in the

---

global stratification the unequal distribution of wealth, power, and prestige on a global basis, resulting in people having vastly different lifestyles and life chances both within and among the nations of the world.

Answers to the Sociology Quiz on Global Wealth and Poverty

1. False.   In 2008, two of the ten richest people were U.S. citizens. They were Warren Buffett (Berkshire Hathaway) and Bill Gates (Microsoft) (*Forbes*, 2008).

2. True.   Assets of the 200 richest people are more than the combined income of 41 percent of the world's population (United Nations Development Programme, 2003).

3. True.   The World Bank estimates that more than 1.4 billion people worldwide live below the international poverty line, which is defined as earning less than $1 each day (World Bank, 2005).

4. False.   One of the consequences of extreme poverty is hunger, and millions of people—including six million children under the age of five years—die of hunger-related diseases or chronic malnutrition each year (PBS, 2008).

5. True.   In almost all low-income countries (as well as middle- and high-income countries), poverty is a more chronic problem for women due to sexual discrimination, resulting in a lack of educational and employment opportunities (Hauchler and Kennedy, 1994).

6. False.   Although the number of poor people residing in urban areas is growing rapidly, the majority of people with incomes below the poverty line live in rural areas of the world (United Nations DPCSD, 1997).

7. False.   Some analysts point out the linkages between the World Bank and the transnational corporate sector on both the borrowing and lending ends of its operation. Although the bank is supposedly owned by its members' governments and lends money only to governments, many of its projects involve vast financial dealings with transnational construction companies, consulting firms, and procurement contractors (see Korten, 1996).

8. True.   Although these fuels are inefficient and harmful to health, many low-income people cannot afford appliances, connection charges, and so forth. In some areas, electric hookups are not available (United Nations DPCSD, 1997).

culture of poverty so as to place it on the path of development, but a massive change in the culture of superfluity in order to place it on the path of counterdevelopment. It does not call for a new value system forcing the world's majority to feel shame at their traditionally moderate consumption habits, but for a new value system forcing the world's rich to see the shame and vulgarity of their overconsumption habits, and the double vulgarity of standing on other people's shoulders to achieve those consumption habits. (Lummis, 1992: 50)

As this statement suggests, the increasing interdependence of all the world's nations was largely overlooked or ignored until increasing emphasis was placed on the global marketplace and the global economy. In addition, there are a number of problems inherent in studying global stratification, one of which is what terminology should be used to describe various nations.

Problems in Studying Global Inequality

One of the primary problems encountered by social scientists studying global stratification and social and economic inequality is what terminology should be used to refer to the distribution of resources in various nations. During the past fifty years, major changes have occurred in the way that inequality is addressed by organizations such as the United Nations and the World Bank. Most definitions of inequality are based on comparisons of levels of income or economic development, whereby countries are identified in terms of the "three worlds" or upon their levels of economic development.

The "Three Worlds" Approach

After World War II, the terms "First World," "Second World," and "Third World" were introduced by social analysts to distinguish among nations on the basis of their levels of economic development and the standard of living of their citizens. *First World* nations were said to consist of the rich, industrialized nations that primarily had capitalist economic systems and democratic political systems. The most frequently noted First World nations were the United States, Canada, Japan, Great Britain, Australia, and New Zealand. *Second World* nations were said to be countries with at least a moderate level of economic development and a moderate standard of living. These nations included China, North Korea, Vietnam, Cuba, and portions of the former Soviet Union.

According to social analysts, although the quality of life in Second World nations was not comparable to that of life in the First World, it was far greater than that of people living in the *Third World*—the poorest countries, with little or no industrialization and the lowest standards of living, shortest life expectancies, and highest rates of mortality.

The Levels of Development Approach

Among the most controversial terminology used for describing world poverty and global stratification has been the language of development. Terminology based on levels of development includes concepts such as developed nations, developing nations, less-developed nations, and underdevelopment. Let's look first at the contemporary origins of the idea of "underdevelopment" and "underdeveloped nations."

Following World War II, the concepts of *underdevelopment* and *underdeveloped nations* emerged out of the Marshall Plan (named after U.S. Secretary of State George C. Marshall), which provided massive sums of money in direct aid and loans to

Vast inequalities in income and lifestyle are evident in this photo of slums and nearby higher-priced housing in Bombay, India. What visible patterns of economic inequality exist in the United States?

© Viviane Moos/CORBIS

rebuild the European economic base destroyed during World War II. Given the Marshall Plan's success in rebuilding much of Europe, U.S. political leaders decided that the Southern Hemisphere nations that had recently been released from European colonialism could also benefit from a massive financial infusion and rapid economic development. Leaders of the developed nations argued that urgent problems such as poverty, disease, and famine could be reduced through the transfer of finance, technology, and experience from the developed nations to lesser-developed countries. From this viewpoint, economic development is the primary way to solve the poverty problem: Hadn't economic growth brought the developed nations to their own high standard of living?

Ideas regarding *underdevelopment* were popularized by President Harry S Truman in his 1949 inaugural address. According to Truman, the nations in the Southern Hemisphere were "underdeveloped areas" because of their low gross national product, which today is referred to as *gross national income* (GNI)—a term that refers to all the goods and services produced in a country in a given year, plus the net income earned outside the country by individuals or corporations. If nations could increase their GNI, then social and economic inequality among the citizens within the country could also be reduced. Accordingly, Truman believed that

it was necessary to assist the people of economically underdeveloped areas to raise their *standard of living*, by which he meant material well-being that can be measured by the quality of goods and services that may be purchased by the per capita national income. Thus, an increase in the standard of living meant that a nation was moving toward economic development, which typically included the improved exploitation of natural resources by industrial development.

What has happened to the issue of development since the post–World War II era? After several decades of economic development fostered by organizations such as the United Nations and the World Bank, it became apparent by the 1970s that improving a country's GNI did not tend to reduce the poverty of the poorest people in that country. In fact, global poverty and inequality were increasing, and the initial optimism of a speedy end to underdevelopment faded.

Why did inequality increase even with greater economic development? Some analysts in the developed nations began to link growing social and economic inequality on a global basis to relatively high rates of population growth taking place in the underdeveloped nations. Organizations such as the United Nations and the World Health Organization stepped up their efforts to provide family planning services to the populations so that they could control their own

Global inequality is most striking in nations that are sometimes referred to as "Third World" or "underdeveloped." In this Wenzhou, China, shoe factory, workers produce shoes only for export to other countries.

fertility. More recently, however, population researchers have become aware that issues such as population growth, economic development, and environmental problems must be seen as interdependent concerns. This changing perception culminated in the U.N. Conference on Environment and Development in Rio de Janeiro, Brazil (the "Earth Summit"), in 1992; as a result, terms such as *underdevelopment* have largely been dropped in favor of measurements such as sustainable development, and economies are now classified by their levels of income.

Classification of Economies by Income

Today, the World Bank (2005) focuses on three development themes: people, the environment, and the economy. Because the World Bank's primary business is providing loans and policy advice to low- and middle-income member countries, it classifies nations into three economic categories: *low-income economies* (a GNI per capita of $875 or less in 2005), *middle-income economies* (a GNI per capita between $876 and $10,725 in 2005), and *high-income economies* (a GNI per capita of more than $10,725 in 2005).

Low-Income Economies

About half the world's population lives in the sixty-one low-income economies, where most people engage in agricultural pursuits, reside in nonurban areas, and are impoverished (World Bank, 2005). As shown in ▶ Map 8.1, low-income economies are primarily found in countries in Asia and Africa, where half of the world's population resides.

Among those most affected by poverty in low-income economies are women and children. Mayra Buvinić, who has served as chief of the women in development program unit at the Inter-American

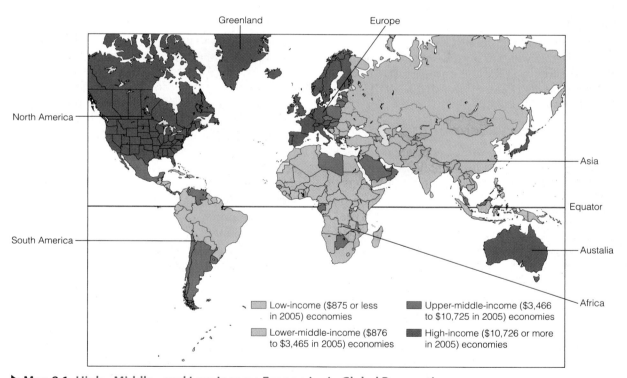

▶ **Map 8.1** **High-, Middle-, and Low-Income Economies in Global Perspective**
How does the United States compare with other nations with regard to income?

Source: World Bank, 2005.

Development Bank, describes the plight of one Nigerian woman as an example:

> On the outskirts of Ibadan, Nigeria, Ade cultivates a small, sparsely planted plot with a baby on her back and other visibly undernourished children nearby. Her efforts to grow an improved soybean variety, which could have improved her children's diet, failed because she lacked the extra time to tend the new crop, did not have a spouse who would help her, and could not afford hired labor. (Buvinić, 1997: 38)

According to Buvinić, Ade's life is typical of many women worldwide who face obstacles to increasing their economic power because they do not have the time to invest in the additional work that could bring in more income. Also, many poor women worldwide also do not have access to commercial credit and have been trained only in traditionally female skills that produce low wages. These factors have contributed to the *global feminization of poverty*, whereby women around the world tend to be more impoverished than men. Despite the fact that women have made some gains in terms of well-being, the income gap between men and women continues to grow wider in the low-income, developing nations as well as in the high-income, developed nations such as the United States (Buvinić, 1997).

Middle-Income Economies

About one-third of the world's population resides in the ninety-three nations with middle-income economies (a GNI per capita between $876 and $10,725 in 2005). The World Bank divides middle-income economies into lower-middle-income ($876 to $3,465) and upper-middle-income ($3,466 to $10,725). Countries classified as lower-middle-income include the Latin American nations of Bolivia, Columbia, Guatemala, and El Salvador. However, even though these countries are referred to as "middle-income," more than half of the people residing in countries such as Bolivia, Guatemala, and Honduras live in poverty, defined as below $1 per day in purchasing power (World Bank, 2005). In recent years, millions of people have migrated from the world's poorest nations in hopes of finding better economic conditions elsewhere. However, many of these migrants remain impoverished as they primarily move between lower- and middle-income economies (see Box 8.2, Sociology in Global Perspective).

Other lower-middle-income economies include Russia, Romania, and Kazakhstan. These nations had centrally planned (i.e., socialist) economies until dramatic political and economic changes occurred in the late 1980s and early 1990s. Since then, these nations have been going through a transition to a market economy. According to the World Bank (2005), some nations have been more successful than others in implementing key elements of change and bringing about a higher standard of living for their citizens. Among other factors, high rates of inflation, the growing gap between the rich and the poor, low life-expectancy rates, and homeless children have been visible signs of problems in the transition toward a free-market economy in countries such as Russia.

As compared with lower-middle-income economies, nations having upper-middle-income economies typically have a somewhat higher standard of living and export diverse goods and services, ranging from manufactured goods to raw materials and fuels. Nations with upper-middle-income economies include Argentina, Chile, Hungary, Mexico, and Saudi Arabia. Although these nations are referred to as middle-income economies, many of them have extremely high levels of indebtedness, leaving them with few resources for fighting poverty. According to one analyst, "A major cause of debt accumulation was investment in ill-considered, ill-conceived projects, many involving bloated capital costs and healthy doses of graft" (George, 1993: 88). Additional debts also accrued through military spending, even though a sizable portion of some populations is living in hunger and misery (George, 1993). The World Bank sought to set up funds to reduce the debts of some nations that have debts in excess of 200 to 250 percent of their annual export earnings. However, some high-income nations believed that such an action on the part of the World Bank might set a dangerous precedent, so the debt-reduction process was limited in scope (Lewis, 1996).

As middle-income nations have been required to make payments on their debts, the requisite structural adjustments have necessitated that the countries make spending cuts in areas that formerly

Box 8.2 Sociology in Global Perspective

Marginal Migration: Moving to a Less-Poor Nation

We are forced to come back here—not because we like it, but because we are poor. When we cross the border, we are a little better off. We are able to buy shoes and maybe a chicken.

> —Anes Moises explaining why he and many other Haitian migrants face continual hardships so that they can live and work in the Dominican Republic (qtd. in DeParle, 2007: A1)

In the Dominican Republic, Anes Moises works in the banana fields and earns six times as much money as he would in his homeland of Haiti. For Anes and the more than 74 million "south-to-south" migrants—people who have moved from one developing country to another—even marginal gains in money and quality of life are important. As one journalist described the living conditions of Haitian migrants in the Dominican Republic, "The scrap-wood shanties on a muddy hillside are a poor man's promised land" (DeParle, 2007: A1). Although we hear much more about the 82 million migrants who have moved "south to north"—from a lower-income nation to a country with a high-income economy—the lived experiences of "south-to-south" migrants are also important in understanding global wealth and poverty in the twenty-first century (Ratha and Shaw, 2007).

What are the typical characteristics of people who move from one poor nation to a slightly less poor one? Do their efforts make a difference in their economic status? Recent studies have found that south-to-south migrants are often poorer than individuals who migrate from lower-income to high-income nations. South-to-south migrants are also more likely to travel without proper documentation and be more vulnerable to unscrupulous people and to apprehension by law enforcement officials than are south-to-north migrants (Ratha and Shaw, 2007). Many south-to-south migrants send money back home to extremely poor family members who reside in remote rural areas. Some analysts estimate that the money sent by these migrants has a significant economic impact on the lives of people in the poorest nations of the world. For example, Haitians residing in the Dominican Republic typically send about $135 million a year to relatives back home (DeParle, 2007: A16). Ironically, numerous jobs are available for Haitians in the Dominican Republic because many Dominicans have migrated to the United States (south-to-north migration) in hopes of finding better jobs and higher wages. According to a recent *New York Times* article, similar patterns exist across many nations as some individuals migrate to high-income economies while others move from a very poor country to a slightly less poor one:

> Nicaraguans build Costa Rican buildings. Paraguayans pick Argentine crops. Nepalis dig Indian mines. Indonesians clean Malaysian homes. Farm hands from Burkina Faso tend the fields in Ivory Coast. Some save for the more expensive journeys north, while others find the move from one poor land to another all they will ever afford. With rich countries tightening their borders, migration within the developing world is likely to grow. (DeParle, 2007: A16)

A comparison of the 2007 average per capita income of several nations shows why south-to-south migration will no doubt continue in the future (based on *New York Times*, 2007b: A16):

- Haiti ($480) to the Dominican Republic ($2,850)
- Nicaragua ($1,000) to El Salvador ($2,540)
- Guatemala ($2,640) to Mexico ($7,870)
- Columbia ($2,740) to Panama ($4,890)

Reflect & Analyze

How do these figures compare with what migrants might gain from moving to the United States or other nations with high-income economies? Does this information provide us with new insights on the nature and extent of global stratification and inequality as they affect people living in the United States? Why or why not?

helped some of the poor, including subsidized food, education, and health care. These structural adjustments have been required in agreements with lending agencies such as the World Bank and the International Monetary Fund. Thus, there is a high rate of poverty in many countries that are classified as middle-income economies.

High-Income Economies

High-income economies are found in fifty-four nations, including the United States, Canada, Japan, Australia, Portugal, Ireland, Israel, Italy, Norway, and Germany. According to the World Bank, people in high-income economies typically have a higher standard of living than those in low- and middle-income economies. Nations with high-income economies continue to dominate the world economy, despite the fact that shifts in the global marketplace have affected some workers who have found themselves without work due to *capital flight*—the movement of jobs and economic resources from one nation to another—and *deindustrialization*—the closing of plants and factories because of their obsolescence or the fact that workers in other nations are being hired to do the work more cheaply.

Some of the nations that have been in the high-income category do not have as high a rate of annual economic growth as the newly industrializing nations, particularly in East Asia and Latin America,

that are still in the process of development. However, the only significant group of middle- and lower-income economies to close the gap with the high-income, industrialized economies over the past few decades has been the nations of East Asia. China has experienced a 270-percent increase in per capita income over the past twenty years.

Despite economic growth, the East Asian region remains home to approximately 350 million poor people. And from all signs, it appears that nations such as India will continue to have an extremely large population and rapid annual growth rates. (The Concept Quick Review describes economies classified by income.)

Measuring Global Wealth and Poverty

On a global basis, measuring wealth and poverty is a difficult task because of conceptual problems and problems in acquiring comparable data from various nations. Although gross national income continues to be one of the most widely used measures of national income, in recent years the United Nations and the World Bank have begun to use the gross domestic product (GDP). The *gross domestic product* is all the goods and services produced *within* a country's economy during a given year. Unlike the

© Dell Inc. for Business Wire via Getty Images

Where are these Dell Computer service technicians working? Dell's headquarters is located in Round Rock, Texas (near Austin), so you might assume that these employees are working there or somewhere else in the United States. However, as part of the global work force in high-income nations, the people you see here are employed at the Dell Enterprise Command Center in Limerick, Ireland.

CONCEPT QUICK REVIEW

Classification of Economies by Income

	Low-Income Economies	Middle-Income Economies	High-Income Economies
Previous categorization	Third World, underdeveloped	Second World, developing	First World, developed
2005 per capita income (GNI)	$875 or less	$876–$10,725	$10,726 or more
Type of economy	Largely agricultural	Diverse, from agricultural to manufacturing	Information-based and postindustrial

GNI, the GDP does not include any income earned by individuals or corporations if the revenue comes from sources outside of the country. For example, using GDP as a measure of economic growth, a World Bank report concluded that nations such as China, India, and Indonesia were well on their way to becoming "economic powerhouses" in the next twenty-five years. However, some scholars criticize the assumption that economic development always benefits low-income nations and that poor nations *can* and *should* play "catch-up" with the wealthy, industrialized nations (Lummis, 1992).

Absolute, Relative, and Subjective Poverty

How is poverty defined on a global basis? Isn't it more a matter of comparison than an absolute standard? According to social scientists, defining poverty involves more than comparisons of personal or household income; it also involves social judgments made by researchers. From this point of view, *absolute poverty*—previously defined as a condition in which people do not have the means to secure the most basic necessities of life—would be measured by comparing personal or household income or expenses with the cost of buying a given quantity of goods and services. The World Bank has defined absolute poverty as living on less than a dollar a day. Similarly, *relative poverty*—which exists when people may be able to afford basic necessities but are still unable to maintain an average standard of living—would be measured by comparing one person's income with the incomes of others. Finally, *subjective poverty* would be measured by comparing the actual income against the income earner's expectations and perceptions. However, for low-income nations in a state of economic transition, data on income and levels of consumption are typically difficult to obtain and are often ambiguous when they are available. Defining levels of poverty involves several dimensions: (1) how many people are poor, (2) how far below the poverty line people's incomes fall, and (3) how long they have been poor (is the poverty temporary or long term?) (World Bank, 2005).

The Gini Coefficient and Global Quality-of-Life Issues

The World Bank uses as its measure of income inequality what is known as the *Gini coefficient,* which ranges from zero (meaning that everyone has the same income) to 100 (one person receives all the income). Using this measure, the World Bank (2005) has concluded that inequality has increased in nations such as Bulgaria, the Baltic countries, and the Slavic countries of the former Soviet Union to levels similar to those in the less-equal industrial market economies, such as the United States. In fact, inequality in Russia rose sharply in the 1990s: By 1998, the top 20 percent of the population received a much larger percentage of the total income than it had in 1993. Stark contrasts also exist in countries such as India, where abject poverty still exists side by side with lavish opulence in Calcutta. Note the sharp contrast in lifestyle on one Calcutta street:

> On one side is the Tollygunge Club, 40 hectares of landscaped serenity with an 18-hole golf course, a driving range, riding stables, tennis courts, and two covered swimming pools. . . . On the other side stands the M.R. Bangur Hospital, a sooty building with a morgue [that sometimes takes] the corpses of paupers who die in Calcutta's streets. (Watson, 1997: F1)

This street is symbolic of the sharp chasm that divides the 11 million people who live in Calcutta. In fact, some analysts believe that the scale of poverty in South and East Asia is most visible in the heavily populated states of India and China. It is estimated that 652 million of Asia's people are poor and that 448 million of them live in India alone, where there are more than twice as many poor people as in sub-Saharan Africa (Watson, 1997). However, similar disparities between the rich and the poor can be seen in nations throughout the world.

Global Poverty and Human Development Issues

Income disparities are not the only factor that defines poverty and its effect on people. Although the average income per person in lower-income countries has doubled in the past thirty years and for many years economic growth has been seen as the primary way to achieve development in low-income economies, the United Nations since the 1970s has more actively focused on human development as a crucial factor in fighting poverty. In 1990 the United Nations Development Program introduced the Human Development Index (HDI), establishing three new criteria—in addition to GDP—for measuring the level of development in a country: life expectancy, education, and living standards. According to the United Nations, human development is the process of "expanding choices that people have in life, to lead a life to its full potential and in dignity, through expanding capabilities and through people taking action themselves to improve their lives"

(United Nations Development Programme, 2002). ▶ Figure 8.2 compares indicators such as gross domestic product, life expectancy, and adult literacy in regions around the world. The World Bank calculates that 37 percent of the population in low- and middle-income countries (1.6 billion people) lack the essentials of well-being, whereas only 21 percent (900 million people) are "income poor," as defined by the bank's poverty line. Women account for most of the "extra" 700 million poor people (Buvinić, 1997: 40).

Life Expectancy

Although some advances have been made in middle- and low-income countries regarding life expectancy, major problems still exist. On the plus side, average life expectancy has increased by about a third in the past three decades and is now more than 70 years in 87 countries (United Nations Development Programme, 2003). Although no country has reached a life expectancy for men of 80 years, 19 countries have now reached an expectancy of 80 years or more for women. On a less positive note, the average life expectancy at birth of people in middle-income countries remains about 12 years less than that of people in high-income countries. Moreover, the life expectancy of people in low-income nations is as much as 23 years less than that of people in high-income nations. Especially striking are the differences in life expectancies in high-income economies and low-income economies such as sub-Saharan Africa, where estimated life expectancy has dropped significantly in Uganda (from 48 to 43 for women and from 46 to 43 for men) over the past two decades.

One major cause of shorter life expectancy in low-income nations is the high rate of infant mortality. The infant mortality rate (deaths per thousand live births) is more than eight times higher in low-income countries than in high-income countries (World Bank, 2005). Low-income countries typically have higher rates of illness and disease, and they do not have adequate health care facilities. Malnutrition is a common problem among children, many of whom are underweight, stunted, and have anemia—a nutritional deficiency with serious consequences for child mortality. Consider

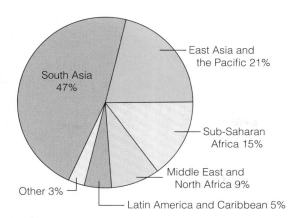

a. **Distribution of the world's illiterate population, age fifteen and older, by region.**

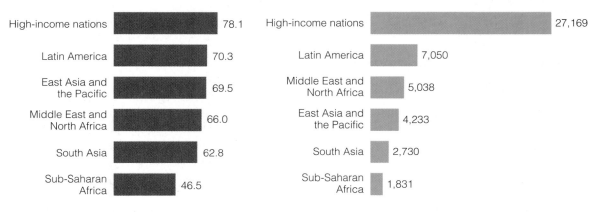

High-income nations	78.1
Latin America	70.3
East Asia and the Pacific	69.5
Middle East and North Africa	66.0
South Asia	62.8
Sub-Saharan Africa	46.5

b. **Life expectancy in years for persons born in 2001.**

High-income nations	27,169
Latin America	7,050
Middle East and North Africa	5,038
East Asia and the Pacific	4,233
South Asia	2,730
Sub-Saharan Africa	1,831

c. **Per capita gross domestic product in U.S. dollars.**

▶ **Figure 8.2 Indicators of Human Development**

Source: World Bank, 2005.

this journalist's description of a child she saw in Haiti:

> Like any baby, Wisly Dorvil is easy to love. Unlike others, this 13-month-old is hard to hold.
>
> That's because his 10-pound frame is so fragile that even the most minimal of movements can dislocate his shoulders.
>
> As lifeless as a rag doll, Dorvil is starving. He has large, brown eyes and a feeble smile, but a stomach so tender that he suffers from ongoing bouts of vomiting and diarrhea.

> Fortunately, though, Dorvil recently came to the attention of U.S. aid workers. With round-the-clock feeding, he is expected to survive.
>
> Others are not so lucky. (Emling, 1997a: A17)

Among adults and children alike, life expectancies are strongly affected by hunger and malnutrition. It is estimated that people in the United States spend more than $5 billion each year on diet products to lower their calorie consumption, whereas the world's poorest 600 million people suffer from chronic malnutrition, and over 40 million people die each year from hunger-related diseases (Kidron and Segal, 1995). To put this figure in perspective, the number of people worldwide dying from hunger-related diseases is the equivalent of more than 300 jumbo-jet crashes per day with no survivors, and half the passengers children (Kidron and Segal, 1995).

Health

Health is defined in the constitution of the World Health Organization as "a state of complete physical, mental and social well-being and not merely the absence of disease or infirmity." Many people in low-income nations are far from having physical, mental, and social well-being. In fact, about 17 million people die each year from diarrhea, malaria, tuberculosis, and other infectious and parasitic illnesses (Hauchler and Kennedy, 1994). According to the World Health Organization, infectious diseases are far from under control in many nations due to such factors as unsanitary or overcrowded living conditions. Despite the possible eradication of diseases such as poliomyelitis, leprosy, guinea-worm disease, and neonatal tetanus in the near future, at least thirty new diseases—for which there is no treatment or vaccine—have recently emerged.

Some middle-income countries are experiencing rapid growth in degenerative diseases such as cancer and coronary heart disease, and many more deaths are expected from smoking-related diseases. Despite the decrease in tobacco smoking in high-income countries, there has been an increase in per capita consumption of tobacco in low- and middle-income countries, many of which have been targeted for free samples and promotional advertising by U.S. tobacco companies (United Nations Development Programme, 2000).

Education and Literacy

According to the Human Development Report (United Nations Development Programme, 2003), education is fundamental to reducing both individual and national poverty. As a result, school enrollment is used as one measure of human development. Although school enrollment has increased at the

© Jenny Matthews/Alamy

In an effort to reduce poverty, some nations have developed adult literacy programs so that people can gain an education that will help lift them out of poverty. In regions such as Solomuna, Eritrea, women's literacy is a particularly crucial issue.

primary and secondary levels in about two-thirds of the lower-income regions, enrollment is not always a good measure of educational achievement because many students drop out of school during their elementary-school years. Many of the children who do not attend school work long hours, under hazardous conditions, for very low wages.

What is literacy, and why is it important for human development? The United Nations Educational, Scientific and Cultural Organization (UNESCO) defines a literate person as "someone who can, with understanding, both read and write a short, simple statement on their everyday life" (United Nations, 1997: 89). Based on this definition, people who can write only with figures, their name, or a memorized phrase are not considered literate. The adult literacy rate in the low-income countries is about half that of the high-income countries, and for women the rate is even lower (United Nations, 1997). Women constitute about two-thirds of those who are illiterate: There are approximately 74 literate women for every 100 literate men (United Nations, 1997). Literacy is crucial for women because it has been closely linked to decreases in fertility, improved child health, and increased earnings potential.

Persistent Gaps in Human Development

Some middle- and lower-income countries have made progress in certain indicators of human development. The gap between some richer and middle- or lower-income nations has narrowed significantly for life expectancy, adult literacy, and daily calorie supply; however, the overall picture for the world's poorest people remains dismal. The gap between the poorest nations and the middle-income nations has continued to widen. Poverty, food shortages, hunger, and rapidly growing populations are pressing problems for at least 1.3 billion people, most of them women and children living in a state of absolute poverty. Although more women around the globe have paid employment than in the past, more and more women are still finding themselves in poverty because of increases in single-person and single-parent households headed by women and the fact that low-wage work is often the only source of livelihood available to them.

Theories of Global Inequality

Why is the majority of the world's population growing richer while the poorest 20 percent—more than one billion people—are so poor that they are effectively excluded from even a moderate standard of living? Social scientists have developed a variety of theories that view the causes and consequences of global inequality somewhat differently. We will examine the development approach and modernization theory, dependency theory, world systems theory, and the new international division of labor theory.

Development and Modernization Theory

According to some social scientists, global wealth and poverty are linked to the level of industrialization and economic development in a given society. Although the process by which a nation industrializes may vary somewhat, industrialization almost

inevitably brings with it a higher standard of living in a nation and some degree of social mobility for individual participants in the society. Specifically, the traditional caste system becomes obsolete as industrialization progresses. Family status, race/ethnicity, and gender are said to become less significant in industrialized nations than in agrarian-based societies. As societies industrialize, they also urbanize as workers locate their residences near factories, offices, and other places of work. Consequently, urban values and folkways overshadow the beliefs and practices of the rural areas. Analysts using a development framework typically view industrialization and economic development as essential steps that nations must go through in order to reduce poverty and increase life chances for their citizens.

Earlier in the chapter, we discussed the post–World War II Marshall Plan, under which massive financial aid was provided to the European nations to help rebuild infrastructure lost in the war. Based on the success of this infusion of cash in bringing about modernization, President Truman and many other politicians and leaders in the business community believed that it should be possible to help so-called underdeveloped nations modernize in the same manner.

The most widely known development theory is *modernization theory*—a perspective that links global inequality to different levels of economic development and suggests that low-income economies can move to middle- and high-income economies by achieving self-sustained economic growth. According to modernization theory, the low-income, less-developed nations can improve their standard of living only with a period of intensive economic growth and accompanying changes in people's beliefs, values, and attitudes toward work. As a result of modernization, the values of people in developing countries supposedly become more similar to those of people in high-income

modernization theory a perspective that links global inequality to different levels of economic development and suggests that low-income economies can move to middle- and high-income economies by achieving self-sustained economic growth.

nations. The number of hours that people work at their jobs each week is one measure of the extent to which individuals subscribe to the *work ethic,* a core value widely believed to be of great significance in the modernization process.

Perhaps the best-known modernization theory is that of Walt W. Rostow (1971, 1978), who, as an economic advisor to U.S. President John F. Kennedy, was highly instrumental in shaping U.S. foreign policy toward Latin America in the 1960s. To Rostow, one of the largest barriers to development in low-income nations was the traditional cultural values held by people, particularly beliefs that are fatalistic, such as viewing extreme hardship and economic deprivation as inevitable and unavoidable facts of life. In cases of fatalism, people do not see any need to work in order to improve their lot in life: If it is predetermined for them, why bother? Based on modernization theory, poverty can be attributed to people's cultural failings, which are further reinforced by governmental policies interfering with the smooth operation of the economy.

Rostow suggested that all countries go through four stages of economic development, with identical content, regardless of when these nations started the process of industrialization. He compared the stages of economic development to an airplane ride. The first stage is the *traditional stage,* in which very little social change takes place, and people do not think much about changing their current circumstances. According to Rostow, societies in this stage are slow to change because the people hold a fatalistic value system, do not subscribe to the work ethic, and save very little money. The second stage is the *take-off stage*—a period of economic growth accompanied by a growing belief in individualism, competition, and achievement. During this stage, people start to look toward the future, to save and invest money, and to discard traditional values. According to Rostow's modernization theory, the development of capitalism is essential for the transformation from a traditional, simple society to a modern, complex one. With the financial help and advice of the high-income countries, low-income countries will eventually be able to "fly" and enter the third stage of economic development. In the third stage, the country moves toward *technological maturity.* At this point, the country will improve its technology,

reinvest in new industries, and embrace the beliefs, values, and social institutions of the high-income, developed nations. In the fourth and final stage, the country reaches the phase of *high mass consumption* and a correspondingly high standard of living.

Modernization theory has had both its advocates and its critics. According to proponents of this approach, studies have supported the assertion that economic development occurs more rapidly in a capitalist economy. In fact, the countries that have been most successful in moving from low- to middle-income status typically have been those that are most centrally involved in the global capitalist economy. For example, the nations of East Asia have successfully made the transition from low-income to higher-income economies through factors such as a high rate of savings, an aggressive work ethic among employers and employees, and the fostering of a market economy.

Critics of modernization theory point out that it tends to be Eurocentric in its analysis of low-income countries, which it implicitly labels as backward (see Evans and Stephens, 1988). In particular, modernization theory does not take into account the possibility that all nations do not industrialize in the same manner. In contrast, some analysts have suggested that modernization of low-income nations today will require novel policies, sequences, and ideologies that are not accounted for by Rostow's approach.

Which sociological perspective is most closely associated with the development approach? Modernization theory is based on a market-oriented perspective which assumes that "pure" capitalism is good and that the best economic outcomes occur when governments follow the policy of laissez-faire (or hands-off) business, giving capitalists the opportunity to make the "best" economic decisions, unfettered by government restraints or cumbersome rules and regulations (see Chapter 13, "Politics and the Economy in Global Perspective"). In today's global economy, however, many analysts believe that national governments are no longer central corporate decision makers and that transnational corporations determine global economic expansion and contraction. Therefore, corporate decisions to relocate manufacturing processes around the world make the rules and regulations of any one nation irrelevant and national boundaries

Poverty and war continue to devastate low-income countries such as Afghanistan. Many low-income countries receive aid from industrialized nations through initiatives such as the World Food Program. Modernization theory links global inequality to levels of economic development, but factors such as war and internal conflict also greatly contribute to patterns of global inequality.

© Chris Hondros/Getty Images

obsolete (Gereffi, 1994). Just as modernization theory most closely approximates a functionalist approach to explaining inequality, dependency theory, world systems theory, and the new international division of labor theory are perspectives rooted in the conflict approach. All four of these approaches are depicted in ▶ Figure 8.3.

Dependency Theory

Dependency theory **states that global poverty can at least partially be attributed to the fact that the low-income countries have been exploited by the high-income countries.** Analyzing events as part of a particular historical process—the expansion of global capitalism—dependency theorists see the greed of the rich countries as a source of increasing impoverishment of the poorer nations and their people. Dependency theory disputes the notion of the development approach, and modernization theory specifically, that economic growth is the key to meeting important human needs in societies. In contrast, the poorer nations are trapped in a cycle of structural dependency on the richer nations due to

their need for infusions of foreign capital and external markets for their raw materials, making it impossible for the poorer nations to pursue their own economic and human development agendas. For this reason, dependency theorists believe that countries such as Brazil, Nigeria, India, and Kenya cannot reach the sustained economic growth patterns of the more-advanced capitalist economies.

Dependency theory has been most often applied to the newly industrializing countries (NICs) of Latin America, whereas scholars examining the NICs of East Asia found that dependency theory had little or no relevance to economic growth and development in that part of the world. Therefore, dependency theory had to be expanded to encompass transnational economic linkages that affect developing countries, including foreign aid, foreign

---

dependency theory the belief that global poverty can at least partially be attributed to the fact that the low-income countries have been exploited by the high-income countries.

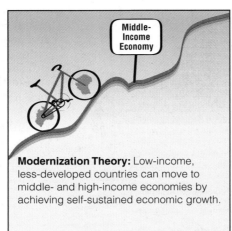

Modernization Theory: Low-income, less-developed countries can move to middle- and high-income economies by achieving self-sustained economic growth.

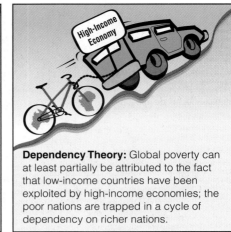

Dependency Theory: Global poverty can at least partially be attributed to the fact that low-income countries have been exploited by high-income economies; the poor nations are trapped in a cycle of dependency on richer nations.

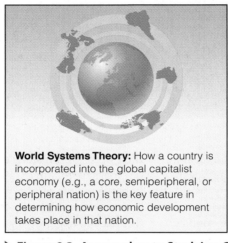

World Systems Theory: How a country is incorporated into the global capitalist economy (e.g., a core, semiperipheral, or peripheral nation) is the key feature in determining how economic development takes place in that nation.

The New International Division of Labor Theory: Commodity production is split into fragments, each of which can be moved (e.g., by a transnational corporation) to whichever part of the world can provide the best combination of capital and labor.

▶ **Figure 8.3 Approaches to Studying Global Inequality**
What causes global inequality? Social scientists have developed a variety of explanations, including the four theories shown here.

trade, foreign direct investment, and foreign loans. On the one hand, in Latin America and sub-Saharan Africa, transnational linkages such as foreign aid, investments by transnational corporations, foreign debt, and export trade have been significant impediments to development within a country. On the other hand, East Asian countries such as Taiwan, South Korea, and Singapore have historically also had high rates of dependency on foreign aid, foreign trade, and interdependence with transnational corporations but have still experienced high rates of economic growth despite dependency. According

to the sociologist Gary Gereffi (1994), differences in outcome are probably associated with differences in the timing and sequencing of a nation's relationship with external entities such as foreign governments and transnational corporations.

Dependency theory makes a positive contribution to our understanding of global poverty by noting that "underdevelopment" is not necessarily the cause of inequality. Rather, it points out that exploitation not only of one country by another but of countries by transnational corporations may limit or retard economic growth and human develop-

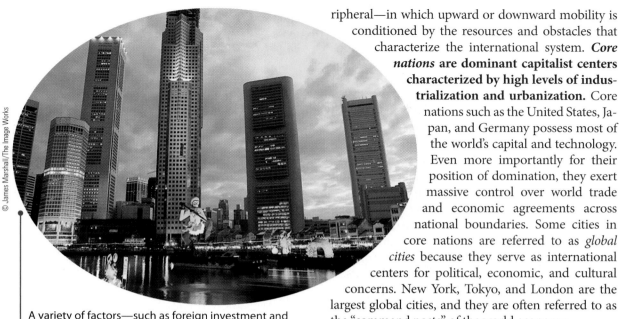

A variety of factors—such as foreign investment and the presence of transnational corporations—have contributed to the economic growth of nations such as Singapore.

ment in some nations. However, what remains unexplained is how East Asia and India had successful "dependency management" whereas many Latin American countries did not (Gereffi, 1994).

World Systems Theory

World systems theory suggests that what exists under capitalism is a truly global system that is held together by economic ties. From this approach, global inequality does not emerge solely as a result of the exploitation of one country by another. Instead, economic domination involves a complex world system in which the industrialized, high-income nations benefit from other nations and exploit their citizens. This theory is most closely associated with the sociologist Immanuel Wallerstein (1979, 1984), who believed that a country's mode of incorporation into the capitalist work economy is the key feature in determining how economic development takes place in that nation. According to *world systems theory*, the capitalist world economy is a global system divided into a hierarchy of three major types of nations—core, semiperipheral, and pe-

ripheral—in which upward or downward mobility is conditioned by the resources and obstacles that characterize the international system. **Core nations are dominant capitalist centers characterized by high levels of industrialization and urbanization.** Core nations such as the United States, Japan, and Germany possess most of the world's capital and technology. Even more importantly for their position of domination, they exert massive control over world trade and economic agreements across national boundaries. Some cities in core nations are referred to as *global cities* because they serve as international centers for political, economic, and cultural concerns. New York, Tokyo, and London are the largest global cities, and they are often referred to as the "command posts" of the world economy.

Semiperipheral nations **are more developed than peripheral nations but less developed than core nations.** Nations in this category typically provide labor and raw materials to core nations within the world system. These nations constitute a midpoint between the core and peripheral nations that promotes the stability and legitimacy of the three-tiered world economy. These nations include South Korea and Taiwan in East Asia, Mexico and Brazil in Latin America, India in South Asia, and Nigeria and South Africa in Africa. Only two global cities are located in semiperipheral nations: São Paulo, Brazil, which is the center of the Brazilian economy, and Singapore, which is the economic center of a multicountry region in Southeast Asia. According to Wallerstein, semiperipheral nations exploit peripheral nations, just as the core nations exploit both the semiperipheral and the peripheral nations.

core nations according to world systems theory, dominant capitalist centers characterized by high levels of industrialization and urbanization.

semiperipheral nations according to world systems theory, nations that are more developed than peripheral nations but less developed than core nations.

Most low-income countries in Africa, South America, and the Caribbean are *peripheral nations*—**nations that are dependent on core nations for capital, have little or no industrialization (other than what may be brought in by core nations), and have uneven patterns of urbanization.** According to Wallerstein (1979, 1984), the wealthy in peripheral nations benefit from the labor of poor workers and from their own economic relations with core-nation capitalists, whom they uphold in order to maintain their own wealth and position. At a global level, uneven economic growth results from capital investment by core nations; disparity between the rich and the poor within the major cities in these nations is increased in the process. The U.S./Mexican border is an example of disparity and urban growth: Transnational corporations have built *maquiladora* plants so that goods can be assembled by low-wage workers to keep production costs down. ▶ Figure 8.4 describes this process. In 2001, there were almost 3,800 maquiladora plants in Mexico, employing 1.3 million peo-

ple, but a number of these plants have subsequently been moved to other nations where the wages that employees earn are even lower.

Not all social analysts agree with Wallerstein's perspective on the hierarchical position of nations in the global economy. However, most scholars acknowledge that nations throughout the world are influenced by a relatively small number of cities and transnational corporations that have prompted a shift from an international to a more global economy (see Knox and Taylor, 1995; Wilson, 1997). Even Wallerstein (1991) acknowledges that world systems theory is an "incomplete, unfinished critique" of long-term, large-scale social change that influences global inequality.

The New International Division of Labor Theory

Although the term *world trade* has long implied that there is a division of labor between societies, the nature and extent of this division have recently

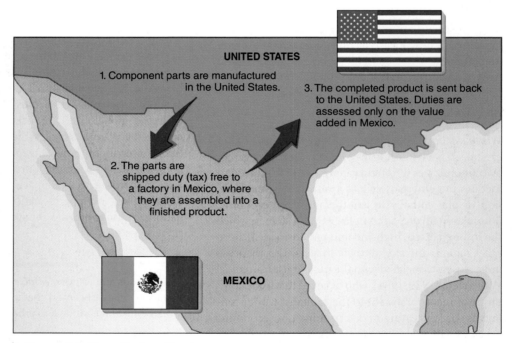

▶ **Figure 8.4 Maquiladora Plants**
Here is the process by which transnational corporations establish plants in Mexico so that profits can be increased by using low-wage workers there to assemble products that are then brought into the United States for sale.

been reassessed based on the changing nature of the world economy. According to the *new international division of labor theory,* commodity production is being split into fragments that can be assigned to whichever part of the world can provide the most profitable combination of capital and labor. Consequently, the new international division of labor has changed the pattern of geographic specialization between countries, whereby high-income countries have now become dependent on low-income countries for labor. The low-income countries provide transnational corporations with a situation in which they can pay lower wages and taxes and face fewer regulations regarding workplace conditions and environmental protection (Waters, 1995). Overall, a global manufacturing system has emerged in which transnational corporations establish labor-intensive, assembly-oriented export production, ranging from textiles and clothing to technologically sophisticated exports such as computers, in middle- and lower-income nations (Gereffi, 1994). At the same time, manufacturing technologies are shifting from the large-scale, mass-production assembly lines of the past toward a more flexible production process involving microelectronic technologies. Even service industries—such as processing insurance claims forms—that were formerly thought to be less mobile have become exportable through electronic transmission and the Internet. The global nature of these activities has been referred to as *global commodity chains,* a complex pattern of international labor and production processes that results in a finished commodity ready for sale in the marketplace.

Some commodity chains are producer-driven, whereas others are buyer-driven. *Producer-driven commodity chains* is the term used to describe industries in which transnational corporations play a central part in controlling the production process. Industries that produce automobiles, computers, and other capital- and technology-intensive products are typically producer-driven. In contrast, *buyer-driven commodity chains* is the term used to refer to industries in which large retailers, brand-name merchandisers, and trading companies set up decentralized production networks in various middle- and low-income countries. This type of chain is most common in labor-intensive, consumer-goods industries such as toys, garments, and footwear

(Gereffi, 1994). Athletic footwear companies such as Nike and Reebok and clothing companies like The Gap and Liz Claiborne are examples of the buyer-driven model. Because these products tend to be labor intensive at the manufacturing stage, the typical factory system is very competitive and globally decentralized. Workers in buyer-driven commodity chains are often exploited by low wages, long hours, and poor working conditions. In fact, most workers cannot afford the products they make. Tini Heyun Alwi, who works on the assembly line of the shoe factory in Indonesia that makes Reebok sneakers, is an example: "I think maybe I could work for a month and still not be able to buy one pair" (qtd. in Goodman, 1996: F1). Because Tini earns only 2,600 Indonesian rupiah ($1.28) per day working a ten-hour shift six days a week, her monthly income would fall short of the retail price of the athletic shoes (Goodman, 1996). Sociologist Gary Gereffi (1994: 225) explains the problem with studying the new global patterns as follows:

> The difficulty may lie in the fact that today we face a situation where (1) the political unit is *national,* (2) industrial production is *regional,* and (3) capital movements are *international.* The rise of Japan and the East Asian [newly industrializing countries] in the 1960s and 1970s is the flip side of the "deindustrialization" that occurred in the United States and much of Europe. Declining industries in North America have been the growth industries in East Asia.

As other analysts suggest, these changes have been a mixed bag for people residing in these countries. For example, Indonesia has been able to woo foreign business into the country, but workers have experienced poverty despite working full time in factories making such consumer goods as Nike tennis shoes (Gargan, 1996). As employers feel pressure from workers to raise wages, clashes erupt between

peripheral nations according to world systems theory, nations that are dependent on core nations for capital, have little or no industrialization (other than what may be brought in by core nations), and have uneven patterns of urbanization.

Sociology *Works!*

Why *Place* Matters in Global Poverty

We're deadly poor. We grow just enough food for ourselves to eat, with no surplus grain. We don't have to pay the grain tax anymore, but our lives aren't much better.

—Zhou Zhiwen, a woman who lives in Yangmiao, China, describing what rural poverty is like in her village (qtd. in French, 2008: YT4)

Zhou Zhiwen lives in an area of China that has largely been untouched by the economic boom in her country. Even with the recent abolition of agricultural taxes for people who are impoverished, local villagers such as Zhou continue to live in abject poverty. As more people have risen out of poverty in China's urban centers in recent decades, poverty in the rural areas, mountainous regions, and deserts remains severe. According to some villagers, the central government is "out of touch with rural realities in places like this," and officials have made little effort to take care of the rural poor (French, 2008). China's poverty is widespread, and the income gap between rural and urban residents has widened over the past three decades (IFAD, 2002). Many people live close to, or below, the minimum standard for poverty: Approximately 350 million people in China live below the international poverty line of $1 per day (World Bank, 2007).

For many years, sociologists studying poverty have focused on differences in rural and urban poverty. Throughout the world, they have found that *place* does matter when it comes to finding the deepest pockets of poverty (Rural Policy Research Institute, 2004). Where people live strongly influences how much money they will make, and income inequalities are important indicators of the life chances of entire families. In some developing countries, the rural poor rely primarily on agriculture, fishing, forestry, and sometimes small-scale industries and services for their livelihood (Khan, 2001). When they are unable to derive sufficient economic resources from these endeavors, little else is available for them. Some migrate to urban centers in hopes of finding new opportunities, but many remain behind, living in grinding poverty. When sociologists speak of "place," they are referring to such things as an area's natural environment, which includes its climate, natural resources, and degree of isolation (Rural Policy Research Institute, 2004). Place also involves the economic structure in the area, such as the extent to which adequate amounts of food can be raised to meet people's needs, or whether an individual can earn sufficient money to purchase food.

Can rural poverty be reduced? According to some social policy analysts, broad economic stability, competitive markets, and public investment in *physical* and *social* infrastructure are important prerequisites for a reduction in rural poverty in developing nations (Khan, 2001). From this perspective, a major reduction in rural poverty in China will occur only if people have access to land and credit, education, health care, support services, and entitlements to food through well-designed public works programs and other transfer mechanisms.

Reflect & Analyze

Whether changes that reduce poverty will occur in China's future remains to be seen, but the sociological premise that *place* matters in regard to poverty remains a valid assumption in helping us explain global poverty and inequality. Can you apply this idea to rural and urban areas with which you are familiar in the United States? Why are issues such as this important to each of us even if we do not live in a rural area and have no personal experience with poverty?

the workers and managers or owners. Similarly, the governments in these countries fear that rising wages and labor strife will drive away the businesses, sometimes leaving behind workers who have no other hopes for employment and become more impoverished than they previously were. Moreover, in situations where government officials were benefiting from the presence of the companies, they also become losers if the workers rebel against their pay or working conditions.

Although most discussions of the new international division of labor focus on changes occurring in the lives of people residing in industrialized urban areas of developing nations, millions of people continue to live in grinding poverty in rural regions of these countries (see "Sociology Works!").

Global Inequality in the Future

As we have seen, social inequality is vast both within and among the countries of the world. Even in high-income nations where wealth is highly concentrated, many poor people coexist with the affluent. In middle- and low-income countries, there are small pockets of wealth in the midst of poverty and despair.

What are the future prospects for greater equality across and within nations? Not all social scientists agree on the answer to this question. Depending on the theoretical framework they apply in studying global inequality, social analysts may describe either an optimistic or a pessimistic scenario for the future. Moreover, some analysts highlight the human rights issues embedded in global inequality, whereas others focus primarily on an economic framework.

In some regions, persistent and growing poverty continues to undermine human development and future possibilities for socioeconomic change. Gross inequality has high financial and quality-of-life costs to people, even among those who are not the poorest of the poor. In the future, continued population growth, urbanization, environmental degradation, and violent conflict threaten even the meager living conditions of those residing in low-income nations. From this approach, the future looks dim not only for people in low-income and middle-income countries but also for those in high-income countries, who will see their quality of life diminish as natural resources are depleted, the environment is polluted, and high rates of immigration and global political unrest threaten the high standard of living that many people have previously enjoyed. According to some social analysts, transnational corporations and financial institutions such as the World Bank and the International Monetary Fund will further solidify and control a globalized economy, which will transfer the power to make significant choices to these organizations

and away from the people and their governments. As a result, further loss of resources and means of livelihood will affect people and countries around the globe.

As a result of global corporate domination, there could be a leveling out of average income around the world, with wages falling in the high-income countries and wages increasing significantly in low- and middle-income countries. If this pessimistic scenario occurs, there is likely to be greater polarization of the rich and the poor and more potential for ethnic and national conflicts over such issues as worsening environmental degradation and who has the right to natural resources. For example, pulp-and-paper companies in Indonesia, along with palm oil plantation owners, have continued clearing land for crops by burning off vast tracts of jungle, producing high levels of smog and pollution across seven Southeast Asian nations and creating havoc for millions of people (Mydans, 1997).

On the other hand, a more optimistic scenario is also possible. With modern technology and worldwide economic growth, it might be possible to

© Scott Olson/Getty Images

The current trend of using grain-based ethanol to power cars and trucks has proved to be a financial boon to some U.S. farmers but has had another effect as well: The price of grain has increased worldwide, making it even harder for people from developing countries to procure enough food to eat.

Box 8.3 You Can Make a Difference

Global Networking to Reduce World Hunger and Poverty

We, the people of the world, will mobilize the forces of transnational civil society behind a widely shared agenda that binds our many social movements in pursuit of just, sustainable, and participatory human societies. In so doing we are forging our own instruments and processes for redefining the nature and meaning of human progress and for transforming those institutions that no longer respond to our needs. We welcome to our cause all people who share our commitment to peaceful and democratic change in the interest of our living planet and the human societies it sustains.

> —International NGO Forum, United Nations Conference on Environment and Development, Rio de Janeiro, Brazil, June 12, 1992 (qtd. in Korten, 1996: 333)

If everyone lit just one little candle, what a bright world this would be.

> —line from the 1950s theme song for Bishop Fulton J. Sheen's television series *Life Is Worth Living* (Sheen, 1995: 245)

© AP Images/Sandra Boulanger

Willie Colon, a Puerto Rican salsa star and spokesperson for CARE, visited this Bolivian classroom as part of that international relief organization's project to reduce women's poverty.

---

reduce absolute poverty and to increase people's opportunities. Among the trends cited by the Human Development Report (United Nations Development Programme, 2003) that have the potential to bring about more sustainable patterns of development are the socioeconomic progress made in many low- and middle-income countries over the past thirty years as technological, social, and environmental improvements have occurred. For example, technological innovation continues to improve living standards for some people. Fertility rates are declining in some regions (but remain high in others, where there remains grave cause for concern about the availability of adequate natural resources for the future). Finally, health and education may continue to improve in lower-income countries. According to the Human Development Report (United Nations Development Programme, 2003), healthy, educated populations are crucial for the future in order to reduce global poverty. The education of women is of primary importance in the future if global inequality is to be reduced. As one analyst stated, "If you educate a boy, you educate a human being. If you educate a girl, you educate generations" (Buvinić, 1997: 49). All aspects of schooling and training are crucial for the future, including agricultural extension services in rural areas to help women farmers in regions such as western Kenya produce more crops to feed their families. As we saw earlier in the chapter, easier access to water can make a crucial difference in people's lives. Mayra

When many of us think about problems such as world poverty, we tend to see ourselves as powerless to bring about change in so vast an issue. However, a recurring message from social activists and religious leaders is that each person can contribute something to the betterment of other people and sometimes the entire world.

An initial way for each of us to become involved is to become more informed about global issues and to learn how we can contribute time and resources to organizations seeking to address social issues such as illiteracy and hunger. We can also find out about meetings and activities of organizations and participate in online discussion forums where we can express our opinions, ask questions, share information, and interact with other people interested in topics such as international relief and development. At first, it may not feel like you are doing much to address global problems; however, information and education are the first steps to promoting greater understanding of social problems and of the world's people, whether they reside in high-, middle-, or low-income countries and regardless of their individual socioeconomic position. Likewise, it is important to help our own nation's children understand that they can make a difference in ending hunger in the United States and other nations.

Would you like to function as a catalyst for change? You can learn how to proceed by gathering information from organizations that seek to reduce problems such as poverty and to provide forums for interacting with other people. Here are a few starting points for your search:

- CARE International is a confederation of 10 national members in North America, Europe, Japan, and Australia. CARE assists the world's poor in their efforts to achieve social and economic well-being. Its work reaches 25 million people in 53 nations in Africa, Asia, Latin America, and Eastern Europe. Programs include emergency relief, education, health and population, children's health, reproductive health, water and sanitation, small economic activity development, agriculture, community development, and environment. Contact CARE at 151 Ellis Street, NE, Atlanta, GA 30303-2439. On the Internet: **http://www.care.org**

Other organizations fighting world hunger and health problems include the following:

- World Hunger Year ("WHY"): **http://www.worldhungeryear.org**

- "Kids Can Make a Difference," an innovative guide developed by WHY: **http://www.kidscanmakeadifference.org**

- World Health Organization: **http://www.who.int**

Buvinić, of the Inter-American Development Bank, puts global poverty in perspective for people living in high-income countries by pointing out that their problems are our problems. She provides the following example:

Reina is a former guerilla fighter in El Salvador who is being taught how to bake bread under a post-civil war reconstruction program. But as she says, "the only thing I have is this training and I don't want to be a baker. I have other dreams for my life."

Once upon a time, women like Reina . . . only migrated [to the United States] to follow or find a husband. This is no longer the case. It is likely that Reina, with few opportunities in her own country, will sooner or later join the rising number of female migrants who leave families and children behind to seek better paying work in the United States and other industrial countries. Wisely spent foreign aid can give Reina the chance to realize her dreams in her *own* country. (Buvinić, 1997: 38, 52)

From this viewpoint, we can enjoy prosperity only by ensuring that other people have the opportunity to survive and thrive in their own surroundings (see Box 8.3). The problems associated with global poverty are therefore of interest to a wide-ranging set of countries and people.

Chapter Review

• What is global stratification, and how does it contribute to economic inequality?

Global stratification refers to the unequal distribution of wealth, power, and prestige on a global basis, which results in people having vastly different lifestyles and life chances both within and among the nations of the world. Today, the income gap between the richest and the poorest 20 percent of the world population continues to widen, and within some nations the poorest one-fifth of the population has an income that is only a slight fraction of the overall average per capita income for that country.

• How are global poverty and human development related?

Income disparities are not the only factor that defines poverty and its effect on people. The United Nations' Human Development Index measures the level of development in a country through indicators such as life expectancy, infant mortality rate, proportion of underweight children under age five (a measure of nourishment and health), and adult literacy rate for low-income, middle-income, and high-income countries.

• What is modernization theory?

Modernization theory is a perspective that links global inequality to different levels of economic development and suggests that low-income economies can move to middle- and high-income economies by achieving self-sustained economic growth.

• How does dependency theory differ from modernization theory?

Dependency theory states that global poverty can at least partially be attributed to the fact that the low-income countries have been exploited by the high-income countries. Whereas modernization theory focuses on how societies can reduce inequality through industrialization and economic development, dependency theorists see the greed of the rich countries as a source of increasing impoverishment of the poorer nations and their people.

• What is world systems theory, and how does it view the global economy?

According to world systems theory, the capitalist world economy is a global system divided into a hierarchy of three major types of nations: Core nations are dominant capitalist centers characterized by high levels of industrialization and urbanization, semiperipheral nations are more developed than peripheral nations but less developed than core nations, and peripheral nations are those countries that are dependent on core nations for capital, have little or no industrialization (other than what may be brought in by core nations), and have uneven patterns of urbanization.

• What is the new international division of labor theory?

The new international division of labor theory is based on the assumption that commodity production is split into fragments that can be assigned to whichever part of the world can provide the most profitable combination of capital and labor. This division of labor has changed the pattern of geographic specialization between countries, whereby high-income countries have become dependent on low-income countries for labor. The low-income countries provide transnational corporations with a situation in which they can pay lower wages and taxes, and face fewer regulations regarding workplace conditions and environmental protection.

www.cengage.com/login

Register for a Student eResource account to maximize your study time online using CengageNOW. First take the system's diagnostic pre-test, and then follow the personalized study plan that is created for you to help you review this chapter. The study plan will

- help you identify areas on which you should concentrate;
- provide interactive exercises to help you master the chapter concepts; and
- provide a post-test to confirm you are ready to move on to the next chapter.

Key Terms

core nations 271

dependency theory 269

global stratification 254

modernization theory 267

peripheral nations 272

semiperipheral nations 271

Questions for Critical Thinking

1. You have decided to study global wealth and poverty. How would you approach your study? What research methods would provide the best data for analysis? What might you find if you compared your research data with popular presentations—such as films and advertising—of everyday life in low- and middle-income countries?

2. How would you compare the lives of poor people living in the low-income nations of the world with those in central cities and rural areas of the United States? In what ways are their lives similar? In what ways are they different?

3. Should U.S. foreign policy include provisions for reducing poverty in other nations of the world? Should U.S. domestic policy include provisions for reducing poverty in the United States? How are these issues similar? How are they different?

4. Using the theories discussed in this chapter, devise a plan to alleviate global poverty. Assume that you have the necessary wealth, political power, and other resources necessary to reduce the problem. Share your plan with others in your class, and create a consolidated plan that represents the best ideas and suggestions presented.

The Kendall Companion Website

www.cengage.com/sociology/kendall

Supplement your review of this chapter by going to the text's companion website, where you can take tutorial quizzes, use flash cards to master key terms, follow live links to useful websites, and explore the other study and research resources you'll find there, such as a comprehensive interactive sociology timeline, GSS Data, and Census 2000 information, much of it presented visually in maps.

The problem is [Jeff Kent] doesn't know how to deal with African-American people. I think that's what's causing everything. It's a pattern of things that have been said off the cuff that I don't interpret as funny. It may be funny to him, but it's not funny to Milton Bradley. But I don't take offense to that because we all joke about race in [the locker room].

Me being an African American is the most important thing to me, more important than baseball. White people never want to see race with anything. But there's race involved in baseball. That's why there's less than 9 percent African-American representation in the game. I'm one of the few African-Americans that starts [for the Los Angeles Dodgers].

Growing up in [Los Angeles], I know how to deal with all types of people, and I do it on an everyday basis. But some people don't deal with all different types of people

Sporting events may bring out the best or the worst in some people. Major League Baseball player Milton Bradley accused teammate Jeff Kent of making racially insensitive remarks about Bradley, creating much controversy among fans and the media. Do we have a responsibility to be thoughtful about what we say to and about others?

every day, and therefore don't know how to handle situations when they arise.

—Major League Baseball player Milton Bradley (then with the Los Angeles Dodgers and now with the Texas Rangers) accusing teammate Jeff Kent of making racially insensitive remarks about Bradley's performance (qtd. in *ABC News,* 2005)

In response to Bradley's comments, Jeff Kent told sports reporters the following:

[Milton Bradley] can go ahead and say those types of things, and it comes from an incident that he still doesn't get. And that's a shame. If you think that I've got a problem with African-Americans, then go talk to Dusty Baker. Go talk to Dave Winfield, who took me under his wing. Go talk to Joe Carter—all the guys that I idolized in this game and all the veteran players who taught me how to play this game [referring to various African-American players]. That's a shame, and I take offense to that. . . . (qtd. in *ABC News,* 2005).

Many people think verbal feuds between athletes such as Milton Bradley and Jeff Kent are nothing more than sensationalist "jock talk" hyped by the media to draw larger audiences and generate more gate revenues. Clearly, Bradley thought that Kent had made racially disparaging remarks about his capabilities as a baseball player, and he was not pleased. However, Kent believed that he had done no such thing and that Bradley was the problem. Eventually, Kent remained with the Dodgers, and

Chapter Focus Question

How is sports a reflection of racial and ethnic relations in the United States and other nations?

Bradley joined the Oakland Athletics, telling reporters that he wanted to "go with the flow" (ESPN. com, 2006).

Ironically, when public attention and the media spotlight are focused on fights over allegedly racist comments made by star athletes or other celebrities, the debate typically descends into a "food fight" that buries other pressing issues of racial and ethnic inequality in sports and other areas of social life. For example, members of the media largely overlooked Bradley's attempt to call attention to the shrinking percentage of African Americans in professional baseball. In 2006, 8.4 percent of all MLB players were African American as compared with 27 percent in 1975 (Blum, 2007; Zirin, 2005). Although changes in the composition of some teams have occurred, five MLB teams (the Boston Red Sox, Baltimore Orioles, Houston Astros, Atlanta Braves, and Colorado Rockies) had no African American players on their active rosters in 2005 (Zirin, 2005).

Although racism is not always the key issue in disputes between athletes such as Bradley and Kent, their comments show the strong racial–ethnic overtones of many contentious arguments in the United States and other nations. To understand the dynamics of contemporary race and ethnicity, we must look deeper than people's inflammatory comments as sometimes reported in the media. We must be aware not only of "in your face" statements and actions that may constitute racism but also of the many subtle forms of discrimination that affect many people and perpetuate racial and ethnic injustice. In this chapter, sports is used as an example of the effects of race and ethnicity on people's lives because this area of social life shows us how, for more than a century, people who have been singled out for negative treatment on the basis of their perceived race or ethnicity have sought to overcome prejudice and discrimination through their determination in endeavors that can be judged based on objective standards ("What's the final score?") rather than subjective standards ("Do I like this person based on characteristics such as race or ethnicity?"). Before reading on, test your knowledge about race, ethnicity, and sports by taking the quiz in Box 9.1.

Race and Ethnicity

What is race? Some people think it refers to skin color (the Caucasian "race"); others use it to refer to a religion (the Jewish "race"), nationality (the British "race"), or the entire human species (the human "race") (Marger, 2009). Popular usages of race have been based on the assumption that a race is a grouping or classification based on *genetic* variations in physical appearance, particularly skin color. However, social scientists and biologists dispute the idea that biological race is a meaningful concept. In fact, the idea of race has little meaning in a biological sense because of the enormous amount of interbreeding that has taken place within the human population. For these reasons, sociologists sometimes place "race" in quotation marks to show that categorizing individuals and population groups on biological characteristics is neither accurate nor based on valid distinctions between the genetic makeup of differently identified "races." Today, sociologists emphasize that race is a *socially constructed reality,* not a biological one. From this approach, the social significance that people accord to race is more significant than any biological differences that might exist among people who are placed in arbitrary categories.

A *race* **is a category of people who have been singled out as inferior or superior, often on the basis of real or alleged physical characteristics such as skin color, hair texture, eye shape, or other subjectively selected attributes** (Feagin and Feagin, 2003). Categories of people frequently thought of as racial groups include Native Americans, Mexican Americans, African Americans, and Asian Americans.

As compared with race, *ethnicity* defines individuals who are believed to share common characteristics that differentiate them from the other collectivities in a society. An *ethnic group* **is a collection of people distinguished, by others or by themselves, primarily on the basis of cultural or nationality characteristics** (Feagin and Feagin, 2003). Examples of ethnic groups include Jewish Americans, Irish Americans, and Italian Americans. Ethnic groups share five main characteristics: (1) *unique cultural traits,* such as language, clothing, holidays,

Box 9.1 Sociology and Everyday Life

How Much Do You Know About Race, Ethnicity, and Sports?

True	False	
T	F	1. Because sports are competitive and fans, coaches, and players want to win, the color of the players has not been a factor, only their performance.
T	F	2. In the late 1800s and early 1900s, boxing provided social mobility for some Irish, Jewish, and Italian immigrants.
T	F	3. African Americans who competed in boxing matches in the late 1800s often had to agree to lose before they could obtain a match.
T	F	4. Racially linked genetic traits explain many of the differences among athletes.
T	F	5. All racial and ethnic groups have viewed sports as a means to become a part of the mainstream.
T	F	6. Until recently, the positions of quarterback and kicker in the National Football League have been held almost exclusively by white players.
T	F	7. In recent years, players of color have moved into coaching, management, and ownership positions in professional sports.
T	F	8. The odds are good that many outstanding high school and college athletes will make the pros if they do not get injured.
T	F	9. Racism and sexism appear to be on the decline in sports in the United States.

Answers on page 284.

or religious practices; (2) *a sense of community;* (3) *a feeling of ethnocentrism;* (4) *ascribed membership from birth;* and (5) *territoriality,* or the tendency to occupy a distinct geographic area (such as Little Italy or Little Havana) by choice and/or for self-protection. Although some people do not identify with any ethnic group, others participate in social interaction with individuals in their ethnic group and feel a sense of common identity based on cultural characteristics such as language, religion, or politics. However, ethnic groups are not only influenced by their own past history but also by patterns of ethnic domination and subordination in societies.

The Social Significance of Race and Ethnicity

Race and ethnicity take on great social significance because how people act in regard to these terms drastically affects other people's lives, including what opportunities they have, how they are treated, and even how long they live. According to the sociologists Michael Omi and Howard Winant (1994: 158), race "permeates every institution, every relationship, and every individual" in the United States:

> As we . . . compare real estate prices in different neighborhoods, select a radio channel to enjoy while we drive to work, size up a potential client, customer, neighbor, or teacher, stand in line at the unemployment office, or carry out a

race a category of people who have been singled out as inferior or superior, often on the basis of physical characteristics such as skin color, hair texture, and eye shape.

ethnic group a collection of people distinguished, by others or by themselves, primarily on the basis of cultural or nationality characteristics.

Box 9.1 Sociology and Everyday Life

Answers to the Sociology Quiz on Race, Ethnicity, and Sports

1. **False.** Discrimination has been pervasive throughout the history of sports in the United States. For example, African American athletes, regardless of their abilities, were excluded from white teams for many years.

2. **True.** Irish Americans were the first to dominate boxing, followed by Jewish Americans and then Italian Americans. Boxing, like other sports, was a source of social mobility for some immigrants.

3. **True.** Promoters, who often set up boxing matches that pitted fighters by race, assumed that white fans were more likely to buy tickets if the white fighters frequently won.

4. **False.** Although some scholars and journalists have used biological or genetic factors to explain the achievements of athletes, sociologists view these explanations as being based on the inherently racist assumption that people have "natural" abilities (or disabilities) because of their race or ethnicity.

5. **False.** Some racial and ethnic groups—including Chinese Americans and Japanese Americans—have not viewed sports as a means of social mobility.

6. **True.** As late as the 1990s, whites accounted for about 90 percent of the quarterbacks and kickers on NFL teams. However, this changed early in the twenty-first century, and today there are some African Americans playing virtually every position on all professional football teams.

7. **False.** Although more African American players are employed by these teams (especially in basketball), their numbers have not increased significantly as owners or in coaching and management positions.

8. **False.** The odds of becoming a professional athlete are very low. The percentage of high school football players who make it to the pros is estimated at about .14 percent; for basketball, about .06 percent; and for baseball, about .18 percent. The percentages of college athletes who make it to the pros are a little higher: 3.5 percent for football, 2.6 percent for basketball, and 4.3 percent for baseball.

9. **False.** Even as people of color and white women have made gains on collegiate and professional teams, scholars have documented the continuing significance of racial and gender discrimination in sports.

Sources: Based on Coakley, 2004; Messner, Duncan, and Jensen, 1993; and Nelson, 1994.

---

thousand other normal tasks, we are compelled to think racially, to use the racial categories and meaning systems into which we have been socialized. (Omi and Winant, 1994: 158)

Historically, stratification based on race and ethnicity has pervaded all aspects of political, economic, and social life. Consider sports as an example. Throughout the early history of the game of baseball, many African Americans had outstanding skills as players but were categorically excluded from Major League teams because of their skin color. Even after Jackie Robinson broke the "color line" to become the first African American in the Major Leagues in 1947, his experience was marred by racial slurs, hate letters, death threats against his infant son, and assaults on his wife (Ashe, 1988; Peterson, 1992/1970). With some professional athletes from diverse racial–ethnic categories having multi-million-dollar contracts and lucrative endorsement deals, it is easy to assume that racism in sports—as well as in the larger society—is a thing of the past.

However, this *commercialization* of sports does not mean that racial prejudice and discrimination no longer exist (Coakley, 2004).

Racial Classifications and the Meaning of Race

If we examine racial classifications throughout history, we find that in ancient Greece and Rome a person's race was the group to which she or he belonged, associated with an ancestral place and culture. From the Middle Ages until about the eighteenth century, a person's race was based on family and ancestral ties, in the sense of a *line,* or ties to a national group. During the eighteenth century, physical differences such as the darker skin hues of Africans became associated with race, but racial divisions were typically based on differences in religion and cultural tradition rather than on human biology. With the intense (though misguided) efforts that surrounded the attempt to justify black slavery and white dominance in all areas of life during the second half of the nineteenth century, *races* came to be defined as distinct biological categories of people who were not all members of the same family but who shared inherited physical and cultural traits that were alleged to be different from those traits shared by people in other races. Hierarchies of races were established, placing the "white race" at the top, the "black race" at the bottom, and others in between.

Despite much evidence to the contrary, many people still assume that racial purity still exists. Prior to the 2000 census, for example, the true diversity of the U.S. population was not revealed in census data because multiracial individuals were forced to either select a single race as being their "race" or to select the vague category of "other." Professional golfer Tiger Woods is an example of how people often have a mixed racial heritage. Woods describes himself as one-half Asian American (one-fourth Thai and one-fourth Chinese), one-eighth white, one-eighth Native American, and one-fourth African American (White, 1997). Census 2000 made it possible—for the first time—for individuals to classify themselves as being of more than one race.

Over time, categories of official racial classifications may create a sense of group membership or "consciousness of kind" among people within arbitrary racial classifications. When people of European descent were classified as "white," some began to see themselves as different from "nonwhite." Consequently, Jewish, Italian, and Irish immigrants may have felt more a part of the Northern European white mainstream in the late eighteenth and early nineteenth centuries. Today, some Chinese Americans, Japanese Americans, Korean Americans, and Filipino Americans think of themselves as "Asian Americans," whereas many others do not.

Dominant and Subordinate Groups

The terms *majority group* and *minority group* are widely used, but their meanings are less clear as the composition of the U.S. population continues to

© Jeff Greenberg/PhotoEdit

Miami's Little Havana is an ethnic enclave where people participate in social interaction with other individuals in their ethnic group and feel a sense of shared identity. Ethnic enclaves provide economic and psychological support for recent immigrants as well as for those who were born in the United States.

change. Accordingly, many sociologists prefer the terms *dominant* and *subordinate* to identify power relationships that are based on perceived racial, ethnic, or other attributes and identities. To sociologists, a **dominant group is one that is advantaged and has superior resources and rights in a society** (Feagin and Feagin, 2003). In the United States, whites with Northern European ancestry (often referred to as Euro-Americans, white Anglo-Saxon Protestants, or WASPs) have been considered to be the dominant group for many years. A **subordinate group is one whose members, because of physical or cultural characteristics, are disadvantaged and subjected to unequal treatment by the dominant group and who regard themselves as objects of collective discrimination.** Historically, African Americans and other persons of color have been considered to be subordinate-group members, particularly when they are from lower-income categories.

It is important to note that, in the sociological sense, the word *group* as used in these two terms is misleading because people who merely share ascribed racial or ethnic characteristics do not constitute a group. However, the terms *dominant group* and *subordinate group* do give us a way to describe relationships of advantage/disadvantage and power/exploitation that exist in contemporary nations.

© AP Images

Contemporary prejudice and discrimination cannot be understood without taking into account the historical background. School integration in the 1950s was accomplished despite white resistance. Today, integration in education, housing, and many other areas of social life remains a pressing social issue.

Prejudice

Although there are various meanings of the word *prejudice,* sociologists define **prejudice as a negative attitude based on faulty generalizations about members of specific racial, ethnic, or other groups.** The term *prejudice* is from the Latin words *prae* ("before") and *judicium* ("judgment"), which means that people may be biased either for or against members of other groups even before they have had any contact with them. Although prejudice can be either *positive* (bias in favor of a group—often our own) or *negative* (bias against a group—one we deem less worthy than our own), it most often refers to the negative attitudes that people may have about members of other racial or ethnic groups.

Stereotypes

Prejudice is rooted in ethnocentrism and stereotypes. When used in the context of racial and ethnic relations, *ethnocentrism* refers to the tendency to regard one's own culture and group as the standard—and thus superior—whereas all other groups are seen as inferior. Ethnocentrism is maintained and perpetuated by **stereotypes—overgeneralizations about the appearance, behavior, or other characteristics of members of particular categories.** Although stereotypes can be either positive or negative, examples of negative stereotyping abound in sports. Think about the Native American names, images, and mascots used by sports teams such as the Atlanta Braves, Cleveland Indians, and Washington

Redskins. Members of Native American groups have been actively working to eliminate the use of stereotypic mascots (with feathers, buckskins, beads, spears, and "warpaint"), "Indian chants," and gestures (such as the "tomahawk chop"), which they claim trivialize and exploit Native American culture. College and university sports teams with Native American names and logos also remain the subject of controversy in the twenty-first century. According to sociologist Jay Coakley (2004), the use of stereotypes and words such as *redskin* symbolizes a lack of understanding of the culture and heritage of native peoples and is offensive to many Native Americans. Although some people see these names and activities as "innocent fun," others view them as a form of racism.

According to the frustration–aggression hypothesis, members of white supremacy groups such as the Ku Klux Klan often use members of subordinate racial and ethnic groups as scapegoats for societal problems over which they have no control.

Racism

What is racism? *Racism* **is a set of attitudes, beliefs, and practices that is used to justify the superior treatment of one racial or ethnic group and the inferior treatment of another racial or ethnic group.** The world has seen a long history of racism: It can be traced from the earliest civilizations. At various times throughout U.S. history, various categories of people, including the Irish Americans, Italian Americans, Jewish Americans, African Americans, and Latinos/as, have been the objects of racist ideology.

Racism may be overt or subtle. Overt racism is more blatant and may take the form of public statements about the "inferiority" of members of a racial or ethnic group. In sports, for example, calling a player of color a derogatory name, participating in racist chanting during a sporting event, and writing racist graffiti in a team's locker room are all forms of overt racism. These racist actions are blatant, but subtle forms of racism are often hidden from sight and more difficult to prove. Examples of subtle racism in sports include those descriptions of African American athletes which suggest that they have "natural" abilities and are better suited for team

positions requiring speed and agility. By contrast, whites are described as having the intelligence, dependability, and leadership and decision-making

dominant group a group that is advantaged and has superior resources and rights in a society.

subordinate group a group whose members, because of physical or cultural characteristics, are disadvantaged and subjected to unequal treatment by the dominant group and who regard themselves as objects of collective discrimination.

prejudice a negative attitude based on faulty generalizations about members of selected racial and ethnic groups.

stereotypes overgeneralizations about the appearance, behavior, or other characteristics of members of particular categories.

racism a set of attitudes, beliefs, and practices that is used to justify the superior treatment of one racial or ethnic group and the inferior treatment of another racial or ethnic group.

skills needed in positions requiring higher levels of responsibility and control.

Theories of Prejudice

Are some people more prejudiced than others? To answer this question, some theories focus on how individuals may transfer their internal psychological problem onto an external object or person. Others look at factors such as social learning and personality types.

The *frustration–aggression hypothesis* states that people who are frustrated in their efforts to achieve a highly desired goal will respond with a pattern of aggression toward others (Dollard et al., 1939). The object of their aggression becomes the *scapegoat*— **a person or group that is incapable of offering resistance to the hostility or aggression of others** (Marger, 2009). Scapegoats are often used as substitutes for the actual source of the frustration. For example, members of subordinate racial and ethnic groups are often blamed for societal problems (such as unemployment or an economic recession) over which they have no control.

According to some symbolic interactionists, prejudice results from social learning; in other words, it is learned from observing and imitating significant others, such as parents and peers. Initially, children do not have a frame of reference from which to question the prejudices of their relatives and friends. When they are rewarded with smiles or laughs for telling derogatory jokes or making negative comments about outgroup members, children's prejudiced attitudes may be reinforced.

Psychologist Theodor W. Adorno and his colleagues (1950) concluded that highly prejudiced individuals tend to have an *authoritarian personality,* **which is characterized by excessive conformity, submissiveness to authority, intolerance, insecurity, a high level of superstition, and rigid, stereotypic thinking** (Adorno et al., 1950). This type of personality is most likely to develop in a family environment in which dominating parents who are anxious about status use physical discipline but show very little love in raising their children (Adorno et al., 1950). Other scholars have linked prejudiced attitudes to traits such as submissiveness to authority, extreme anger toward outgroups, and

conservative religious and political beliefs (Altemeyer, 1981, 1988; Weigel and Howes, 1985).

Discrimination

Whereas prejudice is an attitude, *discrimination* **involves actions or practices of dominant-group members (or their representatives) that have a harmful impact on members of a subordinate group.** Prejudiced attitudes do not always lead to discriminatory behavior. As shown in ▶ Figure 9.1, the sociologist Robert Merton (1949) identified four combinations of attitudes and responses. Unprejudiced nondiscriminators are not personally prejudiced and do not discriminate against others. For example, two players on a professional sports team may be best friends although they are of different races. Unprejudiced discriminators may have no personal prejudice but still engage in discriminatory behavior because of peer-group pressure or economic, political, or social

	Prejudiced attitude?	Discriminatory behavior?
Unprejudiced nondiscriminator	No	No
Unprejudiced discriminator	No	Yes
Prejudiced nondiscriminator	Yes	No
Prejudiced discriminator	Yes	Yes

▶ **Figure 9.1 Merton's Typology of Prejudice and Discrimination**
Merton's typology shows that some people may be prejudiced but not discriminate against others. Do you think that it is possible for a person to discriminate against some people without holding a prejudiced attitude toward them? Why or why not?

interests. For example, in some sports a coach might feel no prejudice toward African American players but believe that white fans will accept only a certain percentage of people of color on the team. Prejudiced nondiscriminators hold personal prejudices but do not discriminate due to peer pressure, legal demands, or a desire for profits. For example, a coach with prejudiced beliefs may hire an African American player to enhance the team's ability to win (Coakley, 2004). Finally, prejudiced discriminators hold personal prejudices and actively discriminate against others. For example, a baseball umpire who is personally prejudiced against African Americans may intentionally call a play incorrectly based on that prejudice.

Discriminatory actions vary in severity from the use of derogatory labels to violence against individuals and groups. The ultimate form of discrimination occurs when people are considered to be unworthy to live because of their race or ethnicity. ***Genocide* is the deliberate, systematic killing of an entire people or nation.** Examples of genocide include the killing of thousands of Native Americans by white settlers in North America and the extermination of six million European Jews by Nazi Germany. More recently, the term *ethnic cleansing* has been used to define a policy of "cleansing" geographic areas by forcing persons of other races or religions to flee—or die.

Discrimination also varies in how it is carried out. Individuals may act on their own, or they may operate within the context of large-scale organizations and institutions, such as schools, churches, corporations, and governmental agencies. How does individual discrimination differ from institutional discrimination? ***Individual discrimination* consists of one-on-one acts by members of the dominant group that harm members of the subordinate group or their property.** For example, a person may decide not to rent an apartment to someone of a different race. By contrast, ***institutional discrimination* consists of the day-to-day practices of organizations and institutions that have a harmful impact on members of subordinate groups.** For example, a bank might consistently deny loans to people of a certain race. Institutional discrimination is carried out by the individuals who implement the policies and procedures of organizations.

Sociologist Joe R. Feagin has identified four major types of discrimination:

1. *Isolate discrimination* is harmful action intentionally taken by a dominant-group member against a member of a subordinate group. This type of discrimination occurs without the support of other members of the dominant group in the immediate social or community context. For example, a prejudiced judge may give harsher sentences to all African American defendants but may not be supported by the judicial system in that action.

2. *Small-group discrimination* is harmful action intentionally taken by a limited number of dominant-group members against members of subordinate groups. This type of discrimination is not supported by existing norms or other dominant-group members in the immediate social or community context. For example, a small group of white students may deface a professor's office with racist epithets without the support of other students or faculty members.

3. *Direct institutionalized discrimination* is organizationally prescribed or community-prescribed action that intentionally has a differential and

---

scapegoat a person or group that is incapable of offering resistance to the hostility or aggression of others.

authoritarian personality a personality type characterized by excessive conformity, submissiveness to authority, intolerance, insecurity, a high level of superstition, and rigid, stereotypic thinking.

discrimination actions or practices of dominant-group members (or their representatives) that have a harmful effect on members of a subordinate group.

genocide the deliberate, systematic killing of an entire people or nation.

individual discrimination behavior consisting of one-on-one acts by members of the dominant group that harm members of the subordinate group or their property.

institutional discrimination the day-to-day practices of organizations and institutions that have a harmful impact on members of subordinate groups.

negative impact on members of subordinate groups. These actions are routinely carried out by a number of dominant-group members based on the norms of the immediate organization or community (Feagin and Feagin, 2003). Intentional exclusion of people of color from public accommodations is an example of this type of discrimination.

4. *Indirect institutionalized discrimination* refers to practices that have a harmful effect on subordinate-group members even though the organizationally or community-prescribed norms or regulations guiding these actions were initially established with no intent to harm. For example, special education classes were originally intended to provide extra educational opportunities for children with various types of disabilities. However, critics claim that these programs have amounted to racial segregation in many school districts.

Various types of racial and ethnic discrimination call for divergent remedies if we are to reduce discriminatory actions and practices in contemporary social life. Since the 1950s and 1960s, many U.S. sociologists have analyzed the complex relationship between prejudice and discrimination. Some have reached the conclusion that prejudice is difficult, if not seemingly impossible, to eradicate because of the deeply held racist beliefs and attitudes that are often passed on from person to person and from one generation to the next. However, the persistence of prejudicial attitudes and beliefs does not mean that racial and ethnic discrimination should be allowed to flourish until such a time as prejudice is effectively eliminated. From this approach, discrimination must be tackled aggressively through demands for change and through policies that specifically target patterns of discrimination (see "Sociology Works!").

Sociological Perspectives on Race and Ethnic Relations

Symbolic interactionist, functionalist, and conflict analysts examine race and ethnic relations in different ways. Functionalists focus on the macrolevel intergroup processes that occur between members of dominant and subordinate groups in society. Conflict theorists analyze power and economic differ-entials between the dominant group and subordinate groups. Symbolic interactionists examine how microlevel contacts between people may produce either greater racial tolerance or increased levels of hostility.

Symbolic Interactionist Perspectives

What happens when people from different racial and ethnic groups come into contact with one another? In the *contact hypothesis,* symbolic interactionists point out that contact between people from divergent groups should lead to favorable attitudes and behavior when certain factors are present. Members of each group must (1) have equal status, (2) pursue the same goals, (3) cooperate with one another to achieve their goals, and (4) receive positive feedback when they interact with one another in positive, nondiscriminatory ways (Allport, 1958; Coakley, 2004).

What happens when individuals meet someone who does not conform to their existing stereotype? Frequently, they ignore anything that contradicts the stereotype, or they interpret the situation to support their prejudices (Coakley, 2004). For example, a person who does not fit the stereotype may be seen as an exception—"You're not like other [persons of a particular race]."

When a person is seen as conforming to a stereotype, he or she may be treated simply as one of "you people." Former Los Angeles Lakers basketball star Earvin "Magic" Johnson (1992: 31–32) described how he was categorized along with all other African Americans when he was bused to a predominantly white school:

> On the first day of [basketball] practice, my teammates froze me out. Time after time I was wide open, but nobody threw me the ball. At first I thought they just didn't see me. But I woke up after a kid named Danny Parks looked right at me and then took a long jumper. Which he missed.
>
> I was furious, but I didn't say a word. Shortly after that, I grabbed a defensive rebound and took the ball all the way down for a basket. I did it again and a third time, too.
>
> Finally Parks got angry and said, "Hey, pass the [bleeping] ball."

Sociology *Works!*

Attacking Discrimination to Reduce Prejudice?

Question: Do you think it is possible to reduce racial and ethnic prejudice in the United States by attacking discrimination at the societal level?

Answer: Well, I think since it's individuals who hate each other and do mean stuff, they should work it out for themselves. We don't need the government telling us what to do or how to behave. Like, my dad runs a company, and he says that the government should "butt out" of our business.

—"Brian," an introductory sociology student, stating why he believes that the government has "no business" trying to reduce discrimination (author's notes)

Brian's comment is typical of how many people feel about court rulings and government initiatives over the past sixty years that have sought to reduce the corrosive effects of racial and ethnic discrimination in American life. Based on the widely held axiom that "prejudice causes discrimination," many people argue that discrimination can be alleviated only through changing the attitudes of individuals. From this perspective, discrimination will go away over time if people are encouraged to shed their negative attitudes about members of other racial or ethnic groups. Discarding negative stereotypes about other groups and bringing to light the truth about popular myths regarding the superiority of one's own race, ethnic group, or nationality are widely seen as the best ways of bringing about positive social change. Diversity training sessions held in schools and at the workplace are a classic example of this approach.

Since the civil rights era in the 1960s, however, sociologists have demonstrated that discrimination can be tackled up front and now—rather than waiting for prejudice to diminish—so that people in subordinate racial/ethnic categories can gain a greater measure of human dignity and a variety of opportunities that they otherwise would not have.

Establishing social policies and laws to eliminate specific practices of segregation and discrimination in education, employment, housing, health care, law enforcement, and other areas of public life, for example, has served as a significant starting point for social change that has positively affected generations of people of color, women of all racial and ethnic categories, religious minorities, and many others who have lived outside the mainstream of social life. By enacting legislation that prohibits discrimination based on race, ethnic origin, and color, our nation has sought to provide people with greater equality before the law and access to crucial opportunities and social resources. These changes have, over time, reduced the prejudiced attitudes of many people on a wide variety of issues.

Although progress has been made in reducing some aspects of overt prejudice and institutional discrimination, racism clearly is not a thing of the past. However, if previous sociological research tells us anything about the future, it is that we must continue to tackle not only individual prejudices and discriminatory conduct but also the larger, societal patterns of discrimination—embedded in the organizations and institutions of which we are a part—that restrict freedom, opportunities, and quality of life for all people.

Reflect & Analyze

Using sports as an example, let's think about these questions: How might college sporting events serve to reduce prejudice and discrimination on campus and beyond? How might these same events serve to perpetuate negative stereotypes and popular myths about racial and ethnic "differences"? What do you think? (See Box 9.3, "You Can Make a Difference," for additional discussion on this topic.)

That did it. I slammed down the ball and glared at him. Then I exploded. "I *knew* this would happen!" I said. "That's why I didn't want to come to this [bleeping] school in the first place!"

"Oh, yeah? Well, you people are all the same," he said. "You think you're gonna come in here and do whatever you want? Look, hotshot, your job is to get the rebound. Let us do the shooting."

The interaction between Johnson and Parks demonstrates that when people from different racial and ethnic groups come into contact with one another, they may treat one another as stereotypes, not as individuals. Eventually, Johnson and Parks were able to work out most of their differences. "There's nothing like winning to help people get along," Johnson explained (1992: 32). Although we might hope that nothing like this happens today, there is much evidence that covert discrimination occurs in many sports and social settings.

Symbolic interactionist perspectives make us aware of the importance of intergroup contact and the fact that it may either intensify or reduce racial and ethnic stereotyping and prejudice.

Functionalist Perspectives

How do members of subordinate racial and ethnic groups become a part of the dominant group? To answer this question, early functionalists studied immigration and patterns of dominant- and subordinate-group interactions.

Assimilation *Assimilation* **is a process by which members of subordinate racial and ethnic groups become absorbed into the dominant culture.** To some analysts, assimilation is functional because it contributes to the stability of society by minimizing group differences that might otherwise result in hostility and violence.

Assimilation occurs at several distinct levels, including the cultural, structural, biological, and psychological stages. *Cultural assimilation,* or *acculturation,* occurs when members of an ethnic group adopt dominant-group traits, such as language, dress, values, religion, and food preferences. Cultural assimilation in this country initially followed an "Anglo conformity" model; members of subordinate ethnic groups were expected to conform to the culture of the dominant white Anglo-Saxon population (Gordon, 1964). However, members of some groups refused to be assimilated and sought to maintain their unique cultural identity.

Structural assimilation, or *integration,* occurs when members of subordinate racial or ethnic groups gain acceptance in everyday social interac-

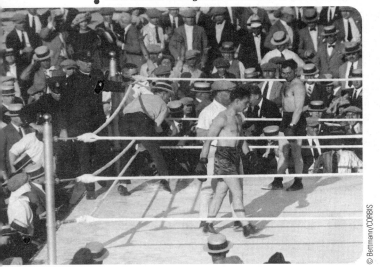

Are sports a source of upward mobility for recent immigrants and ethnic minorities, as was true for some in previous generations? Early-twentieth-century Jewish American and Italian American boxers, for example, not only produced intragroup ethnic pride but also earned a livelihood in such sporting events. And, as this recent NBA match-up shows, U.S. sports now attract immigrants from all over the world.

© Bettmann/CORBIS

© TIM JOHNSON/Reuters/Landov

tion with members of the dominant group. This type of assimilation typically starts in large, impersonal settings such as schools and workplaces, and only later (if at all) results in close friendships and intermarriage. Box 9.2 examines the extent to which television advertisements correctly reflect the extent of structural assimilation in the United States. *Biological assimilation,* or *amalgamation,* occurs when members of one group marry those of other social or ethnic groups.

Psychological assimilation involves a change in racial or ethnic self-identification on the part of an individual. Rejection by the dominant group may prevent psychological assimilation by members of some subordinate racial and ethnic groups, especially those with visible characteristics such as skin color or facial features that differ from those of the dominant group.

Ethnic Pluralism Instead of complete assimilation, many groups share elements of the mainstream culture while remaining culturally distinct from both the dominant group and other social and ethnic groups. ***Ethnic pluralism is the coexistence of a variety of distinct racial and ethnic groups within one society.***

Equalitarian pluralism, or *accommodation,* is a situation in which ethnic groups coexist in equality with one another. Switzerland has been described as a model of equalitarian pluralism; more than six million people with French, German, and Italian cultural heritages peacefully coexist there. *Inequalitarian pluralism,* or *segregation,* exists when specific ethnic groups are set apart from the dominant group and have unequal access to power and privilege. ***Segregation is the spatial and social separation of categories of people by race, ethnicity, class, gender, and/or religion.*** Segregation may be enforced by law. *De jure segregation* refers to laws that systematically enforced the physical and social separation of African Americans in all areas of public life. An example of de jure segregation was the Jim Crow laws, which legalized the separation of the races in public accommodations (such as hotels, restaurants, transportation, hospitals, jails, schools, churches, and cemeteries) in the southern United States after the Civil War (Feagin and Feagin, 2003).

Segregation may also be enforced by custom. *De facto segregation*—racial separation and inequality enforced by custom—is more difficult to document than de jure segregation. For example, residential segregation is still prevalent in many U.S. cities; owners, landlords, real estate agents, and apartment managers often use informal mechanisms to maintain their properties for "whites only." Even middle-class people of color find that racial polarization is fundamental to the residential layout of many cities.

Although functionalist explanations provide a description of how some early white ethnic immigrants assimilated into the cultural mainstream, they do not adequately account for the persistent racial segregation and economic inequality experienced by people of color.

Conflict Perspectives

Conflict theorists focus on economic stratification and access to power in their analyses of race and ethnic relations. Some emphasize the caste-like nature of racial stratification, others analyze class-based discrimination, and still others examine internal colonialism and gendered racism.

The Caste Perspective The caste perspective views racial and ethnic inequality as a permanent feature of U.S. society. According to this approach, the African American experience must be viewed as different from that of other racial or ethnic groups. African Americans were the only group to be subjected to slavery; when slavery was abolished, a caste system was instituted to maintain economic and social inequality between whites and African Americans (Feagin and Feagin, 2003).

The caste system was strengthened by *antimiscegenation laws,* which prohibited sexual intercourse

---

assimilation a process by which members of subordinate racial and ethnic groups become absorbed into the dominant culture.

ethnic pluralism the coexistence of a variety of distinct racial and ethnic groups within one society.

segregation the spatial and social separation of categories of people by race, ethnicity, class, gender, and/or religion.

Box 9.2 Framing Race in the Media

Do Multiracial Scenes in Television Ads Reflect Reality?

Item: Verizon ads feature a fictional interracial family, made up of white and Latino/a members, in seven commercials that (in the words of a company representative) "portray something that is contemporary and realistic."

Item: A Yoplait ad shows a multiracial group of young women who are laughing and comparing the yogurt they are eating to other favorite activities: "This is day-at-the-spa good. This is a-weekend-with-no-boys good."

Item: A Lay's potato chip commercial shows two African American children and two white children—neighbors—looking for a lost softball and consoling themselves with a bag of potato chips. (Texeira, 2005)

These ads suggest that racially mixed families, friendship groups, and neighborhoods are quite common in the United States. According to one journalist, "The ads suggest America's ethnic communities are meshing seamlessly, bonded by a love of yogurt, lipstick and athletic gear" (Texeira, 2005). Sociologically speaking, we might say that these ads employ a particular framing device—*racial harmony framing*—to sell products to diverse audiences. The ads reflect the desire of corporate sponsors to capitalize on the changing racial and ethnic landscape of this country by conveying an inclusive corporate image and showing diverse groups happily involved in one another's daily lives.

Having greater diversity in television advertising and focusing on multiculturalism is a positive development. *Multiculturalism,* as it relates to advertising, refers to those ads that employ diverse casts and give each character an importance and valuable role (Children Now, 1999). For decades, advocacy groups have emphasized the importance of accurately portraying people from various racial and ethnic groups in television shows and advertising. For example, a report from the children's advocacy group Children Now (1999) suggests that

Advertising has the same ability as television programming to impact children's perceptions [about race]. Commercials appear with great repetition, telling us what products will improve our lives and to what we can aspire. By offering roles for kids to admire or reject, advertising can tell children who is important and what they can become. When these pictures involve race and ethnicity, there are important considerations—what do the ads say about being a person of color? What do kids learn from the ads?

How television advertising *frames* race and portrays racial harmony is an important issue because these depictions influence our perceptions of how race and ethnicity are lived out in the real world (see Gallagher, 2003). But how realistic are these TV portrayals?

Although greater representation of diverse racial and ethnic groups is a major step forward on television, a number of media analysts and sociologists question the reality of portrayals that show an "idyllic world of TV commercials [where] Americans increasingly are living together side by side, regardless of race" (Texeira, 2005). As the sociologist Charles Gallagher notes, "The lens through which people learn about other races is absolutely through TV, not through human interaction and contact. Here, we're getting a lens of racial interaction that is far afield from reality." According to Gallagher, ads that extensively show interracial marriages, integrated neighborhoods, and widely diverse friendship groups create a "carefully manufactured racial utopia, a narrative of colorblindness" that does not exist in this country (qtd. in adrants, 2005).

What are the realities of race in the United States? According to the U.S. Census Bureau (Fields, 2004), less than

or marriage between persons of different races. Most states had such laws, which were later expanded to include relationships between whites and Chinese, Japanese, and Filipinos. These laws were not declared unconstitutional until 1967 (Frankenberg, 1993).

Although the caste perspective points out that racial stratification may be permanent because of structural elements such as the law, it has been criticized for not examining the role of class in perpetuating racial inequality.

3 percent of all marriages are interracial, and about 80 percent of white Americans live in neighborhoods that are more than 95 percent white. Similarly, most people in the United States have few close friends from another racial grouping. This is a different picture from the one we receive in much television advertising and programming.

Although there is much to applaud with regard to changes in racial and ethnic framing in the media, it is important for all of us to realize that these media depictions often do not reflect the real world in which we live. In the words of African American studies scholar Jerome Williams, "Despite the progress we've made on civil rights and other things, if you look at the United States in terms of where we live and who our friends are and where we go to church, we live in different worlds" (qtd. in Texeira, 2005).

At the bottom line, showing greater diversity in ads is a step in the right direction. We must be aware of the increasing racial diversity of the U.S. population, but we must also encourage advertisers and other media producers to accurately portray life in this nation instead of creating a fake utopia that contributes to misperceptions about race.

Reflect & Analyze

Which ads and television shows do you think provide viewers with accurate messages about racial diversity and inclusion? Which ads and shows are the weakest (or the most forced) in how they frame multiculturalism in the United States?

Class Perspectives Class perspectives emphasize the role of the capitalist class in racial exploitation. Based on early theories of race relations by the African American scholar W. E. B. Du Bois, the sociologist Oliver C. Cox (1948) suggested that African Americans were enslaved because they were the cheapest and best workers the owners could find for heavy labor in the mines and on plantations. Thus, the profit motive of capitalists, not skin color or racial prejudice, accounts for slavery.

© Shelly Katz/Getty Images

Grinding poverty is a pressing problem for families living along the border between the United States and Mexico. Economic development has been limited in areas where *colonias* such as this one are located, and the wealthy have derived far more benefit than others from recent changes in the global economy.

More recently, sociologists have debated the relative importance of class and race in explaining the unequal life chances of African Americans. Sociologist William Julius Wilson (1996) has suggested that race, cultural factors, social psychological variables, and social class must all be taken into account in examining the life chances of "inner-city residents." His analysis focuses on how class-based economic determinants of social inequality, such as deindustrialization and the decline of the central (inner) city, have affected many African Americans, especially in the Northeast. African Americans were among the most severely affected by the loss of factory jobs because work in the manufacturing sector had previously made upward mobility possible. Wilson (1996) is not suggesting that prejudice and discrimination have been eradicated; rather, he is arguing that they may be less important than class in explaining the current status of African Americans.

How do conflict theorists view the relationship among race, class, and sports? Simply stated, sports reflects the interests of the wealthy and powerful. At all levels, sports exploits athletes (even highly paid ones) in order to gain high levels of profit and prestige for coaches, managers, and owners. Afri-

can American athletes and central-city youths in particular are exploited by the message of rampant consumerism. Many are given the unrealistic expectation that sports can be a ticket out of the ghetto or barrio. If they try hard enough (and wear the right athletic gear), they too can become wealthy and famous.

Internal Colonialism Why do some racial and ethnic groups continue to experience subjugation after many years? According to the sociologist Robert Blauner (1972), groups that have been subjected to internal colonialism remain in subordinate positions longer than groups that voluntarily migrated to the United States. ***Internal colonialism* occurs when members of a racial or ethnic group are conquered or colonized and forcibly placed under the economic and political control of the dominant group.**

In the United States, indigenous populations (including groups known today as Native Americans and Mexican Americans) were colonized by Euro-Americans and others who invaded their lands and conquered them. In the process, indigenous groups lost property, political rights, aspects of their culture, and often their lives. The capitalist class acquired cheap labor and land through this government-sanctioned racial exploitation (Blauner, 1972). The effects of past internal colonialism are reflected today in the number of Native Americans who live on government reservations and in the poverty of Mexican Americans who lost their land and had no right to vote.

The internal colonialism perspective is rooted in the historical foundations of racial and ethnic inequality in the United States. However, it tends to view all voluntary immigrants as having many more opportunities than do members of colonized groups. Thus, this model does not explain the continued exploitation of some immigrant groups, such as the Chinese, Filipinos, Cubans, Vietnamese, and Haitians, and the greater acceptance of others, primarily those from Northern Europe (Cashmore, 1996).

The Split-Labor-Market Theory Who benefits from the exploitation of people of color? Dual- or split-labor-market theory states that white workers and members of the capitalist class both benefit from the exploitation of people of color. *Split labor market* refers to the division of the economy into two areas of employment, a primary sector or upper tier, composed of higher-paid (usually dominant-group) workers in more secure jobs, and a secondary sector or lower tier, composed of lower-paid (often subordinate-group) workers in jobs with little security and hazardous working conditions (Bonacich, 1972, 1976). According to this perspective, white workers in the upper tier may use racial discrimination against nonwhites to protect their positions. These actions most often occur when upper-tier workers feel threatened by lower-tier workers hired by capitalists to reduce labor costs and maximize corporate profits. In the past, immigrants were a source of cheap labor that employers could use to break strikes and keep wages down. Throughout U.S. history, higher-paid workers have responded with racial hostility and joined movements to curtail immigration and thus do away with the source of cheap labor (Marger, 2009).

Proponents of the split-labor-market theory suggest that white workers benefit from racial and ethnic antagonisms. However, these analysts typically do not examine the interactive effects of race, class, and gender in the workplace.

Perspectives on Race and Gender The term *gendered racism* refers to the interactive effect of racism and sexism on the exploitation of women of color. According to the social psychologist Philomena Essed (1991), women's particular position must be explored within each racial or ethnic group because their experiences will not have been the same as men's in each grouping.

All workers are not equally exploited by capitalists. Gender and race or ethnicity are important in this exploitation. Historically, the high-paying primary labor market has been monopolized by white men. Many people of color and white women hold lower-tier jobs. Below that tier is the underground sector of the economy, characterized by illegal or quasi-legal activities such as drug trafficking, prostitution, and working in sweatshops that do not meet minimum wage and safety standards. Many undocumented workers and some white women and people of color attempt to earn a living in this sector (Amott and Matthaei, 1996).

The *theory of racial formation* states that actions of the government substantially define racial and ethnic relations in the United States. Government actions range from race-related legislation to imprisonment of members of groups believed to be a threat to society. Sociologists Michael Omi and Howard Winant (1994) suggest that the U.S. government has shaped the politics of race through actions and policies that cause people to be treated differently because of their race. For example, immigration legislation reflects racial biases. The Naturalization Law of 1790 permitted only white immigrants to qualify for naturalization; the Immigration Act of 1924 favored Northern Europeans and excluded Asians and Southern and Eastern Europeans.

The government's definition of racial realities is periodically challenged by social protest movements of various racial and ethnic groups. When this social rearticulation occurs, people's understanding about race may be restructured somewhat. For example, the African American protest movements of the 1950s and 1960s helped redefine the rights of people of color in the United States.

internal colonialism according to conflict theorists, a practice that occurs when members of a racial or ethnic group are conquered or colonized and forcibly placed under the economic and political control of the dominant group.

split labor market a term used to describe the division of the economy into two areas of employment, a primary sector or upper tier, composed of higher-paid (usually dominant-group) workers in more-secure jobs, and a secondary sector or lower tier, composed of lower-paid (often subordinate-group) workers in jobs with little security and hazardous working conditions.

gendered racism the interactive effect of racism and sexism on the exploitation of women of color.

theory of racial formation the idea that actions of the government substantially define racial and ethnic relations in the United States.

An Alternative Perspective: Critical Race Theory

Emerging out of scholarly law studies on racial and ethnic inequality, critical race theory derives its foundation from the U.S. civil rights tradition. Critical race theory has several major premises, including the belief that racism is such an ingrained feature of U.S. society that it appears to be ordinary and natural to many people (Delgado, 1995). As a result, civil rights legislation and affirmative action laws (formal equality) may remedy some of the more overt, blatant forms of racial injustice but have little effect on subtle, business-as-usual forms of racism that people of color experience as they go about their everyday lives. According to this approach, the best way to document racism and ongoing inequality in society is to listen to the lived experiences of people who have experienced such discrimination. In this way, we can learn what actually happens in regard to racial oppression and the many effects it has on people, including alienation, depression, and certain physical illnesses. Central to this argument is the belief that *interest convergence* is a crucial factor in bringing about social change. According to the legal scholar Derrick Bell, white elites tolerate or encourage racial advances for people of color *only* if the dominant-group members believe that their own self-interest will be served in so doing (cited in Delgado, 1995). From this approach, civil rights laws have typically benefited white Americans as much (or more) as people of color because these laws have been used as mechanisms to ensure that "racial progress occurs at just the right pace: change that is too rapid would be unsettling to society at large; change that is too slow could prove destabilizing" (Delgado, 1995: xiv). The Concept Quick Review outlines the key aspects of each sociological perspective on race and ethnic relations.

CONCEPT QUICK REVIEW

Sociological Perspectives on Race and Ethnic Relations

	Focus	Theory/Hypothesis
Symbolic Interactionist	Microlevel contacts between individuals	Contact hypothesis

Functionalist	Macrolevel intergroup processes	1. Assimilation
		   a. cultural
		   b. biological
		   c. structural
		   d. psychological
		2. Ethnic pluralism
		   a. equalitarian pluralism
		   b. inequalitarian pluralism (segregation)

Conflict	Power/economic differentials between dominant and subordinate groups	1. Caste perspective 2. Class perspective 3. Internal colonialism 4. Split labor market 5. Gendered racism 6. Racial formation
Critical Race Theory	Racism as an ingrained feature of society that affects everyone's daily life	Laws may remedy overt discrimination but have little effect on subtle racism. Interest convergence is required for social change.

Racial and Ethnic Groups in the United States

How do racial and ethnic groups come into contact with one another? How do they adjust to one another and to the dominant group over time? Sociologists have explored these questions extensively; however, a detailed historical account of the unique experiences of each group is beyond the scope of this chapter. Instead, we will look briefly at intergroup contacts. In the process, sports will be used as an example of how members of some groups have attempted to gain upward mobility and become integrated into society.

Native Americans

Native Americans are believed to have migrated to North America from Asia thousands of years ago, as shown on the time line in ▶ Figure 9.2. One of the most widely accepted beliefs about this migration is that the first groups of Mongolians made their way across a natural bridge of land called Beringia into present-day Alaska. From there, they moved to what is now Canada and the northern United States, eventually making their way as far south as the tip of South America (Cashmore, 1996).

As schoolchildren are taught, Spanish explorer Christopher Columbus first encountered the native inhabitants in 1492 and referred to them as "Indians." When European settlers (or invaders) arrived on this continent, the native inhabitants' way of life was changed forever. Experts estimate that approximately two million native inhabitants lived in North America at that time (Cashmore, 1996); however, their numbers had been reduced to less than 240,000 by 1900.

Genocide, Forced Migration, and Forced Assimilation Native Americans have been the victims of genocide and forced migration. Although the

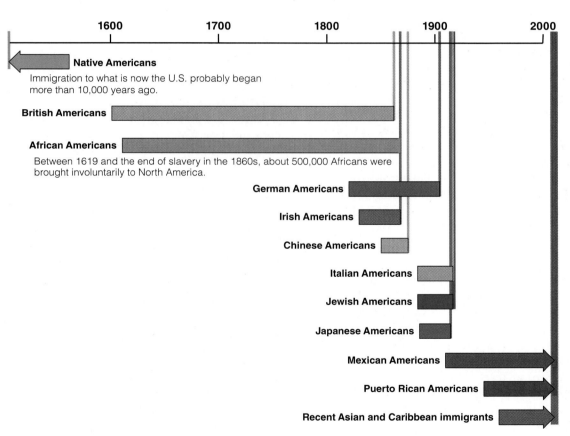

▶ Figure 9.2 **Time Line of Racial and Ethnic Groups in the United States**

United States never had an official policy that set in motion a pattern of deliberate extermination, many Native Americans were either massacred or died from European diseases (such as typhoid, smallpox, and measles) and starvation (Wagner and Stearn, 1945; Cook, 1973). In battle, Native Americans were often no match for the Europeans, who had "modern" weaponry (Amott and Matthaei, 1996). Europeans justified their aggression by stereotyping the Native Americans as "savages" and "heathens" (Takaki, 1993).

After the Revolutionary War, the federal government offered treaties to the Native Americans so that more of their land could be acquired for the growing white population. Scholars note that the government broke treaty after treaty as it engaged in a policy of wholesale removal of indigenous nations in order to clear the land for settlement by Anglo-Saxon "pioneers" (Green, 1977). Entire nations were forced to move in order to accommodate the white settlers. The "Trail of Tears" was one of the most disastrous of the forced migrations. In the coldest part of the winter of 1832, over half of the Cherokee Nation died during or as a result of their forced relocation from the southeastern United States to the Indian Territory in Oklahoma (Thornton, 1984).

Native Americans were subjected to forced assimilation on the reservations after 1871 (Takaki, 1993). Native American children were placed in boarding schools operated by the Bureau of Indian Affairs to hasten their assimilation into the dominant culture. About 98 percent of native lands had been expropriated by 1920 (see McDonnell, 1991).

Native Americans Today Currently, about 4.5 million Native Americans live in the United States, including Aleuts, Inuit (Eskimos), Cherokee, Navajo, Chippewa, Sioux, and more than five hundred other nations of varying sizes and different locales. Most are concentrated in the Southwest, and about one-third live on reservations. Native Americans are the most disadvantaged racial or ethnic group in the United States in terms of income, employment, housing, nutrition, and health. The life chances of Native Americans who live on reservations are

Native Americans have historically had a low rate of college attendance. However, the development of a network of tribal colleges has provided them a local source of upward mobility.

especially limited. They have the highest rates of infant mortality and death by exposure and malnutrition. They also have high rates of suicide, substance abuse, and school violence (Kershaw, 2005). Reservation-based Native American men have an average life expectancy of less than forty-five years; for women, it is less than forty-eight years. Native Americans have had very limited educational opportunities and have a very high rate of unemployment. In recent years, however, a network of tribal colleges has been successful in providing some Native Americans with the education they need to move into the ranks of the skilled working class and beyond (Bordewich, 1996).

In spite of the odds against them, many Native Americans resist oppression. The American Indian Movement, Women of All Red Nations, and other groups have demanded the recovery of Native American lands and reparation for past losses. Native American women have publicized the harmful conditions (including radiation sickness and the forced sterilization of women) that exist on reservations. The American Indian Anti-Defamation Council advocates doing away with the word *tribe* because it demeans Native Americans by equating their level of cultural attainment with "primitivism" or "barbarism."

White Anglo-Saxon Protestants (British Americans)

Whereas Native Americans have been among the most disadvantaged peoples, white Anglo-Saxon Protestants (WASPs) have been the most privileged group in this country. Although many English settlers initially came to North America as indentured servants or as prisoners, they quickly emerged as the dominant group, creating a core culture (including language, laws, and holidays) to which all other groups were expected to adapt. Most of the WASP immigrants arriving from northern Europe were advantaged over later immigrants because they were highly skilled and did not experience high levels of prejudice and discrimination.

Class, Gender, and Wasps Like members of other racial and ethnic groups, not all WASPs are alike. Social class and gender affect their life chances and opportunities. For example, members of the working class and the poor do not have political and economic power; men in the capitalist class do. WASPs constitute the majority of the upper class and maintain cohesion through listings such as the *Social Register* and interactions with one another in elite settings such as private schools and country clubs (Kendall, 2002). However, WASP women do not always have the same rights as the men of their group. Although WASP women have the privilege of a dominant racial position, they do not have the gender-related privileges of men (Amott and Matthaei, 1996).

Wasps and Sports Family background, social class, and gender play an important role in the sports participation of WASPs. Contemporary North American football was invented at the Ivy League colleges and was dominated by young, affluent WASPs who had the time and money to attend college and participate in sports activities. As ◆ Table 9.1 shows, whites are more likely than any other racial or ethnic group to become professional athletes in all sports except football and basketball.

Affluent WASP women participated in intercollegiate women's basketball in the late 1800s, and various other sporting events were used as a means to break free of restrictive codes of femininity (Nelson,

Life chances are extremely limited for Native Americans who live on reservations. Although a few Native Americans early in the twentieth century were well-known athletes, Native Americans today have little opportunity to compete in sports at the college, professional, or Olympic level.

© Kevin Fleming/CORBIS

Native Americans and Sports Early in the twentieth century, Native Americans such as Jim Thorpe gained national visibility as athletes in football, baseball, and track and field. Teams at boarding schools such as the Carlisle Indian Industrial School in Pennsylvania and the Haskell Institute in Kansas were well-known. However, after the first three decades of the twentieth century, Native Americans became much less prominent in sports. Although some Navajo athletes have been very successful in basketball and some Choctaws have excelled in baseball, Native Americans have seldom been able to compete at the college, professional, or Olympic level. Native American scholar Joseph B. Oxendine (2003) attributes the lack of athletic participation to these factors: (1) a reduction in opportunities for developing sports skills, (2) restricted opportunities for participation, and (3) a lessening of Native Americans' interest in competing with and against non-Native Americans.

◆ **Table 9.1 Odds of Becoming a Professional Athlete by Race/Ethnicity and Sport**

Many young people dream of becoming a professional athlete; however, as this table shows, the odds of actually becoming one are very small.

| | Race/Ethnicity | | | |
	White	African American	Latino/a	Asian American
Football	1 in 62,500	1 in 47,600	1 in 2,500,000	1 in 5,000,000
Baseball	1 in 83,300	1 in 333,300	1 in 500,000	1 in 50,000,000
Basketball	1 in 357,100	1 in 153,800	1 in 33,300,000	—
Hockey	1 in 66,700	—	—	—
Golf				
Men's	1 in 312,500	1 in 12,500,000	1 in 33,300,000	1 in 20,000,000
Women's	1 in 526,300	—	1 in 33,300,000	1 in 3,300,000
Tennis				
Men's	1 in 285,700	1 in 2,000,000	1 in 3,300,000	—
Women's	1 in 434,800	1 in 20,000,000	1 in 20,000,000	—

Note: The odds of Native Americans participating in professional football are 1 in 12,500,000; they are not represented in other professional sports.
Source: Leonard and Reyman, 1988: 162–169.

1994). Until recently, however, most women have had little chance for any involvement in college and professional sports.

African Americans

The African American (black) experience has been one uniquely marked by slavery, segregation, and persistent discrimination. There is a lack of consensus about whether *African American* or *black* is the most appropriate term to refer to the 36 million Americans of African descent who live in the United States today. Those who prefer the term *black* point out that it incorporates many African-descent groups living in this country that do not use *African American* as a racial or ethnic self-description. For example, people who trace their origins to Haiti, Puerto Rico, or Jamaica typically identify themselves as "black" but not as "African American" (Cashmore, 1996).

Although the earliest African Americans probably arrived in North America with the Spanish conquerors in the fifteenth century, most historians trace their arrival to about 1619, when the first groups of indentured servants were brought to the colony of Virginia. However, by the 1660s, inden-tured servanthood had turned into full-fledged slavery with the enactment of laws that sanctioned the enslavement of African Americans. Although the initial status of persons of African descent in this country may not have been too different from that of the English indentured servants, all of that changed with the passage of laws turning human beings into property and making slavery a status from which neither individuals nor their children could escape (Franklin, 1980).

Between 1619 and the 1860s, about 500,000 Africans were forcibly brought to North America, primarily to work on southern plantations, and these actions were justified by the devaluation and stereotyping of African Americans. Some analysts believe that the central factor associated with the development of slavery in this country was the plantation system, which was heavily dependent on cheap and dependable manual labor. Slavery was primarily beneficial to the wealthy southern plantation owners, but many of the stereotypes used to justify slavery were eventually institutionalized in southern custom and practice (Wilson, 1978). However, some slaves and whites engaged in active resistance against slavery and its barbaric practices, eventually resulting in slavery being outlawed in the northern states by the late 1700s. Slavery continued in the South until 1863, when it was abolished by the Emancipation Proclamation (Takaki, 1993).

Segregation and Lynching Gaining freedom did not give African Americans equality with whites. African Americans were subjected to many indignities because of race. Through informal practices in the North and *Jim Crow laws* in the South, African Americans experienced segregation in housing, employment, education, and all public accommodations. African Americans who did not stay in their "place" were often the victims of violent attacks and lynch mobs (Franklin, 1980). *Lynching* is a killing carried out by a group of vigilantes seeking revenge for an actual or imagined crime by the victim. Lynchings were used by whites to intimidate African Americans into staying "in their place." It is estimated that as many as 6,000 lynchings occurred between 1892 and 1921 (Feagin and Feagin, 2003). In spite of all odds, many African American women and men resisted oppression and did not give up in their struggle for equality (Amott and Matthaei, 1996).

Discrimination In the twentieth century, the lives of many African Americans were changed by industrialization and two world wars. When factories were built in the northern United States, many African American families left the rural South in hopes of finding jobs and a better life.

During World Wars I and II, African Americans were a vital source of labor in war production industries; however, racial discrimination continued both on and off the job. In World War II, many African Americans fought for their country in segregated units in the military; after the war, they sought—and were denied—equal opportunities in the country for which they had risked their lives.

African Americans began to demand sweeping societal changes in the 1950s. Initially, the Reverend Dr. Martin Luther King, Jr., and the civil rights movement used *civil disobedience*—nonviolent action seeking to change a policy or law by refusing to comply with it—to call attention to racial inequality and to demand greater inclusion of African Americans in all areas of public life. Subsequently, leaders of the Black Power movement, including Malcolm X

and Marcus Garvey, advocated black pride and racial awareness among African Americans. Gradually, racial segregation was outlawed by the courts and the federal government. For example, the Civil Rights Acts of 1964 and 1965 sought to do away with discrimination in education, housing, employment, and health care. Affirmative action programs were instituted in both public-sector and private-sector organizations in an effort to bring about greater opportunities for African Americans and other previously excluded groups. *Affirmative action* refers to policies or procedures that are intended to promote equal opportunity for categories of people deemed to have been previously excluded from equality in education, employment, and other fields on the basis of characteristics such as race or ethnicity. Critics of affirmative action often assert that these policies amount to *reverse discrimination*—a person who is better qualified being denied a position because another person received preferential treatment as a result of affirmative action.

African Americans Today African Americans make up about 13 percent of the U.S. population. Some are descendants of families that have been in this country for many generations; others are recent immigrants from Africa and the Caribbean. Black Haitians make up the largest group of recent Caribbean immigrants; others come from Jamaica and Trinidad and Tobago. Recent African immigrants are primarily from Nigeria, Ethiopia, Ghana, and Kenya. They have been simultaneously "pushed" out of their countries of origin by severe economic and political turmoil and "pulled" by perceived opportunities for a better life in the United States. Recent immigrants are often victimized by the same racism that has plagued African Americans as a people for centuries.

Since the 1960s, many African Americans have made significant gains in politics, education, employment, and income. Between 1964 and 1999, the number of African Americans elected to political office increased from about 100 to almost 9,000 nationwide (Joint Center for Political and Economic Studies, 2000). African Americans won mayoral elections in many major cities that have large African American populations, such as Atlanta, Houston, New Orleans, Philadelphia, and Washington,

© AP Images/Ron Edmonds

In August 2008, Barack Obama made history by becoming the first African American to receive the presidential nomination of a major political party, and on Election Day was voted to become the first African American president of the United States.

D.C. Despite these political gains, African Americans still represent less than 3 percent of all elected officials in the United States.

Some African Americans have made impressive occupational gains and joined the ranks of professionals in the upper middle class. Others have achieved great wealth and fame as entertainers, professional athletes, and entrepreneurs. African Americans head three of the "Fortune 500" list of the nation's largest companies (Daniels, 2002). However, even those who make millions of dollars a year and live in affluent neighborhoods are not always exempt from racial prejudice and discrimination. And although some African Americans have made substantial occupational and educational gains, many more have not. The African American unemployment rate remains twice as high as that of whites.

African Americans and Sports In recent decades, many African Americans have seen sports as a possible source of upward mobility because other means have been unavailable. However, their achievements in sports have often been attributed to "natural ability" and not determination and hard work. Sociologists have rejected such biological explanations for African Americans' success in sports and have focused instead on explanations rooted in the structure of society.

During the slavery era, a few African Americans gained better treatment and, occasionally, freedom by winning boxing matches on which their owners had bet large sums of money (McPherson, Curtis, and Loy, 1989). After emancipation, some African Americans found jobs in horse racing and baseball. For example, fourteen of the fifteen jockeys in the first Kentucky Derby (in 1875) were African Americans. A number of African Americans played on baseball teams; a few played in the Major Leagues until the Jim Crow laws forced them out. Then they formed their own "Negro" baseball and basketball leagues (Peterson, 1992/1970).

Since Jackie Robinson broke baseball's "color line" in 1947, many African American athletes have played collegiate and professional sports. Even now, however, persistent class inequalities between whites and African Americans are reflected in the fact that, until recently, African Americans have primarily excelled in sports (such as basketball or football) that do not require much expensive equipment and specialized facilities in order to develop athletic skills (Coakley, 2004). According to one sports analyst, African Americans typically participate in certain sports and not others because of the *sports opportunity structure*—the availability of facilities, coaching, and competition in the schools and community recreation programs in their area (Phillips, 1993).

As more African Americans have made gains in education and employment, many of them have also made a conscious effort to increase awareness of African culture and to develop a sense of unity, cooperation, and self-determination. The seven-day celebration of Kwanzaa in late December and early January exemplifies this desire to maintain a distinct cultural identity.

Regardless of the sport in which they participate, African American men athletes continue to experience inequalities in assignment of playing positions, rewards and authority structures, and management and ownership opportunities in professional sports (Eitzen and Sage, 1997). For example, at the time of this writing, only 7 of the 32 National Football League head coaches and only 7 of the 119 Division I-A head coaches in college football were African Americans (Gary, 2007). Today, African Americans remain significantly underrepresented in other sports, including hockey, skiing, figure skating, golf, volleyball, softball, swimming, gymnastics, sailing, soccer, bowling, cycling, and tennis (Coakley, 2004).

White Ethnic Americans

The American Dream initially brought many white ethnics to the United States. The term *white ethnic Americans* is applied to a wide diversity of immigrants who trace their origins to Ireland and to Eastern and Southern European countries such as Poland, Italy, Greece, Germany, Yugoslavia, and Russia and other former Soviet republics. Unlike the WASPs, who immigrated primarily from Northern Europe and assumed a dominant cultural posi-

tion in society, white ethnic Americans arrived late in the nineteenth century and early in the twentieth century to find relatively high levels of prejudice and discrimination directed at them by nativist organizations that hoped to curb the entry of non-WASP European immigrants. Because many of the people in white ethnic American categories were not Protestant, they experienced discrimination because they were Catholic, Jewish, or members of other religious bodies, such as the Eastern Orthodox churches (Farley, 2000).

Discrimination Against White Ethnics Many white ethnic immigrants entered the United States between 1830 and 1924. Irish Catholics were among the first to arrive, with more than four million Irish fleeing the potato famine and economic crisis in Ireland and seeking jobs in the United States (Feagin and Feagin, 2003). When they arrived, they found that British Americans controlled the major institutions of society. The next arrivals were Italians who had been recruited for low-wage industrial and construction jobs. British Americans viewed Irish and Italian immigrants as "foreigners": The Irish were stereotyped as ape-like, filthy, bad-tempered, and heavy drinkers; the Italians were depicted as lawless, knife-wielding thugs looking for a fight, "dagos," and "wops" (short for "without papers") (Feagin and Feagin, 2003).

Both Irish Americans and Italian Americans were subjected to institutionalized discrimination in employment. Employment ads read "Help Wanted—No Irish Need Apply" and listed daily wages at $1.30–$1.50 for "whites" and $1.15–$1.25 for "Italians" (Gambino, 1975: 77). In spite of discrimination, white ethnics worked hard to establish themselves in the United States, often establishing mutual self-help organizations and becoming politically active (Mangione and Morreale, 1992).

Between 1880 and 1920, over two million Jewish immigrants arrived in the United States and settled in the Northeast. Jewish Americans differ from other white ethnic groups in that some focus their identity primarily on their religion whereas others define their Jewishness in terms of ethnic group membership (Feagin and Feagin, 2003). In any case, Jews continued to be the victims of *anti-Semitism*— prejudice, hostile attitudes, and discriminatory behavior targeted at Jews. For example, signs in hotels read "No Jews Allowed," and some "help wanted" ads stated "Christians Only" (Levine, 1992: 55). In spite of persistent discrimination, Jewish Americans achieved substantial success in many areas, including business, education, the arts and sciences, law, and medicine.

White Ethnics and Sports Sports provided a pathway to assimilation for many white ethnics. The earliest collegiate football players who were not white Anglo-Saxon Protestants were of Irish, Italian, and Jewish ancestry. Sports participation provided educational opportunities that some white ethnics would not have had otherwise.

Boxing became a way to make a living for white ethnics who did not participate in collegiate sports. Boxing promoters encouraged ethnic rivalries to increase their profits, pitting Italians against Irish or Jews, and whites against African Americans (Levine, 1992; Mangione and Morreale, 1992). Eventually, Italian Americans graduated from boxing into baseball and football. Jewish Americans found that sports lessened the shock of assimilation and gave them an opportunity to refute stereotypes about their physical weaknesses and counter anti-Semitic charges that they were "unfit to become Americans" (Levine, 1992: 272).

Today, assimilation is so complete that little attention is paid to the origins of white ethnic athletes. As former Pittsburgh Steeler running back Franco Harris stated, "I didn't know I was part Italian until I became famous" (qtd. in Mangione and Morreale, 1992: 384).

Asian Americans

The U.S. Census Bureau uses the term *Asian Americans* to designate the many diverse groups with roots in Asia. Chinese and Japanese immigrants were among the earliest Asian Americans. Many Filipinos, Asian Indians, Koreans, Vietnamese, Cambodians, Pakistani, and Indonesians have arrived more recently. Today, Asian Americans belong to the fastest-growing ethnic minority group in the United States.

Chinese Americans The initial wave of Chinese immigration occurred between 1850 and 1880, when more than 200,000 Chinese men were "pushed" from China by harsh economic conditions and "pulled" to the United States by the promise of gold in California and employment opportunities in the construction of transcontinental railroads. Far fewer Chinese women immigrated; however, many of them were brought to the United States against their will and forced into prostitution, where they were treated like slaves (Takaki, 1993).

Chinese Americans were subjected to extreme prejudice and stereotyped as "coolies," "heathens," and "Chinks." Some Asian immigrants were attacked and even lynched by working-class whites who feared that they were losing their jobs to the immigrants. Passage of the Chinese Exclusion Act of 1882 brought Chinese immigration to a halt. The Exclusion Act was not repealed until World War II, when Chinese Americans who were contributing to the war effort by working in defense plants pushed for its repeal (Takaki, 1993). After immigration laws were further relaxed in the 1960s, the second and largest wave of Chinese immigration occurred, with immigrants coming primarily from Hong Kong and Taiwan. These recent immigrants have had more education and workplace skills than earlier arrivals,

© AP Images

Historically, Chinatowns in major U.S. cities have provided a safe haven and an economic enclave for many Asian immigrants. Some contemporary Chinese Americans reside in these neighborhoods, whereas others visit to celebrate cultural diversity and ethnic pride.

and they brought families and capital with them to pursue the American Dream (Chen, 1992).

Today, many Chinese Americans live in large urban enclaves in California, New York, Hawaii, Illinois, and Texas. As a group, they have enjoyed considerable upward mobility. Some own laundries, restaurants, and other businesses; others have professional careers (Chen, 1992). However, many Chinese Americans, particularly recent immigrants, remain in the lower tier of the working class—providing low-wage labor in garment and knitting factories and Chinese restaurants.

Japanese Americans Most of the early Japanese immigrants were men who worked on sugar plantations in the Hawaiian Islands in the 1860s. Like Chinese immigrants, the Japanese American workers were viewed as a threat by white workers, and immigration of Japanese men was curbed in 1908. However, Japanese women were permitted to enter the United States for several years thereafter because of the shortage of women on the West Coast. Although some Japanese women married white men, this practice was stopped by laws prohibiting interracial marriage.

With the exception of the enslavement of African Americans, Japanese Americans experienced one of the most vicious forms of discrimination ever sanctioned by U.S. laws. During World War II, when the United States was at war with Japan, nearly 120,000 Japanese Americans were placed in internment camps, where they remained for more than two years despite the total lack of evidence that they posed a security threat to this country (Takaki, 1993). This action was a direct violation of the citizenship rights of many *Nisei* (second-generation Japanese Americans), who were born in the United States (see Daniels, 1993). Ironically, only Japanese Americans were singled out for such harsh treatment; German Americans avoided this fate even though the United States was also at war with Germany. Four decades later, the U.S. government issued an apology for its actions and eventually paid $20,000 each to some of those who had been placed in internment camps (Daniels, 1993; Takaki, 1993).

Since World War II, many Japanese Americans have been very successful. The median income of Japanese Americans is more than 30 percent above the national average. However, most Japanese Americans (and other Asian Americans) live in states that not only have higher incomes but also higher costs of living than the national average. In addition, many Asian American families have more persons in the paid labor force than do other families (Takaki, 1993).

Korean Americans The first wave of Korean immigrants were male workers who arrived in Hawaii between 1903 and 1910. The second wave came to the U.S. mainland following the Korean War in 1954 and was made up primarily of the wives of servicemen and Korean children who had lost their parents in the war. The third wave arrived after the Immigration Act of 1965 permitted well-educated professionals to migrate to the United States. Korean Americans have helped one another open small businesses by pooling money through the *kye*—an association that grants members money on a rotating basis to gain access to more capital.

Today, many Korean Americans live in California and New York, where there is a concentration of Korean-owned grocery stores, businesses, and churches. Unlike earlier Korean immigrants, more-recent arrivals have come as settlers and have brought their families with them. However, their experiences with other subordinate racial and ethnic groups have not always been harmonious. Ongoing discord has existed between African Americans and Korean Americans in New York and among African Americans, Latinos, and Korean Americans in California.

Filipino Americans Today, Filipino Americans constitute the second largest category of Asian Americans, with over a million population in the United States. To understand the status of Filipino Americans, it is important to look at the complex relationship between the Philippine Islands and the United States government. After Spain lost the Spanish-American War, the United States established colonial rule over the islands, a rule that lasted from 1898 to 1946. Despite control by the United States, Filipinos were not granted U.S. citizenship, but male Filipinos were allowed to migrate to Hawaii and the U.S. mainland to work in agriculture and in fish canneries in Seattle and Alaska. Like other Asian Americans, Filipino

Americans were accused of taking jobs away from white workers and suppressing wages, and Congress restricted Filipino immigration to fifty people per year between the Great Depression and the aftermath of World War II.

The second wave of Filipino immigrants came following the Immigration Act of 1965, when large numbers of physicians, nurses, technical workers, and other professionals moved to the U.S. mainland. Most Filipinos have not had the start-up capital necessary to open their own businesses, and many have been employed in the low-wage sector of the service economy. However, the average household income of Filipino American families is relatively high because about 75 percent of Filipino American women are employed, and nearly half have a four-year college degree (Espiritu, 1995).

Indochinese Americans Indochinese Americans include people from Vietnam, Cambodia, Thailand, and Laos, most of whom have come to the United States in the past three decades. Vietnamese refugees who had the resources to flee at the beginning of the Vietnam War were the first to arrive. Next came Cambodians and lowland Laotians, referred to as "boat people" by the media. Many who tried to immigrate did not survive at sea; others were turned back when they reached this country or were kept in refugee camps for long periods of time. When they arrived in the United States, inflation was high, the country was in a recession, and many native-born citizens feared that they would lose their jobs to these new refugees, who were willing to work very hard for low wages.

Today, many Indochinese Americans are foreign born; about half live in the western states, especially California. Even though most Indochinese immigrants spoke no English when they arrived in this country, some of their children have done very well in school and have been stereotyped as "brains."

Asian Americans and Sports Until recently, Asian Americans received little recognition in sports. However, as women's athletic events, including ice skating and gymnastics, have garnered more

© AP Images/Don Ryan

Increasing numbers of Asian Americans are distinguishing themselves in college and professional athletics. Champion figure skater Michelle Kwan is one of the best-known recent Asian American sports heroes.

media coverage, names of persons such as Kristi Yamaguchi, Michelle Kwan, and Amy Chow have become widely known and often idolized by fans. As one sports analyst stated, "[These athletes] are of Asian descent, but more importantly they are Asian Americans whose actions reflect upon the United States. . . . As role models, particularly for the Asian-American community, they exemplify success, integrity, discipline and a dedicated work ethic" (Shum, 1997).

Latinos/as (Hispanic Americans)

The terms *Latino* (for males), *Latina* (for females), and *Hispanic* are used interchangeably to refer to people who trace their origins to Spanish-speaking Latin America and the Iberian peninsula. However, as racial–ethnic scholars have pointed out, the label *Hispanic* was first used by the U.S. government to designate people of Latin American and Spanish descent living in the United States, and it has not been fully accepted as a source of identity by the more than 42 million Latinos/as who live in the United States today (Oboler, 1995; Romero, 1997). Instead, many of the people who trace their roots to Spanish-speaking countries think of themselves as Mexican Americans, Chicanos/as, Puerto Ricans, Cuban Americans, Salvadorans, Guatemalans, Nicaraguans, Costa Ricans, Argentines, Hondurans, Dominicans, or members of other categories. Many also think of themselves as having a combination of Spanish, African, and Native American ancestry.

Mexican Americans or Chicanos/as Mexican Americans—including both native- and foreign-born people of Mexican origin—are the largest segment (approximately two-thirds) of the Latino/a population in the United States. Most Mexican Americans live in the southwestern region of the United States, although more have moved throughout the United States in recent years.

Immigration from Mexico is the primary vehicle by which the Mexican American population grew in this country. Initially, Mexican-origin workers came to work in agriculture, where they were viewed as a readily available cheap and seasonal labor force. Many initially entered the United States as undocumented workers ("illegal aliens"); however, they were more vulnerable to deportation than other illegal immigrants because of their visibility and the proximity of their country of origin. For more than a century, there has been a "revolving door" between the United States and Mexico that has been open when workers were needed and closed during periods of economic recession and high rates of U.S. unemployment.

Mexican Americans have long been seen as a source of cheap labor, while—ironically—at the same time, they have been stereotyped as lazy and unwilling to work. As has been true of other groups, when white workers viewed Mexican Americans as a threat to their jobs, they demanded that the "illegal aliens" be sent back to Mexico. Consequently, U.S. citizens who happen to be Mexican American have been asked for proof of their citizenship, especially when anti-immigration sentiments are running high. Many Mexican American families have lived in the United States for four or five generations—they have fought in wars, made educational and political gains, and consider themselves to be solid U.S. citizens. Thus, it is a great source of frustration for them to be viewed as illegal immigrants or to be asked "How long have you been in this country?"

Puerto Ricans When Puerto Rico became a territory of the United States in 1917, Puerto Ricans acquired U.S. citizenship and the right to move freely to and from the mainland. In the 1950s, many migrated to the mainland when the Puerto Rican sugar industry collapsed, settling primarily in New York and New Jersey. Although living conditions have improved substantially for some Puerto Ricans, life has been difficult for the many living in poverty in Spanish Harlem and other barrios. Nevertheless, in recent years Puerto Ricans have made dramatic advances in education, the arts, and politics. Increasing numbers have become lawyers, physicians, and college professors (see Rodriguez, 1989).

Cuban Americans Cuban Americans live primarily in the Southeast, especially Florida. As a group, they have fared somewhat better than other Latinos/as because many Cuban immigrants were affluent professionals and businesspeople who fled Cuba after Fidel Castro's 1959 Marxist revolution. This early wave of Cuban immigrants has median incomes well above those of other Latinos/as; however, this group is still below the national average. The second wave of Cuban Americans, arriving in the 1970s, has fared worse. Many had been released from prisons and mental hospitals in Cuba, and their arrival fueled an upsurge in prejudice against all Cuban Americans. The more recent arrivals have developed their own ethnic and economic enclaves in Miami's Little Havana, and many of the earlier immigrants have become mainstream professionals and entrepreneurs.

© AP Images/Kathy Willens

Pedro Martinez of the New York Mets is one of the most visible Latino athletes. Latino and Latina sports figures have gained prominence in a wide variety of sports, including boxing, baseball, golf, and tennis.

Latinos/as and Sports For most of the twentieth century, Latinos have played Major League Baseball. Originally, Cubans, Puerto Ricans, and Venezuelans were selected for their light skin as well as for their skill as players (Hoose, 1989). Today, Latinos represent more than 20 percent of all major leaguers. If not for a 1974 U.S. Labor Department quota limiting how many foreign-born players can play professional baseball, this number might be even larger (Hoose, 1989).

Education is a crucial issue for Latinos/as. Because of past discrimination and unequal educational opportunities, many Latinos/as currently have low levels of educational attainment. Many are unable to attend college or participate in collegiate sports, which is essential for being drafted in professional sports other than baseball. Consequently, the overall number of Latinas/os in college and professional sports is low compared to the rest of the U.S. population who are in this age bracket.

Middle Eastern Americans

Since 1970, many immigrants have arrived in the United States from countries located in the "Middle East," which is the geographic region from Afghani-stan to Libya and includes Arabia, Cyprus, and Asiatic Turkey. Placing people in the "Middle Eastern" American category is somewhat like placing wide diversities of people in the categories of Asian American or Latino/a; some U.S. residents trace their origins to countries such as Bahrain, Egypt, Iran, Iraq, Kuwait, Lebanon, Oman, Qatar, Saudi Arabia, Syria, UAE (United Arab Emirates), and Yemen. Middle Eastern Americans speak a variety of languages and have diverse religious backgrounds: Some are Muslim, some are Coptic Christian, and others are Melkite Catholic. Although some are from working-class families, Lebanese Americans, Syrian Americans, Iranian Americans, and Kuwaiti Americans primarily come from middle- and upper-income family backgrounds. For example, numerous Iranian Americans are scientists, professionals, and entrepreneurs.

In cities across the United States, Muslims have established social, economic, and ethnic enclaves. On the Internet, they have created websites that provide information about Islamic centers, schools, and lists of businesses and services run by those who adhere to Islam, one of the fastest-growing religions in this country. In cities such as Seattle, incorporation into the economic mainstream has been relatively easy for Palestinian immigrants who left their homeland in the 1980s. Some have found well-paid employment with corporations such as Microsoft because they bring educational skills and talents to the information-based economy, including the ability to translate software into Arabic for Middle Eastern markets (Ramirez, 1999). In the United States, Islamic schools and centers often bring together people from a diversity of countries such as Egypt and Pakistan. Many Muslim leaders and parents focus on how to raise children to be good Muslims and good U.S. citizens. However, recent immigrants continue to be torn between establishing roots in the

United States and the continuing divisions and strife that exist in their homelands. Some Middle Eastern Americans experience prejudice and discrimination based on their speech patterns, appearance (such as the *hijabs,* or "head-to-toe covering" that leaves only the face exposed, which many girls and women wear), or the assumption that "all Middle Easterners" are somehow associated with terrorism.

Following the September 11, 2001, attacks on the United States by terrorists whose origins were traced to the Middle East, hate crimes and other forms of discrimination against people who were assumed to be Arabs, Arab Americans, or Muslims escalated in this country. With the passage of the U.S. Patriot Act—a law giving the federal government greater authority to engage in searches and surveillance with less judicial review than previously—in the aftermath of the terrorist attacks, many Arab Americans have expressed concern that this new law could be used to target people who appear to be of Middle Eastern origins.

Middle Eastern Americans and Sports Although more Islamic schools are beginning to focus on sports, particularly for teenage boys, there has been less emphasis on competitive athletics among many Middle Eastern Americans. Based on popular sporting events in their countries of origin, some Middle Eastern Americans play golf or soccer. As well, some Iranian Americans follow the soccer careers of professional players from Iran who now play for German, Austrian, Belgian, and Greek clubs. Keeping up with global sporting events is easy with all-sports television cable channels and websites that provide up-to-the-minute information about players and competitions. Over time, there will probably be greater participation by Middle Eastern American males in competitions such as soccer and golf; however, girls and women in Muslim families are typically not allowed to engage in athletic activities. Although little research has been done on this issue in the United States, one study of Islamic countries in the Middle East found that female athletes face strong cultural opposition to their sports participation (Dupre and Gains, 1997).

Muslims in the United States who wear traditional attire may face prejudice and/or discrimination as they go about their daily lives.

Global Racial and Ethnic Inequality in the Future

Throughout the world, many racial and ethnic groups seek *self-determination*—the right to choose their own way of life. As many nations are currently structured, however, self-determination is impossible.

Worldwide Racial and Ethnic Struggles

The cost of self-determination is the loss of life and property in ethnic warfare. In recent years, the Cold War has given way to dozens of smaller wars over ethnic dominance. In Europe, for example, ethnic

Box 9.3 You Can Make a Difference

Working for Racial Harmony

Suppose that you are talking with several friends about a series of racist incidents at your college. Having studied the sociological imagination, you decide to launch an organization similar to No Time to Hate, which was started at Emory University several years ago to reduce racism on campus. In analyzing racism, your group identifies factors contributing to the problem: (1) divisiveness between different cultural and ethnic communities, (2) persistent lack of trust, (3) the fact that many people never really communicate with one another, (4) the need to bring different voices into the curriculum and college life generally, and (5) the need to learn respect for people from different backgrounds (Loeb, 1994). Your group also develops a set of questions to be answered regarding racism on campus:

- *Encouraging inclusion and acceptance.* Do members of our group reflect the college's racial and ethnic diversity? How much do I know about other people's history and culture? How can I become more tolerant—or accepting—of people who are different from me?

- *Raising consciousness.* What is racism? What causes it? Can people participate in racist language and behavior without realizing what they are doing? What is our college or university doing to reduce racism?

- *Becoming more self-aware.* How much do I know about my own family roots and ethnic background? How do the families and communities in which we grow up affect our perceptions of racial and ethnic relations?

- *Using available resources.* What resources are available for learning more about working to reduce racism? Here are some agencies to contact:

  - ACLU (American Civil Liberties Union), 125 Broad Street, 18th floor, New York, NY 10094. Online: **http://www.aclu.org**

  - ADL (Anti-Defamation League of B'nai B'rith), 823 United Nations Plaza, New York, NY 10017. Online: **http://www.adl.org**

  - NAACP (National Association for the Advancement of Colored People), 4805 Mt. Hope Drive, Baltimore, MD 21215. Online: **http://www.naacp.org**

  - National Council of La Raza, 1111 19th, NW, Suite 1000, Washington, DC 20036. Online: **http://www.nclr.org**

What additional items would you add to the list of problem areas on your campus? How might your group's objective be reached? Over time, many colleges and universities have been changed as a result of involvement by students like you!

---

violence has persisted in Yugoslavia, Spain, Britain (between the Protestant majority and the Catholic minority in Northern Ireland), Romania, Russia, Moldova, and Georgia. Ethnic violence continues in the Middle East, Africa, Asia, and Latin America. Hundreds of thousands have died from warfare, disease, and refugee migration.

Ethnic wars have a high price even for survivors, whose life chances can become bleaker even after the violence subsides. In ethnic conflict between Abkhazians and Georgians in the former Soviet Union, for example, as many as two thousand people have been killed and more than eighty thousand dis-

placed. Ethnic hatred also devastated the province of Kosovo, which is located in Serbia, and brought about the deaths of thousands of ethnic Albanians (Bennahum, 1999).

In the twenty-first century, the struggle between the Israeli government and various Palestinian factions over the future and borders of Palestine continues to make headlines. Discord in this region has heightened tensions among people not only in Israel and Palestine but also in the United States and around the world as deadly clashes continue and political leaders are apparently unable to reach a lasting solution to the decades-long strife.

Growing Racial and Ethnic Diversity in the United States

Racial and ethnic diversity is increasing in the United States. African Americans, Latinos/as, Asian Americans, and Native Americans constitute one-fourth of the U.S. population, whereas whites are a shrinking percentage of the population. Today, white Americans make up 70 percent of the population, in contrast to 80 percent in 1980. It is predicted that by 2056, the roots of the average U.S. resident will be in Africa, Asia, Hispanic countries, the Pacific islands, and the Middle East—not white Europe.

What effect will these changes have on racial and ethnic relations? Several possibilities exist. On the one hand, conflicts may become more overt and confrontational as people continue to use *sincere fictions*—personal beliefs that reflect larger societal mythologies, such as "I am not a racist" or "I have never discriminated against anyone"—even when these are inaccurate perceptions (Feagin and Vera, 1995). Interethnic tensions may increase as competition for education, jobs, and other resources continues to grow.

On the other hand, there is reason for cautious optimism. Throughout U.S. history, members of diverse racial and ethnic groups have struggled to gain the freedom and rights that were previously withheld from them. Today, minority grassroots organizations are pressing for affordable housing, job training, and educational opportunities. As discussed in Box 9.3, movements composed of both whites and people of color continue to oppose racism in everyday life, to seek to heal divisions among racial groups, and to teach children about racial tolerance. Many groups hope not only to affect their own microcosm but also to contribute to worldwide efforts to end racism.

To eliminate racial discrimination, it will be necessary to equalize opportunities in schools and workplaces. As Michael Omi and Howard Winant (1994: 158) have emphasized,

> Today more than ever, opposing racism requires that we notice race, not ignore it, that we afford it the recognition it deserves and the subtlety it embodies. By noticing race we can begin to challenge racism, with its ever-more-absurd reduction of human experience to an essence attributed to all without regard for historical or social context. . . . By noticing race we can develop the political insight and mobilization necessary to make the U.S. a more racially just and egalitarian society.

Chapter Review

• How do race and ethnicity differ?

A race is a category of people who have been singled out as inferior or superior, often on the basis of physical characteristics such as skin color, hair texture, or eye shape. An ethnic group is a collection of people distinguished primarily by cultural or national characteristics, including unique cultural traits, a sense of community, a feeling of ethnocentrism, ascribed membership, and territoriality.

• What are dominant and subordinate groups?

A dominant group is an advantaged group that has superior resources and rights in society. A subordinate group is a disadvantaged group whose members are subjected to unequal treatment by the dominant group. Use of the terms *dominant* and *subordinate* reflects the importance of power in relationships.

• How is prejudice related to discrimination?

Prejudice is a negative attitude often based on stereotypes, which are overgeneralizations about the appearance, behavior, or other characteristics of all members of a group. Discrimination involves actions or practices of dominant-group members that have a harmful effect on members of a subordinate group.

• What are the major psychological explanations of prejudice?

According to the frustration–aggression hypothesis of prejudice, people frustrated in their efforts to achieve a highly desired goal may respond with aggression toward others, who then become scapegoats. Another theory of prejudice focuses on the authoritarian personality, marked by excessive conformity, submissiveness to authority, intolerance, insecurity, superstition, and rigid thinking.

• How do individual discrimination and institutional discrimination differ?

Individual discrimination involves actions by individual members of the dominant group that harm members of subordinate groups or their property. Institutional discrimination involves day-to-day practices of organizations and institutions that have a harmful effect on members of subordinate groups.

• How do sociologists view racial and ethnic group relations?

Symbolic interactionists suggest that increased contact between people from divergent groups should lead to favorable attitudes and behavior when members of each group (1) have equal status, (2) pursue the same goals, (3) cooperate with one another to achieve goals, and (4) receive positive feedback when they interact with one another. Functionalists stress that members of subordinate groups become a part of the mainstream through assimilation, the process by which members of subordinate groups become absorbed into the dominant culture. Conflict theorists focus on economic stratification and access to power in race and ethnic relations. The caste perspective views inequality as a permanent feature of society, whereas class perspectives focus on the link between capitalism and racial exploitation. According to racial formation theory, the actions of the U.S. government substantially define racial and ethnic relations.

• How have the experiences of various racial–ethnic groups differed in the United States?

Native Americans suffered greatly from the actions of European settlers, who seized their lands and made them victims of forced migration and genocide. Today, they lead lives characterized by poverty and lack of opportunity. White Anglo-Saxon Protestants are the most privileged group in the United States, although social class and gender affect their life chances. White ethnic Americans, whose ancestors migrated from southern and eastern European countries, have gradually made their way into the mainstream of U.S. society. Following the abolishment of slavery in 1863, African Americans were still subjected to segregation, discrimination, and lynchings. Despite civil rights legislation and economic and political gains by many African Americans, racial prejudice and discrimination continue to exist. Asian American immigrants as a group have enjoyed considerable upward mobility in U.S. society in recent decades, but many Asian Americans still struggle to survive by working at low-paying jobs and living in urban ethnic enclaves. Although some Latinos/as have made substantial political, economic, and professional gains in U.S. society, as a group they are nevertheless subjected to anti-immigration sentiments. Middle Eastern immigrants to the United States speak a variety of languages and have diverse religious backgrounds. Because they generally come from middle-class backgrounds, they have made inroads into mainstream U.S. society.

www.cengage.com/login

Register for a Student eResource account to maximize your study time online using CengageNOW. First take the system's diagnostic pre-test, and then follow the personalized study plan that is created for you to help you review this chapter. The study plan will

- help you identify areas on which you should concentrate;
- provide interactive exercises to help you master the chapter concepts; and
- provide a post-test to confirm you are ready to move on to the next chapter.

Key Terms

assimilation 292
authoritarian personality 288
discrimination 288
dominant group 286
ethnic group 282
ethnic pluralism 293
gendered racism 297

genocide 289
individual discrimination 289
institutional discrimination 289
internal colonialism 296
prejudice 286
race 282
racism 287

scapegoat 288
segregation 293
split labor market 297
stereotypes 286
subordinate group 286
theory of racial formation 297

Questions for Critical Thinking

1. Do you consider yourself defined more strongly by your race or by your ethnicity? How so?
2. Given that subordinate groups have some common experiences, why is there such deep conflict between some of these groups?

3. What would need to happen in the United States, both individually and institutionally, for a positive form of ethnic pluralism to flourish in the twenty-first century?

The Kendall Companion Website

www.cengage.com/sociology/kendall

Supplement your review of this chapter by going to the text's companion website, where you can take tutorial quizzes, use flash cards to master key terms, follow live links to useful websites, and explore the other study and research resources you'll find there, such as a comprehensive interactive sociology timeline, GSS Data, and Census 2000 information, much of it presented visually in maps.

chapter 10 Sex and Gender

As I sat in the theater at the Aladdin Hotel on the Strip [in Las Vegas, Nevada], I was enclosed by a sea of crowns. Little girls and teenagers attended the [Miss America] Pageant in droves, many wearing the crowns and sashes that represented their biggest pageant victories.

Sitting in the middle of the cheering section for Miss Kansas, surrounded by cardboard daisies on wooden sticks with Miss Kansas's face in the center and shouts of "You go girl!" I felt as if I were at a political convention or a religious revival. Also marooned in the Miss Kansas section, without any daisies, were

a little girl, her mother (who had twice competed in the Miss Nevada state pageant), and her grandmother, who sat in front of me. The twelve-year-old watched the Pageant with wide eyes the whole night. She wore no crown or sash, but she looked smart in a black velvet dress.

After the talent segment, the girl turned to me and asked, "When

Jennifer Berry, Miss Oklahoma, accepts her crown as the 2006 Miss America at the Aladdin Casino in Las Vegas. What is your opinion of pageants such as this one?

© TIM SHAFFER/Reuters/Landov

I'm in the Miss America Pageant, I want to play the piano and the saxophone for my talent. I can switch back and forth. Do you think that would work?"

"Well, it would certainly be different," I replied.

"Good, then that would help me win."

"So, you really want to be Miss America someday?"

The girl nodded her head, her face solemn. Before replying, I paused. "Well, you can do that. But, you know, there are so many other things to do besides being a beauty queen."

The little girl did not hear me. She was rapturously watching as Miss Oklahoma was crowned Miss America 2006. All the other girls in the audience, those with crowns and those without, stood together, mouthing the words to the famous theme song as the new Miss America was serenaded by the voice of the great Pageant emcee, Bert Perks, who died in 1992: *"There she is, Miss America, there she is, your ideal. . . ."*

—Hilary Levey (2007: 72), a graduate student in sociology at Princeton University, describes her thoughts upon attending a Miss America Pageant. Although Ms. Levey never competed in a beauty pageant, her mother was named Miss America (from Michigan) in 1970.

Chapter Focus Question

How do expectations about female and male appearance, and especially weight, reflect gender inequality?

Many little girls are similar to the one that Hilary Levey encountered at the Miss America Pageant: They have their hearts set on being chosen as the winner of a beauty and/or talent competition such as Miss America or Miss USA. Tens of thousands of beauty pageants—ranging from local beach bikini pageants to international scholarship competitions—are held annually. Two competitions—the Miss America Scholarship program and the Miss USA pageant—involve more than 7,500 local and regional pageants across the country each year (Banet-Weiser, 1999).

Of course, all pageants are not identical. For example, organizers of the Miss America pageant claim that their competition focuses on both talent and beauty and that it exists "to provide personal and professional opportunities for young women and to promote their voices in culture, politics and the community" (Miss America, 2008). By contrast, Miss USA, which is part of the Miss Universe system and partly owned by Donald Trump, originated as a "bathing beauty" competition that was sponsored by a swimwear company. Today, Miss USA and its younger counterpart, Miss Teen USA, continue to look for female models who look outstanding in swimsuits and evening gowns, and who can promote a variety array of products ranging from suntan lotion to flashy diamonds (Angelotti, 2006). Regardless of somewhat different stated goals, these talent and beauty competitions are really about physical beauty and appearance.

For this reason, competitions such as Miss America, Miss USA, Miss Universe, and Miss Teen USA have been the subject of both praise and criticism for the ways in which they portray girls and young women. Some individuals believe that beauty pageants are good for women because they encourage individual achievement and promote self-confidence. Pageant winners are often praised for being positive role models for young women, particularly if the title holder remains scandal-free during the year of her reign. However, some critics of beauty pageants claim that these events promote an unrealistic beauty ideal that is not attainable for most people and that is not necessarily desirable in the real world (see Banet-Weiser, 1999). Other critics believe that pageants are degrading to women because the contestants are ranked "like prize horses" and given a sash to put around their neck (Corsbie-Massay, 2005: 1). Some feminist analysts argue that beauty pageants objectify women (Watson and Martin, 2004).

What is objectification? *Objectification* is the process whereby some people treat other individuals as if they were objects or things, not human be-

◆ **Table 10.1 The Objectification of Women**

General Aspects of Objectification	Objectification Based on Cultural Preoccupation with "Looks"
Women are responded to primarily as "females," whereas their personal qualities and accomplishments are of secondary importance.	Women are often seen as the objects of sexual attraction, not full human beings—for example, when they are stared at.
Women are seen as "all alike."	Women are seen by some as depersonalized body parts—for example, "a piece of ass."
Women are seen as being subordinate and passive, so things can easily be "done to a woman"—for example, discrimination, harassment, and violence.	Depersonalized female sexuality is used for cultural and economic purposes—such as in the media, advertising, the fashion and cosmetics industries, and pornography.
Women are seen as easily ignored or trivialized.	Women are seen as being "decorative" and status-conferring objects to be bought (sometimes collected) and displayed by men and sometimes by other women.
	Women are evaluated according to prevailing, narrow "beauty" standards and often feel pressure to conform to appearance norms.

Source: Schur, 1983.

How Much Do You Know About Body Image and Gender?

True	False	
T	F	1. Most people have an accurate perception of their physical appearance.
T	F	2. Recent studies show that up to 95 percent of men express dissatisfaction with some aspect of their bodies.
T	F	3. Many young girls and women believe that being even slightly "overweight" makes them less "feminine."
T	F	4. Physical attractiveness is a more central part of self-concept for women than for men.
T	F	5. Contestants in beauty pageants such as Miss America have remained about the same in body size throughout the history of these competitions.
T	F	6. Thinness has always been the "ideal" body image for women.
T	F	7. Women bodybuilders have gained full acceptance in society.
T	F	8. The media play a significant role in shaping societal perceptions about the ideal female body.

Answers on page 320.

ings. For example, we objectify women—or men—when we judge them strictly on the basis of their physical appearance rather than on their individual qualities, attributes, or actions (Schur, 1983). Although men may be objectified in some societies, the objectification of girls and women is widespread and particularly common in the United States and many other nations (see ◆ Table 10.1). In regard to beauty pageants, organizers seek to deflect this criticism by providing contestants with an opportunity to talk about themselves and their interests or to answer questions that supposedly will show that they are intelligent and knowledgeable about current events. At the end of each pageant, however, the winner's physical attractiveness is most often highlighted rather than the true substance of her life (Angelotti, 2006). Although some people think of the Miss America Pageant and similar competitions as a vestige of the past, many women and men in the twenty-first century are strongly influenced by the images that these competitions project regarding beauty, body image, race/ethnicity, identity, and consumerism (Watson and Martin, 2004).

Some differences between men and women are biological in nature; however, many differences between the sexes are socially constructed. Studying sociology makes us aware of differences that relate to gender (a social concept) as well as differences that are based on a person's biological makeup, or sex. In this chapter, we examine the issue of gender: what it is and how it affects us. Before reading on, test your knowledge about body image and gender by taking the quiz in Box 10.1.

Sex: The Biological Dimension

Whereas the word *gender* is often used to refer to the distinctive qualities of men and women (masculinity and femininity) that are culturally created, *sex* **refers to the biological and anatomical differences between females and males.** At the core of these differences is the chromosomal information transmitted at the moment a child is conceived. The mother contributes an X chromosome and the father

sex the biological and anatomical differences between females and males.

Box 10.1 Sociology and Everyday Life

Answers to the Sociology Quiz on Body Image and Gender

1. False. Many people do not have a very accurate perception of their bodies. For example, many girls and women think of themselves as "fat" when they are not. Some boys and men believe that they need a well-developed chest and arm muscles, broad shoulders, and a narrow waist.

2. True. In recent studies, up to 95 percent of men believed they needed to improve some aspect of their bodies.

3. True. More than half of all adult women in the United States are currently dieting, and over three-fourths of normal-weight women think they are "too fat." Recently, very young girls have developed similar concerns.

4. True. Women have been socialized to believe that being physically attractive is very important. Studies have found that weight and body shape are the central determinants of women's perception of their physical attractiveness.

5. False. Contestants in Miss America and other national beauty pageants decreased in body size, becoming much thinner and less curvaceous, during the 1980s and 1990s. Today, more emphasis is place on being physically fit, but winning contestants typically have much lower body weight than the average woman of their height and age.

6. False. The "ideal" body image for women has changed a number of times. A positive view of body fat has prevailed for most of human history; however, in the twentieth century in the United States, this view gave way to "fat aversion."

7. False. Although bodybuilding among women has gained some degree of acceptance, women bodybuilders are still expected to be very "feminine" and not to overdevelop themselves.

8. True. Women in the United States are bombarded by advertising, television programs, and films containing images of women that typically represent an ideal that most real women cannot attain.

Sources: Based on Fallon, Katzman, and Wooley, 1994; Kilbourne, 1999; Seid, 1994; and Turner, 1997.

---

either an X (which produces a female embryo) or a Y (which produces a male embryo). At birth, male and female infants are distinguished by ***primary sex characteristics:* the genitalia used in the reproductive process.** At puberty, an increased production of hormones results in the development of *secondary sex characteristics:* **the physical traits (other than reproductive organs) that identify an individual's sex.** For women, these include larger breasts, wider hips, and narrower shoulders; a layer of fatty tissue throughout the body; and menstruation. For men, they include development of enlarged genitals, a deeper voice, greater height, a more muscular build, and more body and facial hair.

Hermaphrodites/Transsexuals

Sex is not always clear-cut. Occasionally, a hormone imbalance before birth produces a ***hermaphrodite*—a person in whom sexual differentiation is ambiguous or incomplete.** A hermaphrodite (sometimes called an *intersexed person*) tends to have some combination of male and female genitalia. In one case, for example, a chromosomally normal (XY) male was born with a penis just one centimeter long and a urinary opening similar to that of a female. Some people may be genetically of one sex but have a gender identity of the other. That is true for a ***transsexual,* a person in whom the sex-**

related structures of the brain that define gender identity are opposite from the physical sex organs of the person's body. Consequently, transsexuals often feel that they are the opposite sex from that of their sex organs. Some transsexuals are aware of this conflict between gender identity and physical sex as early as the preschool years. Some transsexuals take hormone treatments or have a sex change operation to alter their genitalia in order to achieve a body congruent with their sense of sexual identity. Many transsexuals who receive hormone treatments or undergo surgical procedures go on to lead lives that they view as being compatible with their true sexual identity.

Western societies acknowledge the existence of only two sexes; some other societies recognize three—men, women, and *berdaches* (or *hijras* or *xaniths*): biological males who behave, dress, work, and are treated in most respects as women. The closest approximation of a third sex in Western societies is a ***transvestite,* a male who lives as a woman or a female who lives as a man but does not alter the genitalia.** Although transvestites are not treated as a third sex, they often "pass" for members of that sex because their appearance and mannerisms fall within the range of what is expected from members of the other sex.

Transsexuality may occur in conjunction with homosexuality, but this is frequently not the case. Some researchers believe that both transsexuality and homosexuality have a common prenatal cause such as a critically timed hormonal release due to stress in the mother or the presence of certain hormone-mimicking chemicals during critical steps of fetal development. Researchers continue to examine this issue and debate the origins of transsexuality and homosexuality.

Sexual Orientation

Sexual orientation **refers to an individual's preference for emotional–sexual relationships with members of the opposite sex (heterosexuality), the same sex (homosexuality), or both (bisexuality)** (Lips, 2001). Some scholars believe that sexual orientation is rooted in biological factors that are present at birth; others believe that sexuality has both biological and social components and is not preordained at birth.

The terms *homosexual* and *gay* are most often used in association with males who prefer same-sex relationships; the term *lesbian* is used in association with females who prefer same-sex relationships. Heterosexual individuals, who prefer opposite-sex relationships, are sometimes referred to as *straight*. However, it is important to note that heterosexual people are much less likely to be labeled by their sexual orientation than are people who are gay, lesbian, or bisexual.

What criteria do social scientists use to classify individuals as gay, lesbian, or homosexual? In a definitive study of sexuality in the mid-1990s, researchers at the University of Chicago established three criteria for identifying people as homosexual or bisexual: (1) *sexual attraction* to persons of one's own gender, (2) *sexual involvement* with one or more persons of one's own gender, and (3) *self-identification* as a gay, lesbian, or bisexual (Michael et al., 1994). According to these criteria, then, having engaged in a homosexual act does not necessarily classify a person as homosexual. In fact, many respondents in the University of Chicago study indicated that although they had at least one homosexual encounter when they were younger, they were no longer involved in homosexual conduct and never identified themselves as gay, lesbian, or bisexual.

primary sex characteristics the genitalia used in the reproductive process.

secondary sex characteristics the physical traits (other than reproductive organs) that identify an individual's sex.

hermaphrodite a person in whom sexual differentiation is ambiguous or incomplete.

transsexual a person who believes that he or she was born with the body of the wrong sex.

transvestite a male who lives as a woman or a female who lives as a man but does not alter the genitalia.

sexual orientation a person's preference for emotional–sexual relationships with members of the opposite sex (heterosexuality), the same sex (homosexuality), or both (bisexuality).

Studies have examined how sexual orientation is linked to identity. Sociologist Kristin G. Esterberg (1997) interviewed lesbian and bisexual women to determine how they "perform" lesbian or bisexual identity through daily activities such as choice of clothing and hairstyles, as well as how they use body language and talk. According to Esterberg (1997), some of the women viewed themselves as being "lesbian from birth" whereas others had experienced shifts in their identities, depending on social surroundings, age, and political conditions at specific periods in their lives. Another study looked at gay and bisexual men. Human development scholar Ritch C. Savin-Williams (2004) found that gay/bisexual youths often believe from an early age that they are different from other boys:

> The pattern that most characterized the youths' awareness, interpretation, and affective responses to childhood attractions consisted of an overwhelming desire to be in the company of men. They wanted to touch, smell, see, and hear masculinity. This awareness originated from earliest childhood memories; in this sense, they "always felt gay."

However, most of the boys and young men realized that these feelings were not typical of other males and were uncomfortable when others attempted to make them conform to the established cultural definitions of masculinity, such as showing a great interest in team sports, competition, and aggressive pursuits.

The term *transgendered* refers to individuals whose appearance, behavior, or self-identification does not conform to common social rules of gender expression. Transgenderism is sometimes used to refer to those who cross-dress, to transsexuals, and to others outside mainstream categories. Although some gay and lesbian advocacy groups oppose the concept of transgender as being somewhat meaningless, others applaud the term as one that might help unify diverse categories of people based on sexual identity. Various organizations of gays, lesbians, and transgendered persons have been unified in their desire to reduce hate crimes and other forms of *homophobia*—extreme prejudice directed at gays, lesbians, bisexuals, and others who are perceived as not being heterosexual.

Gender: The Cultural Dimension

Gender refers to the culturally and socially constructed differences between females and males found in the meanings, beliefs, and practices associated with "femininity" and "masculinity." Although biological differences between women and men are very important, in reality most "sex differences" are socially constructed "gender differences." According to sociologists, social and cultural processes, not biological "givens," are most important in defining what females and males are, what they should do, and what sorts of relations do or should exist between them. Sociologist Judith Lorber (1994: 6) summarizes the importance of gender:

> Gender is a human invention, like language, kinship, religion, and technology; like them, gender organizes human social life in culturally patterned ways. Gender organizes social relations in everyday life as well as in the major social structures, such as social class and the hierarchies of bureaucratic organizations.

Virtually everything social in our lives is *gendered:* People continually distinguish between males and females and evaluate them differentially. Gender is an integral part of the daily experiences of both women and men (Kimmel and Messner, 2004).

A microlevel analysis of gender focuses on how individuals learn gender roles and acquire a gender identity. *Gender role* refers to the attitudes, behavior, and activities that are socially defined as appropriate for each sex and are learned through the socialization process (Lips, 2001). For example, in U.S. society, males are traditionally expected to demonstrate aggressiveness and toughness whereas females are expected to be passive and nurturing. *Gender identity* is a person's perception of the self as female or male. Typically established between eighteen months and three years of age, gender identity is a powerful aspect of our self-concept (Lips, 2001). Although this identity is an individual perception, it is developed through interaction with others. As a result, most people form a gender identity that matches their biological sex: Most biologi-

© Alex Farnsworth/The Image Works

© MOHAMMED AMEEN/Reuters/Landov

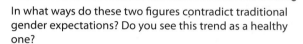

In what ways do these two figures contradict traditional gender expectations? Do you see this trend as a healthy one?

cal females think of themselves as female, and most biological males think of themselves as male. Body consciousness is a part of gender identity. *Body consciousness* **is how a person perceives and feels about his or her body;** it also includes an awareness of social conditions in society that contribute to this self-knowledge (Thompson, 1994). Consider, for example, these comments by Steve Michalik, a former Mr. Universe:

> I was small and weak, and my brother Anthony was big and graceful, and my old man made no bones about loving him and hating me. . . . The minute I walked in from school, it was, "You worthless little s--t, what are you doing home so early?" His favorite way to torture me was to tell me he was going to put me in a home. We'd be driving along in Brooklyn somewhere, and

homophobia extreme prejudice directed at gays, lesbians, bisexuals, and others who are perceived as not being heterosexual.

gender the culturally and socially constructed differences between females and males found in the meanings, beliefs, and practices associated with "femininity" and "masculinity."

gender role the attitudes, behavior, and activities that are socially defined as appropriate for each sex and are learned through the socialization process.

gender identity a person's perception of the self as female or male.

body consciousness a term that describes how a person perceives and feels about his or her body.

we'd pass a building with iron bars on the windows, and he'd stop the car and say to me, "Get out. This is the home we're putting you in." I'd be standing there sobbing on the curb—I was maybe eight or nine at the time. (qtd. in Klein, 1993: 273)

As we grow up, we become aware, as Michalik did, that the physical shape of our bodies subjects us to the approval or disapproval of others. Being small and weak may be considered positive attributes for women, but they are considered negative characteristics for "true men."

A macrolevel analysis of gender examines structural features, external to the individual, that perpetuate gender inequality. These structures have been referred to as *gendered institutions,* meaning that gender is one of the major ways by which social life is organized in all sectors of society. Gender is embedded in the images, ideas, and language of a society and is used as a means to divide up work, allocate resources, and distribute power. For example, every society uses gender to assign certain tasks—ranging from child rearing to warfare—to females and to males, and differentially rewards those who perform these duties.

These institutions are reinforced by a *gender belief system,* which includes all the ideas regarding masculine and feminine attributes that are held to be valid in a society. This belief system is legitimated by religion, science, law, and other societal values (Lorber, 2005). For example, gendered belief systems may change over time as gender roles change. Many fathers take care of young children today, and there is a much greater acceptance of this change in roles. However, popular stereotypes about men and women, as well as cultural norms about gender-appropriate appearance and behavior, serve to reinforce gendered institutions in society.

The Social Significance of Gender

Gender is a social construction with important consequences in everyday life. Just as stereotypes regarding race/ethnicity have built-in notions of superiority and inferiority, gender stereotypes hold that men and women are inherently different in attributes, behavior, and aspirations. Stereotypes define men as strong, rational, dominant, independent, and less concerned with their appearance. Women are stereotyped as weak, emotional, nurturing, dependent, and anxious about their appearance.

The social significance of gender stereotypes is illustrated by eating problems. The three most common eating problems are anorexia, bulimia, and obesity. With *anorexia,* a person has lost at least 25 percent of body weight due to a compulsive fear of becoming fat (Lott, 1994). With *bulimia,* a person binges by consuming large quantities of food and then purges the food by induced vomiting, excessive exercise, laxatives, or fasting. With *obesity,* individuals are 20 percent or more above their desirable weight, as established by the medical profession. For a 5-foot-4-inch woman, that is about twenty-five pounds; for a 5-foot-10-inch man, it is about thirty pounds (Burros, 1994: 1).

Sociologist Becky W. Thompson argues that, based on stereotypes, the primary victims of eating problems are presumed to be white, middle-class, heterosexual women. However, such problems also exist among women of color, working-class women, lesbians, and some men. According to Thompson, explanations regarding the relationship between gender and eating problems must take into account a complex array of social factors, including gender socialization and women's responses to problems such as racism and emotional, physical, and sexual abuse (Thompson, 1994; see also Wooley, 1994).

Bodybuilding is another gendered experience. *Bodybuilding* is the process of deliberately cultivating an increase in the mass and strength of the skeletal muscles by means of lifting and pushing weights. In the past, bodybuilding was predominantly a male activity; musculature connoted power, domination, and virility. Today, however, an increasing number of women engage in this activity. As gendered experiences, eating problems and bodybuilding have more in common than we might think. Women's studies scholar Susan Bordo (2004) has noted that the anorexic body and the muscled body are not opposites, but instead are both united against the common enemy of soft, flabby flesh. In other words, the *body* may be objectified both through compulsive dieting and compulsive bodybuilding.

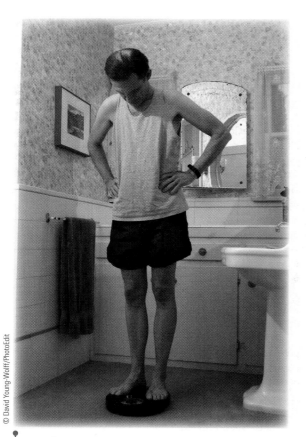

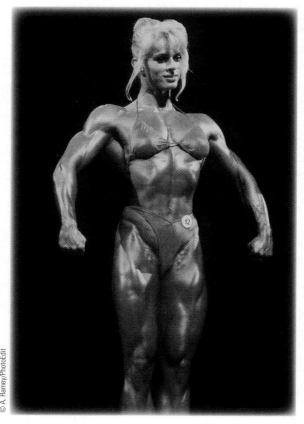

Not all anorectics are women, and not all bodybuilders are men. However, Susan Bordo argues that these two issues are manifestations of the same desire: to avoid having soft, flabby flesh.

Sexism

Sexism **is the subordination of one sex, usually female, based on the assumed superiority of the other sex.** Sexism directed at women has three components: (1) negative attitudes toward women; (2) stereotypical beliefs that reinforce, complement, or justify the prejudice; and (3) discrimination—acts that exclude, distance, or keep women separate (Lott, 1994).

Like racism, sexism is used to justify discriminatory treatment. When women participate in what are considered gender-inappropriate endeavors in the workplace, at home, or in leisure activities, they often find that they are the targets of prejudice and discrimination. Obvious mani-

festations of sexism are found in the undervaluing of women's work and in hiring and promotion practices that effectively exclude women from an organization or confine them to the bottom of the organizational hierarchy. Even today, some women who enter nontraditional occupations (such as firefighting and welding) or professions (such as dentistry, architecture, or investment banking) encounter hurdles that men do not face (see "Sociology Works!").

sexism the subordination of one sex, usually female, based on the assumed superiority of the other sex.

Sociology *Works!*

Institutional Discrimination: Women in a Locker-Room Culture

News Item: Morgan Stanley, a bank holding company, to pay $46 million in a case charging systemic sex discrimination

Responses: "I'm so happy that there's a settlement that's good for everybody," said lead plaintiff Allison Schieffelin, a former bond trader who was awarded a settlement of $12 million (qtd. in Shell, 2007).

"It sends a message to employers everywhere that allegations of discrimination need to be taken seriously," said Elizabeth Grossman, an Equal Employment Opportunity Commission lawyer (qtd. in Shell, 2007).

For decades, sociologists have called attention to the fact that both individual discrimination and institutional discrimination—based on sex, race/ethnicity, and other devalued characteristics and attributes of subordinate-group members—are widespread (Benokraitis and Feagin, 1994). Previous sex-discrimination lawsuits typically involved details about the crude behavior of male employees toward women, or pornography in the workplace, or male bosses' demands that women employees accompany them to strip clubs and other objectionable locations as part of their work-related duties (Anderson, 2007). However, the Morgan Stanley case involved none of these issues and instead focused on women's opportunities for training, gaining new clients, promotion, and pay equity—all factors related to institutional discrimination, previously defined (in Chapter 9) as the day-to-day-practices of organizations and institutions that have a harmful impact on members of subordinate groups.

In the case of Morgan Stanley, the Wall Street investment bank, female employees were passed over for new clients, paid less than male employees, and often passed over for promotion. Although Morgan Stanley denied all charges, the firm agreed to pay at least $46 million to settle this class-action suit filed by eight current and former female brokers. A primary issue in the case involved the way in which accounts were distributed in the firm's retail branches. According to the lawsuit, a "locker-room culture" prevailed in which accounts were often given to golf buddies based on a "power ranking" system that dictated how the accounts of retiring and departing brokers were to be distributed and how new accounts were to be assigned. A new system put in place after the lawsuit will automate the distribution of accounts, brokers will received written information about the distribution of accounts, and data will be more readily available to individuals who believe they are being discriminated against (Anderson, 2007).

Sociological theorizing and research have increased public awareness that the day-to-day practices of organizations and institutions may have a negative and differential effect on individuals who have historically been excluded from workplace settings such as prestigious Wall Street firms, which traditionally have been dominated by white males. Although many gains have been made through legislation and litigation to reduce institutional discrimination, recent cases such as the Morgan Stanley sex-bias lawsuit demonstrate that much remains to be done before women truly have equal opportunities in the workplace.

Reflect & Analyze

Why is it important for all employees to feel that they are being treated fairly at work? Are some employment settings more resistant to change than others? What do you think?

Sexism is interwoven with *patriarchy*—a **hierarchical system of social organization in which cultural, political, and economic structures are controlled by men.** By contrast, *matriarchy* **is a hierarchical system of social organization in which cultural, political, and economic structures are controlled by women;** however, few (if any) societies have been organized in this manner. Patriarchy is reflected in the way that men may think of their position as men as a given whereas women may deliberate on what their position in society should be (see Box 10.2 for an example). As the sociologist Virginia Cyrus (1993: 6) explains, "Under patriarchy, men are seen as 'natural' heads of households,

◆ Table 10.2 Technoeconomic Bases of Society

	Hunting and Gathering	Horticultural and Pastoral	Agrarian	Industrial	Postindustrial
Change from Prior Society	—	Use of hand tools, such as digging stick and hoe	Use of animal-drawn plows and equipment	Invention of steam engine	Invention of computer and development of "high-tech" society
Economic Characteristics	Hunting game, gathering roots and berries	Planting crops, domestication of animals for food	Labor-intensive farming	Mechanized production of goods	Information and service economy
Control of Surplus	None	Men begin to control societies	Men who own land or herds	Men who own means of production	Corporate shareholders and high-tech entrepreneurs
Women's Status	Relative equality	Decreasing in move to pastoralism	Low	Low	Varies by class, race, and age

Source: Adapted from Lorber, 1994: 140.

Presidential candidates, corporate executives, college presidents, etc. Women, on the other hand, are men's subordinates, playing such supportive roles as housewife, mother, nurse, and secretary." Gender inequality and a division of labor based on male dominance are nearly universal, as we will see in the following discussion on the origins of gender-based stratification.

Gender Stratification in Historical and Contemporary Perspective

How do tasks in a society come to be defined as "men's work" or "women's work"? Three factors are important in determining the gendered division of labor in a society: (1) the type of subsistence base, (2) the supply of and demand for labor, and (3) the extent to which women's child-rearing activities are compatible with certain types of work. *Subsistence* refers to the means by which a society gains the basic necessities of life, including food, shelter, and clothing. The three factors vary according to a society's *technoeconomic base*—the level of technology and the organization of the economy in a given society. Five such bases have been identified: hunting and gathering societies, horticultural and pastoral societies, agrarian societies, industrial societies, and postindustrial societies, as shown in ◆ Table 10.2.

Hunting and Gathering Societies

The earliest known division of labor between women and men is in hunting and gathering societies. While the men hunt for wild game, women gather roots and berries. A relatively equitable relationship exists because neither sex has the ability to provide all the food necessary for survival. When wild game is nearby, both men and women may hunt. When it is far away, hunting becomes incompatible with child rearing (which women tend to do because they breast-feed their young), and women

patriarchy a hierarchical system of social organization in which cultural, political, and economic structures are controlled by men.

matriarchy a hierarchical system of social organization in which cultural, political, and economic structures are controlled by women.

Box 10.2 Sociology and Global Perspective

The Rise of Islamic Feminism in the Middle East?

I would like for all of the young Muslim girls to be able to relate to Iman, whether they wear the hijab [head scarf] or not. Boys will also enjoy Iman's adventures because she is one tough, smart girl! Iman gets her super powers from having very strong faith in Allah, or God. She solves many of the problems by explaining certain parts of the Koran that relate to the story.

—Rima Khoreibi, an author from Dubai (United Arab Emirates), explaining that she has written a book about an Islamic superhero who is female because she would like to dispel a widely held belief that sexism in her culture is deeply rooted in Islam (see theadventuresofiman.com, 2007; Kristof, 2006)

Although Rima Khoreibi and many others who have written fictional and nonfictional accounts of girls and women living in the Middle East typically do not deny that sexism exists in their region or that sexism is deeply interwoven with patriarchy around the world, they dispute the perception that Islam is inherently misogynistic (possessing hatred or strong prejudice toward women). As defined in this chapter, *patriarchy* is a hierarchical system of social organization in which cultural, political, and economic structures are con-

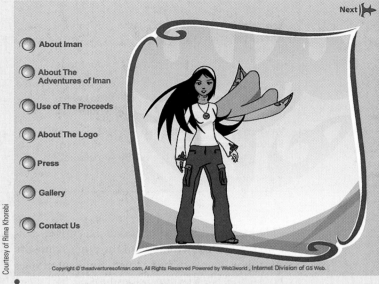

The home page of *The Adventures of Iman.*

are placed at a disadvantage in terms of contributing to the food supply (Lorber, 1994). In most hunting and gathering societies, women are full economic partners with men; relations between them tend to be cooperative and relatively egalitarian (Chafetz, 1984; Bonvillain, 2001). Little social stratification of any kind is found because people do not acquire a food surplus.

Horticultural and Pastoral Societies

In horticultural societies, which first developed ten to twelve thousand years ago, a steady source of food becomes available. People are able to grow their own food because of hand tools, such as the digging stick and the hoe. Women make an important contribution to food production because hoe cultivation is

compatible with child care. A fairly high degree of gender equality exists because neither sex controls the food supply.

When inadequate moisture in an area makes planting crops impossible, *pastoralism*—the domestication of large animals to provide food—develops. Herding is primarily done by men, and women contribute relatively little to subsistence production in such societies. In some herding societies, women have relatively low status; their primary value is their ability to produce male offspring so that the family lineage can be preserved and enough males will exist to protect the group against attack (Nielsen, 1990).

In contemporary horticultural societies, women do most of the farming while men hunt game, clear land, work with arts and crafts, make tools, par-

trolled by men. The influence of religion on patriarchy is a topic of great interest to contemporary scholars, particularly those applying a feminist approach to their explanations of why persistent social inequalities exist between women and men and how these inequalities are greater in some regions of the world than in others.

According to some gender studies specialists, a newer form of feminist thinking is emerging among Muslim women. Often referred to as "feminist Islam" or "Islamic feminism," this approach is based on the belief that greater gender equality may be possible in the Muslim world if the teachings of Islam, as set forth in the Koran (or *Qur' an*)—the Islamic holy book—are followed more closely. Islamic feminism is based on the principle that Muslim women should retain their allegiance to Islam as an essential part of their self-determination and identity but that they should also work to change patriarchal control over the basic Islamic world view (Wadud, 2002). According to the journalist Nicholas D. Kristof (2006), both Islam and evangelical Christianity have been on the rise in recent years because both religions provide "a firm moral code, spiritual reassurance and orderliness to people vexed by chaos and immorality around them, and they offer dignity to the poor."

Islamic feminists believe that the rise of Islam might contribute to greater, rather than less, equality for women. From this perspective, stories about characters such as Iman may help girls and young women realize that they can maintain their deep religious convictions and their head scarf (*hijab*) while, at the same time, working for greater equality for women and more opportunities for themselves. In *The Adventures of Iman,* the female hero always wears a pink scarf around her neck, and she uses the scarf to cover her hair when she is praying to Allah. Iman quotes the Qur' an when she is explaining to others that Muslims are expected to be tolerant, kind, and righteous. For Iman, religion is a form of empowerment, not an extension of patriarchy.

Islamic feminism is quite different from what most people think of as Western feminism (particularly in regard to issues such as the wearing of the *hijab* or the fact that in Saudi Arabia, a woman may own a motor vehicle but may not legally drive it). However, change is clearly under way in many regions of the Middle East and other areas of the world as rapid economic development and urbanization quickly change the lives of many people.

Reflect & Analyze

Why is women's inequality a complex issue to study across nations? What part does culture play in defining the roles of women and men in various societies? How do religious beliefs influence what we think of as "appropriate" or "inappropriate" behaviors for men, women, and children? What do you think?

ticipate in religious and ceremonial activities, and engage in war. A combination of horticultural and pastoral activities is found in some contemporary societies in Asia, Africa, the Middle East, and South America. These societies are characterized by more gender inequality than in hunting and gathering societies but less than in agrarian societies (Bonvillain, 2001).

Agrarian Societies

In agrarian societies, which first developed about eight to ten thousand years ago, gender inequality and male dominance become institutionalized. The most extreme form of gender inequality developed about five thousand years ago in societies in the fertile crescent around the Mediterranean Sea (Lorber, 1994). Agrarian societies rely on agriculture—farming done by animal-drawn or mechanically powered plows and equipment. Because agrarian tasks require more labor and greater physical strength than do horticultural ones, men become more involved in food production. It has been suggested that women are excluded from these tasks because they are viewed as too weak for the work and because child-care responsibilities are considered incompatible with the full-time labor that the tasks require (Nielsen, 1990).

Why does gender inequality increase in agrarian societies? Scholars cannot agree on an answer; some suggest that it results from private ownership of property. When people no longer have to move continually in search of food, they can acquire a surplus. Men gain control over the disposition of

In contemporary societies, women do a wide variety of work and are responsible for many diverse tasks. The women shown here are employed in the agricultural and the postindustrial sectors of the U.S. economy. How might issues of gender inequality differ for these two women? What issues might be the same for both of them?

the surplus and the kinship system, and this control serves men's interests (Lorber, 1994). The importance of producing "legitimate" heirs to inherit the surplus increases significantly, and women's lives become more secluded and restricted as men attempt to ensure the legitimacy of their children. Premarital virginity and marital fidelity are required; indiscretions are punished (Nielsen, 1990). However, some scholars argue that male dominance existed before the private ownership of property (Firestone, 1970; Lerner, 1986).

Male dominance is very strong in agrarian societies. Women are secluded, subordinated, and mutilated as a means of regulating their sexuality and protecting paternity. Most of the world's population currently lives in agrarian societies in various stages of industrialization.

Industrial Societies

An *industrial society* is one in which factory or mechanized production has replaced agriculture as the major form of economic activity. As societies industrialize, the status of women tends to decline further. Industrialization in the United States created a gap between the nonpaid work performed by middle- and upper-class women at home and the paid work that was increasingly performed by men and unmarried girls (Amott and Matthaei, 1996). Husbands were responsible for being "breadwinners"; wives were seen as "homemakers."

This gendered division of labor increased the economic and political subordination of women. It also became a source of discrimination against women of color based on both their race and the fact that many of them had to work in order to survive. In the late 1800s and early 1900s, many African American women were employed as domestic servants in affluent white households (Lorber, 1994).

As people moved from a rural, agricultural lifestyle to an urban existence, body consciousness increased. People who worked in offices often became sedentary and exhibited physical deterioration from their lack of activity. As gymnasiums were built to fight this lack of physical fitness, images of masculinity shifted from the "burly farmer" or "robust workman" to the middle-class man who exercised and lifted weights (Klein, 1993). As industrialization progressed and food became more plentiful, the social symbolism of women's body weight and size also changed, and middle-class women became more preoccupied with body fitness (Bordo, 2004; Seid, 1994). Today, women's bodies (even in bodybuilding programs) are supposed to be "inviting, available, and welcoming" whereas men's bodies should be "self-contained, active, and invasive" (MacSween, 1993: 156).

Postindustrial Societies

Chapter 4 defines *postindustrial societies* as ones in which technology supports a service- and information-based economy. In such societies, the division of labor in paid employment is increasingly based on whether people provide or apply information or are employed in service jobs such as fast-food-restaurant counter help or health care workers. For both women and men in the labor force, formal education is increasingly crucial for economic and social success. However, although some women have moved into entrepreneurial, managerial, and professional occupations, many others have remained in the low-paying service sector, which affords few opportunities for upward advancement.

How do new technologies influence gender relations in the workplace? Although some analysts presumed that technological developments would reduce the boundaries between women's and men's work, researchers have found that the gender stereotyping associated with specific jobs has remained remarkably stable even when the nature of work and the skills required to perform it have been radically transformed. Today, men and women continue to be segregated into different occupations, and this segregation is particularly visible within individual workplaces (as discussed later in the chapter).

How does the division of labor change in families in postindustrial societies? For a variety of reasons, more households are headed by women with no adult male present (see "Census Profiles: Single Mothers with Children Under 18"). Almost one-fourth (23 percent) of all U.S. children live with their mother only (as contrasted with just 5 percent who reside with their father only); among African American children, 48 percent live with their mother only (U.S. Census Bureau, 2007). This means that women in these households truly have a double burden, both from family responsibilities and from the necessity of holding gainful employment in the labor force.

In postindustrial societies such as the United States, approximately 60 percent of adult women are in the labor force, meaning that finding time to care for children, help aging parents, and meet the demands of the workplace will continue to place a heavy burden on women, despite living in an information- and service-oriented economy.

How people accept new technologies and the effect that these technologies have on gender stratification are related to how people are socialized into gender roles. However, gender-based stratification remains rooted in the larger social structures of society, which individuals have little ability to control.

Gender and Socialization

We learn gender-appropriate behavior through the socialization process. Our parents, teachers, friends, and the media all serve as gendered institutions that communicate to us our earliest, and often

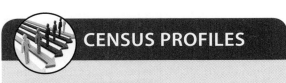

CENSUS PROFILES

Single Mothers with Children Under 18

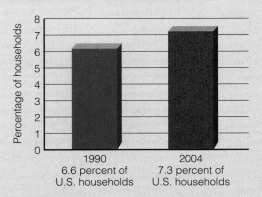

1990	2004
6.6 percent of U.S. households	7.3 percent of U.S. households

As shown above, between 1990 and 2004 the number of U.S. families headed by single mothers increased by about 25 percent, to more than 8.2 million households. For census purposes, a single mother is identified as a woman who is widowed, divorced, separated, or never married and who has children under 18 living at home. Not only does this increase mark a change in the roles of many women, but it may also indicate that "traditional" households are in decline in this country.

Source: U.S. Census Bureau, 2008 (Statistical Abstract).

most lasting, beliefs about the social meanings of being male or female and thinking and behaving in masculine or feminine ways. Some gender roles have changed dramatically in recent years; others have remained largely unchanged over time.

Many parents prefer boys to girls because of stereotypical ideas about the relative importance of males and females to the future of the family and society. Research suggests that social expectations play a major role in this preference. We are socialized to believe that it is important to have a son, especially for a first or only child. For many years, it was assumed that only a male child could support his parents in their later years and carry on the family name. Across cultures, boys are preferred to girls, especially when the number of children that parents can have is limited by law or economic conditions.

For example, in China, which strictly regulates the allowable number of children to one per family, a disproportionate number of female fetuses are aborted, resulting in a shortage of women. However, in the aftermath of the catastrophic earthquakes of early 2008, the government is revising this policy to some degree (see Jacobs, 2008).

Parents and Gender Socialization

From birth, parents act toward children on the basis of the child's sex. Baby boys are perceived to be less fragile than girls and tend to be treated more roughly by their parents. Girl babies are thought to be "cute, sweet, and cuddly" and receive more gentle treatment. Parents strongly influence the gender role development of children by passing on—both overtly and covertly—their own beliefs about gender. When girl babies cry, parents respond to them more quickly, and parents are more prone to talk and sing to girl babies (Wharton, 2004).

Children's toys reflect their parents' gender expectations (Thorne, 1993). Gender-appropriate toys for boys include computer games, trucks and other vehicles, sports equipment, and war toys such as guns and soldiers. Girls' toys include "Barbie" dolls, play makeup, and homemaking items. Parents' choices of toys for their children are not likely to change in the near future. A group of college students in one study were shown slides of toys and asked to decide which ones they would buy for girls and boys. Most said they would buy guns, soldiers, jeeps, carpenter tools, and red bicycles for boys; girls would get baby dolls, dishes, sewing kits, jewelry boxes, and pink bicycles (Fisher-Thompson, 1990).

When children are old enough to help with household chores, they are often assigned different tasks. Maintenance chores (such as mowing the lawn) are assigned to boys, whereas domestic chores (such as shopping, cooking, and clearing the table) are assigned to girls. Chores may also become linked with future occupational choices and personal characteristics. Girls who are responsible for domestic chores such as caring for younger brothers and sisters may learn nurturing behaviors that later translate into employment as a nurse or schoolteacher. Boys may learn about computers and other types of technology that lead to different career options (Wharton, 2004).

Are children's toys a reflection of their own preferences and choices? How do toys reflect gender socialization by parents and other adults?

In the past, most studies of gender socialization focused on white, middle-class families and paid little attention to ethnic differences (Raffaelli and Ontai, 2004). According to earlier studies, children from middle- and upper-income families are less likely to be assigned gender-linked chores than children from lower-income backgrounds. In addition, gender-linked chore assignments occur less frequently in African American families, where both sons and daughters tend to be socialized toward independence, employment, and child care (Bardwell, Cochran, and Walker, 1986; Hale-Benson, 1986). Sociologist Patricia Hill Collins (1991) suggests that African American mothers are less likely to socialize their daughters into roles as subordinates; instead, they are likely to teach them a critical posture that allows them to cope with contradictions.

In contrast, a recent study of gender socialization in U.S. Latino/a families suggests that adolescent females of Mexican, Puerto Rican, Cuban, or other Central or South American descent receive different gender socialization by their parents than do their male siblings (Raffaelli and Ontai, 2004). Latinas are given more stringent curfews and are allowed less interaction with members of the opposite sex than are the adolescent males in their families. Rules for dating, school activities, and part-time jobs are more stringent for the girls because many parents want to protect their daughters and keep them closer to home.

Across classes and racial/ethnic categories, mothers typically play a stronger role in gender socialization of daughters, whereas fathers do more to socialize sons than daughters (McHale, Crouter, and Tucker, 1999).

Peers and Gender Socialization

Peers help children learn prevailing gender-role stereotypes, as well as gender-appropriate and gender-inappropriate behavior (Hibbard and Buhrmester, 1998). During the preschool years, same-sex peers have a powerful effect on how children see their gender roles (Maccoby and Jacklin, 1987); children are more socially acceptable to their peers when they conform to implicit societal norms governing the "appropriate" ways that girls and boys should act in social situations and what prohibitions exist in such cases (Martin, 1989).

Male peer groups place more pressure on boys to do "masculine" things than female peer groups place on girls to do "feminine" things. For example, girls wear jeans and other "boy" clothes, play soccer and softball, and engage in other activities traditionally associated with males. By contrast, if a boy wears a dress, plays hopscotch with girls, and engages in other activities associated with being female, he will be ridiculed by his peers. This distinction between the

© Barbara Campbell

Parents, peers, and the larger society all influence our perceptions about gender-appropriate behavior.

a production of watching her eat lunch and frequently made sounds like grunts or moos (Kolata, 1993). Because peer acceptance is so important for both males and females during their first two decades, such actions can have very harmful consequences for the victims.

As young adults, men and women still receive many gender-related messages from peers. Among college students, for example, peer groups are organized largely around gender relations and play an important role in career choices and the establishment of long-term, intimate relationships. In a study of women college students at two universities (one primarily white, the other predominantly African American), anthropologists Dorothy C. Holland and Margaret A. Eisenhart (1990) found that the peer system propelled women into a world of romance in which their attractiveness to men counted most. Although peers initially did not influence the women's choices of majors and careers, they did influence whether the women continued to pursue their original goals, changed their course of action, or were "derailed."

relative value of boys' and girls' behaviors strengthens the cultural message that masculine activities and behavior are more important and more acceptable (Wood, 1999).

During adolescence, peers are often stronger and more effective agents of gender socialization than adults (Hibbard and Buhrmester, 1998). Peers are thought to be especially important in boys' development of gender identity. Male bonding that occurs during adolescence is believed to reinforce masculine identity (Gaylin, 1992) and to encourage gender-stereotypical attitudes and behavior (Huston, 1985; Martin, 1989). For example, male peers have a tendency to ridicule and bully others about their appearance, size, and weight. Aleta Walker painfully recalls walking down the halls at school when boys would flatten themselves against the lockers and cry, "Wide load!" At lunchtime, the boys made

Teachers, Schools, and Gender Socialization

From kindergarten through college, schools operate as a gendered institution. Teachers provide important messages about gender through both the formal content of classroom assignments and informal interactions with students. Sometimes, gender-related messages from teachers and other students reinforce gender roles that have been taught at home; however, teachers may also contradict parental socialization. During the early years of a child's schooling, teachers' influence is very powerful; many children spend more hours per day with their teachers than they do with their own parents.

According to some researchers, the quantity and quality of teacher–student interactions often vary

Teachers often use competition between boys and girls because they hope to make a learning activity more interesting. Here, a middle-school girl leads other girls against boys in a Spanish translation contest. What are the advantages and disadvantages of gender-based competition in classroom settings?

between the education of girls and that of boys (Wellhousen and Yin, 1997). One of the messages that teachers may communicate to students is that boys are more important than girls. Research spanning the past thirty years shows that unintentional gender bias occurs in virtually all educational settings. *Gender bias* **consists of showing favoritism toward one gender over the other.** Researchers consistently find that teachers devote more time, effort, and attention to boys than to girls (Sadker and Sadker, 1994). Males receive more praise for their contributions and are called on more frequently in class, even when they do not volunteer.

Teacher–student interactions influence not only students' learning but also their self-esteem (Sadker and Sadker, 1985, 1986, 1994). A comprehensive study of gender bias in schools suggested that girls' self-esteem is undermined in school through such experiences as (1) a relative lack of attention from teachers; (2) sexual harassment by male peers; (3) the stereotyping and invisibility of females in textbooks, especially in science and math texts; and (4) test bias based on assumptions about the relative importance of quantitative and visual–spatial ability, as compared with verbal ability, that restricts some girls' chances of being admitted to the most prestigious colleges and being awarded scholarships.

Teachers also influence how students treat one another during school hours. Many teachers use sex segregation as a way to organize students, resulting in unnecessary competition between females and males (Basow, 1992). In addition, teachers may take a "boys will be boys" attitude when girls complain of sexual harassment. Even though sexual harassment is prohibited by law, and teachers and administrators are obligated to investigate such incidents, the complaints may be dealt with superficially. If that happens, the school setting can become a hostile environment rather than a site for learning (Sadker and Sadker, 1994).

Sports and Gender Socialization

Children spend more than half of their nonschool time in play and games, but the type of games played differs with the child's sex. Studies indicate that boys are socialized to participate in highly competitive, rule-oriented games with a larger number of participants than games played by girls. Girls have been socialized to play exclusively with others of their own age, in groups of two or three, in activities such as hopscotch and jump rope that involve a minimum of competitiveness. Other research shows that boys express much more favorable attitudes toward

gender bias behavior that shows favoritism toward one gender over the other.

physical exertion and exercise than girls do. Some analysts believe this difference in attitude is linked to ideas about what is gender-appropriate behavior for boys and girls (Brustad, 1996). For males, competitive sport becomes a means of "constructing a masculine identity, a legitimated outlet for violence and aggression, and an avenue for upward mobility" (Lorber, 1994: 43). Recently, more girls have started to play soccer and softball and to participate in sports formerly regarded as exclusively "male" activities. Girls who go against the grain and participate in masculine play as children are more likely to participate in sports as young women and adults (Greendorfer, 1993; Giuliano, Popp, and Knight, 2000).

Many women athletes believe that they have to manage the contradictory statuses of being both "women" and "athletes." One study found that women college basketball players dealt with this contradiction by dividing their lives into segments. On the basketball court, the women "did athlete": They pushed, shoved, fouled, ran hard, sweated, and cursed. Off the court, they "did woman": After the game, they showered, dressed, applied makeup, and styled their hair, even if they were only getting in a van for a long ride home (Watson, 1987).

© AP Images

In recent years, women have expanded their involvement in professional sports through organizations such as the WNBA; however, most sports remain rigidly divided into female events and male events. Do you think that media coverage of women's and men's college and professional sporting events differs?

Most sports are rigidly divided into female and male events. Assumptions about male and female physiology and athletic capabilities influence the types of sports in which members of each sex are encouraged to participate. For example, women who engage in activities that are assumed to be "masculine" (such as bodybuilding) may either ignore their critics or attempt to redefine the activity or its result as "feminine" or womanly (Klein, 1993).

Mass Media and Gender Socialization

The media, including newspapers, magazines, television, and movies, are powerful sources of gender stereotyping. Although some critics argue that the media simply reflect existing gender roles in society, others point out that the media have a unique ability to shape ideas. Think of the impact that television might have on children if they spend one-third of their waking time watching it, as has been estimated. From children's cartoons to adult shows, television programs are sex-typed, and many are male oriented. More male than female roles are shown, and male characters act strikingly different from female ones. Typically, males are more aggressive, constructive, and direct, and are rewarded for their actions. By contrast, females are depicted as acting deferential toward other people or as manipulating them through helplessness or seductiveness to get their way.

In prime-time television, a number of significant changes in the past three decades have reduced gender stereotyping; however, men still outnumber women as leading characters, and they are often "in charge" in any setting where both men's and women's roles are portrayed. In the popular ABC series *Grey's Anatomy,* for example, the number of women's and men's roles is evenly balanced, but the male characters typically are the top surgeons at the hospital, whereas the female characters are residents, interns, or nurses. In shows with predominantly female characters, such as ABC's *Desperate Housewives,* the women are typically very attractive, thin, and ultimately either hysterical or compliant when dealing with male characters (Stanley, 2004).

Advertising—whether on television and billboards or in magazines and newspapers—can be very persuasive. The intended message is clear to many people: If they embrace traditional notions of masculinity and femininity, their personal and social success is assured; if they purchase the right products and services, they can enhance their appearance and gain power over other people. In commercials, men's roles typically are portrayed differently from women's roles: Men are more likely to be shown working or playing outside the house rather than inside, whereas women are more likely to be doing domestic tasks such as cooking, cleaning, shopping, or taking care of the children. As such, television commercials may act as agents of socialization, showing children and others what women's and men's designated activities are (Kaufman, 1999).

A study by the sociologist Anthony J. Cortese (2004) found that women—regardless of what they were doing in a particular ad—were frequently shown in advertising as being young, beautiful, and seductive. Although such depictions may sell products, they may also have the effect of influencing how we perceive ourselves and others with regard to issues of power and subordination.

Adult Gender Socialization

Gender socialization continues as women and men complete their training or education and join the work force. Men and women are taught the "appropriate" type of conduct for persons of their sex in a particular job or occupation—both by their employers and by coworkers. However, men's socialization usually does not include a measure of whether their work can be successfully combined with having a family; it is often assumed that men can and will do both. Even today, the reason given for women not entering some careers and professions is that this kind of work is not suitable for women because of their physical capabilities or assumed child-care responsibilities.

Different gender socialization may occur as people reach their forties and enter "middle age." A double standard of aging exists that affects women more than men. Often, men are considered to be at the height of their success as their hair turns gray and their face gains a few wrinkles. By contrast, not only do other people in society make middle-aged

women feel as if they are "over the hill," but multimillion-dollar advertising campaigns continually call attention to women's every weakness, every pound gained, and every bit of flabby flesh, wrinkle, or gray hair. Increasingly, both women and men have turned to "miracle" products, and sometimes to cosmetic surgery, to reduce the visible signs of aging.

A knowledge of how we develop a gender-related self-concept and learn to feel, think, and act in feminine or masculine ways is important for an understanding of ourselves. Examining gender socialization makes us aware of the impact of our parents, siblings, teachers, friends, and the media on our perspectives about gender. However, the gender socialization perspective has been criticized on several accounts. Childhood gender-role socialization may not affect people as much as some analysts have suggested. For example, the types of jobs that people take as adults may have less to do with how they were socialized in childhood than with how they are treated in the workplace. From this perspective, women and men will act in ways that bring them the most rewards and produce the fewest

© KAMIL KRZACZYNSKI/Landov

Cosmetic surgery can be dangerous. Dr. Donda West, mother of the rapper Kanye West, died in 2007 after an elective procedure.

punishments (Reskin and Padavic, 2002). Also, gender socialization theories can be used to blame women for their own subordination by not taking into account structural barriers that perpetuate gender inequality. We will now examine a few of those structural forces.

Contemporary Gender Inequality

According to feminist scholars, women experience gender inequality as a result of past and present economic, political, and educational discrimination. Women's position in the U.S. work force reflects the years of subordination that they have experienced in society.

Gendered Division of Paid Work

Where people are located in the occupational structure of the labor market has a major impact on their earnings. The workplace is another example of a gendered institution. In industrialized countries, most jobs are segregated by gender and by race/ethnicity. Sociologist Judith Lorber (1994: 194) gives this example:

> In a workplace in New York City—for instance, a handbag factory—a walk through the various departments might reveal that the owners and managers are white men; their secretaries and bookkeepers are white and Asian women; the order takers and data processors are African American women; the factory hands are [Latinos] cutting pieces and [Latinas] sewing them together; African American men are packing and loading the finished product; and non-English-speaking Eastern European women are cleaning up after everyone. The workplace as a whole seems integrated by race, ethnic group, and gender, but the individual jobs are markedly segregated according to social characteristics.

Lorber notes that in most workplaces, employees are either gender segregated or all of the same gender. *Gender-segregated work* refers to the concentration of women and men in different occupations, jobs, and places of work (Reskin and Padavic, 2002). In 2006, for example, 97 percent of all secretaries in the United States were women; 88 percent of all engineers were men (U.S. Census Bureau, 2007). To eliminate gender-segregated jobs in the United States, more than half of all men or all women workers would have to change occupations. Moreover, women are severely underrepresented at the top of U.S. corporations. Only about 10 percent of the executive jobs at Fortune 500 companies are held by women, and only eight women are the CEO of such a company.

Although the degree of gender segregation in the professional labor market (including physicians, dentists, lawyers, accountants, and managers) has declined since the 1970s, racial–ethnic segregation has remained deeply embedded in the social structure. As the sociologist Elizabeth Higginbotham (1994) points out, for many years African American professional women found themselves limited to employment in certain sectors of the labor market. Although some change has occurred in recent years, women of color are more likely than their white counterparts to be concentrated in public-sector employment (as public schoolteachers, welfare workers, librarians, public defenders, and faculty members at public colleges, for example) rather than in the private sector (for example, in large corporations, major law firms, and private educational institutions). Across all categories of occupations, white women and all people of color are not evenly represented, as shown in ◆ Table 10.3.

Labor market segmentation—the division of jobs into categories with distinct working conditions—results in women having separate and unequal jobs (Amott and Matthaei, 1996; Lorber, 2005). The pay gap between men and women is the best-documented consequence of gender-segregated work (Reskin and Padavic, 2002). Most women work in lower-paying, less-prestigious jobs, with little opportunity for advancement. Because many employers assume that men are the breadwinners, men are expected to make more money than women in order to support their families. For many years, women have been viewed as supplemental wage earners in a male-headed household, regardless of the women's marital status. Consequently, women have not been seen as legitimate workers but mainly

◆ **Table 10.3 Percentage of Work Force Represented by Women, African Americans, and Hispanics in Selected Occupations**

The U.S. Census Bureau accumulates data that show what percentage of the total work force is made up of women, African Americans, and Hispanics. As used in this table, *women* refers to females in all racial–ethnic categories, whereas *African American* and *Hispanic* refer to both women and men. Based on this table, in which occupations are white men most likely to be employed? In which occupations are they least likely to be employed?

	Women	African Americans	Hispanics
All occupations	46.6	11.3	10.9
Managerial and professional specialty (all)	50.0	8.3	5.1
Executive, administrative, and managerial	46.0	7.9	5.6
Professional specialty	53.7	8.6	4.7
Technical, sales, and administrative support (all)	63.7	11.4	9.1
Technicians and related support	53.4	10.3	7.5
Sales occupations	49.4	9.1	8.7
Service occupations (all)	60.4	17.9	16.3
Private household workers	96.2	12.1	32.8
Cleaning and building service workers	46.0	20.7	23.8
Health service assistants, aides, nurses, and attendants	89.2	29.4	11.5
Operators, fabricators, and laborers	23.3	15.6	17.7

Source: U.S. Census Bureau, 2008 (Statistical Abstract).

as wives and mothers (Lorber, 2005). Such thinking has especially harmful consequences for the many women who are the only breadwinner in their family due to divorce, widowhood, unmarried status, or other reasons.

Gender-segregated work affects both men and women. Men are often kept out of certain types of jobs. Those who enter female-dominated occupations often have to justify themselves and prove that they are "real men." They have to fight stereotypes (gay, "wimpy," and passive) about why they are interested in such work (Williams, 2004). Even if these assumptions do not push men out of female-dominated occupations, they affect how the men manage their gender identity at work. For example, men in occupations such as nursing tend to emphasize their masculinity, attempt to distance themselves from female colleagues, and try to move quickly into management and supervisory positions (Williams, 2004).

Occupational gender segregation contributes to stratification in society. Job segregation is structural; it does not occur simply because individual workers have different abilities, motivations, and material

needs. As a result of gender and racial segregation, employers are able to pay many men of color and all women less money, promote them less often, and provide fewer benefits.

Pay Equity (Comparable Worth)

Occupational segregation contributes to a ***pay gap—the disparity between women's and men's earnings.*** It is calculated by dividing women's earnings by men's to yield a percentage, also known as the earnings ratio (Reskin and Padavic, 2002). As ▶ Figure 10.1 shows, women at all levels of educational attainment receive less pay than men with the same levels of education. Across different categories of employment, data continue to demonstrate this disparity. Moreover, the pay gap is even greater for women of color. Although white women in 2006 earned 80 percent as much as white men, African

---

pay gap a term used to describe the disparity between women's and men's earnings.

By Age

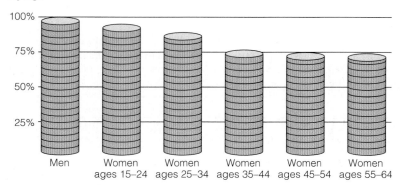

By Racial–Ethnic Group

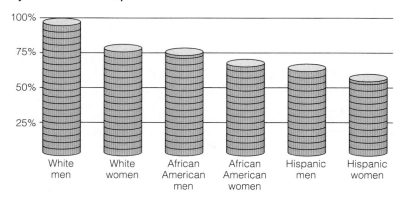

By Occupation

▶ **Figure 10.1 The Wage Gap**

By Age: Although men's average wages vary depending upon age, women's average wages are always lower than those of men in the same age group, and the older women get, the greater the gap. Women ages 16–24 earn 93 cents for every dollar earned by men the same age, but women ages 55–64 earn only 72 cents for every dollar earned by men ages 55–64.

By Racial–Ethnic Group: Across racial–ethnic groupings, men's earnings remain higher than the earnings of women in the same category.

By Occupation: Regardless of occupation, women on average receive lower wages.

Source: U.S. Bureau of Labor Statistics, 2007.

What stereotypes are associated with men in female-oriented positions? With women in male-oriented occupations? Do you think such stereotypes will change in the near future?

American women earned only 68 percent and Latinas 58 percent of what white male workers earned (U.S. Bureau of Labor Statistics, 2007a). The gap increases as a person's salary grows: Whereas 48 percent of people earning $20,000 to $25,000 per year are women, women constitute less than 10 percent of people earning over $200,000 per year (Johnston, 2002). Likewise, among older workers the pay gap between men and women is larger. Women's earnings tend to rise more slowly than men's when they are younger and then drop off when women reach their late thirties and early forties, whereas the wages of men tend to increase as they age (Reskin and Padavic, 2002).

Pay equity or ***comparable worth* is the belief that wages ought to reflect the worth of a job, not the gender or race of the worker.** How can the comparable worth of different kinds of jobs be determined? One way is to compare the actual work of women's and men's jobs and see if there is a disparity in the salaries paid for each. To do this, analysts break a job into components—such as the education, training, and skills required, the extent of responsibility for others' work, and the working conditions—and then allocate points for each (Lorber, 2005). For pay equity to exist, men and women in occupations that receive the same number of points should be paid the same. However, pay equity exists for very few jobs. For example, for every dollar earned by men, women in the same occupation earned 90 cents as office support workers (secretaries, clerks, etc.), 87 cents as restaurant servers, 70 cents as lawyers, and 63 cents as financial managers (U.S. Bureau of Labor Statistics, 2007a).

Paid Work and Family Work

As previously discussed, the first big change in the relationship between family and work occurred with the Industrial Revolution and the rise of capitalism. The cult of domesticity kept many middle- and upper-class women out of the work force during this period. Primarily, working-class and poor women were the ones who had to deal with the work/family conflict. Today, however, the issue spans the entire economic spectrum (Reskin and Padavic, 2002). The typical married woman in the United States combines paid work in the labor force and family work as a homemaker. Although this change has occurred at the

comparable worth (or **pay equity**) the belief that wages ought to reflect the worth of a job, not the gender or race of the worker.

societal level, individual women bear the brunt of the problem.

Even with dramatic changes in women's workforce participation, the sexual division of labor in the family remains essentially unchanged. Most married women now share responsibility for the breadwinner role, yet many men do not accept their share of domestic responsibilities (Reskin and Padavic, 2002). Consequently, many women have a "double day" or "second shift" because of their dual responsibilities for paid and unpaid work (Hochschild, 1989, 2003). Working women have less time to spend on housework; if husbands do not participate in routine domestic chores, some chores simply do not get done or get done less often. Although the income that many women earn is essential to the economic survival of their families, they still must spend part of their earnings on family maintenance, such as day-care centers, fast-food restaurants, and laundries, in an attempt to keep up with their obligations.

Especially in families with young children, domestic responsibilities consume a great deal of time and energy. Although some kinds of housework can be put off, the needs of children often cannot be ignored or delayed. When children are ill or school events cannot be scheduled around work, parents (especially mothers) may experience stressful role conflicts ("Shall I be a good employee or a good mother?"). Many working women care not only for themselves, their husbands, and their children but also for elderly parents or in-laws. Some analysts refer to these women as "the sandwich generation"—caught between the needs of their young children and of their elderly relatives. Many women try to solve their time crunch by forgoing leisure time and sleep. When Arlie Hochschild interviewed working mothers, she found that they talked about sleep "the way a hungry person talks about food" (1989: 9). Perhaps this is one reason that, in later research, Hochschild (1997) learned that some married women with children found more fulfillment at work and that they worked longer hours because they liked work better than facing the pressures of home.

Perspectives on Gender Stratification

Sociological perspectives on gender stratification vary in their approach to examining gender roles and power relationships in society. Some focus on the roles of women and men in the domestic sphere; others note the inequalities arising from a gendered division of labor in the workplace. Still others attempt to integrate both the public and private spheres into their analyses. The Concept Quick Review outlines the key aspects of each sociological perspective on gender socialization.

CONCEPT QUICK REVIEW

Sociological Perspectives on Gender Stratification

Perspective	Focus	Theory/Hypothesis
Functionalist	Macrolevel analysis of women's and men's roles	Traditional gender roles ensure that expressive and instrumental tasks will be performed Human capital model
Conflict	Power and economic differentials between men and women	Unequal political and economic power heightens gender-based social inequalities
Feminist Approaches	Feminism should be embraced to reduce sexism and gender inequality	1. Liberal feminism 2. Radical feminism 3. Socialist feminism 4. Multicultural feminism

Functionalist and Neoclassical Economic Perspectives

As seen earlier, functionalist theory views men and women as having distinct roles that are important for the survival of the family and society. The most basic division of labor is biological: Men are physically stronger, and women are the only ones able to bear and nurse children. Gendered belief systems foster assumptions about appropriate behavior for men and women and may have an impact on the types of work that women and men perform.

The Importance of Traditional Gender Roles

According to functional analysts such as Talcott Parsons (1955), women's roles as nurturers and caregivers are even more pronounced in contemporary industrialized societies. While the husband performs the *instrumental* tasks of providing economic support and making decisions, the wife assumes the *expressive* tasks of providing affection and emotional support for the family. This division of family labor ensures that important societal tasks will be fulfilled; it also provides stability for family members.

This view has been adopted by a number of politically conservative analysts who assert that relationships between men and women are damaged when changes in gender roles occur, and family life suffers as a consequence. From this perspective, the traditional division of labor between men and women is the natural order of the universe.

The Human Capital Model

Functionalist explanations of occupational gender segregation are similar to neoclassical economic perspectives, such as the human capital model (Horan, 1978; Kemp, 1994). According to this model, individuals vary widely in the amount of human capital they bring to the labor market. *Human capital* is acquired by education and job training; it is the source of a person's productivity and can be measured in terms of the return on the investment (wages) and the cost (schooling or training) (Stevenson, 1988; Kemp, 1994).

From this perspective, what individuals earn is the result of their own choices (the kinds of training, education, and experience they accumulate, for example) and of the labor market need (demand) for and availability (supply) of certain kinds of workers at specific points in time. For example, human capital analysts argue that women diminish their human capital when they leave the labor force to engage in childbearing and child-care activities. While women are out of the labor force, their

© Peter Dazeley/Getty Images

According to the human capital model, women may earn less in the labor market because of their child-rearing responsibilities. What other sociological explanations are offered for the lower wages that women receive?

human capital deteriorates from nonuse. When they return to work, women earn lower wages than men because they have fewer years of work experience and have "atrophied human capital" because their education and training may have become obsolete (Kemp, 1994: 70).

Evaluation of Functionalist and Neoclassical Economic Perspectives Although Parsons and other functionalists did not specifically endorse the gendered division of labor, their analysis suggests that it is natural and perhaps inevitable. However, critics argue that problems inherent in traditional gender roles, including the personal role strains of men and women and the social costs to society, are minimized by this approach. For example, men are assumed to be "money machines" for their families when they might prefer to spend more time in child-rearing activities. Also, the woman's place is assumed to be in the home, an assumption that ignores the fact that many women hold jobs due to economic necessity.

In addition, the functionalist approach does not take a critical look at the structure of society (especially the economic inequalities) that makes educational and occupational opportunities more available to some than to others. Furthermore, it fails to examine the underlying power relations between men and women or to consider the fact that the tasks assigned to women and to men are unequally valued by society (Kemp, 1994). Similarly, the human capital model is rooted in the premise that individuals are evaluated based on their human capital in an open, competitive market where education, training, and other job-enhancing characteristics are taken into account. From this perspective, those who make less money (often men of color and all women) have no one to blame but themselves.

Critics note that instead of blaming people for their choices, we must acknowledge other realities. Wage discrimination occurs in two ways: (1) the wages are higher in male-dominated jobs, occupations, and segments of the labor market, regardless of whether women take time for family duties, and (2) in any job, women and people of color will be paid less (Lorber, 1994).

Conflict Perspectives

According to many conflict analysts, the gendered division of labor within families and in the workplace results from male control of and dominance over women and resources. Differentials between men and women may exist in terms of economic, political, physical, and/or interpersonal power. The importance of a male monopoly in any of these arenas depends on the significance of that type of power in a society (Richardson, 1993). In hunting and gathering and horticultural societies, male dominance over women is limited because all members of the society must work in order to survive (Collins, 1971; Nielsen, 1990). In agrarian societies, however, male sexual dominance is at its peak. Male heads of household gain a monopoly not only on physical power but also on economic power, and women become sexual property.

© AP Images/J. Scott Applewhite

Although the demographic makeup of the U.S. Congress has been gradually changing in recent decades, men still dominate it, a fact that the conflict perspective attributes to a very old pattern in human societies.

Although men's ability to use physical power to control women diminishes in industrial societies, men still remain the head of household and control the property. In addition, men gain more power through their predominance in the most highly paid and prestigious occupations and the highest elected offices. By contrast, women have the ability to trade their sexual resources, companionship, and emotional support in the marriage market for men's financial support and social status. As a result, women as a group remain subordinate to men (Collins, 1971; Nielsen, 1990).

All men are not equally privileged; some analysts argue that women and men in the upper classes are more privileged, because of their economic power, than men in lower-class positions and all people of color (Lorber, 1994). In industrialized societies, persons who occupy elite positions in corporations, universities, the mass media, and government or who have great wealth have the most power (Richardson, 1993). Most of these are men, however.

Conflict theorists in the Marxist tradition assert that gender stratification results from private ownership of the means of production; some men not only gain control over property and the distribution of goods but also gain power over women. According to Friedrich Engels and Karl Marx, marriage serves to enforce male dominance. Men of the capitalist class instituted monogamous marriage (a gendered institution) so that they could be certain of the paternity of their offspring, especially sons, whom they wanted to inherit their wealth. Feminist analysts have examined this theory, among others, as they have sought to explain male domination and gender stratification.

Feminist Perspectives

Feminism—the belief that women and men are equal and should be valued equally and have equal rights—is embraced by many men as well as women. It holds in common with men's studies the view that gender is a socially constructed concept that has important consequences in the lives of all people (Craig, 1992). According to the sociologist Ben Agger (1993), men can be feminists and propose feminist theories; both women and men have much in common as they seek to gain a better un-

derstanding of the causes and consequences of gender inequality. Over the past three decades, many different organizations have been formed to advocate causes uniquely affecting women or men and to help people gain a better understanding of gender inequality (see Box 10.3).

Feminist theory seeks to identify ways in which norms, roles, institutions, and internalized expectations limit women's behavior. It also seeks to demonstrate how women's personal control operates even within the constraints of relative lack of power (Stewart, 1994).

Liberal Feminism In liberal feminism, gender equality is equated with equality of opportunity. The roots of women's oppression lie in women's lack of equal civil rights and educational opportunities. Only when these constraints on women's participation are removed will women have the same chance for success as men. This approach notes the importance of gender-role socialization and suggests that changes need to be made in what children learn from their families, teachers, and the media about appropriate masculine and feminine attitudes and behavior. Liberal feminists fight for better child-care options, a woman's right to choose an abortion, and the elimination of sex discrimination in the workplace.

Radical Feminism According to radical feminists, male domination causes all forms of human oppression, including racism and classism (Tong, 1989). Radical feminists often trace the roots of patriarchy to women's childbearing and child-rearing responsibilities, which make them dependent on men (Firestone, 1970; Chafetz, 1984). In the radical feminist view, men's oppression of women is deliberate, and ideological justification for this subordination is provided by other institutions such as the media and religion. For women's condition to improve, radical feminists claim, patriarchy must be abolished. If institutions are currently gendered, alternative institutions—such as women's organizations

feminism the belief that all people—both women and men—are equal and that they should be valued equally and have equal rights.

How Do We "Do Gender" in the Twenty-First Century?

What distinctive ways of acting and feeling are characteristic of women? Of men? For centuries, people have used a male/female dichotomy to answer these questions and, in the process, have identified women's and men's behaviors as opposites in many respects: Men are supposed to be "real men" and meet the normative conception of *masculinity* by being aggressive, independent, and powerful, whereas women are supposed to demonstrate *femininity* by being passive, dependent, and weak.

However, many theorists using a symbolic interactionist perspective suggest that gender is something that we *do* rather than being a set of masculine or feminine traits that resides within the individual. Sociologists Candace West and Don H. Zimmerman (1987) coined the term "doing gender" to refer to the process by which we *socially* create differences between males and females that are *not* based on natural, essential, or biological factors but instead are based on the things we do in our social interactions. According to West and Zimmerman (1987), *accountability* is involved in the process of doing gender: We know that our actions will be evaluated by others based on how well they think we meet the normative conceptions of appropriate attitudes and activities that are expected of people in our sex category (the socially required displays that identify a person as being either male or female).

What is the primary difference in these two approaches? These viewpoints are based on different assumptions. The male/female dichotomy is based on the assumption that women and men have inherently different traits, whereas the concept of *doing gender* is based on the assumption that, through our interactions with others, we produce, reproduce, and sustain the social meanings that are accorded to gender in any specific society at any specific point in time (in other words, those meanings may change from time to time and from place to place). By focusing on gender as an *accomplishment* (rather than something that is previously established), symbolic interactionists emphasize that some people's resistance to existing gendered norms is probable and that social change is possible. Symbolic interactionist theories also make us aware that change is less likely to take place when people feel constrained

to be accountable to others for their behavior as women or men (Fenstermaker and West, 2002).

When you look at the pictures on these two pages, think about how the people in each setting are "doing gender" in everyday life. Are they doing gender based on what they perceive to be the normative expectations of others? Or are they doing gender as they see fit? What do you think?

© Image Source Pink/Getty Images

Gender and Appearance

How do we do gender on a shopping trip? Consider, for example, the differences in clothing shown in this photo. For the man, a casual outing means wearing jeans, sandals, and a partially unbuttoned shirt. For the woman, a casual outing means a "pulled together" outfit with heels, a sexy dress, her hair in place, and carefully positioned sunglasses.

© Ariel Skelley/Getty Images

Gender and Home Life

Although in recent years more men have assumed greater responsibility at home with regard to household tasks and child rearing, women are more likely to be seen as the primary caregivers for children. Does our society still hold women, more than men, accountable for the well-being of children?

© Ed Bock/CORBIS

Gender and Social Interaction

How are the men shown here doing gender after their game of pick-up basketball? Consider, for example, what they are wearing, how they are sitting, and how they are communicating with one another. If three women were in this photo instead of three men, in what ways would their attire, actions, and expressions be different?

© Photodisc/SuperStock

Gender and Careers

In recent decades, more women have become doctors and lawyers than in the past. How has this affected the way that people do gender in settings that reflect their profession? Do professional women look and act more like their male colleagues, or have men changed their appearance and activities at work as a result of having female colleagues?

Gender and Success

When you hear someone referred to as a "wealthy entrepreneur," do you tend to think of a man or a woman? Are the typical trappings of success—such as luxury cars and expensive private airplanes—more associated with how men or how women do gender? Do you believe that this will change in the future?

Reflect & Analyze

1. If you went to a medical clinic for the first time and was told that your physician was "Dr. Smith," would you automatically assume that this person was a man? If so, why?
2. Is your social behavior dependent on gender context? In other words, do you behave differently when around members of your own sex than you do in mixed situations? Come up with a specific example of a situation where you felt the need to adapt your behavior because of social and gender expectations.
3. Do you think that wide differences in how two people look when they go out on a date can affect the outcome of their encounter? Why or why not?

Turning to Video

Watch the ABC video *Are Men Smarter? Sexism in the Classroom* (running time 4:33), available on the Kendall Companion Website and through Cengage Learning eResources accounts. This special report focuses on the experience of leading neurobiologist Ben Barres, who became a female-to-male transsexual at the age of forty. Having been active in his field as both Barbara and Ben, Dr. Barres has gained unique perspective on the biased perceptions of men and women in science. As you watch the video, think about the photographs, commentary, and questions that you encountered in this photo essay. After you've watched the video, consider another question: How do Ben Barres's life experiences and professional accomplishments reflect West and Zimmerman's concept of "doing gender"?

Box 10.3 You Can Make a Difference

Joining Organizations to Overcome Sexism and Gender Inequality

Over the past four decades, many college students—like you—have been actively involved in organizations seeking to dismantle sexism and reduce gender-based inequalities in the United States and other nations. Two people speak about the importance of feminist advocacy for women and men:

I was raised on a pure, unadulterated feminist ethic. . . . [But] the feminism I was raised on was very cerebral. It forced a world full of people to change the way they think about women. I want more than their minds. I want to see them do it. . . . I know that sitting on the sidelines will not get me what I want from my movement. And it is mine. . . . Don't be fooled into thinking that feminism is old-fashioned. The movement is ours and we need it. . . . The next generation is coming. (Neuborne, 1995: 29–35)
   —Ellen Neuborne, a reporter for *USA Today*

For men who are messed up (that is, facing problems related to their emotional lives, sexuality, their place in society, and gender politics—in other words, me and virtually every other man I have ever met) feminism offers the best route to understanding the politics of such personal problems and coming to terms with those problems. (Jensen, 1995: 111)
   —Robert Jensen, a journalism professor at the University of Texas at Austin

If you are interested in joining an organization that deals with the problem of sexism or organizes activities such as Take Back the Night, an annual event that promotes awareness of violence against women, contact your college's student activities office.

Opportunities for involvement in local, state, and national organizations that promote women's and/or men's rights are available. Here are a few examples:

● The National Organization for Women (NOW), 1000 16th St., NW, Suite 700, Washington, DC 20006. (202) 331-0066. NOW works to end gender bias and seeks greater representation of women in all areas of public life. On the Internet, NOW provides links to other feminist resources:
http://www.now.org

● National Organization for Men Against Sexism (NOMAS), 54 Mint Street, San Francisco, CA 94103. NOMAS has a *profeminist stance* that seeks to end sexism and an *affirmative stance* on the rights of gay men and lesbians.

On the Internet:

● *MenWeb (Men's Voices Magazine)* is an e-zine that offers support and advocacy for men. It features articles on the men's movement and a national calendar of men's events, as well as links to other sites:
http://menweb.org

seeking better health care, day care, and shelters for victims of domestic violence and rape—should be developed to meet women's needs.

Socialist Feminism Socialist feminists suggest that women's oppression results from their dual roles as paid *and* unpaid workers in a capitalist economy. In the workplace, women are exploited by capitalism; at home, they are exploited by patriarchy (Kemp, 1994). Women are easily exploited in both sectors; they are paid low wages and have few economic resources. Gendered job segregation is "the primary mechanism in capitalist society that maintains the superiority of men over women, because it enforces lower wages for women in the labor market" (Hartmann, 1976: 139). As a result, women must do domestic labor either to gain a better-paid man's economic support or to stretch their own wages (Lorber, 1994). According to socialist feminists, the only way to achieve gender equality is to eliminate capitalism and develop a socialist economy that would bring equal pay and rights to women.

Multicultural Feminism Recently, academics and activists have been rethinking the experiences of women of color from a feminist perspective. The experiences of African American women and Latinas/Chicanas have been of particular interest to some social analysts. Building on the civil rights and feminist movements of the late 1960s and early 1970s, some contemporary black feminists have focused on the cultural experiences of African American women. A central assumption of this analysis is that race, class, and gender are forces that simultaneously oppress African American women (Hull, Bell-Scott, and Smith, 1982). The effects of these three statuses cannot be adequately explained as "double" or "triple" jeopardy (race + class + gender = a poor African American woman) because these ascribed characteristics are not simply added to one another. Instead, they are multiplicative in nature (race × class × gender); different characteristics may be more significant in one situation than another. For example, a well-to-do white woman (class) may be in a position of privilege when compared to people of color (race) and men from lower socioeconomic positions (class), yet be in a subordinate position as compared with a white man (gender) from the capitalist class (Andersen and Collins, 1998). In order to analyze the complex relationship among these characteristics, the lived experiences of African American women and other previously "silenced people" must be heard and examined within the context of particular historical and social conditions.

Another example of multicultural feminist studies is the work of the psychologist Aida Hurtado (1996), who explored the cultural identification of Latina/Chicana women. According to Hurtado, distinct differences exist between the world views of the white (non-Latina) women who participate in the women's movement and many Chicanas, who have a strong sense of identity with their own communities. Hurtado (1996) suggests that women of color do not possess the "relational privilege" that white women have because of their proximity to white patriarchy through husbands, fathers, sons, and others. Like other multicultural feminists, Hurtado calls for a "politics of inclusion," creating social structures that lead to positive behavior and bring more people into a dialogue about how to improve social life and reduce inequalities.

Evaluation of Conflict and Feminist Perspectives

Conflict and feminist perspectives provide insights into the structural aspects of gender inequality in

© Larry Kolvoord/The Image Works

Latinas have become increasingly involved in social activism for causes that they believe are important. This woman is showing her support for amnesty for undocumented workers in the United States.

society. These approaches emphasize factors external to individuals that contribute to the oppression of white women and people of color; however, they have been criticized for emphasizing the differences between men and women without taking into account the commonalities they share. Feminist approaches have also been criticized for their emphasis on male dominance without a corresponding analysis of the ways in which some men may also be oppressed by patriarchy and capitalism.

Recently, the debate has continued about whether the feminist movement has diminished the well-being of boys and men. Sociologist William J. Goode (1982) suggests that some men have felt "under attack" by women's demands for equality because the men do not see themselves as responsible for societal conditions such as patriarchy and attribute their own achievements to hard work and intelligence,

Author Susan Faludi believes that men and boys are being "stiffed" by current trends in U.S. society, including education. Do you agree?

not to built-in societal patterns of male domination and female subordination.

In *Stiffed: The Betrayal of the American Man,* the author and journalist Susan Faludi (1999) argues that men are not as concerned about the possibility of the feminist movement diminishing their own importance as they are experiencing an identity crisis brought about by the current societal emphasis on wealth, power, fame, and looks (often to the exclusion of significant social values). According to Faludi, men are increasingly aware of their body image and are spending ever-increasing sums of money on men's cosmetics, health and fitness gear and classes, and cigar bars and "gentlemen's clubs." The popularity of her book, among others, suggests that issues of gender inequality and of men's and women's roles in society are far from resolved.

Gender Issues in the Future

Over the past century, women made significant progress in the labor force (Reskin and Padavic, 2002). Laws were passed to prohibit sexual discrimination in the workplace and school. Affirmative action programs helped make women more visible in education, government, and the professional world. More women entered the political arena as candi-

dates instead of as volunteers in the campaign offices of male candidates (Lott, 1994).

Many men joined movements to raise their consciousness, realizing that what is harmful to women may also be harmful to men. For example, women's lower wages in the labor force suppress men's wages as well; in a two-paycheck family, women who are paid less contribute less to the family's finances, thus placing a greater burden on men to earn more money.

In the midst of these changes, however, many gender issues remain unresolved in the twenty-first century. In the labor force, gender segregation and the wage gap are still problems. Although women continue to narrow the pay gap, women earn slightly less than 77 cents on the dollar compared with men (Barakat, 2000), and analysts believe that the narrowing can partly be attributed to the economic boom (for some) of recent decades. Employers have had to look for employees based on merit (rather than race, class, and gender) in order to have the number and types of employees they need to meet the demands of global competition (Barakat, 2000).

Although some employers have implemented family-leave policies, these do not relieve women's domestic burden in the family. Analysts believe that the burden of the "double day" or "second shift" will probably preserve women's inequality at home and in the workplace for another generation (Reskin and Padavic, 2002).

Chapter Review

● How do sex and gender differ?

Sex refers to the biological categories and manifestations of femaleness and maleness; gender refers to the socially constructed differences between females and males. In short, sex is what we (generally) are born with; gender is what we acquire through socialization.

● How do gender roles and gender identity differ from gendered institutions?

Gender role encompasses the attitudes, behaviors, and activities that are socially assigned to each sex and that are learned through socialization. Gender identity is an individual's perception of self as either female or male. By contrast, gendered institutions are those structural features that perpetuate gender inequality.

● How does the nature of work affect gender equity in societies?

In most hunting and gathering societies, fairly equitable relationships exist between women and men because neither sex has the ability to provide all of the food necessary for survival. In horticultural societies, a fair degree of gender equality exists because neither sex controls the food supply. In agrarian societies, male dominance is overt; agrarian tasks require more labor and physical strength, and females are often excluded from these tasks because they are viewed as too weak or too tied to child-rearing activities. In industrialized societies, a gap exists between nonpaid work performed by women at home and paid work performed by men and women. A wage gap also exists between women and men in the marketplace.

● What are the key agents of gender socialization?

Parents, peers, teachers and schools, sports, and the media are agents of socialization that tend to reinforce stereotypes of appropriate gender behavior.

● What causes gender inequality in the United States?

Gender inequality results from economic, political, and educational discrimination against women. In most workplaces, jobs are either gender segregated or

the majority of employees are of the same gender. Although the degree of gender segregation in the professional workplace has declined since the 1970s, racial and ethnic segregation remains deeply embedded.

● How is occupational segregation related to the pay gap?

Many women work in lower-paying, less-prestigious jobs than men. This occupational segregation leads to a disparity, or pay gap, between women's and men's earnings. Even when women are employed in the same job as men, on average they do not receive the same, or comparable, pay.

● How do functionalists and conflict theorists differ in their view of division of labor by gender?

According to functionalist analysts, women's roles as caregivers in contemporary industrialized societies are crucial in ensuring that key societal tasks are fulfilled. While the husband performs the instrumental tasks of economic support and decision making, the wife assumes the expressive tasks of providing affection and emotional support for the family. According to conflict analysts, the gendered division of labor within families and the workplace—particularly in agrarian and industrial societies—results from male control and dominance over women and resources.

www.cengage.com/login

Register for a Student eResource account to maximize your study time online using CengageNOW. First take the system's diagnostic pre-test, and then follow the personalized study plan that is created for you to help you review this chapter. The study plan will

- help you identify areas on which you should concentrate;
- provide interactive exercises to help you master the chapter concepts; and
- provide a post-test to confirm you are ready to move on to the next chapter.

Key Terms

Questions for Critical Thinking

1. Do the media reflect societal attitudes on gender, or do the media determine and teach gender behavior? (As a related activity, watch television for several hours, and list the roles for women and men depicted in programs and those represented in advertising.)
2. Examine the various academic departments at your college. What is the gender breakdown of the faculty in selected departments? What is the gender breakdown of undergraduates and graduate students in those departments? Are there major differences among various academic areas of teaching and study? What hypothesis can you come up with to explain your observations?

The Kendall Companion Website

www.cengage.com/sociology/kendall

Supplement your review of this chapter by going to the text's companion website, where you can take tutorial quizzes, use flash cards to master key terms, follow live links to useful websites, and explore the other study and research resources you'll find there, such as a comprehensive interactive sociology timeline, GSS Data, and Census 2000 information, much of it presented visually in maps.

11 Families and Intimate Relationships

My son began commuting between his two homes at age 4. He traveled with a Hello Kitty suitcase with a pretend lock and key until it embarrassed him. . . . He graduated to a canvas backpack filled with a revolving arsenal of essential stuff: books and journals, plastic vampire teeth, "Star Trek" Micro Machines, a Walkman, CD's, a teddy bear.

The commuter flights between San Francisco and Los Angeles were the only times a parent wasn't lording over him, so he was able to order Coca-Cola, verboten at home. . . . But such benefits were insignificant when contrasted with his preflight nightmares about plane crashes. . . .

Like so many divorcing couples, we divided the china and art and our young son. . . . First he was ferried back and forth between our homes across town, and then, when his mother moved to Los Angeles, across the state. For the eight years since, he has been one of the thousands of American

© Myrleen Cate/Index Stock Imagery/Photolibrary

Changing family patterns and relocations for employment opportunities have made sights such as this increasingly familiar, as children travel alone to various destinations to be reunited with a parent or other family member.

children with two homes, two beds, two sets of clothes and toys and two toothbrushes.

—several years ago, David Sheff (1995: 64) explaining how divorce and joint custody had affected his son, Nick Sheff

Four years later, Nick Sheff stated his own view on long-distance joint custody:

I am 16 now and I still travel back and forth, but it's mostly up to me to decide when. I've chosen to spend more time with my friends at the expense of visits with my mom. When I do go to L.A., it's like my stepdad put it: I have a cameo role in their lives. I say my lines and I'm off. It's painful.

What's the toll of this arrangement? I'm always missing somebody. When I'm in northern California, I miss my mom and stepdad. But when I'm in L.A., I miss hanging out with my friends, my other set of parents and little brothers and sisters. After all these back-and-forth flights, I've learned not to get too emotionally attached. I have to protect myself. . . . No child should be subjected to the hardship of long-distance joint custody. To prevent it, maybe there should be an addition to the marriage vows: Do you promise [that] if you ever have children and wind

Chapter Focus Question

Why are many family-related concerns—such as divorce and child care—viewed primarily as personal problems rather than as social concerns requiring macrolevel solutions?

up divorced, [you will] stay within the same geographical area as your kids? . . . Or how about some common sense? If you move away from your children, *you* have to do the traveling to see them. (Sheff, 1999: 16)

Nick's family is one of millions around the globe experiencing major change as a result of divorce, death, or some other significant event. In this chapter, we examine the increasing complexity and diversity of contemporary families in the United States and other nations. Pressing social issues such as divorce, child care, and new reproductive technologies will be used as examples of how families and intimate relationships continue to change. Before reading on, test your knowledge about some contemporary trends in U.S. family life by taking the quiz in Box 11.1.

Families in Global Perspective

As the nature of family life and work has changed in high-, middle-, and low-income nations, the issue of what constitutes a "family" has been widely debated. For many years, the standard sociological definition of *family* has been a group of people who are related to one another by bonds of blood, marriage, or adoption and who live together, form an economic unit, and bear and raise children. Many people believe that this definition should not be expanded—that social approval should not be extended to other relationships simply because the persons in those relationships wish to consider themselves a family. However, others challenge this definition because it simply does not match the reality of family life in contemporary society (Lamanna and Riedmann, 2009). Today's families include many types of living arrangements and relationships, including single-parent households, unmarried couples, lesbian

and gay couples, and multiple generations (such as grandparent, parent, and child) living in the same household. To accurately reflect these changes in family life, some sociologists believe that we need a more encompassing definition of what constitutes a family. Accordingly, we will define *families* as **relationships in which people live together with commitment, form an economic unit and care for any young, and consider their identity to be significantly attached to the group.** Sexual expression and parent–child relationships are a part of most, but not all, family relationships.

Although families differ widely around the world, they also share certain common concerns in their everyday lives. For example, women and men of all racial–ethnic categories, nationalities, and income levels face problems associated with child care. Various nongovernmental agencies of the United Nations have established international priorities to help families. Suggestions have included the development of "social infrastructures for the care and education of children of working parents in order to reduce their . . . burden" and implementation of flexible working hours so that parents will have more opportunities to spend time with their children (Pietilä and Vickers, 1994: 115).

How do sociologists approach the study of families? In our study of families, we will use our sociological imagination to see how our personal experiences are related to the larger happenings in society. At the microlevel, each of us has a "biography," based on our experience within our family; at the macrolevel, our families are embedded in a specific social context that has a major effect on them. We will examine the institution of the family at both of these levels, starting with family structure and characteristics.

Box 11.1 Sociology and Everyday Life

How Much Do You Know About Contemporary Trends in U.S. Family Life?

True	False	
T	F	1. Today, people in the United States are more inclined to get married than at any time in history.
T	F	2. Most U.S. family households are composed of a married couple with one or more children under age 18.
T	F	3. One in two U.S. preschoolers has a mother in the paid labor force.
T	F	4. Recent studies have found that sexual activity is more satisfying to people who are in sustained relationships such as marriage.
T	F	5. People under age 45 are more likely to cohabit than are people over age 45.
T	F	6. With the strong U.S. economy at the end of the 1990s, the number of children living in extreme poverty went down significantly.
T	F	7. Recent studies have found that adult children of divorced parents are more likely to dissolve their own marriages than they were two decades ago.
T	F	8. Teenage pregnancies have increased dramatically over the past 30 years.
T	F	9. Couples who already have children at the beginning of their marriage are less likely to divorce.
T	F	10. The number of children living with grandparents and with no parent in the home increased by more than 50 percent between 1990 and 2000.

Answers on page 358.

Family Structure and Characteristics

In preindustrial societies, the primary form of social organization is through kinship ties. *Kinship* **refers to a social network of people based on common ancestry, marriage, or adoption.** Through kinship networks, people cooperate so that they can acquire the basic necessities of life, including food and shelter. Kinship systems can also serve as a means by which property is transferred, goods are produced and distributed, and power is allocated.

In industrialized societies, other social institutions fulfill some of the functions previously taken care of by the kinship network. For example, political systems provide structures of social control and authority, and economic systems are responsible for the production and distribution of goods and services. Consequently, families in industrialized so-

cieties serve fewer and more-specialized purposes than do families in preindustrial societies. Contemporary families are responsible primarily for regulating sexual activity, socializing children, and providing affection and companionship for family members.

Families of Orientation and Procreation During our lifetime, many of us will be members of two different types of families—a family of orientation

families relationships in which people live together with commitment, form an economic unit and care for any young, and consider their identity to be significantly attached to the group.

kinship a social network of people based on common ancestry, marriage, or adoption.

Answers to the Sociology Quiz on Contemporary Trends in U.S. Family Life

1. False. Census data show that the marriage rate has gone down by about one-third since 1960. In 1960, there were about 73 marriages per 1,000 unmarried women age 15 and up, whereas today the rate is about 49 per 1,000.

2. False. Less than 25 percent of all family households are composed of married couples with one or more children under age 18.

3. True. One in two preschoolers in the United States has a mother in the paid labor force, which makes affordable, high-quality child care an important issue for many families.

4. True. Recent research indicates that people who are married or are in a sustained relationship with their sexual partner express greater satisfaction with their sexual activity than do those who are not.

5. True. People under age 45 are more likely to cohabit than are older individuals.

6. False. Despite the strong economy prior to 2001, the number of children living in extreme poverty—with incomes below one-half of the poverty line—crept upward by almost 400,000 in the 1990s.

7. False. Adult children of divorced parents are less likely to dissolve their own marriages than they were two decades ago.

8. False. Teenage pregnancies have actually decreased over the past 30 years. However, the percentage of *unmarried* teenage pregnancies has increased dramatically.

9. False. Couples who already have children at the beginning of their marriage are *more* likely to divorce than those who do not.

10. True. The number of children living with grandparents and with no parent in the home grew by 51.5 percent between 1990 and 2000.

Sources: Based on Children's Defense Fund, 2001; Coontz, 1992, 1997; *Fox News*, 1999; and Popenoe and Whitehead, 1999.

---

and a family of procreation. The *family of orientation* **is the family into which a person is born and in which early socialization usually takes place.** Although most people are related to members of their family of orientation by blood ties, those who are adopted have a legal tie that is patterned after a blood relationship. The *family of procreation* **is the family that a person forms by having or adopting children.** Both legal and blood ties are found in most families of procreation. The relationship between a husband and wife is based on legal ties; however, the relationship between a parent and child may be based on either blood ties or legal ties, depending on whether the child has been adopted.

Some sociologists have emphasized that "family of orientation" and "family of procreation" do not encompass all types of contemporary families. Instead, many gay men and lesbians have *families we choose*—social arrangements that include intimate relationships between couples and close familial relationships among other couples and other adults and children. According to the sociologist Judy Root Aulette (1994), "families we choose" include blood ties and legal ties, but they also include *fictive kin*—persons who are not actually related by blood but who are accepted as family members.

Extended and Nuclear Families Sociologists distinguish between extended families and nuclear families based on the number of generations that live within a household. An *extended family* **is a family unit composed of relatives in addition to**

Whereas the relationship between a husband and wife is based on legal ties, relationships between parents and children may be established by either blood ties or legal ties.

parents and children who live in the same household. These families often include grandparents, uncles, aunts, or other relatives who live close to the parents and children, making it possible for family members to share resources. In horticultural and agricultural societies, extended families are extremely important; having a large number of family members participate in food production may be essential for survival. Today, extended-family patterns are found in Latin America, Africa, Asia, and some parts of Eastern and Southern Europe (Busch, 1990). With the advent of industrialization and urbanization, maintaining the extended-family pattern becomes more difficult. Increasingly, young people move from rural to urban areas in search of employment in the industrializing sector of the economy. At that time, some extended families remain, but the nuclear family typically becomes the predominant family form in the society.

A *nuclear family* **is a family composed of one or two parents and their dependent children, all of whom live apart from other relatives.** A traditional definition specifies that a nuclear family is made up of a "couple" and their dependent children; however, this definition became outdated when a significant shift occurred in the family structure. A comparison of Census Bureau data from 1970 and 2003 shows that there has been a significant decline in the percentage of U.S. households comprising a married couple with their own children under eighteen years of age (see "Census Profiles: Household Composition" on next page). Conversely, there has been an increase in the percentage of households in which either a woman or a man lives alone.

Marriage Patterns

Across cultures, families are characterized by different forms of marriage. *Marriage* **is a legally recognized and/or socially approved arrangement between two or more individuals that carries certain rights and obligations and usually involves sexual activity.** In most societies, marriage involves a mutual commitment by each partner, and linkages between two individuals and families are publicly demonstrated.

family of orientation the family into which a person is born and in which early socialization usually takes place.

family of procreation the family that a person forms by having or adopting children.

extended family a family unit composed of relatives in addition to parents and children who live in the same household.

nuclear family a family composed of one or two parents and their dependent children, all of whom live apart from other relatives.

marriage a legally recognized and/or socially approved arrangement between two or more individuals that carries certain rights and obligations and usually involves sexual activity.

CENSUS PROFILES

Household Composition, 1970 and 2003

The Census Bureau asks a representative sample of the U.S. population about their marital status and also asks those individuals questions about other individuals residing in their household. Based on the most recent data, the Census Bureau reports that the percentage distribution of nonfamily and family households has changed substantially during the past three decades. The most noticeable trend is the decline in the number of married-couple households with their own children living with them, which decreased from about 40 percent of all households in 1970 to about 23 percent in 2003. The distribution shown below reflects this significant change in family structure.

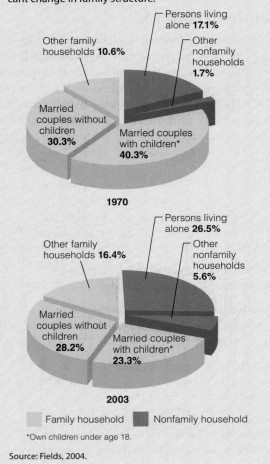

1970

- Persons living alone **17.1%**
- Other nonfamily households **1.7%**
- Married couples with children* **40.3%**
- Married couples without children **30.3%**
- Other family households **10.6%**

2003

- Persons living alone **26.5%**
- Other nonfamily households **5.6%**
- Married couples with children* **23.3%**
- Married couples without children **28.2%**
- Other family households **16.4%**

Family household   Nonfamily household

*Own children under age 18.

Source: Fields, 2004.

In the United States, the only legally sanctioned form of marriage is *monogamy*—**a marriage between two partners, usually a woman and a man.** For some people, marriage is a lifelong commitment that ends only with the death of a partner. For others, marriage is a commitment of indefinite duration. Through a pattern of marriage, divorce, and remarriage, some people practice *serial monogamy*—a succession of marriages in which a person has several spouses over a lifetime but is legally married to only one person at a time.

Polygamy **is the concurrent marriage of a person of one sex with two or more members of the opposite sex.** The most prevalent form of polygamy is *polygyny*—**the concurrent marriage of one man with two or more women.** Polygyny has been practiced in a number of societies, including parts of Europe until the Middle Ages. More recently, some marriages in Islamic societies in Africa and Asia have been polygynous; however, the cost of providing for multiple wives and numerous children makes the practice impossible for all but the wealthiest men. In addition, because roughly equal numbers of women and men live in these areas, this nearly balanced sex ratio tends to limit polygyny.

The second type of polygamy is *polyandry*—**the concurrent marriage of one woman with two or more men.** Polyandry is very rare; when it does occur, it is typically found in societies where men greatly outnumber women because of high rates of female infanticide.

Patterns of Descent and Inheritance

Even though a variety of marital patterns exist across cultures, virtually all forms of marriage establish a system of descent so that kinship can be determined and inheritance rights established. In preindustrial societies, kinship is usually traced through one parent (unilineally). The most common pattern of unilineal descent is *patrilineal descent*—**a system of tracing descent through the father's side of the family.** Patrilineal systems are set up in such a manner that a legitimate son inherits his father's property and sometimes his position upon the father's death. In nations such as India, where boys are seen as permanent patrilineal family members but girls are seen as only temporary family members, girls tend to be considered more expendable than boys (O'Connell, 1994). Recently, some scholars have concluded that cultural and racial nationalism in China is linked to the idea of patrilineal descent being crucial to the modern Chinese national identity (Dikotter, 1996).

Even with the less common pattern of *matrilineal descent*—**a system of tracing descent through the mother's side of the family**—women may not control property. However, inheritance of property and position is usually traced from the maternal uncle (mother's brother) to his nephew (mother's son). In some cases, mothers may pass on their property to daughters.

By contrast, kinship in industrial societies is usually traced through both parents (bilineally). The most common form is *bilateral descent*—**a system of tracing descent through both the mother's and father's sides of the family.** This pattern is used in the United States for the purpose of determining kinship and inheritance rights; however, children typically take the father's last name.

Power and Authority in Families

Descent and inheritance rights are intricately linked with patterns of power and authority in families. The most prevalent forms of familial power and authority are patriarchy, matriarchy, and egalitarianism. A *patriarchal family* is a family structure in which authority is held by the eldest male (usually the father). The male authority figure acts as head of the household and holds power and authority over the women and children, as well as over other males. A *matriarchal family* is a family structure in which authority is held by the eldest female (usually the mother). In this case, the female authority figure acts as head of the household. Although there has been a great deal of discussion about matriarchal families, scholars have found no historical evidence to indicate that true matriarchies ever existed.

The most prevalent pattern of power and authority in families is patriarchy. Across cultures, men are the primary (and often sole) decision makers regarding domestic, economic, and social concerns facing the family. The existence of patriarchy may give men a sense of power over their own lives, but it can also create an atmosphere in which some men feel greater freedom to abuse women and children (O'Connell, 1994).

An *egalitarian family* **is a family structure in which both partners share power and authority equally.** Recently, a trend toward more-egalitarian relationships has been evident in a number of countries as women have sought changes in their legal status and increased educational and employment opportunities. Some degree of economic independence makes it possible for women to delay marriage or to terminate a problematic marriage (O'Connell, 1994). Among gay and lesbian couples, power and authority issues are also important. Although some analysts suggest that legalizing marriage among gay couples would produce more-egalitarian relationships, others argue that such marriages would reproduce the inequalities in power and authority found in conventional heterosexual relationships (see Aulette, 1994).

monogamy a marriage between two partners, usually a woman and a man.

polygamy the concurrent marriage of a person of one sex with two or more members of the opposite sex.

polygyny the concurrent marriage of one man with two or more women.

polyandry the concurrent marriage of one woman with two or more men.

patrilineal descent a system of tracing descent through the father's side of the family.

matrilineal descent a system of tracing descent through the mother's side of the family.

bilateral descent a system of tracing descent through both the mother's and father's sides of the family.

patriarchal family a family structure in which authority is held by the eldest male (usually the father).

matriarchal family a family structure in which authority is held by the eldest female (usually the mother).

egalitarian family a family structure in which both partners share power and authority equally.

Residential Patterns

Residential patterns are interrelated with the authority structure and method of tracing descent in families. *Patrilocal residence* **refers to the custom of a married couple living in the same household (or community) as the husband's family.** Across cultures, patrilocal residency is the most common pattern. One example of contemporary patrilocal residency can be found in al-Barba, a lower-middle-class neighborhood in the Jordanian city of Irbid (McCann, 1997). According to researchers, the high cost of renting an apartment or building a new home has resulted in many sons building their own living quarters onto their parents' home, resulting in multifamily households consisting of an older married couple, their unmarried children, their married sons, and their sons' wives and children.

Few societies have residential patterns known as *matrilocal residence*—**the custom of a married couple living in the same household (or community) as the wife's parents.** In industrialized nations such as the United States, most couples hope to live in a *neolocal residence*—**the custom of a married couple living in their own residence apart from both the husband's and the wife's parents.**

To this point, we have examined a variety of marriage and family patterns found around the world. Even with the diversity of these patterns, most people's behavior is shaped by cultural rules pertaining to endogamy and exogamy. *Endogamy* **is the practice of marrying within one's own group.** In the United States, for example, most people practice endogamy: They marry people who come from the same social class, racial–ethnic group, religious affiliation, and other categories considered important within their own social group. *Exogamy* **is the practice of marrying outside one's own social group or category.** Depending on the circumstances, exogamy may not be noticed at all, or it may result in a person being ridiculed or ostracized by other members of the "in" group. The three most important sources of positive or negative sanctions for intermarriage are the family, the church, and the state. Participants in these social institutions may look unfavorably on the marriage of an in-group member to an "outsider" because of the belief that it diminishes social cohesion in the group (Kalmijn, 1998). However, educational attainment is also a strong indicator of marital choice. Higher education emphasizes individual achievement, and college-educated people may be less likely than others to identify themselves with their social or cultural roots and thus more willing to marry outside their own social group or category if their potential partner shares a similar level of educational attainment (Hwang, Saenz, and Aguirre, 1995; Kalmijn, 1998).

Theoretical Perspectives on Families

The *sociology of family* **is the subdiscipline of sociology that attempts to describe and explain patterns of family life and variations in family structure.** Functionalist perspectives emphasize the functions that families perform at the macrolevel of society, whereas conflict and feminist perspectives focus on families as a primary source of social inequality. Symbolic interactionists examine microlevel interactions that are integral to the roles of different family members.

Functionalist Perspectives

Functionalists emphasize the importance of the family in maintaining the stability of society and the well-being of individuals. According to Emile Durkheim, marriage is a microcosmic replica of the larger society; both marriage and society involve a mental and moral fusion of physically distinct individuals (Lehmann, 1994). Durkheim also believed that a division of labor contributes to greater efficiency in all areas of life—including marriages and families—even though he acknowledged that this division imposes significant limitations on some people.

In the United States, Talcott Parsons was a key figure in developing a functionalist model of the family. According to Parsons (1955), the husband/father fulfills the *instrumental role* (meeting the family's economic needs, making important decisions, and providing leadership), whereas the wife/mother fulfills the *expressive role* (running the household, caring for children, and meeting the emotional needs of family members).

© Jeff Randall/Getty Images

Functionalist theorists believe that families serve a variety of functions that no other social institution can adequately fulfill. In contrast, conflict and feminist theorists believe that families may be a source of conflict over values, goals, and access to resources and power. Children in upper-class families have many advantages and opportunities that are not available to other children.

Contemporary functionalist perspectives on families derive their foundation from Durkheim. Division of labor makes it possible for families to fulfill a number of functions that no other institution can perform as effectively. In advanced industrial societies, families serve four key functions:

1. *Sexual regulation.* Families are expected to regulate the sexual activity of their members and thus control reproduction so that it occurs within specific boundaries. At the macrolevel, incest taboos prohibit sexual contact or marriage between certain relatives. For example, virtually all societies prohibit sexual relations between parents and their children and between brothers and sisters.
2. *Socialization.* Parents and other relatives are responsible for teaching children the necessary knowledge and skills to survive. The smallness and intimacy of families make them best suited for providing children with the initial learning experiences they need.
3. *Economic and psychological support.* Families are responsible for providing economic and psychological support for members. In preindustrial societies, families are economic production units; in industrial societies, the economic security of

families is tied to the workplace and to macrolevel economic systems. In recent years, psychological support and emotional security have been increasingly important functions of the family.
4. *Provision of social status.* Families confer social status and reputation on their members. These statuses include the ascribed statuses with which individuals are born, such as race/ethnicity, nationality, social class, and sometimes religious affiliation. One of the most significant and compelling forms of social placement is the family's class position and the opportunities (or lack thereof) resulting from that position. Examples of class-related opportunities are access to quality health care, higher education, and a safe place to live.

Conflict and Feminist Perspectives

Conflict and feminist analysts view functionalist perspectives on the role of the family in society as idealized and inadequate. Rather than operating harmoniously and for the benefit of all members, families are sources of social inequality and conflict over values, goals, and access to resources and power.

According to some conflict theorists, families in capitalist economies are similar to the work environment of a factory. Women are dominated by

patrilocal residence the custom of a married couple living in the same household (or community) as the husband's family.

matrilocal residence the custom of a married couple living in the same household (or community) as the wife's parents.

neolocal residence the custom of a married couple living in their own residence apart from both the husband's and the wife's parents.

endogamy cultural norms prescribing that people marry within their social group or category.

exogamy cultural norms prescribing that people marry outside their social group or category.

sociology of family the subdiscipline of sociology that attempts to describe and explain patterns of family life and variations in family structure.

men in the home in the same manner that workers are dominated by capitalists and managers in factories (Engels, 1970/1884). Although childbearing and care for family members in the home contribute to capitalism, these activities also reinforce the subordination of women through unpaid (and often devalued) labor. Other conflict analysts are concerned with the effect that class conflict has on the family. The exploitation of the lower classes by the upper classes contributes to family problems such as high rates of divorce and overall family instability.

Some feminist perspectives on inequality in families focus on patriarchy rather than class. From this viewpoint, men's domination over women existed long before capitalism and private ownership of property (Mann, 1994). Women's subordination is rooted in patriarchy and men's control over women's labor power (Hartmann, 1981). According to one scholar, "Male power in our society is expressed in economic terms even if it does not originate in property relations; women's activities in the home have been un-

dervalued at the same time as their labor has been controlled by men" (Mann, 1994: 42). In addition, men have benefited from the privileges they derive from their status as family breadwinners.

Symbolic Interactionist Perspectives

Early symbolic interactionists such as Charles Horton Cooley and George Herbert Mead provided key insights on the roles we play as family members and how we modify or adapt our roles to the expectations of others—especially significant others such as parents, grandparents, siblings, and other relatives. How does the family influence the individual's self-concept and identity? In order to answer questions such as this one, contemporary symbolic interactionists examine the roles of husbands, wives, and children as they act out their own parts and react to the actions of others. From such a perspective, what people think, as well as

© Bob Barkany/Getty Images

Marriage and divorce are complicated processes involving rituals, shared moments of happiness, and then failure and dissolution as couples create a shared reality and then must reestablish individual ones.

© Corbis Super RF/Alamy

© Michael Newman/PhotoEdit

what they say and do, is very important in understanding family dynamics.

According to the sociologists Peter Berger and Hansfried Kellner (1964), interaction between marital partners contributes to a shared reality. Although newlyweds bring separate identities to a marriage, over time they construct a shared reality as a couple. In the process, the partners redefine their past identities to be consistent with new realities. Development of a shared reality is a continuous process, taking place not only in the family but in any group in which the couple participates together. Divorce is the reverse of this process; couples may start with a shared reality and, in the process of uncoupling, gradually develop separate realities (Vaughan, 1985).

Symbolic interactionists explain family relationships in terms of the subjective meanings and everyday interpretations that people give to their lives. As the sociologist Jessie Bernard (1982/1973) pointed out, women and men experience marriage differently. Although the husband may see *his* marriage very positively, the wife may feel less positive about *her* marriage, and vice versa. Researchers have found that husbands and wives may give very different accounts of the same event, and their "two realities" frequently do not coincide (Safilios-Rothschild, 1969).

Postmodernist Perspectives

According to postmodern theories, we have experienced a significant decline in the influence of the family and other social institutions. As people have pursued individual freedom, they have been less inclined to accept the structural constraints imposed on them by institutions. Given this assumption, how might a postmodern perspective view contemporary family life? For example, how might this approach answer the question "How is family life different in the information age"? Social scientist David Elkind (1995) describes the postmodern family as *permeable*—capable of being diffused or invaded in such a manner that the family's original nature is modified or changed. According to Elkind (1995), if the nuclear family is a reflection of the age of modernity, the permeable family reflects the postmodern assumptions of difference and irregularity. This is evident in the fact that the nuclear family is

now only one of many family forms. Similarly, the idea of romantic love under modernity has given way to the idea of consensual love: Individuals agree to have sexual relations with others whom they have no intention of marrying or, if they marry, do not necessarily see the marriage as having permanence. Maternal love has also been transformed into shared parenting, which includes not only mothers and fathers but also caregivers who may either be relatives or nonrelatives (Elkind, 1995).

Urbanity is another characteristic of the postmodern family. The boundaries between the public sphere (the workplace) and the private sphere (the home) are becoming much more open and flexible. In fact, family life may be negatively affected by the decreasing distinction between what is work time and what is family time. As more people are becoming connected "24/7" (twenty-four hours a day, seven days a week), the boss who in the past would not call at 11:30 P.M. may send an e-mail asking for an immediate response to some question that has arisen while the person is away on vacation with family members (Leonard, 1999). According to some postmodern analysts, this is an example of the "power of the new communications technologies to integrate and control labour despite extensive dispersion and decentralization" (Haraway, 1994: 439).

The Concept Quick Review summarizes sociological perspectives on the family. Taken together, these perspectives on the social institution of families reflect various ways in which familial relationships may be viewed in contemporary societies. Now we shift our focus to love, marriage, intimate relationships, and family issues in the United States.

Developing Intimate Relationships and Establishing Families

The United States has been described as a "nation of lovers"; it has been said that we are "in love with love." Why is this so? Perhaps the answer lies in the fact that our ideal culture emphasizes *romantic love,* which refers to a deep emotion, the satisfaction of significant needs, a caring for and acceptance of the

CONCEPT QUICK REVIEW

Theoretical Perspectives on Families

	Focus	Key Points	Perspective on Family Problems
Functionalist	Role of families in maintaining stability of society and individuals' well-being.	In modern societies, families serve the functions of sexual regulation, socialization, economic and psychological support, and provision of social status.	Family problems are related to changes in social institutions such as the economy, religion, education, and law/government.
Conflict/Feminist	Families as sources of conflict and social inequality.	Families both mirror and help perpetuate social inequalities based on class and gender.	Family problems reflect social patterns of dominance and subordination.
Symbolic Interactionist	Family dynamics, including communication patterns and the subjective meanings that people assign to events.	Interactions within families create a shared reality.	How family problems are perceived and defined depends on patterns of communication, the meanings that people give to roles and events, and individuals' interpretations of family interactions.

person we love, and involvement in an intimate relationship (Lamanna and Riedmann, 2009).

Love and Intimacy

In the late nineteenth century, during the Industrial Revolution, people came to view work and home as separate spheres in which different feelings and emotions were appropriate (Coontz, 1992). The public sphere of work—men's sphere—emphasized self-reliance and independence. By contrast, the private sphere of the home—women's sphere—emphasized the giving of services, the exchange of gifts, and love. Accordingly, love and emotions became the domain of women, and work and rationality became the domain of men (Lamanna and Riedmann, 2009). Although the roles of women and men changed dramatically in the twentieth century, women and men may still not share the same perceptions about romantic love today. According to the sociologist Francesca Cancian (1990), women tend to express their feelings verbally whereas men tend to express their love through nonverbal actions, such as running an errand for someone or repairing a child's broken toy.

Love and intimacy are closely intertwined. Intimacy may be psychic ("the sharing of minds"), sexual, or both. Although sexuality is an integral part of many intimate relationships, perceptions about sexual activities vary from one culture to the next and from one time period to another. For example, kissing is found primarily in Western cultures; many African and Asian cultures view kissing negatively (Reinisch, 1990).

For many years, the work of the biologist Alfred C. Kinsey was considered to be the definitive research on human sexuality, even though some of his methodology had serious limitations. More recently, the work of Kinsey and his associates has been superseded by the National Health and Social Life Survey conducted by the National Opinion Research Center at the University of Chicago (see Laumann et al., 1994; Michael et al., 1994). Based on interviews with more than 3,400 men and women aged 18 to 59, this random survey tended to reaffirm the significance of the dominant sexual ideologies. Most respondents reported that they engaged in heterosexual relationships, although 9 percent of the men said they had had at least one homosexual encounter resulting in orgasm. Although 6.2 percent of men and 4.4 percent of women said that they were at least somewhat

In the United States, the notion of romantic love is deeply intertwined with our beliefs about how and why people develop intimate relationships and establish families. Not all societies share this concern with romantic love.

attracted to others of the same gender, only 2.8 percent of men and 1.4 percent of women identified themselves as gay or lesbian. According to the study, persons who engaged in extramarital sex found their activities to be more thrilling than those with a marital partner, but they also felt more guilt. Persons in sustained relationships such as marriage or cohabitation found sexual activity to be the most satisfying emotionally and physically.

Cohabitation and Domestic Partnerships

Attitudes about cohabitation have changed in the past three decades. Until recently, the Census Bureau defined cohabitation as the sharing of a household by one man and one woman who are not related to each other by kinship or marriage, but recently has expanded this to a more inclusive definition. For our purposes, we will define *cohabitation* as **referring to two people who live together, and think of themselves as a couple, without being legally married.** It is not known how many people actually cohabit because the Census Bureau does not ask about emotional or sexual involvement between unmarried individuals sharing living quarters or between gay and lesbian couples.

Based on Census Bureau data, the people who are most likely to cohabit are under age 45, have been married before, or are older individuals who do not want to lose financial benefits (such as retirement benefits) that are contingent upon not remarrying. Among younger people, employed couples are more likely to cohabit than college students are.

Today, some people view cohabitation as a form of "trial marriage." Some people who have cohabited do eventually marry the person with whom they have been living, whereas others do not. A recent study of 11,000 women found that there was a 70 percent marriage rate for women who remained in a cohabiting relationship for at least 5 years. However, of the women in that study who cohabited and then married their partner, 40 percent became divorced within a 10-year period (Bramlett and Mosher, 2001). Whether these findings will be supported by subsequent research remains to be seen. But we do know that studies over the past two decades have supported the proposition that couples who cohabit before marriage do not necessarily have a stable relationship following marriage (Bumpass, Sweet, and Cherlin, 1991; London, 1991).

Among heterosexual couples, many reasons exist for cohabitation; for gay and lesbian couples, however, no alternatives to cohabitation exist in most U.S. states. For that reason, many lesbians and gays seek recognition of their *domestic partnerships*—**household partnerships in which an unmarried couple**

cohabitation a situation in which two people live together, and think of themselves as a couple, without being legally married.

domestic partnerships household partnerships in which an unmarried couple lives together in a committed, sexually intimate relationship and is granted the same rights and benefits as those accorded to married heterosexual couples.

lives together in a committed, sexually intimate relationship and is granted the same rights and benefits as those accorded to married heterosexual couples (Aulette, 1994; Gerstel and Gross, 1995). Benefits such as health and life insurance coverage are extremely important to *all* couples, as Gayle, a lesbian, points out: "It makes me angry that [heterosexuals] get insurance benefits and all the privileges, and Frances [her partner] and I take a beating financially. We both pay our insurance policies, but we don't get the discounts that other people get and that's not fair" (qtd. in Sherman, 1992: 197).

Over the past few years, much controversy has arisen over the legal status of gay and lesbian couples, particularly those who seek to make their relationship a legally binding commitment through marriage. Recently, the California Supreme Court declared that the state's ban on same-sex marriage was unconstitutional and that couples of the same sex could be legally married beginning June 1, 2008. In this decision, the court compared the ban on same-sex marriage with a previous ban on interracial marriage. The Chief Justice quoted from *Perez v. Sharp,* a 1948 California Supreme Court decision that struck down the ban against interracial marriage in that state: "The essence of the right to marry is freedom to join in marriage with the person of one's choice" (qtd. in Liptak, 2008: A10). Will this new law have a long-term effect on the rights of gays and lesbians to be married in California—and perhaps in other states as well? This remains to be seen. Clearly, opponents of same-sex marriage will continue to press for state and national constitutional amendments to restrict marriage to the union of *a man and a woman.* However, significant changes have also occurred in a few states other than California (Liptak, 2008).

Marriage

Why do people get married? Couples get married for a variety of reasons. Some do so because they are "in love," desire companionship and sex, want to have children, feel social pressure, are attempting to escape from a bad situation in their parents' home, or believe that they will have more money or other resources if they get married. These factors notwith-

© GABRIEL BOUYS/AFP/Getty Images

These two couples filed suit in California against the state's ban on same-sex marriage. Here they celebrate the May 2008 decision by the California Supreme Court to rescind that ban. How do you define marriage?

standing, the selection of a marital partner is actually fairly predictable. As previously discussed, most people in the United States tend to choose marriage partners who are similar to themselves. *Homogamy* **refers to the pattern of individuals marrying those who have similar characteristics, such as race/ethnicity, religious background, age, education, or social class.** However, homogamy provides only the general framework within which people select their partners; people are also influenced by other factors. For example, some researchers claim that people want partners whose personalities match their own in significant ways. Thus, people who are outgoing and friendly may be attracted to other people with those same traits. However, other researchers claim that people look for partners whose personality traits differ from but complement their own.

Housework and Child-Care Responsibilities

Today, more than 50 percent of all marriages in the United States are *dual-earner marriages*—**marriages in which both spouses are in the labor force.** Over half of all employed women hold full-time, year-round jobs. Even when their children are very young, most working mothers work full time. For example, in 2004 more than 74 percent of employed mothers with children under age 6 worked full time (U.S. Census Bureau, 2008). Moreover, as Chapter 10 points out, many married women leave their paid employment at the end of the day and then go home to perform hours of housework and child care. Sociologist Arlie Hochschild (1989, 2003) refers to this as the *second shift*—**the domestic work that employed women perform at home after they complete their workday on the job.** Thus, many married women contribute to the economic well-being of their families and also meet many, if not all, of the domestic needs of family members by cooking, cleaning, shopping, taking care of children, and managing household routines. According to Hochschild, the unpaid housework that women do on the second shift amounts to an extra month of work each year. In households with small children or many children, the amount of housework increases (Hartmann, 1981). Across race and class, numerous studies have confirmed that domestic work remains primarily women's work (Gerstel and Gross, 1995).

Hochschild (2003: 28) states that continuing problems regarding the second shift in many families are a sign that the gender revolution has stalled:

> The move of masses of women into the paid workforce has constituted a revolution. But the slower shift in ideas of "manhood," the resistance of sharing work at home, the rigid schedules at work make for a "stall" in this gender revolution. It is a stall in the change of institutional arrangements of which men are the principal keepers.

As Hochschild points out, the second shift remains a problem for many women in dual-earner marriages.

In recent years, more husbands have attempted to share some of the household and child-care responsibilities, especially in families in which the wife's earnings are essential to family finances. Overall, when husbands share some of the household responsibilities, they typically spend much less time in these activities than do their wives. Women and men perform different household tasks, and the deadlines for their work vary widely. Recurring tasks that have specific times for completion (such as bathing a child or cooking a meal) tend to be the women's responsibility; by contrast, men are more likely to do the periodic tasks that have no highly structured schedule (such as mowing the lawn or

Juggling housework, child care, and a job in the paid work force is all part of the average day for many women. Why does sociologist Arlie Hochschild believe that many women work a "second shift"?

changing the oil in the car) (Hochschild, 1989). Some men are also more reluctant to perform undesirable tasks such as scrubbing the toilet or diapering a baby, or to give up leisure pursuits.

homogamy the pattern of individuals marrying those who have similar characteristics, such as race/ethnicity, religious background, age, education, or social class.

dual-earner marriages marriages in which both spouses are in the labor force.

second shift Arlie Hochschild's term for the domestic work that employed women perform at home after they complete their workday on the job.

Couples with more-egalitarian ideas about women's and men's roles tend to share more equally in food preparation, housework, and child care (Wright et al., 1992). For some men, the shift to a more egalitarian household occurs gradually, as Wesley, whose wife works full time, explains:

It was me taking the initiative, and also Connie pushing, saying, "Gee, there's so much that has to be done." At first I said, "But I'm supposed to be the breadwinner," not realizing she's also the breadwinner. I was being a little blind to what was going on, but I got tired of waiting for my wife to come home to start cooking, so one day I surprised the hell out [of] her and myself and the kids, and I had supper waiting on the table for her. (qtd. in Gerson, 1993: 170)

In the United States, millions of parents rely on child care so that they can work and so that their young children can benefit from early educational experiences that will help in their future school endeavors. For millions more parents, after-school care for school-age children is an urgent concern. Nearly five million children are home alone after school each week in this country. The children need productive and safe activities to engage in while their parents are working. Although child care is often unavailable or unaffordable for many parents, those children who are in day care for extended hours often come to think of child-care workers and other caregivers as members of their extended families because they may spend nearly as many hours with them as they do with their own parents. For children of divorced parents and other young people living in single-parent households, the issue of child care is often a pressing concern because of the limited number of available adults and the lack of financial resources.

Child-Related Family Issues and Parenting

Not all couples become parents. Those who decide not to have children often consider themselves to be "child-free," whereas those who do not produce children through no choice of their own may consider themselves "childless."

Deciding to Have Children

Cultural attitudes about having children and about the ideal family size began to change in the United States in the late 1950s. Women, on average, are now having 2.1 children each (see "Sociology Works!"). However, rates of fertility differ across racial and ethnic categories. In 2006, for example, Latinas (Hispanic women) had a total fertility rate of 2.9, which was 50 percent above that of white (non-Hispanic) women (National Center for Health Statistics, 2007). Among Latinas, the highest rate of fertility was found among Mexican American women, whereas Puerto Rican and Cuban American women had relatively lower rates.

Advances in birth control techniques over the past four decades—including the birth control pill and contraceptive patches and shots—now make it possible for people to decide whether or not they want to have children, how many they wish to have, and to determine (at least somewhat) the spacing of their births. However, sociologists suggest that fertility is linked not only to reproductive technologies but also to women's beliefs that they do or do not have other opportunities in society that are viable alternatives to childbearing (Lamanna and Riedmann, 2009).

Today, the concept of reproductive freedom includes both the desire *to have* or *not to have* one or more children. According to the sociologists Leslie King and Madonna Harrington Meyer (1997), many U.S. women spend up to one-half of their life attempting to control their reproductivity. Other analysts have found that women, more often than men, are the first to choose a child-free lifestyle (Seccombe, 1991). However, the desire not to have children often comes in conflict with our society's *pronatalist bias,* which assumes that having children is the norm and can be taken for granted, whereas those who choose not to have children believe they must justify their decision to others (Lamanna and Riedmann, 2009).

However, some couples experience the condition of *involuntary infertility,* whereby they want to have a child but find that they are physically unable to do so. *Infertility* is defined as an inability to conceive after a year of unprotected sexual relations. Today, infertility affects nearly five million U.S. couples, or one in twelve couples in which the wife is between

Sociology *Works!*

Social Factors Influencing Parenting: From the Housing Market to the Baby Nursery

One reason there are so few children in Italy is that housing is so hard to come by. Houses are bigger in the U.S. and generally more available. That may help explain why Americans have more babies.

—Robert Engelman, vice president for programs at the Worldwatch Institute, an environmental research organization, and author of *More: Population, Nature, and What Women Want* (qtd. in Leland, 2008: A12)

Social scientists have long traced a connection between housing and fertility. When homes are scarce or beyond the means of young couples, as in the 1930s, couples delay marriage or have fewer children. This tendency helps account for the relatively dismal birth rates of many developed nations. . . . (Leland, 2008: A12)

For many years, demographers and other sociologists who specialize in population trends have sought to identify how biological and social factors affect fertility rates in various nations. Although biological factors such as general health and levels of nutrition in a region clearly affect the number of children a couple may produce, social factors are also important in determining the estimated number of children a woman will have in her lifetime (see Chapter 15 for additional information). A key social factor is the housing market where a couple lives: The ability to buy a house and having a relatively large home may influence a couple's decision about how many children to have. Early in the 2000s, the housing market in some areas of the United States provided an opportune time for more young couples to purchase their own home. Mortgages were readily available, and it typically did not take much cash (in the way of a down payment) up front to sign the contract and move into one's dream home. The availability of such loans to people who really couldn't afford the

home they were buying has since come back to haunt many underfunded home buyers. By 2006, however, the babies were arriving, and the fertility rate in the United States had grown to an estimated 2.1 children for every woman of child-bearing age, reaching the highest level since the 1970s (Leland, 2008).

Sociological insights on the social aspects of fertility, which at first might appear to be primarily a biological phenomenon, have provided us with new information on why people decide to have children and how many children they might have. However, much remains unknown about the relationship between the housing market and the maternity ward, including how income and feelings of optimism or pessimism about the local and national economy might affect a couple's decisions regarding parenting.

Will the baby boomlet of 2006 continue in the future? According to some social analysts, the boomlet may be short-lived because of a downturn in the economy and the housing market, which has culminated in fewer new homes being built and brought about more foreclosures—two key factors that may discourage couples from either having children or producing larger families.

Reflect & Analyze

How might sociological findings about factors that influence a couple's decision to have children be useful in your community? For example, why is information about the availability of housing and local fertility trends important to school board members and administrators when they are making enrollment projections or deciding where to build a new school in the future?

the ages of fifteen and forty-four (Gabriel, 1996). Research suggests that fertility problems originate in females in approximately 30-40 percent of the cases and with males in about 40 percent of the cases; in the other 20 percent of the cases, the cause is impossible to determine (Gabriel, 1996). A lead-

ing cause of infertility is sexually transmitted diseases, especially those cases that develop into pelvic inflammatory disease (Gold and Richards, 1994). It is estimated that about half of infertile couples who seek treatments such as fertility drugs, artificial insemination, and surgery to unblock fallopian tubes

Box 11.2 **Sociology and Global Perspective**

Wombs-for-Rent: Outsourcing Births to India

Picture four people—three adults and one infant—as they might be shown on *CBS News:* One person in the photo is Karen Kim, a lovely young woman from California, who is cuddling her infant son, Brady. Another person is Karen's husband, Thomas, who lovingly looks on at the mother and child. Brady, the Kims' new son, is the third person in the photo, but who is the fourth person—a woman with long black hair, a red dress, and pearl earrings? Her name is Dr. Nayna Patel, and she is the physician who made it possible for the Kims to become parents because she runs Akanksha Fertility Clinic in Anand, India, where surrogate mothers give birth so that infertile couples such as the Kims can have children. (*CBS News,* 2007)

The notion of extracting resources from the Third World in order to enrich the First World is hardly new. It harks back to imperialism in its most literal form: The nineteenth-century extraction of gold, ivory, and rubber from the Third World. . . . Today, as love and care become the "new gold," the female part of the story has grown in prominence.

—Sociologist Arlie Russell Hochschild (2003: 194) describes what she believes is happening as young women in low-income nations such as the Philippines leave their own children behind to work abroad for long periods of time, taking care of other people's children and households; however, these words certainly can also be applied to nations such as India, where women are serving as surrogate

mothers for infertile couples in the United States, Britain, Taiwan, and beyond. (qtd. in Dolnick, 2008)

The pregnant women at Akanksha Fertility Clinic (where the Kims' son was born) are professional surrogate mothers. Each surrogate must have at least one child of her own before she is allowed to become a surrogate mother. The clinic established this rule based on the assumption that having a child shows potential clients that the surrogate can successfully carry and deliver a baby and that she has other children at home to love and will not be resistant to giving up a newborn that she gestated and birthed (Kohl, 2007).

Why do some infertile couples in the United States, Britain, and elsewhere want to "hire" a woman in India to have their child? Most couples that engage in this practice have made numerous attempts to have a child through in vitro fertilization and other assisted reproductive technologies. If they have been unsuccessful in their efforts, the couple may first attempt to find a surrogate in the United States, but they quickly learn that a gestational surrogate costs more than $50,000, whereas the cost of an Indian surrogate ranges from about $2,250 to $5,000 plus medical expenses (Kohl, 2007; United Press International, 2007). According to Dr. Patel, earning money through surrogacy helps uplift Indian women: It provides money for their household and makes them more independent. For example, the typical woman might earn more for one surrogate pregnancy than she would earn in fifteen years from other kinds of employment (*CBS News,* 2007).

---

can be helped; however, some are unable to conceive despite expensive treatments such as *in vitro fertilization,* which costs as much as $11,000 per attempt (Gabriel, 1996). According to the sociologist Charlene Miall (1986), women who are involuntarily childless engage in "information management" to combat the social stigma associated with childlessness. Their tactics range from avoiding people who make them uncomfortable to revealing their infertility so that others will not think of them as "selfish" for being childless. Some people who are involuntarily child-

less may choose surrogacy or adoption as alternative ways of becoming a parent (see Box 11.2).

Adoption

Adoption is a legal process through which the rights and duties of parenting are transferred from a child's biological and/or legal parents to a new legal parent or parents. This procedure gives the adopted child all the rights of a biological child. In most adoptions, a new birth certificate is issued, and the child has no

A surrogate mother (left) has delivered a baby for Karen Kim (center), with the help of infertility specialist Dr. Nayna Patel (right). This practice, sometimes called "rent-a-womb," remains controversial.

© AP Images/Ajit Solanki

Is there any problem with global "rent-a-womb"? If there is an agreement between a surrogate mother and a couple that badly wants a child, some analysts believe that "offer and acceptance" is nothing more than capitalism at work—where there is a demand (for infants by infertile couples), there will be a supply (from low-income surrogate mothers). However, some ethicists raise troubling questions about the practice of commercial surrogacy: A mother should give birth to her child because it is hers and she loves it, not because she is being paid to give birth to someone else's baby. Other social critics are concerned about the potential mistreatment of low-income women who may be exploited or may suffer long-term emotional damage from functioning as a surrogate (Dunbar, 2007). For the time being, in clinics such as the one in Anand, India, hopeful parents just provide the egg, the sperm, and the money, and all the rest is done for them by the clinic and the surrogates, who live in a spacious house where they are taken care of by maids, cooks, and doctors who want them to remain healthy and happy throughout their pregnancies—after all, it's good business!

Reflect & Analyze

What are your thoughts on surrogacy? Is there any difference between surrogacy when it occurs in high-income nations such as the United States and Britain as compared to situations in which the parents live in a high-income nation and the surrogate mother lives in a lower-income nation? How might we relate the specific issue of outsourced surrogacy to some larger concerns about families and intimate relationships that we have discussed in this chapter?

future contact with the biological parents; however, some states have "right-to-know" laws under which adoptive parents must grant the biological parents visitation rights.

Matching children who are available for adoption with prospective adoptive parents can be difficult. The available children have specific needs, and the prospective parents often set specifications on the type of child they want to adopt. Some adoptions are by relatives of the child; others are by infertile couples (although many fertile couples also adopt). Increasing numbers of gays, lesbians, and people who are single are adopting children. Although thousands of children are available for adoption each year in the United States, many prospective parents seek out children in developing nations such as Romania, South Korea, and India. The primary reason is that the available children in the United States are thought to be "unsuitable." They may have disabilities, or they may be sick, nonwhite (most of the prospective parents are white), or too old. In addition, fewer infants are available for

Megastars Angelina Jolie and Brad Pitt have both produced their own children and adopted others, sometimes in controversial situations.

adoption today than in the past because better means of contraception exist, abortion is more readily available, and more unmarried teenage parents decide to keep their babies.

Ironically, although many couples who would like to have a child are unable to do so, other couples conceive a child without conscious intent. Consider the fact that for the approximately 6.4 million women who become pregnant each year in the United States, about 2.8 million (44 percent) pregnancies are intended whereas about 3.6 million (56 percent) are unintended (Gold and Richards, 1994). As with women and motherhood, some men feel that their fatherhood was planned; others feel that it was thrust upon them (Gerson, 1993). Unplanned pregnancies usually result from failure to use contraceptives or from using contraceptives that do not work. Even with "planned pregnancies," it is difficult to plan exactly when conception will occur and a child will be born.

Teenage Pregnancies

Teenage pregnancies are a popular topic in the media and political discourse, and the United States has the highest rate of teen pregnancy in the Western industrialized world (National Campaign to Prevent Teen Pregnancy, 1997). In 2006 the total number of live births per 1,000 women ages 15 to 17 was 22.0, and for women ages 18 and 19 the number was 73.0, in both instances the first increase since 1991 (National Center for Health Statistics, 2007).

What are the primary reasons for the high rates of teenage pregnancy? At the microlevel, several issues are most important: (1) many sexually active teenagers do not use contraceptives; (2) teenagers—especially those from some low-income families and/or subordinate racial and ethnic groups—may receive little accurate information about the use of, and problems associated with, contraception; (3) some teenage males (due to a double standard based on the myth that sexual promiscuity is acceptable among males but not females) believe that females should be responsible for contraception; and (4) some teenagers view pregnancy as a sign of male prowess or as a way to gain adult status. At the macrolevel, structural factors also contribute to teenage pregnancy rates. Lack of education and employment opportunities in some central-city and rural areas may discourage young people's thoughts of upward mobility that might make early parenting appear less appealing. Likewise, religious and political opposition has resulted in issues relating to reproductive responsibility not being dealt with as openly in the United States as in some other nations. Finally, advertising, films, television programming, magazines, music, and other forms of media often flaunt the idea of being sexually active without showing the possible consequences of such behavior.

Teen pregnancies have been of concern to analysts who suggest that teenage mothers may be less skilled at parenting, are less likely to complete high school than their counterparts without children, and possess few economic and social supports other than their relatives (Maynard, 1996; Moore, Driscoll, and Lindberg, 1998). In addition, the increase in births among unmarried teenagers may have negative long-term consequences for mothers and their children, who have severely limited educational and employment opportunities and a high likelihood of living in poverty. Moreover, the Children's Defense Fund estimates that among those who first gave birth between the ages of fifteen and nineteen, 43 percent will have a second child within three years.

Teenage fathers have largely been left out of the picture. According to the sociologist Brian Robinson (1988), a number of myths exist regarding teenage fathers: (1) they are worldly wise "superstuds" who engage in sexual activity early and often, (2) they are "Don Juans" who sexually exploit unsuspecting females, (3) they have "macho" tendencies because they are psychologically inadequate and need to prove their masculinity, (4) they have few emotional feelings for the women they impregnate, and (5) they are "phantom fathers" who are rarely involved in caring for and rearing their children. However, these myths overlook the fact that some teenage males try to be good fathers.

Single-Parent Households

In recent years, there has been a significant increase in single- or one-parent households due to divorce and to births outside of marriage. Even for a person with a stable income and a network of friends and family to help with child care, raising a child alone can be an emotional and financial burden. Single-parent households headed by women have been stereotyped by some journalists, politicians, and analysts as being problematic for children. About 42 percent of all white children and 86 percent of all African American children will spend part of their childhood living in a household headed by a single mother who is divorced, separated, never married, or widowed (Garfinkel and McLanahan, 1986). According to sociologists Sara McLanahan and Karen Booth (1991), children from mother-only families are more likely than children in two-parent families to have poor academic achievement, higher school absentee and dropout rates, early marriage and parenthood, higher rates of divorce, and more drug and alcohol abuse. Does living in a one-parent family *cause* all of this? Certainly not! Many other factors—including poverty, discrimination, unsafe neighborhoods, and high crime rates—contribute to these problems.

Lesbian mothers and gay fathers are counted in some studies as single parents; however, they often

© Masterfile

Mothers and fathers in single-parent households are confronted with the necessity of meeting most of their children's daily needs without help from others. However, even in two-parent households, children are not guaranteed a happy childhood simply because both parents live in the same household.

share parenting responsibilities with a same-sex partner. Due to homophobia (hatred and fear of homosexuals and lesbians), lesbian mothers and gay fathers are more likely to lose custody to a heterosexual parent in divorce disputes (Falk, 1989; Robson, 1992). In any case, between one million and three million gay men in the United States and Canada are fathers. Some gay men are married natural fathers, others are single gay men, and still others are part of gay couples that have adopted children. Very little research exists on gay fathers; what research does exist tends to show that noncustodial gay fathers try to maintain good relationships with their children (Bozett, 1988).

Single fathers who do not have custody of their children may play a relatively limited role in the lives of those children. Although many remain actively involved in their children's lives, others may become "Disneyland daddies" who take their children to recreational activities and buy them presents for special occasions but have a very small part in their children's day-to-day life. Sometimes, this limited role is by choice, but more often it is caused by workplace demands on time and energy, the location of the ex-wife's residence, and limitations placed on visitation by custody arrangements.

Two-Parent Households

Parenthood in the United States is idealized, especially for women. According to the sociologist Alice Rossi (1992), maternity is the mark of adulthood for women, whether or not they are employed. In contrast, men secure their status as adults by their employment and other activities outside the family (Hoffnung, 1995).

For families in which a couple truly shares parenting, children have two primary caregivers. Some parents share parenting responsibilities by choice; others share out of necessity because both hold full-time jobs. Some studies have found that men's taking an active part in raising the children is beneficial not only for mothers (who then have a little more time for other activities) but also for the men and the children. The men benefit through increased access to children and greater opportunity to be nurturing parents (Coltrane, 1989).

Remaining Single

Some never-married people remain single by choice. Reasons include opportunities for a career (especially for women), the availability of sexual partners without marriage, the belief that the single lifestyle is full of excitement, and the desire for self-sufficiency and freedom to change and experiment (Stein, 1976, 1981). Some scholars have concluded that individuals who prefer to remain single hold more-individualistic values and are less family-oriented than those who choose to marry. Friends and personal growth tend to be valued more highly than marriage and children (Cargan and Melko, 1982; Alwin, Converse, and Martin, 1985).

Other never-married individuals remain single out of necessity. Being single is an economic ne-

cessity for those who cannot afford to marry and set up their own household. Structural changes in the economy have limited the options of many working-class young people. Even some college graduates have found that they cannot earn enough money to set up a household separate from that of their parents.

The proportion of singles varies significantly by racial and ethnic group, as shown in ▶ Figure 11.1. Among persons age 15 and over, 40.8 percent of African Americans have never married, compared with 30.1 percent of Latinos/as, 26.8 percent of Asian and Pacific Islander Americans, and 22.6 percent of whites. Among women age 20 and over, the difference is even more pronounced; almost twice as many African American women in this age category have never married, compared with U.S. women of the same age in general (U.S. Census Bureau, 2008).

Although a number of studies have examined why African Americans remain single, few studies have focused on Latinos/as. Some analysts cite the diversity of experiences among Mexican Americans, Cuban Americans, and Puerto Ricans as the reason for this lack of research. Existing research attributes increased rates of singlehood among Latinos/as to several factors, including the youthful age of the Latino/a population and economic conditions experienced by many young Latinos/as (see Mindel, Habenstein, and Wright, 1988).

Transitions and Problems in Families

Families go through many transitions and experience a wide variety of problems, ranging from high rates of divorce and teen pregnancy to domestic abuse and family violence. These all-too-common experiences highlight two important facts about families: (1) for good or ill, families are central to our existence, and (2) the reality of family life is far more complicated than the idealized image of families found in the media and in many political discussions. Moreover, as people grow older, transitions inevitably occur in family life.

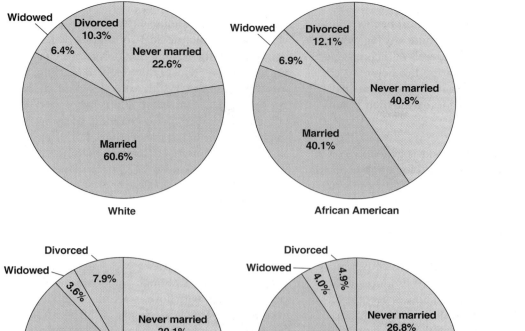

White

African American

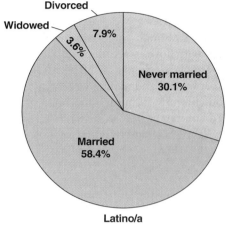

Latino/a

Asian and Pacific Islander American

▶ Figure 11.1 **Marital Status of U.S. Population Age 15 and Over by Race/Ethnicity**

Source: U.S. Census Bureau, 2008.

Family Transitions Based on Age and the Life Course

Throughout our lives, our families play an important role in our individual development. People are assigned different roles and positions based on their age and the family structure in a particular society. Stages in the contemporary life course in industrialized nations typically include infancy and childhood, adolescence and young adulthood, middle adulthood, and late adulthood.

Infancy and Childhood Infancy and childhood are largely creations of industrialized nations. In agricultural societies, children are often a source of physical labor for both their families and the larger society. By contrast, children in industrialized nations are expected to attend school and learn the necessary skills

for future employment rather than perform unskilled labor. In this context, families must provide for extensive periods of time in which their children are economically dependent upon them.

Adolescence and Young Adulthood In contemporary industrialized countries, adolescence roughly spans the teenage years. As compared with preindustrial societies, in which adolescents are seen as adults, adolescents in industrialized nations are treated by their families, teachers, and others as neither children nor adults. Young adulthood, which follows adolescence and lasts to about age 39, is socially significant because during this time people are expected to establish their own families and get a job.

Middle Adulthood The concept of middle adulthood—people between the ages of 40 and 65—did not exist until fairly recently. For most people, middle adulthood represents the time during which their families have the highest levels of income, and they may conclude child rearing but begin an era of grandparenting. Today, more grandparents have partial or full responsibility for rearing their grandchildren.

Late Adulthood Late adulthood is generally considered to begin at age 65—which for some people is "normal" retirement age. *Retirement*—the institutionalized separation of an individual from an occupational position, with continuation of income through a retirement pension based on prior years of service—means the end of a status that has long been a source of income and a means of personal identity. Loss of such status may produce family discord for a period of time as spouses become adjusted to having more time alone with each other. Increasingly, however, older adults are maintaining either full-time or part-time employment outside the home. For those older adults who are financially able, travel and other leisure-time pursuits may become the focus of their activities rather than daily attendance at a job.

Many older adults continue to live in their own homes; however, some may prefer smaller housing units or apartments if they no longer have other

© David Young-Wolff/Getty Images

What can people across generations learn by taking time to talk to each other? Can the learning process flow in both directions?

family members residing in the household. Those who have chronic medical conditions or need the assistance of others may rely on *assisted living* arrangements—a concept that refers to housing, support services, and health care designed to meet the varied needs of older adults who need help with daily activities such as bathing, dressing, food preparation, or taking medications. For others, living with family members remains the preferred approach. Factors involved in living arrangements of older family members include how close they live to other relatives and friends and how frequently they react with one another.

At each stage in the life course, family life differs significantly based on the number of people and the diversity of age categories represented in the social unit at that time. Although some families provide their members with love, warmth, and satisfying emotional experiences, others may be hazardous to the individual's physical and mental well-being. Because of this dichotomy in family life, sociologists have described families as both a "haven in a heartless world" (Lasch, 1977) and a "cradle of violence" (Gelles and Straus, 1988).

Family Violence

Violence between men and women in the home is often called spouse abuse or domestic violence. *Spouse abuse* refers to any intentional act or series of acts—whether physical, emotional, or sexual—that causes injury to a female or male spouse (Wallace, 2002). According to sociologists, the term *spouse abuse* refers not only to people who are married but also to those who are cohabiting or involved in a serious relationship, as well as those individuals who are separated or living apart from their former spouse (Wallace, 2002).

How much do we know about family violence? Women, as compared with men, are more likely to be the victim of violence perpetrated by intimate partners. Recent statistics indicate that women are five times more likely than men to experience such violence and that many of these women live in households with children younger than age twelve. However, we cannot know the true extent of family violence because much of it is not reported to police. For example, it is estimated that only about one-half of the intimate-partner violence against women in the 1990s was reported to police (U.S. Department of Justice, 2000). African American women were more likely than other women to report such violence, which may further skew data about who is most likely to be victimized by a domestic partner (U.S. Department of Justice, 2000).

Although everyone in a household where family violence occurs is harmed psychologically, whether or not they are the victims of violence, children are especially affected by household violence. It is estimated that between three million and ten million children witness some form of domestic violence in their homes each year, and there is evidence to suggest that domestic violence and child maltreatment often take place in the same household (Children's Defense Fund, 2002). According to some experts, domestic violence is an important indicator that child abuse and neglect are also taking place in the household.

In some situations, family violence can be reduced or eliminated through counseling, the removal of one parent from the household, or other steps that are taken either by the family or by social service or law enforcement officials. However, children who witness violence in the home may display certain emotional and behavioral problems that adversely affect their school life and communication with other people. In some families, the problems of family violence are great enough that the children are removed from the household and placed in foster care.

Children in Foster Care

Not all of the children in foster care come from violent homes, but many foster children have been in dysfunctional homes where parents or other relatives lacked the ability to meet the children's daily needs. *Foster care* refers to institutional settings or residences where adults other than a child's own parents or biological relatives serve as caregivers. States provide financial aid to foster parents, and the intent of such programs is that the children will either return to their own families or be adopted by other families. However, this is often not the case for "difficult to place" children, particularly those who are over ten years of age, have illnesses or disabili-

ties, or are perceived as suffering from "behavioral problems." More than 568,000 children are in foster care at any given time (Barovick, 2001). About 60 percent of children in foster care are children of color, with about 42 percent of them being African American—almost three times the percentage of African American children in the total U.S. child population (Children's Defense Fund, 2002). Even when the number of children entering foster care for the first time remains relatively stable, the total number of children in foster care continues to increase because fewer children are leaving foster care and being adopted or placed in permanent homes (Children's Defense Fund, 2002). Many children in foster care have limited prospects for finding a permanent home; however, a few innovative programs have offered hope for children who previously had been moved from one foster care setting to another (see Box 11.3).

Problems in the family contribute to the large numbers of children who are in foster care. Such factors include parents' illness, unemployment, or death; violence or abuse in the family; and high rates of divorce.

Elder Abuse

Abuse and neglect of older persons have received increasing public attention in recent years, due both to the increasing number of older people and to the establishment of more-vocal groups to represent their concerns. *Elder abuse* refers to physical abuse, psychological abuse, financial exploitation, and medical abuse or neglect of people age 65 or older (Hooyman and Kiyak, 2002).

The National Academy of Sciences (2003) estimates that between one and two million people age 65 and older in the United States have been the victims of physical or mental abuse. Just as with violence against children or women, it is difficult to determine how much abuse against older persons occurs. Many victims are understandably reluctant to talk about it. One study indicates that slightly over 2 percent of all older people experience physical abuse (Pillemer and Finkelhor, 1988; Atchley and Barusch, 2004). Although this may appear to be a small percentage, it represents a large number of people. Studies have shown

Box 11.3 You Can Make a Difference

Providing Hope and Help for Children

I take it personally when I see kids mistreated. I just think they need an advocate to fight for them. . . . For me, it's very simple: The kids' needs come first. That's the bottom line at Hope Meadows. We make decisions as if these are our own children, and when you think that way, your decisions are different than if you are just trying to work within a bureaucratic system.

—sociologist Brenda Eheart describing why she founded Hope Meadows (qtd. in Smith, 2001: 22)

After five years of research into the adoptions of older children, the sociologist Brenda Eheart realized that foster families faced many problems when they tried to help children who had been moved from home to home. Thinking that she might be able to make a difference, Eheart developed the plan for Hope Meadows, a community established in 1994 on an abandoned Illinois air force base. Hope Meadows is made up of a three-block-long series of ranch houses that provide multigenerational and multiracial housing for foster children, their temporary families, and older adults who live and work with the children. Older adults who interact with the children receive reduced rent in exchange for at least six hours per week of volunteer work. Foster families that reside at Hope Meadows gain a feeling of community as they work together to help children who have experienced severe abuse or neglect, have been exposed to drugs and numerous foster homes, and often have physical, emotional, and behavioral problems.

Since its commencement, Hope Meadows has been largely successful in helping children get adopted. However, children are not the only beneficiaries of this community: Older residents gain the benefit of interacting with children and feeling that they can *make a positive contribution* to the lives of others (Barovick, 2001). Debbie Calhoun, a foster parent at Hope Meadows, has suggested things that children need the most when they

come there, and we can make a difference by providing the children in our lives with these same things (based on Smith, 2001):

- *Understanding.* We need to gain an awareness of how children feel and why they say and do certain things.
- *Trust.* We need to help children to see us as people they can rely on and believe in as people they can trust.
- *Love.* We must show children that they are loved and that they will still be loved even when they make mistakes.
- *Compassion.* We must show children compassion because they must experience compassion in order to be able to show it to others.
- *Time.* We must give children time to be a part of our lives, and we must also give them time to adjust and to start over when they need to do so.
- *Security.* We must help children to feel secure in their surroundings and to believe that there is stability or permanency in their living arrangements.
- *Praise.* We must tell children when they are doing well and not always be critical of them.
- *Discipline.* We must let children know what behavior is acceptable and what behavior is not, all the while showing them that we love them, even when discipline is necessary.
- *Self-Esteem.* We must help children feel good about themselves.
- *Pride.* We should provide opportunities for children to learn to take pride in their accomplishments and in themselves.

If these suggestions are beneficial for children in foster care settings, then they are certainly useful ideas for each of us to implement in our own families and communities as well. What other ideas would you add to the list? Why?

that elder abuse tends to be concentrated among those over age 75 (Steinmetz, 1987). Most of the victims are white, middle-to-lower-middle-class Protestant women, age 75–85, who suffer some form of impairment (Garbarino, 1989). Sons, fol-lowed by daughters, are the most frequent abusers of older persons.

What causes elder abuse? Scholars do not agree on an answer to this question. Some believe that elder abuse may occur in families because of ageism, which you will recall is prejudice and discrimination against people on the basis of age. Other analysts link elder abuse with the physical impairment of some elderly persons; however, little evidence has been found to support that conclusion. Similarly, the sociologist Karl Pillemer found no support for the common assumption that elderly persons who are dependent on their children are more likely to be abused. To the contrary, the abusers are very likely to be *dependent on the older person* for housing and financial assistance (Pillemer, 1985).

Divorce

Divorce is the legal process of dissolving a marriage that allows former spouses to remarry if they so choose. Most divorces today are granted on the grounds of *irreconcilable differences,* meaning that there has been a breakdown of the marital relationship for which neither partner is specifically blamed. Prior to the passage of more-lenient divorce laws, many states required that the partner seeking the divorce prove misconduct on the part of the other spouse. Under *no-fault divorce laws,* however, proof of "blameworthiness" is generally no longer necessary.

Over the past 100 years, the U.S. divorce rate (number of divorces per 1,000 population) has varied from a low of 0.7 in 1900 to an all-time high of 5.3 in 1981; by 1995, it had stabilized at 4.4 (U.S. Census Bureau, 2008). Although many people believe that marriage should last for a lifetime, others believe that marriage is a commitment that may change over time.

Recent studies have shown that 43 percent of first marriages end in separation or divorce within 15 years (National Centers for Disease Control, 2001). (Figure ▶ 11.2 shows U.S. divorce rates for each state.) Many first marriages do not last even 15 years: One in three first marriages ends within 10 years, and one in five ends within 5 years. The likelihood of divorce goes up with each subsequent marriage in the serial monogamy pattern. When divorces that terminate second or subsequent marriages are taken into account, there are about half as many divorces each year in the United States as there are marriages. These data concern researchers in agencies such as the Centers for Disease Control because separation and divorce often have adverse effects on the health and well-being of both children and adults (National Centers for Disease Control, 2001).

Causes of Divorce Why do divorces occur? As you will recall from Chapter 1, sociologists look for correlations (relationships between two variables) in attempting to answer questions such as this. Existing research has identified a number of factors at both the

One of the most dramatic consequences of divorce affects children, many of whom divide their time between the separate households of their mother and father. Although some divorced fathers spend time and money on activities with their children, the burden of child rearing in single-parent households often falls most heavily on women.

© Tony Freeman/PhotoEdit

United States[c]	Rates per 1,000 population[a,b]						Rates per 1,000 population[a,b]			
	1990	2000	2004	2005			1990	2000	2004	2005
United States[c]	4.7	4.2	3.7	3.6		Missouri	5.1	4.8	3.8	3.6
Alabama	6.1	5.4	4.7	4.9		Montana	5.1	2.4	3.8	3.8
Alaska	5.5	4.4	4.8	5.8		Nebraska	4.0	3.8	3.6	3.4
Arizona	6.9	4.4	4.2	4.1		Nevada	11.4	9.6	6.4	7.7
Arkansas	6.9	6.9	6.3	6.0		New Hampshire	4.7	5.8	3.9	3.3
California	4.3	n.a.	n.a.	n.a.		New Jersey	3.0	3.1	3.0	2.9
Colorado	5.5	n.a.	4.4	4.4		New Mexico	4.9	5.3	4.6	4.6
Connecticut	3.2	2.0	2.9	2.7		New York	3.2	3.4	3.0	2.8
Delaware	4.4	4.2	3.7	3.9		North Carolina	5.1	4.8	4.4	3.8
District of Columbia	4.5	3.0	1.7	2.0		North Dakota	3.6	3.2	2.8	2.4
Florida	6.3	5.3	4.8	4.6		Ohio	4.7	4.4	3.7	3.6
Georgia	5.5	3.9	n.a.	n.a.		Oklahoma	7.7	3.7	n.a.	5.6
Hawaii	4.6	3.9	n.a.	n.a.		Oregon	5.5	5.0	4.1	4.3
Idaho	6.5	5.4	5.1	4.9		Pennsylvania	3.3	3.2	2.5	2.3
Illinois	3.8	3.2	2.6	2.5		Rhode Island	3.7	3.1	3.0	2.9
Indiana	n.a.	n.a.	n.a.	n.a.		South Carolina	4.5	3.7	3.2	2.9
Iowa	3.9	3.3	2.8	2.7		South Dakota	3.7	3.6	3.2	3.0
Kansas	5.0	4.0	3.3	3.1		Tennessee	6.5	6.1	5.0	4.6
Kentucky	5.8	5.4	4.9	4.5		Texas	5.5	4.2	3.6	3.2
Louisiana	n.a.	n.a.	n.a.	n.a.		Utah	5.1	4.5	3.9	4.0
Maine	4.3	4.6	3.6	3.5		Vermont	4.5	8.6	3.9	3.3
Maryland	3.4	3.3	3.1	3.1		Virginia	4.4	4.3	4.0	3.9
Massachusetts	2.8	3.0	2.2	2.2		Washington	5.9	4.7	4.1	4.0
Michigan	4.3	4.0	3.5	3.4		West Virginia	5.3	5.2	4.7	5.1
Minnesota	3.5	3.1	2.8	n.a.		Wisconsin	3.6	3.3	3.1	3.0
Mississippi	5.5	5.2	4.5	4.5		Wyoming	6.6	5.9	5.3	5.3

▶ **Figure 11.2 U.S. Divorce Rates by State, 1990–2005**

[a]Based on total population residing in area; population enumerated as of April 1 for 1990: estimated as of July 1 for all other years.
[b]Includes annulments.
[c]U.S. totals for the number of divorces is an estimate that includes states not reporting (California, Colorado, Indiana, and Louisiana).
Source: U.S. Census Bureau, 2008.

macrolevel and microlevel that make some couples more or less likely to divorce. At the macrolevel, societal factors contributing to higher rates of divorce include changes in social institutions, such as religion and family. Some religions have taken a more lenient attitude toward divorce, and the social stigma associated with divorce has lessened. Further, as we have seen in this chapter, the family institution has undergone a major change that has resulted in less economic and emotional dependency among family members—thus reducing a barrier to divorce.

At the microlevel, a number of factors contribute to a couple's "statistical" likelihood of divorcing. Here are some of the primary social characteristics of those most likely to get divorced:

- Marriage at an early age (59 percent of marriages to brides under 18 end in separation or divorce within 15 years) (National Centers for Disease Control, 2001)
- A short acquaintanceship before marriage
- Disapproval of the marriage by relatives and friends
- Limited economic resources and low wages
- A high school education or less (although deferring marriage to attend college may be more of a factor than education per se)
- Parents who are divorced or have unhappy marriages
- The presence of children (depending on their gender and age) at the start of the marriage

The interrelationship of these and other factors is complicated. For example, the effect of age is intertwined with economic resources; persons from families at the low end of the income scale tend to marry earlier than those at more affluent income levels. Thus, the question becomes whether age itself is a factor or whether economic resources are more closely associated with divorce.

The relationship between divorce and factors such as race, class, and religion is another complex issue. Although African Americans are more likely than whites of European ancestry to get a divorce, other factors—such as income level and discrimination in society—must also be taken into account. Latinos/as share some of the problems faced by African Americans, but their divorce rate is only slightly higher than that of whites of European ancestry. Religion may affect the divorce rate of some people, including many Latinos/as who are Roman Catholic. However, despite the Catholic doctrine that discourages divorce, the rate of Catholic divorces is now approximately equal to that of Protestant divorces.

Consequences of Divorce Divorce may have a dramatic economic and emotional impact on family members. An estimated 60 percent of divorcing couples have one or more children. By age sixteen, about one out of every three white children and two out of every three African American children will experience divorce within their families. As a result, most of them will remain with their mothers and live in a single-parent household for a period of time. In recent years, there has been a debate over whether children who live with their same-sex parent after divorce are better off than their peers who live with an opposite-sex parent. However, sociologists have found virtually no evidence to support the belief that children are better off living with a same-sex parent (Powell and Downey, 1997).

Although divorce decrees provide for parental joint custody of approximately 100,000–200,000 children annually, this arrangement may create unique problems for some children, as the personal narratives of David and Nick Sheff at the beginning of this chapter demonstrate. Furthermore, some children experience more than one divorce during their childhood because one or both of their parents may remarry and subsequently divorce again.

But divorce does not have to be always negative. For some people, divorce may be an opportunity to terminate destructive relationships. For others, it may represent a means to achieve personal growth by managing their lives and social relationships and establishing their own social identity. Still others choose to remarry one or more times.

Remarriage

Most people who divorce get remarried. In recent years, more than 40 percent of all marriages have been between previously married brides and/or grooms. Among individuals who divorce before age thirty-five, about half will remarry within three years of their first divorce (Bramlett and Mosher, 2001). Most divorced people remarry others who have been divorced. However, remarriage rates vary by gender and age. At all ages, a greater proportion of men than women remarry, often relatively soon after the divorce. Among women, the older a woman is at the time of divorce, the lower her likelihood of remarrying. Women who have not graduated from high school and who have young children tend to remarry relatively quickly; by contrast, women with a college degree and without children are less likely to remarry (Bramlett and Mosher, 2001).

© Michelle D. Bridwell/PhotoEdit

Remarriage and blended families create new opportunities and challenges for parents and children alike.

As a result of divorce and remarriage, complex family relationships are often created. Some people become part of stepfamilies or *blended families,* **which consist of a husband and wife, children from previous marriages, and children (if any) from the new marriage.** At least initially, levels of family stress may be fairly high because of rivalry among the children and hostilities directed toward stepparents or babies born into the family. In spite of these problems, however, many blended families succeed. The family that results from divorce and remarriage is typically a complex, binuclear family in which children may have a biological parent and a stepparent, biological siblings and stepsiblings, and an array of other relatives, including aunts, uncles, and cousins.

According to the sociologist Andrew Cherlin (1992), the norms governing divorce and remarriage are ambiguous. Because there are no clear-cut guidelines, people must make decisions about family life (such as whom to invite for a birthday celebration or wedding) based on their beliefs and feelings about the people involved.

Family Issues in the Future

As we have seen, families and intimate relationships changed dramatically during the twentieth century. Some people believe that the family as we know it is doomed. Others believe that a return to traditional family values will save this important social institution and create greater stability in society. However, the sociologist Lillian Rubin (1986: 89) suggests that clinging to a traditional image of families is hypocritical in light of our society's failure to support families: "We are after all, the only advanced industrial nation that has no public policy of support for the family whether with family allowances or decent publicly-sponsored childcare facilities." Some laws even have the effect of hurting children whose families do not fit the traditional model. For example, cutting back on government programs that provide food and medical care for pregnant women and infants will result in seriously ill children rather than model families (Aulette, 1994).

According to the psychologist Bernice Lott (1994: 155), people's perceptions about what constitutes a family will continue to change in the future:

> Persons on whom one can depend for emotional support, who are available in crises and emergencies, or who provide continuing affections, concern, and companionship can be said to make up a family. Members of such a group may live together in the same household or in separate households, alone or with others. They may be related by birth, marriage, or a chosen commitment to one another that has not been legally formalized.

Some of these changes are already becoming evident. For example, many men are attempting to take an active role in raising their children and helping with household chores. Many couples terminate abusive relationships and marriages.

---

blended family a family consisting of a husband and wife, children from previous marriages, and children (if any) from the new marriage.

Chapter Review

• What is the family?

Today, families may be defined as relationships in which people live together with commitment, form an economic unit and care for any young, and consider their identity to be significantly attached to the group.

• How does the family of orientation differ from the family of procreation?

The family of orientation is the family into which a person is born; the family of procreation is the family a person forms by having or adopting children.

• What pattern of marriage is legally sanctioned in the United States?

In the United States, monogamy is the only form of marriage sanctioned by law. Monogamy is a marriage between two partners, usually a woman and a man.

• What are the primary sociological perspectives on families?

Functionalists emphasize the importance of the family in maintaining the stability of society and the well-being of individuals. Conflict and feminist perspectives view the family as a source of social inequality and an arena for conflict. Symbolic interactionists explain family relationships in terms of the subjective meanings and everyday interpretations that people give to their lives. Postmodern analysts view families as being permeable, capable of being diffused or invaded so that the original purpose is modified.

• What are the major stages in the family life course?

The major stages are infancy and childhood, adolescence and young adulthood, middle adulthood, and late adulthood.

• How are families in the United States changing?

Families are changing dramatically in the United States. Cohabitation has increased significantly in the past three decades. With the increase in dual-earner marriages, women have increasingly been burdened by the second shift—the domestic work that employed women perform at home after they complete their workday on the job. Many single-parent families also exist today.

• What is divorce, and what are some of its causes?

Divorce is the legal process of dissolving a marriage. At the macrolevel, changes in social institutions may contribute to an increase in divorce rates; at the microlevel, factors contributing to divorce include age at marriage, length of acquaintanceship, economic resources, education level, and parental marital happiness.

www.cengage.com/login

Register for a Student eResource account to maximize your study time online using CengageNOW. First take the system's diagnostic pre-test, and then follow the personalized study plan that is created for you to help you review this chapter. The study plan will

- help you identify areas on which you should concentrate;
- provide interactive exercises to help you master the chapter concepts; and
- provide a post-test to confirm you are ready to move on to the next chapter.

Key Terms

bilateral descent 361

blended families 384

cohabitation 367

domestic partnerships 367

dual-earner marriages 368

egalitarian family 361

endogamy 362

exogamy 362

extended family 358

families 356

family of orientation 358

family of procreation 358

homogamy 368

kinship 357

marriage 359

matriarchal family 361

matrilineal descent 361

matrilocal residence 362

monogamy 360

neolocal residence 362

nuclear family 359

patriarchal family 361

patrilineal descent 360

patrilocal residence 362

polyandry 360

polygamy 360

polygyny 360

second shift 369

sociology of family 362

Questions for Critical Thinking

1. In your opinion, what constitutes an ideal family? How might functionalist, conflict, feminist, and symbolic interactionist perspectives describe this family?

2. Suppose that you wanted to find out about women's and men's perceptions about love and marriage. What specific issues might you examine? What would be the best way to conduct your research?

3. You have been appointed to a presidential commission on child-care problems in the United States. How to provide high-quality child care at affordable prices is a key issue for the first meeting. What kinds of suggestions would you take to the meeting? How do you think your suggestions should be funded? How does the future look for children in high-, middle-, and low-income families in the United States?

The Kendall Companion Website

www.cengage.com/sociology/kendall

Supplement your review of this chapter by going to the text's companion website, where you can take tutorial quizzes, use flash cards to master key terms, follow live links to useful websites, and explore the other study and research resources you'll find there, such as a comprehensive interactive sociology timeline, GSS Data, and Census 2000 information, much of it presented visually in maps.

12 Education and Religion

We are teaching our children a theory [evolution] that most of us don't believe in. I don't think God creates everything on a day-to-day basis, like the color of the sky. But I do believe he created Adam and Eve—instantly.

—Steve Farrell, a resident of Dover, Pennsylvania, explaining why he approves of the Dover school board's decision to require eighth-grade biology teachers to teach "intelligent design"—an assertion that the universe is so complex that an intelligent, supernatural power must have created it—as an alterative to the theory of evolution (qtd. in Powell, 2004: A1)

I *definitely* would prefer to believe that God created me than that I'm 50th cousin to a silverback ape. What's wrong with wanting our children to hear about all the holes in the theory of evolution?

—Lark Myers, another resident of Dover, who also wants her child to learn about intelligent design at school (qtd. in Powell, 2004: A1)

I believe it is wrong to introduce a non-scientific "explanation" of the origins of life into the science curriculum. This policy was not endorsed by the Dover High School science department. I think this

Function Favored by Natural Selection

© AP Images/Jay Laprete

For many years, people have argued about what should (or should not) be taught in U.S. public schools, including the teaching of creationism or intelligent design as contrasted with evolution. Shown here is Dr. Kenneth Miller, a biology professor, during a discussion of the pros and cons of incorporating the teaching of intelligent design into the Ohio state science curriculum.

policy was approved for religious reasons, not to improve science education for my child.

—Tammy Kitzmiller, one of the eleven parents who filed a lawsuit (*Kitzmiller v. Dover*) against the school board, challenging its controversial decision (qtd. in ACLU, 2005)

People have an impatience about science. They think it's this practical process that explains how everything works, but that's the least interesting part. We understand a lot of the mechanisms of evolution but it's what we *don't* understand that makes it exciting. . . . It's very clear that intelligent design has become a stalking-horse. If these school boards had their druthers, they would teach Noah's flood and the 6,000-year-old design of Earth. My fear is that they are making real headway in the popular imagination.

—Kenneth R. Miller, a university biologist and the author of the biology textbook used in Dover before the school board's decision, explaining why he believes that the teaching of intelligent design in public classrooms is a very bad idea (qtd. in Powell, 2004: A1)

What is all the controversy about? How did a small school district draw so much attention to itself and end up with a district judge ruling that the school board's decision to introduce intelligent design as an alternative to evolution violated the First Amendment to the United States Constitution?

The argument over intelligent design is the latest debate in a lengthy battle over the teaching of creationism versus evolutionism in public schools, and it is only one of many arguments that will continue to take place regarding the appropriate relationship between public education and religion in the United

Chapter Focus Question

Why are education and religion powerful and influential forces in contemporary society?

389

States. More than seventy years ago, for example, evolutionism versus creationism was hotly debated in the famous "Scopes monkey trial," so named because of Charles Darwin's assertion that human beings had evolved from lower primates. In this case, John Thomas Scopes, a substitute high school biology teacher in Tennessee, was found guilty of teaching evolution, which denied the "divine creation of man as taught in the Bible." Although an appeals court later overturned Scopes's conviction and $100 fine (on the grounds that the fine was excessive), teaching evolution in Tennessee's public schools remained illegal until 1967 (Chalfant, Beckley, and Palmer, 1994). By contrast, recent U.S. Supreme Court rulings have looked unfavorably on the teaching of creationism in public schools, based on a provision in the Constitution that requires a "wall of separation" between church (religion) and state (government). Initially, this wall of separation was erected to protect religion from the state, not the state from religion.

Who is to decide what should be taught in U.S. public schools? What is the purpose of education? Of religion? In this chapter, we examine education and religion, two social institutions that have certain commonalities both as institutions and as objects of sociological inquiry. Before reading on, test your knowledge about the impact that religion has had on U.S. education by taking the quiz in Box 12.1.

An Overview of Education and Religion

Education and religion are powerful and influential forces in contemporary societies. Both institutions impart values, beliefs, and knowledge considered essential to the social reproduction of individual personalities and entire cultures (Bourdieu and Passeron, 1990). Education and religion both grapple with issues of societal stability and social change, reflecting society even as they attempt to shape it. Education and religion also share certain commonalities as objects of sociological study; for example, both are socializing institutions. Whereas early socialization is primarily informal and takes place within our families and friendship networks, as we grow older, socialization passes to the more formalized organizations created for the specific purposes of education and religion.

Areas of sociological inquiry that specifically focus on those institutions are (1) the *sociology of education,* which primarily examines formal education or schooling in industrial societies, and (2) the *sociology of religion,* which focuses on religious groups and organizations, on the behavior of individuals within those groups, and on ways in which religion is intertwined with other social institutions (Roberts, 2004). Let's start our examination by looking at sociological perspectives on education.

Sociological Perspectives on Education

Education is the social institution responsible for the systematic transmission of knowledge, skills, and cultural values within a formally organized structure. In all societies, people must acquire certain knowledge and skills in order to survive. In less-developed societies, these skills might include hunting, gathering, fishing, farming, and self-preservation. In contemporary, developed societies, knowledge and skills are often related to the requirements of a highly competitive job market.

Sociologists have divergent perspectives on the purpose of education in contemporary society. Functionalists suggest that education contributes to the maintenance of society and provides people with an opportunity for self-enhancement and upward social mobility. Conflict theorists argue that education perpetuates social inequality and benefits the dominant class at the expense of all others. Symbolic interactionists focus on classroom dynamics and the effect of self-concept on grades and aspirations.

Functionalist Perspectives on Education

Functionalists view education as one of the most important components of society. According to Durkheim, education is the "influence exercised by adult generations on those that are not yet ready

Box 12.1 **Sociology and Everyday Life**

How Much Do You Know About the Impact of Religion on U.S. Education?

True	False	
T	F	1. The U.S. Constitution originally specified that religion should be taught in the public schools.
T	F	2. Virtually all sociologists have advocated the separation of moral teaching from academic subject matter.
T	F	3. The federal government has limited control over how funds are spent by school districts because most of the money comes from the state and local levels.
T	F	4. Parochial schools have decreased in enrollment as interest in religion has waned in the United States.
T	F	5. The number of children from religious backgrounds other than Christianity and Judaism has grown steadily in public schools over the past three decades.
T	F	6. Debates over textbook content focus only on elementary education because of the vulnerability of young children.
T	F	7. More parents are instructing their own children through home schooling because of their concerns about what public schools are (or are not) teaching their children.
T	F	8. The U.S. Congress has the ultimate authority over whether religious education can be included in public school curricula.

Answers on page 392.

for social life" (Durkheim, 1956: 28). Durkheim asserted that moral values are the foundation of a cohesive social order and that schools have the responsibility of teaching a commitment to the common morality.

From this perspective, students must be taught to put the group's needs ahead of their individual desires and aspirations. Contemporary functionalists suggest that education is responsible for teaching U.S. values. According to sociologist Amitai Etzioni (1994: 258–259),

> We ought to teach those values Americans share, for example, that the dignity of all persons ought to be respected, that tolerance is a virtue and discrimination abhorrent, that peaceful resolution of conflicts is superior to violence, that . . . truth telling is morally superior to lying, that democratic government is morally superior to totalitarianism and authoritarianism, that one ought to give a day's work for a day's pay, that saving for one's own and one's country's future is better than

squandering one's income and relying on others to attend to one's future needs.

Etzioni suggests that "shared" values should be transmitted by schools from kindergarten through college. However, not all analysts agree on what those shared values should be or what functions education should serve in contemporary societies. In analyzing the values and functions of education, sociologists using a functionalist framework distinguish between manifest and latent functions. Manifest functions and latent functions of education are compared in Figure ▶ 12.1.

Manifest Functions of Education Some functions of education are *manifest functions*, previously defined as the open, stated, and intended goals or

---

education the social institution responsible for the systematic transmission of knowledge, skills, and cultural values within a formally organized structure.

Box 12.1   **Sociology and Everyday Life**

Answers to the Sociology Quiz on Religion and Education

1. False. Due to the diversity of religious backgrounds of the early settlers, no mention of religion was made in the original Constitution. Even the sole provision that currently exists (the establishment clause of the First Amendment) does not speak directly of the issue of religious learning in public education.

2. False. Obviously, contemporary sociologists hold strong beliefs and opinions on many subjects; however, most of them do not think that it is their role to advocate specific stances on a topic. Early sociologists were less inclined to believe that they had to be "value-free." For example, Durkheim strongly advocated that education should have a moral component and that schools had a responsibility to perpetuate society by teaching a commitment to the common morality.

3. True. Most public school revenue comes from local funding through property taxes and state funding from a variety of sources, including sales taxes, personal income taxes, and, in some states, oil revenues or proceeds from lotteries.

4. False. Just the opposite has happened. As parents have felt that children were not receiving the type of education they desired in public schools, parochial schools have flourished. Christian schools have grown to over five thousand in number; Jewish parochial schools have also grown rapidly over the past decade.

5. True. Although about 86 percent of those age 18 and over in the 48 contiguous states of the United States describe their religion as some Christian denomination, there has still been a significant increase in those who either adhere to no religion (7.5 percent) or who are Jewish, Muslim/Islamic, Unitarian-Universalist, Buddhist, or Hindu.

6. False. Attempts to remove textbooks occur at all levels of schooling. A recent case involved the removal of Chaucer's "The Miller's Tale" and Aristophanes' *Lysistrata* from a high school curriculum.

7. True. Some parents choose home schooling for religious reasons; others embrace it for secular reasons, including fear for their children's safety and concerns about the quality of public schools.

8. False. Ultimately, issues relating to the separation of church and state, including religious instruction in public schools, are constitutional issues that are decided by the U.S. Supreme Court in the absence of a constitutional amendment.

Sources: Based on Ballantine, 2001; Gibbs, 1994; Greenberg and Page, 2002; Johnson, 1994; Kosmin and Lachman, 1993; and Roof, 1993.

---

consequences of activities within an organization or institution. Examples of manifest functions in education include teaching specific subjects, such as science, mathematics, reading, history, and English. Education serves five major manifest functions in society:

1. *Socialization.* From kindergarten through college, schools teach students the student role, specific academic subjects, and political socialization. In primary and secondary schools, students

are taught specific subject matter appropriate to their age, skill level, and previous educational experience. At the college level, students focus on more detailed knowledge of subjects that they have previously studied while also being exposed to new areas of study and research.

2. *Transmission of culture.* Schools transmit cultural norms and values to each new generation and play an active part in the process of assimilation, whereby recent immigrants learn dominant cul-

Manifest functions—open, stated, and intended goals or consequences of activities within an organization or institution. In education, these are:

- socialization
- transmission of culture
- social control
- social placement
- change and innovation

Latent functions—hidden, unstated, and sometimes unintended consequences of activities within an organization. In education, these include:

- matchmaking and production of social networks
- restricting some activities
- creating a generation gap

▶ Figure 12.1
Manifest and Latent Functions of Education

tural values, attitudes, and behavior so that they can be productive members of society.

3. *Social control.* Schools are responsible for teaching values such as discipline, respect, obedience, punctuality, and perseverance. Schools teach conformity by encouraging young people to be good students, conscientious future workers, and law-abiding citizens (see "Sociology Works!").

4. *Social placement.* Schools are responsible for identifying the most qualified people to fill available positions in society. As a result, students are channeled into programs based on individual ability and academic achievement. Graduates receive the appropriate credentials for entry into the paid labor force.

5. *Change and innovation.* Schools are a source of change and innovation. As student populations change over time, new programs are introduced to meet societal needs; for example, sex education, drug education, and multicultural studies have been implemented in some schools to help students learn about pressing social issues. Innovation in the form of new knowledge is required of colleges and universities. Faculty members are encouraged, and sometimes required, to engage

in research and to share the results with students, colleagues, and others.

Latent Functions of Education In addition to manifest functions, all social institutions, including education, have some *latent functions,* which, as you will recall, are the hidden, unstated, and sometimes unintended consequences of activities within an organization or institution. Education serves at least three latent functions:

1. *Restricting some activities.* Early in the twentieth century, all states passed *mandatory education laws* that require children to attend school until they reach a specified age or until they complete a minimum level of formal education. Out of these laws grew one latent function of education, which is to keep students off the streets and out of the full-time job market for a number of years, thus helping keep unemployment within reasonable bounds (Braverman, 1974).

2. *Matchmaking and production of social networks.* Because schools bring together people of similar ages, social class, and race/ethnicity, young people often meet future marriage partners and

Sociology *Works!*

Stopping Bullying: Character Building and Social Norms

Two Viewpoints:

Bullying, particularly in adolescence, is epidemic, not just in the USA but around the world.

— Marvin Berkowitz, professor of character education at the University of Missouri–St. Louis (qtd. in mindoh.com, 2007)

What we've seen consistently is that risk behaviors [and] problems are overestimated, which [means] much of the bullying or violence or substance abuse can continue because the people engaged in that think everybody else is doing it.

— H. Wesley Perkins, a sociology professor and director of the Alcohol Education Project at Hobart and William Smith Colleges in Geneva, New York (qtd. in Teicher, 2006)

What is bullying? By definition, bullying is aggressive behavior that is intentional and that involves an imbalance of power or strength. Bullying is often repeated, and the child who is routinely the object of the bullying usually has a difficult time defending himself or herself (U.S. Department of Health and Human Services, 2008).

How great a problem is bullying in today's schools? What have sociologists learned that might be useful in combating this problem? In regard to the frequency of bullying at school, some studies have found that as many as eight out of ten U.S. students in upper elementary, middle, and high schools who have access to an anonymous online messaging service that provides them with the abil-

ity to communicate the problem to school counselors and administrators report that they have been the victims of bullying (Gary, 2007). Other studies have shown that between 15 to 25 percent of U.S. students are bullied with some frequency and that 15 to 20 percent admit that they bully others with some frequency during the school year (U.S. Department of Health and Human Services, 2008). Although the nature and extent of this problem are unknown, experts are in agreement on one thing: Bullying can and must be addressed and reduced in schools. What methods work best for eliminating negative behaviors and maintaining social control in schools? One method of reducing bullying is rooted in character education, which focuses on the reinforcement of acceptable character traits, such as respect and responsibility. From character education courses and online studies, students learn problem-solving techniques, conflict-resolution approaches, and communication skills to help them interact with one another in a more positive manner.

However, an alternative approach has been suggested by such sociologists as H. Wesley Perkins, who emphasizes that the best practice in bullying prevention and intervention is to focus on the *social environment* of the school. According to Perkins, to reduce bullying it is important to change the climate of the school and social norms regarding bullying. As you will recall from Chapter 2, norms are established rules of behavior or standards of conduct; therefore, a school's social norms must establish standards of conduct that make it "uncool" for one person to bully another but "cool" for individuals to

develop social networks that may last for many years.

3. *Creating a generation gap.* Students may learn information in school that contradicts beliefs held by their parents or their religion. When education conflicts with parental attitudes and beliefs, a generation gap is created if students embrace the newly acquired perspective.

As we have seen, education fulfills both manifest and latent functions in society; however, some aspects of this important social institution may be impaired

and generate problems not only for students but also for others in the larger society (such as lagging educational standards that result in employees coming into the work force with fewer marketable skills).

Dysfunctions of Education Functionalists acknowledge that education has certain dysfunctions. Some analysts argue that U.S. education is not promoting the high-level skills in reading, writing, science, and mathematics that are needed in the workplace and the global economy. For example, mathematics and science education in the United

Sociologist H. Wesley Perkins sees the problem of bullying as an issue involving norms: If bullying becomes uncool, young people will be less likely to take part in it and more likely to help someone who is a victim of this very harmful activity.

bullying which assumes that a person's behavior may be strongly influenced by the *incorrect* perceptions the individual holds about how other members of his or her social group think and act. From this approach, the first step to reducing the problem of bullying is to correct students' misperceptions about how prevalent this type of behavior really is: If students think that bullying is very common in their school and that "everybody is doing it," they may believe that the social norms support such behavior. If this misperception can be corrected, bullying will decrease as students come to see that the "nonbully status" is the social norm (Teicher, 2006).

Emphasizing the social environment of education and its norms is an important sociological contribution to our understanding of both positive and negative actions that take place in schools. It also makes us aware that social institutions fulfill a variety of important functions in society and that one of those functions for education is to maintain social control and to encourage civility among participants, whether or not they subscribe to moral values that teach them to "Love thy neighbor. . . . " Although it is also important to focus on character education for students and to teach them to respect others and have a sense of responsibility for their own actions, all of us have a responsibility for maintaining a social environment in which all individuals are treated with kindness and respect. Bullying isn't just a matter of "kids being kids": It is a serious sociological problem and must be dealt with as such.

help out those who are bullied. Based on his previous research on alcohol use among college students, Perkins developed a social norms approach to preventing

Reflect & Analyze

What do you think about the problem of bullying? How can sociology help us to reduce problems such as this?

States does not compare favorably with that found in many other industrialized countries. In the latest available (2003) Trends in International Mathematics and Science Study (TIMSS), which compares the mathematics and science performance of U.S. students with that of their peers in other nations, U.S. eighth-grade students scored lower than did students in fourteen other nations, including Singapore, the Republic of Korea, and Japan (National Center for Education Statistics, 2005).

Are U.S. schools dysfunctional as a result of scores such as these? Analysts do not agree on what

these score differences mean. For example, Japanese students may outperform their U.S. counterparts because their schools are more structured and teachers focus on drill and practice (Celis, 1994). Educators David C. Berliner and Bruce J. Biddle (1995) believe that data on cross-cultural differences in educational attainment actually involve comparisons of "apples" and "oranges": They found that studies such as this often compare the achievement of eighth-grade Japanese students who have already taken algebra with the achievement of U.S. students who typically take such courses a year or two later.

© Enigma/Alamy

What values are these schoolchildren being taught? Is there a consensus about what today's schools should teach? Why or why not?

Among U.S. students who had already completed an algebra course, most did at least as well as their Japanese counterparts on the mathematics exam (Berliner and Biddle, 1995).

Clearly, test scores are subject to a variety of interpretations; however, for most functionalist analysts, lagging test scores are a sign that dysfunctions exist in the nation's educational system. According to this approach, improvements will occur only when more stringent academic requirements are implemented for students. Dysfunctions may also be reduced by more thorough teacher training and consistent testing of instructors.

Conflict Perspectives on Education

In contrast with the functionalist perspective, conflict theorists argue that schools often perpetuate class, racial–ethnic, and gender inequalities as some groups seek to maintain their privileged position at the expense of others (Ballantine, 2001).

Cultural Capital and Class Reproduction Although many factors—including intelligence, family income, motivation, and previous achievement—are important in determining how much education a person will attain, conflict theorists argue that

access to quality education is closely related to social class. From this approach, education is a vehicle for reproducing existing class relationships. According to the French sociologist Pierre Bourdieu, the school legitimates and reinforces the social elites by engaging in specific practices that uphold the patterns of behavior and the attitudes of the dominant class. Bourdieu asserts that students from diverse class backgrounds come to school with differing different amounts of *cultural capital*—social assets that include values, beliefs, attitudes, and competencies in language and culture (Bourdieu and Passeron, 1990). Cultural capital involves "proper" attitudes toward education, socially approved dress and manners, and knowledge about books, art, music, and other forms of high and popular culture.

Middle- and upper-income parents endow their children with more cultural capital than do working-class and poverty-level parents. Because cultural capital is essential for acquiring an education, children with less cultural capital have fewer opportunities to succeed in school. For example, standardized tests that are used to group students by ability and to assign them to classes often measure students' cultural capital rather than their "natural" intelligence or aptitude. Thus, a circular effect occurs: Students with dominant cultural values are more highly rewarded by the educational system. In turn, the educational system teaches and reinforces those values that sustain the elite's position in society.

Tracking and Social Inequality Closely linked to the issue of cultural capital is how tracking in schools is related to social inequality. *Tracking* refers to the practice of assigning students to specific curriculum groups and courses on the basis of their test scores, previous grades, or other criteria. Conflict theorists believe that tracking seriously affects many students' educational performance and their overall academic accomplishments. In elementary schools, tracking is

Children who are able to visit museums, libraries, and musical events may gain cultural capital that other children do not possess. What is cultural capital? Why is it important in the process of class reproduction?

often referred to *ability grouping* and is based on the assumption that it is easier to teach a group of students who have similar abilities. However, class-based factors also affect which children are most likely to be placed in "high," "middle," or "low" groups, often referred to by such innocuous terms as "Blue Birds," "Red Birds," and "Yellow Birds." This practice is described by Ruben Navarrette, Jr. (1997: 274–275), who tells us about his own experience with tracking:

> One fateful day, in the second grade, my teacher decided to teach her class more efficiently by dividing it into six groups of five students each. Each group was assigned a geometric symbol to differentiate it from the others. There were the Circles. There were the Squares. There were the Triangles and Rectangles.

I remember being a Hexagon. . . . I remember something else, an odd coincidence. The Hexagons were the smartest kids in the class. These distinctions are not lost on a child of seven. . . . Even in the second grade, my classmates and I knew who was smarter than whom. And on the day on which we were assigned our respective shapes, we knew that our teacher knew, too. As Hexagons, we would wait for her to call on us, then answer by hurrying to her with books and pencils in hand. We sat around a table in our "reading group," chattering excitedly to one another and basking in the intoxication of positive learning. We did not notice, did not care to notice, over our shoulders, the frustrated looks on the faces of Circles and Squares and Triangles who sat quietly at their desks, doodling on scratch paper or mumbling to one another. We knew also that, along with our geometric shapes, our books were different and that each group had different amounts of work to do. . . . The Circles had the easiest books and were assigned to read only a few pages at a time. . . . Not surprisingly, the Hexagons had the most difficult books of all, those with the biggest words and the fewest pictures, and we were expected to read the most pages.

The result of all of this education by separation was exactly what the teacher had imagined that it would be: Students could, and did, learn at their own pace without being encumbered by one another. Some learned faster than others. Some, I realized only [later], did not learn at all.

As Navarette suggests, tracking does make it possible for students to work together based on their perceived abilities and at their own pace; however, it also extracts a serious toll from students who are labeled as "underachievers" or "slow learners." Race, class, language, gender, and many other social categories may determine the placement of children in

---

cultural capital Pierre Bourdieu's term for people's social assets, including values, beliefs, attitudes, and competencies in language and culture.

tracking the assignment of students to specific curriculum groups and courses on the basis of their test scores, previous grades, or both.

elementary tracking systems as much or more than their actual academic abilities and interests.

The practice of tracking continues in middle school/junior high and high school. Although schools in some communities bring together students from diverse economic and racial/ethnic backgrounds, the students do not necessarily take the same courses or move on the same academic career paths (Gilbert, 2003). Numerous studies have found that ability grouping and tracking affect students' academic achievements and career choices (Oakes, 1985; Welner and Oakes, 2000). Moreover, some social scientists believe that tracking is one of the most obvious mechanisms through which students of color and those from low-income families receive a diluted academic program, making it much more likely that they will fall even further behind their white, middle-class counterparts (see Miller, 1995). A recent study of Latinas concluded, for example, that school practices such as tracking impose low expectations that create self-fulfilling prophecies for many of these young women (Ginorio and Huston, 2000). Instead of enhancing school performance, tracking systems may result in students dropping out of school or ending up in "dead-end" situations because they have not taken the courses required to go to college.

As Ruben Navarette, Jr., so powerfully describes, school is extremely tedious for underachieving students, who may find themselves "tracked" in such a way as to deny them upward mobility in the future.

The Hidden Curriculum According to conflict theorists, the *hidden curriculum* is the transmission of cultural values and attitudes, such as conformity and obedience to authority, through implied demands found in the rules, routines, and regulations of schools (Snyder, 1971).

Social Class and the Hidden Curriculum Although students from all social classes are subjected to the hidden curriculum, working-class and poverty-level students may be affected the most adversely (Polakow, 1993; Ballantine, 2001; Oakes and Lipton, 2003). When teachers from middle- and upper-middle-class backgrounds instruct students from working and lower-income families, the teachers often have a more structured classroom and a more controlling environment for students. These teachers may also have lower expectations for students' academic achievements. For example, one study of five elementary schools in different communities found significant differences in how knowledge was transmitted to students even though the general curriculum of the school was organized similarly (Anyon, 1980, 1997). Schools for working-class students emphasize procedures and rote memorization without much decision making, choice, or explanation of why something is done a particular way. Schools for middle-class students stress the processes (such as figuring and decision making) involved in getting the right answer. Schools for affluent students focus on creative activities in which students express their own ideas and apply them to the subject under consideration. Schools for students from elite families work to develop students' analytical powers and critical-thinking skills, applying abstract principles to problem solving.

Through the hidden curriculum, schools make working-class and poverty-level students aware that they will be expected to take orders from others, arrive at work punctually, follow bureaucratic rules, and experience high levels of boredom without complaining (Ballantine, 2001). Over time, these students may be disqualified from higher education and barred from obtaining the credentials necessary for well-paid occupations and professions (Bowles and Gintis, 1976). Educational credentials are extremely important in societies that emphasize *credentialism*—a process of social selection in which class advantage and social status are linked to the possession of academic qualifications (Collins, 1979; Marshall, 1998). Credentialism is closely related to *meritocracy*—previously defined as a social system in which status is assumed to be acquired through individual ability and effort (Young, 1994/1958). Persons who acquire the appropriate credentials for a job are assumed to have gained the position through what they know, not who they are or whom they know. According to conflict theorists, the hidden curriculum determines in advance that the most valued credentials will primarily stay in the hands of the elites, so the United States is not actually as meritocratic as some might claim.

Gender Bias and the Hidden Curriculum According to conflict theorists, gender bias is embedded in both the formal and the hidden curricula of schools. Although most girls and young women in the United States have a greater opportunity for education than those living in developing nations, their educational opportunities are not equal to those of boys and young men in their social class (see AAUW, 1995; Orenstein, 1995). For many years, reading materials, classroom activities, and treatment by teachers and peers contributed to a feeling among many girls and young women that they were less important than male students. Over time, this kind of differential treatment undermined females' self-esteem and discouraged them from taking certain courses, such as math and science, which were usually dominated by male teachers and students (Raffalli, 1994).

In recent years, some improvements have taken place in girls' education, as more females have enrolled in advanced placement or honors courses and in academic areas, such as math and science, where they had previously lagged (AAUW, 1998). However, girls are still not enrolled in higher-level science (such as physics) and computer sciences courses in the same numbers as boys. Girls and young women continue to make up only a small percentage of students in computer science and computer design classes, and the gender gap grows even wider from grade eight to grade eleven. Whereas male students are more likely to enroll in courses where they learn how to develop computer programs, female students are more likely to enroll in clerical and data-entry classes (AAUW, 1998). Some researchers find that the hidden curriculum works against young women in that some educators do not provide females with as much information about economic trends and the relationship among curriculum, course-taking choices, and career options as they provide to male students from middle- and upper-income families. Latinas are particularly affected by the hidden curriculum as it relates to gender, race/ethnicity, and language (AAUW, 1998).

Symbolic Interactionist Perspectives on Education

Unlike functionalist analysts, who focus on the functions and dysfunctions of education, and conflict theorists, who focus on the relationship between education and inequality, symbolic interactionists focus on classroom communication patterns and educational practices, such as labeling, that affect students' self-concept and aspirations.

Labeling and the Self-Fulfilling Prophecy Chapter 6 explains that *labeling* is the process whereby a person is identified by others as possessing a specific characteristic or exhibiting a certain pattern of behavior (such as being deviant). According to symbolic interactionists, the process of labeling is directly related to the power and status of those persons who do the labeling and those who are being labeled. In schools, teachers and administrators are empowered to label children in various ways, including grades, written comments on deportment (classroom behavior), and placement in classes. For example, based on standardized test scores or classroom performance, educators label some children

hidden curriculum the transmission of cultural values and attitudes, such as conformity and obedience to authority, through implied demands found in rules, routines, and regulations of schools.

credentialism a process of social selection in which class advantage and social status are linked to the possession of academic qualifications.

as "special ed" or low achievers, whereas others are labeled as average or "gifted and talented." For some students, labeling amounts to a *self-fulfilling prophecy*—previously defined as an unsubstantiated belief or prediction resulting in behavior that makes the originally false belief come true (Merton, 1968). A classic form of labeling and the self-fulfilling prophecy occurs through the use of IQ (intelligence quotient) tests, which claim to measure a person's inherent intelligence, apart from any family or school influences on the individual. In many school systems, IQ tests are used as one criterion in determining student placement in classes and ability groups.

Using Labeling Theory to Examine the IQ Debate

The relationship between IQ testing and labeling theory has been of special interest to sociologists. In the 1960s, two social scientists conducted an experiment in an elementary school during which they intentionally misinformed teachers about the intelligence test scores of students in their classes (Rosenthal and Jacobson, 1968). Despite the fact that the students were randomly selected for the study and had no measurable differences in intelligence, the researchers informed the teachers that some of the students had extremely high IQ test scores, whereas others had average to below-average scores. As the researchers observed, the teachers began to teach "exceptional" students in a different manner from other students. In turn, the "exceptional" students began to outperform their "average" peers and to excel in their classwork. This study called attention to the labeling effect of IQ scores.

Question 2: Consider the following two statements: all farmers who are also ranchers cannot come near town; and most of the ranchers who are also farmers cannot surf. Which of the following statements MUST be true?

○ Most of the farmers who cannot come near town can surf

○ Only some farmers who ranch can surf near town

○ A surfer who ranches and farms cannot surf near town

○ Some ranchers who farm can come to town to learn to surf

○ Any farmer who cannot surf also ranches

IQ tests containing items such as this are often used to place students in ability groups. Such placement can set the course of a person's entire education.

However, experiments such as this also raise other important issues: What if a teacher (as a result of stereotypes based on the relationship between IQ and race) believes that some students of color are less capable of learning? Will that teacher (often without realizing it) treat such students as if they are incapable of learning? In their controversial book *The Bell Curve: Intelligence and Class Structure in American Life,* Richard J. Herrnstein and Charles Murray (1994) argue that intelligence is genetically inherited and that people cannot be "smarter" than they are born to be, regardless of their environment or education. According to Herrnstein and Murray, certain racial–ethnic groups differ in average IQ and are likely to differ in "intelligence genes" as well. For example, they point out that, on average, people living in Asia score higher on IQ tests than white Americans and that African Americans score 15 points lower on average than white Americans. Based on an all-white sample, the authors also concluded that low intelligence leads to social pathology, such as high rates of crime, dropping out of school, and winding up poor. In contrast, high intelligence typically leads to success, and family background plays only a secondary role.

Many scholars disagree with Herrnstein and Murray's research methods and conclusions. Two major flaws found in their approach were as follows: (1) the authors used biased statistics that underestimate the impact of hard-to-measure factors such as family background, and (2) they used scores from the Armed Forces Qualification Test, an exam that depends on the amount of schooling that people have completed. Thus, what the authors claim is immutable intelligence is actually acquired skills (Weinstein, 1997). Despite this refutation, the idea of inherited mental inferiority tends to take on a life of its own when people want to believe that such differences exist. According to researchers, many African American and Mexican American children are placed in special education classes on the basis of IQ scores when the students were not fluent in English and thus could not understand the directions given for the test. Moreover, when children are labeled as "special ed" students or as being *learning disabled,* these terms are social constructions that may lead to stigmatization and become a self-fulfilling prophecy (Carrier, 1986; Coles, 1987).

Labeling students based on IQ scores has been an issue for many decades. Immigrants from southern and eastern Europe—particularly from Italy, Poland, and Russia—who arrived in this country at the beginning of the twentieth century had lower IQ scores on average than did northern European immigrants who had arrived earlier from nations such as Great Britain. For many of the white ethnic students, IQ testing became a self-fulfilling prophecy: Teachers did not expect them to do as well as children from a northern European (WASP) family background and thus did not encourage them or give them an opportunity to overcome language barriers or other educational obstacles. Although many students persisted and achieved an education, the possibility that differences in IQ scores could be attributed to linguistic, cultural, and educational biases in the tests was largely ignored (Feagin and Feagin, 2003). Debates over the possible intellectual inferiority of white ethnic groups are unthinkable today, but arguments pertaining to African Americans and IQ continue to surface.

Problems Within Elementary and Secondary Schools

Education in kindergarten through high school is a microcosm of many of the issues and problems facing the United States. Today, there are almost 15,000 U.S. school districts in what is probably the most decentralized system of public education in any high-income, developed nation of the world. One of the biggest problems in public education today results from unequal funding.

Unequal Funding of Public Schools

Why does unequal funding in public education exist? Most educational funds come from state legislative appropriations and local property taxes. State and local governments contribute about 47 percent *each* toward educational expenses, and the federal government pays the remaining 6 percent, largely for special programs for students who are disadvantaged (e.g., the Head Start program) or have disabilities (Bagby, 1997). Although public school spending increased by 28 percent between 1991 and 1996, there was an 8.4 percent increase in enrollment during that same period, and special education programs received an increasing share of school budgets (*New York Times,* 1997).

Per-capita spending on public and secondary education varies widely from state to state. In part, this is because the local property-tax base has been eroding in central cities as major industries have relocated or gone out of business. Many middle- and upper-income families have moved to suburban areas with their own property-tax base so their children can attend relatively new schools equipped with the latest textbooks and state-of-the-art computers—advantages that schools in central cities and poverty-ridden rural areas lack (see Kozol, 1991; Ballantine, 2001).

In recent years, some states have been held accountable by the courts for unequal funding that

© Justin Sullivan/Getty Images

© Bob Daemmrich/The Image Works

"Rich" schools and "poor" schools are readily identifiable by their buildings and equipment. What are the long-term social consequences of unequal funding for schools?

results in "rich" and "poor" school districts. However, proposals to improve educational funding have been relatively limited in scope. Recently, voucher systems—which would allow students and their families to spend a specified sum of government money to purchase education at the school of their choice—have gained many supporters, who believe that this system might be an answer to some of the problems that plague public education today. However, opponents strongly disagree with this approach, believing that it would harm public education instead of helping it.

Dropping Out

Although there has been a decrease in the overall school dropout rate over the past two decades, about 10 percent of people between the ages of fourteen and twenty-four have left school before earning a high school diploma. (Figure ▶ 12.2 shows which U.S. school districts have the highest graduation rates and which have the lowest.) However, ethnic and class differences are significant in dropout rates. For example, Latinos/as (Hispanics) have the highest dropout rate (24.0 percent), followed by African Americans (12.2 percent), non-Hispanic whites (7.9 percent), and Asian Americans (1.0 percent) (Shin, 2005). The dropout rate also varies by region. In New York City, for example, African American, Puerto Rican, and Italian American youths have the highest dropout rates (Pinderhughes, 1997).

Why are Latinos/as more likely to drop out of high school than are other racial and ethnic groups? First, the category of "Hispanic" or "Latino/a" incorporates a wide diversity of young people—including those who trace their origins to Mexico, Puerto Rico, Haiti, and countries in Central and South America—who may leave school for a variety of reasons. Second, some students may drop out of school partly because their teachers have labeled them as "troublemakers." Some students have been repeatedly expelled from school before they actually become "dropouts." Third, some critics disagree with the use of the term "dropout," believing that Latinos/as have been excluded from meaningful academic programs and discouraged from gaining the education necessary to move into the middle-income categories, and thus have been pushed out of the educational system rather than dropping out of it.

Students who drop out of school may be skeptical about the value of school even while they are still attending because they believe that school will not increase their job opportunities. Upon leaving school, many dropouts have high hopes of making some money and enjoying their newfound freedom. However, these feelings often turn to disappointment when they find that few jobs are available and that they do not meet the minimum educational requirements for any "good" jobs that exist (Pinderhughes, 1997).

Racial Segregation and Resegregation

In many areas of the United States, schools remain racially segregated or have become resegregated after earlier attempts at integration failed. In 1954 the U.S. Supreme Court ruled (in *Brown v. The Board of Education of Topeka, Kansas*) that "separate but equal" segregated schools are unconstitutional because they are inherently unequal. However, five decades later, racial segregation remains a fact of life in education.

▶ Figure 12.2 **Top Five and Bottom Five Graduation Rankings Among the Fifty Largest Public School Districts in the United States**

Source: Reproduced with permission from the June 21, 2006, issue of *The Christian Science Monitor* (www.csmonitor.com). © 2006 The Christian Science Monitor. All rights reserved.

Graduation Rates Among 50 Largest U.S. School Districts

BEST		WORST	
Fairfax County, Va.	82.5%	Detroit	21.7%
Wake County, N.C.	82.2	City of Baltimore, Md.	38.5
Baltimore County, Md.	81.9	New York City	38.9
Montgomery County, Md.	81.5	Milwaukee	43.1
Cypress-Fairbanks, Texas	81.3	Cleveland	43.8

Efforts to bring about *desegregation*—the abolition of legally sanctioned racial–ethnic segregation—or *integration*—the implementation of specific action to change the racial–ethnic and/or class composition of the student body—have failed in many districts throughout the country. Some school districts have bused students across town to achieve racial integration. Others have changed school attendance boundaries or introduced magnet schools with specialized programs such as science or the fine arts to change the racial–ethnic composition of schools. But school segregation does not exist in isolation. Racially segregated housing patterns are associated with the high rate of school segregation experienced by African American, Latina/o, and other students of color. Ethnic enclaves of recent immigrants may also result in a concentration of students in a particular school where they have no opportunity to interact with children from other income levels or family backgrounds.

Even in more-integrated schools, resegregation often occurs at the classroom level (Mickelson and Smith, 1995). Because of past racial discrimination and current socioeconomic inequalities, many children of color are placed in lower-level courses and special education classes. At the same time, non-Latina/o white and Asian American students are more likely to be enrolled in high-achievement courses and programs for the gifted and talented.

School Safety and School Violence

Problems such as bullying, harassment, and high dropout rates are a major concern in education because schools should provide a safe and supportive environment where students and teachers alike can feel secure. Today, officials in schools ranging from the elementary years to two-year colleges and four-year universities are focusing on how to reduce or eliminate bullying, harassment, and violence. In many schools, teachers and counselors are instructed in anger management and peer mediation, and they are encouraged to develop classroom instruction that teaches values such as respect and responsibility (National Education Association,

2007). Some schools create partnerships with local law enforcement agencies and social service organizations to link issues of school safety to larger concerns for safety in the community and the nation. For example, the National Education Association's Safe Schools Program works with other national organizations to advocate for safe schools and communities and to create a positive learning environment for all students.

Clearly, some efforts to make schools a safe haven for students and teachers are paying off. Statistics related to school safety continue to show that U.S. schools are among the safest places for young people. According to the latest "Indicators of School Crime and Safety," jointly released by the National Center for Educational Statistics and the U.S. Department of Justice's Bureau of Justice Statistics, young people are more likely to be victims of violent crime at or near their home, on the streets, at commercial establishments, or at parks than they are at school (National Education Association, 2007). However, these statistics do not keep many people from believing that schools are becoming more dangerous with each passing year and that all schools should have high-tech surveillance equipment to help maintain a safe environment. As a result, many students attend classes in an academic environment that is somewhat similar to a prison. In some schools, all students, faculty, and staff are required to wear a photo ID around their necks or present a school-issued identity card upon entering the school grounds or buildings. Individuals entering school facilities are frequently required to pass through a weapons-scanning metal detector and have their backpacks or other personal items inspected or sent through an X-ray machine. Many schools are equipped with magnetic door locks and have a large staff of security guards who carry walkie-talkies and weapons. Faculty and security guards often use criminal justice vernacular such as "scanning," "holding areas," and "corridor sweeps" to describe their daily activities at the schools (Devine, 1996).

Even with all of these safety measures in place, violence and fear of violence continue to be pressing problems in schools throughout the United States. Each horrendous killing on a school campus further intensifies our fright and heightens our concern about how safe our schools really are. This concern extends from kindergarten through college because violent acts have resulted in numerous elementary and high school deaths in communities such as Jonesboro, Arkansas; Springfield, Oregon; Littleton, Colorado; Santee, California; Red Lake, Minnesota; and an Amish schoolhouse in rural Pennsylvania.

College and university campuses are also not immune to violence and multiple deaths as deranged individuals have engaged in acts of personal terrorism at the expense of students, professors, and other victims. In 1966 Charles Whitman, a University of Texas at Austin student, went to the top of the university's twenty-seven-story tower and began firing shots at people walking around the campus below him. By the time that he was shot and killed by an Austin police officer, Whitman had taken the lives of fifteen people and seriously wounded thirty-one others. People who remember when Whitman went on his rampage sadly relived that experience of grief four decades later on April 16, 2007 (almost exactly eight years after the Columbine High school massacre in Littleton, Colorado, left fifteen people dead at the hands of two high school student gunmen), when Cho Seung-Hui, a Virginia Tech student, shot and killed thirty-two victims and himself on the college campus in Blacksburg, Virginia. Less than one year later, five students were killed on the campus of Northern Illinois University by a twenty-seven-year-old graduate of that school.

The aftermath of these tragedies saw a massive outpouring of public sympathy and a call for greater campus security. As had been true following previous shootings, activist groups offered suggestions on how to make schools safer: Gun control advocates called for greater control over the licensing and ownership of firearms and for heightened police security on college campuses, whereas pro-gun advocates argued that students should be allowed to carry firearms on campus for their own protection (Buchholz, 2007). Neither of these approaches seems sufficient to deal with the real issue of school violence today. Given the problematic nature of discussions about safety and violence in elementary and secondary schools, as well as on college and university campuses, we must reassess at each level of schooling what we as a nation hope to accomplish through the education process and how this can best be achieved. Now let's look more closely at

other opportunities and challenges in today's colleges and universities.

Opportunities and Challenges in Colleges and Universities

Who attends college? What sort of college or university do they attend? For students who complete high school, access to colleges and universities is determined not only by prior academic record but also by the ability to pay. One of the most remarkable success stories over the last fifty years has been the development and rapid growth of community colleges in the United States.

Opportunities and Challenges in Community Colleges

One of the fastest growing areas of U.S. higher education today is the community college; however, the history of two-year colleges goes back more than a century. The nation's first community college was Joliet Junior College in Illinois. Founded as an alternative to four-year and private two-year colleges, JJC still exists today. Like other community colleges, Joliet offers pre-baccalaureate programs for students planning to transfer to a four-year university, occupational education leading directly to employment, adult education and literacy programs, work-force and workplace development services, and support services to help students succeed (Joliet Junior College, 2005). Originally, two-year colleges focused on general liberal arts courses, but during the 1930s these institutions began offering job-training programs. Following World War II, the GI Bill of Rights provided the opportunity for more people to attend college, and in 1948 a presidential commission report called for the establishment of a network of public community colleges that would charge little or no tuition, serve as cultural centers, be comprehensive in their program offerings, and serve the area in which they were located (Vaughan, 2000).

Hundreds of community colleges were opened across the nation during the 1960s, and the number

Courtesy of Joliet Junior College

Courtesy of Miami Dade College, photo by Phil Roche

Joliet Junior College (Illinois) is the oldest two-year college in the United States, having opened its doors in 1901. On the right is a scene from graduation day at the nation's largest two-year school, Miami Dade College (Florida). Today, Joliet, Miami Dade, and others like them fulfill many needs in the competitive world of higher education.

of such institutions has steadily increased since that time as community colleges have responded to the needs of their students and local communities. Community colleges offer a variety of courses, some of which are referred to as "transfer courses," in which students earn credits that are fully transferable to a four-year college or university. Other courses are in technical/occupational programs, which provide formal instruction in fields such as nursing, emergency medical technology, plumbing, carpentry, and computer information technology. Community colleges also pride themselves on offering remedial education for students who need to gain additional background or competence in a subject, as well as courses to benefit those international students who need assistance in learning a new language or developing other skills. Finally, community colleges offer continuing education or lifelong learning courses and work-force development activities in which the schools partner with business and industry to provide skilled workers in their communities.

Did you know that community colleges educate about half of the nation's undergraduates? According to the American Association of Community Colleges (2005), there are a total of 1,166 community colleges (including public and private colleges) in the United States, and these institutions enroll almost 12 million students in credit and noncredit courses. Community college enrollment accounts for 46 percent of all U.S. undergraduates.

Who benefits most from community colleges? Community colleges provide significant educational opportunities to students across lines of income, gender, and race/ethnicity. Because community colleges are more affordable, with an average of about one-half the tuition and fees of the typical four-year college, more students are able to take advantage of the educational opportunities provided in their community. According to the American Association of Community Colleges (2005), almost 40 percent of all community college students receive financial aid to help meet the $1,076 average annual tuition. Women make up a slight majority (58 percent) of community college students, and for working women and mothers of young children, these schools provide a unique opportunity to attend classes on a part-time basis as their schedule permits. Men also benefit from flexible scheduling because they can work part time or full time while enrolled in school. About 62 percent of all community college students are enrolled part time, while 36 percent are full-time students (taking 12 or more credit hours each semester).

One of the greatest challenges facing community colleges today is money. Across the nation, state and local governments struggling to balance their budgets have slashed funding for community colleges. In a number of regions, these cuts have been so severe that schools have been seriously limited in their ability to meet the needs of their students. In some cases, colleges have terminated programs, slashed course offerings, reduced the number of faculty, and eliminated essential student services. Limited resources are one of the major problems that community colleges face, and this problem is shared by many four-year colleges and universities as well.

Opportunities and Challenges in Four-Year Colleges and Universities

More than ten million students attend public or private four-year colleges or universities in the United States. Whereas community colleges award certificates and associate degrees, four-year institutions offer a variety of degrees, including the bachelor's degree, master's degree, and the doctorate, the highest degree awarded. Some also award professional degrees in fields such as law or medicine. According to the Association of American Colleges and Universities, providing a liberal education is the goal of many institutions of higher education. Liberal education is a philosophy of education that aims to empower individuals, liberate the mind from ignorance, and cultivate social responsibility. For this reason, four-year schools typically offer a general education curriculum that gives students exposure to multiple disciplines and ways of knowing, along with more in-depth study (known as a "major") in at least one area of concentration. Having a liberal education provides opportunities for students in that it offers them a diversity of ideas and experiences that will help them not only in a career but in their interpersonal relationships and civic engagements. Today, it is increasingly important for people to acquire education beyond the high school level because the demand for college-educated workers has risen faster than the supply. Average earnings of college graduates are much higher than those of persons with only a high school diploma. However, many challenges are faced by four-year institutions, including the cost of higher education, problems with students completing a degree program in a reasonable time frame, racial and ethnic differences

in enrollment, and lack of faculty diversity. We now turn to the key challenges in higher education.

The Soaring Cost of a College Education

What does a college education cost? Studies by the College Board (a nonprofit organization that provides tests and many other educational services for students, schools, and colleges) have found that a college education is quite expensive and that increases in average yearly tuition for four-year colleges are higher than the overall rate of inflation (Bagby, 1997). Although public institutions such as community colleges and state colleges and universities typically have lower tuition and overall costs—because they are funded primarily by tax dollars—than private colleges have, the cost of attending public institutions has increased dramatically over the past decade. The total number of low-income students has dropped since the 1980s as a result of declining scholarship funds and also because many students must work full time or part time to pay for their education.

According to some social analysts, a college education is a bargain—even at about $90 a day for private schools or $35 for public schools—because for their money students receive instruction, room, board, and other amenities such as athletic facilities and job placement services. However, other analysts believe that the high cost of a college education reproduces the existing class system: Students who lack money may be denied access to higher education, and those who are able to attend college tend to receive different types of education based on their ability to pay. For example, a community college student who receives an associate's degree or completes a certificate program may be prepared for a position in the middle of the occupational status range, such as a dental assistant, computer programmer, or auto mechanic (Gilbert, 2003). In contrast, university graduates with four-year degrees are more likely to find initial employment with firms where they stand a chance of being promoted to high-level management and executive positions. Although higher education may be a source of upward mobility for talented young people from poor families, the U.S. system of higher education is sufficiently stratified

© BRIAN SNYDER/Reuters/Landov

Soaring costs of both public and private institutions of higher education are a pressing problem for today's college students and their parents. What factors have contributed to the higher overall costs of obtaining a college degree?

that it may also reproduce the existing class structure (Gilbert, 2003).

Racial and Ethnic Differences in Enrollment

How does college enrollment differ by race and ethnicity? People of color (who are more likely than the average white student to be from lower-income families) are underrepresented in higher education. However, some increases in minority enrollment have occurred over the past three decades. Latina/o enrollment as a percentage of total college enrollment increased from about 7.7 percent to 10.7 percent between 1995 and 2005 (*Chronicle of Higher Education,* 2007: 15). Even so, Latinos/as are underrepresented in higher education: Today, Latinos/as account for 15 percent of the total U.S. population (U.S. Census Bureau, 2008).

Although African American enrollment increased somewhat between 1990 and 2001, it remains steady today at about 12 percent. Gender differences are evident in African American enrollment: Women accounted for 65 percent of all African American college students in 2005 (*Chronicle of Higher Education,* 2007: 15). Native American enrollment rates have remained stagnant at about 0.9 percent from the 1970s to the 2000s; however, two-year community colleges—known as tribal colleges—on reservations have experienced an increase in Native American student enrollment. Founded to overcome racism experienced by Native American students in traditional four-year colleges and to shrink the high dropout rate among Native American college students, there are now 27 colleges chartered and run by the Native American nations (Holmes, 1997). Unlike other community colleges, however, the tribal colleges receive no funding from state and local governments and, as a result, are chronically short of funds to fulfill their academic mission. But for most Native Americans residing on reservations, these schools provide the best hope for attaining a higher education (Holmes, 1997).

The proportionately low number of people of color enrolled in colleges and universities is reflected in the educational achievement of people age 25 and over, as shown in the "Census Profiles" feature. If we focus on persons who receive doctorate degrees, the underrepresentation of persons of color is even more striking. Minority-group members accounted for only about 19 percent of the doctorates awarded in 2005 (*Chronicle of Higher Education,* 2007: 19).

At all levels of formal education in the United States, some schools are operated by the government (primarily by state and local governments), and others are run by various types of nongovernmental entities. In the private sector, some of the schools either were founded by or are operated by religious organizations. By way of example, some of the best-known private universities in this country were initially run by religious groups that sought to integrate the principles of education with the teaching of the religious and moral beliefs of their group. Consequently, there is an overlap between education and religion in a variety of academic institutions at all levels. To gain a better understanding of how sociologists systematically examine religion in

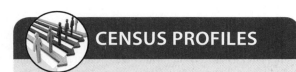

CENSUS PROFILES

Educational Achievement of Persons Age 25 and Over

The Census Bureau asks people to indicate the highest degree or level of schooling they have completed. Sixteen categories, ranging from "no schooling completed" to "doctorate degree," are set forth as responses on the form that is used; however, we are looking only at the categories of high school graduate and above. As shown below, census data reflect that the highest levels of educational attainment are held by Asian Americans, followed by non-Hispanic white respondents. For these statistics to change significantly, greater educational opportunities and more-affordable higher education would need to be readily available to African Americans and Latinos/as, who historically have experienced racial discrimination and inadequately funded public schools with high dropout rates and low high school graduation rates.

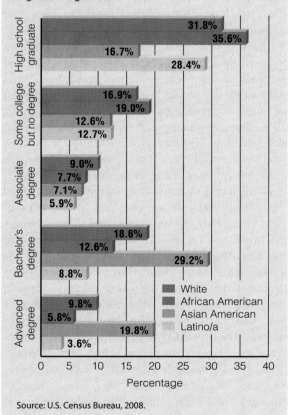

Source: U.S. Census Bureau, 2008.

society, let's first look at religion from a historical perspective.

Religion in Historical Perspective

Religion **is a social institution composed of a unified system of beliefs, symbols, and rituals, based on some sacred or supernatural realm, that guides human behavior, gives meaning to life, and unites believers into a community.** For many people, religious beliefs provide the answers for seemingly unanswerable questions about the meaning of life and death.

Religion and the Meaning of Life

Religion seeks to answer important questions such as why we exist, why people suffer and die, and what happens when we die. Whereas science and medicine typically rely on existing scientific evidence to respond to these questions, religion seeks to explain suffering, death, and injustice in the realm of the sacred. According to Emile Durkheim, *sacred* **refers to those aspects of life that are extraordinary or supernatural**—in other words, those things that are set apart as "holy." People feel a sense of awe, reverence, deep respect, or fear for that which is considered sacred. Across cultures and in different eras, many things have been considered sacred, including invisible gods, spirits, specific animals or trees, altars, crosses, holy books, and special words or songs that only the initiated could speak or sing (Collins, 1982). Those things that people do not set apart as sacred are referred to as *profane*—**the everyday, secular or "worldly" aspects of life** (Collins, 1982). Thus, whereas sacred beliefs are rooted in the holy or supernatural, secular beliefs have their foundation in scientific knowledge or everyday explanations. In the debate between creationists and evolutionists, for example, advocates of creationism view it as a belief founded in sacred (Biblical) teachings, whereas advocates of evolutionism assert that their beliefs are based on provable scientific facts.

In addition to beliefs, religion also comprises symbols and rituals. According to the anthropolo-

One of the rituals in most religions is the identification of a holy place. For Jews, one such place is the Western Wall in Jerusalem. Buddhists might travel to the Mahaparinirvan Stupa, a shrine in India. For Muslims, an example of a holy place is the Blue Mosque in Turkey. And historical Native Americans believed in the ritual of creating burial mounds.

gist Clifford Geertz (1966), religion is a set of cultural symbols that establishes powerful and pervasive moods and motivations to help people interpret the meaning of life and establish a direction for their behavior. People often act out their religious beliefs in the form of *rituals*—symbolic actions that represent religious meanings (McGuire, 2002). Rituals range from songs and prayers to offerings and sacrifices that worship or praise a supernatural being, an ideal, or a set of supernatural principles (Roberts, 2004). Rituals differ from everyday actions in that they have very strictly determined behavior. According to the sociologist Randall Collins (1982: 34), "In rituals, it is the forms that count. Saying prayers, singing a hymn, performing a primitive sacrifice or a dance, marching in a procession, kneeling before

religion a system of beliefs, symbols, and rituals, based on some sacred or supernatural realm, that guides human behavior, gives meaning to life, and unites believers into a community.

sacred those aspects of life that are extraordinary or supernatural.

profane the everyday, secular, or "worldly" aspects of life.

an idol or making the sign of the cross—in these, the action must be done the right way."

Categories of Religion Although it is difficult to establish exactly when religious rituals first began, anthropologists have concluded that all known groups over the past 100,000 years have had some form of religion (Haviland, 2002). Religions have been classified into four main categories based on their dominant belief: simple supernaturalism, animism, theism, and transcendent idealism. In very simple preindustrial societies, religion often takes the form of *simple supernaturalism*—the belief that supernatural forces affect people's lives either positively or negatively. This type of religion does not acknowledge specific gods or supernatural spirits but focuses instead on impersonal forces that may exist in people or natural objects. By contrast, ***animism* is the belief that plants, animals, or other elements of the natural world are endowed with spirits or life forces having an impact on events in society.** Animism is identified with early hunting and gathering societies and with many Native American societies, in which everyday life was not separated from the elements of the natural world (Albanese, 2007).

The third category of religion is *theism*—a belief in a god or gods. Horticultural societies were among the first to practice *monotheism*—a belief in a single, supreme being or god who is responsible for significant events such as the creation of the world. Three of the major world religions—Christianity, Judaism, and Islam—are monotheistic. By contrast, Shinto and a number of indigenous religions of Africa are forms of *polytheism*—a belief in more than one god (see ◆ Table 12.1). The fourth category of religion, transcendent idealism, is *nontheistic* because it does not focus on worship of a god or gods. *Transcendent*

◆ **Table 12.1 Major World Religions**

		Current Followers	Founder/Date	Beliefs
	Christianity	1.7 billion	Jesus—1st century C.E.	Jesus is the Son of God. Through good moral and religious behavior (and/or God's grace), people achieve eternal life with God.
	Islam	1 billion	Muhammad— ca. 600 C.E.	Muhammad received the Qur'an (scriptures) from God. On Judgment Day, believers who have submitted to God's will, as revealed in the Qur'an, will go to an eternal Garden of Eden.
	Hinduism	719 million	No specific founder— ca. 1500 B.C.E.	Brahma (creator), Vishnu (preserver), and Shiva (destroyer) are divine. Union with ultimate reality and escape from eternal reincarnation are achieved through yoga, adherence to scripture, and devotion.
	Buddhism	309 million	Siddhartha Gautama— 500 to 600 B.C.E.	Through meditation and adherence to the Eight-Fold Path (correct thought and behavior), people can free themselves from desire and suffering, escape the cycle of eternal rebirth, and achieve nirvana (enlightenment).
	Judaism	18 million	Abraham, Isaac, and Jacob—ca. 2000 B.C.E.	God's nature and will are revealed in the Torah (Hebrew scripture) and in His intervention in history. God has established a covenant with the people of Israel, who are called to a life of holiness, justice, mercy, and fidelity to God's law.
	Confucianism	5.9 million	K'ung Fu-Tzu (Confucius)—ca. 500 B.C.E.	The sayings of Confucius (collected in the Analects) stress the role of virtue and order in the relationships among individuals, their families, and society.

idealism is a belief in sacred principles of thought and conduct. Principles such as truth, justice, affirmation of life, and tolerance for others are central tenets of transcendent idealists, who seek an elevated state of consciousness in which they can fulfill their true potential.

Religion and Scientific Explanations

During the Industrial Revolution, scientific explanations began to compete with religious views of life. Rapid growth in scientific and technological knowledge gave rise to the idea that science would ultimately answer questions that previously had been in the realm of religion. Many scholars believed that increases in scientific knowledge would result in *secularization*—**the process by which religious beliefs, practices, and institutions lose their significance in sectors of society and culture** (Berger, 1967). Secularization involves a decline of religion in everyday life and a corresponding increase in organizations that are highly bureaucratized, fragmented, and impersonal (Chalfant, Beckley, and Palmer, 1994).

In the United States, some people argue that science and technology have overshadowed religion, but others point to the resurgence of religious beliefs and an unprecedented development of alternative religions in recent years (Kosmin and Lachman, 1993; Roof, 1993; Singer with Lalich, 1995). The issue of whether religion or scientific explanations best explain various aspects of social life (such as when a person becomes a "human being" or when a person dies) is not limited to the teachings of schools and religious organizations: Members of the contemporary media are also key players in the framing of religion and scientific debates, and journalists may influence how we view a number of key issues (see Box 12.2).

Sociological Perspectives on Religion

According to the sociologist Meredith B. McGuire (2002), religion as a social institution is a powerful, deeply felt, and influential force in human society.

Sociologists study the social institution of religion because of the importance religion holds for many people; they also want to know more about the influence of religion on society, and vice versa (McGuire, 2002).

The major sociological perspectives have different outlooks on the relationship between religion and society. Functionalists typically emphasize the ways in which religious beliefs and rituals can bind people together. Conflict explanations suggest that religion can be a source of false consciousness in society. Symbolic interactionists focus on the meanings that people give to religion in their everyday life.

Functionalist Perspectives on Religion

Emile Durkheim was one of the first sociologists to emphasize that religion is essential to the maintenance of society. He suggested that religion is a cultural universal found in all societies because it meets basic human needs and serves important societal functions.

For Durkheim, the central feature of all religions is the presence of sacred beliefs and rituals that bind people together in a collectivity. In his studies of the religion of the Australian aborigines, for example, Durkheim found that each clan had established its own sacred totem, which included kangaroos, trees, rivers, rock formations, and other animals or natural creations. To clan members, their totem was sacred; it symbolized some unique quality of their clan. People developed a feeling of unity by performing ritual dances around their totem, which caused them to abandon individual self-interest. Durkheim suggested that the correct performance of the ritual gives rise to religious conviction. Religious beliefs and rituals are *collective representations*—group-held meanings that express something important

---

animism the belief that plants, animals, or other elements of the natural world are endowed with spirits or life forces having an effect on events in society.

secularization the process by which religious beliefs, practices, and institutions lose their significance in sectors of society and culture.

Box 12.2 Framing Religion in the Media

Shaping the Intersections of Science and Religion

After religious teachers accomplish the refining process indicated, they will surely recognize with joy that true religion has been ennobled and made more profound by scientific knowledge.
> —Albert Einstein, a theoretical physicist known for formulating the theory of relativity (religioustolerance.org, 2005)

Science is almost totally incompatible with religion.
> —Peter Atkins, a chemist at the University of Oxford (religioustolerance.org, 2005)

As the statements by these two well-known scientists show, a lack of consensus exists regarding the relationship between science and religion. However, members of the media are often called upon to write about situations in which science and religion apparently intersect in societies. When these events occur, many print and electronic journalists use specific types of framing to shape their discussion of the intersections of science and religion. Theology professor Jame Schaefer (2005) has identified several ways that the media might approach this topic, and the following discussion is a modification of three of those approaches.

Conflict Framing

According to Schaefer (2005), "The image of conflict between science and religion is conventional in the media today. Coverage of a story is more dramatic when extreme views are highlighted while more subtle positions go unreported." When this type of framing is used by the media, experts are often carefully chosen from widely divergent viewpoints so that extreme viewpoints will be highlighted for readers and viewers. The result is often a form of framing that emphasizes hostility toward religion or toward science and technology. An example of conflict framing is stories about the teaching of evolution versus the teaching of creationism in the classroom.

Conflation Framing

When conflation framing is employed by the media, it becomes difficult to distinguish between religion and science as distinct human endeavors (Schaefer, 2005). Conflation exists when two things are combined into one. An example that Schaefer (2005: 220) provides for this type of framing is the "practice of attributing to God's activity natural phenomena that cannot be explained scientifically." In other words, if there is not a ready scientific explanation for a natural or social phenomenon that occurs, it may be described as supernatural or within the realm of God's doing, not that of human beings. When the media rush to a Catholic church to describe people who are watching a portrait of the Virgin Mary that appears to be weeping or bleeding, some journalists explain this phenomenon with a combination of religious and scientific explanations. For example, the occurrence may be described by onlookers as an act of God, but the weeping or bleeding may also be attributed to environmental factors such as stains on the artwork or the way in which light and shadows at different times of the year affect paintings in the church. Journalists who combine religious and scientific explanations of phenomenon such as the "bleeding Virgin" tend to conflate religion and science for media audiences.

about the group itself (McGuire, 2002). Because of the intertwining of group consciousness and society, functionalists suggest that religion has three important functions in any society:

1. *Meaning and purpose.* Religion offers meaning for the human experience. Some events create a profound sense of loss on both an individual basis (such as injustice, suffering, and the death of a loved one) and a group basis (such as famine, earthquake, economic depression, or subjugation by an enemy). Inequality may cause people to wonder why their own situation is no better than it is. Most religions offer explanations for these concerns. Explanations may differ from one religion to another, yet each tells the individual or group that life is part of a larger system of order in the universe (McGuire, 2002). Some (but not all) religions even offer hope of an afterlife for persons who follow the religion's tenets of moral-

The face on this statue of the Virgin Mary, in Caracas, Venezuela, appears to be bleeding. How journalists report on such a phenomenon is determined by which type of framing that they decide to use.

Contrast Framing

Although media framing based on conflict and media framing based on conflation both suggest that religion and science are interrelated, contrast framing is just the opposite. In the words of Schaefer (2005: 221), the contrast approach views religion and science as "totally indepen-

dent and autonomous ways of knowing. Each is valid only within its clearly defined sphere of inquiry." Contrast framing in the media is based on the assumption that science and religion are too different to be interrelated:

1. Religion and science deal with different types of questions.
2. Religion and science use different languages to explain *reality*.
3. Religion and science tackle different tasks.
4. Religion and science follow different authorities (see also Gilkey, 1993).

According to Schaefer (2005: 221), "Science examines the natural world empirically, while religion addresses the ultimate reality that transcends the empirically known world" (see also Haught, 1995). Today, some people view religion and science as conflicting world views, whereas other individuals view religion and science as indistinguishable.

Does it make a difference how the media frame stories about the intersection of religion and science? How the media frame stories about the intersection of these institutions may influence our views on a number of topics, ranging from the origins of the Earth and human beings to how and when life on Earth might end. Media stories about religion and science may also influence our thinking about moral issues such as what constitutes "right" and "wrong" in our society and who should be allowed to determine policies on key social issues such as abortion, the death penalty, and end-of-life decisions for individuals with a critical illness.

Reflect & Analyze

Do the media affect your views on these issues? If so, how? If not, why not?

ity in this life. Such beliefs help make injustices in this life easier to endure.

2. *Social cohesion and a sense of belonging.* Religious teachings and practices, by emphasizing shared symbolism, help promote social cohesion. An example is the Christian ritual of communion, which not only commemorates a historical event but also allows followers to participate in the unity ("communion") of themselves with other believers (McGuire, 2002). All religions have some form of shared experiences that rekindle the group's consciousness of its own unity.

3. *Social control and support for the government.* All societies attempt to maintain social control through systems of rewards and punishments. Sacred symbols and beliefs establish powerful, pervasive, long-lasting motivations based on the concept of a general order of existence. In other words, if individuals consider themselves to be part of a larger order that holds the ultimate

meaning in life, they will feel bound to one another (and past and future generations) in a way that otherwise might not be possible (McGuire, 2002).

Religion also helps maintain social control in society by conferring supernatural legitimacy on the norms and laws of society. In some societies, social control occurs as a result of direct collusion between the dominant classes and the dominant religious organizations.

In the United States, the separation of church and state reduces religious legitimation of political power. Nevertheless, political leaders often use religion to justify their decisions, stating that they have prayed for guidance in deciding what to do (McGuire, 2002). This informal relationship between religion and the state has been referred to as *civil religion*—the set of beliefs, rituals, and symbols that makes sacred the values of the society and places the nation in the context of the ultimate system of meaning. Civil religion is not tied to any one denomination or religious group; it has an identity all its own. For example, many civil ceremonies in the United States have a marked religious quality. National values are celebrated on "high holy days" such as Memorial Day and the Fourth of July. Political inaugurations and courtroom trials both require people to place their hand on a Bible while swearing to do their duty or tell the truth, as the case may

Throughout recorded history, churches and other religious bodies have provided people with a sense of belonging and of being part of something larger than themselves. Members of this congregation show their unity as they visit with one another.

be. The United States flag is the primary sacred object of our civil religion, and the pledge of allegiance has included the phrase "one nation under God" for many years now. U.S. currency bears the inscription "In God We Trust."

Some critics have attempted to eliminate all vestiges of civil religion from public life. However, sociologist Robert Bellah (1967), who has studied civil religion extensively, argues that civil religion is not the same thing as Christianity; rather, it is limited to affirmations that members of any denomination can accept. As McGuire (2002: 203) explains,

> Civil religion is appropriate to actions in the official public sphere, and Christianity (and other religions) are granted full liberty in the sphere of personal piety and voluntary social action. This division of spheres of relevance is particularly important for countries such as the United States, where religious pluralism is both a valued feature of sociological life and a barrier to achieving a unified perspective for decision making.

However, Bellah's assertion does not resolve the problem for those who do not believe in the existence of God or for those who believe that *true* religion is trivialized by civil religion.

Conflict Perspectives on Religion

Many functionalists view religion, including civil religion, as serving positive functions in society, but some conflict theorists view religion negatively.

Karl Marx on Religion For Marx, *ideologies*—systematic views of the way the world ought to be—are embodied in religious doctrines and political values (Turner, Beeghley, and Powers, 2002). These ideologies also serve to justify the status quo and retard social change. The capitalist class uses religious ideology as a tool of domination to mislead the workers about their true interests. For this reason, Marx wrote his now famous statement that religion is the "opiate of the masses." People become complacent because they have been taught to believe in an afterlife in which they will be rewarded for their suffering and misery in this life. Although these religious teachings soothe the masses' distress, any relief is illusory. Religion unites people under a "false con-

According to Marx and Weber, religion serves to reinforce social stratification in a society. For example, according to Hindu belief, a person's social position in his or her current life is a result of behavior in a former life.

© Devendra M. Singh/AFP/Getty Images

sciousness" that they share common interests with members of the dominant class (Roberts, 2004).

Max Weber on Religion Whereas Marx believed that religion retarded social change, Weber argued just the opposite. For Weber, religion could be a catalyst to produce social change. In *The Protestant Ethic and the Spirit of Capitalism* (1976/1904–1905), Weber asserted that the religious teachings of John Calvin were directly related to the rise of capitalism. Calvin emphasized the doctrine of *predestination*—the belief that, even before they are born, all people are divided into two groups, the saved and the damned, and only God knows who will go to heaven (the elect) and who will go to hell. Because people cannot know whether they will be saved, they tend to look for earthly signs that they are among the elect. According to the Protestant ethic, those who have faith, perform good works, and achieve economic success are more likely to be among the chosen of God. As a result, people work hard, save their money, and do not spend it on worldly frivolity; instead, they reinvest it in their land, equipment, and labor (Chalfant, Beckley, and Palmer, 1994).

The spirit of capitalism grew in the fertile soil of the Protestant ethic. Even as people worked ever harder to prove their religious piety, structural conditions in Europe led to the Industrial Revolution, free markets, and the commercialization of the economy—developments that worked hand in hand with Calvinist religious teachings. From this viewpoint, wealth was an unintended consequence of religious piety and hard work. With the contemporary secularizing influence of wealth, people often think of wealth and material possessions as the major (or only) reason to work. Although it is no longer referred to as the "Protestant ethic," many people still refer to the "work ethic" in somewhat the same manner that Weber did. For example, political and business leaders in the United States often claim that "the work ethic is dead."

Like Marx, Weber was acutely aware that religion could reinforce existing social arrangements, especially the stratification system. The wealthy can use religion to justify their power and privilege: It is a sign of God's approval of their hard work and morality (McGuire, 2002). As for the poor, if they work hard and live a moral life, they will be richly rewarded in another life.

From a conflict perspective, religion tends to promote conflict between groups and societies. According to conflict theorists, conflict may be *between* religious groups (for example, anti-Semitism), *within* a religious group (for example, when a splinter group leaves an existing denomination), or between a religious group and the *larger society* (for example, the conflict over religion in the classroom). Conflict theorists assert that in attempting to provide meaning and purpose in life while at the same time promoting the status quo, religion is used by the dominant classes to impose their own control over society and its resources (McGuire, 2002). Many feminists object to the patriarchal nature of most religions; some advocate a break from traditional religions, whereas others seek to reform religious language, symbols, and rituals to eliminate the elements of patriarchy.

civil religion the set of beliefs, rituals, and symbols that makes sacred the values of the society and places the nation in the context of the ultimate system of meaning.

Symbolic Interactionist Perspectives on Religion

Thus far, we have been looking at religion primarily from a macrolevel perspective. Symbolic interactionists focus their attention on a microlevel analysis that examines the meanings that people give to religion in their everyday life.

Religion as a Reference Group For many people, religion serves as a reference group to help them define themselves. For example, religious symbols have meaning for large bodies of people. The Star of David holds special significance for Jews, just as the crescent moon and star do for Muslims and the cross does for Christians. For individuals as well, a symbol may have a certain meaning beyond that shared by the group. For instance, a symbol given to a child may have special meaning when he or she grows up and faces war or other crises. It may not only remind the adult of a religious belief but also create a feeling of closeness with a relative who is now deceased. It has been said that the symbolism of religion is so very powerful because it "expresses the essential facts of our human existence" (Collins, 1982: 37).

Her Religion and his Religion Early feminist Charlotte Perkins Gilman (1976/1923) believed that "men's religion" taught people to submit and obey rather than to think about and realistically confront situations. Consequently, she asserted, the monopolization of religious thoughts and doctrines by men contributed to intolerance and the subordination of women. Along with other feminist thinkers, Gilman concluded that "God the Mother" images could be useful as a means of encouraging people to be more cooperative and compassionate, rather than competitive and violent.

Not all people interpret religion in the same way. In virtually all religions, women have much less influence in establishing social definitions of appropriate gender roles both within the religious community and in the larger community. Therefore, women and men may belong to the same religious group, but their individual religion will not necessarily be a carbon copy of the group's entire system of beliefs. In fact, according to McGuire (2002), women's versions of a certain religion probably differ markedly from men's versions.

Religious symbolism and language typically create a social definition of the roles of men and women. For example, religious symbolism may depict the higher deities as male and the lower deities as female. Sometimes, females are depicted as negative, or evil, spiritual forces. For instance, the Hindu goddess Kali represents men's eternal battle against the evils of materialism (Daly, 1973). Historically, language has defined women as being nonexistent in the world's major religions. Phrases such as "for all men" in Catholic and Episcopal services gradually have been changed to "for all"; however, some churches retain the traditional liturgy. Although there has been resistance, especially by women, to some of the terms, inclusive language is less common than older male terms for God.

The Concept Quick Review summarizes the major sociological perspectives on education and religion.

Types of Religious Organization

Religious groups vary widely in their organizational structure. Although some groups are large and somewhat bureaucratically organized, others are small and have a relatively informal authority structure. Some require total commitment from their members; others expect members to have only a partial commitment. Sociologists have developed typologies or ideal types of religious organization to enable them to study a wide variety of religious groups. The most common categorization includes four types: ecclesia, church, sect, and cult.

Ecclesia

Ecclesia **is a religious organization that is so integrated into the dominant culture that it claims as its membership all members of a society.** Membership in the ecclesia occurs as a result of being born into the society rather than by any conscious decision on the part of individual members. The linkages between the social institutions of religion and government are often very strong in such societies. Although no true ecclesia exists in the contemporary world, the Anglican church (the official church of England), the Lutheran church in Sweden and

CONCEPT QUICK REVIEW

Sociological Perspectives on Education and Religion

	Education	Religion
Functionalist Perspective	One of the most important components of society: Schools teach students not only content but also to put group needs ahead of the individual's.	Sacred beliefs and rituals bind people together and help maintain social control.
Conflict Perspective	Schools perpetuate class, racial–ethnic, and gender inequalities through what they teach to whom.	Religion may be used to justify the status quo (Marx) or to promote social change (Weber).
Symbolic Interactionist Perspective	Labeling and the self-fulfilling prophecy are an example of how students and teachers affect each other as they interpret their interactions.	Religion may serve as a reference group for many people, but because of race, class, and gender, people may experience it differently.

◆ **Table 12.2 Characteristics of Churches and Sects**

Characteristic	Church	Sect
Organization	Large, bureaucratic organization, led by a professional clergy	Small, faithful group, with high degree of lay participation
Membership	Open to all; members usually from upper and middle classes	Closely guarded membership, usually from lower classes
Type of Worship	Formal, orderly	Informal, spontaneous
Salvation	Granted by God, as administered by the church	Achieved by moral purity
Attitude Toward Other Institutions and Religions	Tolerant	Intolerant

Denmark, the Catholic church in Spain, and Islam in Iran and Pakistan come fairly close.

The Church–Sect Typology

To help explain the different types of religious organizations found in societies, Ernst Troeltsch (1960/1931) and his teacher, Max Weber (1963/1922), developed a typology that distinguishes between the characteristics of churches and sects (see ◆ Table 12.2). Unlike an ecclesia, a church is not considered to be a state religion; however, it may still have a powerful influence on political and economic arrangements in society. A *church* **is a large, bureaucratically organized religious organization that tends to seek accommo-** **dation with the larger society in order to maintain some degree of control over it.** Church membership is largely based on birth; children of church members are typically baptized as infants and become lifelong members of the church. Older children and adults

ecclesia a religious organization that is so integrated into the dominant culture that it claims as its membership all members of a society.

church a large, bureaucratically organized religious organization that tends to seek accommodation with the larger society in order to maintain some degree of control over it.

may choose to join the church, but they are required to go through an extensive training program that culminates in a ceremony similar to the one that infants go through. Leadership is hierarchically arranged, and clergy generally have many years of formal education. Churches have very restrained services that appeal to the intellect rather than the emotions. Religious services are highly ritualized; they are led by clergy who wear robes, enter and exit in a formal processional, administer sacraments, and read services from a prayer book or other standardized liturgical format.

Midway between the church and the sect is a *denomination*—**a large organized religion characterized by accommodation to society but frequently lacking in the ability or intention to dominate society** (Niebuhr, 1929). Denominations have a trained ministry, and although involvement by lay members is encouraged more than in the church, their participation is usually limited to particular activities, such as readings or prayers. Denominations tend to be more tolerant and less likely than churches to expel or excommunicate members. This

◆ **Table 12.3** Major U.S. Denominations That Self-Identify as Christian

Religious Body	Members	Churches
Roman Catholic Church	69,135,000	18,992
Southern Baptist Convention	16,270,000	43,669
United Methodist Church	8,075,000	34,660
Church of Jesus Christ of Latter Day Saints	5,691,000	12,753
Church of God in Christ[a]	5,500,000	15,300
National Baptist Convention, U.S.A.[a]	5,000,000	9,000
Evangelical Lutheran Church in America	4,851,000	10,519
National Baptist Convention of America[a]	3,500,000	(N/A)
Presbyterian Church (U.S.A.)	3,099,000	10,960
Assemblies of God	2,831,000	12,298
African Methodist Episcopal Church[a]	2,500,000	4,174
National Missionary Baptist Convention of America[a]	2,500,000	(N/A)
Progressive National Baptist Convention[a]	2,500,000	2,000
Lutheran Church–Missouri Synod	2,441,000	6,144
Episcopal Church	2,248,000	7,200
Churches of Christ[a]	1,639,000	15,000
Greek Orthodox Church	1,500,000	566
Pentecostal Assemblies of the World[a]	1,500,000	1,750
African Methodist Episcopal Zion Church	1,440,000	3,260
American Baptist Churches in U.S.A.	1,397,000	5,740
United Church of Christ	1,244,000	5,567
Baptist Bible Fellowship, International[a]	1,200,000	4,500
Christian Churches and Churches of Christ[a]	1,072,000	5,579
Jehovah's Witnesses	1,046,000	12,384

[a]Current data not available; prior data used may no longer be comparable.
Source: U.S. Census Bureau, 2008.

form of organization is most likely to thrive in societies characterized by *religious pluralism*—a situation in which many religious groups exist because they have a special appeal to specific segments of the population. Perhaps because of its diversity, the United States has more denominations than any other country. ◆ Table 12.3 shows this diversity.

A *sect* **is a relatively small religious group that has broken away from another religious organization to renew what it views as the original version of the faith.** Unlike churches, sects offer members a more personal religion and an intimate relationship with a supreme being, depicted as taking an active interest in the individual's everyday life. Whereas churches use formalized prayers, often from a prayer book, sects have informal prayers composed at the time they are given. Typically, religious sects appeal to those who might be characterized as lower class, whereas denominations primarily appeal to the middle and upper-middle classes, and churches focus on the upper classes.

According to the church–sect typology, as members of a sect become more successful economically and socially, their religious organization is also likely to focus more on this world and less on the next. If some members of the sect do not achieve financial success, they may feel left behind as other members and the ministers shift their priorities. Eventually, this process will weaken some organizations, and people will split off to create new, less worldly versions of the group that will be more committed to "keeping the faith." Those who defect to form a new religious organization may start another sect or form a cult (Stark and Bainbridge, 1981).

Cults

A *cult* **is a religious group with practices and teachings outside the dominant cultural and religious traditions of a society.** Although many people view cults negatively, some major religions (including Judaism, Islam, and Christianity) and some denominations (such as the Mormons) started as cults. Cult leadership is based on charismatic characteristics of the individual, including an unusual ability to form attachments with other people. An example is the religious movement started by Reverend Sun Myung Moon, a Korean electrical engineer who believed that God had revealed to him that Judgment Day was rapidly approaching. Out of this movement, the Unification church, or "Moonies," grew and flourished, recruiting new members through their personal attachments to present members.

Trends in Religion in the United States

As we have seen throughout this chapter, religion in the United States is very diverse. Pluralism and religious freedom are among the cultural values most widely espoused, and no state church or single denomination predominates. As shown in ◆ Table 12.4, Protestants constitute the largest religious body in the United States, followed by Roman Catholics, Jews, Eastern Churches, and others.

The rise of a new fundamentalism has occurred at the same time that a number of mainline denominations have been losing membership. Whereas "old" fundamentalism usually appealed to people from lower-income, rural, southern backgrounds, the "new" fundamentalism appears to have a much wider following among persons from all socioeconomic levels, geographical areas, and occupations. Some member of the political elite in Washington have vowed to bring religion "back" into schools and public life. "New-right" fundamentalists have been especially critical of *secular humanism*—a belief in the perfectibility of human beings through their own efforts rather than through a belief in God and a religious conversion. According to fundamentalists, "creeping" secular humanism has been most visible in the public schools, which, instead of offering children a fair and balanced picture, are

denomination a large, organized religion characterized by accommodation to society but frequently lacking in ability or intention to dominate society.

sect a relatively small religious group that has broken away from another religious organization to renew what it views as the original version of the faith.

cult a religious group with practices and teachings outside the dominant cultural and religious traditions of a society.

◆ **Table 12.4 U.S. Religious Traditions' Membership**

Religious Tradition	Percentage of All U.S. Adults
Protestants	51.3
Roman Catholics	23.9
Mormons	1.7
Jews	1.7
Jehovah's Witnesses	0.7
Buddhists	0.7
Orthodox Christians	0.6
Muslims	0.6
Hindus	0.4
Other faiths	1.5
Unaffiliated	16.1
Don't know/no answer	0.6

Source: Pew Forum on Religion and Public Life, 2008.

© Bob Daemmrich/The Image Works

Should prayer be permitted in the classroom? On the school grounds? At school athletic events? Given the diversity of beliefs that U.S. people hold, arguments and court cases over activities such as prayer around the school flagpole will no doubt continue in the future.

teaching things that seem to children to prove that their parents' lifestyle and religion are inferior and perhaps irrational (Carter, 1994). The new-right fundamentalists claim that banning the teaching of Christian beliefs in the classroom while teaching things that are contrary to their faith is an infringement on their freedom of religion. Worse yet, they argue, it is the equivalent of establishing an unconstitutional state religion—a secular religion that does not recognize God.

But how might students and teachers who come from very diverse religious and cultural backgrounds feel about religious instruction or organized prayer in public schools? Rick Nelson, a teacher in the Fairfax County, Virginia, public school system, explains his concern about the potential impact of group prayer on students in his classroom:

> I think it really trivializes religion when you try to take such a serious topic with so many different viewpoints and cover it in the public schools. At my school we have teachers and students who are Hindu. They are really devout, but they are not monotheistic. . . . I am not opposed to individual prayer by students. I expect students to pray when I give them a test. They need to do that for my

tests. . . . But when there is group prayer . . . who's going to lead the group? And if I had my Hindu students lead the prayer, I will tell you it will disrupt many of my students and their parents. It will disrupt the mission of my school, unfortunately . . . if my students are caused to participate in a group Hindu prayer. (CNN, 1994)

As we have seen in this chapter, the debate continues over what should be taught and what practices (such as Bible reading and prayer) should be permitted in public schools.

Education and Religion in the Future

Education and religion are powerful and influential forces because these social institutions instill in individuals the values, beliefs, and knowledge that many people believe are essential for the survival of individuals and entire cultures. Because education and religion are both socializing institutions, children and young people are particularly affected by the ideas that teachers and religious leaders impart to them. Through formal education or schooling, people learn about some aspects of the society in which they live and the dominant culture of which they are a part. In religious groups, individuals of all ages learn about the institutionalized beliefs and rituals of specific organizations that have been established to serve people's religious needs and other purposes.

What will be the future of education in the United States? The answer to this question, particularly in regard to public elementary and secondary schools in this country, is linked to the No Child Left Behind Act of 2001. The purpose of this act is to close the achievement gap between rich and poor students. No Child Left Behind represents the most far-reaching educational reform to be implemented since compulsory education laws were passed early in the twentieth century. Under this law, schools are to be held accountable for students' learning, and specific steps are set forth toward producing a more accountable education system. The law requires states to test every student's progress toward meeting specific standards that were established for the end of each grade. School districts must report students' results to demonstrate that they are making progress toward meeting these standards. Schools that close the education gap will receive additional federal dollars, but schools and districts that do not show adequate progress may lose funding and pupils: Parents may be allowed to move their children from low-performing schools to other schools that meet or exceed their district's educational standards.

In 2007, on the sixth anniversary of the No Child Left Behind Act, President George W. Bush and Education Secretary Margaret Spellings issued a statement indicating that across-the-board improvements had occurred in fourth- and eighth-grade reading and math scores nationwide. The report concluded that African American and Hispanic students had made significant gains in closing the achievement gap in areas such as reading and mathematics. However, the Bush administration acknowledged that No Child Left Behind had also produced a series of unanticipated problems and that much remained to be done to improve education in this country. More than 75 specific suggestions on how to improve teaching, learning, and student performance were made in the report (*New York Times*, 2007a). Among the needed changes were the development of rigorous, voluntary national standards that might more effectively prepare students for success in college and the workplace (whitehouse .gov, 2008). Current problems also exist regarding how states collect data and assess both student and school performance. Other problems are related to the quality of teachers and how effectively they instruct their students. For schools to be able to attract outstanding teachers, the best teachers will need to be more adequately rewarded, and incentives must be offered to encourage good instructors to teach in underperforming schools.

At the bottom line, many critics believe that No Child Left Behind has fundamentally misdefined the problems facing education in this country. They state that schools need more money and more incentives, not more testing of students or teachers. One major criticism of this law is that it does not adequately address the profound educational inequalities that are so prevalent in the United States, particularly with high-spending schools outspending low-spending schools by at least three to one in most states (Darling-Hammond, 2007).

Similarly, colleges and universities face many challenges in the 2000s, particularly given the major financial constraints that many institutions face. Institutions of higher education continue to be expected to expand their focus while undergoing strenuous budget cuts coupled with increasing demands to meet the needs of widely diverse student populations. For example, some U.S. universities are expanding their educational operations to emerging nations where demand is high for certain kinds of curricula, such as advanced business and petroleum engineering courses in Qatar and other Middle Eastern countries. One of the major problems, present and future, in colleges and universities is the increasing cost of education for students and the part that higher education plays in maintaining and perpetuating social inequality in the larger society.

As we shift our focus to religion, we ask this: What significance will religion have in the future? Religion will continue to be important in the lives of many people. On the one hand, religion may unify

Box 12.3 You Can Make a Difference

Understanding and Tolerating Religious and Cultural Differences

[The small strip of kente cloth that hangs on the marble altar and the flags of a dozen West Indies nations, including Trinidad, Jamaica, and Barbados,] represent all of the people in this parish. It's an opportunity for us to celebrate everyone's culture. You won't find your typical Episcopal church looking like that.

—Rev. Cannon Peter P. Q. Golden, rector of St. Paul's Episcopal Church in Brooklyn, New York (qtd. in Pierre-Pierre, 1997: A11)

One part of the history of St. Paul's Episcopal Church in Flatbush, Brooklyn, is etched into its huge stained-glass windows, which tell the story of the prominent descendants of English and Dutch settlers who founded the church. However, the church's more recent history is told in the faces of its parishioners—most of whom are Caribbean immigrants from the West Indies who labor as New York City's taxi drivers, factory workers, accountants, and medical professionals (Pierre-Pierre, 1997).

Traveling only a short distance, we find a growing enclave of Muslims in Brooklyn. It is their wish that people will recognize that Islam shares a great deal with Christianity and Judaism, including the fact that all three believe in one God, are rooted in the same part of the world, and share some holy sites (Sengupta, 1997). Some Muslim adherents also hope that more people in this country will learn greater tolerance toward those who have a different religion and celebrate different holidays. For example, in some years the Islamic holy season, Ramadan, falls at about the same time as Christmas and Hanukkah, but many Muslim children and adults are disparaged by their neighbors because they do not celebrate the same holidays as others. As one person observed, "As Muslims, we have to respect all religions" (Sengupta, 1997: A12). Left unsaid was the belief that other people should do likewise. How can each of us—regardless of our race, color, creed, or national origin—help to bring about greater tolerance of religious diversity in this country? Here is one response to this question, from Huston Smith (1991: 389–390), a historian of religion:

Whether religion is, for us, a good word or bad; whether (if on balance it is a good word) we side with a single religious tradition or to some degree open our arms to all: How do we comport ourselves in a pluralistic world that is riven by ideologies, some sacred, some profane? . . . We listen. . . . If one of the [world's religions] claims us, we begin by listening to it. Not uncritically, for new occasions teach new duties and everything finite is flawed in some respects. Still, we listen to it expectantly, knowing that it houses more truth than can be encompassed in a single lifetime.

But we also listen to the faith of others, including the secularists. We listen first because . . . our times require it. The community today can be no single tradition; it is the planet. Daily, the world grows smaller, leaving understanding the only place where peace can find a home. . . . Those who listen work for peace, a peace built not on ecclesiastical or political hegemonies but on understanding and mutual concern.

Perhaps this is how each of us can make a difference—by *learning* more about our own beliefs and about the diverse denominations and world religions represented in the United States and around the globe, and *listening* to what other people have to say about their own beliefs and religious experiences.

Will religious tolerance increase in the United States? In the world? What steps can you take to help make a difference? Below are a few websites that provide more information about the world's religions.

● The Big Religion Chart
  http://www.religionfacts.com/big_religion_chart .htm

● BBC's Religion and Ethics pages
  http://www.bbc.co.uk/religion

● Sacred Destinations
  http://www.sacred-destinations.com

people; on the other, it may result in tensions and confrontations among individuals and groups. However, one thing appears certain: With the influx of recent immigrants and increasing cultural diversity, religious congregations in the United States in the future will be much more diverse and hold a wider variety of beliefs than did the traditional mainline denominations of the past. In the final analysis, the influence of religion may be felt even by those who claim no religious beliefs of their own. As a result, the debate will continue over what religion is, what it should do, and what its relationship to other social institutions (such as education) should be. All of us should work to gain a better understanding of the wide variety of religions found around the world and the many religious organizations within the United States so that we can have a better understanding of the diverse people who make up our nation and world (see Box 12.3).

To conclude with the ideas of the sociologist Lester Kurtz (1995), the task of religious institutions is to tend to spiritual and ethical issues. However, it is difficult to contain the ideas and actions of one social institution within its own domain so that they do not spread into other social institutions. According to Kurtz (1995: 167), "Religion intrudes on all other spheres . . . because its ethics generally apply to all areas of life. A modern society compartmentalizes institutions, but we cannot compartmentalize people." And so it is in the realm of education and religion: Each of these social institutions serves a specific purpose and has its own functions; however, individuals carry their beliefs, values, and attitudes with them as they move from one social institution to another, and this brings about some of the patterns of cooperation and conflict we have examined in this chapter.

Chapter Review

● **How are the social institutions of education and religion similar?**

Education and religion are powerful and influential forces in society. Both institutions impart values, beliefs, and knowledge considered essential to the social reproduction of individual personalities and entire cultures.

● **What is the primary function of education?**

Education is the social institution responsible for the systematic transmission of knowledge, skills, and cultural values within a formally organized structure.

● **What is the functionalist perspective on education?**

According to functionalists, education has both manifest functions (socialization, transmission of culture, social control, social placement, and change and innovation) and latent functions (keeping young people off the streets and out of the job market, matchmaking and producing social networks, and creating a generation gap).

● **What is the conflict perspective on education?**

From a conflict perspective, education is used to perpetuate class, racial–ethnic, and gender inequalities through tracking, ability grouping, and a hidden curriculum that teaches subordinate groups conformity and obedience.

● **What is the symbolic interactionist perspective on education?**

According to symbolic interactionists, education may be a self-fulfilling prophecy for some students, such that these students come to perform up—or down—to the expectations held for them by teachers.

● **What is religion?**

Religion is a social institution composed of a unified system of beliefs, symbols, and rituals, based on some sacred or supernatural realm, that guides human behavior, gives meaning to life, and unites believers into a community.

• What is the functionalist perspective on religion?

According to functionalists, religion has three important functions in any society: (1) providing meaning and purpose to life, (2) promoting social cohesion and a sense of belonging, and (3) providing social control and support for the government.

• What is the conflict perspective on religion?

From a conflict perspective, religion can have negative consequences in that the capitalist class uses religion as a tool of domination to mislead workers about their true interests. However, Max Weber believed that religion could be a catalyst for social change.

• What is the symbolic interactionist perspective on religion?

Symbolic interactionists examine the meanings that people give to religion and the meanings that they attach to religious symbols in their everyday life.

• What are the different types of religious organizations?

Religious organizations can be categorized as ecclesia, churches, denominations, sects, and cults. Some of the world's major religions started off as cults built around a charismatic leader and developed into sects, denominations, and churches. No true ecclesia, or government-sponsored religion, exists in the contemporary world.

www.cengage.com/login

Register for a Student eResource account to maximize your study time online using CengageNOW. First take the system's diagnostic pre-test, and then follow the personalized study plan that is created for you to help you review this chapter. The study plan will

- help you identify areas on which you should concentrate;
- provide interactive exercises to help you master the chapter concepts; and
- provide a post-test to confirm you are ready to move on to the next chapter.

Key Terms

animism 410
church 417
civil religion 414
credentialism 399
cult 419
cultural capital 396

denomination 418
ecclesia 416
education 390
hidden curriculum 398
profane 408

religion 408
sacred 408
sect 419
secularization 411
tracking 396

Questions for Critical Thinking

1. Why does so much controversy exist over what should be taught in U.S. public schools?
2. How are the values and attitudes you learned from your family reflected in your beliefs about education and religion?
3. How would you design a research project to study the effects of civil religion on everyday life? What kind of data would be most accessible?
4. If Durkheim, Marx, and Weber were engaged in a discussion about education and religion, on what topics might they agree? On what topics would they disagree?

The Kendall Companion Website

www.cengage.com/sociology/kendall

Supplement your review of this chapter by going to the text's companion website, where you can take tutorial quizzes, use flash cards to master key terms, follow live links to useful websites, and explore the other study and research resources you'll find there, such as a comprehensive interactive sociology timeline, GSS Data, and Census 2000 information, much of it presented visually in maps.

13 Politics and the Economy in Global Perspective

Do you think that mainstream TV news is boring? Does the idea of sitting down to watch *NewsHour with Jim Lehrer* make you yawn . . . just thinking about it? If so, you're not alone—in fact, this is now the norm among the college-aged demographic. A recent article [notes] that many young adults eschew traditional nightly news for [Jon Stewart's] *The Daily Show*, . . . which proudly bills itself as "the most trusted name in fake news."

Wow. While I'm a huge fan of Comedy Central, I usually tune in to watch something with an inherently comedic purpose— like South Park, Mencia, or the Colbert Report. While Colbert is billed as a blatant mockery of conservative TV pundits, the lines are more blurred when it comes to *The Daily Show*. Yes, it is largely satirical, but I can understand why many might watch the show as a main source of news. While it is based on real

© AP Images/Jason DeCrow

Jon Stewart, host of the extremely popular Comedy Central production *The Daily Show*, parodies mainstream news programming. Many people now watch Stewart's reports instead of watching network news programs. How might comedy news shows affect our perceptions of politics?

news, it is also written by comedy writers, and has ratings in mind—not necessarily the best interests of the American public or young people. . . . Yes, *The Daily Show* is funny. Yes, Jon Stewart is right on the nail with his ironic insight and sarcastic humor. However, it isn't a real news show. . . .

—Katie Stapleton-Paff (2007), a writer for the *Daily of the University of Washington,* describing her concerns about the fact that many people between the ages of eighteen and twenty-five use "comedy" news programs as a prime source of information

I love the mock interviews, but I never go into the show thinking I'm watching real news. What I see is what I get and in the case of *The Daily Show* I see a funny show that makes fun of the day's news, much like Jay Leno [of NBC's *Tonight Show*] does in his monologue every night. . . . Fair and balanced news is extremely hard to come by these days and if students can't get fair and balanced as well as entertaining, they'll just stick with what's entertaining.

—Allen, a blogger, responding to the article by Stapleton-Paff

However comedic the show might be, it is important to understand that fiction and comedy have largely been a critique of mainstream society. Just take a look at George Orwell's *Animal*

Chapter Focus Question

What effect do the media have on the perception of people in the United States and other nations regarding politics and the economy?

Farm. It's clearly fiction and somewhat comedic but it does raise valid points in the critique of society. *The Daily Show* has managed to do the same in a different manner.

—Jeff, another blogger, stating his point of view regarding the article

There are days when I watch *The Daily Show,* and I kind of chuckle. There are days when I laugh out loud. There are days when I stand up and point to the TV and say, "You're damn right!" . . . The stock-in-trade of *The Daily Show* is hypocrisy, exposing hypocrisy. And nobody else has the guts to do it. They really know how to crystallize an issue on all sides, see the silliness everywhere.

—Hub Brown, chair of the Communications Department at the S. I. Newhouse School of Public Communications and associate professor of journalism at Syracuse University, now refers to himself as a *The Daily Show* "convert" after he began watching the show in response to his students' comments that they greatly preferred it to mainstream news programs (qtd. in Smolkin, 2007).

Even as a massive financial crisis descended upon the United States and other nations of the world in the autumn of 2008 and Congress approved government bailouts of more than $700 billion, media audiences were drawn to pseudonews programs and political comedy on television (rather than mainstream media) to learn about the consequences of a burst in the housing bubble, widespread mortgage defaults, massive losses in financial institutions, and just how expensive the government bailouts might be for taxpayers and others. And these media audiences were not alone. In recent years, more television viewers have channel-checked away from network news broadcasts and late-night entertainment such as *The Tonight Show with Jay Leno* and *Late Night with David Letterman* to watch satirical "fake news" shows such as Jon Stewart's *The Daily Show* and *The Colbert Report,* both of which are broadcast on cable television's Comedy Central. For many individuals, these satirical programs have become an important source of news, and studies conducted at Indiana University and elsewhere have concluded that these programs are typically as substantive in their political coverage as the broadcast television networks' nightly newscasts are (see Fox, Koloen, and Sahin, 2007). The extent to which television entertainment and "hard news" have become blurred is related to the extent to which our major social institutions of politics and the economy are both reflective of and influenced by a mass media that constitute yet another powerful social institution in our society and around the world.

In twenty-first-century America, the subject of this chapter—the issue of politics and the global economy—is a hot topic for concerned people because we live in an age of discord regarding many decisions that have been made by our country's political and business leaders. Sociologists are concerned about how the social institutions of politics and the economy operate and how the decisions made in these social arenas affect people's everyday lives. In this chapter, we discuss the intertwining nature of contemporary politics, the economy, and the media. Before reading on, test your knowledge of the media by taking the quiz in Box 13.1.

Politics, Power, and Authority

Politics is the social institution through which power is acquired and exercised by some people and groups. In contemporary societies, the government is the primary political system. **Government is the formal organization that has the legal and political authority to regulate the relationships among members of a society and between the society and those outside its borders.** Some social analysts refer to the government as the **state—the political entity that possesses a legitimate monopoly over the use of force within its territory to achieve its goals.**

Whereas political science focuses primarily on power and its distribution in different types of political systems, *political sociology* is the area of sociology that examines politics and the government. Political sociology primarily focuses on the *social circumstances* of politics and explores how the political arena and its actors are intertwined with social institutions such as the economy, religion, education, and the media. According to the political analyst Michael Parenti (1998: 7–8), many people underrate the significance of politics in daily life:

> Politics is something more than what politicians do when they run for office. Politics is concerned with the struggles that shape social relations within societies and affairs between nations. The taxes and prices we pay and the jobs available to

politics the social institution through which power is acquired and exercised by some people and groups.

government the formal organization that has the legal and political authority to regulate the relationships among members of a society and between the society and those outside its borders.

state the political entity that possesses a legitimate monopoly over the use of force within its territory to achieve its goals.

Box 13.1 Sociology and Everyday Life

Answers to the Sociology Quiz on the Media

1. True. Mass media industries earn about $311 billion annually. Radio and television industries earn about $88 billion, newspapers earn $50 billion, movies and videos earn $73 billion, and book publishing, magazines, recordings, and Internet industries account for the rest.

2. False. Although most media outlets are privately owned, Public Broadcasting Service (PBS) and National Public Radio (NPR) are funded by government support, grants from nonprofit foundations, and donations from viewers and listeners.

3. False. The top ten newspaper chains control about 51 percent of total weekday newspaper circulation in the United States.

4. False. Previous marathon broadcast coverage occurred during the Johnson administration's involvement in the Vietnam War, the Nixon administration's involvement in the Watergate scandal, and the Reagan administration's involvement in Iran-Contra, a congressional investigation of how weapons were illegally supplied to the Nicaraguan Contras.

5. False. News programs such as *NBC Nightly News* typically have 19 minutes of news content and 11 minutes of commercials, promotional announcements, and other non-news items.

6. True. "Soft-money" contributions, which are made outside the limits imposed by federal election laws, have allegedly been used by both parties for campaign-style ads, but leaders of the national political parties claimed that their own ads were about social issues, not candidates.

7. False. Thirty-six percent of journalists identify themselves as Democrats, 18 percent as Republicans, and 46 percent as independents.

8. True. Because TV "reality shows" imitate news stories by reenacting events and interviewing crime victims, it is difficult for some viewers to determine whether they are watching "real" news.

Sources: Based on Accuracy in Media, 2006; Journalism.com, 2004; and U.S. Census Bureau, 2008.

us, the chances that we will live in peace or perish in war, the costs of education and the availability of scholarships, the safety of the airliner or highway we travel on, the quality of the food we eat and the air we breathe, the availability of affordable housing and medical care, the legal protections against racial and sexual discrimination—all the things that directly affect the quality of our lives are influenced in some measure by politics. . . . To say you are not interested in politics, then, is like saying you are not interested in your own well-being.

Using a power–conflict framework for his analysis, Parenti (1998) suggests that the media often distort—either intentionally or unintentionally—the information they provide to citizens. According to Parenti, the media have the power to influence public opinion in a way that favors management over labor, corporations over their critics, affluent whites over subordinate-group members in central cities, political officials over protestors, and free-market capitalism over public-sector development. Parenti's assertion raises an interesting issue about the distribution of power in the United States and other high-income nations: Do the media distort information to suit their own interests?

Power and Authority

Power is the ability of persons or groups to achieve their goals despite opposition from others (Weber, 1968/1922). Through the use of persuasion, author-

Max Weber's three types of global authority are shown here in global perspective. Pope Benedict XVI is an example of traditional authority sanctioned by custom. Mother Teresa exemplifies charismatic authority, for her leadership was based on personal qualities. The U.S. Supreme Court represents rational–legal authority, which depends upon established rules and procedures.

ity, or force, some people are able to get others to acquiesce to their demands. Consequently, power is a *social relationship* that involves both leaders and followers. Power is also a dimension in the structure of social stratification. Persons in positions of power control valuable resources of society—including wealth, status, comfort, and safety—and are able to direct the actions of others while protecting and enhancing the privileged social position of their class (Domhoff, 2002). For example, the sociologist G. William Domhoff (2002) argues that the media tend to reflect "the biases of those with access to them—corporate leaders, government officials, and policy experts." However, although Domhoff believes the media can amplify the message of powerful people and marginalize the concerns of others, he does not think the media are as important as government officials and corporate leaders are in the U.S. power equation.

What about power on a global basis? Although the most basic form of power is physical violence or force, most political leaders do not want to base their power on force alone. Instead, they seek to legitimize their power by turning it into ***authority*—power that people accept as legitimate rather than coercive.**

Ideal Types of Authority

Who is most likely to accept authority as legitimate and adhere to it? People have a greater tendency to accept authority as legitimate if they are economically or politically dependent on those who hold power. They may also accept authority more readily if it reflects their own beliefs and values (Turner, Beeghley, and Powers, 2007). Weber's outline of three *ideal types* of authority—traditional, charismatic, and rational–legal—shows how different bases of legitimacy are tied to a society's economy.

power according to Max Weber, the ability of people or groups to achieve their goals despite opposition from others.

authority power that people accept as legitimate rather than coercive.

Traditional Authority According to Weber, *traditional authority* **is power that is legitimized on the basis of long-standing custom.** In preindustrial societies, the authority of traditional leaders, such as kings, queens, pharaohs, emperors, and religious dignitaries, is usually grounded in religious beliefs and custom. For example, British kings and queens have historically traced their authority from God. Members of subordinate classes obey a traditional leader's edicts out of economic and political dependency and sometimes personal loyalty. However, as societies industrialize, traditional authority is challenged by a more complex division of labor and by the wider diversity of people who now inhabit the area as a result of high immigration rates.

Gender, race, and class relations are closely intertwined with traditional authority. Political scientist Zillah R. Eisenstein (1994) suggests that *racialized patriarchy*—the continual interplay of race and gender—reinforces traditional structures of power in contemporary societies. According to Eisenstein (1994: 2), "Patriarchy differentiates women from men while privileging men. Racism simultaneously differentiates people of color from whites and privileges whiteness. These processes are distinct but intertwined."

Charismatic Authority *Charismatic authority* **is power legitimized on the basis of a leader's exceptional personal qualities or the demonstration of extraordinary insight and accomplishment that inspire loyalty and obedience from followers.** Charismatic leaders may be politicians, soldiers, or entertainers, among others.

Charismatic authority tends to be temporary and relatively unstable; it derives primarily from individual leaders (who may change their minds, leave, or die) and from an administrative structure usually limited to a small number of faithful followers. For this reason, charismatic authority often becomes routinized. The *routinization of charisma* **occurs when charismatic authority is succeeded by a bureaucracy controlled by a rationally established authority or by a combination of traditional and bureaucratic authority** (Turner, Beeghley, and Powers, 2007). According to Weber (1968/1922: 1148), "It is the fate of charisma to recede . . . after it has entered the permanent structures of social action."

Rational–Legal Authority According to Weber, *rational–legal authority* **is power legitimized by law or written rules and regulations.** Rational–legal authority—also known as *bureaucratic authority*—is based on an organizational structure that includes a clearly defined division of labor, hierarchy of authority, formal rules, and impersonality. Power is legitimized by procedures; if leaders obtain their positions in a procedurally correct manner (such as by election or appointment), they have the right to act.

Rational–legal authority is held by elected or appointed government officials and by officers in a formal organization. However, authority is invested in the *office,* not in the *person* who holds the office. For example, although the U.S. Constitution grants rational–legal authority to the office of the presidency, a president who fails to uphold the public trust may be removed from office. In contemporary society, the media may play an important role in bringing to light allegations about presidents or other elected officials. Examples include the media blitzes surrounding the Watergate investigation of the 1970s that led to the resignation of President Richard M. Nixon and the late 1990s political firestorm over campaign fund-raising and the sex scandal involving President Bill Clinton.

In a rational–legal system, the governmental bureaucracy is the apparatus responsible for creating and enforcing rules in the public interest. Weber believed that rational–legal authority was the only means to attain efficient, flexible, and competent regulation under a rule of law (Turner, Beeghley, and Powers, 2007). Weber's three types of authority are summarized in the Concept Quick Review.

Political Systems in Global Perspective

Political systems as we know them today have evolved slowly. In the earliest societies, politics was not an entity separate from other aspects of life. Political institutions first emerged in agrarian societies as they acquired surpluses and developed greater social inequality. Elites took control of politics and used custom or traditional authority to justify their

CONCEPT QUICK REVIEW

Weber's Three Types of Authority

	Description	Examples
Traditional	Legitimized by long-standing custom.	Patrimony (authority resides in traditional leader supported by larger social structures, as in old British monarchy).
	Subject to erosion as traditions weaken.	Patriarchy (rule by men occupying traditional positions of authority, as in the family).
Charismatic	Based on leader's personal qualities.	Napoleon
	Temporary and unstable.	Adolf Hitler
		Martin Luther King, Jr.
		César Chávez
		Mother Teresa
Rational–Legal	Legitimized by rationally established rules and procedures.	Modern British Parliament.
	Authority resides in the office, not the person.	U.S. presidency, Congress, federal bureaucracy.

position. When cities developed circa 3500–3000 B.C.E., the *city-state*—a city whose power extended to adjacent areas—became the center of political power.

Nation-states as we know them began to develop in Europe between the twelfth and fifteenth centuries (see Tilly, 1975). A *nation-state* is a unit of political organization that has recognizable national boundaries and whose citizens possess specific legal rights and obligations. Nation-states emerge as countries develop specific geographic territories and acquire greater ability to defend their borders. Improvements in communication and transportation make it possible for people in a larger geographic area to share a common language and culture. As charismatic and traditional authority are superseded by rational–legal authority, legal standards come to prevail in all areas of life, and the nation-state claims a monopoly over the legitimate use of force (Kennedy, 1993).

Approximately 190 nation-states currently exist throughout the world; today, everyone is born, lives, and dies under the auspices of a nation-state (see Skocpol and Amenta, 1986). Four main types of political systems are found in nation-states: monarchy, authoritarianism, totalitarianism, and democracy.

Monarchy

Monarchy **is a political system in which power resides in one person or family and is passed from generation to generation through lines of inheritance.** Monarchies are most common in agrarian

traditional authority power that is legitimized on the basis of long-standing custom.

charismatic authority power legitimized on the basis of a leader's exceptional personal qualities or the demonstration of extraordinary insight and accomplishment that inspire loyalty and obedience from followers.

routinization of charisma the process by which charismatic authority is succeeded by a bureaucracy controlled by a rationally established authority or by a combination of traditional and bureaucratic authority.

rational–legal authority power legitimized by law or written rules and regulations.

monarchy a political system in which power resides in one person or family and is passed from generation to generation through lines of inheritance.

societies and are associated with traditional authority patterns. However, the relative power of monarchs has varied across nations, depending on religious, political, and economic conditions.

Absolute monarchs claim a hereditary right to rule (based on membership in a noble family) or a divine right to rule (a God-given right to rule that legitimizes the exercise of power). In limited monarchies, rulers depend on powerful members of the nobility to retain their thrones. Unlike absolute monarchs, *limited monarchs* are not considered to be above the law. In *constitutional monarchies,* the royalty serve as symbolic rulers or heads of state while actual authority is held by elected officials in national parliaments. In present-day monarchies such as the United Kingdom, Sweden, Spain, and the Netherlands, members of royal families primarily perform ceremonial functions. In the United Kingdom, for example, the media often focus large amounts of time and attention on the royal family, but concentrate on the personal lives of its members.

Authoritarianism

Authoritarianism **is a political system controlled by rulers who deny popular participation in government.** A few authoritarian regimes have been absolute monarchies whose rulers claimed a hereditary right to their position. Today, Saudi Arabia and Kuwait are examples of authoritarian absolute monarchies. In *dictatorships,* power is gained and held by a single individual. Pure dictatorships are rare; all rulers need the support of the military and the backing of business elites to maintain their position. *Military juntas* result when military officers seize power from the government, as has happened in recent decades in Argentina, Chile, and Haiti. Today, authoritarian regimes exist in Cuba and in the People's Republic of China.

Totalitarianism

Totalitarianism **is a political system in which the state seeks to regulate all aspects of people's public and private lives.** Totalitarianism relies on modern technology to monitor and control people; mass propaganda and electronic surveillance are widely used to influence people's thinking and control their

© Samir Hussein/WireImage/Getty Images

Through its many ups and downs, the British royal family has remained a symbol of Great Britain's monarchy, today headed by Queen Elizabeth. Monarchies typically pass power from generation to generation, and the two young men on the left, Princes William and Harry, represent the future of the royal family's rule.

actions. One example of a totalitarian regime was the National Socialist (Nazi) Party in Germany during World War II; military leaders there sought to control all aspects of national life, not just government operations. Other examples include the former Soviet Union and contemporary Iraq before the fall of Saddam Hussein's regime.

To keep people from rebelling, totalitarian governments enforce conformity. People are denied the right to assemble for political purposes, access to information is strictly controlled, and secret police enforce compliance, creating an environment of constant fear and suspicion.

Many nations do not recognize totalitarian regimes as being the legitimate government for a particular country. Afghanistan in the year 2001 was an example. As the war on terrorism began in the aftermath of the September 11 terrorist attacks on the United States, many people developed a heightened awareness of the Taliban regime, which ruled most of Afghanistan and was engaged in fierce fighting to capture the rest of the country. The Taliban regime maintained absolute control over the Afghan people in most of that country. For example, it required that all Muslims take part in prayer five times each day and that men attend prayer at mosques, where women were forbidden (Marquis, 2001). Taliban leaders claimed that their actions were based on Muslim law and espoused a belief in never-ending *jihad*—a struggle against one's perceived enemies. Although the totalitarian nature of the Taliban regime was difficult for many people, it was particularly oppressive for women, who were viewed by this group as being "biologically, religiously and prophetically" inferior to men (McGeary, 2001: 41). Consequently, this regime made the veil obligatory and banned women from public life. U.S. government officials believed that the Taliban regime was protecting Osama bin Laden, the man thought to have been the mastermind behind numerous terrorist attacks on U.S. citizens and facilities, both on the mainland and abroad. As a totalitarian regime, the Taliban leadership was recognized by only three other governments despite controlling most of Afghanistan.

Once the military action commenced in Afghanistan, most of what U.S. residents knew about the Taliban and about the war on terrorism was based on media accounts and "expert opinions" that were voiced on television. According to the political analyst Michael Parenti (1998), the media play a significant role in framing the information we receive about the political systems of other countries. As you will recall, *framing* refers to how news is packaged, including the amount of exposure given to a story, the positive or negative tone of the story, the headlines and photographs, and the accompanying visual and auditory effects if the story is being broadcast. The war in Afghanistan, like other wars in the past, was typically framed as a fight to save freedom and democracy.

Democracy

***Democracy* is a political system in which the people hold the ruling power either directly or through elected representatives.** The literal meaning of *democracy* is "rule by the people" (from the Greek words *demos,* meaning "the people," and *kratein,* meaning "to rule"). In an ideal-type democracy, people would actively and directly rule themselves. *Direct participatory democracy* requires that citizens be able to meet together regularly to debate and decide the issues of the day. However, if all 303 million people in the United States came together in one place

for a meeting, they would occupy an area of more than seventy square miles, and a single round of five-minute speeches would require more than five thousand years (based on Schattschneider, 1969).

In countries such as the United States, Canada, Australia, and the United Kingdom, people have a voice in the government through ***representative democracy*, whereby citizens elect representatives to serve as bridges between themselves and the government.** The U.S. Constitution requires that each state have two senators and a minimum of one member in the House of Representatives. The current size of the House (435 seats) has not changed since the apportionment following the 1910 census. Therefore, based on Census 2000, those 435 seats were reapportioned based on the increase or decrease in a state's population between 1990 and 2000.

In a representative democracy, elected representatives are supposed to convey the concerns and interests of those they represent, and the government is expected to be responsive to the wishes of the people. Elected officials are held accountable to the people through elections. However, representative democracy is not always equally accessible to all people in a nation. Throughout U.S. history, members of subordinate racial–ethnic groups have been denied full participation in the democratic process. Gender and social class have also limited some people's democratic participation. For example, women have not always had the same rights as men. Full voting rights were not gained by women until ratification of the Nineteenth Amendment in 1920.

---

authoritarianism a political system controlled by rulers who deny popular participation in government.

totalitarianism a political system in which the state seeks to regulate all aspects of people's public and private lives.

democracy a political system in which the people hold the ruling power either directly or through elected representatives.

representative democracy a form of democracy whereby citizens elect representatives to serve as bridges between themselves and the government.

Even representative democracies are not all alike. As compared to the winner-takes-all elections in the United States, which are usually decided by who wins the most votes, the majority of European elections are based on a system of proportional representation, meaning that each party is represented in the national legislature according to the proportion of votes that party received. For example, a party that won 40 percent of the vote would receive 40 seats in a 100-seat legislative body, and a party receiving 20 percent of the votes would receive 20 seats.

Perspectives on Power and Political Systems

Is political power in the United States concentrated in the hands of the few or distributed among the many? Sociologists and political scientists have suggested many different answers to this question; however, two prevalent models of power have emerged: pluralist and elite.

Functionalist Perspectives: The Pluralist Model

The pluralist model is rooted in a functionalist perspective which assumes that people share a consensus on central concerns, such as freedom and protection from harm, and that the government serves important functions no other institution can fulfill. According to Emile Durkheim (1933/1893), the purpose of government is to socialize people to be good citizens, to regulate the economy so that it operates effectively, and to provide necessary services for citizens. Contemporary functionalists state the four main functions as follows: (1) maintaining law and order, (2) planning and directing society, (3) meeting social needs, and (4) handling international relations, including warfare (see ▶ Figure 13.1).

But what happens when people do not agree on specific issues or concerns? Functionalists suggest that divergent viewpoints lead to a system of political pluralism, in which the government functions as an arbiter between competing interests and viewpoints. According to the *pluralist model,* **power in**

▶ Figure 13.1 **Government from a Functionalist Perspective**

From the functionalist perspective, government serves important functions no other institution can fulfill. Contemporary functionalists identify four main functions:
(a) maintaining law and order,
(b) planning and directing society,
(c) meeting social needs, and
(d) handling international relations, including warfare.

political systems is widely dispersed throughout many competing interest groups (Dahl, 1961).

In the pluralist model, the diverse needs of women and men, people of all religions and racial–ethnic backgrounds, and the wealthy, middle class, and poor are met by political leaders who engage in a process of bargaining, accommodation, and compromise. Competition among leadership groups in government, business, labor, education, law, medicine, and consumer organizations, among others, helps prevent abuse of power by any one group. Everyday people can influence public policy by voting in elections, participating in existing special interest groups, or forming new ones to gain access to the political system. In sum, power is widely dispersed, and leadership groups that wield influence on some decisions are not the same groups that may be influential in other decisions (Dye and Zeigler, 2006).

Special Interest Groups *Special interest groups* are political coalitions made up of individuals or groups that share a specific interest they wish to protect or advance with the help of the political system (Greenberg and Page, 2002). Examples of special interest groups include the AFL-CIO (representing the majority of labor unions) and public interest or citizens' groups such as the American Conservative Union and Zero Population Growth.

What purpose do special interest groups serve in the political process? According to some analysts, special interest groups help people advocate their own interests and further their causes. Broad categories of special interest groups include banking, business, education, energy, the environment, health, labor, persons with a disability, religious groups, retired persons, women, and those espousing a specific ideological viewpoint; obviously, many groups overlap in interests and membership. Special interest groups are also referred to as *pressure groups* (because they put pressure on political leaders) or *lobbies*. Lobbies are often referred to in terms of the organization they represent or the single issue on which they focus—for example, the "gun lobby" and the "dairy lobby." The people who are paid to influence legislation on behalf of specific clients are referred to as *lobbyists*.

Over the past four decades, special interest groups have become more involved in "single-issue

Special interest groups help people advocate their interests and further their causes. Advocates may run for public office and gain a wider voice in the political process. An example is former U.S. Senator Ben Nighthorse Campbell of Colorado, a Cheyenne chief, who is shown here walking to the Senate floor after participating in a ground-breaking ceremony for the National Museum of the American Indian in Washington, D.C.

politics," in which political candidates are often supported or rejected solely on the basis of their views on a specific issue—such as abortion, gun control, gay and lesbian rights, or the environment. Single-issue groups derive their strength from the intensity of their beliefs; leaders have little room to compromise on issues.

pluralist model an analysis of political systems that views power as widely dispersed throughout many competing interest groups.

Political Action Committees Funding of lobbying efforts has become more complex in recent years. Reforms in campaign finance laws in the 1970s set limits on direct contributions to political candidates and led to the creation of *political action committees* (PACs)—organizations of special interest groups that solicit contributions from donors and fund campaigns to help elect (or defeat) candidates based on their stances on specific issues.

As the cost of running for political office has skyrocketed, candidates have relied more on PACs for financial assistance. PACs contributed more than $316,800,000 to candidates for the U.S. House and Senate between January 1, 2005, and October 18, 2006, representing 28 percent of all contributions received (Federal Election Commission, 2006). Advertising, staff, direct-mail operations, telephone banks, computers, consultants, travel expenses, office rentals, and other expenses incurred in political campaigns make PAC money vital to candidates.

Some PACs represent the "public interest" and ideological interest groups such as gay rights or the National Rifle Association. Other PACs represent the capitalistic interests of large corporations. Realistically, members of the least-privileged sectors of society are not represented by PACs. As one senator pointed out, "There aren't any Poor PACs or Food Stamp PACs or Nutrition PACs or Medicare PACs" (qtd. in Greenberg and Page, 1993: 240). Critics of pluralism argue that "Big Business" wields such disproportionate power in U.S. politics that it undermines the democratic process (see Lindblom, 1977; Domhoff, 1978).

As an outgrowth of record-setting campaign spending in the 1996 national election, campaign financing abuses were alleged by both Republicans and Democrats in Washington. At the center of the controversy was the issue of "soft-money" contributions, which are made outside the limits imposed by federal election laws. In 2002, Congress passed the McCain–Feingold campaign finance law, prohibiting soft-money contributions in federal elections, and the U.S. Supreme Court upheld the soft-money provisions of that law in 2003. However, the McCain–Feingold law applies only to federal elections and does not bar soft-money contributions in state and local elections.

Conflict Perspectives: Elite Models

Although conflict theorists acknowledge that the government serves a number of important purposes in society, they assert that government exists for the benefit of wealthy or politically powerful elites who use the government to impose their will on the masses. According to the *elite model,* **power in political systems is concentrated in the hands of a small group of elites and the masses are relatively powerless.** The pluralist model and the elite model are compared in ▶ Figure 13.2.

Contemporary elite models are based on the assumption that decisions are made by the elites, who agree on the basic values and goals of society. However, the needs and concerns of the masses are not often given full consideration by those in the elite. According to this approach, power is highly concentrated at the top of a pyramid-shaped social hierarchy, and public policy reflects the values and preferences of the elite, not the preferences of the people (Dye and Zeigler, 2006).

C. Wright Mills and The Power Elite Who makes up the U.S. power elite? According to the sociologist C. Wright Mills (1959a), the *power elite* comprises leaders at the top of business, the executive branch of the federal government, and the military. Of these three, Mills speculated that the "corporate rich" (the highest-paid officers of the biggest corporations) are the most powerful because of their unique ability to parlay the vast economic resources at their disposal into political power (see "Sociology Works!" on pages 440–441). At the middle level of the pyramid, Mills placed the legislative branch of government, special interest groups, and local opinion leaders. The bottom (and widest layer) of the pyramid is occupied by the unorganized masses, who are relatively powerless and are vulnerable to economic and political exploitation.

G. William Domhoff and the Ruling Class Sociologist G. William Domhoff (2002) asserts that, in fact, this nation has a *ruling class*—the corporate rich, who make up less than 1 percent of the U.S. population. Domhoff uses the term *ruling class* to

PLURALIST MODEL

- Decisions are made on behalf of the people by leaders who engage in bargaining, accommodation, and compromise.

- Competition among leadership groups makes abuse of power by any one group difficult.

- Power is widely dispersed, and people can influence public policy by voting.

- Public policy reflects a balance among competing interest groups.

▶ Figure 13.2 **Pluralist and Elite Models**

ELITE MODEL

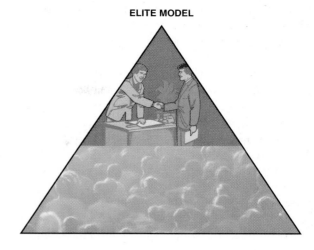

- Decisions are made by a small group of elite people.

- Consensus exists among the elite on the basic values and goals of society.

- Power is highly concentrated at the top of a pyramid-shaped social hierachy.

- Public policy reflects the values and preferences of the elite.

signify a relatively fixed group of privileged people who wield sufficient power to constrain political processes and serve underlying capitalist interests. Although the power elite controls the everyday operation of the political system, who *governs* is less important than who *rules.*

According to Domhoff (2002), the corporate rich influence the political process in three ways. First, they affect the candidate selection process by helping to finance campaigns and providing favors to political candidates. Second, through participation in the special interest process, the corporate rich are able to obtain favors, tax breaks, and favorable regulatory rulings. Finally, the corporate rich gain access to the policy-making process through their appointments to governmental advisory committees, presidential commissions, and other governmental positions.

Power elite models call our attention to a central concern in contemporary U.S. society: the ability of democracy and its ideals to survive in the context of the increasingly concentrated power held by capitalist oligarchies such as the media giants we discuss in this chapter.

political action committees organizations of special interest groups that solicit contributions from donors and fund campaigns to help elect (or defeat) candidates based on their stances on specific issues.

elite model a view of society that sees power in political systems as being concentrated in the hands of a small group of elites whereas the masses are relatively powerless.

Sociology *Works!*

C. Wright Mills's Ideas About the Media: Ahead of His Time?

In a mass society the dominant type of communication is the formal media, and the [various audiences] become mere *media markets:* all those exposed to the contents of a given mass media.

> —In *The Power Elite,* the influential sociologist C. Wright Mills (1959a: 299) provided this early critique of the U.S. mass media and its effect on the general public.

During the mid-twentieth century, when Mills wrote these words, the press in the United States was often described as a "watchdog for the public." It was widely believed that journalists would reliably report what the public needed to know to stay informed, including any government and/or corporate dishonesty or ineptitude (Headley, 1985: 333). However, C. Wright Mills rejected this notion because he believed the media industry lulled individuals into complacency and persuaded them to accept the status quo rather than encouraging the audience member to become a "public" person who formulates his or her own opinion on a given topic. As one contemporary scholar explained, "To recover Mills's 'public man' who is capable of formulating opinions, we would have to salvage whatever is left of him after he is stripped of MTV, CNN, and http://www.,

whose Italian translation, *ragnatele mondiale,* is useful in conveying the sense 'spider-web' and therefore the suspicion that we may be flies" (Cinquemani, 1997: 89).

Do the contemporary media inform people or trap them in a "spiderweb"? Mills hoped that individuals would develop the capacity for critical judgment based on information they received from the media industry; however, he was doubtful that this would occur for two important sociological reasons: (1) media communication involves a limited number of people who communicate *to* a great number of others ("the masses"), and (2) audiences have no effective way of answering back, making mass communication a one-way process.

As part of his legacy for making sociology work in the real world, Mills made an important prediction. According to Mills (1956: 303–304), critical thinking is more likely to occur among individuals under the following circumstances: "If public communications are so organized that there is a chance immediately and effectively to answer back any opinion expressed in public. Opinion formed by such discussion readily finds an outlet in effective action, even against—if necessary—the prevailing system of authority." If media audiences are able to respond immediately and

The U.S. Political System

The U.S. political system is made up of formal elements, such as the legislative process and the duties of the president, and informal elements, such as the role of political parties in the election process. We now turn to an examination of these informal elements, including political parties, political socialization, and voter participation.

Political Parties and Elections

A *political party* is an organization whose purpose is to gain and hold legitimate control of government; it is usually composed of people with simi-

lar attitudes, interests, and socioeconomic status. A political party (1) develops and articulates policy positions, (2) educates voters about issues and simplifies the choices for them, and (3) recruits candidates who agree with those policies, helps those candidates win office, and holds the candidates responsible for implementing the party's policy positions. In carrying out these functions, a party may try to modify the demands of special interests, build a consensus that could win majority support, and provide simple and identifiable choices for the voters on election day. Political parties create a *platform,* a formal statement of the party's political positions on various social and economic issues.

Since the Civil War, the Democratic and Republican parties have dominated the U.S. political system. Although one party may control the presidency

C. Wright Mills believed that mass communication was a one-way street: The audience had no way to react to it and tended to become passive recipients of its content. However, today's digital technology allows people to respond quickly and, through developments such as web logs, to provide news content and opinions of their own. What are the benefits and drawbacks of interactive technologies?

provide their own opinions, media communications may become a two-way street. Technologies and methods of communication—including cell phones, text messaging, e-mail, and websites—now provide us with opportunities for media interactivity through which we can respond to media discourse and make our thoughts known.

Sociology works today in Mills's ideas because he encourages us to think for ourselves and to express our ideas and opinions rather than being passive recipients of information from the media. Mills challenges us to think about the extent to which we rely on *mediated* experiences of others, such as the ideas presented by television commentators or online bloggers, as compared with becoming our own "public person" and responding with thoughts and judgments of our own.

Reflect & Analyze

How might we relate Mills's ideas to the contemporary role of the media industries in shaping political and economic "realities" around the world?

---

for several terms, at some point the voters elect the other party's nominee, and control shifts.

How well do the parties measure up to the ideal-type characteristics of a political party? Although both parties have been successful in getting their candidates elected at various times, they generally do not meet the ideal characteristics for a political party because they do not offer voters clear policy alternatives. Moreover, the two parties are oligarchies, dominated by active elites who hold views that are further from the center of the political spectrum than are those of a majority of members of their party. As a result, voters in primary elections (in which the nominees of political parties for most offices other than president and vice president are chosen) may select nominees whose views are closer to the center of the political spectrum and further away from the party's own platform. Likewise,

party loyalties appear to be declining among voters, who may vote in one party's primary but then cast their ballot in general elections without total loyalty to that party, or cast a "split-ticket" ballot (voting for one party's candidate in one race and another party's candidate in the next one).

Political Participation and Voter Apathy

Why do some people vote and others not? How do people come to think of themselves as being conservative, moderate, or liberal? Key factors include individuals' political socialization and attitudes.

---

political party an organization whose purpose is to gain and hold legitimate control of government.

As gasoline prices soared during the summer of 2008, presidential hopeful Senator John McCain backed an initiative to eliminate federal gas taxes until prices eased. Are political decisions sometimes linked to a person's political aspirations?

***Political socialization* is the process by which people learn political attitudes, values, and behavior.** For young children, the family is the primary agent of political socialization, and children tend to learn and hold many of the same opinions held by their parents. By the time children reach school age, they typically identify with the political party (if any) of their parents (Burnham, 1983). As we grow older, other agents of socialization begin to affect our political beliefs, including our peers, teachers, and the media. Over time, these other agents may cause people's political attitudes and values to change.

In addition to the socialization process, people's socioeconomic status affects their political attitudes, values, and beliefs. For example, individuals who are very poor or are unable to find employment may believe that society has failed them and therefore tend to be indifferent toward the political system (Zipp, 1985; Pinderhughes, 1986). Believing that casting a ballot would make no difference to their circumstances, they do not vote.

Democracy in the United States has been defined as a government "of the people, by the people, and for the people." Accordingly, it would stand to reason that "the people" would actively participate in their government at any or all of four levels: (1) voting, (2) attending and taking part in political meetings, (3) actively participating in political campaigns, and (4) running for and/or holding political office. At most, about 10 percent of the voting-age population in this country participates at a level higher than simply voting, and over the past 40 years, less than half of the voting-age population has voted in nonpresidential elections. Even in presidential elections, voter turnout is often relatively low. Slightly over 54 percent of the voting-age population (age 18 and older) voted in the 2004 presidential election—an election in which the winning candidate (George W. Bush) received only slightly more than one-half of all votes cast.

The United States has one of the lowest percentages of voter turnout of all Western nations. In many of the other Western nations, the average turnout is between 80 and 90 percent of all eligible voters. Why is it that so many eligible voters in this country stay away from the polls? During any election, millions of voting-age persons do not go to the polls due to illness, disability, lack of transportation, nonregistration, or absenteeism. However, these explanations do not account for why many other people do not vote. According to some conservative analysts, people may not vote because they are satisfied with the status quo or because they are apathetic and un-

Although U.S. voting rates are very low—just over half of the eligible population cast a vote in the 2004 presidential election—hotly contested elections will still bring voters out, as shown in the long line of people waiting to vote in Hawaii's 2008 Democratic presidential caucus.

◆ **Table 13.1** Voter Preferences in the 2004 Presidential Election, by Selected Characteristics

		Republican	Democrat
Gender	Men	55%	44%
	Women	48	52
Race/Ethnicity	Whites	58	41
	African Americans	11	88
	Latinos/as	43	56
	Asian Americans	41	58
Age	18–29	45	54
	30–44	53	46
	45–59	51	48
	60 and older	54	46
Sexual Orientation	Gay, lesbian, bisexual	23	77
Education	Did not graduate from high school	49	50
	High school graduate	52	47
	Some college	54	46
	College graduate	52	46
Region	Eastern U.S.	44	56
	Midwest	51	49
	Southern U.S.	57	43
	Western U.S.	49	51
Family Income	Under $15,000	36	63
	$15,000–$29,999	42	57
	$30,000–$49,999	48	50
	Over $50,000	56	43

Note: Due to the presence of other candidates on the ballot, not all rows total 100 percent.
Source: Connelly, 2004.

informed—they lack an understanding of both public issues and the basic processes of government.

By contrast, liberals argue that people stay away from the polls because they feel alienated from politics at all levels of government—federal, state, and local—due to political corruption and influence peddling by special interests and large corporations. Participation in politics is influenced by gender, age, race/ethnicity, and, especially, socioeconomic status (SES). The rate of participation increases as a person's SES increases. ◆ Table 13.1 shows voting preferences in the 2004 presidential election by gender, race/ethnicity, age, sexual orientation, education, region, and income. One explanation for the higher rates of political participation at higher SES levels is that advanced levels of education may give people a better understanding of government processes, a belief that they have more at stake in the political process, and greater economic resources to contribute to the process. Some studies suggest that during their college years, many people develop assumptions about political participation that continue throughout their lives.

Governmental Bureaucracy

When most people think of political power, they overlook one of its major sources—the governmental bureaucracy. Negative feelings about bureaucracy are perhaps strongest when people are describing the "red tape" and "faceless bureaucrats" in government with whom they must deal. But who are these "faceless bureaucrats," and what do they do?

political socialization the process by which people learn political attitudes, values, and behavior.

Roughly 65 percent of the top-echelon positions in the federal government are held by white men (U.S. Office of Personnel Management, 2004).

Because rising to the top of the bureaucracy may take as much as twenty years, some analysts argue that white women and all people of color simply have not held positions in the federal government long enough to reach the top positions. Others argue that the entrenched rules of this bureaucracy, one of the most powerful in the world, weigh against the promotion of those individuals who are not white and male.

The governmental bureaucracy has been able to perpetuate itself and expand because many of its employees have highly specialized knowledge and skills and cannot be replaced easily by "outsiders." In addition, as the United States has grown in size and complexity, public policy is made more by bureaucrats than by elected officials. For example, offices and agencies have been established to create rules, policies, and procedures for dealing with complex issues such as nuclear power, environmental protection, and drug safety; bureaucracies announce an estimated twenty rules or regulations for every one law passed by Congress (Dye and Zeigler, 2006).

The federal budget is the central ingredient in the bureaucracy. Preparing the annual federal budget is a major undertaking for the president and the Office of Management and Budget, one of the most important agencies in Washington. Getting the budget approved by Congress is an even more monumental task; however, even with the highly publicized wrangling over the budget by the president and Congress, the final congressional appropriations are usually within 2–3 percent of the budget originally proposed by the president (Dye and Zeigler, 2006).

In part, this is due to the *iron triangle of power*—a three-way arrangement in which a private interest group (usually a corporation), a congressional committee or subcommittee, and a bureaucratic agency make the final decision on a political issue that is to be decided by that agency. ▶ Figure 13.3 illustrates the alliance among the Defense Department (Pentagon), private military (or defense) contractors, and members of Congress. According to the sociologist Joe Feagin,

> The Iron Triangle has a revolving door of money, influence, and jobs among these three sets of actors, involving trillions of dollars. Military con-

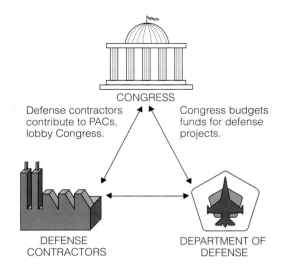

▶ Figure 13.3 **Example of the Iron Triangle of Power**

tractors who receive contracts from the Defense Department serve on the advisory committees that recommend what weapons they believe are needed. Many people move around the triangle from job to job, serving in the military, then in the Defense Department, then in military industries. (Feagin and Feagin, 1994: 405)

The Iron Triangle is also referred to as the ***military–industrial complex—the mutual interdependence of the military establishment and private military contractors.*** According to the sociologist C. Wright Mills (1976), this alliance of economic, military, and political power amounts to a "permanent war economy" or "military economy." However, the economist John Kenneth Galbraith (1985) has argued that the threat of war is good for the economy because government money spent on military preparedness stimulates the private sector of the economy, creates jobs, and encourages consumer spending. For example, at the beginning of the war on terrorism, the Pentagon awarded what was expected to become the largest military contract in U.S. history to Lockheed Martin to build a new generation of supersonic stealth fighter jets. The contract was expected to be worth more than $200 billion, creating thousands of new jobs in an economy that was struggling at the time of the announcement (Dao, 2000). Issues regarding the military–industrial complex are intricately linked to problems in the economy, which we will examine now.

Economic Systems in Global Perspective

The *economy* **is the social institution that ensures the maintenance of society through the production, distribution, and consumption of goods and services.** *Goods* are tangible objects that are necessary (such as food, clothing, and shelter) or desired (such as DVDs and electric toothbrushes). *Services* are intangible activities for which people are willing to pay (such as dry cleaning, a movie, or medical care).

Preindustrial, Industrial, and Postindustrial Economies

In all societies, the specific method of producing goods is related to the technoeconomic base of the society, as discussed in Chapter 10. In each society, people develop an economic system, ranging from simple to very complex, for the sake of survival.

Preindustrial economies include hunting and gathering, horticultural and pastoral, and agrarian societies. Most workers engage in ***primary sector production*—the extraction of raw materials and natural resources from the environment.** These materials and resources are typically consumed or used without much processing. The production units in hunting and gathering societies are small; most goods are produced by family members. The division of labor is by age and by gender (Hodson and Sullivan, 2008). The potential for producing surplus goods increases as people learn to domesticate animals and grow their own food. In horticultural and pastoral societies, the economy becomes distinct from family life. The distribution process becomes more complex with the accumulation of a *surplus* such that some people can engage in activities other than food production. In agrarian societies, production is related primarily to producing food. However, workers have a greater variety of

specialized tasks, such as warlord or priest; for example, warriors are necessary to protect the surplus goods from plunder by outsiders (Hodson and Sullivan, 2008).

Industrial economies result from sweeping changes to the system of production and distribution of goods and services during industrialization. Drawing on new forms of energy (such as steam, gasoline, and electricity) and machine technology, factories proliferate as the primary means of producing goods. Most workers engage in ***secondary sector production*—the processing of raw materials (from the primary sector) into finished goods.** For example, steel workers process metal ore; auto workers then convert the ore into automobiles, trucks, and buses. In industrial economies, work becomes specialized and repetitive, activities become bureaucratically organized, and workers primarily work with machines instead of with one another. With the emergence of mass production, larger surpluses are generated, typically benefiting some people and organizations but not others.

A *postindustrial economy* is based on ***tertiary sector production*—the provision of services rather than goods**—as a primary source of livelihood for workers and profit for owners and corporate shareholders. Tertiary sector production includes a wide range of activities, such as fast-food service, transportation, communication, education, real estate, advertising, sports, and entertainment.

military–industrial complex the mutual interdependence of the military establishment and private military contractors.

economy the social institution that ensures the maintenance of society through the production, distribution, and consumption of goods and services.

primary sector production the sector of the economy that extracts raw materials and natural resources from the environment.

secondary sector production the sector of the economy that processes raw materials (from the primary sector) into finished goods

tertiary sector production the sector of the economy that is involved in the provision of services rather than goods.

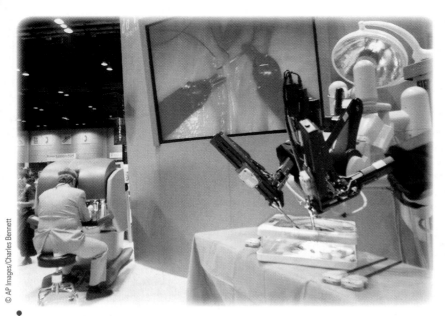

© AP Images/Charles Bennett

The nature of work has changed dramatically over the past century. Not all social scientists agree on whether robots will eventually replace many humans in the work force. What do you think work will be like in the future?

Capitalism and socialism are the principal economic models in industrial and postindustrial countries. As we examine these two models, keep in mind that no society has a purely capitalist or socialist economy.

Capitalism

Capitalism **is an economic system characterized by private ownership of the means of production, from which personal profits can be derived through market competition and without government intervention.** Most of us think of ourselves as "owners" of private property because we own a car, a television, or other possessions. However, most of us are not capitalists; we *spend money* on the things we own rather than *make money* from them. Only a relatively few people own income-producing property from which a profit can be realized by producing and distributing goods and services; everyone else is a consumer. "Ideal" capitalism has four distinctive features: (1) private ownership of the means of production, (2) pursuit of personal profit, (3) competition, and (4) lack of government intervention.

Private Ownership of the Means of Production Capitalist economies are based on the right of individuals to own income-producing property, such as land, water, mines, and factories, and the right to "buy" people's labor. However, under early monopoly capitalism (1890–1940), most ownership shifted from individuals to huge *corporations*—**organizations that have legal powers, such as the ability to enter into contracts and buy and sell property, separate from their individual owners.** In advanced monopoly capitalism (1940–present), ownership and control of major industrial and business sectors have become increasingly concentrated, and many corporations have become more global in scope. *Transnational corporations* **are large corporations that are headquartered in one country but sell and produce goods and services in many countries.** Ten of the largest U.S. transnationals are listed in ◆ Table 13.2.

Pursuit of Personal Profit A tenet of capitalism is the belief that people are free to maximize their individual gain through personal profit; in the process, the entire society will benefit from their activities (Smith, 1976/1776). Economic development is assumed to benefit both capitalists and workers, and the general public also benefits from public expenditures (such as for roads, schools, and parks) made possible through an increase in business tax revenues. However, critics assert that specific individuals and families (not the general public) have been the primary recipients of profits. In early monopoly capitalism, stockholders in companies such as American Tobacco, Westinghouse, and Sears, Roebuck derived massive profits from companies that held near-monopolies on specific goods and services (Hodson and Sullivan, 2008). In advanced (late) monopoly capitalism, profits have become even more concentrated.

◆ Table 13.2 The Largest Nonfinancial U.S. Transnational Corporations
(Ranked by Foreign Revenues)

Corporation	Product	Foreign Revenues ($ Billions)	Total Revenues ($ Billions)	Assets ($ Billions)
Exxon/Mobil	Oil refining	143.0	206.0	149.0
Chevron Texaco	Oil refining	65.0	117.0	77.6
Ford	Motor vehicles	51.7	170.0	283.4
IBM	Office equipment	51.1	88.4	88.3
General Electric	Electronics	49.5	129.8	437.0
General Motors	Motor vehicles	48.2	184.6	303.1
Wal-Mart Stores	Retail sales	32.1	191.3	78.1
Philip Morris	Tobacco products	32.1	63.3	79.0
Hewlett-Packard	Office equipment	27.5	48.9	34.0
Motorola	Telecommunications	21.8	37.6	42.3

Source: United Nations Conference on Trade and Development, 2002.

Competition In theory, competition acts as a balance to excessive profits. When producers vie with one another for customers, they must be able to offer innovative goods and services at competitive prices. However, from the time of early industrial capitalism, the trend has been toward less, rather than more, competition among companies. In early monopoly capitalism, competition was diminished by increasing concentration *within* a particular industry, a classic case being the virtual monopoly on oil held by John D. Rockefeller's Standard Oil Company (Tarbell, 1925/1904; Lundberg, 1988). Today, Microsoft Corporation has been the subject of federal investigations and lawsuits because it so dominates certain areas of the computer software industry that it has virtually no competitors in those areas. In other situations, several companies may dominate certain industries. An ***oligopoly* exists when several companies overwhelmingly control an entire industry.** An example is the music industry, in which four giant companies are behind many of the labels and artists (see ◆ Table 13.3). More specifically, a ***shared monopoly* exists when four or fewer companies supply 50 percent or more of a particular market.** Examples include U.S. automobile manufacturers (referred to as the "Big Three") and cereal companies (three of which control 77 percent of the market).

In advanced monopoly capitalism, mergers also occur *across* industries: Corporations gain near-monopoly control over all aspects of the production

capitalism an economic system characterized by private ownership of the means of production, from which personal profits can be derived through market competition and without government intervention.

corporations large-scale organizations that have legal powers, such as the ability to enter into contracts and buy and sell property, separate from their individual owners.

transnational corporations large corporations that are headquartered in one country but sell and produce goods and services in many countries.

oligopoly a condition existing when several companies overwhelmingly control an entire industry.

shared monopoly a condition that exists when four or fewer companies supply 50 percent or more of a particular market.

◆ **Table 13.3** The Music Industry's Big Four

Company	Country	Leading Artists
Universal Music Group (MCA, Polygram)	France	U2 Jimi Hendrix Sting
Sony BMG Music Entertainment (RCA, Arista, Sony)	Japan	Avril Lavigne Pearl Jam Carlos Santana
EMI Group (Capitol, EMI, Virgin)	United Kingdom	The Beatles Pink Floyd Radiohead
Warner Music Group (Atlantic, Elektra)	United States	Led Zeppelin Jewel Hootie and the Blowfish

and distribution of a product by acquiring both the companies that supply the raw materials and the companies that are the outlets for the product. For example, an oil company may hold leases on the land where the oil is pumped out of the ground, own the plants that convert the oil into gasoline, and own the individual gasoline stations that sell the product to the public.

Corporations with control both within and across industries are often formed by a series of mergers and acquisitions across industries. These corporations are referred to as *conglomerates*—**combinations of businesses in different commercial areas, all of which are owned by one holding company.** Media ownership is a case in point; companies such as Time Warner have extensive holdings in radio and television stations, cable television companies, book publishing firms, and film production and distribution companies, to name only a few. Today, most of our news originates from ten massive conglomerates that dominate the global media market. According to media scholars, three factors triggered the rapid growth of the media conglomerates: (1) aggressive political and economic maneuvering by dominant media firms, (2) introduction of new technologies that increased the cost efficiency of global systems, and (3) policies of the U.S. government and organizations such as the World Bank, International Monetary Fund, and World Trade Organization (McChesney, 1998).

Although media titans such as Rupert Murdoch, head of News Corporation—the conglomerate that owns more than a hundred newspapers worldwide, major movie studios, publishing interests, the Fox TV network, and cable channels—seldom discuss the business dealings or political negotiations involved in mass media and telecommunications megamergers, journalists occasionally provide insights on the thinking of such individuals. For example, the journalist Ken Aulette describes a discussion he had with Rupert Murdoch about the continual growth of Murdoch's media empire:

Does [he] tire of the constant competition to get bigger, to win each war? When does Murdoch say *enough*? I asked him this question at the end of a long night that started with a drink and a stroll over his six acre property in Beverly Hills. . . . After dinner in the dining room of his comfortable Spanish-style home, Murdoch sipped a glass of California Chardonnay and treated the question as something so alien as to be incomprehensible. "You go on," he responded, opaquely. "There is a global village in some sense. You are competing everywhere. . . . And I just enjoy it." (Aulette, 1998: 287)

When questioned on another occasion regarding mergers, joint ventures, and cross-ownership of media firms, Murdoch stated, "We [media and telecommunications firms] can join forces now, or we can kill

Rupert Murdoch, CEO of News Corporation, controls many newspapers, magazines, and television and radio stations, part of a recent trend in media consolidation. Murdoch has stated that "Monopoly is a terrible thing until you have it." Do you agree?

boards of directors of two corporations that are in *direct* competition with each other, a person may serve simultaneously on the board of a financial institution (a bank, for example) and the board of a commercial corporation (a computer manufacturing company or a furniture store chain, for example) that borrows money from the bank. Directors of competing corporations may also serve together on the board of a third corporation that is not in direct competition with the other two. In 1996, former Defense Secretary Frank C. Carlucci sat on the corporate boards of fourteen Fortune 1,000 companies; former Labor Secretary Ann D. McLaughlin sat on eleven; and former Health, Education, and Welfare Secretary Joseph A. Califano, Jr., sat on nine (Dobrzynski, 1996). An example of interlocking directorates is depicted in Figure ▶ 13.4. Interlocking directorates diminish competition by producing interdependence. Individuals who serve on multiple boards are often able to forge cooperative arrangements that benefit their corporations but not necessarily the general public. When several corporations are controlled by the same financial interests, they are more likely to cooperate with one another than to compete (Mintz and Schwartz, 1985).

Lack of Government Intervention Ideally, capitalism works best without government intervention in the marketplace. The policy of *laissez-faire* (les-ay-FARE, which means "leave alone") was advocated by the economist Adam Smith in his 1776 treatise *An Inquiry into the Nature and Causes of the Wealth of Nations*. Smith argued that when people pursue their own selfish interests, they are guided "as if by an invisible hand" to promote the best interests of society (see Smith, 1976/1776). Today, terms such as *market economy* and *free enterprise* are often used, but the underlying assumption is the same: Free market competition, not the government, should regulate prices and wages. However,

each other and then join forces" (McChesney, 1998: 15). Murdoch and other media CEOs acknowledge that two or three companies will eventually dominate the media and telecommunications industry; however, in their view this is not necessarily a bad thing. As Murdoch commented, "Monopoly is a terrible thing until you have it" (McChesney, 1998: 54). Concerned about the degree to which media conglomerates control the news that we receive about politics and the economy, media watchdog groups encourage each of us to keep an eye on the media (see Box 13.2).

Competition is also reduced over the long run by *interlocking corporate directorates*—**members of the board of directors of one corporation who also sit on the board(s) of other corporations.** Although the Clayton Antitrust Act of 1914 made it illegal for a person to sit simultaneously on the

conglomerate a combination of businesses in different commercial areas, all of which are owned by one holding company.

interlocking corporate directorates members of the board of directors of one corporation who also sit on the board(s) of other corporations.

Box 13.2 You Can Make a Difference

Keeping an Eye on the Media

Do we get all of the news that we should about how our government operates and about the pressing social problems of our nation? Consider this list (shown by number) of the top five news stories that some media analysts believe were *not* adequately covered by the U.S. media in recent years:

1. "No Habeas Corpus for 'Any Person'": Passage of the Military Commissions Act, which legalized military roundups and lengthy detention for any person—including a U.S. citizen—who is deemed to be an "enemy of the state."

2. "Bush Moves Toward Martial Law": Passage of the John Warner Defense Authorization Act, which allows the president to station military troops anywhere in the United States and take control of state National Guard units in order to "suppress public disorder."

3. "U.S. Military Control of Africa's Resources": Replaces U.S. military command posts in Africa with a more centralized and intensified military presence for the stated purpose of providing a humanitarian guard in the war on terror. However, critics believe these actions set the stage for the United States to procure and control Africa's oil and its global delivery systems.

4. "Frenzy of Increasingly Destructive Trade Agreements": The United States and the European Union have established numerous trade and investment agreements outside of the auspices of the World Trade Organization that require major concessions on the part of developing nations and offer little in return, particularly for increasingly impoverished local farmers in these regions.

5. "Human Traffic Builds U.S. Embassy in Iraq": Work on the building of a massive, very expensive U.S. embassy in Iraq's Green Zone (which is located "right under the nose of the [U.S.] State Department") is being done by a Kuwait contractor who has repeatedly been accused of using forced labor, which is trafficked from South Asia, to do the work. (Project Censored, 2008)

According to Project Censored, an organization of more than 200 students and professors who produce the annual "Top 25 Censored Stories" list at Sonoma State University's Sociology Department, many important stories are either missing from the news altogether or do not receive the attention they deserve. (To view the entire list, visit Project Censored's website at **http://www.projectcensored.org.**)

What should be the role of the media in keeping us informed? The media are often referred to as the "Fourth Estate" or the "Fourth Branch of the Government" because they are supposed to provide people relevant information on important topics regarding how the government operates in a democratic society. This information can then be used by citizens to decide how they will vote on candidates and issues presented for their approval or disapproval on the election ballot.

The first step in keeping an eye on the news is to become more analytical about the "news" that we do receive. How can we evaluate the information we receive from the media? In *How to Watch TV News,* the media analysts Neil Postman and Steve Powers (1992: 160–168) suggest the following:

1. We should keep in mind that television news shows are called "shows" for a reason. They are not a public service or a public utility.

2. We should never underestimate the power of commercials, which tell us much about our society.

3. We should learn about the economic and political interests of those who run television stations or own a controlling interest in a media conglomerate.

4. We should pay attention to the *language* of newscasts, not just the visual imagery. For example, a *question* may reveal as much about the *questioner* as the person answering the question.

Becoming aware of the media's role in influencing people's opinions about how our government is run is the first step toward becoming an informed participant in the democratic political process.

The second step in keeping an eye on the news is becoming aware of national and international events that should receive more coverage than they do or that might not be reported in a fair and unbiased manner. To form your own opinion about media coverage of the news, you may wish to also visit these sites, which examine media practices:

● FAIR (Fairness and Accuracy in Reporting):
http://www.fair.org

● MediaChannel.org, which states, "As the media watch the world, we watch the media":
http://www.mediachannel.org

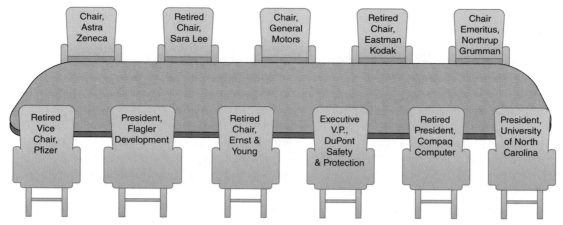

▶ **Figure 13.4 The General Motors Board of Directors**
The 2007 General Motors Board of Directors shows the nature of interlocking directorates.
On the chair representing each of the directors is the name of another entity each director is
connected with, and his or her position with that entity.

Source: General Motors, 2007.

the "ideal" of unregulated markets benefiting all citizens has seldom been realized. Individuals and companies in pursuit of higher profits have run roughshod over weaker competitors, and small businesses have grown into large, monopolistic corporations. Accordingly, government regulations were implemented in an effort to curb the excesses of the marketplace brought about by laissez-faire policies.

However, much of what is referred to as government intervention has been in the form of aid to business. Between 1850 and 1900, corporations received government assistance in the form of public subsidies and protection from competition by tariffs, patents, and trademarks. Government intervention in the 1990s included billions of dollars in subsidies to farmers, tax credits for corporations, and large subsidies or loan guarantees to automakers, aircraft companies, railroads, and others. Overall, most corporations have gained much more than they have lost as a result of government involvement in the economy.

Socialism

Socialism **is an economic system characterized by public ownership of the means of production, the pursuit of collective goals, and centralized deci-**

sion making. Like "pure" capitalism, "pure" socialism does not exist. Karl Marx described socialism as a temporary stage en route to an ideal communist society. Although the terms *socialism* and *communism* are associated with Marx and are often used interchangeably, they are not identical. Marx defined communism as an economic system characterized by common ownership of all economic resources. In the *Communist Manifesto* and *Das Kapital*, he predicted that the working class would become increasingly impoverished and alienated under capitalism. As a result, the workers would become aware of their own class interests, revolt against the capitalists, and overthrow the entire system (see Turner, Beeghley, and Powers, 2007). After the revolution, private property would be abolished, and capital would be controlled by collectives of workers who would own the means of production. The government (previously used to further the interests of the capitalists) would no longer be necessary. People would contribute according to their abilities and

socialism an economic system characterized by public ownership of the means of production, the pursuit of collective goals, and centralized decision making.

receive according to their needs (Marx and Engels, 1967/1848; Marx, 1967/1867). "Ideal" socialism has three distinctive features: (1) public ownership of the means of production, (2) pursuit of collective goals, and (3) centralized decision making.

Public Ownership of the Means of Production

In a truly socialist economy, the means of production are owned and controlled by a collectivity or the state, not by private individuals or corporations. For example, prior to the early 1990s, the state owned all the natural resources and almost all the capital in the Soviet Union. At least in theory, goods were produced to meet the needs of the people. Access to housing and medical care was considered to be a right.

Karl Marx's beliefs about capitalism, socialism, and communism are widely known today, although it has been more than 100 years since his famous proclamation—as emblazoned on his tombstone outside London—"Workers of all lands unite."

Leaders of the Soviet Union and some Eastern European nations decided to abandon government ownership and control of the means of production because the system was unresponsive to the needs of the marketplace and offered no incentive for increased efficiency (Boyes and Melvin, 2002). Since the early 1990s, Russia and other states in the former Soviet Union have attempted to privatize ownership of production. China—previously the world's other major communist economy—announced in 1997 that it would privatize most state industries (Serrill, 1997). In *privatization,* resources are converted from state ownership to private ownership; the government takes an active role in developing, recognizing, and protecting private property rights (Boyes and Melvin, 2002). However, it appears that these measures may have failed in Russia: Shortages once again loom, and many people want to see a transformation to a more socialist form of economy.

Pursuit of Collective Goals Socialism is based on the pursuit of collective goals, rather than on personal profits. Equality in decision making replaces hierarchical relationships (such as between owners and workers or political leaders and citizens). Everyone shares in the goods and services of society—especially necessities such as food, clothing, shelter, and medical care—based on need, not on ability to pay. In reality, however, few societies pursue purely collective goals.

Centralized Decision Making Another tenet of socialism is centralized decision making. In theory, economic decisions are based on the needs of society; the government is responsible for aiding the production and distribution of goods and services. Central planners set wages and prices to ensure that the production process works. When problems such as shortages and unemployment arise, they can be dealt with quickly and effectively by the central government (Boyes and Melvin, 2002).

Mixed Economies

As we have seen, no economy is truly capitalist or socialist; most economies are mixtures of both. A *mixed economy* combines elements of a market economy (capitalism) with elements of a command economy (socialism). Sweden, Great Britain, and France have mixed economies, sometimes referred to as *democratic socialism*—an economic and political system that combines private ownership of some of the means of production, governmental distribution of some essential goods and services, and free elections. For example, government ownership in Sweden is limited primarily to railroads, mineral resources, a public bank, and liquor and tobacco operations (Feagin and Feagin, 1997). Compared with capitalist economies, however, the government in a mixed economy plays a larger role in setting rules, policies, and objectives.

The government is also heavily involved in providing services such as medical care, child care, and transportation. In Sweden, for example, all residents have health insurance, housing subsidies, child allowances, paid parental leave, and day-care subsidies. Recently, some analysts have suggested that the United States has assumed many of the characteristics of a *welfare state,* **a state in which there is extensive government action to provide support and services to the citizens,** as it has attempted to meet the basic needs of older persons, young children, unemployed people, and persons with a disability (Esping-Andersen, 1990).

Work in the Contemporary United States

The economy in the United States and other contemporary societies is partially based on the work (purposeful activity, labor, or toil) that people perform. However, work in high-income nations is highly differentiated and often fragmented because people have many kinds of occupations. Some occupations are referred to as professions.

Professions

Although sociologists do not always agree on exactly which occupations are professions, most of them agree that the term *professionals* includes most doctors, natural scientists, engineers, computer scientists, certified public accountants, economists, social scientists, psychotherapists, lawyers, policy experts of various sorts, professors, at least some journalists and editors, some clergy, and some artists and writers.

Characteristics of Professions *Professions* are **high-status, knowledge-based occupations that have five major characteristics** (Freidson, 1970, 1986; Larson, 1977):

1. *Abstract, specialized knowledge.* Professionals have abstract, specialized knowledge of their field based on formal education and interaction with colleagues. Education provides the credentials, skills, and training that allow professionals to have job opportunities and assume positions of authority within organizations (Brint, 1994).
2. *Autonomy.* Professionals are autonomous in that they can rely on their own judgment in selecting the relevant knowledge or the appropriate technique for dealing with a problem. Consequently, they expect patients, clients, or students to respect that autonomy.
3. *Self-regulation.* In exchange for autonomy, professionals are theoretically self-regulating. All professions have licensing, accreditation, and regulatory associations that set professional standards and that require members to adhere to a code of ethics as a form of public accountability. Realistically, however, many professionals work within large-scale bureaucracies that have rules, policies, and procedures to which everyone must adhere.

---

mixed economy an economic system that combines elements of a market economy (capitalism) with elements of a command economy (socialism).

democratic socialism an economic and political system that combines private ownership of some of the means of production, governmental distribution of some essential goods and services, and free elections.

welfare state a state in which there is extensive government action to provide support and services to the citizens.

professions high-status, knowledge-based occupations.

4. *Authority.* Because of their authority, professionals expect compliance with their directions and advice. Their authority is based on mastery of the body of specialized knowledge and on their profession's autonomy: Professionals do not expect the client to argue about the professional advice rendered.

5. *Altruism.* Ideally, professionals have concern for others, not just their own self-interest. The term *altruism* implies some degree of self-sacrifice whereby professionals go beyond their self-interest or personal comfort so that they can help a patient or client (Hodson and Sullivan, 2008).

Social Reproduction of Professionals Although higher education is one of the primary qualifications for a profession, the emphasis on education gives children whose parents are professionals a disproportionate advantage early in life (Brint, 1994). There is a direct linkage between parental education/income and children's scores on college admissions tests such as the SAT, as shown in ▶ Figure 13.5. In turn, test scores are directly related to students' ability to gain admission to colleges and universities, which serve as springboards to most professions.

Race and gender are also factors in access to the professions. Historically, people of color have been underrepresented. Today, as more persons from underrepresented groups have gained access to professions such as law and medicine, their children have also gained the educational and mentoring opportunities necessary for professional careers. Between 1980 and 1997, there was a 63 percent increase in the number of African Americans, age 25 and above, with at least four years of college education. However, by 2004, only 17.6 percent of African Americans and 12.1 percent of Latinos/as age 25 and above had graduated from college, as compared with 28.2 percent of whites and 49.4 percent of Asian Americans (U.S. Census Bureau, 2006).

Deprofessionalization Certain professions are undergoing a process of *deprofessionalization,* in which some of the characteristics of a profession are

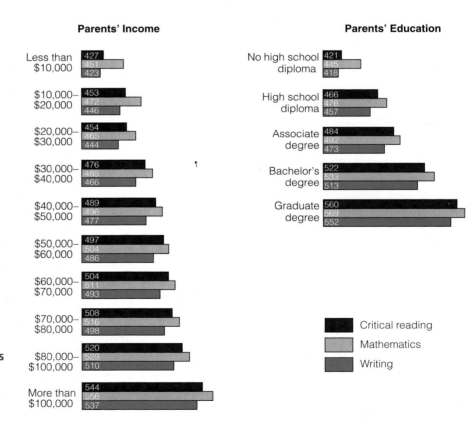

▶ Figure 13.5 **SAT Scores by Parents' Income and Education, 2007**

Source: College Board, 2007.

eliminated (Hodson and Sullivan, 2008). Occupations such as pharmacist have already been *deskilled,* as Nino Guidici explains:

> In the old days [people] took druggists as doctors. . . . All we do [now] is count pills. Count out twelve on the counter, put 'em in here, count out twelve more. . . . Doctors used to write out their own formulas and we made most of these things. Most of the work is now done in the laboratory. The real druggist is found in the manufacturing firms. They're the factory workers and they're the pharmacists. We just get the name of the drugs and the number and the directions. It's a lot easier. (qtd. in Terkel, 1990/1972)

However, colleges of pharmacy in many universities have fought against deprofessionalization by upgrading the degrees awarded to pharmacy graduates from the traditional B.S. in pharmacy to a Ph.D. This upgrading of degrees has also occurred over the past two decades in law schools, where the Bachelor of Laws (LL.B.) has been changed to the Juris Doctor (J.D.) degree.

Other Occupations

Occupations **are categories of jobs that involve similar activities at different work sites** (Reskin and Padavic, 2002). More than 500 different occupational categories and 31,000 occupation titles, ranging from motion picture cartoonist to drop-hammer operator, are currently listed by the U.S. Census Bureau. Historically, occupations have been classified as blue collar and white collar. Blue-collar workers were primarily factory and craft workers who did manual labor; white-collar workers were office workers and professionals. However, contemporary workers in the service sector do not easily fit into either of these categories; neither do the so-called pink-collar workers, primarily women, who are employed in occupations such as preschool teacher, dental assistant, secretary, and clerk (Hodson and Sullivan, 2008).

Sociologists establish broad occupational categories by distinguishing between employment in the primary labor market and in the secondary labor market. The *primary labor market* **consists of high-paying jobs with good benefits that have** some degree of security and the possibility of future advancement. **By contrast, the** *secondary labor market* **consists of low-paying jobs with few benefits and very little job security or possibility for future advancement** (Bonacich, 1972).

Upper-Tier Jobs: Managers and Supervisors

Managers are essential in contemporary bureaucracies, where work is highly specialized and authority structures are hierarchical. Workers at each level of the hierarchy take orders from their immediate superiors and perhaps give orders to a few subordinates. Upper-level managers are typically responsible for coordination of activities and control of workers. The *span of control,* or the number of workers a manager supervises, is affected by the organizational structure and by technology. Some analysts believe that hierarchical organization is necessary to coordinate the activities of a large number of people; others suggest that it produces apathy and alienation among workers (Blauner, 1964).

Lower-Tier and Marginal Jobs Positions in the lower tier of the service sector are part of the secondary labor market, characterized by low wages, little job security, few chances for advancement, higher unemployment rates, and very limited (if any) unemployment benefits. Typical lower-tier positions include janitor, waitress, messenger, sales clerk, typist, file clerk, migrant laborer, and textile worker. Large numbers of young people, people of color, recent immigrants, and white women are employed in this sector (Callaghan and Hartmann, 1991).

---

occupations categories of jobs that involve similar activities at different work sites.

primary labor market the sector of the labor market that consists of high-paying jobs with good benefits that have some degree of security and the possibility of future advancement.

secondary labor market the sector of the labor market that consists of low-paying jobs with few benefits and very little job security or possibility for future advancement.

Marginal jobs differ from the employment norms of the society in which they are located; examples in the U.S. labor market include personal service and private household workers. Marginal jobs are frequently not covered by government work regulations—such as minimum standards of pay, working conditions, and safety standards—or do not offer sufficient hours of work each week to provide a living (Hodson and Sullivan, 2008). Today, more than 11 million U.S. workers are employed in personal service industries such as eating and drinking places, hotels, laundries, beauty shops, and household service, primarily maid service (Hodson and Sullivan, 2008). Occupational segregation by race and gender is clearly visible in marginal jobs in personal service industries. Private household workers, including launderers, cooks, maids, housekeepers, gardeners, babysitters, and nannies, frequently must travel long distances to get to work, and many rely on public transportation, which can add hours to their workday. Moreover, household work is marginal: It lacks regularity, stability, and adequacy. The jobs are excluded from most labor legislation, employers often pay cash in order to avoid payroll taxes and Social Security, and the jobs typically provide no insurance or retirement benefits (Hodson and Sullivan, 2008).

Contingent Work

Contingent work **is part-time work, temporary work, or subcontracted work that offers advantages to employers but that can be detrimental to the welfare of workers.** Contingent work is found in every segment of the work force, including colleges and universities, where tenure-track positions are more scarce than in the past and a series of one-year, nontenure-track appointments at the lecturer or instructor level has become a means of livelihood for many professionals. The federal government is part of this trend, as is private enterprise. In the health care field, physicians, nurses, and other workers are increasingly employed through temporary agencies.

Employers benefit by hiring workers on a part-time or temporary basis; they are able to cut costs, maximize profits, and have workers available only when they need them. Temporary workers are the fastest-growing segment of the contingent work force, and agencies that "place" them have increased dramatically in number in the last decade. The agencies provide

© Peter Hvizdak/The Image Works

Occupational segregation by race and gender is clearly visible in personal service industries, such as restaurants and fast-food chains. Women and people of color are disproportionately represented in marginal jobs such as waitperson or fast-food server—jobs that typically do not meet societal norms for benefits and security.

workers on a contract basis to employers for an hourly fee; workers are paid a portion of this fee.

Subcontracted work is another form of contingent work that often cuts employers' costs at the expense of workers. Instead of employing a large work force, many companies have significantly reduced the size of their payrolls and benefit plans by *subcontracting*—**an agreement in which a corporation contracts with other (usually smaller) firms to provide specialized components, products, or services to the larger corporation.** Hiring and paying workers become the responsibility of the subcontractor, not of the larger corporation. The employment practices of some subcontractors are unethical, if not illegal.

The Underground Economy

Some social analysts make a distinction between the legitimate and the underground economies in the United States. For the most part, the occupations previously described in this chapter operate within the *legitimate economy*: Taxes on income are paid by employers and employees, and individuals who hold jobs requiring a specialized license (such as craftspeople or taxi drivers) possess the appropriate credentials for their work. By contrast, the *underground economy* is made up of a wide variety of activities through which people make money that they do not report to the government, and in some cases, their endeavors may involve criminal behavior (Venkatesh, 2006). Sometimes referred to as the "shadow economy," the underground economy is made up of workers who are paid "off the books," which means that they are paid in cash, their earnings are not reported, and no taxes are paid. Lawful jobs, such as nannies, construction workers, and landscape/yard workers, are often part of the shadow economy because workers and bosses make under-the-table deals so that both can gain through the transaction: Employers pay less for workers' services, and workers have more money to take home than if they paid taxes on their earnings. According to some financial analysts, the underground economy is increasing rapidly because of the growing number of low-wage, undocumented immigrant workers who come to the United States hoping for a better life for themselves and their families (McTague, 2005). Fear of apprehension by immigration authorities is enough to drive many undocumented workers into the underground economy.

The underground economy also involves trade in lawful goods that are sold "off the books" because no taxes are paid on the sales. In his study of the underground economy of the urban poor, the sociologist Sudhir Venkatesh (2006: 9) explains how these transactions take place:

> Most underground exchanges are short-term efforts to make a buck, but they can nevertheless follow strict patterns. Individuals know where to meet one another to trade off the books; there are usually particular places where this trading occurs and particular people who are known to be involved. People will have a rough idea of prices or rates of barter and trade before initiating the exchange. And it is not difficult to predict when conflict may arise; nor are people entirely unaware of the means for addressing disputes over quality, pricing, and service. In other words, while there are endless reasons to participate in the underground (or to stop doing so), there are always rules to be obeyed, codes to be followed, and likely consequences of actions.

The selling of goods in the underground economy may have increased in recent years due to the popularity of online commerce. Although individual sellers are responsible for paying taxes on items sold on websites such as eBay, it is unknown how many of the sellers actually report income from their sales. According to one way of thinking, operating a business in the underground economy reveals capitalism at its best because it shows how the "free market" might work if there were no government intervention. However, from another perspective, selling goods or services in the underground economy borders on—or moves into—criminal behavior. Some estimates of the output of the underground economy place it between $970 million and $1 trillion, and analysts believe that increasing numbers of people work or sell goods in this shadow economy (McTague, 2005). For some individuals, the underground economy offers the only alternative to unemployment, particularly in poor communities where people often feel alienated from the wider world and believe that they must use shady means to survive (see Venkatesh, 2006).

Unemployment

There are three major types of unemployment—cyclical, seasonal, and structural. *Cyclical unemployment* occurs as a result of lower rates of production during recessions in the business cycle; a recession is a decline in an economy's total production that lasts at least six months. Although massive layoffs initially occur, some of the workers will eventually be rehired, largely depending on the length and severity of the recession. *Seasonal unemployment* results from shifts in the demand for workers based on conditions such as the weather (in agriculture, the construction

---

marginal jobs jobs that differ from the employment norms of the society in which they are located.

contingent work part-time work, temporary work, or subcontracted work that offers advantages to employers but that can be detrimental to the welfare of workers.

subcontracting an agreement in which a corporation contracts with other (usually smaller) firms to provide specialized components, products, or services to the larger corporation.

industry, and tourism) or the season (holidays and summer vacations). Both of these types of unemployment tend to be relatively temporary.

By contrast, structural unemployment may be permanent. *Structural unemployment* arises because the skills demanded by employers do not match the skills of the unemployed or because the unemployed do not live where the jobs are located (McEachern, 2003). This type of unemployment often occurs when a number of plants in the same industry are closed or when new technology makes certain jobs obsolete. Structural unemployment often results from capital flight—the investment of capital in foreign facilities, as previously discussed. Today, many workers fear losing their jobs, exhausting their unemployment benefits (if any), and still not being able to find another job.

The *unemployment rate* **is the percentage of unemployed persons in the labor force actively seeking jobs.** In early 2004 the U.S. unemployment rate was 5.6 percent, and for specific categories ranged from 4.9 percent for white Americans to 10.5 percent for African Americans. By 2008, the overall rate had decreased to 4.9 percent. Like other types of "official" statistics, however, unemployment rates may be misleading. Individuals who become discouraged in their attempt to find work and are no longer actively seeking employment are not counted as unemployed.

Labor Unions and Worker Activism

In their individual and collective struggles to improve their work environment and gain some measure of control over their work-related activities, workers have used a number of methods to resist workplace alienation. Many have joined labor unions to gain strength through collective action.

Labor Unions U.S. labor unions came into being in the mid-nineteenth century. Unions have been credited with gaining an eight-hour workday, a five-day workweek, health and retirement benefits, sick leave and unemployment insurance, and workplace health and safety standards for many employees. Most of these gains have occurred through *collective bargaining*—negotiations between employers and labor union leaders on behalf of workers. In some cases, union leaders have called strikes to force employers to accept the union's position on wages and benefits. While on strike, workers may picket in front of the workplace to gain media attention, to fend off "scabs" (nonunion workers) who might take over their jobs, and in some cases to discourage customers from purchasing products made or sold by their employer. In recent years, strike activity has diminished significantly because many workers have feared losing their jobs. In 2006, only 20 strikes, or work stoppages, involving more than 1,000 workers were reported, as compared with significantly higher numbers in the 1960s and 1970s (U.S. Census Bureau, 2008). The number of workers involved in the actions declined from a peak of more than 2.5 million in 1971 to 70,000 in 2006.

Although the overall *number* of union members in the United States has increased since the 1960s, primarily as a result of the growth of public employee unions (such as the American Federation of Teachers), the *proportion* of all employees who are union members has declined. Today, about 17 percent of all U.S. employees belong to unions or employee associations. Part of the decline in union participation has been attributed to the structural shift away from manufacturing and manual work and toward the service sector, where employees have been less able to unionize.

© AP Images/Mark Lennihan

During economic downturns and periods of high unemployment, good jobs become highly prized. These people are waiting for interviews at Con Edison, a large public utility in New York.

Seeking to improve economic and social opportunities for farmworkers, the late César Chávez held rallies and engaged in other protest activities in an effort to better the workers' lives.

Absenteeism and Sabotage Absenteeism is one means by which workers resist working conditions they consider to be oppressive. Other workers use sabotage to bring about informal work stoppages. The phrase "throwing a monkey wrench in the gears" originated with the practice of workers "losing" a tool in assembly-line machinery. Sabotage of this sort effectively brought the assembly line to a halt until the now-defective piece of machinery could be repaired. Although most workers do not sabotage machinery, a significant number do resist what they perceive to be oppression from supervisors and employers.

Studies of worker resistance are very important to our understanding of how people deal with the social organization of work. Previously, workers (especially women) have been portrayed as passive "victims" of their work environment. However, this is not always the case for either women or men; many workers resist problematic situations and demand better wages and work conditions.

Employment Opportunities for Persons with a Disability

An estimated 49.7 million persons in the United States have one or more physical or mental disabilities that differentially affect their opportunities for employment. In 1990 the United States became the first nation to formally address the issue of equality for persons with a disability. When Congress passed the Americans with Disabilities Act (ADA), this law established "a clear and comprehensive prohibition of discrimination on the basis of disability." Combined with previous disability rights laws (such as those that provide for the elimination of architectural barriers from new, federally funded buildings and for the maximum integration of schoolchildren with disabilities), the ADA is a legal mandate for the full equality of people with disabilities.

Despite this law, about two-thirds of working-age persons with a disability are unemployed today (U.S. Census Bureau, 2008). Most persons with a disability believe they could work if they were offered the opportunity. However, even when persons with a disability are able to find jobs, they earn less than persons without a disability (Yelin, 1992). On the average, workers with a disability make 85 percent (for men) and 70 percent (for women) of what their coworkers without disabilities earn, and the gap is growing (U.S. Census Bureau, 2008). Studies have shown a thirty-year overall decline in the economic condition of persons with disabilities (Yelin, 1992; Burkhauser, Haveman, and Wolfe, 1993). The problem has been particularly severe for African Americans and Latinos/as with disabilities (Kirkpatrick, 1994). Among Latinos/as with disabilities, only 10 percent are employed full time; those who work full time earn 73 percent of what white (non-Latino) persons with disabilities earn.

What does it cost to "mainstream" persons with disabilities? A survey of personnel directors and other executives responsible for making hiring decisions for their companies found that the average cost of workplace modifications to accommodate employees with a disability was less than $500 (Heldrich Center for Workforce Development, 2003). Other accommodations cost about $1,000, with the largest single cost being $14,500 each for special Braille computer displays for visually impaired employees (Noble, 1995). As employment opportunities change in the global economy, it is important that employers use common sense and a creative approach when thinking about how persons with disabilities can fit into the marketplace.

unemployment rate the percentage of unemployed persons in the labor force actively seeking jobs.

Politics and the Economy in the Future

Will the government of the United States and other nation-states become obsolete in the future? According to analysts, there has been some erosion of the powers of governments in high-income nations such as the United States, along with a corresponding increase in transnational politics and the economy. Trade agreements are an example: Consider the North American Free Trade Agreement (NAFTA), which unites the United States, Canada, Mexico, and portions of South America in an economic alliance; the European Union (EU), which binds many European nations together in a political and economic partnership; and the General Agreement on Tariffs and Trade (GATT), which brings together the 124 nations in the World Trade Organization.

Do such agreements signal the end of individual governments? Despite trade alliances, many analysts believe that nation-states remain of great significance. According to scholar Paul Kennedy (1993), individual nations will remain the primary locus of identity for most people. Regardless of who their employer is and what they do for a living, individuals pay taxes to the state, are subject to its laws, serve (if need be) in its armed forces, and can travel abroad only by having its passport. Therefore, as new challenges—such as terrorism, war, and economic instability—increase, most people in democracies will turn to their own governments and demand solutions.

What is the future of democracy? Today, millions of people around the globe remain committed to what they believe are the "ideals of democracy." For example, recent U.S. immigrants who have escaped ethnic conflicts and poverty in their own nations often seek a life of dignity in a society they believe to be governed by democratic principles. Similarly, millions of people in Eastern Europe, the former Soviet Union, Latin America, and China have put their lives on the line for democracy, with varying degrees of success. Ironically, people who believe most in the ideals of democracy and freedom may be those who have experienced the least of it, either in other nations or in our own. In the United States, for example, many working-class and poor people, white women, people of color, and members of religious minorities who have experienced prejudice

The fate of high-income nations is increasingly linked to the globalization of markets. Dramatic changes in stock markets in nations such as Japan and the United States create reverberations that are felt in financial markets around the world.

and discrimination still believe that a reenvisioning of democratic ideals is possible.

Politics and the economy are so intertwined in the United States and on a global basis that many social scientists speak of the two as a single entity, the *political economy*. As we have seen, both political power and economic power are concentrated in the hands of a few in the "new" global economy. Most social analysts predict that transnational corporations will become even more significant in the global economy of this century. As they continue to compete for world market shares, these corporations will become even less aligned with the values of any one nation. However, those who advocate increased globalization typically focus on its potential impact on high-income nations, not the effect it may have on the 80 percent of the world's population that resides in lower-income nations. Persons in lower-income nations may become increasingly baffled and resentful when they are bombarded with media images of affluence and consumption that are not attainable by them.

What effect will global political and economic changes have on the average worker? In recent years, the average worker in the United States and other high-income nations has benefited from global competition and economic growth more than the average worker in industrializing nations. For example, the average citizen of Switzerland has an income several hundred times that of a resident of Ethiopia. More than a billion of the world's people live in abject poverty; for many, this means attempting

Box 13.3 Sociology and Social Policy

Does Globalization Change the Nature of Social Policy?

In 1492 Christopher Columbus set sail for India, going west. He had the Nina, the Pinta and the Santa Maria. He never did find India, but he called the people he met "Indians" and came home and reported to his king and queen: "The world is round." I set off for India 512 years later. I knew just which direction I was going. I went east. I had Lufthansa business class, and I came home and reported only to my wife and only in a whisper: "The world is flat."

> —author and newspaper columnist Thomas L. Friedman (2005a) describing how the global economy is changing all areas of economic, political, and social life, including how we formulate social policy

What does Thomas Friedman mean when he states that "the world is flat"? As discussed in Friedman's (2005b) best-selling book, *The World Is Flat: A Brief History of the Twenty-First Century,* "flat" means "level" or "connected" because (in his opinion) there is a more level global playing field in business (and almost any other endeavor) in the twenty-first century. According to Friedman, global telecommunications and the lowering of many trade and political barriers have brought about a new global era driven by *individuals,* not just by major corporations or giant trade organizations such as the World Bank. These individuals include entrepreneurs who create start-up ventures around the world and computer freelancers whose work knows no boundaries (based on the idea of the older nation-state borders) when it comes to the transfer of information. Other factors that Friedman also believes have contributed to the "flattening" of the world include the fall of the Berlin Wall, the emergence of Netscape, the streaming of the supply chain (Wal-Mart, for example), and the organization of information on the Internet by Google and Yahoo (Zakaria, 2005). Worldwide, many freelancers and business entrepreneurs are not in the United States: They reside in nations such as India and China, where it is now possible to do more than merely compete in low-wage manufacturing and routine information labor (such as workers in call centers), but also in the top levels of research and design work.

If Friedman's assertions are correct that the world has become flat and the United States is losing dominance in the global political and economic arena, where does this leave us in regard to social policy? Can we do something to stem the flow of jobs out of this country that has resulted from outsourcing and the shift of some information technologies from the United States to India, China, and other emerging nations?

Friedman suggests that if the United States is to remain competitive in the global economy, we must not continue to do things as they have previously been done. He believes we should have a thoughtful national discussion about what globalization means in all of our lives. As Friedman (2005a) states, "When it comes to responding to the challenges of the flat world, there is no help line we can call. We have to dig into ourselves. We in America have all the basic economic and educational tools to do that. But we have not been improving those tools as much as we should."

Because there has been a shift from large-scale corporate players in the global economy to individual entrepreneurs and freelancers, we must look to each *individual* in our country to see how we can best play the economic game in the twenty-first century. In addition to nationwide policies, Friedman (2005a) believes that our social policy regarding globalization must begin at home; children and young adults must be encouraged to rise to the economic challenge that faces them:

> We need to get going immediately. It takes 15 years to train a good engineer, because, ladies and gentlemen, this really is rocket science. So parents, throw away the Game Boy, turn off the television and get your kids to work. There is no sugar-coating this: in a flat world, every individual is going to have to run a little faster if he or she wants to advance his or her standard of living. When I was growing up, my parents used to say to me, "Tom, finish your dinner—people in China are starving." But after sailing to the edges of the flat world for a year, I am now telling my own daughters, "Girls, finish your homework—people in China and India are starving for your jobs."

When social policy becomes personal (as Friedman believes), are we willing to engage in the changes it requires?

Reflect & Analyze

Are Friedman's assumptions about the changing world order accurate? What do you think? What other arguments might be presented?

to survive on less than $365 a year. However, many factors contribute to the complexity of predicting how globalization will affect the United States and other nations. Our nation's social policies need to be crafted with an understanding that the global political economy is rapidly changing (see Box 13.3).

In the twenty-first century, terrorism and war have influenced how many people think about politics and the economy within and across na-tions. Acts of terrorism extract a massive toll by producing rampant fear, widespread loss of human life, and extensive destruction of property. For the first time, people in the United States have been confronted with threats of terrorism and problems associated with war that have been facts of life in some other countries for many years, and we have become more aware of the vulnerability of all human beings.

Chapter Review

• What are the three types of authority?

Max Weber identified three types of authority: char-ismatic, traditional, and rational–legal. Charismatic authority is power based on a leader's personal qualities. Traditional authority is based on long-standing custom. Rational–legal authority is based on law or written rules and regulations, as found in contemporary bureaucracies.

• What are the main types of political systems?

The main types of political systems are monarchies, authoritarian systems, totalitarian systems, and dem-ocratic systems.

• How do pluralist and power elite perspec-tives view power in the United States?

According to the pluralist (functionalist) model, power is widely dispersed throughout many com-peting interest groups. People influence policy by voting, joining special interest groups and political action campaigns, and forming new groups. Ac-cording to the elite (conflict) model, power is con-centrated in a small group of elites, whereas the masses are relatively powerless.

• Who are the power elite, and why are they important?

According to C. Wright Mills, the power elite is composed of influential business leaders, key gov-ernment leaders, and the military. The elites possess greater resources than the masses, and public policy reflects their preferences.

• What is the primary function of the economy?

The economy is the social institution that ensures the maintenance of society through the production, dis-tribution, and consumption of goods and services.

• How do the major contemporary economic systems differ?

Capitalism, socialism, and mixed economies are the main systems in industrialized countries. Capital-ism is characterized by ownership of the means of production, pursuit of personal profit, competition, and limited government intervention. Socialism is characterized by public ownership of the means of production, the pursuit of collective goals, and cen-tralized decision making. In mixed economies, ele-ments of a capitalist, market economy are combined with elements of a command, socialist economy. These mixed economies are often referred to as democratic socialism.

• What are the characteristics of professions?

Professions are high-status, knowledge-based occu-pations characterized by abstract, specialized knowl-edge; autonomy; authority over clients and subordi-nate occupational groups; and a degree of altruism.

• What is contingent work?

Contingent work is part-time work, temporary work, or subcontracted work that offers advantages to employers but may be detrimental to workers. Through the use of contingent workers, employers are able to cut costs and maximize profits, but work-ers have little or no job security.

www.cengage.com/login

Register for a Student eResource account to maximize your study time online using CengageNOW. First take the system's diagnostic pre-test, and then follow the personalized study plan that is created for you to help you review this chapter. The study plan will

- help you identify areas on which you should concentrate;
- provide interactive exercises to help you master the chapter concepts; and
- provide a post-test to confirm you are ready to move on to the next chapter.

Key Terms

authoritarianism 434
authority 431
capitalism 446
charismatic authority 432
conglomerates 448
contingent work 456
corporations 446
democracy 435
democratic socialism 453
economy 445
elite model 438
government 429
interlocking corporate directorates 449
marginal jobs 456

military–industrial complex 444
mixed economy 452
monarchy 433
occupations 455
oligopoly 447
pluralist model 436
political action committees 438
political party 440
political socialization 442
politics 429
power 430
primary labor market 455
primary sector production 445
professions 453
rational–legal authority 432

representative democracy 435
routinization of charisma 432
secondary labor market 455
secondary sector production 445
shared monopoly 447
socialism 451
state 429
subcontracting 456
tertiary sector production 445
totalitarianism 434
traditional authority 432
transnational corporations 446
unemployment rate 458
welfare state 453

Questions for Critical Thinking

1. Who is ultimately responsible for decisions and policies that are made in a democracy such as the United States: the people or their elected representatives?
2. How would you design a research project that studies the relationship between campaign contributions to elected representatives and their subsequent voting records? What would be your hypothesis? What kinds of data would you need to gather? How would you gather accurate data?

The Kendall Companion Website

www.cengage.com/sociology/kendall

Supplement your review of this chapter by going to the text's companion website, where you can take tutorial quizzes, use flash cards to master key terms, follow live links to useful websites, and explore the other study and research resources you'll find there, such as a comprehensive interactive sociology timeline, GSS Data, and Census 2000 information, much of it presented visually in maps.

14 Health, Health Care, and Disability

Medicine is, I have found, a strange and in many ways disturbing business. The stakes are high, the liberties taken tremendous. We drug people, put needles and tubes into them, manipulate their chemistry, biology, and physics, lay them unconscious and open their bodies up to the world. We do so out of an abiding confidence in our know-how as a profession. What you find when you get in close, however—close enough to see the furrowed brows, the doubts and missteps, the failures as well as the successes—is how messy, uncertain, and also surprising medicine turns out to be.

The thing that still startles me is how fundamentally human an endeavor it is. Usually, when we think about medicine and its remarkable abilities, what comes to mind is the science and all it has given us to fight sickness and misery: the tests, the machines, the drugs, the procedures. And without question, these

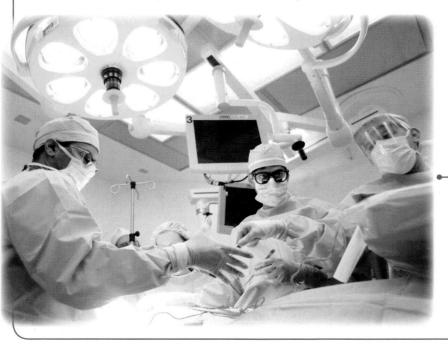

Dr. Atul Gawande (center) has written movingly about the differences between people's expectations of physicians and the medical establishment and the realities that they find in health care today. Sociologists study these contradictions to better understand a very complex and important part of U.S. social life.

are at the center of virtually everything medicine achieves. But we rarely see how it all actually works. You have a cough that won't go away—and then? It's not science you call upon but a doctor. A doctor with good days and bad days. A doctor with a weird laugh and a bad haircut. A doctor with three other patients to see and, inevitably, gaps in what he knows and skills he's still trying to learn. . . . We look for medicine to be an orderly field of knowledge and procedure. But it is not. It is an imperfect science, an enterprise of constantly changing knowledge, uncertain information, fallible individuals, and at the same time lives on the line. There is science in what we do, yes, but also habit, intuition, and sometimes plain old guessing. The gap between what we know and what we aim for persists. And this gap complicates everything we do.

—Atul Gawande, M.D. (2002: 4, 5, 7), a surgeon at Brigham and Women's Hospital in Boston, was a surgical resident when he wrote these words describing how he feels about the power and the limits of medicine.

The everyday life of a doctor like Atul Gawande is filled with its high and low points: Some patients benefit from medical treatments they receive from physicians, whereas others have sustained injuries or developed illnesses that are too severe or are beyond the scope of current knowledge and practice in the health care system to be successfully resolved. Physicians are human beings just like the patients they treat; however, much more is expected of them because of the availability of health care in the United

Chapter Focus Question

Why are health, health care, and disability significant concerns not only for individuals but also for entire societies?

States and other high-income nations and because the dominant role of doctors in modern high-tech medicine has led many individuals to believe that virtually anything should be possible when it comes to one's health and longevity. However, this assumption is often not an accurate reflection of how health, illness, and health care actually work.

In this chapter, we will explore the dynamics of health, health care, and disability from a sociological perspective, as well as look at issues through the eyes of those who have experienced medical problems. Before reading on, test your knowledge about health, illness, and health care by taking the quiz in Box 14.1.

What does the concept of health mean to you? At one time, health was considered to be simply the absence of disease. However, the World Health Organization (2003: 7) defines **health** as **a state of complete physical, mental, and social well-being.** According to this definition, health involves not only the absence of disease, but also a positive sense of wellness. In other words, health is a multidimensional phenomenon: It includes physical, social, and psychological factors.

What is illness? Illness refers to an interference with health; like health, illness is socially defined and may change over time and between cultures. For example, in the United States and Canada, obesity is viewed as unhealthy, whereas in other times and places, obesity has signaled that a person was prosperous and healthy. A disease, by comparison, is an objective reality: It is a particular destructive process in the body, with specific causes and characteristic symptoms. There are specific medical criteria for identifying a disease.

What happens when a person is perceived to have an illness or disease? Healing involves both personal and institutional responses to perceived illness and disease. One aspect of institutional healing is health care and the health care delivery system in a society. **Health care is any activity intended to improve health.** When people experience illness, they often seek medical attention in hopes of having their health restored. A vital part of health care is **medicine—an institutionalized system for the scientific diagnosis, treatment, and prevention of illness.**

Health in Global Perspective

Studying health and health care issues around the world offers insights on illness and how political and economic forces shape health care in nations. Disparities in health are glaringly apparent between high-income and low-income nations when we examine factors such as the prevalence of life-threatening diseases, rates of life expectancy and infant mortality, and access to health services. In regard to global health, for example, the number of people infected with HIV/AIDS more than doubled between 1990 and 2000 (from fewer than 15 million to more than 34 million). AIDS has cut life expectancy by 5 years in Nigeria, 18 years in Kenya, and 33 years in Zimbabwe (U.S. Census Bureau, 2008). *Life expectancy* **refers to an estimate of the average lifetime of people born in a specific year.** AIDS results in higher mortality rates in childhood and young adulthood, stages in the life course when mortality is otherwise low. However, AIDS is not the only disease reducing life expectancy in some nations. Most deaths in low- and middle-income nations are linked to infectious and parasitic diseases that are now rare in high-income, industrialized nations. Among these diseases are tuberculosis, polio, measles, diphtheria, meningitis, hepatitis, malaria, and leprosy. Although it is estimated that only 13 percent of U.S. citizens and 9 percent of Canadians will die prior to age 60, health experts estimate that more than 1.5 billion people around the world will die prior to age 60. This is particularly true in low-income nations such as Zambia, where 80 percent of the people are not expected to see their sixtieth birthday.

The *infant mortality rate* **is the number of deaths of infants under 1 year of age per 1,000 live births in a given year.** The infant mortality rate in some low-income nations is staggering: 261 infants under 1 year of age die per 1,000 live births in Angola, 257 die in Sierra Leone, and 239 die in Niger (World Health Organization, 2004a). In fact, almost 14 percent of all children born in low-income nations die before they reach their first birthday. The World Health Organization (2004a) estimates that two-thirds of those infants die during the *first*

Box 14.1 Sociology and Everyday Life

How Much Do You Know About Health, Illness, and Health Care?

True	False	
T	F	1. Some social scientists view sickness as a special form of deviant behavior.
T	F	2. The field of epidemiology focuses primarily on how individuals acquire disease and bodily injury.
T	F	3. The primary reason that African Americans have shorter life expectancies than whites is the high rate of violence in central cities and the rural South.
T	F	4. Native Americans have shown dramatic improvement in their overall health level since the 1950s.
T	F	5. Health care in most high-income, developed nations is organized on a fee-for-service basis as it is in the United States.
T	F	6. The medical–industrial complex has operated in the United States with virtually no regulation, and allegations of health care fraud have largely been overlooked by federal and state governments.
T	F	7. Media coverage of chronic depression and other mental conditions focuses almost exclusively on these problems as "women's illnesses."
T	F	8. It is extremely costly for employees to "mainstream" persons with disabilities in the workplace.

Answers on page 468.

month of life. A child born in Latin America or Asia can expect to live between 7 and 13 fewer years, on average, than one born in North America or Western Europe (Epidemiological Network for Latin America and the Caribbean, 2000). There are many reasons for these differences in life expectancy and infant mortality. Many people in low-income countries have insufficient or contaminated food; lack access to pure, safe water; and do not have adequate sewage and refuse disposal. Added to these hazards is a lack of information about how to maintain good health. Many of these nations also lack qualified physicians and health care facilities with up-to-date equipment and medical procedures.

Nevertheless, tremendous progress has been made in saving the lives of children and adults over the past 15 years. Life expectancy at birth has risen to more than 70 years in 84 countries, up from only 55 countries in 1990. Life expectancy in low-income nations increased on average from 53 to 62 years, and

mortality of children under 5 years of age dropped from 149 to 85 per 1,000 live births. Although this increase has been attributed to a number of factors, an especially important advance has been the development of a safe water supply. The percentage of the world's population with access to safe water nearly

health a state of complete physical, mental, and social well-being.

health care any activity intended to improve health.

medicine an institutionalized system for the scientific diagnosis, treatment, and prevention of illness.

life expectancy an estimate of the average lifetime of people born in a specific year.

infant mortality rate the number of deaths of infants under 1 year of age per 1,000 live births in a given year.

Box 14.1 Sociology and Everyday Life

Answers to the Sociology Quiz on Health, Illness, and Health Care

1. True. Some social scientists view sickness as a special form of deviant behavior. However, it is not equivalent to other forms of deviance such as crime or violent behavior. Unlike many who are defined as criminal, the sick are provided with therapeutic care so that their health will be restored and they can fulfill their roles in society (Weiss and Lonnquist, 2003).

2. False. The primary focus of the epidemiologist is on the health problems of social aggregates or large groups of people, not on individuals as such (Cockerham, 2004).

3. False. The lower life expectancy of African Americans as a category is due to a higher prevalence of life-threatening illnesses, such as cancer, heart disease, hypertension, and AIDS. However, it should be noted that African American males do have the highest death rates from homicide of any racial–ethnic category in the United States (Cockerham, 2004).

4. True. Native Americans (including American Indians and native Alaskans) as a category have had significant improvement in health in recent decades. Some analysts attribute this change to better nutrition and health care services. However, other analysts point out that Native Americans continue to have high rates of mortality from diabetes, alcohol-related illnesses, and suicide (Cockerham, 2004).

5. False. The United States is one of only two high-income, developed nations that do not have some form of universal health coverage. In the United States, health care has traditionally been purchased by the patient. In most other high-income nations, health care is provided or purchased by the government (Cockerham, 2004).

6. False. In the mid-to-late 1990s, government investigations focused on rising health care payments and allegations of fraud in the health care delivery system. Thus far, billing frauds have been found in Medicare and Medicaid payments to physicians, hospitals, nursing homes, home health agencies, medical labs, and medical equipment manufacturers (Findlay, 1997).

7. False. Until recently, chronic depression and other mental conditions were most often depicted as "female" problems. However, in the late 1990s, male celebrities such as Mike Wallace, the news correspondent who is coeditor of *60 Minutes,* have made the general public more aware of male depression.

8. False. Although disability expenditures nationwide may be costly, individual employers often find that they can accommodate the workplace needs of a worker with a disability for costs ranging from zero to several thousand dollars, thus opening up new opportunities for people previously excluded from certain types of jobs and careers.

doubled between 1990 and 2000 (United Nations Development Programme, 2003).

Will improvements in health around the world continue to occur? Organizations such as the United Nations argue that both public-sector and private-sector initiatives will be required to improve global health conditions. For example, a United Nations report states that in the era of globalization and dominance by transnational corporations, "money talks louder than need" when "cosmetic drugs and slow-ripening tomatoes come higher on the list [of priorities] than a vaccine against malaria or drought-resistant crops for marginal lands" (United Nations Development Programme, 1999: 68).

Recently, pressing questions have arisen about the availability of new technologies and lifesaving drugs around the world. An example is the problem of providing access to drugs in countries with high rates

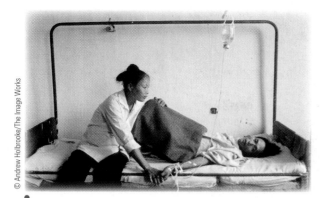

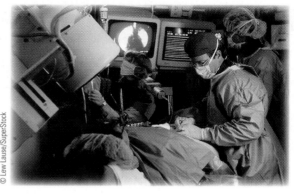

Access to quality health care is much greater for some people than for others. The factors that are involved vary not only for people within one nation but also across the nations of the world.

of HIV/AIDS. Many people cannot afford to pay for drugs, such as the three-drug combination therapy that prolongs the life of many AIDS patients. Transnational pharmaceutical companies fear that if they provide their name-brand drugs at a lower price in low-income countries, that might undercut their major sales base in high-income countries if those drugs become available as generic products (which are less costly and can be made by more than one manufacturer) or are reimported into the high-income countries at a reduced price. The companies claim that they need the money generated from sales of their name-brand drugs in order to fund research on other products that will reduce suffering and sometimes prolong human life. For this reason, pharmaceutical companies have increasingly marketed their prescription drugs to patients through the media, particularly television and print advertisements (see Box 14.2). Pharmaceutical companies that hold the patents on various drugs see their products as something that needs to be protected by law, whereas people in human relief agencies around the world are concerned about the fact that one-third of the world's population does not have access to essential medicines and that, even worse, this figure rises to one-half in the poorest parts of Africa and Asia (United Nations Development Programme, 2003). If we are to see a significant improvement in life expectancy and health among people in all of the nations of the world, improvements are needed in the availability to them of new medical technologies and lifesaving drugs.

How about improvements in health and health care within one nation? Is there a positive relation-

ship between the amount of money that a society spends on health care and the overall physical, mental, and social well-being of its people? Not necessarily. If there were such a relationship, people in the United States would be among the healthiest and most physically fit people in the world. We spend more than one trillion dollars—the equivalent of $3,925 per person—on health care each year, and the amount has almost doubled in the past ten years (Anell and Willis, 2000). But by comparing health care expenditures in Sweden and the United States with infant mortality rates in these two countries, we can see that large expenditures for health care do not always produce better health care for individuals. Sweden spends an average of $1,701 per person on health care and has an infant mortality rate of 3.5; by contrast, the United States has an infant mortality rate of 6.5 (Anell and Willis, 2000; U.S. Census Bureau, 2008). Similarly, a child born in Sweden in 2005 had a life expectancy of 78.4 years, whereas a child born in the United States that same year had a life expectancy of 77.7 years (U.S. Census Bureau, 2008).

Health in the United States

Even if we limit our discussion (for the moment) to people in the United States, why are some of us healthier than others? Is it biology—our genes—that accounts for this difference? Does the environment

Box 14.2 Framing Health Issues in the Media

It's Right for You! The Framing of Drug Ads

I thought I could get myself through this on my own....I started seeing signs developing within me I should have paid attention to. I pushed and pushed and pushed, until I pushed myself right into a corner....

—"Mary" explaining how depression affected her life before her doctor prescribed Cymbalta, a drug for major depressive disorder and generalized anxiety disorder

You know when you feel the weight of sadness. You may feel exhausted, hopeless, anxious. Whatever you do you feel lonely and do not enjoy the things you once loved. Things just don't feel like they used to. These are some symptoms of depression, a serious medical condition affecting over 20 million Americans.... You just shouldn't have to feel this way anymore.... When you know more about what's wrong, you can help make it right.

—ad for Zoloft, an antidepression drug

These television and Internet ads for prescription anti-depression drugs are only two of thousands of pharmaceutical advertisements that constantly bombard us. Drug advertising aimed at consumers accounts for nearly $5 billion in annual advertising dollars (Freudenheim, 2007). Most drug ads use *sympathetic and intuitive framing* to help television viewers, newspaper and magazine readers, and Internet users believe that they are not alone if they have high cholesterol, feelings of depression, patterns of sexual dysfunction, or whatever else may be bothering them. In many ads, a "real person" gives his or her testimonial about

how a particular drug has helped with his or her problem. The undertone of the person's statement is to offer a *sympathetic* message to viewers ("I know what you're going through because I've been there myself!"). Then a narrator or voiceover (a person who is heard but not seen) makes several understanding (*intuitive*) statements that suggest to individuals in the media audience, "We know you don't want to discuss this with anyone, but we can help." Ads such as this offer a simple solution (in the form of a pill) to help people solve their problems if they will only follow the narrator's advice: "Ask your doctor if [the product] is right for you."

What are the strengths and limitations of framing prescription drug ads in this manner and addressing consumers directly rather than pitching the products only to the physicians who must prescribe the drugs? An obvious strength of direct advertising is that patients become aware of newer products and learn about certain illnesses or conditions of which they were unaware. However, major limitations exist when it comes to direct advertising of such products. Many drug ads do not tell the whole story: Some play down the fact that other drugs, or a nondrug option, may be more effective in treating the condition than the pill being marketed. Other ads encourage people who are not good candidates for a particular drug to insist that their physician prescribe it for them anyway.

Although more prescription drug ads today inform media audiences about possible side effects than those

within which we live have an effect? How about our own individual lifestyle?

Social Epidemiology

The field of social epidemiology attempts to answer questions such as these. **Social epidemiology is the study of the causes and distribution of health, disease, and impairment throughout a population** (Weiss and Lonnquist, 2009). Typically, the target of the investigation is disease agents, the environment, and the human host. *Disease agents* include biologi-

cal agents such as insects, bacteria, and viruses that carry or cause disease; nutrient agents such as fats and carbohydrates; chemical agents such as gases and pollutants in the air; and physical agents such as temperature, humidity, and radiation. The *environment* includes the physical (geography and climate), biological (presence or absence of known disease agents), and social (socioeconomic status, occupation, and location of home) environments. The human *host* takes into account demographic factors (age, sex, and race/ethnicity), physical condition, habits and customs, and lifestyle (Weiss and

in the past, the narrators usually read the list of negative effects very quickly and then—in a highly reassuring voice—state that most side effects are extremely rare. One example is an ad for Mirapex, the restless-leg-syndrome drug, which states, "Tell your doctor . . . if

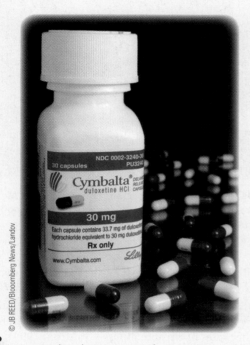

© JB REED/Bloomberg News/Landov

Increasingly, advertisements for prescription drugs are aimed at consumers, although they cannot purchase such products until a doctor prescribes them. Do you agree with this approach to informing patients? Why or why not?

you experience increased gambling, sexual, or other intense urges." Comments such as this may produce laughter from viewers rather than giving them a reason to seriously consider the possible negative consequences of certain prescription medications.

In the 2000s, the U.S. Congress and various governmental agencies continue to ponder the future of prescription drug ads that directly target consumers, but little is likely to come of these inquiries because spending on consumer drug advertising is so high and pressure from the pharmaceutical industry to continue these lucrative ads is so great. However, Mo Rocca (2007), a commentator on CBS News' *Sunday Morning* show, offered the following humorous, short-term solution to dealing with the side effects of TV drug ads:

> If there's one side effect to all these ads, it's heightened anxiety. . . . Maybe the best solution is to tune out the ads. Or at least ask your doctor if watching them is right for you. And in no instance should you watch them while you drink or operate heavy machinery. Otherwise, you're likely to experience nausea, dizziness . . . and an uncontrollable urge to throw your TV out the window.

Reflect & Analyze

On a more serious sociological note, how might functionalist, conflict, and symbolic interactionist theorists differ in their explanation of these ads? As a sociology student, what effect (if any) do you think prescription drug ads in the media have on people you know?

Lonnquist, 2009). Let's look briefly at some of these factors.

Age Rates of illness and death are highest among the old and the young. Mortality rates drop shortly after birth and begin to rise significantly during middle age. After age 65, rates of chronic diseases and mortality increase rapidly. *Chronic diseases* **are illnesses that are long term or lifelong and that develop gradually or are present from birth;** in contrast, *acute diseases* **are illnesses that strike suddenly and cause dramatic**

social epidemiology the study of the causes and distribution of health, disease, and impairment throughout a population.

chronic diseases illnesses that are long term or lifelong and that develop gradually or are present from birth.

acute diseases illnesses that strike suddenly and cause dramatic incapacitation and sometimes death.

incapacitation and sometimes death (Weitz, 2004).

Two of the most common sources of chronic disease and premature death are tobacco use, which increases mortality among both smokers and people who breathe the tobacco smoke of others, and alcohol abuse, both of which are discussed later in this chapter. The fact that rates of chronic diseases increase rapidly after age 65 has obvious implications not only for people reaching that age (and their families) but also for society. The Census Bureau projects that about 20 percent of the U.S. population will be at least age 65 by the year 2050 and that the population of persons aged 85 and over will have tripled from about 4 million (1.5 percent) in 2000 to about 12 million (5 percent). The cost of caring for many of these people—especially those who must be institutionalized—will increase in at least direct proportion to their numbers.

Sex Prior to the twentieth century, women had lower life expectancies than men because of high mortality rates during pregnancy and childbirth. Preventive measures have greatly reduced this cause of female mortality, and women now live longer than men. For babies born in the United States in 2004, for example, life expectancy at birth was 75.2 years for males and 80.4 years for females. Females have a slight biological advantage over males in this regard from the beginning of life, as can be seen in the fact that they have lower mortality rates both in the prenatal stage and in the first month of life (Weiss and Lonnquist, 2009). However, the sociologist Ingrid Waldron (1994) notes that gender roles and gender socialization also contribute to the difference in life expectancy. Men are more likely to work in dangerous occupations such as commercial fishing, mining, construction, and public safety/firefighting. As a result of gender roles, males may be more likely than females to engage in risky behavior such as drinking alcohol, smoking cigarettes (there is more social pressure on women not to smoke), using drugs, driving dangerously, and engaging in fights. Finally, women are more likely to use the health care system, with the result that health problems are identified and treated earlier (while there is a better chance of a successful outcome), whereas many men are more reluctant to consult doctors.

© chuck kuhn photography/Getty Images

Occupation and life expectancy may be related. Men are overrepresented in high-risk jobs, such as long-haul trucking, that may affect their life expectancies.

Because women on average live longer than men, it is easy to jump to the conclusion that they are healthier than men. However, although men at all ages have higher rates of fatal diseases, women have higher rates of chronic illness (Waldron, 1994).

Race/Ethnicity and Social Class Although race/ethnicity and social class are related to issues of health and mortality, recent research tends to suggest that income and factors such as the neighborhood in which a person lives may be more significant than race or ethnicity with respect to these issues. How is it possible that the neighborhood you live in may significantly affect your risk of dying during the next year? According to a study by the Stanford Center for Research in Disease Prevention (Winkleby and Cubbin, 2003), people have a higher survival rate if they live in better-educated or wealthier neighborhoods than if the neighborhood is low-income and has low levels of education. Among the reasons researchers believe that neighborhoods make a difference are the availability (or lack thereof) of safe areas to exercise, grocery stores with nutritious foods, and access to transportation, education, and good jobs. Many low-income neigh-

Can your neighborhood be bad for your health? According to recent research, it can indeed, especially if it predominantly contains fast-food restaurants, liquor stores, and similarly unhealthy lifestyle options.

However, although Latinas/os are more likely than non-Latino/a whites to live below the poverty line, they have lower death rates from heart disease, cancer, accidents, and suicide, and an overall lower death rate. One explanation may be dietary factors and the strong family life and support networks found in many Latina/o families (Weiss and Lonnquist, 2009). Obviously, more research is needed on this point, for the answer might be beneficial to all people.

Lifestyle Factors

As noted previously, social epidemiologists also examine lifestyle choices as a factor in health, disease, and impairment. We will examine three lifestyle factors as they relate to health: drugs, sexually transmitted diseases, and diet and exercise.

Drug Use and Abuse What is a drug? There are many different definitions, but for our purposes, a *drug* is any substance—other than food and water—that, when taken into the body, alters its functioning in some way. Drugs are used for either therapeutic or recreational purposes. *Therapeutic* use occurs when a person takes a drug for a specific purpose such as reducing a fever or controlling a cough. In contrast, *recreational* drug use occurs when a person takes a drug for no purpose other than achieving a pleasurable feeling or psychological state. Alcohol and tobacco are examples of drugs that are primarily used for recreational purposes; their use by people over a fixed age (which varies from time to time and place to place) is lawful. Other drugs—such as some antianxiety drugs and tranquilizers—may be used legally only if prescribed by a physician for therapeutic use but are frequently used illegally for recreational purposes.

borhoods are characterized by fast-food restaurants, liquor stores, and other facilities that do not afford residents healthy options.

As discussed in prior chapters, people of color are more likely to have incomes below the poverty line, and the poorest people typically receive less preventive care and less optimal management of chronic diseases than do other people. People living in central cities, where there are high levels of poverty and crime, or in remote rural areas generally have greater difficulty in getting health care because most doctors prefer to locate their practice in a "safe" area, particularly one with a patient base that will produce a high income. Although rural Americans make up 20 percent of the U.S. population, only 9 percent of the nation's physicians practice in rural areas, and fewer specialists such as cardiologists are available in these areas (Ricketts, 1999).

Another factor is occupation. People with lower incomes are more likely to be employed in jobs that expose them to danger and illness—working in the construction industry or around heavy equipment in a factory, for example, or holding a job as a convenience store clerk or other position that exposes a person to the risk of armed robbery. Finally, people of color and poor people are more likely to live in areas that contain environmental hazards.

drug any substance—other than food and water—that, when taken into the body, alters its functioning in some way.

Alcohol The use of alcohol is considered an accepted part of the dominant culture in the United States. Adults in this country consume an average of 2.4 gallons of wine, 21.3 gallons of beer, and 1.4 gallons of liquor a year (U.S. Census Bureau, 2008). In fact, adults consume more beer on average than milk or coffee. However, these statistics overlook the fact that among people who drink, 10 percent account for roughly half the total alcohol consumption in this country (Levinthal, 2002).

Although the negative short-term effects of alcohol are usually overcome, chronic heavy drinking or alcoholism can cause permanent damage to the brain or other parts of the body (Fishbein and Pease, 1996). For alcoholics, the long-term negative health effects include *nutritional deficiencies* resulting from poor eating habits (chronic heavy drinking contributes to high caloric consumption but low nutritional intake); *cardiovascular problems* such as inflammation and enlargement of the heart muscle, high blood pressure, and stroke; and eventually to *alcoholic cirrhosis*—a progressive development of scar tissue that chokes off blood vessels in the liver and destroys liver cells by interfering with their use of oxygen (Levinthal, 2002). Alcoholic cirrhosis is the ninth most frequent cause of death in the United States. The social consequences of heavy drinking are not always limited to the person doing the drinking. For example, abuse of alcohol and other drugs by a pregnant woman can damage her unborn fetus.

Nicotine (Tobacco) The nicotine in tobacco is a toxic, dependency-producing psychoactive drug that is more addictive than heroin. It is classified as a stimulant because it stimulates central nervous system receptors and activates them to release adrenaline, which raises blood pressure, speeds up the heartbeat, and gives the user a temporary sense of alertness. Although the overall proportion of smokers in the general population has declined somewhat since the 1964 Surgeon General warning that smoking is linked to cancer and other serious diseases, tobacco is still responsible for about one in every five deaths in this country (Akers, 1992). (▶ Figure 14.1 displays the nations and U.S. states that had instituted full or partial bans on public smoking as of 2007.) Even people who never light

up a cigarette are harmed by *environmental tobacco smoke*—the smoke in the air inhaled by nonsmokers as a result of other people's tobacco smoking (Levinthal, 2002). Researchers have found that environmental smoke is especially hazardous for nonsmokers who carpool or work with heavy smokers.

Illegal Drugs Marijuana is the most extensively used illegal drug in the United States. About one-third of all people over age twelve have tried marijuana at least once. Although most marijuana users are between the ages of eighteen and twenty-five, use by teenagers has more than doubled during the past decade. High doses of marijuana smoked during pregnancy can disrupt the development of a fetus and result in congenital abnormalities and neurological disturbances (Fishbein and Pease, 1996). Furthermore, some studies have found an increased risk of cancer and other lung problems associated with marijuana because its smokers are believed to inhale more deeply than tobacco users.

Another widely used illegal drug is cocaine: About 23 million people over the age of twelve in the United States report that they have used cocaine at least once, and about one million acknowledge having used it during the past month (Substance

© Michael Newman/PhotoEdit

Despite a variety of warnings from the U.S. Surgeon General about the potentially harmful effects of smoking, many people continue to light up cigarettes. Even those who do not smoke may be affected by environmental tobacco smoke.

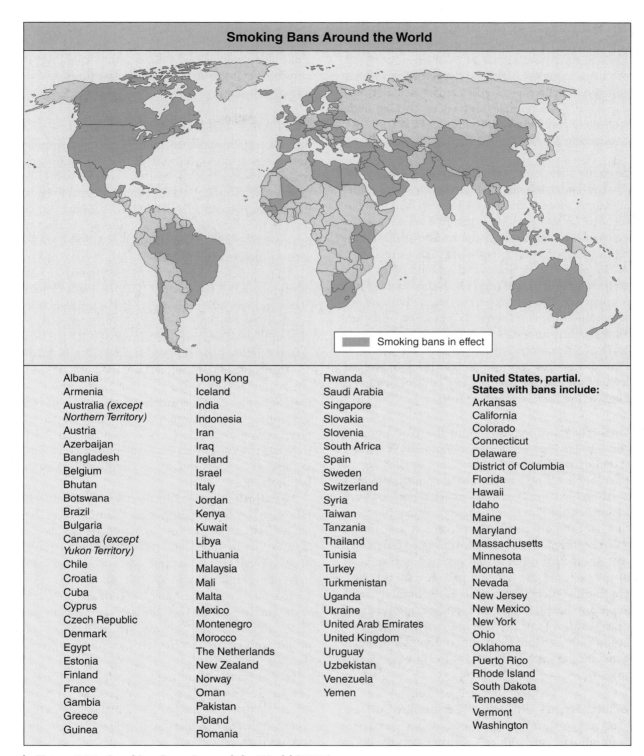

Smoking Bans Around the World

Smoking bans in effect

			United States, partial.
Albania	Hong Kong	Rwanda	**States with bans include:**
Armenia	Iceland	Saudi Arabia	Arkansas
Australia *(except Northern Territory)*	India	Singapore	California
	Indonesia	Slovakia	Colorado
Austria	Iran	Slovenia	Connecticut
Azerbaijan	Iraq	South Africa	Delaware
Bangladesh	Ireland	Spain	District of Columbia
Belgium	Israel	Sweden	Florida
Bhutan	Italy	Switzerland	Hawaii
Botswana	Jordan	Syria	Idaho
Brazil	Kenya	Taiwan	Maine
Bulgaria	Kuwait	Tanzania	Maryland
Canada *(except Yukon Territory)*	Libya	Thailand	Massachusetts
	Lithuania	Tunisia	Minnesota
Chile	Malaysia	Turkey	Montana
Croatia	Mali	Turkmenistan	Nevada
Cuba	Malta	Uganda	New Jersey
Cyprus	Mexico	Ukraine	New Mexico
Czech Republic	Montenegro	United Arab Emirates	New York
Denmark	Morocco	United Kingdom	Ohio
Egypt	The Netherlands	Uruguay	Oklahoma
Estonia	New Zealand	Uzbekistan	Puerto Rico
Finland	Norway	Venezuela	Rhode Island
France	Oman	Yemen	South Dakota
Gambia	Pakistan		Tennessee
Greece	Poland		Vermont
Guinea	Romania		Washington

▶ Figure 14.1 **Smoking Bans Around the World (2007)**

Source: Matt Ray, *Environmental Health Perspectives*, Vol. 115, No. 8, August 2007.

Abuse and Mental Health Services Administration, 2000). People who use cocaine over extended periods of time have higher rates of infection, heart problems, internal bleeding, hypertension, stroke, and other neurological and cardiovascular disorders than do nonusers. Intravenous cocaine users who share contaminated needles are also at risk for contracting AIDS.

Sexually Transmitted Diseases The circumstances under which a person engages in sexual activity is another lifestyle choice with health implications. Although most people find sexual activity enjoyable, it can result in the transmission of certain *sexually transmitted diseases* (STDs), including AIDS, gonorrhea, syphilis, and genital herpes. Prior to 1960, the incidence of STDs in this country had been reduced sharply by barrier-type contraceptives (e.g., condoms) and the use of penicillin as a cure. However, in the 1960s and 1970s the number of cases of STDs increased rapidly with the introduction of the birth control pill, which led to women having more sexual partners and couples being less likely to use barrier contraceptives.

Gonorrhea and Syphilis Until the 1960s, gonorrhea (today the second-most-common STD) and syphilis (which can be acquired not only by sexual intercourse but also by kissing or coming into intimate bodily contact with an infected person) were the principal STDs in this country. Today, however, they constitute less than 15 percent of all cases of STDs reported in U.S. clinics. Untreated gonorrhea may spread from the sexual organs to other parts of the body, among other things negatively affecting fertility; it can also spread to the brain or heart and cause death. Untreated syphilis can, over time, cause cardiovascular problems, brain damage, or even death. Penicillin can cure most cases of either gonorrhea or syphilis as long as the disease has not spread.

Genital Herpes This sexually transmitted disease produces a painful rash on the genitals. Genital herpes cannot be cured: Once the virus enters the body, it stays there for the rest of a person's life, regardless of treatment. However, the earlier that treatment is received, the more likely it is that the severity of the symptoms will be reduced. About 40 percent of persons infected with genital herpes have only a first attack of symptoms of the disease; the remaining 60 percent may have attacks four or five times a year for several years.

AIDS AIDS (acquired immunodeficiency syndrome), which is caused by HIV (human immunodeficiency virus), is among the most significant health problems that this nation—and the world—faces today. Although AIDS almost inevitably ends in death, no one actually dies *of* AIDS. Rather, AIDS reduces the body's ability to fight diseases, making a person vulnerable to many diseases—such as pneumonia—that result in death.

AIDS was first identified in 1981, and the total number of AIDS-related deaths in the United States through 1985 was only 12,493; however, the numbers rose rapidly and precipitously after that. The number of *new* reported cases of AIDS in this country in calendar year 1993 was 103,533; in 2004, 15,798 people in the United States died of AIDS-related diseases (U.S. Department of Health and Human Services, 2006). Fortunately, awareness of the syndrome and how it can be acquired has begun to produce some results: The 15,798 deaths in 2004 were less than one-third the number of such deaths in 1995 (U.S. Department of Health and Human Services, 2006).

Worldwide, however, the number of people with HIV or AIDS is increasing at an alarming rate. The United Nations Joint Programme on HIV/Aids estimates that in 2005, more than 40 million people had HIV/AIDS. Of all new cases worldwide, 14 percent are children, and children under the age of 15 account for about 570,000 AIDS-related deaths each year. More than two-thirds of the people with HIV/AIDS live in sub-Saharan Africa; 14 percent of people with HIV/AIDS live in South and Southeast Asia.

HIV is transmitted through unprotected (or inadequately protected) sexual intercourse with an infected partner (either male or female), by sharing a contaminated hypodermic needle with someone who is infected, by exposure to blood or blood products (usually from a transfusion), and by an infected woman who passes the virus on to her child during pregnancy, childbirth, or breast feeding.

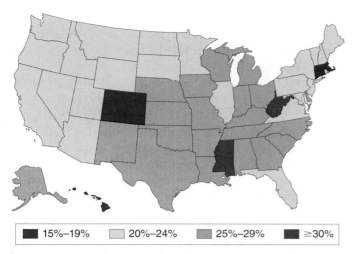

▶ Figure 14.2 **Obesity in the United States**

Source: Centers for Disease Control and Prevention, 2006.

■ 15%–19%   □ 20%–24%   ▨ 25%–29%   ■ ≥30%

Staying Healthy: Diet and Exercise Lifestyle choices also include positive actions such as a healthy diet and good exercise. Over the past several decades, many people in the United States have begun to improve their dietary habits. More individuals now eat larger amounts of vegetables, fruits, and cereals, and substitute unsaturated fats and oils for saturated fats. These changes have contributed to a marked decrease in the incidence of heart disease and of some types of cancer. However, recent studies have raised concern that a significant percentage of children and adults in this country are overweight or obese to an extent that may decrease their life expectancy. ▶ Figure 14.2 maps obesity rates across the United States.

Exercise is another factor. Regular exercise (at least three times a week) keeps the heart, lungs, muscles, and bones in good health and slows the aging process.

Health Care in the United States

Understanding health care as it exists in the United States today requires a brief examination of its history. During the nineteenth century, people became doctors in this country either through apprenticeships, purchasing a mail-order diploma, complet-ing high school and attending a series of lectures, or obtaining bachelor's and M.D. degrees and studying abroad for a number of years. At that time, medical schools were largely proprietary institutions, and their officials were often more interested in acquiring students than in enforcing standards. Medical school graduates were largely poor and frustrated because of the overabundance of doctors and quasi-medical practitioners, so doctors became highly competitive and anxious to limit the number of new practitioners. The obvious way to accomplish this was to reduce the number of medical schools and set up licensing laws to eliminate unqualified or ir-regular practitioners (Kendall, 1980).

The Rise of Scientific Medicine and Professionalism

Although medicine had been previously viewed more as an art than as a science, several significant discoveries during the nineteenth century in areas such as bacteriology and anesthesiology began to give medicine increasing credibility as a science (Nuland, 1997). At the same time that these discoveries were occurring, the ideology of science was being advocated in all areas of life, and people came to believe that almost any task could be done better if the appropriate scientific methods were used. To make medicine in the United States more scientific (and more profitable), the Carnegie Foundation (at the request of the American Medical Association and the forerunner of the Association of American Medical Colleges) commissioned an official study of medical education. The "Flexner report" that resulted from this study has been described as the catalyst of modern medical education but has also been criticized for its lack of objectivity.

The Flexner Report To conduct his study, Abraham Flexner met with the leading faculty at the Johns Hopkins University School of Medicine to develop a model of how medical education should take place; he next visited each of the 155 medical schools then in existence, comparing them with the model. Included in the model was the belief that a medical school should be a full-time, research-oriented, laboratory facility that devoted all of its energies to teaching and research, not to the practice of medicine (Kendall, 1980). It should employ "laboratory men" to train students in the "science" of medi-

cine, and the students should then apply the principles they had learned in the sciences to the illnesses of patients (Brown, 1979). Only a few of the schools Flexner visited were deemed to be equipped to teach scientific medicine; nonetheless, his model became the standard for the profession (Duffy, 1976).

As a result of the Flexner report (1910), all but two of the African American medical schools then in existence were closed, and only one of the medical schools for women survived. As a result, white women and people of color were largely excluded from medical education for the first half of the twentieth century. Until the civil rights movement and the women's movement of the 1960s and 1970s, virtually all physicians were white, male, and upper- or upper-middle class.

The Professionalization of Medicine Despite its adverse effect on people of color and women who might desire a career in medicine, the Flexner re-

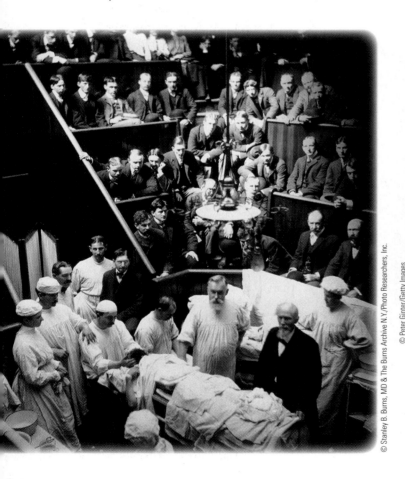

© Stanley B. Burns, MD & The Burns Archive N.Y./Photo Researchers, Inc.

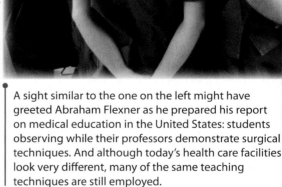

© Peter Ginter/Getty Images

A sight similar to the one on the left might have greeted Abraham Flexner as he prepared his report on medical education in the United States: students observing while their professors demonstrate surgical techniques. And although today's health care facilities look very different, many of the same teaching techniques are still employed.

port did help professionalize medicine. When we compare post-Flexner medicine with the characteristics of professions (see Chapter 13), we find that it meets those characteristics:

1. *Abstract, specialized knowledge.* Physicians undergo a rigorous education that results in a theoretical understanding of health, illness, and medicine. This education provides them with the credentials, skills, and training associated with being a professional.
2. *Autonomy.* Physicians are autonomous and (except as discussed subsequently in this chapter) rely on their own judgment in selecting the appropriate technique for dealing with a problem. They expect patients to respect that autonomy.
3. *Self-regulation.* Theoretically, physicians are self-regulating. They have licensing, accreditation, and regulatory boards and associations that set professional standards and require members to adhere to a code of ethics as a form of public accountability.
4. *Authority.* Because of their authority, physicians expect compliance with their directions and advice. They do not expect clients to argue about the advice rendered (or the price to be charged).
5. *Altruism.* Physicians perform a valuable service for society rather than acting solely in their own self-interest. Many physicians go beyond their self-interest or personal comfort so that they can help a patient.

However, with professionalization, licensed medical doctors gained control over the entire medical establishment, a situation that has continued until the present and—despite current efforts at cost control by insurance companies and others—may continue into the future.

Medicine Today

Throughout its history in the United States, medical care has been on a *fee-for-service* basis: Patients are billed individually for each service they receive, including treatment by doctors, laboratory work, hospital visits, prescriptions, and other health-related expenses. Fee for service is an expensive way to deliver health care because there are few restrictions on the fees charged by doctors, hospitals, and other medical providers.

There are both good sides and bad sides to the fee-for-service approach. The good side is that in the "true spirit" of capitalism, coupled with the hard work and scholarship of many people, this approach has resulted in remarkable advances in medicine. The bad side of fee-for-service medicine is its inequality of distribution. In effect, the United States has a two-tier system of medical care. Those who can afford it are able to get top-notch medical treatment. And where they receive it may not be much like the hospitals that most of us have visited:

> Every afternoon, between three and five, high above New York's Fifth Avenue, the usual quiet of Eleven West is broken by the soft rustle of white linen cloths and the clink of silver and china as high tea is served room by room. . . . Down the hall, a concierge waits to take your dinner order, provide a video from a list of over 950 titles, arrange for a manicure or massage, or send up that magazine or best-seller that you wanted to read. No, this is not a hitherto unknown outpost of the Four Seasons or the Ritz, but a 19-room wing of the Mt. Sinai Medical Center, one of the nation's leading hospitals. . . .
>
> Here at Mt. Sinai and a few other top hospitals . . . sheets are 250-count cotton, the bathrooms are marble and stocked with toiletries, and there is ample room to accommodate a nice seating group of leather wing chairs and a brocade sofa. And here no call button is pressed in vain. Hospital personnel not only come when summoned but are eagerly waiting to cater to your every need. . . .
>
> On these select floors, multi-tiered food service carts and their clattering, plastic trays are gone. "Room service" is in full force. Meals are presented, often course by course, by bow-tied, black-jacketed waiters from rolling, linen-covered tables. (Winik, 1997)

However, the charges for rooms on floors such as those described above may run from $250 to $1,000 per night more than the cost of the standard private room (Winik, 1997). Obviously, this sort of medical care is not within the budget of most of us. Yet the cost of health care per person in the United States rose from $141 in 1960 to $6,270 (almost forty-five times as much) in 2005 (U.S. Census Bureau, 2008) and is still increasing. ▶ Figure 14.3 reflects recent cost increases.

Paying for Medical Care in the United States

The United States and the Union of South Africa are the only developed nations without some form of universal health coverage for all citizens. Before we examine the health care systems of several other nations, however, let's look more closely at our own system.

Private Health Insurance Part of the reason that the cost of fee-for-service health care in the United States escalated rapidly beginning in the 1960s was the expansion of medical insurance programs at that time. Third-party providers (public and private insurers) began picking up large portions of doctor and hospital bills for insured patients. With third-party fee-for-service payment, patients pay premiums into a fund that in turn pays doctors and hospitals for each treatment the patient receives. According to the medical sociologist Paul Starr (1982), third-party fee for service is the main reason for medical inflation because it gives doctors and hospitals an incentive to increase medical services. In other words, the more services they provide, the more fees they charge and the more money they make. Patients have no incentive to limit their visits to doctors or hospitals because they have already paid their premiums and feel entitled to medical care, regardless of the cost.

Public Health Insurance The United States has two nationwide public health insurance programs, Medicare and Medicaid. Medicare is a program for persons age 65 or older who are covered by Social Security or who are eligible and "buy into" the program by paying a monthly premium (Atchley and Barusch, 2004). Medicare pays part of the health care costs of these people. Medicaid, a jointly funded federal–state–local program, was established to make health care more available to the poor. However, both the Medicaid program and the Medicare program are in financial difficulty. These two programs cost $627 billion annually and account for almost one-fourth (23 percent) of all federal spending. Medicare and Medicaid are growing more rapidly than the U.S. economy and the revenues that are used to finance them. As a result, the Congressional Budget Office estimates that the costs of these

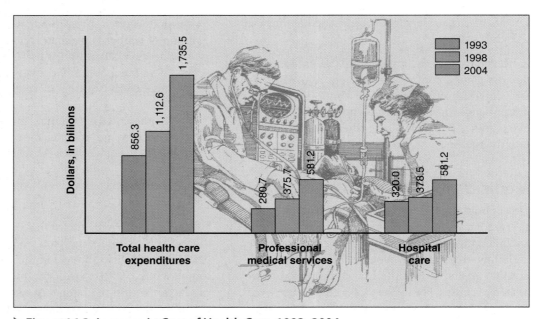

▶ **Figure 14.3 Increase in Cost of Health Care, 1993–2004**
Source: Health Affairs, 2005.

programs will double in the next decade. Medicare is particularly endangered because the number of older Americans who rely on this program to pay for their health care is rising dramatically. As the same time, health care spending is also growing rapidly because of the increased use of new medical technologies (Pear, 2008).

Health Maintenance Organizations (HMOS) Created in an effort to provide workers with health coverage by keeping costs down, *health maintenance organizations (HMOs)* **provide, for a set monthly fee, total care with an emphasis on prevention to avoid costly treatment later.** The doctors do not work on a fee-for-service basis, and patients are encouraged to get regular checkups and to practice good health practices (e.g., exercise and eat right). As long as patients use only the doctors and hospitals that are affiliated with their HMO, they pay no fees, or only small co-payments, beyond their insurance premiums. Seeing HMOs as a potential source of high profits because of their emphasis on preventing serious illness, many for-profit corporations moved into the HMO business in the 1980s (Anders, 1996). However, research shows that preventive care is good for the individual's health but does not necessarily lower total costs.

Recent concerns about physicians being used as gatekeepers who might prevent some patients from obtaining referrals to specialists or from getting needed treatment have resulted in changes in the policies of some HMOs, which now allow patients to visit health care providers outside an HMO's network or to receive other previously unauthorized services by paying a higher co-payment. However, critics charge that those HMOs whose primary-care physicians are paid on a capitation basis—meaning that they receive only a fixed amount per patient that they see, regardless of how long they spend with that patient—in effect encourage doctors to undertreat patients (Weitz, 2004).

Managed Care Another approach to controlling health care costs in the United States is known as *managed care*—**any system of cost containment that closely monitors and controls health care providers' decisions about medical procedures, diagnostic tests, and other services that should be**

provided to patients (Weitz, 2004). In most managed care programs, patients choose a primary-care physician from a list of participating doctors. When patients need medical services, they must first contact the primary-care physician; if a specialist is needed for treatment, the primary-care physician refers the patient to a specialist who participates in the program. Doctors must get approval before they perform certain procedures or admit a patient to a hospital; if they fail to obtain such advance approval, the insurance company has the right to refuse to pay for the treatment or hospital stay.

The Uninsured and the Underinsured Despite public and private insurance programs, about one-third of all U.S. citizens are without health insurance or had difficulty getting or paying for medical care at some time in the last year. As shown on ▶ Map 14.1, the number of people not covered by health insurance varies from state to state. Of the people not covered by health insurance, 8.7 million are children (DeNavas-Walt, Proctor, and Smith, 2007). An estimated 47.0 million people in the United States had no health insurance in 2006—approximately 15.8 percent of the nation's population (DeNavas-Walt, Proctor, and Smith, 2007). The working poor constitute a substantial portion of this category, and it is estimated that 20 million of the uninsured hold full-time jobs. They make too little to afford health insurance but too much to qualify for Medicaid, and their employers do not provide health insurance coverage. What happens when they need medical treatment? "They do without," explains Ray Hanley, medical services director for the Arkansas Department of Human Services (qtd. in Kilborn, 1997: A10).

---

health maintenance organizations (HMOs) companies that provide, for a set monthly fee, total care with an emphasis on prevention to avoid costly treatment later.

managed care any system of cost containment that closely monitors and controls health care providers' decisions about medical procedures, diagnostic tests, and other services that should be provided to patients.

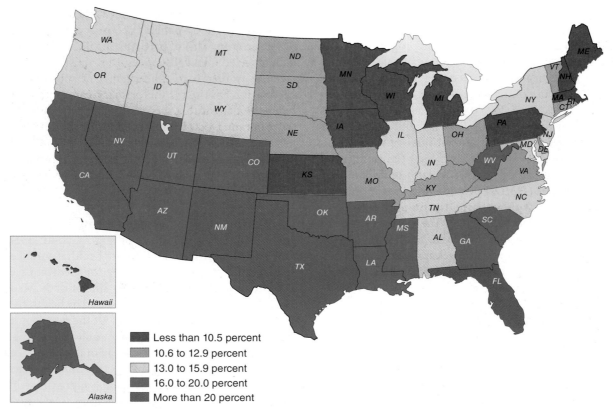

Less than 10.5 percent
10.6 to 12.9 percent
13.0 to 15.9 percent
16.0 to 20.0 percent
More than 20 percent

▶ **Map 14.1** **Persons Not Covered by Health Insurance, by State**
Source: U.S. Census Bureau, 2008.

Paying for Medical Care in Other Nations

Other industrialized and industrializing countries do not leave their citizens in the situation in which some people in the United States find themselves. Let's examine how other nations pay for health care.

Canada Prior to the 1960s, Canada's health care system was similar to that of the United States today. However, in 1962 the government of the province of Saskatchewan implemented a health insurance plan despite opposition from doctors, who went on strike to protest the program. The strike was not successful, as the vast majority of citizens supported the government, which maintained health services by importing doctors from Great Britain. The Saskatchewan program proved itself viable in the years following the strike, and by 1972 all Canadian provinces and terri-

tories had coverage for medical and hospital services (Kendall, Linden, and Murray, 2008). As a result, Canada has a ***universal health care system***—**a health care system in which all citizens receive medical services paid for by tax revenues.** In Canada, these revenues are supplemented by insurance premiums paid by all taxpaying citizens.

One major advantage of the Canadian system over that of the United States is a significant reduction in administrative costs. Whereas more than 20 percent of the U.S. health care dollar represents administrative costs, in Canada the corresponding figure is 10 percent (Weiss and Lonnquist, 2009). However, the system is not without its critics, who claim that it is costly and often wasteful. For example, Canadians are allowed unlimited trips to the doctor, and doctors can increase their income by ordering extensive tests and repeat visits, just as in the United States (Kendall, Linden, and Murray, 2008).

The Canadian health care system does not constitute what is referred to as *socialized medicine*—**a health care system in which the government owns the medical care facilities and employs the physicians.** Rather, Canada has maintained the private nature of the medical profession. Although the government pays most health care costs, the physicians are not government employees and have much greater autonomy than physicians in the health care system in Great Britain.

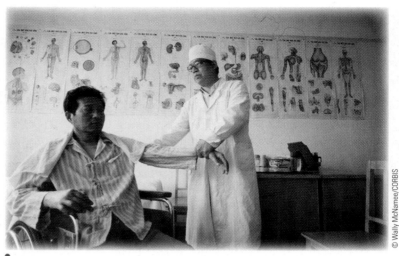

© Wally McNamee/CORBIS

Many physicians in China are trained in both Western and traditional Chinese medicine. Here, a doctor in Tianjin, China, performs pressure therapy to help an aching patient.

Great Britain In 1946, Great Britain passed the National Health Service Act, which provided for all health care services to be available at no charge to the entire population. Although physicians work out of offices or clinics—as in the United States or Canada—the government sets health care policies, raises funds and controls the medical care budget, owns health care facilities, and directly employs physicians and other health care personnel (Weiss and Lonnquist, 2009). Unlike the Canadian model, the health care system in Great Britain *does* constitute socialized medicine. Physicians receive capitation payments from the government: a fixed annual fee for each patient in their practice regardless of how many times they see the patient or how many procedures they perform. They also receive supplemental payments for each low-income or elderly patient in their practice, to compensate for the extra time such patients may require; bonus payments if they meet targets for providing preventive services, such as immunizations against disease; and financial incentives if they practice in medically underserved areas (Weitz, 2004). Physicians may accept private patients, but such patients rarely constitute more than a small fraction of a physician's practice; hospitals reserve a small number of beds for private patients (Weiss and Lonnquist, 2009). Why would anyone want to be a private patient who pays for her or his own care or hospital bed? The answer is primarily found in the desire to avoid the long waits ("queues") that the general population encounters and the fact that private patients can enter the hospital for surgery at times convenient to the consumer rather than wait upon the convenience of the system (Gill, 1994).

China After a lengthy civil war, in 1949 the Communist Party won control of mainland China but found itself in charge of a vast nation with a population of one billion people, most of whom lived in poverty and misery. Malnutrition was prevalent, life expectancies were short, and infant and maternal mortality rates were high. In the cities, only the elite could afford medical care; in the rural areas where most of the population resided, Western-style health care barely existed (Weitz, 2004). With a lack of both financial resources and trained health care personnel, China needed to adopt innovative strategies in order to improve the health of its populace. One such policy was developing a large number of *physician extenders* and sending them out into the cities and rural areas to educate the public regarding

universal health care system a health care system in which all citizens receive medical services paid for by tax revenues.

socialized medicine a health care system in which the government owns the medical care facilities and employs the physicians.

health and health care and to treat illness and disease. Referred to as *street doctors* in urban areas and *barefoot doctors* in the countryside, these individuals had little formal training and worked under the supervision of trained physicians (Weitz, 2004).

Over the past four decades, Chinese medical training has become more rigorous, and supervision has increased. All doctors receive training in both Western and traditional Chinese medicine. Doctors who work in hospitals receive a salary; all other doctors now work on a fee-for-service basis. In urban areas, the cost of health care is paid by employers; however, 78 percent of the population lives in rural areas, where most work on family farms and are expected to pay for their own health care. The cost of health care generally remains low, but the cost of hospital care has risen; accordingly, many Chinese—if they can afford it—purchase health care insurance to cover the cost of hospitalization (Weitz, 2004). As a low-income country, China spends only 5 percent of its gross domestic product on health care, but the health of its citizens is only slightly below that of most industrialized nations.

Regardless of which system of delivering medical care that a nation may have, the health care providers and the general population of the nation are having to face new issues that arise as a result of new technology.

Social Implications of Advanced Medical Technology

Advances in medical technology are occurring at a speed that is almost unbelievable; however, sociologists and other social scientists have identified specific social implications of some of the new technologies (see Weiss and Lonnquist, 2009):

1. *The new technologies create options for people and for society, but options that alter human relationships.* An example is the ability of medical personnel to sustain a life that in earlier times would have ended as the result of disease or an accident. Although this can be beneficial, technologically advanced equipment that can sustain life after consciousness is lost and there is no likelihood that the person will recover can create a difficult decision for the family of that person if he or she

has not left a *living will*—a document stating the person's wishes regarding the medical circumstances under which his or her life should be terminated. Federal law requires all hospitals and other medical facilities to honor the terms of a living will. Recent media coverage of individuals whose lives have been prolonged by new medical technologies has made more people aware of some end-of-life issues.

2. *The new technologies increase the cost of medical care.* For example, the computerized axial tomography (CT or CAT) scanner—which combines a computer with X-rays that are passed through the body at different angles—produces clear images of the interior of the body that are invaluable in investigating disease. However, the cost of such a scanner is around $1 million. Magnetic resonance imaging (MRI) equipment that allows pictures to be taken of internal organs ranges in cost from $1 million to $2.5 million. Can the United States afford such equipment in every hospital for every patient? The money available for health care is not unlimited, and when it is spent on high-tech equipment and treatment, it is being reallocated from other health care programs that might be of greater assistance to more people.

3. *The new technologies raise provocative questions about the very nature of life.* In Chapter 11, we briefly discuss in vitro fertilization—a form of assisted reproductive technology. But during 1997, Dr. Ian Williams and his associates in Scotland took in vitro fertilization a step further: They cloned a lamb (that they named Dolly) from the DNA of an adult sheep. Subsequently, scientists have cloned other animals in the same manner, raising a number of profound questions: If scientists can duplicate mammals from adult DNA, is it possible to clone perfect (whatever that may be) human beings instead of taking a chance on a child that is born to a couple? If it is possible, would it be ethical? For example, if—as discussed earlier in this text—most parents prefer a boy over a girl if they are going to have only one child, would the world suddenly have substantially more boys than girls? Would everyone start to look alike, eliminating diversity? If a child were born other than through cloning and had some sort of biological defect, would the child have the right to sue his or her parents for negligence?

However, at the same time that high-tech medicine is becoming a major part of overall health care, many people are turning to holistic medicine and alternative healing practices.

Holistic Medicine and Alternative Medicine

When examining the subject of medicine, it is easy to think only in terms of conventional (or mainstream) medical treatment. By contrast, **holistic medicine is an approach to health care that focuses on prevention of illness and disease and is aimed at treating the whole person—body and mind—rather than just the part or parts in which symptoms occur.** Under this approach, it is important that people not look solely to medicine and doctors for their health, but rather that people engage in health-promoting behavior. Likewise, medical professionals must not only treat illness and disease but also work with the patient to promote a healthy lifestyle and self-image.

Many practitioners of *alternative medicine*—healing practices inconsistent with dominant medical practice—take a holistic approach, and today many people are turning to alternative medicine

© Royalty Free/CORBIS

Use of herbal therapies is a form of alternative medicine that is increasing in popularity in the United States. How does this approach to health care differ from a more traditional medical approach?

either in addition to or in lieu of traditional medicine. However, many medical doctors are opposed to alternative medicine. In understanding the medical establishment's reaction to alternative medicine, it is important to keep in mind the philosophy of scientific medicine—that medicine is a science, not an art. Thus, to the extent to which alternative medicine is "nonscientific," it must be quackery and therefore something that is undoubtedly worthless and possibly harmful. Undoubtedly, self-interest is also involved in mainstream medicine's reaction to alternative medicine: If the public can be persuaded that scientific medicine is the only legitimate healing practice, fewer health care dollars will be spent on a form of medical treatment that is (at least to some extent) in competition with the medical establishment (Weiss and Lonnquist, 2009). But if all forms of alternative medicine (including chiropractic, massage, and spiritual) are taken into account, people spend more money on unconventional therapies than they do for all hospitalizations (Weiss and Lonnquist, 2009).

Sociological Perspectives on Health and Medicine

Functionalist, conflict, symbolic interactionist, and postmodernist perspectives focus on different aspects of health and medicine; each provides us with significant insights on the problems associated with these pressing social concerns.

A Functionalist Perspective: The Sick Role

According to the functionalist approach, if society is to function as a stable system, it is important for people to be healthy and to contribute to their society. Consequently, sickness is viewed as a form of

holistic medicine an approach to health care that focuses on prevention of illness and disease and is aimed at treating the whole person—body and mind—rather than just the part or parts in which symptoms occur.

deviant behavior that must be controlled by society. This view was initially set forth by the sociologist Talcott Parsons (1951) in his concept of the *sick role*—**the set of patterned expectations that defines the norms and values appropriate for individuals who are sick and for those who interact with them.** According to Parsons, the sick role has four primary characteristics:

1. People who are sick are not responsible for their condition. It is assumed that being sick is not a deliberate and knowing choice of the sick person.
2. People who assume the sick role are temporarily exempt from their normal roles and obligations. For example, people with illnesses are typically not expected to go to school or work.
3. People who are sick must want to get well. The sick role is considered to be a temporary role that people must relinquish as soon as their condition improves sufficiently. Those who do not return to their regular activities in a timely fashion may be labeled as hypochondriacs or malingerers.
4. People who are sick must seek competent help from a medical professional to hasten their recovery.

As these characteristics show, Parsons believed that illness is dysfunctional for both individuals and the larger society. Those who assume the sick role are unable to fulfill their necessary social roles, such as being parents or employees. Similarly, people who are ill lose days from their productive roles in society, thus weakening the ability of groups and organizations to fulfill their functions.

According to Parsons, it is important for the society to maintain social control over people who enter the sick role. Physicians are empowered to determine who may enter this role and when patients are ready to exit it. Because physicians spend many years in training and have specialized knowledge about illness and its treatment, they are certified by the society to be "gatekeepers" of the sick role. When patients seek the advice of a physician, they enter into the patient–physician relationship, which does not contain equal power for both parties. The patient is expected to follow the "doctor's orders" by adhering to a treatment regime, recovering from the malady, and returning to a normal routine as soon as possible.

What are the major strengths and weaknesses of Parsons's model and, more generally, of the functionalist view of health and illness? Parsons's analysis of the sick role was pathbreaking when it was introduced. Some social analysts believe that Parsons made a major contribution to our knowledge of how society explains illness-related behavior and how physicians have attained their gatekeeper status. In contrast, other analysts believe that the sick-role model does not take into account racial–ethnic, class, and gender variations in the ways that people view illness and interpret this role. For example, this model does not take into account the fact that many individuals in the working class may choose not to accept the sick role unless they are seriously ill—because they cannot afford to miss time from work and lose a portion of their earnings. Moreover, people without health insurance may not have the option of assuming the sick role.

A Conflict Perspective: Inequalities in Health and Health Care

Unlike the functionalist approach, conflict theory emphasizes the political, economic, and social forces that affect health and the health care delivery system. Among the issues of concern to conflict theorists are the ability of all people to obtain health care; how race, class, and gender inequalities affect health and health care; power relationships between doctors and other health care workers; the dominance of the medical model of health care; and the role of profit in the health care system.

Who is responsible for problems in the U.S. health care system? According to many conflict theorists, problems in U.S. health care delivery are rooted in the capitalist economy, which views medicine as a commodity that is produced and sold by the medical–industrial complex. The *medical–industrial complex* **encompasses both local physicians and hospitals as well as global health-related industries such as insurance companies and pharmaceutical and medical supply companies** (Relman, 1992).

The United States is one of the few industrialized nations that relies almost exclusively on the medical–industrial complex for health care delivery and does not have universal health coverage, which provides some level of access to medical treatment for all people. Consequently, access to high-quality medical care is linked to people's ability to pay and to their position within the class structure. Those who are affluent or have good medical insurance may receive high-quality, state-of-the-art care in the medical–industrial complex because of its elaborate technologies and treatments. However, people below the poverty level and those just above it have greater difficulty gaining access to medical care. Referred to as the *medically indigent,* these individuals do not earn enough to afford private medical care but earn just enough money to keep them from qualifying for Medicaid (Weiss and Lonnquist, 2009). In the profit-oriented capitalist economy, these individuals are said to "fall between the cracks" in the health care system.

Who benefits from the existing structure of medicine? According to conflict theorists, physicians—who hold a legal monopoly over medicine—benefit from the existing structure because they can charge inflated fees. Similarly, clinics, pharmacies, laboratories, hospitals, supply manufacturers, insurance companies, and many other corporations derive excessive profits from the existing system of payment in medicine. In recent years, large drug companies and profit-making hospital corporations have come to occupy a larger and larger part of health care delivery. As a result, medical costs have risen rapidly, and the federal government and many insurance companies have placed pressure for cost containment on other players in the medical–industrial complex (Tilly and Tilly, 1998).

Conflict theorists increase our awareness of inequalities of race, class, and gender as these statuses influence people's access to health care. They also inform us about the problems associated with health care becoming "big business." However, some analysts believe that the conflict approach is unduly pessimistic about the gains that have been made in health status and longevity—gains that are at least partially due to large investments in research and treatment by the medical–industrial complex.

A Symbolic Interactionist Perspective: The Social Construction of Illness

Symbolic interactionists attempt to understand the specific meanings and causes that we attribute to particular events. In studying health, symbolic

interactionists focus on the meanings that social actors give their illness or disease and how these meanings affect people's self-concept and relationships with others. According to symbolic interactionists, we socially construct "health" and "illness" and how both should be treated. For example, some people explain disease by blaming it on those who are ill. If we attribute cancer to the acts of a person, we can assume that we will be immune to that disease if we do not engage in the same behavior. Nonsmokers who learn that a lung cancer victim had a two-pack-a-day habit feel comforted that they are unlikely to suffer the same fate. Similarly, victims of AIDS are often blamed for promiscuous sexual conduct or intravenous drug use, regardless of how they contracted HIV. In this case, the social definition of the illness leads to the stigmatization of individuals who suffer from the disease.

Although biological characteristics provide objective criteria for determining medical conditions such as heart disease, tuberculosis, or cancer, there is also a subjective component to how illness is defined. This subjective component is very important when we look at conditions such as childhood hyperactivity, mental illness, alcoholism, drug abuse, cigarette smoking, and overeating, all of which have been medicalized. The term *medicalization* **refers to the process whereby nonmedical problems become defined and treated as illnesses or disorders.** Medicalization may occur on three levels: (1) the conceptual level (e.g., the use of medical terminology to define the problem), (2) the institutional level (e.g., physicians are supervisors of treatment

sick role the set of patterned expectations that defines the norms and values appropriate for individuals who are sick and for those who interact with them.

medical–industrial complex local physicians, local hospitals, and global health-related industries such as insurance companies and pharmaceutical and medical supply companies that deliver health care today.

medicalization the process whereby nonmedical problems become defined and treated as illnesses or disorders.

and gatekeepers to applying for benefits), and (3) the interactional level (e.g., when physicians treat patients' conditions as medical problems). For example, the sociologists Deborah Findlay and Leslie Miller (1994: 277) explain how gambling has been medicalized:

> Habitual gambling . . . has been regarded by a minority as a sin, and by most as a leisure pursuit—perhaps wasteful but a pastime nevertheless. Lately, however, we have seen gambling described as a psychological illness—"compulsive gambling." It is in the process of being medicalized. The consequences of this shift in discourse (that is, in the way of thinking and talking) about gambling are considerable for doctors, who now have in gamblers a new market for their services or "treatment"; perhaps for gambling halls, which may find themselves subject to new regulations, insofar as they are deemed to contribute to the "disease"; and not least, for gamblers themselves, who are no longer treated as sinners or wastrels, but as patients, with claims on our sympathy, and to our medical insurance plans as well.

Sociologists often refer to this form of medicalization as the *medicalization of deviance* because it gives physicians and other medical professionals greater authority to determine what should be considered "normal" and "acceptable" behavior and to establish the appropriate mechanisms for controlling "deviant behaviors."

According to symbolic interactionists, medicalization is a two-way process: Just as conditions can be medicalized, so can they be demedicalized. **Demedicalization refers to the process whereby a problem ceases to be defined as an illness or a disorder.** Examples include the removal of certain behaviors (such as homosexuality) from the list of mental disorders compiled by the American Psychiatric Association and the deinstitutionalization of mental health patients. The process of demedicalization also continues in women's health as advocates seek to redefine childbirth and menopause as natural processes rather than as illnesses (Conrad, 1996).

In addition to how health and illness are defined, symbolic interactionists examine how doctors and patients interact in health care settings (see "Soci-

© Thinkstock/Jupiterimages

Is gambling a moral issue or a medical one? According to sociologists, the recent trend toward viewing compulsive gambling as a health care issue is an example of the medicalization of deviance.

ology Works!" on page 490). Some physicians may hesitate to communicate certain kinds of medical information to patients, such as why they are prescribing certain medications or what side effects or drug interactions may occur (Kendall, 2004).

Symbolic interactionist perspectives on health and health care provide us with new insights on the social construction of illness and how health and illness cannot be strictly determined by medical criteria. Symbolic interactionists also make us aware of the importance of communication between physicians and patients, including factors that may reduce effective medical treatment for some individuals. However, these approaches have been criticized for suggesting that few objective medical criteria exist for many illnesses and for overemphasizing microlevel issues without giving adequate recognition to macrolevel issues such as the effects on health care of managed care, health maintenance organizations, and for-profit hospital chains.

A Postmodernist Perspective: The Clinical Gaze

As previously discussed in Chapter 6, some postmodern perspectives focus on how the powerful exert control over less powerful individuals. Through

CONCEPT QUICK REVIEW

Sociological Perspectives on Health and Medicine

A functionalist perspective: the sick role	People who are sick are temporarily exempt from normal obligations but must want to get well and seek competent help.
A conflict perspective: inequalities in health and health care	Problems in health care are rooted in the capitalist system, exemplified by the medical–industrial complex.
A symbolic interactionist perspective: the social construction of illness	People socially construct both "health" and "illness," and how both should be treated.
A postmodernist perspective: the clinical gaze	Doctors gain power through observing patients to gather information, thus appearing to speak "wisely."

the use of tightly controlled knowledge and specialized norms and values, institutions such as clinics, hospitals, and other branches of the medical establishment are able to categorize and treat people in ways that are not available to people outside the medical environment.

How, for example, did doctors gain so much power over patients? How did the medical establishment achieve the level of prestige it possesses in contemporary societies? In *The Birth of the Clinic* (1994/1963), social theorist Michel Foucault asserted that doctors gain power through the *clinical* (or "observing") *gaze,* which they use to gather information. In earlier times, doctors developed the clinical gaze through their observation of patients; as the doctors began to diagnose and treat medical conditions, they also started to speak "wisely" about everything. As a result, other people started to believe that doctors could "penetrate illusion and see . . . the hidden truth" (Shawver, 1998). Consequently, truth in medicine—as in all other areas of social life—is a social construction.

According to Foucault, the prestige of the medical establishment was further enhanced when it became possible to categorize all illnesses within a definitive network of disease classification under which physicians can claim that they know why patients are sick. Moreover, the invention of new tests made it necessary for physicians to gaze upon the naked body, to listen to the human heart with an instrument, and to run tests on the patient's body

fluids. Patients who objected were criticized by the doctors for their "false modesty" and "excessive restraint" (Foucault, 1994/1963: 163). As the new rules allowed for the patient to be touched and prodded, the myth of the doctor's diagnostic wisdom was further enhanced, and "medical gestures, words, gazes took on a philosophical density that had formerly belonged only to mathematical thought" (Foucault, 1994/1963: 199). For Foucault, the formation of clinical medicine was merely one of the more-visible ways in which the fundamental structures of human experience have changed throughout history. According to postmodernists, medical knowledge is a form of power that is not always used for the benefit of patients and the general public.

This chapter's Concept Quick Review summarizes the major sociological perspectives on health and medicine.

Disability

What is a disability? There are many different definitions. In business and government, disability is often defined in terms of work—for instance, "an inability

---

demedicalization the process whereby a problem ceases to be defined as an illness or a disorder.

Sociology *Works!*

Sociology Sheds Light on the Physician–Patient Relationship

DOCTOR: What's the problem?

(Chair noise).

PATIENT: . . . had since last Monday evening so it's a week of sore throat.

DOCTOR: . . . hm . . . hm . . .

PATIENT: . . . which turned into a cold . . . and then a cough.

DOCTOR: A cold you mean what? Stuffy nose?

PATIENT: uh stuffy nose yeah not a chest . . . cold . . .

DOCTOR: . . . hm . . . hm . . . And a cough.

PATIENT: And a cough . . . which is the most irritating aspect. . . .

DOCTOR: Okay. Uh, any fever?

PATIENT: Not that I know of . . . I took it a couple of times in the beginning but haven't felt like—

DOCTOR: How bout your ears? . . . (Mishler, 2005: 322)

In this brief excerpt from the transcript of a discussion between a doctor and patient, the patient responds to the doctor's request for information by telling him what her symptoms are and how they began to change when her illness "turned into a cold . . . and then a cough." According to sociologists who study the social organization of health care, this transcript indicates that the physician wants the patient to continue speaking when he makes sounds such as "hm . . . hm," but he also wants to remain in control of the conversation. When the patient mentions that she has a "cold," for example, the doctor asks for further clarification of her specific symptoms so that he can determine, efficiently and in a short period of time, what kind of cold she has and what the treatment plan should be (Mishler, 2005). This is one brief passage from many pages of medical transcripts that researchers have used to study what they refer to as the struggle between the voice of medicine and the voice of the lifeworld (Mishler, 1984, 2005).

What are the voices of medicine and of the lifeworld? In this context, the voice of medicine refers to the technical, scientific attitude adopted by many doctors in their communication with patients. This type of discourse is generally abstract, neutral, and somewhat distant. By contrast, the voice of the lifeworld refers to the natural, everyday attitudes that are expressed by patients when they talk to their physician in the hope of gaining additional insight on their medical condition. Some sociologists believe that a constant struggle exists between these two "voices" in the doctor–patient relationship and that this struggle affects the outcome of each medical encounter. The voice of medicine makes it difficult for patients to believe that their concerns are being heard when the physician is visibly in a hurry, does not listen, interrupts frequently, and/or talks down to the patient.

To minimize the voice of medicine, some sociologists advocate *therapeutic communication* between doctors and patients. In therapeutic communication, (1) the physician engages in full and open communication with the patient and feels free to ask questions about psychosocial as well as physical conditions, (2) the patient provides full and open information to the physician and feels free to ask questions and seek clarifications, and (3) a genuine rapport develops between physician and patient (Weiss and Lonnquist, 2009). According to some social analysts, it takes a doctor no additional time to use a positive communication style that conveys friendliness, empathy, genuineness and candor, an openness to conversation, and a nonjudgmental attitude toward the patient. Such a communication style certainly helps physicians establish positive relationships with their patients. Of course, patients should also try to communicate in a positive manner with physicians, people whom the patients hope will be able to help them remain healthy or help them resolve an existing medical problem (Weiss and Lonnquist, 2009).

Reflect & Analyze

Sociologists who study the social organization of medicine will continue to look for new insights on the physician–patient relationship in the future. What other sociological perspectives do you believe might be useful in explaining the dynamics of doctor–patient communications or other social interactions (such as between physicians and other health professionals) that routinely take place within the health care system?

to engage in gainful employment." Medical professionals tend to define it in terms of organically based impairments—the problem being entirely within the body. However, not all disabilities are visible to others or necessarily limit people physically. ***Disability* refers to a reduced ability to perform tasks one would normally do at a given stage of life and that may result in stigmatization or discrimination against the person with disabilities.** In other words, the notion of disability is based not only on physical conditions but also on social attitudes and the social and physical environments in which people live. In an elevator, for example, the buttons may be beyond the reach of persons using a wheelchair. In this context, disability derives from the fact that certain things have been made inaccessible to some people (Weitz, 2004). According to disability rights advocates, disability must be thought of in terms of how society causes or contributes to the problem—not in terms of what is "wrong" with the person with a disability.

An estimated 49.7 million persons in the United States have one or more physical or mental disabilities. This number continues to increase for several

David Paterson, the governor of New York, is legally blind. According to disability-rights advocates, this combination is rare: People with disabilities have higher rates of underemployment and unemployment, and few attain executive positions.

reasons. First, with advances in medical technology, many people who once would have died from an accident or illness now survive, although with an impairment. Second, as more people live longer, they are more likely to experience diseases (such as arthritis) that may have disabling consequences. Third, persons born with serious disabilities are more likely to survive infancy because of medical technology. However, less than 15 percent of persons with a disability today were born with it; accidents, disease, and war account for most disabilities in this country.

Although anyone can become disabled, some people are more likely to be or to become disabled than others. African Americans have higher rates of disability than whites, especially more serious disabilities; persons with lower incomes also have higher rates of disability (Weitz, 2004). However, "disability knows no socioeconomic boundaries. You can become disabled from your mother's poor nutrition or from falling off your polo pony," says Patrisha Wright, a spokesperson for the Disability Rights Education and Defense Fund (qtd. in Shapiro, 1993: 10).

For persons with chronic illness and disability, life expectancy may take on a different meaning. Knowing that they will likely not live out the full life expectancy for persons in their age cohort, they may come to "treasure each moment," as did the late James Keller, a well-respected former college baseball coach:

In December 1992, I found out I have Lou Gehrig's disease—amyotrophic lateral sclerosis, or ALS. I learned that this disease destroys every muscle in the body, that there's no known cure or treatment and that the average life expectancy for people with ALS is two to five years after diagnosis.

Those are hard facts to accept. Even today, nearly two years after my diagnosis, I see myself as 42-year-old career athlete who has always been blessed with excellent health. Though not an hour goes by in which I don't see or hear in my mind that phrase "two to five years," I still can't

disability a physical or health condition that stigmatizes or causes discrimination.

quite believe it. Maybe my resistance to those words is exactly what gives me the strength to live with them and the will to make the best of every day in every way. (Keller, 1994)

As Keller's comments illustrate, disease and disability are intricately linked.

Environment, lifestyle, and working conditions may all contribute to either temporary or chronic disability. For example, air pollution in automobile-clogged cities leads to a higher incidence of chronic respiratory disease and lung damage, which may result in severe disability for some people. Eating certain types of food and smoking cigarettes increase the risk for coronary and cardiovascular diseases. In contemporary industrial societies, workers in the second tier of the labor market (primarily recent immigrants, white women, and people of color) are at the greatest risk for certain health hazards and disabilities. Employees in data processing and service-oriented jobs may also be affected by work-related disabilities. The extensive use of computers has been shown to harm some workers' vision; to produce joint problems such as arthritis, low-back pain, and carpal tunnel syndrome; and to place employees under high levels of stress that may result in neuroses and other mental health problems (Albrecht, 1992). As shown in ◆ Table 14.1, about one out of five people in the United States (20.8 percent) has a "chronic health condition which, given the physical, attitudinal, and financial barriers built into the social system, makes it difficult to perform one or more activities generally considered appropriate for persons of their age" (Weitz, 2004).

Can a person in a wheelchair have equal access to education, employment, and housing? If public transportation is not accessible to those in wheelchairs, the answer is certainly no. As disability rights activist Mark Johnson put it, "Black people fought for the right to ride in the front of the bus. We're fighting for the right to get on the bus" (qtd. in Shapiro, 1993: 128).

Many disability rights advocates argue that persons with a disability have been kept out of the mainstream of society. They have been denied equal opportunities in education by being consigned to special education classes or special schools. For example, people who grow up deaf are often viewed as disabled; however,

◆ Table 14.1 Percentage of U.S. Population with Disabilities

Characteristic	Percentage[a]
With a disability	20.8
Severe	13.7
Not severe	7.0
Has difficulty or is unable to:	
See words and letters	3.5
Hear normal conversation	3.5
Have speech understood	1.2
Lift or carry ten pounds	6.9
Use stairs	9.2
Walk	9.4
Has difficulty or needs assistance with:	
Getting around inside the house	1.7
Getting in/out of bed or a chair	2.5
Taking a bath or shower	2.2
Dressing	1.7
Eating	0.8
Getting to or using the toilet	1.1
Has difficulty and needs assistance with:	
Going outside the home alone	4.0
Managing money and bills	2.2
Preparing meals	2.3
Doing light housework	3.1
Using the telephone	1.3

[a]Percentage of persons age 15 and older.
Source: Steinmetz, 2006.

many members of the deaf community instead view themselves as a "linguistic minority" that is part of a unique culture (Lane, 1992; Cohen, 1994). They believe that they have been restricted from entry into schools and the work force not due to their own limitations, but by societal barriers.

Living with disabilities is a long-term process. For infants born with certain types of congenital (present at birth) problems, their disability first acquires social significance for their parents and caregivers. In a study of children with disabilities in Israel, the sociologist Meira Weiss (1994) challenged the assumption that parents automatically bond with infants, especially those born with visible disabilities. She found that an infant's appearance may determine how parents will view the child. Parents are more likely to be bothered by external, openly vis-

Double amputee Oscar Pistorius uses carbon fiber prosthetics to allow him to race in international events. Many observers found it ironic when Pistorius had to fight for his chance to qualify for the 2008 Olympics after some of his competitors argued that the prosthetics gave him an unfair advantage.

© ALESSANDRO BIANCHI/Reuters/Landov

Among persons who acquire disabilities through disease or accidents later in life, the social significance of their disability can be seen in how they initially respond to their symptoms and diagnosis, how they view the immediate situation and their future, and how the illness and disability affect their lives. When confronted with a disability, most people adopt one of two strategies—avoidance or vigilance. Those who use the avoidance strategy deny their condition in order to maintain hopeful images of the future and elude depression; for example, some individuals refuse to participate in rehabilitation following a traumatic injury because they want to pretend that the disability does not exist. By contrast, those using the vigilance strategy actively seek knowledge and treatment so that they can respond appropriately to the changes in their bodies (Weitz, 2004).

Sociological Perspectives on Disability

How do sociologists view disability? Those using the functionalist framework often apply Parsons's sick-role model, which is referred to as the *medical model* of disability. According to the medical model, people with disabilities become, in effect, chronic patients under the supervision of doctors and other medical personnel, subject to a doctor's orders or a program's rules, and not to their own judgment (Shapiro, 1993). From this perspective, disability is deviance. The deviance framework is also apparent in some symbolic interactionist perspectives. According to symbolic interactionists, people with a disability experience *role ambiguity* because many people equate disability with deviance (Murphy et al., 1988). By labeling individuals with a disability as "deviant," other people can avoid them or treat them as outsiders. Society marginalizes people with a disability because they have lost old roles and statuses and are labeled as "disabled" persons. According to the sociologist Eliot Freidson (1965), how the people are labeled results from three factors: (1) their degree of responsibility for

ible disabilities than by internal or disguised ones; some parents are more willing to consent to or even demand the death of an "appearance-impaired" child (Weiss, 1994). According to Weiss, children born with internal (concealed) disabilities are at least initially more acceptable to parents because they do not violate the parents' perceived body images of their children. Weiss's study provides insight into the social significance that people attach to congenital disabilities.

their impairment, (2) the apparent seriousness of their condition, and (3) the perceived legitimacy of the condition. Freidson concluded that the definitions of and expectations for people with a disability are socially constructed factors.

Finally, from a conflict perspective, persons with a disability are members of a subordinate group in conflict with persons in positions of power in the government, in the health care industry, and in the rehabilitation business, all of whom are trying to control their destinies (Albrecht, 1992). Those in positions of power have created policies and artificial barriers that keep people with disabilities in a subservient position (Asch, 1986; Hahn, 1987). Moreover, in a capitalist economy, disabilities are big business. When people with disabilities are defined as a social problem and public funds are spent to purchase goods and services for them, rehabilitation becomes a commodity that can be bought and sold by the medical–industrial complex (Albrecht, 1992). From this perspective, persons with a disability are objectified. They have an economic value as consumers of goods and services that will allegedly make them "better" people. Many persons with a disability endure the same struggle for resources faced by people of color, women, and older persons. Individuals who hold more than one of these ascribed statuses, combined with experiencing disability, are doubly or triply oppressed by capitalism.

Social Inequalities Based on Disability

People with visible disabilities are often the objects of prejudice and discrimination, which interfere with their everyday life. For example, Marylou Breslin, executive director of the Disability Rights Education and Defense Fund, was wearing a businesswoman's suit, sitting at the airport in her battery-powered wheelchair, and drinking a cup of coffee while waiting for a plane. A woman walked by and plunked a coin in the coffee cup that Breslin held in her hand, splashing coffee on Breslin's blouse (Shapiro, 1993). Why did the woman drop the coin in Breslin's cup? The answer to this question is found in stereotypes built on lack of knowledge about or exaggeration of the characteristics of people with a disability.

Some stereotypes project the image that persons with disabilities are deformed individuals who may also be horrible deviants. For example, slasher movies such as the *Nightmare on Elm Street* series show a villain who was turned into a hateful, sadistic killer because of disfigurement resulting from a fire. Somewhat lighter fare such as the *Batman* movies depict villains as individuals with disabilities: the Joker, disfigured by a fall into a vat of acid, and the Penguin, born with flippers instead of arms (Shapiro, 1993). Other stereotypes show persons with disabilities as pathetic individuals to be pitied. Fund-raising activities by many charitable organizations—such as "poster child" campaigns showing a photograph of a friendly-looking child with a visible disability—sometimes contribute to this perception. Even apparently positive stereotypes become harmful to people with a disability. An example is what some disabled persons refer to as "supercrips"—people with severe disabilities who seem to excel despite the impairment and who receive widespread press coverage in the process. Disability rights advocates note that such stereotypes do not reflect the day-to-day reality of most persons with disabilities, who must struggle constantly with smaller challenges (Shapiro, 1993).

Today, many working-age persons with a disability in the United States are unemployed (see "Census Profiles: Disability and Employment Status"). Most of them believe that they could and would work if offered the opportunity. However, even when persons with a severe disability are able to find jobs, they typically earn less than persons without a disability. On average, workers with a severe disability make 59 percent of what their coworkers without disabilities earn (U.S. Census Bureau, 2005a). The problem has been particularly severe for African Americans and Latinos/as with disabilities. Among Latinos/as with a severe disability, only 26 percent are employed; those who work earn 80 percent of what white (non-Latino) persons with a severe disability earn.

Employment, poverty, and disability are related. On the one hand, people may become economically disadvantaged as a result of chronic illness or disability. On the other hand, poor people are less likely to be educated and more likely to be malnourished and have inadequate access to health care—all of which contribute to the risk of chronic illness, physical and

Disability and Employment Status

Each month, the Census Bureau conducts a survey of about 50,000 representative households, gathering the primary source of information on the labor force characteristics of the U.S. population. This monthly survey provides information about employment, unemployment, earnings, hours of work, and other data, including the labor force participation of persons with a work disability. As you can see from the figure set forth below, for the population ages 25 to 64 (the period during which people are most likely to be employed), less than 25 percent of persons with a disability are employed, compared with more than 80 percent of persons without a disability who are employed. Is employment among persons with a disability related primarily to their disability status, or do other factors—such as prejudice or lack of willingness to make the necessary accommodations that would allow such persons to hold a job—play a significant part in the high rate of unemployment for persons with a disability?

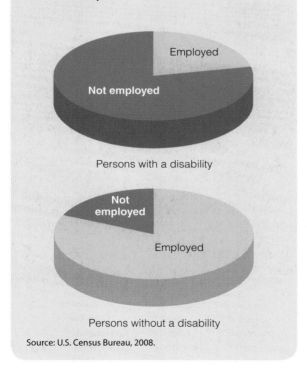

Persons with a disability

Persons without a disability

Source: U.S. Census Bureau, 2008.

mental disability, and the inability to participate in the labor force. In addition, the type of employment available to people with limited resources increases their chances of becoming disabled. They may work in hazardous places such as mines, factory assembly lines, and chemical plants, or in the construction industry, where the chance of becoming seriously disabled is much higher (Albrecht, 1992).

Generalizations about the relationship between disability and income are difficult to make for at least three reasons. First, most research on disability is organized around specific conditions or impairments, making it problematic to reach conclusions about how much income a person loses because of a disability (DeJong, Batavia, and Griss, 1989). Second, generalizations about entire categories of people (such as whites, African Americans, or Latinos/as) tend to be inaccurate because of differences *within* each group. For example, the sociologist Ronald Angel (1984) found significant differences in the effect of disabilities on Latinos depending on whether they were Mexican American, Puerto Rican, or Cuban American. Third is the problem of determining which factor occurred first. For example, some of the problems for Puerto Ricans with disabilities are tied to their overall position of economic disadvantage. Although for both males and females, disability is associated with lower earnings and higher rates of unemployment, a disability has a stronger negative effect on women's labor force participation than it does on men's. Compared to men, women with disabilities are overrepresented as clerical and service workers and underrepresented as managers and administrators. Women with disabilities are also much less likely to be covered by pension and health plans than are men (Russo and Jansen, 1988).

Health Care in the Future

A central question regarding the future of health care is the role that advanced technologies will play in medical diagnosis and treatment. Technology is a major stimulus for social change, and the health care systems in high-income nations such as the

Box 14.3 You Can Make a Difference

Joining the Fight Against Illness and Disease!

Each day trained volunteers nurture female patients in hospitals and long-term care facilities by administering complimentary makeup and moisturizer. Among the benefits to patients receiving this caring social interaction and gentle touch are higher self-esteem and shorter recovery times. (Fiffer and Fiffer, 1994: 120)

Beverly Barnes did not intend to start the program now known as Patient Pride when she went to the hospital to visit a friend who was recuperating from surgery. However, by the time she left the hospital, having helped her friend put on makeup and freshen her appearance, Barnes realized how much seemingly small acts of kindness can mean to people who are ill or recuperating from surgery. Eventually, Patient Pride, Inc., became a nonprofit corporation that relies on volunteers to visit patients, and new chapters

servicing hospitals and long-term health care facilities have opened throughout the country. Recently, the program has sought to include male patients as well as women.

Patient Pride is only one of many examples of how everyday people can make a difference in the lives of people experiencing illness or disability. Here are some Internet sources that you can check out to learn more about volunteering and participating in the fight against diseases such as cancer and cardiovascular disease:

- The American Cancer Society's home page (which has links to various volunteer activities):
  http://www.cancer.org

- The American Heart Association's home page:
  http://www.amhrt.org

United States reflect the rapid rate of technological innovation that has occurred in the last few decades. In the future, advanced health care technologies will no doubt provide even more accurate and quicker diagnosis, improved treatment techniques, and increased life expectancy. However, technology alone cannot solve many of the problems confronting us in health and health care delivery. In fact, some aspects of technological innovation may be dysfunctional for individuals and society. As we have seen, some technological "advances" raise new ethical concerns, such as the moral and legal issues surrounding the cloning of human life. Some "advances" may also fail: A new prescription drug may be found to cause side-effects that are more serious than the illness that it was supposed to remedy.

Whether advanced technology succeeds or fails in some areas, it will probably continue to increase the cost of health care in the future. As a result, the gap between the rich and the poor in the United States will contribute to inequalities of access to vital medical services. On a global basis, new technologies may lower the death rate in some low-

© AP Images/Jennifer Graylock/JGRYL, Graylock

Michael Moore's 2007 documentary *Sicko* addresses inequities in the U.S. health care system. These advocates chose the film's New York premiere to present their argument for universal health care in the United States.

income counties, but it will primarily be the wealthy in those nations who will have access to the level of health care that many people in higher-income countries take for granted.

The organization of U.S. health care will continue to change in the future. Despite recent legal challenges, many managed care companies continue their cost-containment strategies and intervene in medical decisions formerly reserved for physicians and other health care professionals. Consequently, managed care and new technologies will continue

to change the traditional doctor–patient relationship, even in how people communicate with one another.

Finally, to a degree, health care in the future will be up to each of us. What measures will we take to safeguard ourselves against illness and disorders? How can we help others who are the victims of acute and chronic diseases or disabilities? Although we cannot change global or national health problems, there are some small (but not insignificant) things we can do to make a difference (see Box 14.3).

Chapter Review

• What is health, and why are sociologists interested in studying health and medicine?

According to the World Health Organization, health is a state of complete physical, mental, and social well-being. In other words, health is not only a biological issue but also a social issue. For this reason, sociologists are interested in studying health and medicine. As a social institution, medicine is one of the most important components of quality of life.

• How is health care paid for in the United States?

Throughout most of the past hundred years, medical care in the United States has been paid for on a fee-for-service basis. The term *fee for service* means that patients are billed individually for each service they receive. This approach to paying for medical services is expensive because few restrictions are placed on the fees that doctors, hospitals, and other medical providers can charge patients. Recently, there have been efforts at cost containment, and HMOs and managed care have produced both positive and negative results in the contemporary practice of medicine. Health maintenance organizations (HMOs) provide, for a set monthly fee, total care with an emphasis on prevention to avoid costly treatment later. Managed care refers to any system of cost containment that closely monitors and controls health care providers' decisions about medical

procedures, diagnostic tests, and other services that should be provided to patients.

• What is the functionalist perspective on health and illness?

According to the functionalist approach, if society is to function as a stable system, it is important for people to be healthy and to contribute to their society. Consequently, sickness is viewed as a form of deviant behavior that must be controlled by society. Sociologist Talcott Parsons (1951) described the sick role—the set of patterned expectations that defines the norms and values appropriate for individuals who are sick and for those who interact with them. Although individuals are given permission to not perform their usual activities for a period of time, they are expected to seek medical attention and get well as soon as possible so that they can go about their normal routine.

• What is the conflict perspective on health and illness?

Conflict theory tends to emphasize the political, economic, and social forces that affect health and the health care delivery system. Among these issues are the ability of all people to obtain health care; how race, class, and gender inequalities affect health and health care; power relations between doctors and other health care workers; the dominance of the medical model of health care; and the role of profit in the health care system.

● **What is the symbolic interactionist perspective on health and illness?**

In studying health, symbolic interactionists focus on the fact that the meaning that social actors give their illness or disease will affect their self-concept and their relationships with others. According to symbolic interactionists, we socially construct "health" and "illness" and how both should be treated. Symbolic interactionists also examine medicalization—the process whereby nonmedical problems become defined and treated as illnesses or disorders.

● **What is the postmodernist perspective on health and illness?**

Postmodern theorists such as Michel Foucault argue that doctors and the medical establishment have gained control over illness and patients at least partly because of the physicians' clinical gaze, which replaces all other systems of knowledge. The myth of

the wise doctor was also supported by the development of disease classification systems and new tests.

www.cengage.com/login

Register for a Student eResource account to maximize your study time online using CengageNOW. First take the system's diagnostic pre-test, and then follow the personalized study plan that is created for you to help you review this chapter. The study plan will

● help you identify areas on which you should concentrate;
● provide interactive exercises to help you master the chapter concepts; and
● provide a post-test to confirm you are ready to move on to the next chapter.

Key Terms

acute diseases 471

chronic diseases 471

demedicalization 488

disability 491

drug 473

health 466

health care 466

health maintenance organization (HMO) 481

holistic medicine 485

infant mortality rate 466

life expectancy 466

managed care 481

medical–industrial complex 486

medicalization 487

medicine 466

sick role 486

social epidemiology 470

socialized medicine 483

universal health care system 482

Questions for Critical Thinking

1. Why is it important to explain the social, as well as the biological, aspects of health and illness in societies?
2. In what ways are race, class, and gender intertwined with physical and mental disorders?
3. How would functionalists, conflict theorists, and symbolic interactionists suggest that health

care delivery might be improved in the United States?
4. Based on this chapter, how do you think illness and disability will be handled in the United States in the near future? Are there things that we can learn from other nations regarding the delivery of health care? Why or why not?

The Kendall Companion Website

www.cengage.com/sociology/kendall

Supplement your review of this chapter by going to the text's companion website, where you can take tutorial quizzes, use flash cards to master key terms, follow live links to useful websites, and explore the other study and research resources you'll find there, such as a comprehensive interactive sociology timeline, GSS Data, and Census 2000 information, much of it presented visually in maps.

15 Population and Urbanization

The tomato-farming region of Immokalee, Florida:

It's a very bad thing because we're working very hard here and there's no support from the government. We're only working. We're not committing a sin.

—Rigoberto Morales expressing his frustration upon learning that Congress is doing little to help him become a U.S. citizen after he has spent eight long years working in the hot fields of Florida picking tomatoes and sending most of his earnings to his family in Mexico (qtd. in Goodnough and Steinhauer, 2006: A35)

We live here in fear. We fear Immigration will come, and many people just don't go out.

—Antonia Fuentes, a Mexican farmworker who has picked tomatoes for two years and would welcome even a guest worker program that would permit her to stay in the United States for a set period of time (qtd. in Goodnough and Steinhauer, 2006: A35)

A migrant shelter in Nogales, Mexico:

We want to try our luck [in the United States]. We can't go back to Michoacan [a state in Mexico] because there is no future there. My brothers said there is plenty of work [in the

© A. Ramey/PhotoEdit

Many undocumented workers seek to enter the United States so that they can work and obtain a better quality of life for themselves and their families. However, the aspirations of these workers often run contrary to the mandate of law enforcement officials to patrol the border and keep out individuals who seek to enter the country illegally.

United States], and that it looks like they will start giving [work] permits.

—Edith Mondragon explaining why she and her husband paid a smuggler in a failed attempt to get across the Mexican border into the United States (qtd. in CNN.com, 2006)

It's hard to cross. But it's harder to see your children have little to eat.

—Raul Gonzalez, who turned himself in to U.S. authorities after he was robbed and his feet started bleeding from walking for five days to get across the border into the United States, stating the common theme of most immigrants: We want a better life for ourselves and our children (qtd. in CNN.com, 2006)

Washington, D.C., immigration rally:

We want to be legal. We want to live without hiding, without fear. We have to speak so that our voices are listened to and we are taken into account.

—Ruben Arita, a construction worker from Honduras, explaining why he participated in a rally to push for legal status for undocumented workers residing in the United States (qtd. in Swarns, 2006: A1)

Immigrants are coming together in a way that we have never seen before, and it's going to keep going. . . . This is a movement. We're sending a strong message that we are people of dignity. All that we want is to have a shot at the American dream.

—Jaime Contreras, who entered the United States illegally from El Salvador but is now a U.S. citizen, discussing why he would like to see a major change in current immigration and citizenship laws (qtd. in Swarns, 2006: A1)

Chapter Focus Question

What effect does migration have on cities and on shifts in the global population?

The first decade of the twenty-first century has resonated with pleas from undocumented workers and leaders of various immigration reform organizations asking for change in U.S. immigration and citizenship laws for millions of people who dream of U.S. citizenship (Kilgannon, 2006). However, even as tens of thousands of demonstrators have taken to the streets to proclaim "We Are America" and to ask for legal status and citizenship for millions of illegal immigrants, many political leaders and ordinary citizens fear that immigration will become the downfall of this nation. One member of Congress stated that the United States will become "the biggest magnet ever. It would be like a dinner bell, 'Come one, come all'" (qtd. in Hulse and Swarns, 2006: A12).

In the United States, as well as in many other nations that have experienced high rates of immigration, questions of permanent residency and citizenship for immigrant workers have stirred up extensive debates and sometimes produced hostility or violence toward people who are labeled as "illegal immigrants" or "outsiders." According to recent estimates, between 11.5 and 12 million unauthorized migrants are living in the United States, and two-thirds (66 percent) of this population has been in the country for ten years or less. In fact, about 40 percent of the unauthorized population (about 4.4 million people) has been in the United States for five years or less (Passel,

2006). This sharp upswing in unauthorized immigration has produced a complex immigration debate, involving many pressing economic, political, and social issues, and has divided people in this country. By contrast, other factors associated with population growth and urban change, such as the number of live births annually as compared with the number of deaths, have received very little recent attention.

In this chapter, we explore the dynamics of population and urbanization with a focus on how migration, and particularly immigration, affect growth and change in societies such as ours. Before reading on, text your knowledge about current U.S. immigration issues by taking the quiz in Box 15.1.

Demography: The Study of Population

Although population growth has slowed in the United States, the world's population of 6.6 billion in 2008 is increasing by almost 78 million people per year as a result of the larger number of births than deaths worldwide (see ▶ Figure 15.1). Between 2000 and 2030, almost all of the world's 1.4 percent annual population growth will occur in low-income countries in Africa, Asia, and Latin America (Population Reference Bureau, 2001).

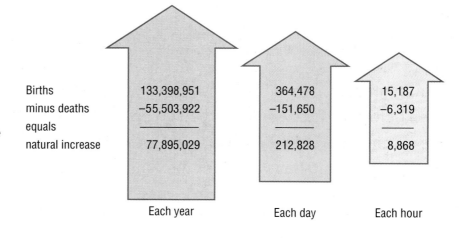

▶ Figure 15.1 Growth in the World's Population
Every single day, the world's population increases by more than 200,000 people.

Source: U.S. Census Bureau, 2008.

	Each year	Each day	Each hour
Births	133,398,951	364,478	15,187
minus deaths	−55,503,922	−151,650	−6,319
equals natural increase	77,895,029	212,828	8,868

How Much Do You Know About U.S. Immigration?

True	False	
T	F	1. All "unauthorized immigrants" in the United States entered the country illegally.
T	F	2. Nearly two-thirds of the children living in unauthorized immigrant families in the United States are U.S. citizens by birth.
T	F	3. Adult males make up a larger percentage of the unauthorized immigrant population than adult females.
T	F	4. Unauthorized immigrants from Mexico and Latin America represent slightly less than 50 percent of the unauthorized population in the United States.
T	F	5. During the past ten years, more unauthorized immigrants came to the United States from Europe and Canada than from Asia.
T	F	6. Unauthorized male immigrants are more likely to be employed than are males who are either legal immigrants or native-born.
T	F	7. The percentage of unauthorized immigrant workers in white-collar occupations has risen substantially in the 2000s.
T	F	8. In the 2000s, most undocumented immigrant workers have settled in six states (California, New York, Florida, Texas, New Jersey, and Illinois).

Answers on page 504.

What causes the population to grow rapidly in some nations? This question is of interest to scholars who specialize in the study of *demography*—a subfield of sociology that examines population size, composition, and distribution. Many sociological studies use demographic analysis as a component of the research design because all aspects of social life are affected by demography. For example, an important relationship exists between population size and the availability of food, water, energy, and housing. Population size, composition, and distribution are also connected to issues such as poverty, racial and ethnic diversity, shifts in the age structure of society, and concerns about environmental degradation (Weeks, 2008).

Increases or decreases in population can have a powerful impact on the social, economic, and political structures of societies. As used by demographers, a *population* is a group of people who live in a specified geographic area. Changes in populations occur as a result of three processes: fertility (births), mortality (deaths), and migration.

Fertility

Fertility **is the actual level of childbearing for an individual or a population.** The level of fertility in a society is based on biological and social factors, the primary biological factor being the number of women of childbearing age (usually between ages 15 and 45). Other biological factors affecting fertility include the general health and level of nutrition of women of childbearing age. Social factors influencing the level of fertility include the roles available to women in a society and prevalent viewpoints regarding what constitutes the "ideal" family size.

Based on biological capability alone, most women could produce twenty or more children during

demography a subfield of sociology that examines population size, composition, and distribution.

fertility the actual level of childbearing for an individual or a population.

Answers to the Sociology Quiz on U.S. Immigration

1. False. Although the term "unauthorized immigrant" refers to a U.S. resident who is not a citizen of this country, who has not been admitted for permanent residence, or who does not have an authorized temporary status that permits longer-term residence and work, some "unauthorized immigrants" originally entered the country with valid visas but overstayed their visas' expiration or otherwise violated the terms of their admission.

2. True. Based on the latest available estimates, about 3.1 million children, or 64 percent of all children living in unauthorized immigrant families, are U.S. citizens by birth.

3. True. Estimates suggest that about 58 percent of the unauthorized adult immigrant population is composed of males, whereas females make up about 42 percent of the adult unauthorized population.

4. False. Unauthorized immigrants from Mexico and Latin America account for almost 80 percent of the unauthorized population in the United States. Most unauthorized immigrants come from Mexico, and the Pew Hispanic Center estimates that about 80 to 85 percent of all Mexican immigrants residing in the United States for less than 10 years are undocumented.

5. False. During the past ten years, unauthorized immigrants from South and East Asia made up about 13 percent of the undocumented population, whereas immigrants from Europe and Canada accounted for about 6 percent of all illegal U.S. immigration.

6. True. Undocumented male workers are often younger than legal immigrant or native-born males, and they are also less likely to attend college. As a result, 94 percent of unauthorized male immigrants are in the work force, as compared with 86 percent of male legal immigrants and 83 percent of native-born adult males.

7. False. Unauthorized immigrant workers continue to be underrepresented in white-collar occupations such as management, business, and professional occupations and overrepresented in occupational categories (such as farming, cleaning, construction, and food preparations) that typically require less education and have no licensing requirement.

8. False. Although most immigrant workers previously settled in one of these six states, many more are moving to—or initially arriving at—destinations throughout various regions of the United States, including states such as North Carolina, Georgia, Nebraska, and Idaho.

Source: Based on Passel, 2006 (based on the March 2005 Current Population Survey).

their childbearing years. *Fecundity* **is the potential number of children who could be born if every woman reproduced at her maximum biological capacity.** Fertility rates are not as high as fecundity rates because people's biological capabilities are limited by social factors such as practicing voluntary abstinence and refraining from sexual intercourse until an older age, as well as by contraception, voluntary sterilization, abortion, and infanticide. Additional social factors affecting fertility include significant changes in the number of available partners for sex and/or marriage (as a result of war, for example), increases in the number of women of childbearing age in the work force, and high rates of unemployment. In some countries, governmental policies also affect the fertility rate. For example, if China's two-decades-old policy of allowing only one child per family in order to limit population growth stays in place, it will result in that country's population starting to decline in 2042, according to United Nations projections (Beech, 2001).

◆ **Table 15.1 The Ten Leading Causes of Death in the United States, 1900 and 2000**

Cause of Death—1900	Rank	Cause of Death—2000
Influenza/pneumonia	1	Heart disease
Tuberculosis	2	Cancer
Stomach/intestinal disease	3	Stroke
Heart disease	4	Chronic lung disease
Cerebral hemorrhage	5	Accidents
Kidney disease	6	Pneumonia and influenza
Accidents	7	Diabetes
Cancer	8	HIV
Diseases in early infancy	9	Suicide
Diphtheria	10	Homicide

Sources: Hostetler, 1994; Cockerham, 1995; and U.S. Census Bureau, 2006.

The most basic measure of fertility is the *crude birth rate*—**the number of live births per 1,000 people in a population in a given year.** In 2006 the crude birth rate in the United States was 14.1 per 1,000, as compared with an all-time high rate of 27 per 1,000 in 1947 (following World War II). This measure is referred to as a "crude" birth rate because it is based on the entire population and is not "refined" to incorporate significant variables affecting fertility, such as age, marital status, religion, and race/ethnicity.

In most areas of the world, women are having fewer children. Women who have six or seven children tend to live in agricultural regions of the world, where children's labor is essential to the family's economic survival and child mortality rates are very high. For example, Uganda has a crude birth rate of 49.1 per 1,000, as compared with 14.1 per 1,000 in the United States (U.S. Census Bureau, 2008). However, in Uganda and some other African nations, families need to have many children in order to ensure that one or two will live to adulthood due to high rates of poverty, malnutrition, and disease.

Mortality

The primary cause of world population growth in recent years has been a decline in *mortality*—**the incidence of death in a population.** The simplest measure of mortality is the *crude death rate*—**the number of deaths per 1,000 people in a population in a given year.** In 2006 the U.S. crude death rate was 8.3 per 1,000 (U.S. Census Bureau, 2008). In high-income, developed nations, mortality rates have declined dramatically as diseases such as malaria, polio, cholera, tetanus, typhoid, and measles have been virtually eliminated by vaccinations and improved sanitation and personal hygiene (Weeks, 2008). Just as smallpox appeared to be eradicated, however, HIV/AIDS rapidly rose to surpass the 30 percent fatality rate of smallpox. (The ten leading causes of death are shown in ◆ Table 15.1.) In low-income, less-developed nations, infectious diseases remain the leading cause of death; in some areas, mortality rates are increasing rapidly as a result of

fecundity the potential number of children who could be born if every woman reproduced at her maximum biological capacity.

crude birth rate the number of live births per 1,000 people in a population in a given year.

mortality the incidence of death in a population.

crude death rate the number of deaths per 1,000 people in a population in a given year.

HIV/AIDS. Children under age fifteen constitute a growing number of those who are infected with HIV.

But many children do not survive long enough to contract communicable diseases. On a global basis, large numbers of newborn infants do not live to see their first birthday. The measure of these deaths is referred to as the *infant mortality rate,* which is defined in Chapter 14 as the number of deaths of infants under 1 year of age per 1,000 live births in a given year. The infant mortality rate is an important reflection of a society's level of preventive (prenatal) medical care, maternal nutrition, childbirth procedures, and neonatal care for infants. Differential levels of access to these services are reflected in the divergent infant mortality rates for African Americans and whites. In 2004, for example, the U.S. mortality rate for white infants was 5.7 per 1,000 live births, as compared with 13.8 per 1,000 live births for African American infants (U.S. Census Bureau, 2008).

As discussed in Chapter 14, *life expectancy* is an estimate of the average lifetime in years of people born in a specific year. For persons born in the United States in 2006, for example, life expectancy at birth was 77.8 years, as compared with 82.0 years in Japan and 50 years or less in the African nations of Ethiopia, Mozambique, Nigeria, and Uganda. Life expectancy varies by sex; for instance, females born in the United States in 2004 could expect to live about 80.4 years as compared with 75.2 years for males. Life expectancy also varies by race; for example, African American men have a life expectancy at

birth of about 69.5 years, compared to 75.7 years for white males (U.S. Census Bureau, 2008).

Migration

Migration **is the movement of people from one geographic area to another for the purpose of changing residency.** Migration affects the size and distribution of the population in a given area. *Distribution* refers to the physical location of people throughout a geographic area. In the United States, people are not evenly distributed throughout the country; many of us live in densely populated areas. *Density* is the number of people living in a specific geographic area. In urbanized areas, density may be measured by the number of people who live per room, per block, or per square mile.

Migration may be either international (movement between two nations) or internal (movement within national boundaries). Internal migration has occurred throughout U.S. history and has significantly changed the distribution of the population over time.

Migration involves two types of movement: immigration and emigration. *Immigration* is the movement of people into a geographic area to take up residency. Each year, about one million people enter the United States, primarily from Latin America and Asia. Over the last three decades, there has been a tenfold increase in the number of adult Mexican immigrants living in the United States. Today, these immigrants alone account for more than eight million people in this country, and their children make

Political unrest, violence, and war are "push" factors that encourage people to leave their country of origin. By contrast, job opportunities, such as construction work in the United States, are a major "pull" factor for people from low-income countries.

507

CHAPTER 15 • POPULATION AND URBANIZATION

up between 10 and 20 percent of the U.S. child population, with these rates being even higher in states such as Texas and California (Crosnoe, 2006). Working with immigrant families and their children has become an important concern (see Box 15.2).

Immigration rates are not an accurate reflection of the actual number of immigrants who enter a country. The Department of Homeland Security records only legal immigration based on entry visas and change-of-immigration-status forms. Similarly, few records are maintained regarding *emigration*—the movement of people out of a geographic area to take up residency elsewhere. To determine the net migration in a geographic area, the number of people leaving that area to take up permanent or semipermanent residence elsewhere (emigrants) is subtracted from the number of people entering that area to take up residence there (immigrants), unless more people are moving out of the area than into it, in which case the mathematical process is reversed.

People migrate either voluntarily or involuntarily. *Pull* factors at the international level, such as a democratic government, religious freedom, employment opportunities, or a more temperate climate, may draw voluntary immigrants into a nation. Within nations, people from large cities may be pulled to rural areas by lower crime rates, more space, and a lower cost of living. People such as Antonia Fuentes, whose decision to migrate to the United States is described at the beginning of this chapter, are drawn by pull factors such as greater economic opportunities at their destination and are pushed by factors such as low wages and few employment opportunities in their previous place of residence. *Push* factors at the international level, such as political unrest, violence, war, famine, plagues, and natural disasters, may encourage people to leave one area and relocate elsewhere. Push factors in regional U.S. migration include unemployment, harsh weather conditions, a high cost of living, inadequate school systems, and high crime rates.

Involuntary, or forced, migration usually occurs as a result of political oppression, such as when Jews fled Nazi Germany in the 1930s or when Afghans left their country to escape oppression there in the late 1990s. Slavery is the most striking example of involuntary migration; for example, the 10–20 million Africans forcibly transported to the Western Hemisphere prior to 1800 did not come by choice (see Chapter 9).

Population Composition

Changes in fertility, mortality, and migration affect the *population composition*—**the biological and social characteristics of a population, including age, sex, race, marital status, education, occupation, income, and size of household.**

One measure of population composition is the *sex ratio*—**the number of males for every hundred females in a given population.** A sex ratio of 100 indicates an equal number of males and females in the population. If the number is greater than 100, there are more males than females; if it is less than 100, there are more females than males. In the United States, the estimated sex ratio for 2006 was 97.1, which means there were about 97 males per 100 females. Although approximately 124 males are conceived for every 100 females, male fetuses miscarry at a higher rate. From birth to age 14, the sex ratio is 105; in the age 40–44 category, however, the ratio shifts to 99.3, and from this point on, women outnumber men. By age 65, the sex ratio is about 84.6—that is, there are 84 men for every 100 women.

For demographers, sex and age are significant population characteristics; they are key indicators of fertility and mortality rates. The age distribution of a population has a direct bearing on the demand for schooling, health, employment, housing, and pensions. The current distribution of a population can be depicted in a *population pyramid*—**a graphic representation of the distribution of a population by sex and age.**

migration the movement of people from one geographic area to another for the purpose of changing residency.

population composition the biological and social characteristics of a population, including age, sex, race, marital status, education, occupation, income, and size of household.

sex ratio a term used by demographers to denote the number of males for every hundred females in a given population.

population pyramid a graphic representation of the distribution of a population by sex and age.

Immigration and the Changing Face(s) of the United States

Throughout U.S. history, immigration has had a profound effect on our nation. Chances are very good that almost all of us can trace our heritage and our family roots to one or more nations where our ancestors lived before coming to the United States. Immigration has also been a controversial topic at times throughout our nation's history, just as it is at the start of the twenty-first century.

When we look at the faces of the people around our country today, we see a wide diversity of human beings, most of whom are seeking to live their lives together positively and peacefully. Demographers and other social science researchers are interested in studying how people become part of the mainstream of a country to which they have immigrated while still maintaining their own unique cultural identity, and why other immigrants do not want to become part of that mainstream.

As you view the pictures on these pages, think about how you and other members of your family came to view yourselves as Americans—residents of the United States of America—and what this means to you and to them in terms of what you think and do on a daily basis. Doing so helps us gain a better understanding of some of the sociological issues relating to immigration.

© AP Images/Robert F. Bukaty

At a march to show support for the Somali community in Lewiston, Maine, marchers urged people to "Love Thy Neighbor!" rather than to attempt to keep new arrivals from other countries from moving into the city.

Immigrants and Educational Opportunities

For recent immigrants, education is a key that may open the door to greater involvement in one's new country. Some educational experiences may be more long term (such as completing a college degree), whereas others serve more immediate needs, such as the acculturation and parenting class that these recent Hmong immigrants have completed to help them adjust to their new life in the United States.

Politics and Immigration

Voting and political participation help people have a voice in the U.S. democratic process. Here, four languages are used to encourage residents of South Boston to vote on "Super Tuesday" in 2008. How does politics influence our thinking about immigration? Does immigration influence our thinking about politics? Can our opinions change over time?

Immigration and the Changing Face of the Media

As recent immigrants to the United States reach out to find media sources that reflect their culture and interests, executives at many media outlets strive to reach the large number of young people (typically between the ages of 18 and 24) who represent the future of the larger ethnic categories in this country. Latinos and Latinas, for example, are a key target audience for both Hispanic and so-called mainstream media. On the Spanish-language cable network program *Acceso Maximo,* for instance, viewers vote for their favorite videos and artists by text messaging or voting online.

Immigrants and Employment

Many people often view the worlds *immigrants* and *workers* as being almost interchangeable because the primary purpose of much immigration is either to find work or for employers to have a larger pool of low-paid workers from which to hire new employees. Immigrant workers in the United States hold many jobs, ranging from agricultural and gardening positions to high-tech and health-care-related professions. As with many of our ancestors, these California landscape workers hope that their earnings will help their children have a more secure future in this country.

Reflect & Analyze

1. When did your ancestors immigrate to the United States? Or, if you are a Native American, what is the history of your group? Do some research to find out about your ancestors.
2. Do you believe that non-English-speaking immigrants to this country should be expected—or required—to read and speak English at school, work, and other public places? Why or why not?
3. Should undocumented workers have access to government services, such as medical clinics for low-income patients? Should the minor children of undocumented workers have this type of access? Why or why not?

Turning to Video

Watch the ABC video *Minutemen Patrol the Border* (running time 2:13), available on the Kendall Companion Website and through Cengage Learning eResources accounts. Following President George W. Bush's controversial efforts in 2005 to combat illegal immigration, this news report provides a look at the Minutemen, private citizens who live usually along U.S. borders and patrol for illegal immigrants, as well as communities where illegal immigration is an unexpected source of conflict. As you watch the video, think about the photographs, commentary, and questions that you encountered in this photo essay. After you've watched the video, consider another question: To what degree is racism involved in some U.S. citizens' anti–illegal immigrant feelings and actions?

Windows Media Player

RUN FOR THE BORDER
ILLEGAL ALIEN CRISIS?

00:00:39   Playing: 277 Kbps

Box 15.2 You Can Make a Difference

Creating a Vital Link Between College Students and Immigrant Children

The program that I volunteered for is the Youth Tutoring Program. This program is designed to help refugee, immigrant, and underserved minority children get additional help on homework or class assignments. The program also offers activities to help children succeed in school. Its goal is to help the children develop effective problem-solving skills and create supportive peer networks.

My experience with the program and especially with the kids [has] been wonderful and truly rewarding. I love the children there, and I enjoy teaching them as well as building a relationship with them. . . . The students have taught me the importance of a strong and caring friendship.

> —Sovanny That (2007), who was completing her doctor of pharmacy degree when she wrote these words, explaining why she feels so strongly about making a contribution through her volunteer work with immigrant and refugee children

Sovanny That's volunteer work took place with the Refugee Women's Alliance (ReWA), a nonprofit, multiethnic organization that provides services for newly arrived immigrant families in the state of Washington. Across the nation, volunteers provide much-needed services for recent immigrants and their children who may need assistance not only in learning about life in the United States but also in developing basic learning skills for school. Adjusting to a new environment is difficult for adults, but it can be even more challenging for young children who may not understand the reasons why their families have moved to faraway places.

Like Sovanny, thousands of other college students seek to make a difference in the lives of immigrant children through organized volunteer activities at their schools. At Northwestern University in Evanston, Illinois, for example, the Northwestern Community Development Corps (NCDC) is a student-run organization that engages students in community development in the greater Chicago area. According to the group's mission statement, "We promote civic engagement through direct service, social awareness, and advocacy" (NCDC, 2007). Among the projects offered by NCDC is the Centro Romero After School program, where college volunteers tutor seven- to twelve-year-old children from immigrant or refugee families in math, reading, and spelling. After the homework is completed, the volunteers enjoy playing games with the children. Tutoring programs for immigrant students exist nationwide in various public schools and colleges, and most college students who volunteer in these programs find their activities to be both rewarding and a learning experience as the children begin to thrive through their efforts.

Depending on the area of the country in which you live, children from immigrant and refugee families may trace their roots to Mexico, Central or South America, Somalia, Ethiopia, or Eastern Europe, among other places. By serving in such a program, volunteers may not only help the children gain a foothold in this country but may also help dispel a pervasive myth that immigrant students are a drain on the American educational system. Volunteers may help other people see that these students are a valuable resource in our schools because they provide firsthand knowledge of the larger world in which we live. As the sociologist Robert Crosnoe (2006) suggested in his recent book on how to help Mexican immigrant children succeed in school (and this would apply to other racial/ethnic groups as well), recent immigrants should be seen not as a *threat* to our nation but as a potential *resource*. Consequently, we should make every effort to see that the American Dream is a realistic goal, and not just an ideology, for new arrivals as well as for everyone else residing in this country.

Does your college or university have a volunteer coordinator who might provide you with the names of programs where you could put your skills to work helping immigrant students or other people who can benefit from your knowledge and expertise?

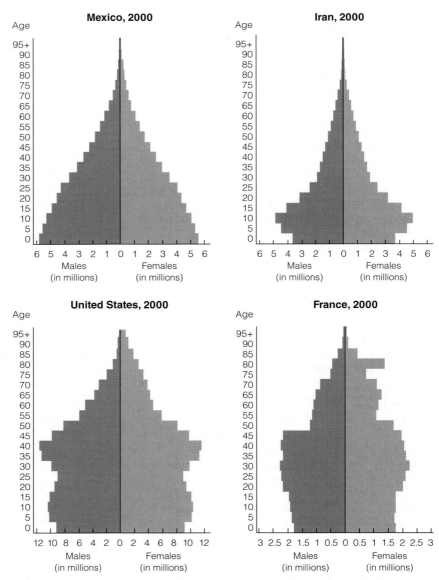

▶ **Figure 15.2 Population Pyramids for Mexico, Iran, the United States, and France**

Sources: Weeks, 2005; UNAIDS/WHO, 2000.

Population pyramids are a series of bar graphs divided into five-year age cohorts: The left side of the pyramid shows the number or percentage of males in each age bracket; the right side provides the same information for females. The age/sex distribution in the United States and other high-income nations does not have the ap-pearance of a classic pyramid, but rather is more rectan-gular or barrel-shaped. By contrast, low-income nations, such as Mexico and Iran, which have high fertility and mortality rates, do fit the classic population pyramid. ▶ Figure 15.2 compares the demographic composition of Mexico, Iran, the United States, and France.

Population Growth in Global Context

What are the consequences of global population growth? Scholars do not agree on the answer to this question. Some biologists have warned that Earth is a finite ecosystem that cannot support the 10 billion people predicted by 2050; however, some economists have emphasized that free-market capitalism is capable of developing innovative ways to solve such problems. The debate is not a new one; for several centuries, strong opinions have been voiced about the effects of population growth on human welfare.

The Malthusian Perspective

English clergyman and economist Thomas Robert Malthus (1766–1834) was one of the first scholars to systematically study the effects of population. Displeased with societal changes brought about by the Industrial Revolution in England, Malthus (1965/1798: 7) anonymously published *An Essay on the Principle of Population, As It Affects the Future Improvement of Society,* in which he argued that "the power of population is infinitely greater than the power of the earth to produce subsistence [food] for man."

According to Malthus, the population, if left unchecked, would exceed the available food supply. He argued that the population would increase in a geometric (exponential) progression (2, 4, 8, 16 . . .) while the food supply would increase only by an arithmetic progression (1, 2, 3, 4 . . .). In other words, a *doubling effect* occurs: Two parents can have four children, sixteen grandchildren, and so on, but food production increases by only one acre at a time. Thus, population growth inevitably surpasses the food supply, and the lack of food ultimately ends population growth and perhaps eliminates the existing population (Weeks, 2008). Even in a best-case scenario, overpopulation results in poverty.

However, Malthus suggested that this disaster might be averted by either positive or preventive checks on population. *Positive checks* are mortality risks such as famine, disease, and war; *preventive checks* are limits to fertility. For Malthus, the only acceptable preventive check was *moral restraint;* people should practice sexual abstinence before marriage and postpone marriage as long as possible in order to have only a few children.

Malthus has had a lasting impact on the field of population studies. Most demographers refer to his dire predictions when they examine the relationship between fertility and subsistence needs. Overpopulation is still a daunting problem that capitalism and technological advances thus far have not solved, especially in middle- and low-income nations with rapidly growing populations and very limited resources.

The Marxist Perspective

Among those who attacked the ideas of Malthus were Karl Marx and Frederick Engels. According to Marx and Engels, the food supply is not threatened by overpopulation; technologically, it is possible to produce the food and other goods needed to meet the demands of a growing population. Marx and Engels viewed poverty as a consequence of exploitation of workers by the owners of the means of production. For example, they argued that England had poverty because the capitalists skimmed off some of the workers' wages as profits. The labor of the working classes was used by capitalists to earn profits, which, in turn, were used to purchase machinery that could replace the workers, rather than supply food for all. From this perspective, overpopulation occurs because capitalists desire to have a surplus of workers (an industrial reserve army) so as to suppress wages and force workers concerned about losing their livelihoods to be more productive.

According to some contemporary economists, the greatest crisis today facing low-income nations is capital shortage, not food shortage. Through technological advances, agricultural production has reached the level at which it can meet the food needs of the world if food is distributed efficiently. Capital shortage refers to the lack of adequate money or property to maintain a business; it is a problem because the physical capital of the past no longer meets the needs of modern economic development. In the past, self-contained rural economies survived on local labor, using local materials to produce the capital needed for other laborers. For example, in a typical village a carpenter made the loom needed by the weaver to make cloth. Today, in the global economy, the one-to-one exchange between the carpenter and the weaver is lost. With an antiquated, locally made loom, the weaver cannot compete against electronically controlled, mass-produced looms. Therefore, the village must purchase capital from the outside, using its own meager financial resources. In the process, the complementary relationship between labor and capital is lost; modern technology brings with it steep costs and results in village noncompetitiveness and underemployment (see Keyfitz, 1994).

Marx and Engels made a significant contribution to the study of demography by suggesting that poverty, not overpopulation, is the most important issue with regard to food supply in a capitalist economy. Although Marx and Engels offer an interesting counterpoint to Malthus, some scholars argue that the Marxist perspective is self-limiting because it attributes the population problem solely to capitalism. In actuality, nations with socialist economies also have demographic trends similar to those in capitalist societies.

The Neo-Malthusian Perspective

More recently, *neo-Malthusians* (or "new Malthusians") have reemphasized the dangers of overpopulation. To neo-Malthusians, Earth is "a dying planet" with too many people and too little food, compounded by environmental degradation. Overpopulation and rapid population growth result in global environmental problems ranging from global warming and rain-forest destruction to famine and vulnerability to epidemics (Ehrlich, Ehrlich, and Daily, 1995). Unless significant changes are made, including improving the status of women, reducing racism and religious prejudice, reforming the agriculture system, and shrinking the growing gap between rich and poor, the consequences will be dire (Ehrlich, Ehrlich, and Daily, 1995).

Early neo-Malthusians published birth control handbooks, and widespread acceptance of birth control eventually reduced the connection between people's sexual conduct and fertility (Weeks, 2008). Later neo-Malthusians have encouraged people to be part of the solution to the problem of overpopulation by having only one or two children in order to bring about *zero population growth*—**the point at which no population increase occurs from year to year** because the number of births plus immigrants is equal to the number of deaths plus emigrants (Weeks, 2008).

Demographic Transition Theory

Some scholars who disagree with the neo-Malthusian viewpoint suggest that the theory of demographic transition offers a more accurate picture of future population growth. *Demographic transition* **is the process by which some societies have moved from high birth and death rates to relatively low birth and death rates as a result of technological development.** Demographic transition is linked to four stages of economic development (see ▶ Figure 15.3):

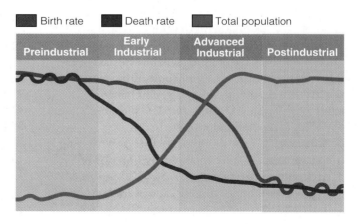

▶ Figure 15.3 **The Demographic Transition**

- *Stage 1: Preindustrial societies.* Little population growth occurs because high birth rates are offset by high death rates. Food shortages, poor sanitation, and lack of adequate medical care contribute to high rates of infant and child mortality.
- *Stage 2: Early industrialization.* Significant population growth occurs because birth rates are relatively high whereas death rates decline. Improvements in health, sanitation, and nutrition produce a substantial decline in infant mortality rates. Overpopulation is likely to occur because more people are alive than the society has the ability to support.
- *Stage 3: Advanced industrialization and urbanization.* Very little population growth occurs because both birth rates and death rates are low. The birth rate declines as couples control their fertility through contraceptives and become less likely to adhere to religious directives against their use. Children are not viewed as an economic asset; they consume income rather than produce it. Societies in this stage attain zero population growth, but the actual number of births per year may still rise due to an increased number of women of childbearing age.
- *Stage 4: Postindustrialization.* Birth rates continue to decline as more women gain full-time employment and the cost of raising children continues to increase. The population grows very slowly, if at all, because the decrease in birth rates is coupled with a stable death rate.

Debate continues as to whether this evolutionary model accurately explains the stages of population growth in all societies. Advocates note that demographic transition theory highlights the relationship between technological development and population growth, thus making Malthus's predictions obsolete. Scholars also point out that demographic transitions occur at a faster rate in now-low-income nations than they previously did in the nations that are already developed. For example, nations in the process of development have higher birth rates and death rates than the now-developed societies did when they were going through the transition. The death rates declined in the now-developed nations as a result of internal economic development—not, as is the case today, through improved methods of disease control (Weeks, 2008). Critics suggest that this theory best explains development in Western societies.

Other Perspectives on Population Change

In recent decades, other scholars have continued to develop theories about how and why changes in population growth patterns occur. Some have studied the relationship between economic development and a decline in fertility; others have focused on the process of *secularization*—the decline in the significance of the sacred in daily life—and how a change from believing that otherworldly powers are responsible for one's life to a sense of responsibility for one's own well-being is linked to a decline in fertility. Based on this premise, some analysts argue that the processes of industrialization and economic development are typically accompanied by secularization, but the relationship between these factors is complex when it comes to changes in fertility.

Shifting from the macrolevel to the microlevel, education and social psychological factors also play into the decisions that individuals make about how many children to have. Family planning information is more readily available to people with more years of formal education and may cause them to engage in decision making in accord with *rational choice theory,* which is based on the assumption that people make decisions based on a calculated cost–benefit analysis ("What do I gain and lose from a specific action?"). In low-income countries or other settings in which children are identified as an economic resource for their parents throughout life, fertility rates are higher than in higher-income countries. However, as modernization and urbanization occur in such societies, the positive economic effects of having more children may be offset by the cost of

---

zero population growth the point at which no population increase occurs from year to year.

demographic transition the process by which some societies have moved from high birth rates and death rates to relatively low birth rates and death rates as a result of technological development.

caring for those children and the lowered economic advantage gained from having children in an industrialized nation.

As demographers have reformulated the demographic transition theory, they have highlighted additional factors that are likely to be causes of fertility decline, and they have suggested that demographic transition is not just one process, but rather a set of intertwined transitions. One is the epidemiological transition—the shift from deaths at younger ages due to acute, communicable diseases. Another is the fertility transition—the shift from natural fertility to controlled fertility, resulting in a decrease in the fertility rate. Other transitions include the migration transition, the urban transition, the age transition, and the family and household transition, which occur as a result of lower fertility, longer life, an older age structure, and predominantly urban residence.

As the demographer John R. Weeks (2008) points out, we can best understand demographic events and behavior by studying the context of global change to determine how factors such as political change, economic development, and perhaps the process of "westernization" may influence population growth and patterns of migration.

A Brief Glimpse at International Migration Theories

Why do people relocate from one nation to another? Several major theories have been developed in an attempt to explain international migration. The *neoclassical economic approach* assumes that migration patterns occur based on geographic differences in the supply of and demand for labor. The United States and other high-income countries that have had growing economies and a limited supply of workers for certain types of jobs have paid higher wages than are available in areas with a less-developed economy and a large labor force. As a result, people move to gain higher wages and sometimes better living conditions. They may also take jobs in other countries so that they can send money to their families in their country of origin (see Box 15.3). For

example, it is estimated that Mexican workers in the United States send about half of what they earn to their families across the border (Ferriss, 2001), an amount that may total nearly $12 billion per year in good economic times.

Unlike the neoclassical explanation of migration, which focuses on individual decision making, the *new households economics of migration approach* emphasizes the part that entire families or households play in the migration process. From this approach, the previous example of Mexican workers' temporary migration to the United States would be examined not only from the perspective of the individual worker but also in terms of what the entire family gains from the process of having one or more migrant family members work in another country. By having a diversity of family income (originating from more than one source), the family is cushioned from the economic woes of the nation that most of the family members think of as "home."

Two conflict perspectives on migration add to our knowledge of why people migrate. Split-labor-market theory (as previously discussed in Chapter 9) suggests that immigrants from low-income countries are often recruited for secondary-labor-market positions: dead-end jobs with low wages, unstable employment, and sometimes hazardous working conditions. By contrast, migrants from higher-income countries may migrate for primary-sector employment—jobs in which well-educated workers are paid high wages and receive benefits such as health insurance and a retirement plan. The global migration of some high-tech workers is an example of this process, whereas the migration of farm workers and construction helpers is an example of secondary-labor-market migration.

Finally, world systems theory (discussed later in this chapter) views migration as linked to the problems caused by capitalist development around the world (Massey et al., 1993). As the natural resources, land, and work force in low-income countries with little or no industrialization have come under the influence of international markets, there has been a corresponding flow of migrants from those nations to the highly industrialized, high-income countries, especially those with which the poorer nations have had the most economic, political, or military contact.

Box 15.3 Framing Immigration in the Media

Media Framing and Public Opinion

Inside his little Western wear store tucked in a corner of East Riverside Drive, Francisco Javier Aceves can't help but feel a kinship with the angular young men who come in to buy jeans, cowboy boots, phone cards and cell phones. As sure as regular payday, they come in also to wire money to their families back home in Mexico, in places such as Veracruz, Tabasco, Chiapas and Oaxaca.

"Sometimes they come three or four in a car," Aceves said about his customers. "Sometimes they just start lining up to wire money. . . ." The flow of money repeats and repeats, affirming emotional bonds between people separated by a border. . . . It is part of a growing phenomenon. Every month, Mexican immigrants working in the United States in mostly low-paying jobs send more than $1 billion to their families back home, more than Mexico earns from tourism or foreign investment. (Castillo, 2003: J1)

This newspaper article about Mexican immigrants who send money to their families in Mexico is an example of sympathetic framing by the news media. Sympathetic framing refers to news writing that focuses on the human interest side of a story and shows that the individuals involved are caring people who are representative of a larger population.

In stark contrast to sympathetic framing of such stories are those news reports that employ negative framing to describe recent immigrants from countries such as Mexico. Negative framing describes immigrants as nothing more than cheap labor that benefits employers in this country who do not believe in paying a living wage. Articles focusing on the problematic aspects of illegal immigrants or of "guest worker" programs usually emphasize that such workers suppress the wages of other low-income employees because they are willing to work for less money. Negative framing also emphasizes that these workers bring in (or give birth to) millions of children who speak little English, contributing to the decline of public education.

Negative framing of the issue of immigrant labor is not new in the United States. For many years, "immigrant, foreign labor" has been described as a threat to the livelihood of other workers and as a menace to public safety. In 1904 the *San Francisco Chronicle* carried lengthy articles describing how Japanese laborers were taking jobs away from U.S. workers, reflecting a pattern of media reporting that continues to be a topic in the twenty-first century (Puette, 1992).

In the early 2000s, when President George W. Bush introduced his "Guest Worker Proposal," both sympathetic and negative media framing of immigrant workers ensued. Some articles emphasized the importance of immigrant workers' contributions to the United States. Other reports highlighted the problematic aspects of guest worker programs, arguing that immigration undermines the future of the country.

How the media frame stories about immigrant workers may influence social policy. Will we close our borders to immigration? Will we develop new programs to allow limited entry of immigrant workers? Not only are these important legal and social policy questions, but they are also issues that journalists and other news analysts must face as they frame their stories on immigrants and the billions of dollars that are sent to other countries in exchange for the labor of these workers.

Perhaps the American Dream now transcends the borders of this country. Francisco Javier Aceves (the store proprietor) described to one reporter how the young men who came into his shop have their own dreams of making money, building a home in Mexico one piece at a time, and then going back: "They tell me this. They'll say, 'With this, I'm going to put the roof on my house.'" He paused, smiling slowly, "They're excited!" (Castillo, 2003: J4).

Reflect & Analyze

Since the recent decline in the U.S. housing market, the number of residential construction jobs has been drastically reduced. Would such an event have any effect on the "pull" factors that attract immigrants to this country? Why or why not?

After flows of migration commence, the pattern may continue because potential migrants have personal ties with relatives and friends who now live in the country of destination and can serve as a source of stability when the potential migrants relocate to the new country. Known as *network theory*, this approach suggests that once migration has commenced, it takes on a life of its own and that the migration pattern which ensues may be different from the original *push* or *pull* factors that produced the earlier migration. Another approach, *institutional theory*, suggests that migration may be fostered by groups—such as humanitarian aid organizations relocating refugees or smugglers bringing people into a country illegally—and that the actions of these groups may produce a larger stream of migrants than would otherwise be the case.

As you can see from these diverse approaches to explaining contemporary patterns of migration, the reasons that people migrate are numerous and complex, involving processes occurring at the individual, family, and societal levels.

Urbanization in Global Perspective

Urban sociology is a subfield of sociology that examines social relationships and political and economic structures in the city. According to urban sociologists, a *city* is a relatively dense and permanent settlement of people who secure their livelihood primarily through nonagricultural activities. Although cities have existed for thousands of years, only about 3 percent of the world's population lived in cities 200 years ago, as compared with almost 50 percent today. Current estimates suggest that two out of every three people around the world will live in urban areas by 2050 (United Nations, 2000). Thus, the process of urbanization continues on a global basis.

Emergence and Evolution of the City

Cities are a relatively recent innovation when compared with the length of human existence. The earliest humans are believed to have emerged

An increasing proportion of the world's population lives in cities. How is this scene in Lagos, Nigeria, similar to and different from major U.S. cities?

© AP Images/George Osodi

anywhere from 40,000 to 1,000,000 years ago, and permanent human settlements are believed to have begun first about 8000 B.C.E. However, some scholars date the development of the first city between 3500 and 3100 B.C.E., depending largely on whether a formal writing system is considered as a requisite for city life (Sjoberg, 1965; Weeks, 2008; Flanagan, 2002).

According to the sociologist Gideon Sjoberg (1965), three preconditions must be present in order for a city to develop:

1. *A favorable physical environment,* including climate and soil favorable to the development of plant and animal life and an adequate water supply to sustain both.
2. *An advanced technology* (for that era) that could produce a social surplus in both agricultural and nonagricultural goods.

3. *A well-developed social organization,* including a power structure, in order to provide social stability to the economic system.

Based on these prerequisites, Sjoberg places the first cities in the Middle Eastern region of Mesopotamia or in areas immediately adjacent to it at about 3500 B.C.E. However, not all scholars concur; some place the earliest city in Jericho (located in present-day Jordan) at about 8000 B.C.E. with a population of about 600 people (see Kenyon, 1957).

The earliest cities were not large by today's standards. The population of the larger Mesopotamian centers was between 5,000 and 10,000 (Sjoberg, 1965). The population of ancient Babylon (probably founded around 2200 B.C.E.) may have grown as large as 50,000 people; Athens may have held 80,000 people (Weeks, 2008). Four to five thousand years ago, cities with at least 50,000 people existed in the Middle East (in what today is Iraq and Egypt) and Asia (in what today is Pakistan and China), as well as in Europe. About 3,500 years ago, cities began to reach this size in Central and South America.

Preindustrial Cities

The largest preindustrial city was Rome; by 100 C.E., it may have had a population of 650,000 (Chandler and Fox, 1974). With the fall of the Roman Empire in 476 C.E., the nature of European cities changed. Seeking protection and survival, those persons who lived in urban settings typically did so in walled cities containing no more than 25,000 people. For the next 600 years, the urban population continued to live in walled enclaves as competing warlords battled for power and territory during the "dark ages." Slowly, as trade increased, cities began to tear down their walls.

Preindustrial cities were limited in size by a number of factors. For one thing, crowded housing conditions and a lack of adequate sewage facilities increased the hazards from plagues and fires, and death rates were high. For another, food supplies were limited. In order to generate food for each city resident, at least fifty farmers had to work in the fields (Davis, 1949), and animal power was the only means of bringing food to the city. Once foodstuffs arrived in the city, there was no effective way to preserve them. Finally, migration to the city

was difficult. Many people were in serf, slave, and caste systems whereby they were bound to the land. Those able to escape such restrictions still faced several weeks of travel to reach the city, thus making it physically and financially impossible for many people to become city dwellers.

In spite of these problems, many preindustrial cities had a sense of *community*—**a set of social relationships operating within given spatial boundaries or locations that provides people with a sense of identity and a feeling of belonging.** The cities were full of people from all walks of life, both rich and poor, and they felt a high degree of social integration. You will recall that Ferdinand Tönnies (1940/1887) described such a community as *Gemeinschaft*—a society in which social relationships are based on personal bonds of friendship and kinship and on intergenerational stability, such that people have a commitment to the entire group and feel a sense of togetherness. By contrast, industrial cities were characterized by Tönnies as *Gesellschaft*—societies exhibiting impersonal and specialized relationships, with little long-term commitment to the group or consensus on values (see Chapter 4). In *Gesellschaft* societies, even neighbors are "strangers" who perceive that they have little in common with one another.

Industrial Cities

The Industrial Revolution changed the nature of the city. Factories sprang up rapidly as production shifted from the primary, agricultural sector to the secondary, manufacturing sector. With the advent of factories came many new employment opportunities not available to people in rural areas. Emergent technology, including new forms of transportation and agricultural production, made it easier for people to leave the countryside and move to the city. Between 1700 and 1900, the population of many European cities mushroomed. For example, the population of London increased from 550,000

---

community a set of social relationships operating within given spatial boundaries or locations that provides people with a sense of identity and a feeling of belonging.

to almost 6.5 million. Although the Industrial Revolution did not start in the United States until the mid-nineteenth century, the effect was similar. Between 1870 and 1910, for example, the population of New York City grew by 500 percent. In fact, New York City became the first U.S. *metropolis*—**one or more central cities and their surrounding suburbs that dominate the economic and cultural life of a region.** Nations, such as Japan and Russia, that became industrialized after England and the United States experienced a delayed pattern of urbanization, but this process moved quickly once it commenced in those countries.

Postindustrial Cities

Since the 1950s, postindustrial cities have emerged in nations such as the United States as their economies have gradually shifted from secondary (manufacturing) production to tertiary (service and information-processing) production. Postindustrial cities increasingly rely on an economic structure that is based on scientific knowledge rather than industrial production, and as a result, a class of professionals and technicians grows in size and influence. Postindustrial cities are dominated by "light" industry, such as software manufacturing; informa-

Despite an increase in telecommuting and more diverse employment opportunities in the high-tech economy, our highways have grown increasingly congested. Can we implement measures to reduce the problems of urban congestion and environmental pollution, or will these problems grow worse with each passing year?

During the Industrial Era, people not only moved from the countryside into cities, but some people also moved from the cities to the suburbs after transportation became available to make getting from home to work and back again an easier process.

tion-processing services, such as airline and hotel reservation services; educational complexes; medical centers; convention and entertainment centers; and retail trade centers and shopping malls. Most families do not live close to a central business district. Technological advances in communication and transportation make it possible for middle- and upper-income individuals and families to have more work options and to live greater distances from the workplace; however, these options are not often available to people of color and those at the lower end of the class structure.

On a global basis, cities such as New York, London, and Tokyo appear to fit the model of the postindustrial city (see Sassen, 2001). These cities have experienced a rapid growth in knowledge-based industries such as financial services. London, Tokyo, and New York have—at least until recently—experienced an increase in the number of highly paid professional jobs, and more workers have been in high-income categories. Many people have benefited for a number of years from these high incomes and have created a lifestyle that is based on materialism and the gentrification of urban spaces. Meanwhile, those persons outside the growing professional categories have seen their own quality of life further deteriorate and their job opportunities become increasingly restricted to secondary labor markets in their respective "global" cities.

Perspectives on Urbanization and the Growth of Cities

Urban sociology follows in the tradition of early European sociological perspectives that compared social life with biological organisms or ecological processes. For example, Auguste Comte pointed out that cities are the "real organs" that make a society function. Emile Durkheim applied natural ecology to his analysis of *mechanical solidarity,* characterized by a simple division of labor and shared religious beliefs such as are found in small, agrarian societies, and *organic solidarity,* characterized by interdependence based on the elaborate division of labor found in large, urban societies (see Chapter 4). These early analyses became the foundation for ecological models/functionalist perspectives in urban sociology.

Functionalist Perspectives: Ecological Models

Functionalists examine the interrelations among the parts that make up the whole; therefore, in studying the growth of cities, they emphasize the life cycle of urban growth. Like the social philosophers and sociologists before him, the University of Chicago sociologist Robert Park (1915) based his analysis of the city on **human ecology—the study of the relationship between people and their physical environment.** According to Park (1936), economic competition produces certain regularities in land-use patterns and population distributions. Applying Park's idea to the study of urban land-use patterns, the sociologist Ernest W. Burgess (1925) developed the concentric zone model, an ideal construct that attempted to explain why some cities expand radially from a central business core.

The Concentric Zone Model Burgess's *concentric zone model* is a description of the process of urban growth that views the city as a series of circular areas or zones, each characterized by a different type of land use, that developed from a central core (see ▶ Figure 15.4a). *Zone 1* is the central business district and cultural center. In *Zone 2,* houses formerly occupied by wealthy families are divided into rooms and rented to recent immigrants and poor persons; this zone also contains light manufacturing and marginal businesses (such as secondhand stores, pawnshops, and taverns). *Zone 3* contains working-class residences and shops and ethnic enclaves. *Zone 4* comprises homes for affluent families, single-family residences of white-collar workers, and shopping centers. *Zone 5* is a ring of small cities and towns populated by persons who commute to the central city to work and by wealthy people living on estates.

Two important ecological processes are involved in the concentric zone theory: invasion and succession. **Invasion is the process by which a new category of people or type of land use arrives in an area previously occupied by another group or type of land use** (McKenzie, 1925). For example, Burgess noted that recent immigrants and low-income individuals "invaded" Zone 2, formerly occupied by wealthy families. **Succession is the process by which a new category of people or type of land use gradually predominates in an area formerly dominated by another group or activity** (McKenzie, 1925). In Zone 2, for example, when some of the single-family residences were sold and subsequently divided into multiple housing units, the remaining single-family owners moved out because the "old" neighborhood had changed. As a result of their move, the process of invasion was complete and succession had occurred.

Invasion and succession theoretically operate in an outward movement: Those who are unable

---

metropolis one or more central cities and their surrounding suburbs that dominate the economic and cultural life of a region.

human ecology the study of the relationship between people and their physical environment.

invasion the process by which a new category of people or type of land use arrives in an area previously occupied by another group or land use.

succession the process by which a new category of people or type of land use gradually predominates in an area formerly dominated by another group or activity.

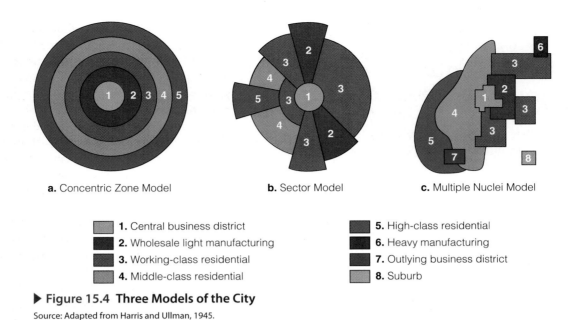

a. Concentric Zone Model   **b.** Sector Model   **c.** Multiple Nuclei Model

1. Central business district
2. Wholesale light manufacturing
3. Working-class residential
4. Middle-class residential

5. High-class residential
6. Heavy manufacturing
7. Outlying business district
8. Suburb

▶ **Figure 15.4 Three Models of the City**
Source: Adapted from Harris and Ullman, 1945.

to "move out" of the inner rings are those without upward social mobility, so the central zone ends up being primarily occupied by the poorest residents—except when gentrification occurs. *Gentrification* **is the process by which members of the middle and upper-middle classes, especially whites, move into the central-city area and renovate existing properties.** Centrally located, naturally attractive areas are the most likely candidates for gentrification. To urban ecologists, gentrification is the solution to revitalizing the central city. To conflict theorists, however, gentrification creates additional hardships for the poor by depleting the amount of affordable housing available and by "pushing" them out of the area (Flanagan, 2002).

The concentric zone model demonstrates how economic and political forces play an important part in the location of groups and activities, and it shows how a large urban area can have internal differentiation (Gottdiener, 1985). However, the model is most applicable to older cities that experienced high levels of immigration early in the twentieth century and to a few midwestern cities such as St. Louis (Queen and Carpenter, 1953). No city, including Chicago (on which the model is based), entirely conforms to this model.

The Sector Model In an attempt to examine a wider range of settings, urban ecologist Homer Hoyt (1939) studied the configuration of 142 cities. Hoyt's *sector model* emphasizes the significance of terrain and the importance of transportation routes in the layout of cities. According to Hoyt, residences of a particular type and value tend to grow outward from the center of the city in wedge-shaped sectors, with the more-expensive residential neighborhoods located along the higher ground near lakes and rivers or along certain streets that stretch in one direction or another from the downtown area (see Figure 15.4b). By contrast, industrial areas tend to be located along river valleys and railroad lines. Middle-class residential zones exist on either side of the wealthier neighborhoods. Finally, lower-class residential areas occupy the remaining space, bordering the central business area and the industrial areas. Hoyt (1939) concluded that the sector model applied to cities such as Seattle, Minneapolis, San Francisco, Charleston (South Carolina), and Richmond (Virginia).

The Multiple Nuclei Model According to the *multiple nuclei model* developed by urban ecologists Chauncey Harris and Edward Ullman (1945), cities do not have one center from which all growth radi-

According to conflict theorists, exploitation by the capitalist class impoverishes poor whites and low-income minority-group members. Increasing rates of homelessness have made scenes such as this a recurring sight in many cities.

ates, but rather have numerous centers of development based on specific urban needs or activities (see Figure 15.4c). As cities began to grow rapidly, they annexed formerly outlying and independent townships that had been communities in their own right. In addition to the central business district, other nuclei developed around entities such as an educational institution, a medical complex, or a government center. Residential neighborhoods may exist close to or far away from these nuclei. A wealthy residential enclave may be located near a high-priced shopping center, for instance, whereas less-expensive housing must locate closer to industrial and transitional areas of town. This model may be applicable to cities such as Boston. However, critics suggest that it does not provide insights about the uniformity of land-use patterns among cities and relies on an after-the-fact explanation of why certain entities are located where they are (Flanagan, 2002).

Contemporary Urban Ecology Urban ecologist Amos Hawley (1950) revitalized the ecological tradition by linking it more closely with functionalism. According to Hawley, urban areas are complex and expanding social systems in which growth patterns are based on advances in transportation and communication. For example, commuter railways and automobiles led to the decentralization of city life and the movement of industry from the central city to the suburbs (Hawley, 1981).

Other urban ecologists have continued to refine the methodology used to study the urban environment. *Social area analysis* examines urban populations in terms of economic status, family status, and ethnic classification (Shevky and Bell, 1966). For example, middle- and upper-middle-class parents with school-aged children tend to cluster together in "social areas" with a "good" school district; young single professionals may prefer to cluster in the central city for entertainment and nightlife.

The influence of human ecology on the field of urban sociology is still very strong today (see Frisbie and Kasarda, 1988). Contemporary research on European and North American urban patterns is often based on the assumption that spatial arrangements in cities conform to a common, most efficient design (Flanagan, 2002). However, some critics have noted that ecological models do not take into account the influence of powerful political and economic elites on the development process in urban areas (Feagin and Parker, 1990).

Conflict Perspectives: Political Economy Models

Conflict theorists argue that cities do not grow or decline by chance. Rather, they are the product of specific decisions made by members of the capitalist

---

gentrification the process by which members of the middle and upper-middle classes, especially whites, move into a central-city area and renovate existing properties.

class and political elites. These far-reaching decisions regarding land use and urban development benefit the members of some groups at the expense of others (see Castells, 1977/1972). Karl Marx suggested that cities are the arenas in which the intertwined processes of class conflict and capital accumulation take place; class consciousness and worker revolt are more likely to develop when workers are concentrated in urban areas (Flanagan, 2002).

According to the sociologists Joe R. Feagin and Robert Parker (1990), three major themes prevail in political economy models of urban growth. First, both economic *and* political factors affect patterns of urban growth and decline. Economic factors include capitalistic investments in production, workers, workplaces, land, and buildings. Political factors include governmental protection of the right to own and dispose of privately held property as owners see fit and the role of government officials in promoting the interests of business elites and large corporations.

Second, urban space has both an exchange value and a use value. *Exchange value* refers to the profits that industrialists, developers, bankers, and others make from buying, selling, and developing land and buildings. By contrast, *use value* is the utility of space, land, and buildings for everyday life, family life, and neighborhood life. In other words, land has purposes other than simply for generating profits—for example, for homes, open spaces, and recreational areas (see ▶ Figure 15.5). Today, class conflict exists over the use of urban space, as is evident in battles over rental costs, safety, and development of large-scale projects (see Tabb and Sawers, 1984).

Third, both structure and agency are important in understanding how urban development takes place. *Structure* refers to institutions such as state bureaucracies and capital investment circuits that are involved in the urban development process. *Agency* refers to human actors, including developers, busi-

Exchange Value

Profits from buying, selling, and developing urban land

Use Value

Utility of urban land, space, and buildings for everyday personal and community life

Examples of who profits

- Industrialists
- Developers
- Bankers
- Tax collectors

Examples of uses

- Affordable housing
- Open spaces
- Recreational areas
- Public services

▶ Figure 15.5 **The Value of Urban Space**

ness elites, and activists protesting development, who are involved in decisions about land use.

Capitalism and Urban Growth in the United States According to political economy models, urban growth is influenced by capital investment decisions, power and resource inequality, class and class conflict, and government subsidy programs. Members of the capitalist class choose corporate locations, decide on sites for shopping centers and factories, and spread the population that can afford to purchase homes into sprawling suburbs located exactly where the capitalists think they should be located (Feagin and Parker, 1990).

Today, a few hundred financial institutions and developers finance and construct most major and many smaller urban development projects around the country, including skyscrapers, shopping malls, and suburban housing projects. These decision makers set limits on the individual choices of the ordinary citizen with regard to real estate, just as they do with regard to other choices (Feagin and Parker, 1990). They can make housing more affordable or totally unaffordable for many people. Ultimately, their motivation rests not in benefiting the community but in making a profit; the cities that they produce reflect this mindset.

One of the major results of these urban development practices is *uneven development*—the tendency of some neighborhoods, cities, or regions to grow and prosper whereas others stagnate and decline (Perry and Watkins, 1977). Conflict theorists argue that uneven development reflects inequalities of wealth and power in society. The problem not only affects areas in a state of decline but also produces external costs, even in "boom" areas, that are paid by the entire community. Among these costs are increased pollution, increased traffic congestion, and rising rates of crime and violence. According to the sociologist Mark Gottdiener (1985: 214), these costs are "intrinsic to the very core of capitalism, and those who profit the most from development are not called upon to remedy its side effects."

Gender Regimes in Cities Feminist perspectives have only recently been incorporated in urban studies (Garber and Turner, 1995). From this perspective, urbanization reflects the workings not only of the political economy but also of patriarchy. According to the sociologist Lynn M. Appleton (1995), different kinds of cities have different *gender regimes*—prevailing ideologies of how women and men should think, feel, and act; how access to social positions and control of resources should be managed; and how relationships between men and women should be conducted. The higher density and greater diversity found in central cities such as New York City serve as a challenge to the private patriarchy found in the home and workplace in lower-density, homogeneous areas such as suburbs and rural areas. *Private patriarchy* is based on a strongly gendered division of labor in the home, gender-segregated paid employment, and women's dependence on men's income. At the same time, cities may foster *public patriarchy* in the form of women's increasing dependence on paid work and the state for income and their decreasing emotional interdependence with men. At this point, gender often intersects with class and race as a form of oppression because lower-income women of color often reside in central cities. Public patriarchy may be perpetuated by cities through policies that limit women's access to paid work and public transportation. However, such cities may also be a forum for challenging patriarchy; all residents who differ in marital status, paternity, sexual orientation, class, and/or race/ethnicity tend to live close to one another and may hold a common belief that both public and private patriarchy should be eliminated (Appleton, 1995).

Symbolic Interactionist Perspectives: The Experience of City Life

Symbolic interactionists examine the *experience* of urban life. How does city life affect the people who live in a city? Some analysts answer this question positively; others are cynical about the effects of urban living on the individual.

Simmel's View of City Life According to the German sociologist Georg Simmel (1950/1902–1917), urban life is highly stimulating, and it shapes people's thoughts and actions. Urban residents are influenced by the quick pace of the city and the pervasiveness of economic relations in everyday life. Due to the intensity of urban life, people have no choice but to become somewhat insensitive to events and individuals around them. Many urban residents avoid emotional involvement with one another and try to ignore events taking place around them. Urbanites feel wary toward other people because most interactions in the city are economic rather than social. Simmel suggests that attributes such as punctuality and exactness are rewarded but that friendliness and warmth in interpersonal relations are viewed as personal weaknesses. Some people act reserved to cloak their deeper feelings of distrust or dislike toward others. However, Simmel did not view city life as completely negative; he also pointed out that urban living could have a liberating effect on people because they had opportunities for individualism and autonomy (Flanagan, 2002).

Urbanism as a Way of Life Based on Simmel's observations on social relations in the city, the early Chicago School sociologist Louis Wirth (1938) suggested that urbanism is a "way of life." *Urbanism* refers to the distinctive social and psychological patterns of life typically found in the city. According to Wirth, the size, density, and heterogeneity of urban populations typically result in an elaborate division of labor and in spatial segregation of people by race/ethnicity, social class, religion, and/or lifestyle. In the city, primary-group ties are largely replaced by secondary relationships; social interaction is fragmented, impersonal, and often superficial. Even though people gain some degree of freedom and privacy by living in the city, they pay a price for their autonomy, losing the group support and reassurance that come from primary-group ties.

From Wirth's perspective, people who live in urban areas are alienated, powerless, and lonely. A sense of community is obliterated and replaced by the "mass society"—a large-scale, highly institutionalized society in which individuality is supplanted by mass messages, faceless bureaucrats, and corporate interests.

Gans's Urban Villagers In contrast to Wirth's gloomy assessment of urban life, the sociologist Herbert Gans (1982/1962) suggested that not everyone experiences the city in the same way. Based on research in the west end of Boston in the late 1950s,

© Jeff Greenberg/PhotoEdit

Festive occasions such as this street fair provide opportunities for urban villagers to mingle with others, enjoying entertainment and social interaction.

Gans concluded that many residents develop strong loyalties and a sense of community in central-city areas that outsiders may view negatively. According to Gans, there are five major categories of adaptation among urban dwellers. *Cosmopolites* are students, artists, writers, musicians, entertainers, and professionals who choose to live in the city because they want to be close to its cultural facilities. *Unmarried people and childless couples* live in the city because they want to be close to work and entertainment. *Ethnic villagers* live in ethnically segregated neighborhoods; some are recent immigrants who feel most comfortable within their own group. The *deprived* are poor individuals with dim future prospects; they have very limited education and few, if any, other resources. The *trapped* are urban dwellers who can find no escape from the city; this group in-

Sociology Works!

Herbert Gans and Twenty-First-Century Urban Villagers

I moved to Austin because it's a high-tech city with a small-town feel. Kinda my own "urban village" where I can cycle around town when I want but still own a nice car to go out in. I chose my neighborhood because it's centrally located to downtown eating and live entertainment. Austin calls itself the "Live Music Capital of the World," and I have plenty of opportunities to hear the music I like. Overall, I'd say that I'm compatible with Austin, and Austin's compatible with me.

—"Brad," a twenty-four-year-old college graduate, explaining why he chose to become an "urban villager" in Austin, Texas (author's files, 2007)

In the more than five decades since the urban sociologist Herbert J. Gans examined life in Boston's west end and identified five major categories of adaptation among urban dwellers, we continue to find that many residents think of themselves as living in an "urban village" despite the differences in high-rise buildings and the smaller settings in which the Boston west enders lived. Today, many younger urban residents think of themselves as having strong loyalties to a specific segment of their community with which they share interests and common experiences. Although many contemporary studies of urban life have emphasized the problems of poverty, crime, racial and ethnic discrimination, inadequate health care, and poor schools in our nation's major cities, it is useful for us to examine positive aspects of urban life as well. We can also gain important insights from examining the experiences of middle- and upper-income residents of our cities.

Herbert Gans believed that the people he referred to as *cosmopolites* chose to live in the city so that they could be close to cultural facilities, while he thought that *unmarried people and childless couples* chose to live there because they wanted to be close to work and entertainment. Among some affluent residents in high-tech cities such as Austin, Texas, married people and families with children have increasingly joined the ranks of individuals who live in or near the city's downtown area. According to contemporary urban villagers such as "Brad," they can find other people who are like themselves, who participate in activities they enjoy, and who support one another much like the members of an extended family when they need friendship or assistance. From this perspective, Gans's ideas about the urban village work because they show us that an important way to understand city life is through the experiences of people who live there. In the final analysis, of course, all people—including lower-income individuals, who have been further disadvantaged or even displaced by gentrification, and the poor and homeless—must be included in any thorough sociological examination of city life in the twenty-first century. However, "urban villagers," as coined by Gans, has staying power as a concept because it encourages us to look at urban life as a kaleidoscope of diversity that includes the wealthy and the merely affluent, as well as those who are "just getting by" or who are poor, because they live near to one another as contemporary urban dwellers.

Reflect & Analyze

Can you identify categories of urban villagers in a city with which you are familiar? To what extent do people live in certain areas of the city based on personal choice? What factors appear to be beyond their control?

cludes persons left behind by the process of invasion and succession, downwardly mobile individuals who have lost their former position in society, older persons who have nowhere else to go, and individuals addicted to alcohol or other drugs.

However, this urban village in Boston's west end was demolished so that high-rise apartments could be constructed. In the aftermath, Gans was unimpressed with the impersonal lives of people who lived in the modernistic towers as compared with the richness and diversity of individuals' lives in the urban village. Gans concluded that the city is a pleasure and a challenge for some urban dwellers and an urban nightmare for others (see "Sociology Works!").

Gender and City Life In their everyday lives, do women and men experience city life differently? According to the scholar Elizabeth Wilson (1991), some men view the city as *sexual space* in which women, based on their sexual desirability and accessibility, are categorized as prostitutes, lesbians, temptresses, or virtuous women in need of protection. Wilson suggests that more-affluent, dominant-group women are more likely to be viewed as virtuous women in need of protection by their own men or police officers. Cities offer a paradox for women: On the one hand, cities offer more freedom than is found in comparatively isolated rural, suburban, and domestic settings; on the other, women may be in greater physical danger in the city. For Wilson, the answer to women's vulnerability in the city is not found in offering protection for them, but rather in changing people's perceptions that they can treat women as sexual objects because of the impersonality of city life (Wilson, 1991).

Cities and Persons with a Disability Chapter 14 describes how disability rights advocates believe that structural barriers create a "disabling" environment for many people, particularly in large urban settings. Many cities have made their streets and sidewalks more user-friendly for persons in wheelchairs and individuals with visual disability by constructing concrete ramps with slide-proof surfaces at intersections or installing traffic lights with sounds designating when to "Walk." However, both urban and rural areas have a long way to go before many persons with disabilities will have access to the things they need to become productive members of the community: educational and employment opportunities. Because some persons with disabilities cannot navigate the streets and sidewalks of their communities or face obstacles getting into buildings that marginally, at best, meet the accessibility standards of the Americans with Disabilities Act, many persons with a disability are unemployed.

Political scientist Harlan Hahn (1997: 177–178) traces the problem of lack of access to the beginnings of industrialism:

The rise of industrialism produced extensive changes in the lives of disabled as well as nondisabled people. As factories replaced private dwellings as the primary sites of production, routines and architectural configurations were standardized to suit nondisabled workers. Both the design of worksites and of the products that were manufactured gave virtually no attention to the needs of people with disabilities. As a result, patterns of aversion and avoidance toward disabled persons were embedded in the construction of commodities, landscapes, and buildings that would remain for centuries. . . .

The social and economic changes fostered by industrialization may have been exacerbated by the accompanying process of urbanization. As workers increasingly moved from farms and rural villages to live near the institutions of mass production, the character of community life appeared to shift perceptibly. Deviant or atypical personal characteristics that may have gradually become familiar in a small community seemed bizarre or disturbing in an urban milieu.

As Hahn's statement suggests, historical patterns in the dynamics of industrial capitalism contributed to discrimination against persons with disabilities, and this legacy remains evident in contemporary cities. Structural barriers are further intensified when other people do not respond favorably toward persons with disabilities. Scholar and disability rights advocate Sally French (1999: 25–26), who is visually disabled, describes her own experience:

I have lived in the same house for 16 years and yet I cannot recognize my neighbors. I know nothing about them at all; which children belong to whom, who has come and gone, who is old or young, ill or well, black or white. . . . On moving to my present house I informed several neighbors that, because of my inability to recognize them, I would doubtless pass them by in the street without greeting them. One neighbor, who had previously seen me striding confidently down the road, refused to believe me, but the others said they understood and would talk to me if our paths crossed. For the first couple of weeks it worked and I was surprised how often we met, but after that their greetings rapidly decreased and then ceased altogether. Why this happened I am not sure, but I suspect that my lack of recognition strained the interaction and limited the social reward they received from the encounter.

CONCEPT QUICK REVIEW

Perspectives on Urbanism and the Growth of Cities

Functionalist Perspectives: Ecological Models	Concentric zone model	Due to invasion, succession, and gentrification, cities are a series of circular zones, each characterized by a particular land use.
	Sector model	Cities consist of wedge-shaped sectors, based on terrain and transportation routes, with the most-expensive areas occupying the best terrain.
	Multiple nuclei model	Cities have more than one center of development, based on specific needs and activities.
Conflict Perspectives: Political Economy Models	Capitalism and urban growth	Members of the capitalist class choose locations for skyscrapers and housing projects, limiting individual choices by others.
	Gender regimes in cities	Different cities have different prevailing ideologies regarding access to social positions and resources for men and women.
	Global patterns of growth	Capital investment decisions by core nations result in uneven growth in peripheral and semiperipheral nations.
Symbolic Interactionist Perspectives: The Experience of City Life	Simmel's view of city life	Due to the intensity of city life, people become somewhat insensitive to individuals and events around them.
	Urbanism as a way of life	The size, density, and heterogeneity of urban population result in elaborate division of labor and space.
	Gans's urban villagers	Five categories of adaptation occur among urban dwellers, ranging from cosmopolites to trapped city dwellers.
	Gender and city life	Cities offer women a paradox: more freedom than in more isolated areas, yet greater potential danger.

This chapter's Concept Quick Review examines the multiple perspectives on urban growth and urban living.

Problems in Global Cities

As we have seen, although people have lived in cities for thousands of years, the time is rapidly approaching when more people worldwide will live in or near a city than live in a rural area. In the middle-income and low-income regions of the world, Latin America is becoming the most urbanized: Four megacities—Mexico City, Buenos Aires, Lima, and Santiago—already contain more than half of the region's population and continue to grow rapidly. Within the next ten years, Rio de Janeiro and São Paulo are expected to have a combined population of about 40 million people living in a 350-mile-long megalopolis. By 2015, New York City will be the only U.S. city among the world's 10 most populous (▶ Figure 15.6 shows current populations).

Natural increases in population (higher birth rates than death rates) account for two-thirds of new urban growth, and rural-to-urban migration accounts for the rest. Some people move from rural areas to urban areas because they have been displaced from their land. Others move because they

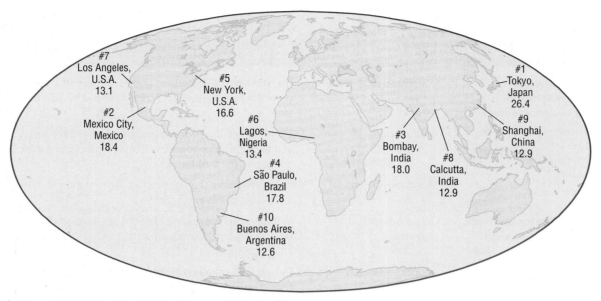

▶ **Figure 15.6 The World's Ten Largest Metropolises**

Note: 2000 population in millions.
Source: Population Reference Bureau, 2001.

are looking for a better life. No matter what the reason, migration has caused rapid growth in cities in sub-Saharan Africa, India, Algeria, and Egypt. At the same time that the population is growing rapidly, the amount of farmland available for growing crops to feed people is decreasing. In Egypt, for example, land that was previously used for growing crops is now used for petroleum refineries, food-processing plants, and other factories (Kaplan, 1996).

Rapid global population growth in Latin America and other regions is producing a wide variety of urban problems, including overcrowding, environmental pollution, and the disappearance of farmland. In fact, many cities in middle- and low-income nations are quickly reaching the point at which food, housing, and basic public services are available to only a limited segment of the population (Crossette, 1996). With urban populations growing at a rate of 170,000 people per day, cities such as Cairo, Lagos, Dhaka, Beijing, and São Paulo are likely to soon have acute water shortages; Mexico City is already experiencing a chronic water shortage (*New York Times*, 1996).

As global urbanization has increased over the past three decades, differences in urban areas based on economic development at the national level have become apparent. Some cities in what Immanuel Wallerstein's (1984) world systems theory describes as core nations (see Chapter 8) are referred to as *global cities*—interconnected urban areas that are centers of political, economic, and cultural activity. New York, Tokyo, and London are generally considered the largest global cities. These cities are the sites of new and innovative product development and marketing, and they are often the "command posts" for the world economy (Sassen, 2001). But economic prosperity is not shared equally by all of the people in the core-nation global cities. Sometimes the living conditions of workers in low-wage service sector jobs or in assembly production jobs more closely resemble the living conditions of workers in semiperipheral nations than they resemble the conditions of middle-class workers in their own country.

Most African countries and many countries in South America and the Caribbean are *peripheral* nations, previously defined as nations that depend on core nations for capital, have little or no industrialization (other than what may be brought in by core nations), and have uneven patterns of urbanization. According to Wallerstein (1984), the wealthy in peripheral nations support the exploitation of poor workers by core-nation capitalists in return for maintaining their own wealth and position. Poverty is thus perpetuated, and the problems worsen because of the unprecedented population growth in these countries.

In regard to the semiperipheral nations, only two cities are considered to be global cities: São Paulo, Brazil, the center of the Brazilian economy, and Singapore, the economic center of a multicountry region in Southeast Asia (Friedmann, 1995). Like peripheral nations, semiperipheral nations—such as India, Iran, and Mexico—are confronted with unprecedented population growth. What is the outlook for cities in the United States?

Urban Problems in the United States

Even the most optimistic of observers tends to agree that cities in the United States have problems brought on by years of neglect and deterioration. As we have seen in previous chapters, poverty, crime, racism, sexism, homelessness, inadequate public school systems, alcoholism and other drug abuse, gangs and guns, and other social problems are most visible and acute in urban settings. Issues of urban growth and development are intertwined with many of these problems.

Divided Interests: Cities, Suburbs, and Beyond

Since World War II, a dramatic population shift has occurred in this country as thousands of families have moved from cities to suburbs. Even though some people lived in suburban areas prior to the twentieth century, it took the involvement of the federal government and large-scale development to spur the dramatic shift that began in the 1950s (Palen, 1995). According to urban historian Kenneth T. Jackson (1985), postwar suburban growth was fueled by aggressive land developers, inexpensive real estate and construction methods, better transportation, abundant energy, government subsidies, and racial stress in the cities. However, the sociologist J. John Palen (1995) suggests that the Baby Boom following World War II and the liberal-

ization of lending policies by federal agencies such as the Veterans Administration (VA) and the Federal Housing Authority (FHA) were significant factors in mass suburbanization.

Regardless of its causes, mass suburbanization has created a territorial division of interests between cities and suburban areas (Flanagan, 2002). Although many suburbanites rely on urban centers for their employment, entertainment, and other services, they pay their property taxes to suburban governments and school districts. Some affluent suburbs have state-of-the-art school districts, police and fire departments, libraries, and infrastructures (such as roads, sewers, and water treatment plants). By contrast, central-city services and school districts languish for lack of funds. Affluent families living in "gentrified" properties typically send their children to elite private schools, whereas the children of poor families living in racially segregated public housing projects attend underfunded (and often substandard) public schools.

Race, Class, and Suburbs The intertwining impact of race and class is visible in the division between central cities and suburbs. About 41 percent of central-city residents are persons of color, although they constitute a substantially smaller portion of the nation's population; just 27 percent of all African Americans live in suburbs. For most African American suburbanites, class is more important than race in determining one's neighbors. According to Vincent Lane, chairman of the Chicago Housing Authority, "Suburbanization isn't about race now; it's about class. Nobody wants to be around poor people, because of all the problems that go along with poor people: poor schools, unsafe streets, gangs" (qtd. in De Witt, 1994: A12).

Nationally, most suburbs are predominantly white, and many upper-middle- and upper-class suburbs remain virtually white. For example, only 5 percent of the population in northern Fulton County (adjoining Atlanta, Georgia) is African American. Likewise, in Plano (adjoining Dallas, Texas), nearly nine out of ten students in the public schools are white, whereas the majority of students in the Dallas Independent School District are African American, Latina/o, or Asian American. In the suburbs, people of color (especially African Americans) often become resegregated (see

Feagin and Sikes, 1994). An example is Chicago, which remains one of the most-segregated metropolitan areas in the country in spite of its fair-housing ordinance. African Americans who have fled the high crime of Chicago's South Side primarily reside in nearby suburbs such as Country Club Hill and Chicago Heights, whereas suburban Asian Americans are most likely to live in Skokie and Naperville and suburban Latinos/as to reside in Maywood, Hillside, and Bellwood (De Witt, 1994). Similarly, suburban Latinas/os are highly concentrated in eight metropolitan areas in California, Texas, and Florida; by far the largest such racial–ethnic concentration is found in the Los Angeles–Long Beach metropolitan area, with over 1.7 million Latinas/os. Like other groups, affluent Latinas/os live in affluent suburbs, whereas poorer Latinas/os remain segregated in less desirable central-city areas (Palen, 1995).

Some analysts argue that the location of one's residence is a matter of personal choice. However, other analysts suggest that residential segregation reflects discriminatory practices by landlords,

homeowners, and white realtors and their agents, who engage in *steering* people of color to different neighborhoods than those shown to their white counterparts. Lending practices of banks (including the *redlining* of certain properties so that acquiring a loan is virtually impossible) and the behavior of neighbors further intensify these problems (see Feagin and Sikes, 1994). In a study of suburban property taxes, the sociologist Andrew A. Beveridge found that African American homeowners are taxed more than whites on comparable homes in 58 percent of the suburban regions and 30 percent of the cities (cited in Schemo, 1994). Some analysts suggest that African Americans are more likely to move to suburbs with declining tax bases because they have limited finances, because they are steered there by real estate agents, or because white flight occurs as African American homeowners move in, leaving a heavier tax burden for the newcomers and those who remain behind. Longer-term residents may not see their property reassessed or

© THOM BAUR/AFP/Getty Images

© AP Images/Mark Duncan

Affluent gated communities and enclaves of million-dollar homes stand in sharp contrast to low-income housing when we see them on the urban landscape. What sociological theories help us describe the disparity of lifestyles and life chances shown in these two settings?

their taxes go up for some period of time; in some cases, reassessment does not occur until the house is sold (Schemo, 1994).

Beyond the Suburbs: Edge Cities New urban fringes (referred to as *edge cities*) have been springing up beyond central cities and suburbs in recent years (Garreau, 1991). The Massachusetts Turnpike corridor west of Boston and the Perimeter area north and east of Atlanta are examples. Edge cities initially develop as residential areas; then retail establishments and office parks move into the area, creating the unincorporated edge city. Commuters from the edge city are able to travel around (rather than in and out of) the metropolitan region's center and can avoid its rush-hour traffic quagmires. Edge cities may not have a governing body or correspond to municipal boundaries; however, they drain taxes from central cities and older suburbs. Many businesses and industries have moved physical plants and tax dollars to these areas: Land is cheaper, and utility rates and property taxes are lower.

Lower taxes are a contributing factor to another recent development in the United States—the growth of Sunbelt cities in the southern and western states. In the 1970s, millions of people moved from the north-central and northeastern states to this area. Four reasons are generally given for this population shift: (1) more jobs and higher wages; (2) lower taxes; (3) pork-barrel programs that funneled federal money into projects in the Sunbelt, creating jobs and encouraging industry; and (4) easier transition to new industry, because most industry in the northern states was based on heavy manufacturing rather than high technology.

The Continuing Fiscal Crises of the Cities

The largest cities in the United States have faced periodic fiscal crises for many years. In the twenty-first century, these crises have been intensified by higher employee health care and pension costs, declining revenues, and increased expenditures for public safety and homeland security. A 2004 survey of 328 cities found that 80 percent of those cities were less able to meet their financial needs than in the previous year (Fecht, 2004). As a result of declining rev-

enues and increased costs, many cities have cut back on spending in areas other than public safety. According to John DeStefano, president of the National League of Cities, this financial crisis is hurting not only the cities themselves but also the people who live there: "Under-funded public schools, smaller police forces, deteriorating transportation systems, expensive health care, sprawl—these are [factors] that increasingly subvert our American ideal. . . ." (qtd. in Fecht, 2004).

Even if the U.S. economy improves significantly in the near future, analysts believe that the positive effects of such a rebound will not improve the budgetary problems of our cities and towns for a number of years.

Rural Community Issues in the United States

Although most people think of the United States as highly urbanized, about 25 percent of the U.S. population resides in rural areas, identified as communities of 2,500 people or less by the U.S. Census Bureau. Sociologists typically identify *rural communities* as small, sparsely settled areas that have a relatively homogeneous population of people who primarily engage in agriculture (Johnson, 2000). However, rural communities today are more diverse than this definition suggests.

Unlike the standard migration patterns from rural to urban places in the past, recently more people have moved from large urban areas and suburbs into rural areas. Many of those leaving urban areas today want to escape the high cost of living, crime, traffic congestion, and environmental pollution that make daily life difficult. Technological advances make it easier for people to move to outlying rural areas and still be connected to urban centers if they need to be. The proliferation of computers, cell phones, commuter airlines, and highway systems has made previously remote areas seem much more accessible to many people. However, many recent immigrants to rural areas do not face some traditional problems experienced by long-term rural residents, particularly farmers, small-business owners, teachers, doctors, and medical personnel in these rural communities.

For many people in rural areas who have made their livelihood through farming and other agricultural endeavors, recent decades have been very difficult, both financially and emotionally. Rural crises such as droughts, crop failures, and the loss of small businesses in the community have had a negative effect on many adults and their children. Like their urban counterparts, rural families have experienced problems of divorce, alcoholism, abuse, and other crises, but these issues have sometimes been exacerbated by such events as the loss of the family farm or business (Pitzer, 2003). Because home is also the center of work in farming families, the loss of the farm may also mean the loss of family and social life, and the loss of things dear to children such as their 4-H projects—often an animal that a child raises to show and sell (Pitzer, 2003). Some rural children and adolescents are also subject to injuries associated with farm work, such as livestock kicks or crushing, falling out of a tractor or pickup, and operating machinery designed for adults, that are not typically experienced by their urban counterparts (Schutske, 2002).

Economic opportunities are limited in many rural areas, and average salaries are typically lower than in urban areas, based on the assumption that a family can live on less money in rural communities than in cities. An example is rural teachers, who earn substantially less than their urban and suburban counterparts. Some rural areas have lost many teachers and administrators to higher-paying districts in other cities.

Although many of the problems we have examined in this book are intensified in rural areas, one of the most pressing is the availability of health services and doctors. Recently, some medical schools have established clinics and practices in outlying rural regions of the states in which they are located in an effort to increase the number of physicians available to rural residents. Typically, physicians who have just started to practice medicine have chosen to work in large urban centers with accessible high-tech medical facilities. Because of the pressing time constraints of tending to patients with life-threatening problems, the availability of community clinics and hospitals in rural areas may be a life-or-death matter for some residents. Loss of these facilities can have a devastating effect on people's health and life chances.

In addition to the movement of some urban dwellers to rural areas, two other factors have changed the face of rural America in some regions. One is the proliferation of superstores, such as Wal-Mart, PetSmart, Lowes, and Home Depot. In some cases, these superstores have effectively put small businesses such as hardware stores and pet shops out of business because local merchants cannot meet the prices established by these large-volume discount chains. The development of superstores and outlet malls along the rural highways of this country has raised new concerns about environmental issues such as air and water pollution, and has brought about new questions regarding whether these stores benefit the rural communities where they are located.

A second factor that has changed the face of some rural areas (and is sometimes related to the growth of mega-stores and outlet malls) is an increase in tourism in rural America (Brown, 2003). According to a recent study, about 87 million people (nearly two-thirds of all U.S. adults) have taken a trip to a rural destination, usually for leisure purposes, over the past few years. Tourism produces jobs; however, many of the positions are for food servers, retail clerks, and hospitality workers, which are often low-paying, seasonal jobs that have few benefits. Tourism may improve a community's tax base, but this does not occur when the outlet malls, hotels, and fast-food restaurants are located outside of the rural community's taxing authority, as frequently occurs when developers decide where to locate malls and other tourist amenities.

Population and Urbanization in the Future

In the future, rapid global population growth is inevitable. Although death rates have declined in many low-income nations, there has not been a corresponding decrease in birth rates. Between 1985 and 2025, 93 percent of all global population growth will have occurred in Africa, Asia, and Latin America; 83 percent of the world's population will live in those regions by 2025. Perhaps even more amazing is the fact that in the five-year span between 1995

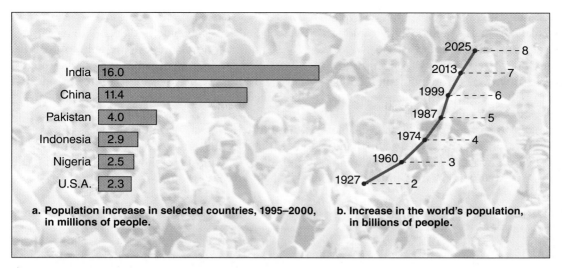

India 16.0
China 11.4
Pakistan 4.0
Indonesia 2.9
Nigeria 2.5
U.S.A. 2.3

a. Population increase in selected countries, 1995–2000, in millions of people.

2025 — 8
2013 — 7
1999 — 6
1987 — 5
1974 — 4
1960 — 3
1927 — 2

b. Increase in the world's population, in billions of people.

▶ Figure 15.7 Growth of the World's Population

Source: United Nations Population Division, 1999.

and 2000, 21 percent of the entire world's population increase occurred in two countries: China and India. ▶ Figure 15.7a shows the net annual additions to the populations of six countries during that five-year period. Figure 15.7b shows the growth of the world's population from 1927 to 1999 and the expected growth to eight billion by the year 2025.

In the future, low-income countries will have an increasing number of poor people. While the world's population will *double,* the urban population will *triple* as people migrate from rural to urban areas in search of food, water, and jobs.

One of the many effects of urbanization is greater exposure of people to the media. Increasing numbers of poor people in less-developed nations will see images from the developed world that are beamed globally by news networks such as CNN. As social analyst John L. Petersen (1994: 119) notes, "For the first time in history, the poor are beginning to understand how relatively poor they are compared to the rich nations. They see, in detail, how the rest of the world lives and feel their increasing disenfranchisement."

Infants born today will be teenagers in the year 2020. By then, in a worst-case scenario, central cities and nearby suburbs in the United States will have experienced bankruptcy exacerbated by sporadic race- and class-oriented violence. The infrastructure will be be-

yond the possibility of repair. Families and businesses with the ability to do so will have long since moved to "new cities," where they will inevitably diminish the quality of life that they originally sought there. Areas that we currently think of as being relatively free from such problems will be characterized by depletion of natural resources and greater air and water pollution (see Ehrlich, Ehrlich, and Daily, 1995).

By contrast, in a best-case scenario, the problems brought about by rapid population growth in low-income nations will be remedied by new technologies that make goods readily available to people. International trade agreements are removing trade barriers and making it possible for all nations to fully engage in global trade. People in low-income nations will benefit by gaining jobs and opportunities to purchase goods at lower prices. Of course, the opposite may also occur: People may be exploited as inexpensive labor, and their country's natural resources may be depleted as transnational corporations buy up raw materials without contributing to the long-term economic stability of the nation.

In the United States, a best-case scenario for the future might include improvements in how tax revenues are collected and spent. Some analysts have suggested that regional governments must be developed. Regional governments would be responsible for water, wastewater (sewage), transportation, schools, parks, hospitals, and other public services over a wider area. Revenues would be shared among central cities, affluent suburbs, and edge cities based on the assumption that everyone will benefit if the quality of life is improved throughout the region. With regard to pollution in urban areas, some futurists predict that environmental activism will increase dramatically as people see irreversible changes in the atmosphere and experience firsthand the effects of environmental hazards and pollution on their health and well-being. Environmental ac-

tivism is discussed in Chapter 16 ("Collective Behavior, Social Movements, and Social Change").

At the macrolevel, we may be able to do little about population and urbanization; at the microlevel, however, we may be able to exercise some degree of control over our communities and our lives. In both cases, analysts suggest that as we approach the future, we must "leave the old ways and invent new ones" (Petersen, 1994: 340). What aspects of our "old ways" do you think we should discard? Can you help invent new ones?

Chapter Review

● What are the processes that produce population changes?

Populations change as the result of fertility (births), mortality (deaths), and migration.

● What is the Malthusian perspective?

Over two hundred years ago, Thomas Malthus warned that overpopulation would result in poverty, starvation, and other major problems that would limit the size of the population. According to Malthus, the population would increase geometrically while the food supply would increase only arithmetically, resulting in a critical food shortage and poverty.

● What are the views of Karl Marx and the neo-Malthusians on overpopulation?

According to Karl Marx, poverty is the result of capitalist greed, not overpopulation. More recently, neo-Malthusians have reemphasized the dangers of overpopulation and encouraged zero population growth—the point at which no population increase occurs from year to year.

● What are the stages in demographic transition theory?

Demographic transition theory links population growth to four stages of economic development: (1) the preindustrial stage, with high birth rates and death rates; (2) early industrialization, with relatively high birth rates and a decline in death rates; (3) advanced industrialization and urbanization, with low birth rates and death rates; and (4) postin-

dustrialization, with additional decreases in the birth rate coupled with a stable death rate.

● How do preindustrial cities differ from industrial and postindustrial cities?

Because of their limited size, preindustrial cities tend to provide a sense of community and a feeling of belonging. The Industrial Revolution changed the size and nature of the city; people began to live close to factories and to one another, resulting in overcrowding and poor sanitation. In postindustrial cities, some people live and work in suburbs or outlying edge cities.

● What are the three functionalist models of urban growth?

Functionalists view urban growth in terms of ecological models. The concentric zone model sees the city as a series of circular areas, each characterized by a different type of land use. The sector model describes urban growth in terms of terrain and transportation routes. The multiple nuclei model views cities as having numerous centers of development from which growth radiates.

● What is the political economy model/conflict perspective on urban growth?

According to political economy models/conflict perspectives, urban growth is influenced by capital investment decisions, power and resource inequality, class and class conflict, and government subsidy programs. At the global level, capitalism also influences the development of cities in core, peripheral, and semiperipheral nations.

● **How do symbolic interactionists view urban life?**

Symbolic interactionist perspectives focus on how people experience urban life. Some analysts view the urban experience positively; others believe that urban dwellers become insensitive to events and to people around them.

● **How did the U.S. population change during the second half of the twentieth century?**

During the 1950s, a dramatic population shift occurred in the United States as people moved from cities to suburbs; almost 80 percent of the U.S. population lives in urban areas today. Edge cities have also developed beyond central cities and suburbs, first with residential areas and then with retail establishments and office parks.

www.cengage.com/login

Register for a Student eResource account to maximize your study time online using CengageNOW. First take the system's diagnostic pre-test, and then follow the personalized study plan that is created for you to help you review this chapter. The study plan will

● help you identify areas on which you should concentrate;
● provide interactive exercises to help you master the chapter concepts; and
● provide a post-test to confirm you are ready to move on to the next chapter.

Key Terms

community 519
crude birth rate 505
crude death rate 505
demographic transition 514
demography 503
fecundity 504

fertility 503
gentrification 522
human ecology 521
invasion 521
metropolis 520
migration 506

mortality 505
population composition 507
population pyramid 507
sex ratio 507
succession 521
zero population growth 514

Questions for Critical Thinking

1. What impact might a high rate of immigration have on culture and personal identity in the United States?
2. If you were designing a study of growth patterns for the city where you live (or one you know well), which theoretical model(s) would provide the most useful framework for your analysis?
3. What do you think everyday life in U.S. cities, suburbs, and rural areas will be like in 2020? Where would you prefer to live? What does your answer reflect about the future of U.S. cities?

The Kendall Companion Website

www.cengage.com/sociology/kendall

Supplement your review of this chapter by going to the text's companion website, where you can take tutorial quizzes, use flash cards to master key terms, follow live links to useful websites, and explore the other study and research resources you'll find there, such as a comprehensive interactive sociology timeline, GSS Data, and Census 2000 information, much of it presented visually in maps.

16 Collective Behavior, Social Movements, and Social Change

One of the by-products of having grown up alongside the Houston Ship Channel was very nearly becoming desensitized to the vast amounts of pollutants the oil and chemical industry poured into East Houston's air and waterways. I once fell in the Ship Channel while working on a crew that built launching pads for a new supertanker. The resulting kidney infection took nine months to heal. I urinated blood for three weeks. No one can tell me that the current state of global consumerism does not impact the world's climate adversely. To [people] who pooh pooh the notion of global warming, I say this: Go take a swim in the Houston Ship Channel.

—Grammy-winning songwriter and recording artist Rodney Crowell explaining why he joined the virtual march against global warming (qtd. in StopGlobalWarming.org, 2006)

When I circled the moon and looked back at Earth, my outlook on life and my viewpoint on Earth changed. You don't see Las Vegas, Boston or even

Environmental activism is a powerful type of social movement that seeks to call public attention to pressing social concerns such as global warming. Activists often stage public events such as the one shown here in an attempt to gain the attention of political leaders and everyday citizens.

© Lu Mingxiang/Xinhua/Landov

New York. You don't see boundaries or people. No whites, blacks, French, Greeks, Christians or Jews. The Earth looks completely uninhabited, and you know that on Spaceship Earth, there live over six billion astronauts, all seeking the same things from life.

When viewed in total, Earth is a spaceship just like Apollo. We are all the crew of Spaceship Earth; and just like Apollo, the crew must learn to live and work together. We must learn to manage the resources of this world with new imagination. The future is up to you.

—Jim Lovell, a retired NASA astronaut, describing how his experience in space gave him a new perspective on environmental problems such as global warming (qtd. in StopGlobalWarming.org, 2006)

Along with many other well-known persons, Rodney Crowell and Jim Lovell are expressing their concerns about global warming and encouraging others to engage in collective behavior to try to save the planet from serious harm. The term *global warming* refers to a process that occurs when carbon dioxide (produced by cars, factories, and power plants, among other sources) stays in the atmosphere and acts like a blanket that holds in the heat. Over time, global warming results in higher temperatures, rises in sea levels, and catastrophic weather such as powerful hurricanes (Revkin, 2006). However, global warming is not a major concern for most people, who, when asked on surveys to identify the most important problems facing the United States, put at the top of their list the war in Iraq, the economy and jobs, immigration, terrorism, health care, and energy prices. Even when respondents mention environmental issues as a concern, global warming is placed far down the list after the pollution of rivers, lakes, reservoirs, and drinking water; air pollution; and toxic waste, according to a recent Gallup survey (*New York Times*, 2006: WK14). The message of the environmental movement is that we must act collectively and immediately to reduce

Chapter Focus Question

Can collective behavior and social movements make people aware of important social issues such as global warming?

global warming before havoc comes to the Earth: Social change is essential. Sociologists define *social change* **as the alteration, modification, or transformation of public policy, culture, or social institutions over time.** Social change is usually brought about by collective behavior and social movements.

In this chapter, we will examine collective behavior, social movements, and social change from a sociological perspective. We will use environmental activism as an example of how people may use social movements as a form of mass mobilization and social transformation (Buechler, 2000). Before reading on, test your knowledge about collective behavior and environmental issues by taking the quiz in Box 16.1.

Collective Behavior

Collective behavior **is voluntary, often spontaneous activity that is engaged in by a large number of people and typically violates dominant-group norms and values.** Unlike the *organizational behavior* found in corporations and voluntary associations (such as labor unions and environmental organizations), collective behavior lacks an official division of labor, hierarchy of authority, and established rules and procedures. Unlike *institutional behavior* (in education, religion, or politics, for example), it lacks institutionalized norms to govern behavior. Collective behavior can take various forms, including crowds, mobs, riots, panics, fads, fashions, and public opinion.

According to the sociologist Steven M. Buechler (2000), early sociologists studied collective behavior because they lived in a world that was responding to the processes of modernization, including urbanization, industrialization, and the proletarianization of workers. Contemporary forms of collective behavior, particularly social protests, are variations on the themes that originated during the transition from feudalism to capitalism and the rise of modernity in Europe (Buechler, 2000). Today, some forms of collective behavior and social movements are directed toward public issues such as air pollution, water pollution, and the exploitation of workers in

global sweatshops by transnational corporations (see Shaw, 1999).

Conditions for Collective Behavior

Collective behavior occurs as a result of some common influence or stimulus that produces a response from a collectivity. A *collectivity* is a number of people who act together and may mutually transcend, bypass, or subvert established institutional patterns and structures. Three major factors contribute to the likelihood that collective behavior will occur: (1) structural factors that increase the chances of people responding in a particular way, (2) timing, and (3) a breakdown in social control mechanisms and a corresponding feeling of normlessness (McPhail, 1991; Turner and Killian, 1993).

A common stimulus is an important factor in collective behavior. For example, the publication of *Silent Spring* (1962) by former Fish and Wildlife Service biologist Rachel Carson is credited with triggering collective behavior directed at demanding a clean environment and questioning how much power large corporations should have. Carson described the dangers of pesticides such as DDT, which was then being promoted by the chemical industry as the miracle that could give the United States the unchallenged position as food supplier to the world (Cronin and Kennedy, 1999). Carson's activism has been described in this way:

> Carson was not a wild-eyed reformer intent on bringing the industrial age to a grinding halt. She wasn't even opposed to pesticides per se. She was a careful scientist and brilliant writer whose painstaking research on pesticides proved that the "miraculous" bursts of agricultural productivity had long-term costs undisclosed in the chemical industry's exaggerated puffery. Americans were losing things—their health, many birds and fishes, and the purity of their waterways—that they should value more than modest savings at the grocery store. (Cronin and Kennedy, 1999: 151)

Timing is another significant factor in bringing about collective behavior. For example, in the 1960s smog had started staining the skies in this country; in Europe, birds and fish were dying from environmental pollution; and oil spills from tankers were

How Much Do You Know About Collective Behavior and Environmental Issues?

True	False	
T	F	1. Scientists are forecasting a global warming of between 2 and 11 degrees Fahrenheit over the next century.
T	F	2. The environmental movement in the United States started in the 1960s.
T	F	3. People who hold strong attitudes regarding the environment are very likely to be involved in social movements to protect the environment.
T	F	4. Environmental groups may engage in civil disobedience or use symbolic gestures to call attention to their issue.
T	F	5. People are most likely to believe rumors when no other information is readily available on a topic.
T	F	6. Influencing public opinion is a very important activity for many social movements.
T	F	7. Most social movements in the United States seek to improve society by changing some specific aspect of the social structure.
T	F	8. Sociologists have found that people in a community respond very similarly to natural disasters and to disasters caused by technological failures.

Answers on page 542.

provoking public outrage worldwide (Cronin and Kennedy, 1999). People in this country were ready to acknowledge that problems existed. By writing *Silent Spring*, Carson made people aware of the hazards of chemicals in their foods and the destruction of wildlife. However, that is not all she produced: As a consequence of her careful research and writing, she also produced anger in people at a time when they were beginning to wonder if they were being deceived by the very industries that they had entrusted with their lives and their resources. Once aroused to action, many people began demanding an honest, comprehensive accounting of where pollution was occurring and how it might be endangering public health and environmental resources. Public outcries also led to investigations in courts and legislatures throughout the United States as people began to demand legal recognition of the right to a clean environment (Cronin and Kennedy, 1999).

A breakdown in social control mechanisms has been a powerful force in triggering collective behavior regarding environmental protection and degradation. During the 1970s, people in the "Love Canal" area of Niagara Falls, New York, became aware that their neighborhood and their children's school had been built over a canal where tons of poisonous waste had been dumped by a chemical company between 1930 and 1950. After the company closed the site, covered it with soil, and sold it (for $1) to the city of Niagara Falls, homes and a school were built on the sixteen-acre site. Over the next two decades, an oily black substance began oozing into the homes in the area and killing the trees and grass on the lots; schoolchildren reported mysterious illnesses and feelings of malaise.

social change the alteration, modification, or transformation of public policy, culture, or social institutions over time.

collective behavior voluntary, often spontaneous activity that is engaged in by a large number of people and typically violates dominant-group norms and values.

Box 16.1 Sociology and Everyday Life

Answers to the Sociology Quiz on Collective Behavior and Environmental Issues

1. True. Global surface temperatures have increased about 0.4 degrees Fahrenheit during the past 25 years, and scientists believe that this trend will grow more pronounced during the next 100 years.

2. False. The environmental movement in the United States is the result of more than 100 years of collective action. The first environmental organization, the American Forestry Association (now American Forests), originated in 1875.

3. False. Since the 1980s, public opinion polls have shown that the majority of people in the United States have favorable attitudes regarding protection of the environment and banning nuclear weapons; however, far fewer individuals are actually involved in collective action to further these causes.

4. True. Environmental groups have held sit-ins, marches, boycotts, and strikes, which sometimes take the form of civil disobedience. Others have hanged political leaders in effigy or held officials hostage. Still others have dressed as grizzly bears to block traffic in Yellowstone National Park or created a symbolic "crack" (made of plastic) on the Glen Canyon Dam on the Colorado River to denounce development in the area.

5. True. Rumors are most likely to emerge and circulate when people have very little information on a topic that is important to them. For example, rumors abound in times of technological disasters, when people are fearful and often willing to believe a worst-case scenario.

6. True. Many social movements, including grassroots environmental activism, attempt to influence public opinion so that local decision makers will feel obliged to correct a specific problem through changes in public policy.

7. True. Most social movements are reform movements that focus on improving society by changing some specific aspect of the social structure. Examples include environmental movements and the disability rights movement.

8. False. Most sociological studies have found that people respond differently to natural disasters, which usually occur very suddenly, than to technological disasters, which may occur gradually. One of the major differences is the communal bonding that tends to occur following natural disasters, as compared with the extreme social conflict that may follow technological disasters.

Sources: Based on Adams, 1991; Gamson, 1990; Hynes, 1990; Worster, 1985; and Young, 1990.

Tests indicated that the dump site contained more than two hundred different chemicals, many of which could cause cancer or other serious health problems. Upon learning this information, Lois Gibbs, a mother of one of the schoolchildren, began a grassroots campaign to force government officials to relocate community members injured by seepages from the chemical dump. The collective behavior of neighborhood volunteers was not only successful in eventually bringing about social change but also inspired others to engage in collective behavior regarding environmental problems in their communities.

Dynamics of Collective Behavior

To better understand the dynamics of collective behavior, let's briefly examine several questions. First, how do people come to transcend, bypass, or subvert established institutional patterns and structures? Some environmental activists have found that they cannot get their point across unless they go outside

The Love Canal area of Niagara Falls, New York, has been the site of protests and other forms of collective behavior because of hazardous environmental pollution. Original protests in the 1970s, demanding a cleanup of the site, were followed in the 1990s by new protests, this time over the proposed resettlement of the area.

established institutional patterns and organizations. For example, Lois Gibbs and other Love Canal residents initially tried to work within established means through the school administration and state health officials to clean up the problem. However, they quickly learned that their problems were not being solved through "official" channels. As the problem appeared to grow worse, organizational responses became more defensive and obscure. Accordingly, some residents began acting outside of established norms by holding protests and strikes (Gibbs, 1982). Some situations are more conducive to collective behavior than others. When people can communicate quickly and easily with one another, spontaneous behavior is more likely (Turner and Killian, 1993). When people are gathered together in one general location (whether lining the streets or assembled in a massive stadium), they are more likely to respond to a common stimulus.

Second, how do people's actions compare with their attitudes? People's attitudes (as expressed in public opinion surveys, for instance) are not always reflected in their political and social behavior. Issues pertaining to the environment are no exception. For example, people may indicate in survey research that they believe the quality of the environment is very important, but the same people may not turn out on election day to support propositions that protect the environment or candidates who promise to focus on environmental issues. Likewise, individuals who indicate on a questionnaire that they are concerned about increases in ground-level ozone—the primary component of urban smog—often drive single-occupant, oversized vehicles which government studies have shown to be "gas guzzlers" that contribute to lowered air quality in urban areas. As a result, smog levels increase, contributing to human respiratory problems and dramatically reduced agricultural crop yields (Voynick, 1999).

Third, why do people act collectively rather than singly? As the sociologists Ralph H. Turner and Lewis M. Killian (1993: 12) note, people believe that there is strength in numbers: "the rhythmic stamping of feet by hundreds of concert-goers in unison is different from isolated, individual cries of 'bravo.'" Likewise, people may act as a collectivity when they believe it is the only way to fight those with greater power and resources. Collective behavior is not just the sum total of a large number of individuals acting at the same time; rather, it reflects people's joint response to some common influence or stimulus.

Distinctions Regarding Collective Behavior

People engaging in collective behavior may be divided into crowds and masses. A ***crowd* is a relatively large number of people who are in one another's immediate vicinity** (Lofland, 1993). Examples of crowds include the audience in a movie theater or people at a pep rally for a sporting event. By contrast, a ***mass* is a number of people who share an interest in a specific idea or issue but who are not in one another's immediate vicinity** (Lofland,

---

crowd a relatively large number of people who are in one another's immediate vicinity.

mass a number of people who share an interest in a specific idea or issue but who are not in one another's immediate vicinity.

1993). An example is the popularity of blogging on the Internet. A *blog,* which is short for "web log," is an online journal maintained by an individual who frequently records entries that are maintained in a chronological order. People who self-publish blogs are widely diverse in their interests. Some may include poetry, diary entries, or discussions of such activities as body piercings. However, others express their beliefs about social issues such as the environment, terrorism, and war, and their concerns about the future. Readers often share a common interest with the blogger on the topics the person is writing about, but these individuals have never met—and probably will never meet—each other in a face-to-face encounter.

Collective behavior may also be distinguished by the dominant emotion expressed. According to the sociologist John Lofland (1993: 72), the *dominant emotion* refers to the "publicly expressed feeling perceived by participants and observers as the most prominent in an episode of collective behavior." Lofland suggests that fear, hostility, and joy are three fundamental emotions found in collective behavior; however, grief, disgust, surprise, or shame may also predominate in some forms of collective behavior.

Types of Crowd Behavior

When we think of a crowd, many of us think of *aggregates,* previously defined as a collection of people who happen to be in the same place at the same time but who share little else in common. However, the presence of a relatively large number of people in the same location does not necessarily produce collective behavior. Sociologist Herbert Blumer (1946) developed a typology in which crowds are divided into four categories: casual, conventional, expressive, and acting. Other scholars have added a fifth category, protest crowds.

Casual and Conventional Crowds *Casual crowds* are relatively large gatherings of people who happen to be in the same place at the same time; if they interact at all, it is only briefly. People in a shopping mall or a subway car are examples of casual crowds. Other than sharing a momentary interest, such as a clown's performance or a small child's fall, a casual crowd has nothing in common. The casual crowd

plays no active part in the event—such as the child's fall—which likely would have occurred whether or not the crowd was present; the crowd simply observes.

Conventional crowds are made up of people who come together for a scheduled event and thus share a common focus. Examples include religious services, graduation ceremonies, concerts, and college lectures. Each of these events has preestablished schedules and norms. Because these events occur regularly, interaction among participants is much more likely; in turn, the events would not occur without the crowd, which is essential to the event.

Expressive and Acting Crowds *Expressive crowds* provide opportunities for the expression of some strong emotion (such as joy, excitement, or grief). People release their pent-up emotions in conjunction with other persons experiencing similar emotions. Examples include worshippers at religious revival services; mourners lining the streets when a celebrity, public official, or religious leader has died; and revelers assembled at Mardi Gras or on New Year's Eve at Times Square in New York.

Acting crowds are collectivities so intensely focused on a specific purpose or object that they may erupt into violent or destructive behavior. Mobs, riots, and panics are examples of acting crowds, but casual and conventional crowds may become acting crowds under some circumstances. A **mob is a highly emotional crowd whose members engage in, or are ready to engage in, violence against a specific target—a person, a category of people, or physical property.** Mob behavior in this country has included lynchings, fire bombings, effigy hangings, and hate crimes. Mob violence tends to dissipate relatively quickly once a target has been injured, killed, or destroyed. Sometimes, actions such as an effigy hanging are used symbolically by groups that are not otherwise violent. For example, Lois Gibbs and other Love Canal residents called attention to their problems with the chemical dump site by staging a protest in which they "burned in effigy" the governor and the health commissioner to emphasize their displeasure with the lack of response from these public officials (A. Levine, 1982).

Compared with mob actions, riots may be of somewhat longer duration. A **riot is violent crowd**

Crowds of people come together for a variety of reasons. The people pictured here wanted to be near the front of the line to purchase an iPhone on the first day the new device became available. How does a crowd such as this differ from other types of crowds?

behavior that is fueled by deep-seated emotions but not directed at one specific target. Riots are often triggered by fear, anger, and hostility; however, not all riots are caused by deep-seated hostility and hatred—people may be expressing joy and exuberance when rioting occurs. Examples include celebrations after sports victories such as those that occurred in Montreal, Canada, following a Stanley Cup win and in Vancouver following a playoff victory (Kendall, Lothian Murray, and Linden, 2004).

A *panic* is a form of crowd behavior that occurs when a large number of people react to a real or perceived threat with strong emotions and self-destructive behavior. The most common type of panic occurs when people seek to escape from a perceived danger, fearing that few (if any) of them will be able to get away from that danger. Panics can also arise in response to events that people believe are beyond their control—such as a major disruption in the economy. Although panics are relatively rare, they receive massive media coverage because they provoke strong feelings of fear in readers and viewers, and the number of casualties may be large.

Protest Crowds Sociologists Clark McPhail and Ronald T. Wohlstein (1983) added protest crowds to the four types of crowds identified by Blumer. *Protest crowds* engage in activities intended to achieve specific political goals. Examples include sit-ins, marches, boycotts, blockades, and strikes. Some protests take the form of *civil disobedience—nonviolent action that seeks to change a policy or law by refusing to comply with it.* Acts of civil disobedience may become violent, as in a confrontation between protesters and police officers; in this case, a protest crowd becomes an *acting crowd*. In the 1960s, African American students and sympathetic whites used sit-ins to call attention to racial injustice and demand social change (see Morris, 1981; McAdam, 1982). Some of these protests can escalate into violent confrontations even when violence was not the intent of the organizers.

Explanations of Crowd Behavior

What causes people to act collectively? How do they determine what types of action to take? One of the earliest theorists to provide an answer to these questions was Gustave Le Bon, a French scholar who focused on crowd psychology in his contagion theory.

Contagion Theory *Contagion theory* focuses on the social–psychological aspects of collective behavior; it attempts to explain how moods, attitudes, and behavior are communicated rapidly and why they are accepted by others (Turner and Killian, 1993). Le Bon (1841–1931) argued that people are more likely to engage in antisocial behavior in a crowd because they are anonymous and feel invulnerable. Le Bon (1960/1895) suggested that a crowd takes on a life

---

mob a highly emotional crowd whose members engage in, or are ready to engage in, violence against a specific target—a person, a category of people, or physical property.

riot violent crowd behavior that is fueled by deep-seated emotions but is not directed at one specific target.

panic a form of crowd behavior that occurs when a large number of people react to a real or perceived threat with strong emotions and self-destructive behavior.

civil disobedience nonviolent action that seeks to change a policy or law by refusing to comply with it.

of its own that is larger than the beliefs or actions of any one person. Because of its anonymity, the crowd transforms individuals from rational beings into a single organism with a collective mind. In essence, Le Bon asserted that emotions such as fear and hate are contagious in crowds because people experience a decline in personal responsibility; they will do things as a collectivity that they would never do when acting alone.

Le Bon's theory is still used by many people to explain crowd behavior. However, critics argue that the "collective mind" has not been documented by systematic studies.

Social Unrest and Circular Reaction
Sociologist Robert E. Park was the first U.S. sociologist to investigate crowd behavior. Park believed that Le Bon's analysis of collective behavior lacked several important elements. Intrigued that people could break away from the powerful hold of culture and their established routines to develop a new social order, Park added the concepts of social unrest and circular reaction to contagion theory. According to Park, social unrest is transmitted by a process of *circular reaction*—the interactive communication between persons such that the discontent of one person is communicated to another, who, in turn, reflects the discontent back to the first person (Park and Burgess, 1921).

Convergence Theory
Convergence theory focuses on the shared emotions, goals, and beliefs that many people may bring to crowd behavior. Because of their individual characteristics, many people have a predisposition to participate in certain types of activities (Turner and Killian, 1993). From this perspective, people with similar attributes find a collectivity of like-minded persons with whom they can express their underlying personal tendencies. Although people may reveal their "true selves" in crowds, their behavior is not irrational; it is highly predictable to those who share similar emotions or beliefs.

Convergence theory has been applied to a wide array of conduct, from lynch mobs to environmental movements. In social psychologist Hadley Cantril's (1941) study of one lynching, he found that the participants shared certain common attributes: They were poor and working-class whites who felt that their status was threatened by the presence of successful African Americans. Consequently, the characteristics of these individuals made them susceptible to joining a lynch mob even if they did not know the target of the lynching.

Convergence theory adds to our understanding of certain types of collective behavior by pointing out how individuals may have certain attributes—such as racial hatred or fear of environmental problems that directly threaten them—that initially

© David Young-Wolff/PhotoEdit

Convergence theory is based on the assumption that crowd behavior involves shared emotions, goals, and beliefs, such as the importance of protecting the environment. An example is the Earth Day events that brought together these children carrying this banner to foster environmental causes.

bring them together. However, this theory does not explain how the attitudes and characteristics of individuals who take some collective action differ from those who do not.

Emergent Norm Theory Unlike contagion and convergence theories, *emergent norm theory* emphasizes the importance of social norms in shaping crowd behavior. Drawing on the symbolic interactionist perspective, the sociologists Ralph Turner and Lewis Killian (1993: 12) asserted that crowds develop their own definition of a situation and establish norms for behavior that fit the occasion:

> Some shared redefinition of right and wrong in a situation supplies the justification and coordinates the action in collective behavior. People do what they would not otherwise have done when they panic collectively, when they riot, when they engage in civil disobedience, or when they launch terrorist campaigns, because they find social support for the view that what they are doing is the right thing to do in the situation.

According to Turner and Killian (1993: 13), emergent norms occur when people define a new situation as highly unusual or see a long-standing situation in a new light.

Sociologists using the emergent norm approach seek to determine how individuals in a given collectivity develop an understanding of what is going on, how they construe these activities, and what type of norms are involved. For example, in a study of audience participation, the sociologist Steven E. Clayman (1993) found that members of an audience listening to a speech applaud promptly and independently but wait to coordinate their booing with other people; they do not wish to "boo" alone.

Some emergent norms are permissive—that is, they give people a shared conviction that they may disregard ordinary rules such as waiting in line, taking turns, or treating a speaker courteously. Collective activity such as mass looting may be defined (by participants) as taking what rightfully belongs to them and punishing those who have been exploitative. For example, following the Los Angeles riots of 1992, some analysts argued that Korean Americans were targets of rioters because they were viewed by Latinos/as and African Americans as "callous and greedy invaders" who became wealthy at the expense of members of other racial–ethnic groups (Cho, 1993). Thus, rioters who used this rationalization could view looting and burning as a means of "paying back" Korean Americans or of gaining property (such as TV sets and microwave ovens) from those who had already taken from them. Once a crowd reaches some agreement on the norms, the collectivity is supposed to adhere to them. If crowd members develop a norm that condones looting or vandalizing property, they will proceed to cheer for those who conform and ridicule those who are unwilling to abide by the collectivity's new norms.

Emergent norm theory points out that crowds are not irrational. Rather, new norms are developed in a rational way to fit the immediate situation. However, critics note that proponents of this perspective fail to specify exactly what constitutes a norm, how new ones emerge, and how they are so quickly disseminated and accepted by a wide variety of participants. One variation of this theory suggests that no single dominant norm is accepted by everyone in a crowd; instead, norms are specific to the various categories of actors rather than to the collectivity as a whole (Snow, Zurcher, and Peters, 1981). For example, in a study of football victory celebrations, the sociologists David A. Snow, Louis A. Zurcher, and Robert Peters (1981) found that each week, behavioral patterns were changed in the postgame revelry, with some being modified, some added, and some deleted.

Mass Behavior

Not all collective behavior takes place in face-to-face collectivities. *Mass behavior* **is collective behavior that takes place when people (who often are geographically separated from one another) respond to the same event in much the same way.** For people to respond in the same way, they typically have common sources of information that provoke their collective behavior. The most frequent

mass behavior collective behavior that takes place when people (who often are geographically separated from one another) respond to the same event in much the same way.

types of mass behavior are rumors, gossip, mass hysteria, public opinion, fashions, and fads. Under some circumstances, social movements constitute a form of mass behavior. However, we will examine social movements separately because they differ in some important ways from other types of dispersed collectivities.

Rumors and Gossip *Rumors* **are unsubstantiated reports on an issue or subject** (Rosnow and Fine, 1976). Whereas a rumor may spread through an assembled collectivity, rumors may also be transmitted among people who are dispersed geographically, including people posting messages on the Internet or talking by cell phone. Although rumors may initially contain a kernel of truth, they may be modified as they spread to serve the interests of those repeating them. Rumors thrive when tensions are high and when little authentic information is available on an issue of great concern. For example, when the blackout of August 2003 occurred, leaving fifty million people in eight states and parts of Canada without electricity, the earliest rumors about the power failures reflected the turbulent times in which we live. One of the first rumors that began to spread was that the blackout was an act of terrorism; television broadcasters and public officials quickly tried to deflect this rumor for fear that people might panic and be injured as they sought to leave their workplaces in cities such as New York and make their way home. Fortunately, most people acted responsibly, and the riots and looting that took place during prior blackouts in New York City did not recur.

People are willing to give rumors credence when no opposing information is available. Environmental issues are similar. For example, when residents of Love Canal waited for information from health department officials about their exposure to the toxic chemicals and from the government about possible relocation at state expense to another area, new waves of rumors spread through the community daily. By the time a meeting was called by health department officials to provide homeowners with the results of air-sample tests for hazardous chemicals (such as chloroform and benzene) performed on their homes, already fearful residents were ready to believe the worst, as Lois Gibbs (1982: 25) describes:

© Jonathan Fickies/Getty Images

When unexpected events such as the massive 2003 power outage in the United States and Canada occur, people frequently rely on rumors to help them know what is going on. Getting accurate information out quickly helped prevent people from panicking as tens of thousands of workers in Manhattan sought to get home any way they could while electricity was unavailable in the city.

Next to the names [of residents] were some numbers. But the numbers had no meaning. People stood there looking at the numbers, knowing nothing of what they meant but suspecting the worst.

One woman, divorced and with three sick children, looked at the piece of paper with numbers and started crying hysterically: "No wonder my children are sick. Am I going to die? What's going to happen to my children?" No one could answer. . . .

The night was very warm and humid, and the air was stagnant. On a night like that, the smell of Love Canal is hard to describe. It's all around you. It's as though it were about to envelop you and smother you. By now, we were outside, standing in the parking lot. The woman's panic caught on, starting a chain reaction. Soon, many people there were hysterical.

Once a rumor begins to circulate, it seldom stops unless compelling information comes to the forefront that either proves the rumor false or makes it obsolete.

In industrialized societies with sophisticated technology, rumors come from a wide variety of sources and may be difficult to trace. Print media (newspapers and magazines) and electronic media (radio and television), fax machines, cellular networks, satellite systems, and the Internet aid the rapid movement of rumors around the globe. In addition, modern communications technology makes anonymity much easier. In a split second, messages (both factual and fictitious) can be disseminated to thousands of people through e-mail, computerized bulletin boards, and Internet newsgroups.

Whereas rumors deal with an issue or a subject, **gossip refers to rumors about the personal lives of individuals.** Charles Horton Cooley (1963/1909) viewed gossip as something that spread among a

Although a spokesperson for CBS Radio stated to listeners that they were hearing a dramatization of a novel, the 1938 presentation of H. G. Wells's *The War of the Worlds,* as presented by Orson Welles and his Mercury Theatre, terrified untold numbers of people. Here Welles talks to interviewers the day after the event caused a nationwide panic.

small group of individuals who personally knew the person who was the object of the rumor. Today, this is frequently not the case; many people enjoy gossiping about people whom they have never met. Tabloid newspapers and magazines such as the *National Enquirer* and *People,* and television "news" programs that purport to provide "inside" information on the lives of celebrities, are sources of contemporary gossip, much of which has not been checked for authenticity.

Mass Hysteria and Panic *Mass hysteria* is a form of dispersed collective behavior that occurs when a large number of people react with strong emotions and self-destructive behavior to a real or perceived threat. Does mass hysteria actually occur? Although the term has been widely used, many sociologists believe that this behavior is best described as a panic with a dispersed audience.

An example of mass hysteria or a panic with a widely dispersed audience was actor Orson Welles's 1938 Halloween eve radio dramatization of H. G. Wells's science fiction classic *The War of the Worlds.* A CBS radio dance music program was interrupted suddenly by a news bulletin informing the audience that Martians had landed in New Jersey and were in the process of conquering Earth. Some listeners became extremely frightened even though an announcer had indicated before, during, and after the performance that the broadcast was a fictitious dramatization. According to some reports, as many as one million of the estimated ten million listeners believed that this astonishing event had occurred. Thousands were reported to have hidden in their storm cellars or to have gotten in their cars so that they could flee from the Martians (see Brown, 1954). In actuality, the program probably did not generate mass hysteria, but rather a panic among gullible listeners. Others switched stations to determine if the same "news" was being broadcast elsewhere. When they discovered that it was not, they merely laughed at the joke being played on listeners by CBS. In

rumor an unsubstantiated report on an issue or subject.

gossip rumors about the personal lives of individuals.

1988, on the fiftieth anniversary of the broadcast, a Portuguese radio station rebroadcast the program; once again, a panic ensued.

Fads and Fashions As you will recall from Chapter 2, a *fad* is a temporary but widely copied activity enthusiastically followed by large numbers of people. Fads can be embraced by widely dispersed collectivities; news networks such as CNN and Internet websites may bring the latest fad to the attention of audiences around the world.

Unlike fads, fashions tend to be longer lasting. In Chapter 2, *fashion* is defined as a currently valued style of behavior, thinking, or appearance. Fashion also applies to art, music, drama, literature, architecture, interior design, and automobiles, among other things. However, most sociological research on fashion has focused on clothing, especially women's apparel (Davis, 1992).

In preindustrial societies, clothing styles remained relatively unchanged. With the advent of industrialization, items of apparel became readily available at low prices because of mass production. Fashion became more important as people embraced the "modern" way of life and as advertising encouraged "conspicuous consumption."

Georg Simmel, Thorstein Veblen, and Pierre Bourdieu have all viewed fashion as a means of status differentiation among members of different social classes. Simmel (1957/1904) suggested a classic "trickle-down" theory (although he did not use those exact words) to describe the process by which members of the lower classes emulate the fashions of the upper class. As the fashions descend through the status hierarchy, they are watered down and "vulgarized" so that they are no longer recognizable to members of the upper class, who then regard them as unfashionable and in bad taste (Davis, 1992). Veblen (1967/1899) asserted that fashion serves mainly to institutionalize conspicuous consumption among the wealthy. Almost eighty years later, Bourdieu (1984) similarly (but more subtly) suggested that "matters of taste," including fashion sensibility, constitute a large share of the "cultural capital" possessed by members of the dominant class.

Herbert Blumer (1969) disagreed with the trickle-down approach, arguing that "collective se-

Collective behavior occurs in many different forms, including flash mobs, a phenomenon observed in cities across the world. This public pillow fight took place in Seattle in 2008. Typically, organizers arrange such an event on the Internet and through text messaging so that they can gather a large crowd at a location for a short time. Have you participated in a similar, relatively spontaneous group activity?

lection" best explains fashion. Blumer suggested that people in the middle and lower classes follow fashion because it is *fashion*, not because they desire to emulate members of the elite class. Blumer thus shifts the focus on fashion to collective mood, tastes, and choices: "Tastes are themselves a product of experience. . . . They are formed in the context of social interaction, responding to the definitions and affirmation given by others. People thrown into areas of common interaction and having similar runs of experience develop common tastes" (qtd. in Davis, 1992: 116). Perhaps one of the best refutations of the trickle-down approach is the way in which fashion today often originates among people in the lower social classes and is mimicked by the elites. In the mid-1990s, the so-called grunge look was a prime example of this.

Public Opinion *Public opinion* **consists of the attitudes and beliefs communicated by ordinary citizens to decision makers** (Greenberg and Page, 2002). It is measured through polls and surveys, which use research methods such as interviews and questionnaires, as described in Chapter 1. Many people are not interested in all aspects of public policy but are concerned about issues they believe are relevant to themselves. Even on a single topic, public opinion will vary widely based on race/ethnicity, religion, region, social class, education level, gender, age, and so on.

Scholars who examine public opinion are interested in the extent to which the public's attitudes are communicated to decision makers and the effect (if any) that public opinion has on policy making (Turner and Killian, 1993). Some political scientists argue that public opinion has a substantial effect on decisions at all levels of government (see Greenberg and Page, 2002); others strongly disagree.

Today, people attempt to influence elites, and vice versa. Consequently, a two-way process occurs with the dissemination of *propaganda*—**information provided by individuals or groups that have a vested interest in furthering their own cause or damaging an opposing one.** Although many of us think of propaganda in negative terms, the information provided can be correct and can have a positive effect on decision making.

In recent decades, grassroots environmental activists (including the Love Canal residents) have attempted to influence public opinion. In a study of public opinion on environmental issues, the sociologist Riley E. Dunlap (1992) found that public awareness of the seriousness of environmental problems and public support for environmental protection increased dramatically between the late 1960s and the 1990s. However, it is less clear that public opinion translates into action by either decision makers in government and industry or by individuals (such as a willingness to adopt a more ecologically sound lifestyle).

Initially, most grassroots environmental activists attempt to influence public opinion so that local decision makers will feel the necessity of correcting a specific problem through changes in public policy. Although activists usually do not start out seeking broader social change, they often move in that direction when they become aware of how widespread the problem is in the larger society or on a global basis. One of two types of social movements often develops at this point—one focuses on NIMBY ("not in my backyard"), whereas the other focuses on NIABY ("not in anyone's backyard") (Freudenberg and Steinsapir, 1992).

Social Movements

Although collective behavior is short-lived and relatively unorganized, social movements are longer lasting, are more organized, and have specific goals or purposes. A *social movement* **is an organized group that acts consciously to promote or resist change through collective action** (Goldberg, 1991). Because social movements have not become institutionalized and are outside the political mainstream, they offer "outsiders" an opportunity to have their voices heard.

Social movements are more likely to develop in industrialized societies than in preindustrial societies, where acceptance of traditional beliefs and practices makes such movements unlikely. Diversity and a lack of consensus (hallmarks of industrialized nations) contribute to demands for social change, and people who participate in social movements typically lack power and other resources to bring about change without engaging in collective action. Social movements are most likely to spring up when people come to see their personal troubles as public issues that cannot be solved without a collective response.

Social movements make democracy more available to excluded groups (see Greenberg and Page, 2002). Historically, people in the United States have worked at the grassroots level to bring about changes even when elites sought to discourage activism (Adams, 1991). For example, the civil rights movement brought into its ranks African Americans in the South who had never before been allowed to participate in politics (see Killian, 1984). The women's suffrage movement gave voice to women who

public opinion the attitudes and beliefs communicated by ordinary citizens to decision makers.

propaganda information provided by individuals or groups that have a vested interest in furthering their own cause or damaging an opposing one.

social movement an organized group that acts consciously to promote or resist change through collective action.

© AP Images

Martin Luther King, Jr., a leader of the civil rights movement in the 1950s and 1960s, advocated nonviolent protests that sometimes took the form of civil disobedience. Here he marches alongside his wife, Coretta Scott King, who for many years took over Dr. King's activities after he was assassinated.

had been denied the right to vote (Rosenthal et al., 1985). Similarly, a grassroots environmental movement gave the working-class residents of Love Canal a way to "fight city hall" and Hooker Chemicals, as Lois Gibbs (1982: 38–40) explains:

People were pretty upset. They were talking and stirring each other up. I was afraid there would be violence. We had a meeting at my house to try to put everything together [and] decided to form a homeowners' association. We got out the word as best we could and told everyone to come to the Frontier Fire Hall on 102d Street. . . . The firehouse was packed with people, and more were outside. . . .

I was elected president. . . . I took over the meeting but I was scared to death. It was only the second time in my life I had been in front of a microphone or a crowd. . . . We set four goals right at the beginning—(1) get all the residents within the Love Canal area who wanted to be evacuated, evacuated and relocated, especially during the construction and repair of the canal; (2) do something about propping up property values; (3) get the canal fixed properly; and (4) have air

sampling and soil and water testing done throughout the whole area, so we could tell how far the contamination had spread. . . .

Most social movements rely on volunteers like Lois Gibbs to carry out the work. Traditionally, women have been strongly represented in both the membership and the leadership of many grassroots movements (A. Levine, 1982; Freudenberg and Steinsapir, 1992).

The Love Canal activists set the stage for other movements that have grappled with the kind of issues that the sociologist Kai Erikson (1994) refers to as a "new species of trouble." Erikson describes the "new species" as environmental problems that "contaminate rather than merely damage . . . they pollute, befoul, taint, rather than just create wreckage . . . they penetrate human tissue indirectly rather than just wound the surfaces by assaults of a more straightforward kind. . . . And the evidence is growing that they scare human beings in new and special ways, that they elicit an uncanny fear in us" (Erikson, 1991: 15). The chaos that Erikson (1994: 141) describes is the result of technological disasters—"meaning everything that can go wrong when systems fail, humans err, designs prove faulty, engines misfire, and so on." A recent example of such a disaster occurred in Japan, where more than 300,000 residents living within six miles of the nuclear plant at Tokaimura were told to stay indoors in the aftermath of three workers' mishandling of stainless-steel pails full of uranium, causing the worst nuclear accident in Japan's history (Larimer, 1999). Although no lives were immediately lost, workers in the plant soaked up potentially lethal doses of radiation, some of which also leaked from the plant into the community.

Social movements provide people who otherwise would not have the resources to enter the game of politics a chance to do so. We are most familiar with those movements that develop around public policy issues considered newsworthy by the media, ranging from abortion and women's rights to gun control and environmental justice. However, a number of other types of social movements exist as well.

Types of Social Movements

Social movements are difficult to classify; however, sociologists distinguish among movements on the basis of their *goals* and the *amount of change* they seek to produce (Aberle, 1966; Blumer, 1974). Some movements seek to change people, whereas others seek to change society.

Reform Movements Grassroots environmental movements are an example of *reform movements,* which seek to improve society by changing some specific aspect of the social structure. Members of reform movements usually work within the existing system to attempt to change existing public policy so that it more adequately reflects their own value system. Examples of reform movements (in addition to the environmental movement) include labor movements, animal rights movements, antinuclear movements, Mothers Against Drunk Driving, and the disability rights movement.

Sociologist Lory Britt (1993) suggested that some movements arise specifically to alter social responses to and definitions of stigmatized attributes. From this perspective, social movements may bring about changes in societal attitudes and practices while at the same time causing changes in participants' social emotions. For example, the civil rights and gay rights movements helped replace shame with pride (Britt, 1993). Such a sense of pride may extend beyond current members of a reform movement. The late César Chávez, organizer of a Mexican American farmworkers' movement that developed into the United Farm Workers Union, noted that the "consciousness and pride raised by our union is alive and thriving inside millions of young Hispanics who will never work on a farm!" (qtd. in Ayala, 1993: E4).

Revolutionary Movements Movements seeking to bring about a total change in society are referred to as *revolutionary movements.* These movements usually do not attempt to work within the existing system; rather, they aim to remake the system by replacing existing institutions with new ones. Revolutionary movements range from utopian groups seeking to establish an ideal society to radical terrorists who use fear tactics to intimidate those with whom they disagree ideologically.

Movements based on terrorism often use tactics such as bombings, kidnappings, hostage taking, hijackings, and assassinations (White, 2003). Over the past thirty years, a number of movements in the United States have engaged in terrorist activities or supported a policy of violence. However, the terrorist attacks in New York City and Washington, D.C., on September 11, 2001, and the events that followed those attacks proved to all of us that terrorism within this country can originate from the activities of revolutionary terrorists from outside the country as well.

Religious Movements Social movements that seek to produce radical change in individuals are typically based on spiritual or supernatural belief systems. Also referred to as *expressive movements, religious movements* are concerned with renovating or renewing people through "inner change." Fundamentalist religious groups seeking to convert nonbelievers to their belief system are an example of this type of movement. Some religious movements are *millenarian*—that is, they forecast that "the end is near" and assert that an immediate change in behavior is imperative. Relatively new religious movements in industrialized Western societies have included Hare Krishnas, the Unification Church, Scientology, and the Divine Light Mission, all of which tend to appeal to the psychological and social needs of young people seeking meaning in life that mainstream religions have not provided for them.

Alternative Movements Movements that seek limited change in some aspect of people's behavior are referred to as *alternative movements.* For example, early in the twentieth century the Women's Christian Temperance Union attempted to get people to abstain from drinking alcoholic beverages. Some analysts place "therapeutic social movements" such as Alcoholics Anonymous in this category; however, others do not, due to their belief that people must change their lives completely in order to overcome alcohol abuse (see Blumberg, 1977). More recently, a variety of "New Age" movements have directed people's behavior by emphasizing spiritual consciousness combined with a belief in reincarnation and astrology. Such practices as vegetarianism, meditation, and holistic medicine are

Yoga has become an increasingly popular activity in recent years as many people have turned to alternative social movements derived from Asian traditions.

often included in the self-improvement category. Beginning in the 1990s, some alternative movements have included the practice of yoga (usually without its traditional background in the Hindu religion) as a means by which the self can be liberated and union can be achieved with the supreme spirit or universal soul.

Resistance Movements Also referred to as *regressive movements, resistance movements* seek to prevent change or to undo change that has already occurred. Virtually all of the social movements previously discussed face resistance from one or more reactive movements that hold opposing viewpoints and want to foster public policies that reflect their own beliefs. Examples of resistance movements are groups organized since the 1950s to oppose school integration, civil rights and affirmative action legislation, and domestic partnership initiatives. However, perhaps the most widely known resistance movement includes many who label themselves "pro-life" advocates—such as Operation Rescue, which seeks to close abortion clinics and make abortion illegal under all circumstances (Gray, 1993; Van Biema,

1993). Protests by some radical antiabortion groups have grown violent, resulting in the death of several doctors and clinic workers, and creating fear among health professionals and patients seeking abortions (Belkin, 1994).

Stages in Social Movements

Do all social movements go through similar stages? Not necessarily, but there appear to be identifiable stages in virtually all movements that succeed beyond their initial phase of development.

In the *preliminary* (or *incipiency*) *stage,* widespread unrest is present as people begin to become aware of a problem. At this stage, leaders emerge to agitate others into taking action. In the *coalescence stage,* people begin to organize and to publicize the problem. At this stage, some movements become formally organized at local and regional levels. In the *institutionalization* (or *bureaucratization*) *stage,* an organizational structure develops, and a paid staff (rather than volunteers) begins to lead the group. When the movement reaches this stage, the initial zeal and idealism of members may diminish as administrators take over management of the organization. Early grassroots supporters may become disillusioned and drop out; they may also start another movement to address some as-yet-unsolved aspect of the original problem. For example, some national environmental organizations—such as the Sierra Club, the National Audubon Society, and the National Parks and Conservation Association—that started as grassroots conservation movements are currently viewed by many people as being unresponsive to local environmental problems (Cable and Cable, 1995). As a result, new movements have arisen.

Social Movement Theories

What conditions are most likely to produce social movements? Why are people drawn to these movements? Sociologists have developed several theories to answer these questions.

Relative Deprivation Theory

According to relative deprivation theory, people who are satisfied with their present condition are less likely to seek social change. Social movements arise as a response to people's perception that they have been deprived of their "fair share" (Rose, 1982). Thus, people who suffer relative deprivation are more likely to feel that change is necessary and to join a social movement in order to bring about that change. *Relative deprivation* refers to the discontent that people may feel when they compare their achievements with those of similarly situated persons and find that they have less than they think they deserve (Orum and Orum, 1968). Karl Marx captured the idea of relative deprivation in this description: "A house may be large or small; as long as the surrounding houses are small it satisfies all social demands for a dwelling. But let a palace arise beside the little house, and it shrinks from a little house to a hut" (qtd. in Ladd, 1966: 24). Movements based on relative deprivation are most likely to occur when an upswing in the standard of living is followed by a period of decline, such that people have *unfulfilled rising expectations*—newly raised hopes of a better lifestyle that are not fulfilled as rapidly as the people expected or are not realized at all.

Although most of us can relate to relative deprivation theory, it does not fully account for why people experience social discontent but fail to join a social movement. Even though discontent and feelings of deprivation may be necessary to produce certain types of social movements, they are not sufficient to bring movements into existence. In fact, the sociologist Anthony Orum (1974) found the best predictor of participation in a social movement to be prior organizational membership and involvement in other political activities.

Value-Added Theory

The value-added theory developed by sociologist Neil Smelser (1963) is based on the assumption that certain conditions are necessary for the development of a social movement. Smelser called his theory the "value-added" approach based on the concept (borrowed from the field of economics) that each step in the production process adds something to the finished product. For example, in the process of converting iron ore into automobiles, each stage "adds value" to the final product (Smelser, 1963). Similarly, Smelser asserted, six conditions are necessary and sufficient to produce social movements when they combine or interact in a particular situation:

1. *Structural conduciveness.* People must become aware of a significant problem and have the opportunity to engage in collective action. According to Smelser, movements are more likely to occur when a person, class, or agency can be singled out as the source of the problem; when channels for expressing grievances either are not available or fail; and when the aggrieved have a chance to communicate among themselves.
2. *Structural strain.* When a society or community is unable to meet people's expectations that something should be done about a problem, strain occurs in the system. The ensuing tension and conflict contribute to the development of a social movement based on people's belief that the problem would not exist if authorities had done what they were supposed to do.
3. *Spread of a generalized belief.* For a movement to develop, there must be a clear statement of the problem and a shared view of its cause, effects, and possible solution.
4. *Precipitating factors.* To reinforce the existing generalized belief, an inciting incident or dramatic event must occur. With regard to technological disasters, some (including Love Canal) gradually emerge from a long-standing environmental threat whereas others (including the Three Mile Island nuclear power plant) involve a suddenly imposed problem.
5. *Mobilization for action.* At this stage, leaders emerge to organize others and give them a sense of direction.
6. *Social control factors.* If there is a high level of social control on the part of law enforcement officials, political leaders, and others, it becomes more difficult to develop a social movement or engage in certain types of collective action.

Value-added theory takes into account the complexity of social movements and makes it possible to test

Smelser's assertions regarding the necessary and sufficient conditions that produce such movements. However, critics note that the approach is rooted in the functionalist tradition and views structural strains as disruptive to society.

Resource Mobilization Theory

Smelser's value-added theory tends to underemphasize the importance of resources in social movements. By contrast, *resource mobilization theory* focuses on the ability of members of a social movement to acquire resources and mobilize people in order to advance their cause (Oberschall, 1973; McCarthy and Zald, 1977). Resources include money, people's time and skills, access to the media, and material goods such as property and equipment. Assistance from outsiders is essential for social movements. For example, reform movements are more likely to succeed when they gain the support of political and economic elites (Oberschall, 1973).

Resource mobilization theory is based on the assumption that participants in social movements are rational people. According to the sociologist Charles Tilly (1973, 1978), movements are formed and dissolved, mobilized and deactivated, based on rational decisions about the goals of the group, available resources, and the cost of mobilization and collective action. Resource mobilization theory also assumes that participants must have some degree of economic and political resources to make the movement a success. In other words, widespread discontent alone cannot produce a social movement; adequate resources and motivated people are essential to any concerted social action (Aminzade, 1973; Gamson, 1990). Based on an analysis of fifty-three U.S. social protest groups (ranging from labor unions to peace movements) between 1800 and 1945, the sociologist William Gamson (1990) concluded that the organization and tactics of a movement strongly influence its chances of success. However, critics note that this theory fails to account for social changes brought about by groups with limited resources.

At the beginning of the twenty-first century, scholars continue to modify resource mobilization theory and to develop new approaches for investigating the diversity of movements (see Buechler, 2000). For example, emerging perspectives based on resource mobilization theory emphasize the ideology and legitimacy of movements as well as material resources (see Zald and McCarthy, 1987; McAdam, McCarthy, and Zald, 1988). Additional perspectives are also needed on social movements in other nations to determine how activists in those countries acquire resources and mobilize people to advance causes such as environmental protection (see Box 16.2 on page 558).

Social Constructionist Theory: Frame Analysis

Recent theories based on a symbolic interactionist perspective focus on the importance of the symbolic presentation of a problem to both participants

How is the issue of immigration framed in these photos? Research based on frame analysis often investigates how social issues are framed and what names they are given.

and the general public (see Snow et al., 1986; Capek, 1993). Social constructionist theory is based on the assumption that a social movement is an interactive, symbolically defined, and negotiated process that involves participants, opponents, and bystanders (Buechler, 2000).

Research based on this perspective often investigates how problems are framed and what names they are given. This approach reflects the influence of the sociologist Erving Goffman's *Frame Analysis* (1974), in which he suggests that our interpretation of the particulars of events and activities is dependent on the framework from which we perceive them. According to Goffman (1974: 10), the purpose of frame analysis is "to try to isolate some of the basic frameworks of understanding available in our society for making sense out of events and to analyze the special vulnerabilities to which these frames of reference are subject." In other words, various "realities" may be simultaneously occurring among participants engaged in the same set of activities. Sociologist Steven M. Buechler (2000: 41) explains the relationship between frame analysis and social movement theory:

> Framing means focusing attention on some bounded phenomenon by imparting meaning and significance to elements within the frame and setting them apart from what is outside the frame. In the context of social movements, framing refers to the interactive, collective ways that movement actors assign meanings to their activities in the conduct of social movement activism. The concept of framing is designed for discussing the social construction of grievances as a fluid and variable process of social interaction—and hence a much more important explanatory tool than resource mobilization theory has maintained.

Sociologists have identified at least three ways in which grievances are framed. First, *diagnostic framing* identifies a problem and attributes blame or causality to some group or entity so that the social movement has a target for its actions. Second, *prognostic framing* pinpoints possible solutions or remedies, based on the target previously identified. Third, *motivational framing* provides a vocabulary of motives that compel people to take action (Benford, 1993; Snow and Benford, 1988). When

successful framing occurs, the individual's vague dissatisfactions are turned into well-defined grievances, and people are compelled to join the movement in an effort to reduce or eliminate those grievances (Buechler, 2000).

Beyond motivational framing, additional frame alignment processes are necessary in order to supply a continuing sense of urgency to the movement. *Frame alignment* is the linking together of interpretive orientations of individuals and social movement organizations so that there is congruence between individuals' interests, beliefs, and values and the movement's ideologies, goals, and activities (Snow et al., 1986). Four distinct frame alignment processes occur in social movements: (1) *frame bridging* is the process by which movement organizations reach individuals who already share the same world view as the organization, (2) *frame amplification* occurs when movements appeal to deeply held values and beliefs in the general population and link those to movement issues so that people's preexisting value commitments serve as a "hook" that can be used to recruit them, (3) *frame extension* occurs when movements enlarge the boundaries of an initial frame to incorporate other issues that appear to be of importance to potential participants, and (4) *frame transformation* refers to the process whereby the creation and maintenance of new values, beliefs, and meanings induce movement participation by redefining activities and events in such a manner that people believe they must become involved in collective action (Buechler, 2000). Some or all of these frame alignment processes are used by social movements as they seek to define grievances and recruit participants.

Frame analysis provides new insights on how social movements emerge and grow when people are faced with problems such as technological disasters, about which greater ambiguity typically exists, and when people are attempting to "name" the problems associated with things such as nuclear or chemical contamination. However, frame analysis has been criticized for its "ideational biases" (McAdam, 1996). According to the sociologist Doug McAdam (1996), frame analyses of social movements have looked almost exclusively at ideas and their formal expression, whereas little attention has been paid to other significant factors such as movement tactics, mobilizing structures, and changing political

Box 16.2 Sociology in
Global Perspective

China: A Nation of Environmental Woes and Emergent Social Activism

News Bulletin:

Estimated number of premature deaths in China each year that are caused by pollution:

- Outdoor Air Pollution: 350,000 to 400,000 people
- Indoor Air Pollution: 300,000 people
- Water Pollution: 60,000 people
  —World Bank data (Kahn and Yardley, 2008: A1)

China is frequently in the international news these days because of its rapid industrial growth and swift rise as a global economic power. Accompanying this nation's double-digit growth rate, however, has been an unprecedented pollution problem. According to some social analysts, "China is choking on its own success" (Kahn and Yardley, 2008: A1). Although the economy is on an upward swing, much of this growth is related to a vast expansion of industry and rapid patterns of urbanization. For this kind of growth to be possible, staggering amounts of energy are needed, and China derives almost all of its energy from coal, one of the dirtiest sources of energy.

If this is China's problem, why should those of us who live in the United States be concerned? For humanitarian reasons, we must be concerned about the effects of deadly pollution on the residents of China. But we must

also be concerned about the effects of such environmental degradation because "What happens in China *does not stay in China.*" China's pollution problems are not just national problems; they are global problems. According to the *Journal of Geophysical Research*, "Sulfur dioxide and nitrogen oxides spewed by China's coal-fired power plants fall as acid rain on Seoul, South Korea, and Tokyo. Much of the particulate pollution over Los Angeles originates in China" (qtd. in Kahn and Yardley, 2008: A6). Yes, that is correct: Some of the pollution found in Los Angeles, California, may be attributed to what happens in China!

Can anything be done about the problem? Are activists and environmental movements trying to bring about environmental conservation in China? Some environmental activists are indeed attempting to highlight the causes and consequences of the various forms of pollution that are assaulting their nation. For example, environmental activist Wu Lihong repeatedly warned public officials in Wuxi, China, that pollution was strangling Lake Tai, but little attention was paid to his concerns until after the city was forced to shut off its drinking water because a deadly algae bloom was growing rapidly on the lake. Environmental researchers partly attributed this algae bloom to heavy pollution in the area. However, rather than praising Wu Lihong for his efforts to mobilize people and

opportunities that influence the signifying work of movements. In this context, "political opportunity" means government structure, public policy, and political conditions that set the boundaries for change and political action. These boundaries are crucial variables in explaining why various social movements have different outcomes (Meyer and Staggenborg, 1996; Gotham, 1999).

New Social Movement Theory

New social movement theory looks at a diverse array of collective actions and the manner in which those actions are based on politics, ideology, and culture. It also incorporates factors of identity, including race, class, gender, and sexuality, as sources of collective action and social movements. Examples of "new social movements" include ecofeminism and environmental justice movements.

Ecofeminism emerged in the late 1970s and early 1980s out of the feminist, peace, and ecology movements. Prompted by the near-meltdown at the Three Mile Island nuclear power plant, ecofeminists established World Women in Defense of the Environment. *Ecofeminism* is based on the belief that patriarchy is a root cause of environmental problems. According to ecofeminists, patriarchy not only results in the domination of women by men but also contributes to a belief that nature is to be possessed and dominated, rather than treated as a partner

Although some people believe that the Chinese government sends mixed messages about care for the environment, China has joined other countries in an effort to limit the use of plastic shopping bags. Other alternatives are shown here.

raise awareness of the problem, public officials had him arrested on blackmail and extortion charges, claiming that he demanded money from businesses by threatening that he would expose them for illegal pollution (Bodeen, 2007). Other social movement organizers in China have also found that they risk arrest and prosecution if they publicize their concerns and try to gather resources and mobilize others for environmental causes. As a result,

some organizers are hesitant to take action because they fear the consequences.

Indeed, it appears that "China is choking on its own success" and that social movements so far have made few, if any, inroads on addressing the problem. According to the environmental researcher Wang Jinnan, "It is a very awkward situation for the country because our greatest achievement is also our biggest burden. There is pressure for change, but many people refuse to accept that we need a new approach so soon" (qtd. in Kahn and Yardley, 2008: A1, A6).

What will the future hold for environmental protection in China? According to resource mobilization theory, widespread discontent alone cannot produce a social movement: Adequate resources and motivated people are essential for any concerned social action. Some analysts believe that environmental leaders in China will eventually be able to mobilize people for change because more-affluent Chinese residents are becoming very concerned about quality-of-life issues and because cell phones, the Internet, and other methods of rapid communications are making it possible for people to organize quickly and demand governmental intervention on pressing problems such as this one.

Reflect & Analyze

Do you believe that environmental movements in China might be organized like the most successful ones in the United States? Why or why not?

(see Ortner, 1974; Merchant, 1983, 1992; Mies and Shiva, 1993).

Another "new social movement" focuses on environmental justice and the intersection of race and class in the environmental struggle (see "Sociology Works!"). Sociologist Stella M. Capek (1993) investigated a contaminated landfill in the Carver Terrace neighborhood of Texarkana, Texas, and found that residents were able to mobilize for change and win a federal buyout and relocation by symbolically linking their issue to a larger *environmental justice* framework. Since the 1980s, the emerging environmental justice movement has focused on the issue of *environmental racism*—the belief that a disproportionate number of hazardous

facilities (including industries such as waste disposal/treatment and chemical plants) are placed in low-income areas populated primarily by people of color (Bullard and Wright, 1992). These areas have been left out of most of the environmental cleanup that has taken place in the last two decades (Schneider, 1993). Capek concludes that linking Carver Terrace with environmental justice led to it

environmental racism the belief that a disproportionate number of hazardous facilities (including industries such as waste disposal/treatment and chemical plants) are placed in low-income areas populated primarily by people of color.

© Andrew Lichtenstein/The Image Works

Referred to as "Cancer Alley," this area of Baton Rouge, Louisiana, is home to a predominantly African American population and also to many refineries that heavily pollute the region. Sociologists suggest that environmental racism is a significant problem in the United States and other nations. What do you think?

being designated as a cleanup site. She also views this as an important turning point in new social movements: "Carver Terrace is significant not only as a federal buyout and relocation of a minority community, but also as a marker of the emergence of environmental racism as a major new component of environmental social movements in the United States" (Capek, 1993: 21).

Sociologist Steven M. Buechler (2000) has argued that theories pertaining to twenty-first-century social movements should be oriented toward the structural, macrolevel contexts in which movements arise. These theories should incorporate both political and cultural dimensions of social activism:

> Social movements are historical products of the age of modernity. They arose as part of a sweeping social, political, and intellectual change that led a significant number of people to view society as a social construction that was susceptible to social reconstruction through concerted collective effort. Thus, from their inception, social movements have had a dual focus. Reflecting the political, they have always involved some form of challenge to prevailing forms of author-

ity. Reflecting the cultural, they have always operated as symbolic laboratories in which reflexive actors pose questions of meaning, purpose, identity, and change. (Buechler, 2000: 211)

This chapter's Concept Quick Review summarizes the main theories of social movements.

As we have seen, social movements may be an important source of social change. Throughout this text, we have examined a variety of social problems that have been the focus of one or more social movements during the past hundred years. In the process of bringing about change, most movements initially develop innovative ways to get their ideas across to decision makers and the public. Some have been successful in achieving their goals; others have not. As the historian Robert A. Goldberg (1991) has suggested, gains made by social movements may be fragile, acceptance brief, and benefits minimal and easily lost. For this reason, many groups focus on preserving their gains while simultaneously fighting for those that they believe they still desire.

Social Change in the Future

In this chapter, we have focused on collective behavior and social movements as potential forces for social change in contemporary societies. A number of other factors also contribute to social change, including the physical environment, population trends, technological development, and social institutions.

The Physical Environment and Change

Changes in the physical environment often produce changes in the lives of people; in turn, people can make dramatic changes in the physical environment,

Sociology *Works!*

Fine-Tuning Theories and Data Gathering on Environmental Racism

Throughout *Sociology in Our Times,* we have examined social theories that help us understand the interplay of factors such as race, class, and gender on the everyday lives of millions of people. In the "Sociology Works!" feature, we have focused on specific theories and how applications of those theories can help us understand the world and sometimes make it a better place in which to live.

In this chapter, we have looked at the work of new social movement theorists who have demonstrated the intersection of environmental justice with race and class: the belief that hazardous-waste treatment, storage, and disposal facilities are more likely to be located near low-income, nonwhite neighborhoods than to higher-income, predominantly white neighborhoods. This is an important issue because of the potential health risks that such sites may pose for people who live nearby. However, critics have scoffed at the suggestion that race- or class-based discrimination is involved in decisions about where hazardous-waste-materials facilities are located. Can more accurate data be gathered to help determine the nature and extent to which environmental racism exists?

During the 1980s and 1990s, the most frequently used method employed in national-level studies documenting the location of waste sites and other polluting industrial facilities was referred to as "unit-hazard coincidence" methodology. Based on this approach, researchers selected a predefined geographic unit (such as certain ZIP code areas or census tracts). Then they identified subsets of the units (areas located within a specific ZIP code or census tract) that had, or did not have, the hazard present. The researchers then compared the demographic characteristics of people living within each of the subsets to see if a larger minority population was present near the hazardous facility (see Mohai and Saha, 2007). Unit-hazard coincidence methodology assumes that the people who live within the predefined geographic units included in a study are located closer to the hazard than those individuals who do not live in those geographic units (Mohai and Saha, 2007). The problem with this approach is that the hazardous site is usually not located at the center of the ZIP code or census tract and that the geographic area being examined may be large or small, making it difficult to know for sure the racial and class characteristics of the people who live the closest to the waste facility.

In recent years, sociologists and other social scientists have begun to use other methods such as GIS (a computer system for capturing, storing, checking, integrating, manipulating, analyzing, and displaying data related to positions on the Earth's surface) to more adequately determine the distance between environmentally hazardous sites and nearby populations. By using distance-based methods to control for proximity around environmentally hazardous sites, those researchers have demonstrated that nonwhites, who made up about 25 percent of the nation's population in 1990, constituted over 40 percent of the population living within one mile of hazardous-waste facilities, meaning that racial disparities in the distribution of hazardous sites are much greater than what had been previously reported. According to social scientists Paul Mohai and Robin Saha (2007: 343), "We [find that] these disparities persist even when controlling for economic and sociopolitical variables, suggesting that factors uniquely associated with race, such as racial targeting, housing discrimination, or other race-related factors, are associated with the location of the nation's hazardous waste facilities."

Sociological theories and research pertaining to environmental racism have raised public awareness that the location of hazardous facilities is not purely coincidental in communities throughout our nation. Clearly, proximity to hazardous sites is related to the cost of the land on which the facilities are located, but the issue of proximity based on the racial/ethnic composition of residents raises an even more challenging social and ethical dilemma. But it is also clear that the vast quantity of data available today—and the methods for obtaining that data—make it possible for us to fine-tune previous theories and obtain a better understanding of the social world in which we live.

Reflect & Analyze

New technology can cause problems for society—for example, nuclear weapons—but it can also improve people's lives. Can you think of another way that current technology could be used to help correct a social problem in your community?

CONCEPT QUICK REVIEW

Social Movement Theories

	Key Components
Relative Deprivation	People who are discontent when they compare their achievements with those of others consider themselves relatively deprived and join social movements in order to get what they view as their "fair share," especially when there is an upswing in the economy followed by a decline.
Value-Added	Certain conditions are necessary for a social movement to develop: (1) structural conduciveness, such that people are aware of a problem and have the opportunity to engage in collective action; (2) structural strain, such that society or the community cannot meet people's expectations for taking care of the problem; (3) growth and spread of a generalized belief as to causes and effects of and possible solutions to the problem; (4) precipitating factors, or events that reinforce the beliefs; (5) mobilization of participants for action; and (6) social control factors, such that society comes to allow the movement to take action.
Resource Mobilization	A variety of resources (money, members, access to media, and material goods such as equipment) are necessary for a social movement; people participate only when they feel the movement has access to these resources.
Social Construction Theory: Frame Analysis	Based on the assumption that social movements are an interactive, symbolically defined, and negotiated process involving participants, opponents, and bystanders, frame analysis is used to determine how people assign meaning to activities and processes in social movements.
New Social Movement	The focus is on sources of social movements, including politics, ideology, and culture. Race, class, gender, sexuality, and other sources of identity are also factors in movements such as ecofeminism and environmental justice.

over which we have only limited control. Throughout history, natural disasters have taken their toll on individuals and societies. Major natural disasters—including hurricanes, floods, and tornados—can devastate an entire population. In September 2005, the United States experienced the worst natural disaster in its history when Hurricane Katrina left a wide path of death and destruction through Louisiana, Mississippi, and Alabama. However, damage from the hurricane itself was just the beginning of how the physical environment abruptly changed, how this disaster altered the lives of millions of people, and how it raised serious questions about our national priorities and the future of the environment. People who did not lose family members or suffer extensive property loss in this disaster may still have experienced trauma that will remain with them in the future. Even comparatively "small" natural disasters change the lives of many people. As the sociologist Kai Erikson (1976, 1994) has suggested, the trauma that people experience from disasters may outweigh the actual loss of physical property—memories of such events can haunt people for many years.

Some natural disasters are exacerbated by human decisions. For example, floods are viewed as natural disasters, but excessive development may contribute to a flood's severity. As office buildings, shopping malls, industrial plants, residential areas, and highways are developed, less land remains as groundcover to absorb rainfall. When heavier-than-usual rains occur, flooding becomes inevitable; some regions of the United States—such as in and around New Orleans—have remained under water for days or even weeks in recent years. Clearly, humans cannot control the rain, but human decisions can worsen the consequences.

The destruction of large sections of New Orleans by Hurricane Katrina and the flooding that followed is an example of how human decisions may worsen the consequences of a natural disaster. If Hurricane Katrina's first wave was the storm itself, the second wave was a *man-made disaster* resulting in part from human decisions relating to planning and budgetary priorities, allocation of funds for maintaining infrastructure, and the importance of emergency preparedness. For many years it was widely known that New Orleans had a great risk of taking a direct hit from a major hurricane and that, in a worse-case scenario, the city might be flooded, badly damaged, and portions of the city rendered uninhabitable. However, despite the city's unique topography (some sections are located below sea level) and its vulnerability to harsh storms, these concerns simply were not a top priority in urban planning or allocation of resources. Weak components of the infrastructure also contributed to the city's problems when water pumps and several levees on Lake Pontchartrain failed, allowing millions of gallons of polluted water to pour out onto the city's already-flooded streets, thereby forcing residents to leave their homes. *Infrastructure* refers to a framework of systems, such as transportation and utilities, that makes it possible to have specific land uses (commercial, residential, and recreational, for example) and a built environment (buildings, houses, and highways) that support people's daily activities and the nation's economy. It takes money and commitment to make sure that the components of the infrastructure remain strong so that cities can withstand natural disasters and other catastrophes.

Hurricane Katrina was a massive lesson in sociology, bringing to the public's attention so many of the issues discussed in this text. By way of example, many of the people most affected by Hurricane Katrina were low-income African Americans whose residences were located in the low-lying areas of the city; the homes of many of the wealthier white residents of the city were located on higher ground and were spared the brunt of the flooding. Apparently, no adequate disaster evacuation plans had been developed for individuals without vehicles or sufficient money to leave on their own.

Both the evacuation process before the hurricane and the relief efforts after the hurricane and flooding have been widely criticized because of the length of time it took for political leaders, the military, and officials in other governmental agencies to mobilize and actively begin to rescue victims, care for the ill and dying, and bring order to the city—again, a lesson in sociology. In essence, this natural disaster

Photo by Jocelyn Augustino/FEMA Photo Library

© AP Images/Bill Haber

Natural disasters such as Hurricane Katrina not only can produce flooding and devastation in cities such as New Orleans and in less populated areas, but they also may make us acutely aware of vast racial and economic inequalities that persist in the United States and other nations. A significant part of the recovery work in New Orleans was done by volunteers.

brought profound changes to the lives of many individuals, and it also revealed how much remains to be done if we are to overcome deep divisions of race and class that affect everyone's identity, life chances, and opportunities. Finally, this dramatic change in the physical environment (the damage and destruction caused directly or indirectly by the hurricane) of the region revealed to our nation that it potentially lacks preparedness for massive disasters such as terrorist attacks, hurricanes, and floods.

Because flooding is one of the many problems that face the world today, it may seem strange that one of the major concerns in the twenty-first century is the availability of water. Many experts warn that water is a finite resource that is necessary for both human survival and the production of goods. However, water is being wasted and polluted, and the supply of *potable* (drinkable) water is limited. People contribute to changes in the Earth's physical condition. Through soil erosion and other degradation of grazing land, often at the hands of people, an estimated 24 billion tons of topsoil is lost annually. As people clear forests to create farmland and pastures and to acquire lumber and firewood, the Earth's tree cover continues to diminish. As millions of people drive motor vehicles, the amount of carbon dioxide in the environment continues to rise each year, contributing to global warming.

Just as people contribute to changes in the physical environment, human activities must also be adapted to changes in the environment. For example, we are being warned to stay out of the sunlight because of increases in ultraviolet rays, a cause of skin cancer, as a result of the accelerating depletion of the ozone layer. If the ozone warnings are accurate, the change in the physical environment will dramatically affect those who work or spend their leisure time outside.

Population and Change

Changes in population size, distribution, and composition affect the culture and social structure of a society and change the relationships among nations. As discussed in Chapter 15, the countries experiencing the most rapid increases in population have a less-developed infrastructure to deal with those changes. How will nations of the world deal with population growth as the global population contin-

ues to move toward seven billion? Only time will provide a response to this question.

In the United States, a shift in population distribution from central cities to suburban and exurban areas has produced other dramatic changes. Central cities have experienced a shrinking tax base as middle-income and upper-middle-income residents and businesses have moved to suburban and outlying areas. As a result, schools and public services have declined in many areas, leaving those people with the greatest needs with the fewest public resources and essential services. The changing composition of the U.S. population has resulted in children from more diverse cultural backgrounds entering school, producing a demand for new programs and changes in curricula. An increase in the birthrate has created a need for more child care; an increase in the older population has created a need for services such as medical care, placing greater demands on programs such as Social Security.

As we have seen in previous chapters, population growth and the movement of people to urban areas have brought profound changes to many regions and intensified existing social problems. Among other factors, growth in the global population is one of the most significant driving forces behind environmental concerns such as the availability and use of natural resources.

As the very large Baby Boom generation enters retirement age, the health care needs of people in this age bracket will strain the capabilities of federal programs such as Medicare.

Technology and Change

Technology is an important force for change; in some ways, technological development has made our lives much easier. Advances in communication and transportation have made instantaneous worldwide communication possible but have also brought old belief systems and the status quo into question as never before. Today, we are increasingly moving information instead of people—and doing it almost instantly. Advances in science and medicine have made significant changes in people's lives in high-income countries.

Scientific advances will continue to affect our lives, from the foods we eat to our reproductive capabilities. Genetically engineered plants have been developed and marketed in recent years, and biochemists are creating potatoes, rice, and cassava with the same protein value as meat (Petersen, 1994). Advances in medicine have made it possible for those formerly unable to have children to procreate; women well beyond menopause are now able to become pregnant with the assistance of medical technology. Advances in medicine have also increased the human lifespan, especially for white and middle- or upper-class individuals in high-income nations; medical advances have also contributed to the declining death rate in low-income nations, where birth rates have not yet been curbed.

Just as technology has brought about improvements in the quality and length of life for many, it has also created the potential for new disasters, ranging from global warfare to localized technological disasters at toxic waste sites. As the sociologist William Ogburn (1966) suggested, when a change in the material culture occurs in society, a period of *cultural lag* follows in which the nonmaterial (ideological) culture has not yet caught up with material development. The rate of technological advance at the level of material culture today is mind-boggling. Many of us can never hope to understand technological advances in the areas of artificial intelligence, holography, virtual reality, biotechnology, cold fusion, and robotics.

One of the ironies of twenty-first-century high technology is the increased vulnerability that results from the increasing complexity of such systems. As futurist John L. Petersen (1994: 70) notes, "The more complex a system becomes, the more likely the

© AP Images/Steve Yeater

Would you like to eat a sandwich containing meat from a cloned animal? California State Senator Carole Migden has advocated the passage of legislation that would require clear labeling of such content. According to Migden, cloning has not yet been perfected and not enough data exist to indicate whether cloned meat could have harmful effects on humans who consume it.

chance of system failure. There are unknown secondary effects and particularly vulnerable nodes." He also asserts that most of the world's population will not participate in the technological revolution that is occurring in high-income nations.

Social Institutions and Change

Many changes occurred in the family, religion, education, the economy, and the political system during the twentieth century and early in the twenty-first century. As we saw in Chapter 11, the size and composition of families in the United States changed with the dramatic increase in the number of single-person and single-parent households. Changes in families produced changes in the socialization of children, many of whom spend large amounts of time in front of a television set or in child-care facilities outside their own homes. Although some political and religious leaders advocate a return to "traditional" family life, numerous scholars argue that such families never worked quite as well as some might wish to believe.

Public education changed dramatically in the United States during the twentieth century. This country was one of the first to provide "universal" education for students regardless of their ability to pay. As a result, at least until recently, the United States has had one of the most highly educated populations in the world. Today, the United States still has one of the best public education systems in the world for the top 15 percent of the students, but it badly fails the bottom 25 percent. As the nature of the economy changes, schools almost inevitably will have to change, if for no other reason than demands from leaders in business and industry for an educated work force that allows U.S. companies to compete in a global economic environment.

A new concept of world security is emerging, requiring the cooperation of high-income countries in halting the proliferation of weapons of mass destruction and combating terrorism. That concept also requires the cooperation of *all* nations in reducing the plight of the poorest people in the low-income countries of the world.

Although we have examined changes in the physical environment, population, technology, and social institutions separately, they all operate together in a complex relationship, sometimes producing large, unanticipated consequences. As we move further into the twenty-first century, we need new ways of conceptualizing social life at both the macrolevel and the microlevel. The sociological imagination helps us think about how personal troubles—regardless of our race, class, gender, age, sexual orientation, or physical abilities and disabilities—are intertwined with the public issues of our society and the global community of which we are a part. As one analyst noted regarding Lois Gibbs and Love Canal,

> If Love Canal has taught Lois Gibbs—and the rest of us—anything, it is that ordinary people become very smart very quickly when their lives are threatened. They become adept at detecting absurdity, even when it is concealed in bureaucratese and scientific jargon. Lois Gibbs learned that one cannot always rely on government to act in the best interests of ordinary citizens—at least, not without considerable prodding. She

© Jeff Greenberg/PhotoEdit

Pollution of lakes, rivers, and other bodies of water has an adverse effect on food supplies, air quality, and the entire environment. What influence does a "business as usual" approach have on environmental quality in your area?

determined that she would prod them until her objectives were attained. She led one of the most successful, single-purpose grass roots efforts of our time. (M. Levine, 1982: xv)

Taking care of the environment is an example of something that government and each of us as individuals can do to help (see Box 16.3). And it is vitally important that we all do everything that we can in order to protect the environment.

A Few Final Thoughts

In this text, we have covered a substantial amount of material, examined different perspectives on a wide variety of social issues, and suggested different methods by which to deal with them. The purpose of this text is not to encourage you to take any particular point of view; rather, it is to allow you to understand different viewpoints and ways in which they may be helpful to you and to society in dealing with the issues of the twenty-first century. Possessing that understanding, we can hope that the future will be something we can all look forward to—producing a better way of life, not only in this country but worldwide as well.

Box 16.3 You Can Make a Difference

It's Now or Never: The Imperative of Taking Action Against Global Warming

It's pretty gut-wrenching. People say climate change is something for your kids to worry about. No. It's now!
—Canadian forestry scientist Allan Carroll emphasizing the importance of acting now to reduce global warming (qtd. in Struck, 2006)

A recurring theme of environmental social movements is that problems associated with the environment are *everyone's problem* and that change must take place before we run out of time. Clearly, we need massive initiatives implemented by elected officials and corporate leaders to reduce carbon emissions, reduce our dependency on fossil fuels, and "get serious" about dealing with the global warming problem. However, at the individual level, we each can take some steps to help reduce the amount of *greenhouse gases*—any gas that absorbs infrared radiation, such as carbon dioxide, methane, nitrous oxide, ozone, and hydro-fluorocarbons—that we put into the atmosphere. Here are a few suggestions from experts:

- Walk, take mass transit, or ride a bike rather than driving alone in your car.
- Promote community carpooling and the creation of bike lanes.
- Buy a car that is fuel efficient (for every gallon of gas that is burned, twenty pounds of carbon dioxide go into the atmosphere).
- Purchase appliances, computers, televisions, and sound systems that are energy efficient as shown by the ENERGY STAR®awarded by the Environmental Protection Agency.
- Have an energy audit performed by the your local electric or gas utility, and make changes in your apartment or house.
- Turn off lights, televisions, and computers when you are not using them.
- Recycle cans, bottles, plastic bags, newspapers, and other items (recycling helps diminish waste disposal problems by reducing the amount of waste hauled off to landfills).
- Become actively involved in social movements and engage in political activism to put pressure on government officials to encourage industry to protect the health of our environment by reducing carbon emissions (based on Union of Concerned Scientists, 2006, and EPA, 2006).

A famous statement by the author Mark Twain is often quoted: "Everybody talks about the weather, but nobody ever does anything about it." In the case of global warming, if nobody does anything to stop it, we may face dire consequences now, and future generations truly may be imperiled. For additional information on global warming and how you can get involved to bring about social change, visit these websites:

http://www.stopglobalwarming.org

http://www.greenhousenet.org

Chapter Review

● **What is the relationship between social change and collective behavior?**

Social change—the alteration, modification, or transformation of public policy, culture, or social institutions over time—is usually brought about by collective behavior, which is defined as relatively spontaneous, unstructured activity that typically violates established social norms.

● **When is collective behavior likely to occur?**

Collective behavior occurs when some common influence or stimulus produces a response from a relatively large number of people.

● What is a crowd?

A crowd is a relatively large number of people in one another's immediate presence. Sociologist Herbert Blumer divided crowds into four categories: (1) casual crowds, (2) conventional crowds, (3) expressive crowds, and (4) acting crowds (including mobs, riots, and panics). A fifth type of crowd is a protest crowd.

● What causes crowd behavior?

Social scientists have developed several theories to explain crowd behavior. Contagion theory asserts that a crowd takes on a life of its own as people are transformed from rational beings into part of an organism that acts on its own. A variation on this is social unrest and circular reaction—people express their discontent to others, who communicate back similar feelings, resulting in a conscious effort to engage in the crowd's behavior. Convergence theory asserts that people with similar attributes find other like-minded persons with whom they can release underlying personal tendencies. Emergent norm theory asserts that as a crowd develops, it comes up with its own norms that replace more conventional norms of behavior.

● What are the primary forms of mass behavior?

Mass behavior is collective behavior that occurs when people respond to the same event in the same way even if they are not in geographic proximity to one another. Rumors, gossip, mass hysteria, fads and fashions, and public opinion are forms of mass behavior.

● What are the major types of social movements, and what are their goals?

A social movement is an organized group that acts consciously to promote or resist change through collective action. Reform, revolutionary, religious, and alternative movements are the major types identified by sociologists. Reform movements seek to improve society by changing some specific aspect of the social structure. Revolutionary movements seek to bring about a total change in society—sometimes by the use of terrorism. Religious movements seek to produce radical change in individuals based on spiritual or supernatural belief systems. Alternative movements seek limited change of some aspect of people's behavior. Resistance movements seek to prevent change or to undo change that has already occurred.

● How do social movements develop?

Social movements typically go through three stages: (1) a preliminary stage (unrest results from a perceived problem), (2) coalescence (people begin to organize), and (3) institutionalization (an organization is developed, and paid staff replaces volunteers in leadership positions).

● How do relative deprivation theory, value-added theory, and resource mobilization theory explain social movements?

Relative deprivation theory asserts that if people are discontented when they compare their accomplishments with those of others similarly situated, they are more likely to join a social movement than are people who are relatively content with their status. According to value-added theory, six conditions are required for a social movement: (1) a perceived problem, (2) a perception that the authorities are not resolving the problem, (3) a spread of the belief to an adequate number of people, (4) a precipitating incident, (5) mobilization of other people by leaders, and (6) a lack of social control. By contrast, resource mobilization theory asserts that successful social movements can occur only when they gain the support of political and economic elites, who provide access to the resources necessary to maintain the movement.

● What is the primary focus of research based on frame analysis and new social movement theory?

Research based on frame analysis often highlights the social construction of grievances through the process of social interaction. Various types of framing occur as problems are identified, remedies are sought, and people feel compelled to take action. Like frame analysis, new social movement theory has been used in research that looks at technological disasters and cases of environmental racism.

Register for a Student eResource account to maximize your study time online using CengageNOW. First take the system's diagnostic pre-test, and then follow the personalized study plan that is created for you to help you review this chapter. The study plan will

- help you identify areas on which you should concentrate;
- provide interactive exercises to help you master the chapter concepts; and
- provide a post-test to confirm you are ready to move on to the next chapter.

Key Terms

civil disobedience 545
collective behavior 540
crowd 543
environmental racism 559
gossip 549

mass 543
mass behavior 547
mob 544
panic 545
propaganda 551

public opinion 550
riot 544
rumors 548
social change 540
social movement 551

Questions for Critical Thinking

1. What types of collective behavior in the United States do you believe are influenced by inequalities based on race/ethnicity, class, gender, age, or disabilities? Why?
2. Which of the four explanations of crowd behavior (contagion theory, social unrest and circular reaction, convergence theory, and emergent norm theory) do you believe best explains crowd behavior? Why?
3. In the text, the Love Canal environmental movement is analyzed in terms of the value-added theory. How would you analyze that movement under (a) the relative deprivation theory and (b) the resource mobilization theory?
4. Using the sociological imagination that you have gained in this course, what are some positive steps that you believe might be taken in the United States to make our society a better place for everyone? What types of collective behavior and/or social movements might be required in order to take those steps?

The Kendall Companion Website

Supplement your review of this chapter by going to the text's companion website, where you can take tutorial quizzes, use flash cards to master key terms, follow live links to useful websites, and explore the other study and research resources you'll find there, such as a comprehensive interactive sociology timeline, GSS Data, and Census 2000 information, much of it presented visually in maps.

Glossary

absolute poverty a level of economic deprivation that exists when people do not have the means to secure the most basic necessities of life.

achieved status a social position that a person assumes voluntarily as a result of personal choice, merit, or direct effort.

acute diseases illnesses that strike suddenly and cause dramatic incapacitation and sometimes death.

ageism prejudice and discrimination against people on the basis of age, particularly against older people.

agents of socialization the persons, groups, or institutions that teach us what we need to know in order to participate in society.

aggregate a collection of people who happen to be in the same place at the same time but share little else in common.

alienation a feeling of powerlessness and estrangement from other people and from oneself.

animism the belief that plants, animals, or other elements of the natural world are endowed with spirits or life forces having an effect on events in society.

anomie Emile Durkheim's designation for a condition in which social control becomes ineffective as a result of the loss of shared values and of a sense of purpose in society.

anticipatory socialization the process by which knowledge and skills are learned for future roles.

ascribed status a social position conferred at birth or received involuntarily later in life based on attributes over which the individual has little or no control, such as race/ethnicity, age, and gender.

assimilation a process by which members of subordinate racial and ethnic groups become absorbed into the dominant culture.

authoritarian leaders people who make all major group decisions and assign tasks to members.

authoritarian personality a personality type characterized by excessive conformity, submissiveness to authority, intolerance, insecurity, a high level of superstition, and rigid, stereotypic thinking.

authoritarianism a political system controlled by rulers who deny popular participation in government.

authority power that people accept as legitimate rather than coercive.

beliefs the mental acceptance or conviction that certain things are true or real.

bilateral descent a system of tracing descent through both the mother's and father's sides of the family.

blended family a family consisting of a husband and wife, children from previous marriages, and children (if any) from the new marriage.

body consciousness a term that describes how a person perceives and feels about his or her body.

bureaucracy an organizational model characterized by a hierarchy of authority, a clear division of labor, explicit rules and procedures, and impersonality in personnel matters.

bureaucratic personality a psychological construct that describes those workers who are more concerned with following correct procedures than they are with getting the job done correctly.

capitalism an economic system characterized by private ownership of the means of production, from which personal profits can be derived through market competition and without government intervention.

capitalist class (or **bourgeoisie**) Karl Marx's term for the class that consists of those who own and control the means of production.

caste system a system of social inequality in which people's status is permanently determined at birth based on their parents' ascribed characteristics.

category a number of people who may never have met one another but share a similar characteristic, such as education level, age, race, or gender.

charismatic authority power legitimized on the basis of a leader's exceptional personal qualities or the demonstration of extraordinary insight and accomplishment that inspire loyalty and obedience from followers.

chronic diseases illnesses that are long term or lifelong and that develop gradually or are present from birth.

church a large, bureaucratically organized religious organization that tends to seek accommodation with the larger society in order to maintain some degree of control over it.

civil disobedience nonviolent action that seeks to change a policy or law by refusing to comply with it.

civil religion the set of beliefs, rituals, and symbols that makes sacred the values of the society and places the nation in the context of the ultimate system of meaning.

class conflict Karl Marx's term for the struggle between the capitalist class and the working class.

class system a type of stratification based on the ownership and control of resources and on the type of work that people do.

cohabitation a situation in which two people live together, and think of themselves as a couple, without being legally married.

collective behavior voluntary, often spontaneous activity that is engaged in by a large number of people and typically violates dominant-group norms and values.

community a set of social relationships operating within given spatial boundaries or locations that provides people with a sense of identity and a feeling of belonging.

comparable worth (or **pay equity**) the belief that wages ought to reflect the worth of a job, not the gender or race of the worker.

conflict perspectives the sociological approach that views groups in society as engaged in a continuous power struggle for control of scarce resources.

conformity the process of maintaining or changing behavior to comply with the norms established by a society, subculture, or other group.

conglomerate a combination of businesses in different commercial areas, all of which are owned by one holding company.

content analysis the systematic examination of cultural artifacts or various forms of communication to extract thematic data and draw conclusions about social life.

contingent work part-time work, temporary work, or subcontracted work that offers advantages to employers but that can be detrimental to the welfare of workers.

control group in an experiment, the group that contains the subjects who are not exposed to the independent variable.

core nations according to world systems theory, dominant capitalist centers characterized by high levels of industrialization and urbanization.

corporate crime illegal acts committed by corporate employees on behalf of the corporation and with its support.

corporations large-scale organizations that have legal powers, such as the ability to enter into contracts and buy and sell property, separate from their individual owners.

correlation a relationship that exists when two variables are associated more frequently than could be expected by chance.

counterculture a group that strongly rejects dominant societal values and norms and seeks alternative lifestyles.

credentialism a process of social selection in which class advantage and social status are linked to the possession of academic qualifications.

crime behavior that violates criminal law and is punishable with fines, jail terms, and other sanctions.

criminal justice system the more than 55,000 local, state, and federal agencies that enforce laws, adjudicate crimes, and treat and rehabilitate criminals.

criminology the systematic study of crime and the criminal justice system, including the police, courts, and prisons.

crowd a relatively large number of people who are in one another's immediate vicinity.

crude birth rate the number of live births per 1,000 people in a population in a given year.

crude death rate the number of deaths per 1,000 people in a population in a given year.

cult a religious group with practices and teachings outside the dominant cultural and religious traditions of a society.

cultural capital Pierre Bourdieu's term for people's social assets, including values, beliefs, attitudes, and competencies in language and culture.

cultural imperialism the extensive infusion of one nation's culture into other nations.

cultural lag William Ogburn's term for a gap between the technical development of a society (material culture) and its moral and legal institutions (nonmaterial culture).

cultural relativism the belief that the behaviors and customs of any culture must be viewed and analyzed by the culture's own standards.

cultural universals customs and practices that occur across all societies.

culture the knowledge, language, values, customs, and material objects that are passed from person to person and from one generation to the next in a human group or society.

culture shock the disorientation that people feel when they encounter cultures radically different from their own and believe they cannot depend on their own taken-for-granted assumptions about life.

demedicalization the process whereby a problem ceases to be defined as an illness or a disorder.

democracy a political system in which the people hold the ruling power either directly or through elected representatives.

democratic leaders leaders who encourage group discussion and decision making through consensus building.

democratic socialism an economic and political system that combines private ownership of some of the means of production, governmental distribution of some essential goods and services, and free elections.

demographic transition the process by which some societies have moved from high birth rates and death rates to relatively low birth rates and death rates as a result of technological development.

demography a subfield of sociology that examines population size, composition, and distribution.

denomination a large, organized religion characterized by accommodation to society but frequently lacking in ability or intention to dominate society.

dependency theory the belief that global poverty can at least partially be attributed to the fact that the low-income countries have been exploited by the high-income countries.

dependent variable in an experiment, the variable assumed to be caused by the independent variable(s).

deviance any behavior, belief, or condition that violates significant social norms in the society or group in which it occurs.

differential association theory the proposition that individuals have a greater tendency to deviate from societal norms when they frequently associate with persons who are more favorable toward deviance than conformity.

disability a physical or health condition that stigmatizes or causes discrimination.

discrimination actions or practices of dominant-group members (or their representatives) that have a harmful effect on members of a subordinate group.

division of labor how the various tasks of a society are divided up and performed.

domestic partnerships household partnerships in which an unmarried couple lives together in a committed, sexually intimate relationship and is granted the same rights and benefits as those accorded to married heterosexual couples.

dominant group a group that is advantaged and has superior resources and rights in a society.

dramaturgical analysis the study of social interaction that compares everyday life to a theatrical presentation.

drug any substance—other than food and water—that, when taken into the body, alters its functioning in some way.

dual-earner marriages marriages in which both spouses are in the labor force.

dyad a group composed of two members.

ecclesia a religious organization that is so integrated into the dominant culture that it claims as its membership all members of a society.

economy the social institution that ensures the maintenance of society through the production, distribution, and consumption of goods and services.

education the social institution responsible for the systematic transmission of knowledge, skills, and cultural values within a formally organized structure.

egalitarian family a family structure in which both partners share power and authority equally.

ego according to Sigmund Freud, the rational, reality-oriented component of personality that imposes restrictions on the innate pleasure-seeking drives of the id.

elite model a view of society that sees power in political systems as being concentrated in the hands of a small group of elites whereas the masses are relatively powerless.

endogamy cultural norms prescribing that people marry within their social group or category.

environmental racism the belief that a disproportionate number of hazardous facilities (including industries such as waste disposal/treatment and chemical plants) are placed in low-income areas populated primarily by people of color.

ethnic group a collection of people distinguished, by others or by themselves, primarily on the basis of cultural or nationality characteristics.

ethnic pluralism the coexistence of a variety of distinct racial and ethnic groups within one society.

ethnocentrism the practice of judging all other cultures by one's own culture.

ethnography a detailed study of the life and activities of a group of people by researchers who may live with that group over a period of years.

ethnomethodology the study of the commonsense knowledge that people use to understand the situations in which they find themselves.

exogamy cultural norms prescribing that people marry outside their social group or category.

experiment a research method involving a carefully designed situation in which the researcher studies the impact of certain variables on subjects' attitudes or behavior.

experimental group in an experiment, the group that contains the subjects who are exposed to an independent variable (the experimental condition) to study its effect on them.

expressive leadership an approach to leadership that provides emotional support for members.

extended family a family unit composed of relatives in addition to parents and children who live in the same household.

face-saving behavior Erving Goffman's term for the strategies we use to rescue our performance when we experience a potential or actual loss of face.

families relationships in which people live together with commitment, form an economic unit and care for any young, and consider their identity to be significantly attached to the group.

family of orientation the family into which a person is born and in which early socialization usually takes place.

family of procreation the family that a person forms by having or adopting children.

fecundity the potential number of children who could be born if every woman reproduced at her maximum biological capacity.

feminism the belief that all people—both women and men—are equal and that they should be valued equally and have equal rights.

feminization of poverty the trend in which women are disproportionately represented among individuals living in poverty.

fertility the actual level of childbearing for an individual or a population.

folkways informal norms or everyday customs that may be violated without serious consequences within a particular culture.

formal organization a highly structured group formed for the purpose of completing certain tasks or achieving specific goals.

functionalist perspectives the sociological approach that views society as a stable, orderly system.

Gemeinschaft (guh-MINE-shoft) a traditional society in which social relationships are based on personal bonds of friendship and kinship and on intergenerational stability.

gender the culturally and socially constructed differences between females and males found in the meanings, beliefs, and practices associated with "femininity" and "masculinity."

gender bias behavior that shows favoritism toward one gender over the other.

gender identity a person's perception of the self as female or male.

gender role the attitudes, behavior, and activities that are socially defined as appropriate for each sex and are learned through the socialization process.

gender socialization the aspect of socialization that contains specific messages and practices concerning the nature of being female or male in a specific group or society.

gendered racism the interactive effect of racism and sexism on the exploitation of women of color.

generalized other George Herbert Mead's term for the child's awareness of the demands and expectations of the society as a whole or of the child's subculture.

genocide the deliberate, systematic killing of an entire people or nation.

gentrification the process by which members of the middle and upper-middle classes, especially whites, move into a central-city area and renovate existing properties.

Gesellschaft (guh-ZELL-shoft) a large, urban society in which social bonds are based on impersonal and specialized relationships, with little long-term commitment to the group or consensus on values.

global stratification the unequal distribution of wealth, power, and prestige on a global basis, resulting in people having vastly different lifestyles and life chances both within and among the nations of the world.

goal displacement a process that occurs in organizations when the rules become an end in themselves rather than a means to an end, and organizational survival becomes more important than achievement of goals.

gossip rumors about the personal lives of individuals.

government the formal organization that has the legal and political authority to regulate the relationships among members of a society and between the society and those outside its borders.

groupthink the process by which members of a cohesive group arrive at a decision that many individual members privately believe is unwise.

health a state of complete physical, mental, and social well-being.

health care any activity intended to improve health.

health maintenance organizations (HMOs) companies that provide, for a set monthly fee, total care with an emphasis on prevention to avoid costly treatment later.

hermaphrodite a person in whom sexual differentiation is ambiguous or incomplete.

hidden curriculum the transmission of cultural values and attitudes, such as conformity and obedience to authority, through implied demands

found in rules, routines, and regulations of schools.

high culture classical music, opera, ballet, live theater, and other activities usually patronized by elite audiences.

high-income countries (sometimes referred to as **industrial countries**) nations with highly industrialized economies; technologically advanced industrial, administrative, and service occupations; and relatively high levels of national and personal income.

holistic medicine an approach to health care that focuses on prevention of illness and disease and is aimed at treating the whole person—body and mind—rather than just the part or parts in which symptoms occur.

homogamy the pattern of individuals marrying those who have similar characteristics, such as race/ethnicity, religious background, age, education, or social class.

homophobia extreme prejudice directed at gays, lesbians, bisexuals, and others who are perceived as not being heterosexual.

human ecology the study of the relationship between people and their physical environment.

hypothesis a statement of the expected relationship between two or more variables.

id Sigmund Freud's term for the component of personality that includes all of the individual's basic biological drives and needs that demand immediate gratification.

ideal type an abstract model that describes the recurring characteristics of some phenomenon (such as bureaucracy).

illegitimate opportunity structures circumstances that provide an opportunity for people to acquire through illegitimate activities what they cannot achieve through legitimate channels.

impression management (presentation of self) Erving Goffman's term for people's efforts to present themselves to others in ways that are most favorable to their own interests or image.

income the economic gain derived from wages, salaries, income transfers (governmental aid), and ownership of property.

independent variable in an experiment, the variable assumed to be the cause of the relationship between variables.

individual discrimination behavior consisting of one-on-one acts by members of the dominant group that harm members of the subordinate group or their property.

industrial society a society based on technology that mechanizes production.

industrialization the process by which societies are transformed from dependence on agriculture and handmade products to an emphasis on manufacturing and related industries.

infant mortality rate the number of deaths of infants under 1 year of age per 1,000 live births in a given year.

informal side of a bureaucracy those aspects of participants' day-to-day activities and interactions that ignore, bypass, or do not correspond with the official rules and procedures of the bureaucracy.

ingroup a group to which a person belongs and with which the person feels a sense of identity.

institutional discrimination the day-to-day practices of organizations and institutions that have a harmful impact on members of subordinate groups.

instrumental leadership goal- or task-oriented leadership.

intergenerational mobility the social movement (upward or downward) experienced by family members from one generation to the next.

interlocking corporate directorates members of the board of directors of one corporation who also sit on the board(s) of other corporations.

internal colonialism according to conflict theorists, a practice that occurs when members of a racial or ethnic group are conquered or colonized and forcibly placed under the economic and political control of the dominant group.

interview a research method using a data-collection encounter in which an interviewer asks the respondent questions and records the answers.

intragenerational mobility the social movement (upward or downward) of individuals within their own lifetime.

invasion the process by which a new category of people or type of land use arrives in an area previously occupied by another group or land use.

iron law of oligarchy according to Robert Michels, the tendency of bureaucracies to be ruled by a few people.

job deskilling a reduction in the proficiency needed to perform a specific job that leads to a corresponding reduction in the wages for that job.

juvenile delinquency a violation of law or the commission of a status offense by young people.

kinship a social network of people based on common ancestry, marriage, or adoption.

labeling theory the proposition that deviants are those people who have been successfully labeled as such by others.

laissez-faire leaders leaders who are only minimally involved in decision making and who encourage group members to make their own decisions.

language a set of symbols that expresses ideas and enables people to think and communicate with one another.

latent functions unintended functions that are hidden and remain unacknowledged by participants.

laws formal, standardized norms that have been enacted by legislatures and are enforced by formal sanctions.

life chances Max Weber's term for the extent to which individuals have access to important societal resources such as food, clothing, shelter, education, and health care.

life expectancy an estimate of the average lifetime of people born in a specific year.

looking-glass self Charles Horton Cooley's term for the way in which a person's sense of self is derived from the perceptions of others.

low-income countries (sometimes referred to as **underdeveloped countries**) nations with little industrialization and low levels of national and personal income.

macrolevel analysis an approach that examines whole societies, large-scale social structures, and social systems.

managed care any system of cost containment that closely monitors and controls health care providers' decisions about medical procedures, diagnostic tests, and other services that should be provided to patients.

manifest functions functions that are intended and/or overtly recognized by the participants in a social unit.

marginal jobs jobs that differ from the employment norms of the society in which they are located.

marriage a legally recognized and/or socially approved arrangement between two or more individuals that carries certain rights and obligations and usually involves sexual activity.

mass a number of people who share an interest in a specific idea or issue but who are not in one another's immediate vicinity.

mass behavior collective behavior that takes place when people (who often are geographically separated from one another) respond to the same event in much the same way.

mass media large-scale organizations that use print or electronic means (such as radio, television, film, and the Internet) to communicate with large numbers of people.

master status the most important status that a person occupies.

material culture a component of culture that consists of the physical or tangible creations (such as clothing, shelter, and art) that members of a society make, use, and share.

matriarchal family a family structure in which authority is held by the eldest female (usually the mother).

matriarchy a hierarchical system of social organization in which cultural, political, and economic structures are controlled by women.

matrilineal descent a system of tracing descent through the mother's side of the family.

matrilocal residence the custom of a married couple living in the same household (or community) as the wife's parents.

mechanical solidarity Emile Durkheim's term for the social cohesion in preindustrial societies, in which there is minimal division of labor and people feel united by shared values and common social bonds.

medical–industrial complex local physicians, local hospitals, and global health-related industries such as insurance companies and pharmaceutical and medical supply companies that deliver health care today.

medicalization the process whereby nonmedical problems become defined and treated as illnesses or disorders.

medicine an institutionalized system for the scientific diagnosis, treatment, and prevention of illness.

meritocracy a hierarchy in which all positions are rewarded based on people's ability and credentials.

metropolis one or more central cities and their surrounding suburbs that dominate the economic and cultural life of a region.

microlevel analysis sociological theory and research that focus on small groups rather than on large-scale social structures.

middle-income countries (sometimes referred to as **developing countries**) nations with industrializing economies, particularly in urban areas, and moderate levels of national and personal income.

migration the movement of people from one geographic area to another for the purpose of changing residency.

military–industrial complex the mutual interdependence of the military establishment and private military contractors.

mixed economy an economic system that combines elements of a market economy (capitalism) with elements of a command economy (socialism).

mob a highly emotional crowd whose members engage in, or are ready to engage in, violence against a specific target—a person, a category of people, or physical property.

modernization theory a perspective that links global inequality to different levels of economic development and suggests that low-income economies can move to middle- and

high-income economies by achieving self-sustained economic growth.

monarchy a political system in which power resides in one person or family and is passed from generation to generation through lines of inheritance.

monogamy a marriage between two partners, usually a woman and a man.

mores strongly held norms with moral and ethical connotations that may not be violated without serious consequences in a particular culture.

mortality the incidence of death in a population.

neolocal residence the custom of a married couple living in their own residence apart from both the husband's and the wife's parents.

nonmaterial culture a component of culture that consists of the abstract or intangible human creations of society (such as attitudes, beliefs, and values) that influence people's behavior.

nonverbal communication the transfer of information between persons without the use of speech.

norms established rules of behavior or standards of conduct.

nuclear family a family composed of one or two parents and their dependent children, all of whom live apart from other relatives.

occupational (white-collar) crime illegal activities committed by people in the course of their employment or financial affairs.

occupations categories of jobs that involve similar activities at different work sites.

official poverty line the federal income standard that is based on what is considered to be the minimum amount of money required for living at a subsistence level.

oligopoly a condition existing when several companies overwhelmingly control an entire industry.

organic solidarity Emile Durkheim's term for the social cohesion found in industrial societies, in which people perform very specialized tasks and feel united by their mutual dependence.

organized crime a business operation that supplies illegal goods and services for profit.

outgroup a group to which a person does not belong and toward which the person may feel a sense of competitiveness or hostility.

panic a form of crowd behavior that occurs when a large number of people react to a real or perceived threat with strong emotions and self-destructive behavior.

participant observation a research method in which researchers collect data while being part of the activities of the group being studied.

patriarchal family a family structure in which authority is held by the eldest male (usually the father).

patriarchy a hierarchical system of social organization in which cultural, political, and economic structures are controlled by men.

patrilineal descent a system of tracing descent through the father's side of the family.

patrilocal residence the custom of a married couple living in the same household (or community) as the husband's family.

pay gap a term used to describe the disparity between women's and men's earnings.

peer group a group of people who are linked by common interests, equal social position, and (usually) similar age.

peripheral nations according to world systems theory, nations that are dependent on core nations for capital, have little or no industrialization (other than what may be brought in by core nations), and have uneven patterns of urbanization.

personal space the immediate area surrounding a person that the person claims as private.

pink-collar occupations relatively low-paying, nonmanual, semiskilled positions primarily held by women, such as day-care workers, checkout clerks, cashiers, and waitpersons.

pluralist model an analysis of political systems that views power as widely dispersed throughout many competing interest groups.

political action committees organizations of special interest groups that solicit contributions from donors and fund campaigns to help elect (or defeat) candidates based on their stances on specific issues.

political crime illegal or unethical acts involving the usurpation of power by government officials, or illegal/unethical acts perpetrated against the government by outsiders seeking to make a political statement, undermine the government, or overthrow it.

political party an organization whose purpose is to gain and hold legitimate control of government.

political socialization the process by which people learn political attitudes, values, and behavior.

politics the social institution through which power is acquired and exercised by some people and groups.

polyandry the concurrent marriage of one woman with two or more men.

polygamy the concurrent marriage of a person of one sex with two or more members of the opposite sex.

polygyny the concurrent marriage of one man with two or more women.

popular culture the component of culture that consists of activities, products, and services that are assumed to appeal primarily to members of the middle and working classes.

population composition the biological and social characteristics of a population, including age, sex, race, marital status, education, occupation, income, and size of household.

population pyramid a graphic representation of the distribution of a population by sex and age.

positivism a term describing Auguste Comte's belief that the world can best be understood through scientific inquiry.

postindustrial society a society in which technology supports a service- and information-based economy.

postmodern perspectives the sociological approach that attempts to explain social life in modern societies that are characterized by postindustrialization, consumerism, and global communications.

power according to Max Weber, the ability of people or groups to achieve their goals despite opposition from others.

prejudice a negative attitude based on faulty generalizations about members of selected racial and ethnic groups.

prestige the respect or regard with which a person or status position is regarded by others.

primary deviance the initial act of rule-breaking.

primary group Charles Horton Cooley's term for a small, less specialized group in which members engage in face-to-face, emotion-based interactions over an extended period of time.

primary labor market the sector of the labor market that consists of high-paying jobs with good benefits that have some degree of security and the possibility of future advancement.

primary sector production the sector of the economy that extracts raw materials and natural resources from the environment.

primary sex characteristics the genitalia used in the reproductive process.

profane the everyday, secular, or "worldly" aspects of life.

professions high-status, knowledge-based occupations.

propaganda information provided by individuals or groups that have a vested interest in furthering their own cause or damaging an opposing one.

property crimes burglary (breaking into private property to commit a serious crime), motor vehicle theft, larceny-theft (theft of property worth $50 or more), and arson.

public opinion the attitudes and beliefs communicated by ordinary citizens to decision makers.

punishment any action designed to deprive a person of things of value (including liberty) because of some offense the person is thought to have committed.

qualitative research sociological research methods that use interpretive description (words) rather than statistics (numbers) to analyze underlying meanings and patterns of social relationships.

quantitative research sociological research methods that are based on the goal of scientific objectivity and that focus on data that can be measured numerically.

race a category of people who have been singled out as inferior or superior, often on the basis of physical characteristics such as skin color, hair texture, and eye shape.

racial socialization the aspect of socialization that contains specific messages and practices concerning the nature of one's racial or ethnic status.

racism a set of attitudes, beliefs, and practices that is used to justify the superior treatment of one racial or ethnic group and the inferior treatment of another racial or ethnic group.

rationality the process by which traditional methods of social organization, characterized by informality and spontaneity, are gradually replaced by efficiently administered formal rules and procedures.

rational–legal authority power legitimized by law or written rules and regulations.

reference group a group that strongly influences a person's behavior and social attitudes, regardless of whether that individual is an actual member.

relative poverty a condition that exists when people may be able to afford basic necessities but are still unable to maintain an average standard of living.

reliability in sociological research, the extent to which a study or research instrument yields consistent results when applied to different individuals at one time or to the same individuals over time.

religion a system of beliefs, symbols, and rituals, based on some sacred or supernatural realm, that guides human behavior, gives meaning to life, and unites believers into a community.

representative democracy a form of democracy whereby citizens elect representatives to serve as bridges between themselves and the government.

research methods specific strategies or techniques for systematically conducting research.

resocialization the process of learning a new and different set of attitudes, values, and behaviors from those in one's background and experience.

riot violent crowd behavior that is fueled by deep-seated emotions but is not directed at one specific target.

role a set of behavioral expectations associated with a given status.

role conflict a situation in which incompatible role demands are placed on a person by two or more statuses held at the same time.

role exit a situation in which people disengage from social roles that have been central to their self-identity.

role expectation a group's or society's definition of the way that a specific role *ought* to be played.

role performance how a person *actually* plays a role.

role strain a condition that occurs when incompatible demands are built into a single status that a person occupies.

role-taking the process by which a person mentally assumes the role of another person in order to understand the world from that person's point of view.

routinization of charisma the process by which charismatic authority is succeeded by a bureaucracy controlled by a rationally established authority or by a combination of traditional and bureaucratic authority.

rumor an unsubstantiated report on an issue or subject.

sacred those aspects of life that are extraordinary or supernatural.

sanctions rewards for appropriate behavior or penalties for inappropriate behavior.

Sapir–Whorf hypothesis the proposition that language shapes the view of reality of its speakers.

scapegoat a person or group that is incapable of offering resistance to the hostility or aggression of others.

second shift Arlie Hochschild's term for the domestic work that employed women perform at home after they complete their workday on the job.

secondary analysis a research method in which researchers use ex-isting material and analyze data that were originally collected by others.

secondary deviance the process that occurs when a person who has been labeled a deviant accepts that new identity and continues the deviant behavior.

secondary group a larger, more specialized group in which members engage in more-impersonal, goal-oriented relationships for a limited period of time.

secondary labor market the sector of the labor market that consists of low-paying jobs with few benefits and very little job security or possibility for future advancement.

secondary sector production the sector of the economy that processes raw materials (from the primary sector) into finished goods.

secondary sex characteristics the physical traits (other than reproductive organs) that identify an individual's sex.

sect a relatively small religious group that has broken away from another religious organization to renew what it views as the original version of the faith.

secularization the process by which religious beliefs, practices, and institutions lose their significance in sectors of society and culture.

segregation the spatial and social separation of categories of people by race, ethnicity, class, gender, and/or religion.

self-concept the totality of our beliefs and feelings about ourselves.

self-fulfilling prophecy the situation in which a false belief or prediction produces behavior that makes the originally false belief come true.

semiperipheral nations according to world systems theory, nations that are more developed than peripheral nations but less developed than core nations.

sex the biological and anatomical differences between females and males.

sex ratio a term used by demographers to denote the number of males for every hundred females in a given population.

sexism the subordination of one sex, usually female, based on the assumed superiority of the other sex.

sexual orientation a person's preference for emotional–sexual relationships with members of the opposite sex (heterosexuality), the same sex (homosexuality), or both (bisexuality).

shared monopoly a condition that exists when four or fewer companies supply 50 percent or more of a particular market.

sick role the set of patterned expectations that defines the norms and values appropriate for individuals who are sick and for those who interact with them.

significant others those persons whose care, affection, and approval are especially desired and who are most important in the development of the self.

slavery an extreme form of stratification in which some people are owned by others.

small group a collectivity small enough for all members to be acquainted with one another and to interact simultaneously.

social bond theory the proposition that the probability of deviant behavior increases when a person's ties to society are weakened or broken.

social change the alteration, modification, or transformation of public policy, culture, or social institutions over time.

social construction of reality the process by which our perception of reality is shaped largely by the subjective meaning that we give to an experience.

social control systematic practices developed by social groups to encourage conformity to norms, rules, and laws and to discourage deviance.

social Darwinism Herbert Spencer's belief that those species of animals, including human beings, best adapted to their environment survive and prosper, whereas those poorly adapted die out.

social devaluation a situation in which a person or group is considered to have less social value than other individuals or groups.

social epidemiology the study of the causes and distribution of health, disease, and impairment throughout a population.

social facts Emile Durkheim's term for patterned ways of acting, thinking, and feeling that exist *outside* any one individual but that exert social control over each person.

social group a group that consists of two or more people who interact frequently and share a common identity and a feeling of interdependence.

social institution a set of organized beliefs and rules that establishes how a society will attempt to meet its basic social needs.

social interaction the process by which people act toward or respond to other people; the foundation for all relationships and groups in society.

social mobility the movement of individuals or groups from one level in a stratification system to another.

social movement an organized group that acts consciously to promote or resist change through collective action.

social stratification the hierarchical arrangement of large social groups based on their control over basic resources.

social structure the stable pattern of social relationships that exists within a particular group or society.

socialism an economic system characterized by public ownership of the means of production, the pursuit of collective goals, and centralized decision making.

socialization the lifelong process of social interaction through which individuals acquire a self-identity and the physical, mental, and social skills needed for survival in society.

socialized medicine a health care system in which the government owns the medical care facilities and employs the physicians.

society a large social grouping that shares the same geographical territory and is subject to the same political authority and dominant cultural expectations.

sociobiology the systematic study of how biology affects social behavior.

socioeconomic status (SES)   a combined measure that, in order to determine class location, attempts to classify individuals, families, or households in terms of factors such as income, occupation, and education.

sociological imagination   C. Wright Mills's term for the ability to see the relationship between individual experiences and the larger society.

sociology   the systematic study of human society and social interaction.

sociology of family   the subdiscipline of sociology that attempts to describe and explain patterns of family life and variations in family structure.

split labor market   a term used to describe the division of the economy into two areas of employment, a primary sector or upper tier, composed of higher-paid (usually dominant-group) workers in more-secure jobs, and a secondary sector or lower tier, composed of lower-paid (often subordinate-group) workers in jobs with little security and hazardous working conditions.

state   the political entity that possesses a legitimate monopoly over the use of force within its territory to achieve its goals.

status   a socially defined position in a group or society characterized by certain expectations, rights, and duties.

status set   all the statuses that a person occupies at a given time.

status symbol   a material sign that informs others of a person's specific status.

stereotypes   overgeneralizations about the appearance, behavior, or other characteristics of members of particular categories.

strain theory   the proposition that people feel strain when they are exposed to cultural goals that they are unable to obtain because they do not have access to culturally approved means of achieving those goals.

subcontracting   an agreement in which a corporation contracts with other (usually smaller) firms to provide specialized components, products, or services to the larger corporation.

subculture   a group of people who share a distinctive set of cultural beliefs and behaviors that differs in some significant way from that of the larger society.

subordinate group   a group whose members, because of physical or cultural characteristics, are disadvantaged and subjected to unequal treatment by the dominant group and who regard themselves as objects of collective discrimination.

succession   the process by which a new category of people or type of land use gradually predominates in an area formerly dominated by another group or activity.

superego   Sigmund Freud's term for the conscience, consisting of the moral and ethical aspects of personality.

survey   a poll in which the researcher gathers facts or attempts to determine the relationships among facts.

symbol   anything that meaningfully represents something else.

symbolic interactionist perspectives   the sociological approach that views society as the sum of the interactions of individuals and groups.

taboos   mores so strong that their violation is considered to be extremely offensive and even unmentionable.

technology   the knowledge, techniques, and tools that allow people to transform resources into a usable form and the knowledge and skills required to use what is developed.

terrorism   the calculated unlawful use of physical force or threats of violence against persons or property in order to intimidate or coerce a government, organization, or individual for the purpose of gaining some political, religious, economic, or social objective.

tertiary deviance   deviance that occurs when a person who has been labeled a deviant seeks to normalize the behavior by relabeling it as nondeviant.

tertiary sector production   the sector of the economy that is involved in the provision of services rather than goods.

theory   a set of logically interrelated statements that attempts to describe, explain, and (occasionally) predict social events.

theory of racial formation   the idea that actions of the government substantially define racial and ethnic relations in the United States.

total institution   Erving Goffman's term for a place where people are isolated from the rest of society for a set period of time and come under the control of the officials who run the institution.

totalitarianism   a political system in which the state seeks to regulate all aspects of people's public and private lives.

tracking   the assignment of students to specific curriculum groups and courses on the basis of their test scores, previous grades, or both.

traditional authority   power that is legitimized on the basis of long-standing custom.

transnational corporations   large corporations that are headquartered in one country but sell and produce goods and services in many countries.

transsexual   a person who believes that he or she was born with the body of the wrong sex.

transvestite   a male who lives as a woman or a female who lives as a man but does not alter the genitalia.

triad   a group composed of three members.

underclass   those who are poor, seldom employed, and caught in long-term deprivation that results from low levels of education and income and high rates of unemployment.

unemployment rate   the percentage of unemployed persons in the labor force actively seeking jobs.

universal health care system   a health care system in which all citizens receive medical services paid for by tax revenues.

urbanization   the process by which an increasing proportion of a population lives in cities rather than in rural areas.

validity   in sociological research, the extent to which a study or research instrument accurately measures what it is supposed to measure.

value contradictions   values that conflict with one another or are mutually exclusive.

values   collective ideas about what is right or wrong, good or bad, and desirable or undesirable in a particular culture.

variable   in sociological research, any concept with measurable traits or characteristics that can change or vary from one person, time, situation, or society to another.

victimless crimes   crimes involving a willing exchange of illegal goods or services among adults.

violent crime   actions—murder, forcible rape, robbery, and aggravated assault—involving force or the threat of force against others.

wealth   the value of all of a person's or family's economic assets, including income, personal property, and income-producing property.

welfare state   a state in which there is extensive government action to provide support and services to the citizens.

working class   (or **proletariat**) those who must sell their labor to the owners in order to earn enough money to survive.

zero population growth   the point at which no population increase occurs from year to year.

References

AAUW (American Association of University Women). 1995. *How Schools Shortchange Girls/ The AAUW Report: A Study of Major Findings on Girls and Education.* New York: Marlowe.

—. 1998. *Gender Gap: Where Schools Still Fail Our Children.* Washington, DC: American Association of University Women. Retrieved Aug. 7, 2003. Online: http://aauw.org/research/girls_education/gg.cfm

ABC News. 2005. "Milton Bradley Accuses Jeff Kent of Racism." Retrieved Feb. 26, 2006. Online: http://abcnews.go.com/Sports/wireStory?id=1063252

Aberle, D. F., A. K. Cohen, A. K. Davis, M. J. Leng, Jr., and F. N. Sutton. 1950. "The Functional Prerequisites of Society." *Ethics,* 60 (January): 100–111.

Aberle, David F. 1966. *The Peyote Religion Among the Navaho.* Chicago: Aldine.

Accuracy in Media. 2006. "Media Monitor." Retrieved Feb. 23, 2008. Online: http://www.aim.org/media-monitor/print/affirmative-action-for-conservatives

ACLU. 2005. "Pennsylvania Parents File First-Ever Challenge to 'Intelligent Design' Instruction in Public Schools." Retrieved Mar. 18, 2006. Online: http://www.aclu.org//religion/schools/16372prs20041214.html

Adams, Tom. 1991. *Grass Roots: How Ordinary People Are Changing America.* New York: Citadel.

Adler, Jerry. 1999. "The Truth About High School." *Newsweek* (May 10): 56–58.

Adler, Patricia A., and Peter Adler. 1998. *Peer Power: Preadolescent Culture and Identity.* New Brunswick, NJ: Rutgers University Press.

—. 2003. *Constructions of Deviance: Social Power, Context, and Interaction* (4th ed.). Belmont, CA: Wadsworth.

Adorno, Theodor W., Else Frenkel-Brunswick, Daniel J. Levinson, and R. Nevitt Sanford. 1950. *The Authoritarian Personality.* New York: Harper & Row.

adrants. 2005. "Reality Is for Losers." Retrieved Apr. 9, 2005. Online: http://www.adrants.com/2005/02/reality-is-for-losers.php

Agger, Ben. 1993. *Gender, Culture, and Power: Toward a Feminist Postmodern Critical Theory.* Westport, CT: Praeger.

Agnew, Robert. 1985. "Social Control Theory and Delinquency: A Longitudinal Test." *Criminology,* 23: 47–61.

Aiello, John R., and S. E. Jones. 1971. "Field Study of Proxemic Behavior of Young School Children in Three Subcultural Groups." *Journal of Personality and Social Psychology,* 19: 351–356.

Akers, Ronald. 1992. *Drugs, Alcohol, and Society: Social Structure, Process, and Policy.* Belmont, CA: Wadsworth.

Akers, Ronald L. 1998. *Social Learning and Social Structure: A General Theory of Crime and Deviance.* Boston: Northeastern University Press.

Albanese, Catherine L. 2007. *America, Religions and Religion* (4th ed.). Belmont, CA: Wadsworth.

Albas, Daniel, and Cheryl Albas. 1988. "Aces and Bombers: The Post-Exam Impression Management Strategies of Students." *Symbolic Interaction,* 11 (Fall): 289–302.

Albrecht, Gary L., 1992. *The Disability Business: Rehabilitation in America.* Newbury Park, CA: Sage.

Allport, Gordon. 1958. *The Nature of Prejudice* (abridged ed.). New York: Doubleday/Anchor.

Alonso, Alejandro A. 1998. "Urban Graffiti on the City Landscape." Paper presented at the Western Geography Graduate Conference, San Diego State University (Feb. 14). Retrieved Mar. 10, 2007. Online: http://www.streetgangs.com/academic/alonsograffiti.pdf

Altemeyer, Bob. 1981. *Right-Wing Authoritarianism.* Winnipeg: University of Manitoba Press.

—. 1988. *Enemies of Freedom: Understanding Right-Wing Authoritarianism.* San Francisco: Jossey-Bass.

Alwin, Duane, Philip Converse, and Steven Martin. 1985. "Living Arrangements and Social Integration." *Journal of Marriage and the Family,* 47: 319–334.

American Academy of Child and Adolescent Psychiatry. 1997. "Children and Watching TV." Retrieved June 22, 1999. Online: http://www.aacap.org/factsfam/tv.htm

American Association of Community Colleges. 2005. "Community College Facts." Retrieved Apr. 21, 2005. Online: http://www.aacc.nche.edu

American Association of Suicidology. 2005. "Media Guidelines for Suicide Reporting." Retrieved Jan. 8, 2006. Online: http://www.211bigbend.org/hotlines/suicide/media.htm

—. 2006. "AAS Fact Sheets." Retrieved Jan. 28, 2007. Online: http://www.suicidology.org/displaycommon.cfm?an=1&subarticlenbr=185

American Sociological Association. 1997. *Code of Ethics.* Washington, DC: American Sociological Association (orig. pub. 1971).

Aminzade, Ronald. 1973. "Revolution and Collective Political Violence: The Case of the Working Class of Marseille, France, 1830–1871." Working Paper #86, Center for Research on Social Organization. Ann Arbor: University of Michigan, October 1973.

Amott, Teresa, and Julie Matthaei. 1996. *Race, Gender, and Work: A Multicultural Economic History of Women in the United States* (rev. ed.). Boston: South End.

Anders, George. 1996. *Health Against Wealth: HMOs and the Breakdown of Medical Trust.* Boston: Houghton Mifflin.

Andersen, Margaret L., and Patricia Hill Collins (Eds.). 1998. *Race, Class, and Gender: An Anthology* (3rd ed.). Belmont, CA: Wadsworth.

Anderson, Elijah. 1990. *Streetwise: Race, Class, and Change in an Urban Community.* Chicago: University of Chicago Press.

Anderson, Jenny. 2007. "Morgan Stanley to Settle Sex Bias Suit." *New York Times* (Apr. 25): C1, C18.

Anell, Anders, and Michael Willis. 2000. "International Comparison of Health Care Systems Using Resource Profiles." Geneva, Switzerland: World Health Organization. Retrieved Aug. 7, 2003. Online: http://www.who.int/docstore/bulletin/pdf/2000/issue6/bu0585.pdf

Angel, Ronald. 1984. "The Costs of Disability for Hispanic Males." *Social Science Quarterly,* 65: 426–443.

Angelotti, Amanda. 2006. "Confessions of a Beauty Pageant Drop-Out." *Campus Progress: A Project of Center for American Progress* (Jan. 25). Retrieved Jan. 27, 2008. Online: http://www.campusprogress/org/features/727/confessions-of-a-beauty-pageant-drop-out

Angier, Natalie. 1993. "'Stopit!' She Said. 'Nomore!'" *New York Times Book Review* (Apr. 25): 12.

Anyon, Jean. 1980. "Social Class and the Hidden Curriculum of Work." *Journal of Education,* 162: 67–92.

—. 1997. *Ghetto Schooling: A Political Economy of Urban Educational Reform.* New York: Teachers College Press.

APA Online. 2000. "Psychiatric Effects of Violence." Public Information: APA Fact Sheet Series. Washington, DC: American Psychological Association. Retrieved Apr. 5, 2000. Online: http://www.psych.org/psych/htdocs/public_info/media_violence.html

Appleton, Lynn M. 1995. "The Gender Regimes in American Cities." In Judith A. Garber and Robyne S. Turner (Eds.), *Gender in Urban Research.* Thousand Oaks, CA: Sage, pp. 44–59.

Arenson, Karen W. 2004. "Suicide of N.Y.U. Student, 19, Brings Sadness and Questions." *New York Times* (Mar. 10). Retrieved Jan. 8, 2006. Online: http://query.nytimes.com/gst/fullpage.html?sec=health&res=9904EFDF153EF933A25750C0A9629C8B63

Arnold, Regina A. 1990. "Processes of Victimization and Criminalization of Black Women." *Social Justice,* 17 (3): 153–166.

Asch, Adrienne. 1986. "Will Populism Empower Disabled People?" In Harry G. Boyle and Frank Reissman (Eds.), *The New Populism: The Power of Empowerment.* Philadelphia: Temple University Press, pp. 213–228.

Asch, Solomon E. 1955. "Opinions and Social Pressure." *Scientific American,* 193 (5): 31–35.

—. 1956. "Studies of Independence and Conformity: A Minority of One Against a Unanimous Majority." *Psychological Monographs,* 70 (9) (Whole No. 416).

Ashe, Arthur R., Jr. 1988. *A Hard Road to Glory: A History of the African-American Athlete.* New York: Warner.

Atchley, Robert C., and Amanda Barusch. 2004. *Social Forces and Aging: An Introduction to Social Gerontology* (10th ed.). Belmont, CA: Wadsworth.

Aulette, Judy Root. 1994. *Changing Families.* Belmont, CA: Wadsworth.

Aulette, Ken. 1998. *The Highwaymen: Warriors of the Information Superhighway.* San Diego, CA: Harvest/Harcourt.

Axtell, Roger E. 1991. *Gestures: The Do's and Taboos of Body Language Around the World.* New York: Wiley.

Ayala, Elaine. 1993. "Unfinished Work: Cesar Chavez Is Dead but His Struggle Has Been Reborn, Followers Say." *Austin American-Statesman* (Sept. 6): E1, E4. Quoted from a speech by Cesar Chavez to the Commonwealth Club of San Francisco, Nov. 9, 1984.

Babbie, Earl. 2004. *The Practice of Social Research* (10th ed.). Belmont, CA: Wadsworth.

Bagby, Meredith (Ed.). 1997. *Annual Report of the United States of America 1997.* New York: McGraw-Hill.

Baker, Robert. 1993. "'Pricks' and 'Chicks': A Plea for 'Persons.'" In Anne Minas (Ed.), *Gender Basics: Feminist Perspectives on Women and Men.* Belmont, CA: Wadsworth, pp. 66–68.

Ballantine, Jeanne H. 2001. *The Sociology of Education: A Systematic Analysis* (5th ed.). Englewood Cliffs, NJ: Prentice Hall.

Bane, Mary Jo. 1986. "Household Composition and Poverty: Which Comes First?" In Sheldon H. Danziger and Daniel H. Weinberg (Eds.), *Fighting Poverty: What Works and What Doesn't.* Cambridge, MA: Harvard University Press.

Bane, Mary Jo, and David T. Ellwood. 1994. *Welfare Realities: From Rhetoric to Reform.* Cambridge, MA: Harvard University Press.

Banet-Weiser, Sarah. 1999. *The Most Beautiful Girl in the World: Beauty Pageants and National Identity.* Berkeley: University of California Press.

Barakat, Matthew. 2000. "Survey: Women's Salaries Beat Men's in Some Fields." *Austin American-Statesman* (July 4): D1, D3.

Bardwell, Jill R., Samuel W. Cochran, and Sharon Walker. 1986. "Relationship of Parental Education, Race, and Gender to Sex Role Stereotyping in Five-Year-Old Kindergartners." *Sex Roles,* 15: 275–281.

Barlow, Hugh D., and David Kauzlarich. 2002. *Introduction to Criminology* (8th ed.). Upper Saddle River, NJ: Prentice Hall.

Barnett, Harold. 1979. "Wealth, Crime and Capital Accumulation." *Contemporary Crises,* 3: 171–186.

Baron, Dennis. 1986. *Grammar and Gender.* New Haven, CT: Yale University Press.

Barovick, Harriet. 2001. "Hope in the Heartland." *Time* (special edition, July): G1–G3.

Barrymore, Drew, with Todd Gold. 1994. *Little Girl Lost.* New York: Pocket. In Jay David (Ed.), *The Family Secret: An Anthology.* New York: Morrow, pp. 171–199.

Basow, Susan A. 1992. *Gender Stereotypes and Roles* (3rd ed.). Pacific Grove, CA: Brooks/Cole.

Baudrillard, Jean. 1983. *Simulations.* New York: Semiotext.

Bauman, Zygmunt. 1992. *Imitations of Postmodernity.* London: Routledge.

Baxter, J. 1970. "Interpersonal Spacing in Natural Settings." *Sociology,* 36 (3): 444–456.

Beck, Ulrich. 1992. *Risk Society: Towards a New Modernity.* London: Sage.

Becker, Howard S. 1963. *Outsiders: Studies in the Sociology of Deviance.* New York: Free Press.

Beech, Hannah. 2001. "China's Lifestyle Choice." *Time* (Aug. 6): 32.

Beeghley, Leonard. 2000. *The Structure of Social Stratification in the United States* (3rd ed.). Boston: Allyn & Bacon.

Belkin, Lisa. 1994. "Kill for Life?" *New York Times Magazine* (Oct. 30): 47–51, 62–64, 76, 80.

Bell, Inge Powell. 1989. "The Double Standard: Age." In Jo Freeman (Ed.), *Women: A Feminist Perspective* (4th ed.). Mountain View, CA: Mayfield, pp. 236–244.

Bellah, Robert N. 1967. "Civil Religion." *Daedalus,* 96: 1–21.

Benford, Robert D. 1993. "'You Could Be the Hundredth Monkey': Collective Action Frames and Vocabularies of Motive Within the Nuclear Disarmament Movement." *Sociological Quarterly,* 34: 195–216.

Benjamin, Lois. 1991. *The Black Elite: Facing the Color Line in the Twilight of the Twentieth Century.* Chicago: Nelson-Hall.

Bennahum, David S. 1999. "For Kosovars, an On-Line Phone Directory of a People in Exile." *New York Times* (July 15): D7.

Benokraitis, Nijole V., and Joe R. Feagin. 1994. *Modern Sexism: Blatant, Subtle and Covert Discrimination* (2nd ed.). Englewood Cliffs, NJ: Prentice Hall.

Berger, Peter. 1963. *Invitation to Sociology: A Humanistic Perspective.* New York: Anchor.

—. 1967. *The Sacred Canopy: Elements of a Sociological Theory of Religion.* New York: Doubleday.

Berger, Peter, and Hansfried Kellner. 1964. "Marriage and the Construction of Reality." *Diogenes,* 46: 1–32.

Berger, Peter, and Thomas Luckmann. 1967. *The Social Construction of Reality: A Treatise in the Sociology of Knowledge.* Garden City, NY: Anchor.

Berliner, David C., and Bruce J. Biddle. 1995. *The Manufactured Crisis: Myths, Fraud, and the Attack on America's Public Schools.* Reading, MA: Addison-Wesley.

Bernard, Jessie. 1982. *The Future of Marriage.* New Haven, CT: Yale University Press (orig. pub. 1973).

Better Health Channel. 2007. "Food and Celebrations." Retrieved Feb. 11, 2007. Online: http://www.betterhealth.vic.gov.au

Biblarz, Arturo, R. Michael Brown, Dolores Noonan Biblarz, Mary Pilgram, and Brent F. Baldree. 1991. "Media Influence on Attitudes Toward Suicide." *Suicide and Life-Threatening Behavior,* 21 (4): 374–385.

Blau, Peter M., and Marshall W. Meyer. 1987. *Bureaucracy in Modern Society* (3rd ed.). New York: Random House.

Blauner, Robert. 1964. *Alienation and Freedom.* Chicago: University of Chicago Press.

—. 1972. *Racial Oppression in America.* New York: Harper & Row.

Block, Fred, Anna C. Korteweg, and Kerry Woodward, with Zach Schiller and Imrul Mazid. 2008. "The Compassion Gap in American Poverty Policy." In Jeff Goodwin and James M. Jasper (Eds.), *The Contexts Reader.* New York: Norton, pp. 166–175. Originally appeared in *Contexts,* a journal published by the American Sociological Association (Spring 2006).

Bluestone, Barry, and Bennett Harrison. 1982. *The Deindustrialization of America.* New York: Basic.

Blum, Ronald. 2007. "Only 8.4 Percent of MLB Players Black." Retrieved Jan. 26, 2008. Online: http://www.foxnews.com/wires/2007Mar29/0,4670,BBOMLBDiversity,00.html

Blumberg, Leonard. 1977. "The Ideology of a Therapeutic Social Movement: Alcoholics Anonymous." *Journal of Studies on Alcohol,* 38: 2122–2143.

Blumer, Herbert G. 1946. "Collective Behavior." In Alfred McClung Lee (Ed.), *A New Outline of the Principles of Sociology.* New York: Barnes & Noble, pp. 167–219.

—. 1969. *Symbolic Interactionism: Perspective and Method.* Englewood Cliffs, NJ: Prentice Hall.

—. 1974. "Social Movements." In R. Serge Denisoff (Ed.), *The Sociology of Dissent.* New York: Harcourt, pp. 74–90.

Bodeen, Christopher. 2007. "China's Woes Vindicate Whistle-Blower, But Court Case Against Him Continues." *International Business Times* (June 9). Retrieved Mar. 8, 2008. Online: http://www.ibtimes.com/articles/20070609/china-whistleblower-wu-lihong_all.htm

Bonacich, Edna. 1972. "A Theory of Ethnic Antagonism: The Split Labor Market." *American Sociological Review,* 37: 547–549.

—. 1976. "Advanced Capitalism and Black–White Relations in the United States: A Split Labor Market Interpretation." *American Sociological Review,* 41: 34–51.

Bonvillain, Nancy. 2001. *Women & Men: Cultural Constructs of Gender* (3rd ed.). Upper Saddle River, NJ: Prentice Hall.

Bordewich, Fergus M. 1996. *Killing the White Man's Indian.* New York: Anchor/Doubleday.

Bordo, Susan. 2004. *Unbearable Weight: Feminism, Western Culture, and the Body* (10th anniversary edition). Berkeley: University of California Press.

Bourdieu, Pierre. 1984. *Distinction: A Social Critique of the Judgement of Taste.* Trans. Richard Nice. Cambridge, MA: Harvard University Press.

Bourdieu, Pierre, and Jean-Claude Passeron. 1990. *Reproduction in Education, Society and Culture.* Newbury Park, CA: Sage.

Bowles, Samuel, and Herbert Gintis. 1976. *Schooling in Capitalist America: Education and the Contradictions of Economic Life.* New York: Basic.

Boyes, William, and Michael Melvin. 2002. *Economics* (5th ed.). Boston: Houghton Mifflin.

Bozett, Frederick. 1988. "Gay Fatherhood." In Phyllis Bronstein and Carolyn Pape Cowan (Eds.), *Fatherhood Today: Men's Changing Role in the Family.* New York: Wiley, pp. 60–71.

Bramlett, Matthew D., and William D. Mosher. 2001. "First Marriage Dissolution, Divorce, and Remarriage: United States." DHHS publication no. 2001–1250 01–0384 (5/01). Hyattsville, MD: Department of Health and Human Services.

Brandl, Steven, Meghan Stroshine, and James Frank. 2001. "Who Are the Complaint-Prone Officers? An Examination of the Relationship Between Police Officers' Attributes, Arrest Activity, Assignment, and Citizens' Complaints About Excessive Force." *Journal of Criminal Justice,* 29: 521–529.

Braverman, Harry. 1974. *Labor and Monopoly Capital.* New York: Monthly Review Press.

Breault, K. D. 1986. "Suicide in America: A Test of Durkheim's Theory of Religious and Family Integration, 1933–1980." *American Journal of Sociology,* 92 (3): 628–656.

Brint, Steven. 1994. *In an Age of Experts: The Changing Role of Professionals in Politics and Public Life.* Princeton, NJ: Princeton University Press.

Britt, Lory. 1993. "From Shame to Pride: Social Movements and Individual Affect." Paper presented at the 88th annual meeting of the American Sociological Association, Miami, August.

Broder, John M. 2003. "Debris Is Now Leading Suspect in Shuttle Catastrophe." *New York Times* (Feb. 4): A1–A25.

Brooks-Gunn, Jeanne. 1986. "The Relationship of Maternal Beliefs About Sex Typing to Maternal and Young Children's Behavior." *Sex Roles,* 14: 21–35.

Brown, Dennis M. 2003. "Rural Tourism: An Annotated Bibliography." United States Department of Agriculture. Retrieved Aug. 9, 2003. Online: http://www.nal.usda.gov/ric/ricpubs/rural_tourism.html#summary

Brown, E. Richard. 1979. *Rockefeller Medicine Men.* Berkeley: University of California Press.

Brown, Robert W. 1954. "Mass Phenomena." In Gardner Lindzey (Ed.), *Handbook of Social Psychology* (vol. 2). Reading, MA: Addison-Wesley, pp. 833–873.

Brustad, Robert J. 1996. "Attraction to Physical Activity in Urban Schoolchildren: Parental Socialization and Gender Influence." *Research Quarterly for Exercise and Sport,* 67: 316–324.

Buchholz, Brad. 2007. "Gunshots Shatter Serenity of Sacred Sanctuary." *Austin American-Statesman* (Apr. 22): G1, G4.

Bucks, Brian K., Arthur B. Kennickell, and Kevin B. Moore. 2006. "Recent Changes in U.S. Family Finances: Evidence from the 2001 and 2004 Survey of Consumer Finances." Federal Reserve Board. Retrieved Feb. 25, 2006. Online: http://www.federalreserve.gov/pubs/bulletin/2006/financesurvey.pdf

Buechler, Steven M. 2000. *Social Movements in Advanced Capitalism: The Political Economy and Cultural Construction of Social Activism.* New York: Oxford University Press.

Bullard, Robert B., and Beverly H. Wright. 1992. "The Quest for Environmental Equity: Mobilizing the African-American Community for Social Change." In Riley E. Dunlap and Angela G. Mertig (Eds.), *American Environmentalism: The U.S. Environmental Movement, 1970–1990.* New York: Taylor & Francis, pp. 39–49.

Bumpass, Larry, James E. Sweet, and Andrew J. Cherlin. 1991. "The Role of Cohabitation in Declining Rates of Marriage." *Journal of Marriage and the Family,* 53: 913–927.

Burgess, Ernest W. 1925. "The Growth of the City." In Robert E. Park and Ernest W. Burgess (Eds.), *The City.* Chicago: University of Chicago Press, pp. 47–62.

Burkhauser, Richard V., Robert H. Haveman, and Barbara L. Wolfe. 1993. "How People with Disabilities Fare When Public Policies Change." *Journal of Policy Analysis and Management,* 12: 251–269.

Burnham, Walter Dean. 1983. *Democracy in the Making: American Government and Politics.* Englewood Cliffs, NJ: Prentice Hall.

Burros, Marian. 1994. "Despite Awareness of Risks, More in U.S. Are Getting Fat." *New York Times* (July 17): 1, 8.

—. 1996. "Eating Well: A Law to Encourage Sharing in a Land of Plenty." *New York Times* (Dec. 11): B6.

Busch, Ruth C. 1990. *Family Systems: Comparative Study of the Family.* New York: Lang.

Butler, John S. 1991. *Entrepreneurship and Self-Help Among Black Americans.* New York: SUNY Press.

Buvinić, Mayra. 1997. "Women in Poverty: A New Global Underclass." *Foreign Policy* (Fall): 38–53.

Cable, Sherry, and Charles Cable. 1995. *Environmental Problems, Grassroots Solutions: The Politics of Grassroots Environmental Conflict.* New York: St. Martin's.

Callaghan, Polly, and Heidi Hartmann. 1991. *Contingent Work.* Washington, DC: Economic Policy Institute.

Campbell, Anne. 1984. *The Girls in the Gang* (2nd ed.). Cambridge, MA: Basil Blackwell.

Campus Progress. 2006. "Five Minutes with: Morgan Spurlock." Retrieved Mar. 10, 2007. Online: http://www.campusprogress.org/features/336/five-minutes-with-morgan-spurlock

Cancian, Francesca M. 1990. "The Feminization of Love." In C. Carlson (Ed.), *Perspectives on the Family: History, Class, and Feminism.* Belmont, CA: Wadsworth, pp. 171–185.

—. 1992. "Feminist Science: Methodologies That Challenge Inequality." *Gender & Society,* 6 (4): 623–642.

Canetto, Silvia Sara. 1992. "She Died for Love and He for Glory: Gender Myths of Suicidal Behavior." *OMEGA,* 26 (1): 1–17.

Canter, R. J., and S. S. Ageton. 1984. "The Epidemiology of Adolescent Sex-Role Attitudes." *Sex Roles,* 11: 657–676.

Cantor, Muriel G. 1980. *Prime-Time Television: Content and Control.* Newbury Park, CA: Sage.

—. 1987. "Popular Culture and the Portrayal of Women: Content and Control." In Beth B. Hess and Myra Marx Ferree (Eds.), *Analyzing Gender: A Handbook of Social Science Research.* Newbury Park, CA: Sage, pp. 190–214.

Cantril, Hadley. 1941. *The Psychology of Social Movements.* New York: Wiley.

Capek, Stella M. 1993. "The 'Environmental Justice' Frame: A Conceptual Discussion and Application." *Social Problems,* 40 (1): 5–23.

Cargan, Leonard, and Matthew Melko. 1982. *Singles: Myths and Realities.* Newbury Park, CA: Sage.

Carrier, James G. 1986. *Social Class and the Construction of Inequality in American Education.* New York: Greenwood.

Carson, Rachel. 1962. *Silent Spring.* Boston: Houghton Mifflin.

Carter, Stephen L. 1994. *The Culture of Disbelief: How American Law and Politics Trivializes Religious Devotion.* New York: Anchor/Doubleday.

Cashmore, E. Ellis. 1996. *Dictionary of Race and Ethnic Relations* (4th ed.). London: Routledge.

Castells, Manuel. 1977. *The Urban Question.* London: Edward Arnold (orig. pub. 1972 as *La Question Urbaine,* Paris).

—. 1998. *End of Millennium.* Malden, MA: Blackwell.

Castillo, Juan. 2003. "U.S. Payday Is Something to Write Home About." *Austin American-Statesman* (Dec. 14): J1, J4.

Cavender, Gray. 1995. "Alternative Theory: Labeling and Critical Perspectives." In Joseph F. Sheley (Ed.), *Criminology: A Contemporary Handbook* (2nd ed.). Belmont, CA: Wadsworth, pp. 349–371.

CBS News. 2000. "Russian Mafia's Worldwide Grip" (July 21). Retrieved Apr. 23, 2005. Online: http://www.cbsnews.com/stories/2000/07/21/world/217683.shtml

—. 2007. "Outsourced 'Wombs-for-Rent' in India." Retrieved Feb. 9, 2008. Online: http://www.cbsnews.com/stories/2007/12/31/health/main3658750.shtml

Celis, William, III. 1994. "Nations Envied for Schools Share Americans' Worries." *New York Times* (July 13): B4.

Center for Problem-Oriented Policing. 2002. "Graffiti." Retrieved Mar. 10, 2007. Online: http://www.popcenter.org/problems/problem-graffiti.htm

Chafetz, Janet Saltzman. 1984. *Sex and Advantage: A Comparative, Macro-Structural Theory of Sex Stratification.* Totowa, NJ: Rowman & Allanheld.

Chagnon, Napoleon A. 1992. *Yanomamo: The Last Days of Eden.* New York: Harcourt (rev. from 4th ed., *Yanomamo: The Fierce People,* published by Holt, Rinehart & Winston).

Chalfant, H. Paul, Robert E. Beckley, and C. Eddie Palmer. 1994. *Religion in Contemporary Society* (3rd ed.). Itasca, IL: Peacock.

Chambliss, William J. 1973. "The Saints and the Roughnecks." *Society,* 11: 24–31.

Chandler, Tertius, and Gerald Fox. 1974. *3000 Years of Urban History.* New York: Academic Press.

Chen, Hsiang-shui. 1992. *Chinatown No More: Taiwan Immigrants in Contemporary New York.* Ithaca, NY: Cornell University Press.

Cherlin, Andrew J. 1992. *Marriage, Divorce, Remarriage.* Cambridge, MA: Harvard University Press.

Chesney-Lind, Meda. 1997. *The Female Offender.* Thousand Oaks, CA: Sage.

Children Now. 1999. "Children, Race and Advertising." Childrennow.org. Retrieved Mar. 27, 2005. Online: http://www.childrennow.org/media/medianow/mnwinter1999.html

Children's Defense Fund. 2001. *The State of America's Children: Yearbook 2001.* Boston: Beacon.

—. 2002. *The State of Children in America's Union: A 2002 Action Guide to Leave No Child Behind.* Retrieved June 29, 2003. Online: http://www.childrensdefense.org/pdf/minigreenbook.pdf

Cho, Sumi K. 1993. "Korean Americans vs. African Americans: Conflict and Construction." In Robert Gooding-Williams (Ed.), *Reading Rodney King, Reading Urban Uprising.* New York: Routledge, pp. 196–211.

Chronicle of Higher Education. 2007. "Almanac Issue, 2007–8" (Aug. 31).

Cinquemani, Anthony M. 1997. "C. Wright Mills, Media, and Mass: Only More So." *Educational Change* (Spring): 88–90.

City of Farmers Branch. 2006. "Resolution No. 2006–130." Retrieved Feb. 17, 2007. Online: http://www.ci.farmers-branch.tx.us/Communication/Resolution%202006-130.html

Clayman, Steven E. 1993. "Booing: The Anatomy of a Disaffiliative Response." *American Sociological Review,* 58 (1): 110–131.

Clinard, Marshall B., and Peter C. Yeager. 1980. *Corporate Crime.* New York: Free Press.

Clines, Francis X. 1996. "A Chef's Training Program That Feeds Hope as Well as Hunger." *New York Times* (Dec. 11): B1, B6.

Cloward, Richard A., and Lloyd E. Ohlin. 1960. *Delinquency and Opportunity: A Theory of Delinquent Gangs.* New York: Free Press.

CNN. 1994. "Both Sides: School Prayer." (Nov. 26).

CNN.com. 2006. "Immigrant Hopefuls Gather Near U.S. Border" (Apr. 12). Retrieved Apr. 29, 2006. Online: http://www.cnn.com/2006/WORLD/americas/04/12/mexico.border.ap/index.html

Coakley, Jay J. 2004. *Sport in Society: Issues and Controversies* (8th ed.). New York: McGraw-Hill.

Cockerham, William C. 1995. *Medical Sociology* (6th ed.). Englewood Cliffs, NJ: Prentice Hall.

—. 2004. *Medical Sociology* (9th ed.). Upper Saddle River, NJ: Prentice Hall.
Cohen, Adam. 1999. "A Curse of Cliques." *Time* (May 3): 44–45.
Cohen, Leah Hager. 1994. *Train Go Sorry: Inside a Deaf World.* Boston: Houghton Mifflin.
Cohen, Robin A., and Michael E. Martinez. 2007. "Health Insurance Coverage: Early Release of Estimates from the National Health Interview Survey, 2006." National Center for Health Statistics. Retrieved Jan. 12, 2008. Online: http://www.cdc.gov/nchs/data/nhis/earlyrelease/insur200706.pdf
Cole, David. 2000. *No Equal Justice: Race and Class in the American Criminal Justice System.* New York: New Press.
Cole, George F., and Christopher E. Smith. 2004. *The American System of Criminal Justice* (10th ed.). Belmont, CA: Wadsworth.
Coleman, Richard P., and Lee Rainwater. 1978. *Social Standing in America: New Dimensions of Class.* New York: Basic.
Coles, Gerald. 1987. *The Learning Mystique: A Critical Look at "Learning Disabilities."* New York: Pantheon.
College Board. 2007. "2007 College-Bound Seniors: Total Group Profile Report." Retrieved Oct. 8, 2007. Online: http://www.collegeboard.com/prod_downloads/about/news_info/cbsenior/yr2007/national-report.pdf
Collins, Patricia Hill. 1990. *Black Feminist Thought: Knowledge, Consciousness, and the Politics of Empowerment.* London: HarperCollins Academic.
—. 1991. "The Meaning of Motherhood in Black Culture." In Robert Staples (Ed.), *The Black Family: Essays and Studies.* Belmont, CA: Wadsworth, pp. 169–178. Orig. pub. in *SAGE: A Scholarly Journal on Black Women,* 4 (Fall 1987): 3–10.
—. 1998. *Fighting Words: Black Women and the Search for Justice.* Minneapolis: University of Minnesota Press.
Collins, Randall. 1971. "A Conflict Theory of Sexual Stratification." *Social Problems,* 19 (1): 3–21.
—. 1979. *The Credential Society: An Historical Sociology of Education.* New York: Academic Press.
—. 1982. *Sociological Insight: An Introduction to Non-Obvious Sociology.* New York: Oxford University Press.
Colt, George Howe. 1991. *The Enigma of Suicide.* New York: Summit.
Coltrane, Scott. 1989. "Household Labor and the Routine Production of Gender." *Social Problems,* 36: 473–490.
Condry, Sandra McConnell, John C. Condry, Jr., and Lee Wolfram Pogatshnik. 1983. "Sex Differences: A Study of the Ear of the Beholder." *Sex Roles,* 9: 697–704.
Connelly, Marjorie. 2004. "How Americans Voted: A Political Portrait." *New York Times* (Nov. 7): WK4.
Conrad, Peter. 1996. "Medicalization and Social Control." In Phil Brown (Ed.), *Perspectives in Medical Sociology* (2nd ed.). Prospect Heights, IL: Waveland, pp. 137–162.
Cook, Sherburn F. 1973. "The Significance of Disease in the Extinction of the New England Indians." *Human Biology,* 45: 485–508.
Cookson, Peter W., Jr., and Caroline Hodges Persell. 1985. *Preparing for Power: America's Elite Boarding Schools.* New York: Basic.
Cooley, Charles Horton. 1963. *Social Organization: A Study of the Larger Mind.* New York: Schocken (orig. pub. 1909).
—. 1998. "The Social Self—the Meaning of 'I.'" In Hans-Joachim Schubert (Ed.), *On Self and Social Organization—Charles Horton Cooley.* Chicago: University of Chicago Press, pp. 155–175. Reprinted from Charles Horton Cooley, *Human Nature and the Social Order.* New York: Schocken, 1902.
Coontz, Stephanie. 1992. *The Way We Never Were: American Families and the Nostalgia Trap.* New York: Basic.
—. 1997. *The Way We Really Are: Coming to Terms with America's Changing Families.* New York: Basic.
Corsaro, William A. 1985. *Friendship and Peer Culture in the Early Years.* Norwood, NJ: Ablex.
—. 1992. "Interpretive Reproduction in Children's Peer Cultures." *Social Psychology Quarterly,* 55 (2): 160–177.
—. 1997. *Sociology of Childhood.* Thousand Oaks, CA: Pine Forge.
Corsbie-Massay, Charisse L. 2005. "Beauty Pageants and Television: A Perfect Marriage." Retrieved Jan. 27, 2008. Online: http://alum.mit.edu/www/charisse
Cortese, Anthony J. 2004. *Provocateur: Images of Women and Minorities in Advertising* (2nd ed.). Latham, MD: Rowman and Littlefield.
Coser, Lewis A. 1956. *The Functions of Social Conflict.* Glencoe, IL: Free Press.
Cox, Oliver C. 1948. *Caste, Class, and Race.* Garden City, NY: Doubleday.
Craig, Steve. 1992. "Considering Men and the Media." In Steve Craig (Ed.), *Men, Masculinity, and the Media.* Newbury Park, CA: Sage, pp. 1–7.
Cressey, Donald. 1969. *Theft of the Nation.* New York: Harper & Row.
Crinnion, Walter J. 1995. "Are Organic Foods Really Healthier for You?" Retrieved Feb. 17, 2007. Online: http://lookwayup.com/free/organic.htm
Cronin, John, and Robert F. Kennedy, Jr. 1999. *The Riverkeepers: Two Activists Fight to Reclaim Our Environment as a Basic Human Right.* New York: Touchstone.
Crosnoe, Robert. 2006. *Mexican Roots, American Schools: Helping Mexican Immigrant Children Succeed.* Stanford, CA: Stanford University Press.
Crossette, Barbara. 1996. "Hope, and Pragmatism, for U.S. Cities Conference." *New York Times* (June 3): A3.
Currie, Elliott. 1998. *Crime and Punishment in America.* New York: Metropolitan.
Curtiss, Susan. 1977. *Genie: A Psycholinguistic Study of a Modern Day "Wild Child."* New York: Academic Press.
Cyrus, Virginia. 1993. *Experiencing Race, Class, and Gender in the United States.* Mountain View, CA: Mayfield.
Dahl, Robert A. 1961. *Who Governs?* New Haven, CT: Yale University Press.
Dahrendorf, Ralf. 1959. *Class and Class Conflict in an Industrial Society.* Stanford, CA: Stanford University Press.
Dailytroll.com. 2006. "Food as Bonding." Retrieved Dec. 29, 2007. Online: http://dailytroll.com/?p=890
Daly, Kathleen, and Meda Chesney-Lind. 1988. "Feminism and Criminology." *Justice Quarterly,* 5: 497–533.
Daly, Mary. 1973. *Beyond God the Father.* Boston: Beacon.
Daniels, Cora. 2002. "Most Powerful Black Executives." Retrieved July 13, 2002. Online: http://www.fortune.com/indext.jhtml?channel=print_article.jhtml&doc_id=208625
Daniels, Roger. 1993. *Prisoners Without Trial: Japanese-Americans in World War II.* New York: Hill & Wang.
Danziger, Sheldon, and Peter Gottschalk. 1995. *America Unequal.* Cambridge, MA: Harvard University Press.
Dao, James. 2000. "Lockheed Wins $200 Billion Deal for Fighter Jet." *New York Times* (Oct. 27): A1, A9.
Darley, John M., and Thomas R. Shultz. 1990. "Moral Rules: Their Content and Acquisition." *Annual Review of Psychology,* 41: 525–556.
Darling-Hammond, Linda. 2007. "Evaluating 'No Child Left Behind.'" *The Nation* (May 21). Retrieved Feb. 11, 2008. Online: http://www.thenation.com/doc/20070521/darling-hammond
Dart, Bob. 1999. "Kids Get More Screen Time Than School Time." *Austin American-Statesman* (June 28): A1, A5.
Davis, Fred. 1992. *Fashion, Culture, and Identity.* Chicago: University of Chicago Press.
Davis, Kingsley. 1940. "Extreme Social Isolation of a Child." *American Journal of Sociology,* 45 (4): 554–565.
—. 1949. *Human Society.* New York: Macmillan.
Davis, Kingsley, and Wilbert Moore. 1945. "Some Principles of Stratification." *American Sociological Review,* 7 (April): 242–249.
Davis, Sampson, George Jenkins, and Rameck Hunt (with Lisa Frazier Page). 2003. *The Pact: Three Young Men Make a Promise and Fulfill a Dream.* New York: Riverhead.
Day, Jennifer Cheeseman, Alex Janus, and Jessica Davis. 2005. "Computer and Internet Use in the United States: 2003." U.S. Census Bureau, Current Population Reports, P23-208. Washington, DC: U.S. Government Printing Office.
Dean, L. M., F. N. Willis, and J. N. la Rocco. 1976. "Invasion of Personal Space as a Function of Age, Sex and Race." *Psychological Reports,* 38 (3) (pt. 1): 959–965.
Death Penalty Information Center. 2007. "Facts About the Death Penalty, March 15, 2007." Retrieved Mar. 31, 2007. Online: http://www.deathpenaltyinfo.org/FactSheet.pdf
Deegan, Mary Jo. 1988. *Jane Addams and the Men of the Chicago School, 1892–1918.* New Brunswick, NJ: Transaction.
Degher, Douglas, and Gerald Hughes. 1991. "The Identity Change Process: A Field Study of Obesity." *Deviant Behavior,* 12: 385–402.
DeJong, Gerben, Andrew I. Batavia, and Robert Griss. 1989. "America's Neglected Health Minority: Working-Age Persons with Disabilities." *Milbank Quarterly,* 67 (suppl. 2): 311–351.
Delgado, Richard. 1995. "Introduction." In Richard Delgado (Ed.), *Critical Race Theory: The Cutting Edge.* Philadelphia: Temple University Press, pp. xiii–xvi.
DeNavas-Walt, Carmen, Robert W. Cleveland, and Bruce H. Webster, Jr. 2003. "Income in the United States: 2002." U.S. Census Bureau, Current Population Reports, P60–221. Washington, DC: U.S. Government Printing Office.
DeNavas-Walt, Carmen, Bernadette D. Proctor, and Robert J. Mills. 2004. "Income, Poverty, and Health Insurance Coverage in the United States: 2003." U.S. Census Bureau, Current Population Reports, P60–226. Washington, DC: U.S. Government Printing Office.
DeNavas-Walt, Carmen, Bernadette D. Proctor, and Jessica Smith. 2007. "Income, Poverty, and Health Insurance Coverage in the United States: 2006." U.S. Census Bureau, Current Population Reports, P60-233. Retrieved Jan. 12, 2008. Online: http://www.census.gov/prod/2007pubs/p60-233.pdf
DeParle, Jason. 2007. "A Global Trek to Poor Nations, from Poorer Ones." *New York Times* (Dec. 27): A1, A16.
Derber, Charles. 1983. *The Pursuit of Attention: Power and Individualism in Everyday Life.* New York: Oxford University Press.
Devine, John. 1996. *Maximum Security: The Culture of Violence in Inner-City Schools.* Chicago: University of Chicago Press.
De Witt, Karen. 1994. "Wave of Suburban Growth Is Being Fed by Minorities." *New York Times* (Aug. 15): A1, A12.
Dikotter, Frank. 1996. "Culture, 'Race' and Nation: The Formation of Identity in Twentieth Century China." *Journal of International Affairs* (Winter): 590–605.
DiMicco, Joan Morris, and David R. Millen. 2007. "Identity Management: Multiple Presentations of Self in Facebook." Retrieved Jan. 5, 2008. Online: http://www.joandimicco.com/pubs/dimicco-millen-group07.pdf
Dobrzynski, Judith H. 1996. "When Directors Play Musical Chairs." *New York Times* (Nov. 17): F1, F8, F9.
Dollard, John, Neal E. Miller, Leonard W. Doob, O. H. Mowrer, and Robert R. Sears. 1939. *Frustration and Aggression.* New Haven, CT: Yale University Press.
Dolnick, Sam. 2008. "Surrogate Business Makes Birth the Latest Job Outsourced to India." *Austin American-Statesman* (Jan. 1): A10.
Domhoff, G. William. 1978. *The Powers That Be: Processes of Ruling Class Domination in America.* New York: Random House.
—. 1983. *Who Rules America Now? A View for the '80s.* Englewood Cliffs, NJ: Prentice Hall.
—. 2002. *Who Rules America? Power and Politics* (4th ed.). New York: McGraw-Hill.
Driskell, Robyn Bateman, and Larry Lyon. 2002. "Are Virtual Communities True Communities? Examining the Environments and Elements of Community." *City & Community* 1 (4): 1–18.
Du Bois, W. E. B. 1967. *The Philadelphia Negro: A Social Study.* New York: Schocken (orig. pub. 1899).
Dubowitz, Howard, Maureen Black, Raymond H. Starr, Jr., and Susan Zuravin. 1993. "A Conceptual Definition of Child Neglect." *Criminal Justice and Behavior,* 20 (1): 8–26.
Duffy, John. 1976. *The Healers.* New York: McGraw-Hill.
Dunbar, Polly. 2007. "Wombs to Rent: Childless British Couples Pay Indian Women to Carry Their Babies." *The Daily Mail* (Dec. 8). Retrieved Feb. 9, 2008. Online: http://www.dailymail.co.uk/pages/live/articles/news/worldnews.html?in_article_id=500601
Dunlap, Riley E. 1992. "Trends in Public Opinion Toward Environmental Issues: 1965–1990." In Riley E. Dunlap and Angela G. Mertig (Eds.), *American Environmentalism: The U.S. Environmental Movement, 1970–1990.* New York: Taylor & Francis, pp. 89–113.
Dupre, Roslyn, and Paul Gains. 1997. "Fundamental Differences." *Women's Sports & Fitness* (October): 63–68.
Durkheim, Emile. 1933. *The Division of Labor in Society.* Trans. George Simpson. New York: Free Press (orig. pub. 1893).
—. 1956. *Education and Sociology.* Trans. Sherwood D. Fox. Glencoe, IL: Free Press.
—. 1964a. *The Rules of Sociological Method.* Trans. Sarah A. Solovay and John H. Mueller. New York: Free Press (orig. pub. 1895).
—. 1964b. *Suicide.* Trans. John A. Sparkling and George Simpson. New York: Free Press (orig. pub. 1897).
Dye, Thomas R., and Harmon Zeigler. 2006. *The Irony of Democracy: An Uncommon Introduction to American Politics* (13th ed.). Belmont, CA: Wadsworth.
Early, Kevin E. 1992. *Religion and Suicide in the African-American Community.* Westport, CT: Greenwood.
Ebaugh, Helen Rose Fuchs. 1988. *Becoming an EX: The Process of Role Exit.* Chicago: University of Chicago Press.
Egley, Arlen, Jr. 2000. "Highlights of the 1999 National Youth Gang Survey." Washington, DC: U.S. Department of Justice, Office of Justice Programs.
Ehrenreich, Barbara. 2001. *Nickel and Dimed: On (Not) Getting by in America.* New York: Metropolitan.
Ehrlich, Paul R., Anne H. Ehrlich, and Gretchen C. Daily. 1995. *The Stork and the Plow: The Equity Answer to the Human Dilemma.* New Haven, CT: Yale University Press.
Eighner, Lars. 1993. *Travels with Lizbeth.* New York: St. Martin's.
Eisenstein, Zillah R. 1994. *The Color of Gender: Reimaging Democracy.* Berkeley: University of California Press.
Eisler, Benita. 1983. *Class Act: America's Last Dirty Secret.* New York: Franklin Watts.
Eitzen, D. Stanley, and George H. Sage. 1997. *The Sociology of North American Sport* (6th ed.). Dubuque, IA: Brown.
Elkin, Frederick, and Gerald Handel. 1989. *The Child and Society: The Process of Socialization* (5th ed.). New York: Random House.
Elkind, David. 1995. "School and Family in the Postmodern World." *Phi Delta Kappan* (September): 8–21.

Emling, Shelley. 1997a. "Haiti Held in Grip of Another Drought." *Austin American-Statesman* (Sept. 19): A17, A18.

—. 1997b. "In Haiti, It's Resort vs. Reality." *Austin American-Statesman* (Sept. 27): A17, A19.

Engels, Friedrich. 1970. *The Origins of the Family, Private Property, and the State.* New York: International (orig. pub. 1884).

Engerman, Stanley L. 1995. "The Extent of Slavery and Freedom Throughout the World as a Whole and in Major Subareas." In Julian L. Simon (Ed.), *The State of Humanity.* Cambridge, MA: Blackwell, pp. 171–177.

EPA. 2006. "Global Warming: We Can Make a Difference!" Retrieved May 6, 2006. Online: http://www.epa.gov/globalwarming/kids/difference.html

Epidemiological Network for Latin America and the Caribbean. 2000. "HIV and AIDS in the Americas: An Epidemic with Many Faces." Retrieved Nov. 23, 2001. Online: http://www.census.gov/ipc/www/hivaidinamerica.pdf

Epstein, Cynthia Fuchs. 1988. *Deceptive Distinctions: Sex, Gender, and the Social Order.* New Haven, CT: Yale University Press.

Erikson, Kai T. 1962. "Notes on the Sociology of Deviance." *Social Problems*, 9: 307–314.

—. 1964. "Notes on the Sociology of Deviance." In Howard S. Becker (Ed.), *The Other Side: Perspectives on Deviance.* New York: Free Press, pp. 9–21.

—. 1976. *Everything in Its Path: Destruction of Community in the Buffalo Creek Flood.* New York: Simon & Schuster.

—. 1991. "A New Species of Trouble." In Stephen Robert Couch and J. Stephen Kroll-Smith (Eds.), *Communities at Risk: Collective Responses to Technological Hazards.* New York: Land, pp. 11–29.

—. 1994. *A New Species of Trouble: Explorations in Disaster, Trauma, and Community.* New York: Norton.

Esping-Andersen, Gosta. 1990. *The Three Worlds of Welfare Capitalism.* Cambridge, MA: Polity.

Espiritu, Yen Le. 1995. *Filipino American Lives.* Philadelphia: Temple University Press.

ESPN.com. 2006. "A's Bradley: 'I'm Not Going to Cause Problems.'" Retrieved Feb. 26, 2006. Online: http://sports.espn.go.com/mlb/news/story?id=2344878

Essed, Philomena. 1991. *Understanding Everyday Racism.* Newbury Park, CA: Sage.

Esterberg, Kristin G. 1997. *Lesbian and Bisexual Identities: Constructing Communities, Constructing Self.* Philadelphia: Temple University Press.

Etzioni, Amitai. 1975. *A Comparative Analysis of Complex Organizations: On Power, Involvement, and Their Correlates* (rev. ed.). New York: Free Press.

—. 1994. *The Spirit of Community: The Reinvention of American Society.* New York: Touchstone.

Evans, Peter B., and John D. Stephens. 1988. "Development and the World Economy." In Neil J. Smelser (Ed.), *Handbook of Sociology.* Newbury Park, CA: Sage, pp. 739–773.

Fagan, Kevin. 2003. "Shame of the City: Homeless Island." *San Francisco Chronicle* (Nov. 30). Retrieved Apr. 11, 2004. Online: http://www.sfgate.com

Falk, Patricia. 1989. "Lesbian Mothers: Psychological Assumptions in Family Law." *American Psychologist*, 44: 941–947.

Fallon, Patricia, Melanie A. Katzman, and Susan C. Wooley. 1994. *Feminist Perspectives on Eating Disorders.* New York: Guilford.

Faludi, Susan. 1999. *Stiffed: The Betrayal of the American Man.* New York: Morrow.

Farb, Peter. 1973. *Word Play: What Happens When People Talk.* New York: Knopf.

Farley, John E. 2000. *Majority–Minority Relations* (4th ed.). Englewood Cliffs, NJ: Prentice Hall.

Feagin, Joe R. 1991. "The Continuing Significance of Race: Antiblack Discrimination in Public Places." *American Sociological Review*, 56 (February): 101–116.

Feagin, Joe R., and Clairece Booher Feagin. 1994. *Social Problems: A Critical Power–Conflict Perspective* (4th ed.). Englewood Cliffs, NJ: Prentice Hall.

—. 1997. *Social Problems: A Critical Power–Conflict Perspective* (5th ed.). Englewood Cliffs, NJ: Prentice Hall.

—. 2003. *Racial and Ethnic Relations* (7th ed.). Upper Saddle River, NJ: Prentice Hall.

—. 2008. *Racial and Ethnic Relations* (8th ed.). Upper Saddle River, NJ: Prentice Hall.

Feagin, Joe R., Anthony M. Orum, and Gideon Sjoberg (Eds.). 1991. *A Case for the Case Study.* Chapel Hill: University of North Carolina Press.

Feagin, Joe R., and Robert Parker. 1990. *Building American Cities: The Urban Real Estate Game* (2nd ed.). Englewood Cliffs, NJ: Prentice Hall.

Feagin, Joe R., and Melvin P. Sikes. 1994. *Living with Racism: The Black Middle-Class Experience.* Boston: Beacon.

Feagin, Joe R., and Hernán Vera. 1995. *White Racism: The Basics.* New York: Routledge.

Fecht, Josh. 2004. "U.S. Cities Cut Civic Services and Staff to Confront Financial Crisis." Retrieved June 12, 2004. Online: http://www.citymayors.com/report/usfiscal_crisis.html

Federal Bureau of Investigation (FBI). 2007. *Crime in the United States: 2006.* Retrieved Jan. 4, 2008. Online: http://www.fbi.gov/ucr/cius2006

Federal Election Commission. 2006. "Congressional Campaigns Spend $966 Million Through Mid October." Retrieved Mar. 24, 2007. Online: http://www.fec.gov/press/press2006/20061102can/20061102can.html

Ferriss, Susan. 2001. "Cold Spell: U.S. Recession Chills Mexico's Economic Hot Spot." *Austin American-Statesman* (Nov. 25): E1, E4 (based on data from the Bank of Mexico).

Fields, Jason. 2004. *America's Families and Living Arrangements: 2003.* Current Population Reports, P20–553. Washington, DC: U.S. Census Bureau.

Fiffer, Steve, and Sharon Sloan Fiffer. 1994. *50 Ways to Help Your Community.* New York: Mainstream/Doubleday.

Findlay, Deborah A., and Leslie J. Miller. 1994. "Through Medical Eyes: The Medicalization of Women's Bodies and Women's Lives." In B. Singh Bolaria and Harley D. Dickinson (Eds.), *Health, Illness, and Health Care in Canada* (2nd ed.). Toronto: Harcourt, pp. 276–306.

Findlay, Steven. 1997. "Health Care Industry Feeling a Little Discomfort." *USA Today* (Aug. 6): 4B.

Fine, Michelle, and Lois Weis. 1998. *The Unknown City: The Lives of Poor and Working-Class Young People.* Boston: Beacon.

Finley, M. I. 1980. *Ancient Slavery and Modern Ideology.* New York: Viking.

Firestone, Shulamith. 1970. *The Dialectic of Sex.* New York: Morrow.

Fishbein, Diana H., and Susan E. Pease. 1996. *The Dynamics of Drug Abuse.* Boston: Allyn & Bacon.

Fisher-Thompson, Donna. 1990. "Adult Sex-Typing of Children's Toys." *Sex Roles*, 23: 291–303.

Fjellman, Stephen M. 1992. *Vinyl Leaves: Walt Disney World & America.* Boulder, CO: Westview.

Flanagan, William G. 2002. *Urban Sociology: Images and Structures* (4th ed.). Boston: Allyn & Bacon.

Flexner, Abraham. 1910. *Medical Education in the United States and Canada.* New York: Carnegie Foundation.

Florida, Richard, and Martin Kenney. 1991. "Transplanted Organizations: The Transfer of Japanese Industrial Organization to the U.S." *American Sociological Review*, 56 (3): 381–398.

Foderaro, Lisa W. 2007. "Child Wants Cellphone; Reception Is Mixed." *New York Times* (Mar. 29): E1–E2.

Fong-Torres, Ben. 2007. "Hungry Heart." *New York Times Book Review* (Feb. 4): 11.

Forbes. 2007. "The World's Richest People." *Forbes* (Mar. 26): 104–208.

—. 2008. "The World's Billionaires." Retrieved Mar. 19, 2008. Online: http://www.forbes.com/2008/03/05/richest-people-billionaires-billionaires08-cx_lk_0305billie_land.html

Foucault, Michel. 1979. *Discipline and Punish: The Birth of the Prison.* New York: Vintage.

—. 1994. *The Birth of the Clinic: An Archeology of Medical Perception.* New York: Vintage (orig. pub. 1963).

Fox, Julia R., Glory Koloen, and Volkan Sahin. 2007. "No Joke: A Comparison of Substance in *The Daily Show with Jon Stewart* and Broadcast Network Television Coverage of the 2004 Presidential Election Campaign." *Journal of Broadcast & Electronic Media*, 51(2): 213–227.

Fox News. 1999. "Divorce Rates for Children of Broken Families on Decline." Retrieved Sept. 18, 1999. Online: http://www.foxnews.com/js_index.sml?content=/news/national/0812/d_ap_0812_11.sml

Frankenberg, Ruth. 1993. *White Women, Race Matters: The Social Construction of Whiteness.* Minneapolis: University of Minnesota Press.

Franklin, John Hope. 1980. *From Slavery to Freedom: A History of Negro Americans.* New York: Vintage.

Freidson, Eliot. 1965. "Disability as Social Deviance." In Marvin B. Sussman (Ed.), *Sociology and Rehabilitation.* Washington, DC: American Sociology Association, pp. 71–99.

—. 1970. *Profession of Medicine.* New York: Dodd, Mead.

—. 1986. *Professional Powers.* Chicago: University of Chicago Press.

French, Howard W. 2008. "Lives of Grinding Poverty, Untouched by China's Boom." *New York Times* (Jan. 13): YT4.

French, Sally. 1999. "The Wind Gets in My Way." In Mairian Corker and Sally French (Eds.), *Disability Discourse.* Buckingham, England: Open University Press, pp. 21–27.

Freud, Sigmund. 1924. *A General Introduction to Psychoanalysis* (2nd ed.). New York: Boni & Liveright.

Freudenberg, Nicholas, and Carl Steinsapir. 1992. "Not in Our Backyards: The Grassroots Environmental Movement." In Riley E. Dunlap and Angela G. Mertig (Eds.), *American Environmentalism: The U.S. Environmental Movement, 1970–1990.* New York: Taylor & Francis, pp. 27–37.

Freudenheim, Milt. 2007. "Showdown Looms in Congress Over Drug Advertising on TV." *New York Times* (Jan. 22): A1.

Friedman, Robert I. 2000. *Red Mafiya: How the Russian Mob Has Invaded America.* New York: Little, Brown.

Friedman, Thomas L. 2005a. "It's a Flat World, After All." *New York Times Magazine* (Apr. 3): 33ff.

—. 2005b. *The World Is Flat: A Brief History of the Twenty-First Century.* New York: Farrar, Straus & Giroux.

Friedmann, John. 1995. "The World City Hypothesis." In Paul L. Knox and Peter J. Taylor (Eds.), *World Cities in a World-System.* Cambridge, England: Cambridge University Press, pp. 317–331.

Frisbie, W. Parker, and John D. Kasarda. 1988. "Spatial Processes." In Neil Smelser (Ed.), *The Handbook of Sociology.* Newbury Park, CA: Sage, pp. 629–666.

Gabriel, Trip. 1996. "High-Tech Pregnancies Test Hope's Limits." *New York Times* (Jan. 7): 1, 10–11.

Gaines, Donna. 1991. *Teenage Wasteland: Suburbia's Dead-End Kids.* New York: Harper-Perennial.

Galbraith, John Kenneth. 1985. *The New Industrial State* (4th ed.). Boston: Houghton Mifflin.

Gallagher, Charles A. 2003. "Miscounting Race: Explaining Whites' Misperceptions of Racial Group Size." *Sociological Perspectives*, 46 (3): 381–396.

Gambino, Richard. 1975. *Blood of My Blood.* New York: Doubleday/Anchor.

Gamson, William. 1990. *The Strategy of Social Protest* (2nd ed.). Belmont, CA: Wadsworth.

Gandara, Ricardo. 1995. "*Dichos de la Vida:* Homespun Proverbs Link Hispanic Culture's Past with the Present." *Austin American-Statesman* (Jan. 21): E1, E10.

Gans, Herbert. 1974. *Popular Culture and High Culture: An Analysis and Evaluation of Tastes.* New York: Basic.

—. 1982. *The Urban Villagers: Group and Class in the Life of Italian Americans* (updated and expanded ed.; orig. pub. 1962). New York: Free Press.

Garbarino, James. 1989. "The Incidence and Prevalence of Child Maltreatment." In L. Ohlin and M. Tonry (Eds.), *Family Violence.* Chicago: University of Chicago Press, pp. 219–261.

Garber, Judith A., and Robyne S. Turner. 1995. "Introduction." In Judith A. Garber and Robyne S. Turner (Eds.), *Gender in Urban Research.* Thousand Oaks, CA: Sage, pp. x–xxvi.

Gardner, Carol Brooks. 1989. "Analyzing Gender in Public Places: Rethinking Goffman's Vision of Everyday Life." *American Sociologist*, 20 (Spring): 42–56.

Garfinkel, Harold. 1967. *Studies in Ethnomethodology.* Englewood Cliffs, NJ: Prentice Hall.

Garfinkel, Irwin, and Sara S. McLanahan. 1986. *Single Mothers and Their Children: A New American Dilemma.* Washington, DC: Urban Institute Press.

Gargan, Edward A. 1996. "An Indonesian Asset Is Also a Liability." *New York Times* (Mar. 16): 17, 18.

Garreau, Joel. 1991. *Edge City: Life on the New Frontier.* New York: Doubleday.

Gary, Keahn. 2007. "New Report Reveals Top Ten Problems Facing U.S. Students." Retrieved Feb. 11, 2008. Online: http://www.nbc26.com/news/trends/8000952.html

Gawande, Atul. 2002. *Complications: A Surgeon's Notes on an Imperfect Science.* New York: Picador.

Gaylin, Willard. 1992. *The Male Ego.* New York: Viking/Penguin.

Geertz, Clifford. 1966. "Religion as a Cultural System." In Michael Banton (Ed.), *Anthropological Approaches to the Study of Religion.* London: Tavistock, pp. 1–46.

Gelles, Richard J., and Murray A. Straus. 1988. *Intimate Violence: The Definitive Study of the Causes and Consequences of Abuse in the American Family.* New York: Simon & Schuster.

General Motors. 2007. "The Company: Investor Information—Board of Directors." Retrieved Apr. 7, 2007. Online: http://www.gm.com/company/investor_information/corp_gov/board.html#board

George, Susan. 1993. "A Fate Worse Than Debt." In William Dan Perdue (Ed.), *Systemic Crisis: Problems in Society, Politics, and World Order.* Fort Worth: Harcourt, pp. 85–96.

Gerbner, George, Larry Gross, Michael Morton, and Nancy Signorielli. 1987. "Charting the Mainstream: Television's Contributions to Political Orientations." In Donald Lazere (Ed.), *American Media and Mass Culture: Left Perspectives.* Berkeley: University of California Press, pp. 441–464.

Gereffi, Gary. 1994. "The International Economy and Economic Development." In Neil J. Smelser and Richard Swedberg (Eds.), *The Handbook of Economic Sociology.* Princeton, NJ: Princeton University Press, pp. 206–233.

Gerson, Kathleen. 1993. *No Man's Land: Men's Changing Commitment to Family and Work.* New York: Basic.

Gerstel, Naomi, and Harriet Engel Gross. 1995. "Gender and Families in the United States: The Reality of Economic Dependence." In Jo Freeman (Ed.), *Women: A Feminist Perspective* (5th ed.). Mountain View, CA: Mayfield, pp. 92–127.

Gerth, Hans H., and C. Wright Mills. 1946. *From Max Weber: Essays in Sociology.* New York: Oxford University Press.

Gibbs, Lois Marie, as told to Murray Levine. 1982. *Love Canal: My Story.* Albany: SUNY Press.

Gibbs, Nancy. 1994. "Home Sweet School." *Time* (Oct. 31): 62–63.

Giffin, Mary, and Carol Felsenthal. 1983. *A Cry for Help.* New York: Doubleday.

Gilbert, Dennis. 2003. *The American Class Structure in an Age of Growing Inequality* (6th ed.). Belmont, CA: Wadsworth.

Gilkey, Langdon. 1993. "Theories in Science and Religion." In James Huchingson (Ed.), *Religion and the Natural Sciences: The Range of Engagement*. Fort Worth, TX: Harcourt, pp. 61–65.

Gill, Derek. 1994. "A National Health Service: Principles and Practice." In Peter Conrad and Rochelle Kern (Eds.), *The Sociology of Health and Illness* (4th ed.). New York: St. Martin's, pp. 480–494.

Gilligan, Carol. 1982. *In a Different Voice: Psychological Theory and Women's Development*. Cambridge, MA: Harvard University Press.

Gilman, Charlotte Perkins. 1976. *His Religion and Hers*. Westport, CT: Hyperion (orig. pub. 1923).

Gilmore, David D. 1990. *Manhood in the Making: Cultural Concepts of Masculinity*. New Haven, CT: Yale University Press.

Ginorio, Angela, and Michelle Huston. 2000. *¡Sí Puede! Yes, We Can: Latinas in School*. Washington, DC: American Association of University Women.

Giuliano, Traci A., Kathryn E. Popp, and Jennifer L. Knight. 2000. "Footballs Versus Barbies: Childhood Play Activities as Predictors of Sports Participation by Women." *Sex Roles*, 42: 159–181.

Glanz, James, and Edward Wong. 2003. "'97 Report Warned of Foam Damaging Tiles." *New York Times* (Feb. 4): A1–A26.

Glastris, Paul. 1990. "The New Way to Get Rich." *U.S. News & World Report* (May 7): 26–36.

Goffman, Erving. 1956. "The Nature of Deference and Demeanor." *American Anthropologist*, 58: 473–502.

—. 1959. *The Presentation of Self in Everyday Life*. Garden City, NY: Doubleday.

—. 1961a. *Asylums: Essays on the Social Situation of Mental Patients and Other Inmates*. Chicago: Aldine.

—. 1961b. *Encounters: Two Studies in the Sociology of Interaction*. London: Routledge and Kegan Paul.

—. 1963a. *Behavior in Public Places: Notes on the Social Structure of Gatherings*. New York: Free Press.

—. 1963b. *Stigma: Notes on the Management of Spoiled Identity*. Englewood Cliffs, NJ: Prentice Hall.

—. 1967. *Interaction Ritual: Essays on Face to Face Behavior*. Garden City, NY: Anchor.

—. 1974. *Frame Analysis: An Essay on the Organization of Experience*. Boston: Northeastern University Press.

Gold, Rachel Benson, and Cory L. Richards. 1994. "Securing American Women's Reproductive Health." In Cynthia Costello and Anne J. Stone (Eds.), *The American Woman 1994–95*. New York: Norton, pp. 197–222.

Goldberg, Robert A. 1991. *Grassroots Resistance: Social Movements in Twentieth Century America*. Belmont, CA: Wadsworth.

Goode, Erich. 1996. "The Stigma of Obesity." In Erich Goode (Ed.), *Social Deviance*. Boston: Allyn & Bacon, pp. 332–340.

Goode, William J. 1960. "A Theory of Role Strain." *American Sociological Review*, 25: 483–496.

—. 1982. "Why Men Resist." In Barrie Thorne with Marilyn Yalom (Eds.), *Rethinking the Family: Some Feminist Questions*. New York: Longman, pp. 131–150.

Goodman, Peter S. 1996. "The High Cost of Sneakers." *Austin American-Statesman* (July 7): F1, F6.

Goodnough, Abby, and Jennifer Steinhauer. 2006. "Senate's Failure to Agree on Immigration Plan Angers Workers and Employers Alike." *New York Times* (Apr. 9): A35.

Gordon, Milton. 1964. *Assimilation in American Life: The Role of Race, Religion, and National Origins*. New York: Oxford University Press.

Gotham, Kevin Fox. 1999. "Political Opportunity, Community Identity, and the Emergence of a Local Anti-Expressway Movement." *Social Problems*, 46: 332–354.

Gottdiener, Mark. 1985. *The Social Production of Urban Space*. Austin: University of Texas Press.

Gray, Gary Norris. 2007. "The NCAA's Shabby Record with Black Football Coaches." Retrieved Jan. 26, 2008. Online: http://www.blackathlete.net/artman2/publish/Commentary_1/The_NCAA_s_Shabby_Record_With_Black_Football_Coaches.shtml

Gray, Paul. 1993. "Camp for Crusaders." *Time* (Apr. 19): 40.

Green, Donald E. 1977. *The Politics of Indian Removal: Creek Government and Society in Crisis*. Lincoln: University of Nebraska Press.

Greenberg, Edward S., and Benjamin I. Page. 1993. *The Struggle for Democracy*. New York: HarperCollins.

—. 2002. *The Struggle for Democracy* (5th ed.). Boston: Allyn & Bacon.

Greendorfer, Susan L. 1993. "Gender Role Stereotypes and Early Childhood Socialization." *Psychology of Women Quarterly*, 18: 85–104.

Guha, Ramachandra. 2004. "The Sociology of Suicide." Retrieved Dec. 20, 2007. Online: http://www.indiatogether.org/2004/aug/rgh-suicide.htm

Hahn, Harlan. 1987. "Civil Rights for Disabled Americans: The Foundation of a Political Agenda." In Alan Gartner and Tom Joe (Eds.), *Images of the Disabled, Disabling Images*. New York: Praeger, pp. 181–203.

—. 1997. "Advertising the Acceptably Employable Image." In Lennard J. Davis (Ed.), *The Disability Studies Reader*. New York: Routledge, pp. 172–186.

Haines, Valerie A. 1997. "Spencer and His Critics." In Charles Camic (Ed.), *Reclaiming the Sociological Classics: The State of the Scholarship*. Malden, MA: Blackwell, pp. 81–111.

Halberstadt, Amy G., and Martha B. Saitta. 1987. "Gender, Nonverbal Behavior, and Perceived Dominance: A Test of the Theory." *Journal of Personality and Social Psychology*, 53: 257–272.

Hale-Benson, Janice E. 1986. *Black Children: Their Roots, Culture and Learning Styles* (rev. ed.). Provo, UT: Brigham Young University Press.

Hall, Edward. 1966. *The Hidden Dimension*. New York: Anchor/Doubleday.

Haraway, Donna. 1994. "A Cyborgo Manifesto: Science, Technology, and Socialist-Feminism in the Late Twentieth Century." In Anne C. Herrmann and Abigail J. Stewart (Eds.), *Theorizing Feminism: Parallel Trends in the Humanities and Social Sciences*. Boulder, CO: Westview, pp. 424–457.

Harlow, Harry F., and Margaret Kuenne Harlow. 1962. "Social Deprivation in Monkeys." *Scientific American*, 207 (5): 137–146.

—. 1977. "Effects of Various Mother–Infant Relationships on Rhesus Monkey Behaviors." In Brian M. Foss (Ed.), *Determinants of Infant Behavior* (vol. 4). London: Methuen, pp. 15–36.

Harrington, Michael. 1985. *The New American Poverty*. New York: Viking/Penguin.

Harris, Anthony, and James W. Shaw. 2000. "Looking for Patterns: Race, Class, and Crime." In Joseph F. Sheley (Ed.), *Criminology: A Contemporary Handbook* (3rd ed.). Belmont, CA: Wadsworth, pp. 128–163.

Harris, Chauncey D., and Edward L. Ullman. 1945. "The Nature of Cities." *Annals of the Academy of Political and Social Sciences* (November): 7–17.

Harris, Marvin. 1974. *Cows, Pigs, Wars, and Witches*. New York: Random House.

—. 1985. *Good to Eat: Riddles of Food and Culture*. New York: Simon & Schuster.

Harrison, Algea O., Melvin N. Wilson, Charles J. Pine, Samuel Q. Chan, and Raymond Buriel. 1990. "Family Ecologies of Ethnic Minority Children." *Child Development*, 61 (2): 347–362.

Hartmann, Heidi. 1976. "Capitalism, Patriarchy, and Job Segregation by Sex." *Signs: Journal of Women in Culture and Society*, 1 (Spring): 137–169.

—. 1981. "The Unhappy Marriage of Marxism and Feminism." In Lydia Sargent (Ed.), *Women and Revolution*. Boston: South End.

Haseler, Stephen. 2000. *The Super-Rich: The Unjust New World of Global Capitalism*. New York: St. Martin's.

Hattori, James. 2006. "Marathoner Runs Race Against Hunger." MSNBC.com (Jan. 20).

Hauchler, Ingomar, and Paul M. Kennedy (Eds.). 1994. *Global Trends: The World Almanac of Development and Peace*. New York: Continuum.

Haught, John F. 1995. *Science & Religion: From Conflict to Conversation*. New York: Paulist.

Haviland, William A. 1993. *Cultural Anthropology* (7th ed.). Orlando, FL: Harcourt.

—. 2002. *Cultural Anthropology* (11th ed.). Belmont, CA: Wadsworth.

Hawley, Amos. 1950. *Human Ecology*. New York: Ronald.

—. 1981. *Urban Society* (2nd ed.). New York: Wiley.

Headley, Bernard. 1985. "The Atlanta Establishment and the Atlanta Tragedy." *Phylon*, 46(4): 333–340.

Healey, Joseph F. 2002. *Race, Ethnicity, Gender, and Class: The Sociology of Group Conflict and Change* (3rd ed.). Thousand Oaks, CA: Pine Forge.

Health Affairs. 2005. "U.S. Health Spending Projections for 2004–2014." Retrieved May 5, 2006. Online: http://content.healthaffairs.org/cgi/content/full/hlthaff.w5.74/DC1

Heldrich Center for Workforce Development. 2003. "Work Trends Survey of Employers About People with Disabilities." Retrieved Aug. 23, 2003. Online: http://www.heldrich.rutgers.edu

Henley, Nancy. 1977. *Body Politics: Power, Sex, and Nonverbal Communication*. Englewood Cliffs, NJ: Prentice Hall.

Heritage, John. 1984. *Garfinkel and Ethnomethodology*. Cambridge, MA: Polity.

Herrnstein, Richard J., and Charles Murray. 1994. *The Bell Curve: Intelligence and Class Structure in American Life*. New York: Free Press.

Heshka, Stanley, and Yona Nelson. 1972. "Interpersonal Speaking Distances as a Function of Age, Sex, and Relationship." *Sociometry*, 35 (4): 491–498.

Hibbard, David R., and Duane Buhrmester. 1998. "The Role of Peers in the Socialization of Gender-Related Social Interaction Styles." *Sex Roles*, 39: 185–203.

Higginbotham, Elizabeth. 1994. "Black Professional Women: Job Ceilings and Employment Sectors." In Maxine Baca Zinn and Bonnie Thornton Dill (Eds.), *Women of Color in U.S. Society*. Philadelphia: Temple University Press, pp. 113–131.

Hirschi, Travis. 1969. *Causes of Delinquency*. Berkeley: University of California Press.

Hochschild, Arlie Russell. 1983. *The Managed Heart: Commercialization of Human Feeling*. Berkeley: University of California Press.

—. 1997. *The Time Bind: When Work Becomes Home and Home Becomes Work*. New York: Metropolitan.

—. 2003. *The Commercialization of Intimate Life: Notes from Home and Work*. Berkeley: University of California Press.

Hochschild, Arlie Russell, with Ann Machung. 1989. *The Second Shift: Working Parents and the Revolution at Home*. New York: Viking/Penguin.

Hochschild, Jennifer L. 1995. *Facing Up to the American Dream: Race, Class, and the Soul of the Nation*. Princeton, NJ: Princeton University Press.

Hodson, Randy, and Robert E. Parker. 1988. "Work in High Technology Settings: A Review of the Empirical Literature." *Research in the Sociology of Work*, 4: 1–29.

Hodson, Randy, and Teresa A. Sullivan. 2008. *The Social Organization of Work* (4th ed.). Belmont, CA: Wadsworth.

Hoffnung, Michele. 1995. "Motherhood: Contemporary Conflict for Women." In Jo Freeman (Ed.), *Women: A Feminist Perspective* (5th ed.). Mountain View, CA: Mayfield, pp. 162–181.

Holland, Dorothy C., and Margaret A. Eisenhart. 1990. *Educated in Romance: Women, Achievement, and College Culture*. Chicago: University of Chicago Press.

Holmes, Steven A. 1997. "Bringing Hope and Education to the Reservation." *New York Times* Education Life Supplement (Aug. 3): 28–29, 34–35.

Hoose, Phillip M. 1989. *Necessities: Racial Barriers in American Sports*. New York: Random House.

Hooyman, Nancy R. R., and H. Asuman Kiyak. 2002. *Social Gerontology: A Multidisciplinary Approach* (6th ed.). Boston: Allyn & Bacon.

Horan, Patrick M. 1978. "Is Status Attainment Research Atheoretical?" *American Sociological Review*, 43: 534–541.

Horovitz, Brude. 2006. "More University Students Call for Organic, 'Sustainable' Food." *USA Today* (Sept. 26). Retrieved Feb. 17, 2007. Online: http://www.usatoday.com/money/industries/food/2006-09-26-college-food-usat_x.htm

Hostetler, A. J. 1994. "U.S. Death Rate Falls to Lowest Level Ever Despite Rise in AIDS." *Austin American-Statesman* (Dec. 16): A4.

Hoyt, Homer. 1939. *The Structure and Growth of Residential Neighborhoods in American Cities*. Washington, DC: Federal Housing Administration.

Hughes, Everett C. 1945. "Dilemmas and Contradictions of Status." *American Journal of Sociology*, 50: 353–359.

Hull, Gloria T., Patricia Bell-Scott, and Barbara Smith. 1982. *All the Women Are White, All the Blacks Are Men, But Some of Us Are Brave*. Old Westbury, NY: Feminist.

Hulse, Carl, and Rachel L. Swarns. 2006. "Conservatives Stand Firm on Immigration." *New York Times* (Mar. 31): A12.

Hurst, Charles E. 2007. *Social Inequality: Forms, Causes, and Consequences* (6th ed.). Boston: Allyn & Bacon.

Hurtado, Aida. 1996. *The Color of Privilege: Three Blasphemies on Race and Feminism*. Ann Arbor: University of Michigan Press.

Huston, Aletha C. 1985. "The Development of Sex Typing: Themes from Recent Research." *Developmental Review*, 5: 2–17.

Huyssen, Andreas. 1984. *After the Great Divide*. Bloomington: Indiana University Press.

Hwang, S. S., R. Saenz, and B. E. Aguirre. 1995. "The SES-Selectivity of Interracially Married Asians." *International Migration Review*, 29: 469–491.

Hynes, H. Patricia. 1990. *Earth Right: Every Citizen's Guide*. Rocklin, CA: Prima.

Idaho Association of Soil Conservation Districts. 2004. "Organic Pest Control." Retrieved Feb. 17, 2007. Online: http://www.oneplan.org/Crop/OrganicPestCtrl.shtml

IFAD. 2002. "IFAD in China: The Rural Poor Speak." Rome, Italy: International Fund for Agricultural Development. Retrieved Jan. 22, 2008. Online: http://www.ifad.org/media/success/China.pdf

Inciardi, James A., Ruth Horowitz, and Anne E. Pottieger. 1993. *Street Kids, Street Drugs, Street Crime: An Examination of Drug Use and Serious Delinquency in Miami*. Belmont, CA: Wadsworth.

International Monetary Fund. 1992. *World Economic Outlook*. Washington, DC: International Development Fund.

Ishikawa, Kaoru. 1984. "Quality Control in Japan." In Naoto Sasaki and David Hutchins (Eds.), *The Japanese Approach to Product Quality: Its Applicability to the West*. Oxford: Permagon, pp. 1–5.

Jackson, Kenneth T. 1985. *Crabgrass Frontier: The Suburbanization of the United States*. New York: Oxford University Press.

Jacobs, Andrew. 2008. "One-Child Policy Lifted for Quake Victims' Parents." *New York Times* (May 27): A1.

Jameson, Fredric. 1984. "Postmodernism, or, The Cultural Logic of Late Capitalism." *New Left Review*, 146: 59–92.

Janis, Irving. 1972. *Victims of Groupthink*. Boston: Houghton Mifflin.
—-. 1989. *Crucial Decisions: Leadership in Policymaking and Crisis Management*. New York: Free Press.
Jankowski, Martin Sanchez. 1991. *Islands in the Street: Gangs and American Urban Society*. Berkeley: University of California Press.
Jaramillo, P. T., and Jesse T. Zapata. 1987. "Roles and Alliances Within Mexican-American and Anglo Families." *Journal of Marriage and the Family*, 49 (November): 727–735.
Jary, David, and Julia Jary. 1991. *The Harper Collins Dictionary of Sociology*. New York: HarperPerennial.
Jensen, Robert. 1995. "Men's Lives and Feminist Theory." *Race, Gender & Class*, 2 (2): 111–125.
Jewell, K. Sue. 1993. *From Mammy to Miss America and Beyond: Cultural Images and the Shaping of US Social Policy*. New York: Routledge.
Johnson, Allan G. 2000. *The Blackwell Dictionary of Sociology* (2nd ed.). Malden, MA: Blackwell.
Johnson, Claudia. 1994. *Stifled Laughter: One Woman's Story About Fighting Censorship*. Golden, CO: Fulcrum.
Johnson, Earvin "Magic," with William Novak. 1992. *My Life*. New York: Fawcett Crest.
Johnston, David Cay. 2002. "As Salary Grows, So Does a Gender Gap." *New York Times* (May 12): BU8.
Joint Center for Political and Economic Studies. 2000. "Joint Center Releases 1999 Count of Black Elected Officials." Retrieved Sept. 25, 2002. Online: http://www.jointcenter.org/pressrel/2000_beo.htm
Joliet Junior College. 2005. "History of Joliet Junior College." Retrieved Apr. 21, 2005. Online: http://www.jjc.edu/campus/history
Journalism.org. 2004. "The State of the News Media 2004." Retrieved Feb. 23, 2008. Online: http://www.stateofthenewsmedia.org/2007/index.asp
Kabagarama, Daisy. 1993. *Breaking the Ice: A Guide to Understanding People from Other Cultures*. Boston: Allyn & Bacon.
Kahn, Joseph, and Jim Yardley. 2008. "As China Roars, Pollution Reaches Deadly Extremes." *New York Times* (Aug. 26): A1, A6–7.
Kalmijn, Matthijs. 1998. "Intermarriage and Homogamy: Causes, Patterns, Trends." *Annual Review of Sociology*, 24: 395–422.
Kanter, Rosabeth Moss. 1983. *The Change Masters: Innovation and Entrepreneurship in the American Corporation*. New York: Simon & Schuster.
—. 1985. "All That Is Entrepreneural Is Not Gold." *Wall Street Journal* (July 22): 18.
—. 1993. *Men and Women of the Corporation*. New York: Basic (orig. pub. 1977).
Kaplan, Robert D. 1996. "Cities of Despair." *New York Times* (June 6): A19.
Kaspar, Anne S. 1986. "Consciousness Reevaluated: Interpretive Theory and Feminist Scholarship." *Sociological Inquiry*, 56 (1): 30–49.
Katz, Michael B. 1989. *The Undeserving Poor: From the War on Poverty to the War on Welfare*. New York: Pantheon.
Katzer, Jeffrey, Kenneth H. Cook, and Wayne W. Crouch. 1991. *Evaluating Information: A Guide for Users of Social Science Research*. New York: McGraw-Hill.
Kaufman, Gayle. 1999. "The Portrayal of Men's Family Roles in Television Commercials." *Sex Roles*, 313: 439–451.
Keller, James. 1994. "'I Treasure Each Moment.'" *Parade* (Sept. 4): 4–5.
Kemp, Alice Abel. 1994. *Women's Work: Degraded and Devalued*. Englewood Cliffs, NJ: Prentice Hall.
Kendall, Diana. 1980. Square Pegs in Round Holes: Non-Traditional Students in Medical Schools. Unpublished doctoral dissertation, Department of Sociology, the University of Texas at Austin.
—. 2002. *The Power of Good Deeds: Privileged Women and the Social Reproduction of the Upper Class*. Lanham, MD: Rowman & Littlefield.
—. 2004. *Social Problems in a Diverse Society* (3rd ed.). Boston: Allyn & Bacon.
—. 2005. *Framing Class: Media Representations of Wealth and Poverty in America*. Lanham, MD: Rowman & Littlefield.
—. 2008. *Members Only: Elite Clubs and the Process of Exclusion*. Lanham, MD: Rowman & Littlefield.
Kendall, Diana, Rick Linden, and Jane Murray. 2008. *Sociology in Our Times: The Essentials* (4th Canadian edition). Scarborough, Ontario: Nelson Thomson Learning.
Kendall, Diana, Jane Lothian Murray, and Rick Linden. 2004. *Sociology in Our Times* (3rd Canadian edition). Scarborough, Ontario: Nelson Thomson Learning.
Kennedy, Paul. 1993. *Preparing for the Twenty-First Century*. New York: Random House.
Kenyon, Kathleen. 1957. *Digging Up Jericho*. London: Benn.
Kerbo, Harold R. 2000. *Social Stratification and Inequality: Class Conflict in Historical, Comparative, and Global Perspective* (4th ed.). New York: McGraw-Hill.
Kershaw, Sarah. 2005. "Crisis of Indian Children Intensifies as Families Fail." *New York Times* (Apr. 5): A13.
Keyfitz, Nathan. 1994. "The Scientific Debate: Is Population Growth a Problem? An Interview with Nathan Keyfitz." *Harvard International Review* (Fall): 10–11, 74.
Khan, Mahmood Hasan. 2001. "Rural Poverty in Developing Countries: Implications for Public Policy: Economic Issues No. 26." Washington, DC: International Monetary Fund. Retrieved Jan. 21, 2008. Online: http://www.imf.org/external/pubs/ft/issues/issues26/index.htm
Kidron, Michael, and Ronald Segal. 1995. *The State of the World Atlas*. New York: Penguin.
Kilborn, Peter T. 1997. "Illness Is Turning into Financial Catastrophe for More of the Uninsured." *New York Times* (Aug. 1): A10.
Kilbourne, Jean. 1999. *Deadly Persuasion: The Addictive Power of Advertising*. New York: Simon & Schuster.
Kilgannon, Corey. 2006. "3 Perspectives on Immigration, from the Inside: The Immigrant; Dreaming of U.S. Citizenship." *New York Times* (Apr. 16): A23.
Killian, Lewis. 1984. "Organization, Rationality, and Spontaneity in the Civil Rights Movement." *American Sociological Review*, 49: 770–783.
Kim, Ryan. 2006. "The World's a Cell-Phone Stage." *San Francisco Chronicle* (Feb. 27). Retrieved Mar. 30, 2007. Online: http://http://www.sfgate.com/cgi-bin/article.cgi?file=/chronicle/archive/2006/02/27/BUG2IHECTO1.DTL&type=printable
Kimmel, Michael S., and Michael A. Messner. 2004. *Men's Lives* (6th ed.). Boston: Allyn & Bacon.
King, Gary, Robert O. Keohane, and Sidney Verba. 1994. *Designing Social Inquiry: Scientific Inference in Qualitative Research*. Princeton, NJ: Princeton University Press.
King, Leslie, and Madonna Harrington Meyer. 1997. "The Politics of Reproductive Benefits: U.S. Insurance Coverage of Contraceptive and Infertility Treatments." *Gender and Society*, 11 (1): 8–30.
Kirkpatrick, P. 1994. "Triple Jeopardy: Disability, Race and Poverty in America." *Poverty and Race*, 3: 1–8.
Kitsuse, John I. 1980. "Coming Out All Over: Deviance and the Politics of Social Problems." *Social Problems*, 28: 1–13.
Klein, Alan M. 1993. *Little Big Men: Bodybuilding Subculture and Gender Construction*. Albany: SUNY Press.
Knox, Paul L., and Peter J. Taylor (Eds.). 1995. *World Cities in a World-System*. Cambridge, England: Cambridge University Press.
Knudsen, Dean D. 1992. *Child Maltreatment: Emerging Perspectives*. Dix Hills, NY: General Hall.
Kohl, Beth. 2007. "On Indian Surrogates." *The Huffington Post* (Oct. 30). Retrieved Feb. 9, 2008. Online: http://www.huffingtonpost.com/beth-kohl/on-indian-surrogates_b_70425.html
Kohlberg, Lawrence. 1969. "Stage and Sequence: The Cognitive–Developmental Approach to Socialization." In David A. Goslin (Ed.), *Handbook of Socialization Theory and Research*. Chicago: Rand McNally, pp. 347–480.
—. 1981. *The Philosophy of Moral Development: Moral Stages and the Idea of Justice*, vol. 1: *Essays on Moral Development*. San Francisco: Harper & Row.
Kohn, Alfie. 2001. "Five Reasons to Stop Saying 'Good Job!'" Retrieved Jan. 3, 2008. Online: http://www.alfiekohn.org/parenting/gj.htm
Kohn, Melvin L. 1977. *Class and Conformity: A Study in Values* (2nd ed.). Homewood, IL: Dorsey.
Kohn, Melvin L., Atsushi Naoi, Carrie Schoenbach, Carmi Schooler, and Kazimierz M. Slomczynski. 1990. "Position in the Class Structure and Psychological Functioning in the United States, Japan, and Poland." *American Journal of Sociology*, 95: 964–1008.
Kolata, Gina. 1993. "Fear of Fatness: Living Large in a Slimfast World." *Austin American-Statesman* (Jan. 3): C1, C6.
Korsmeyer, Carolyn. 1981. "The Hidden Joke: Generic Uses of Masculine Terminology." In Mary Vetterling-Braggin (Ed.), *Sexist Language: A Modern Philosophical Analysis*. Totowa, NJ: Littlefield, Adams, pp. 116–131.
Korten, David C. 1996. *When Corporations Rule the World*. West Hartford, CT: Kumarian.
Kosmin, Barry A., and Seymour P. Lachman. 1993. *One Nation Under God: Religion in Contemporary American Society*. New York: Crown.
Kozol, Jonathan. 1991. *Savage Inequalities: Children in America's Schools*. New York: Crown.
Kranning, Antoinette, and Lee Ehman. 1999. "Help! I'm Lost in Cyberspace." *Social Education*, 63(3): 152–156. Retrieved Mar. 4, 2007. Online: http://www.indiana.edu/~leeehman/mystery.pdf
Kristof, Nicholas D. 2006. "Looking for Islam's Luthers." *New York Times* (Oct. 15): A22.
Kroloff, Charles A. 1993. *54 Ways You Can Help the Homeless*. Southport, CT: Hugh Lauter Levin Associates; and West Orange, NJ: Behrman.
Kurtz, Lester. 1995. *Gods in the Global Village: The World's Religions in Sociological Perspective*. Thousand Oaks, CA: Pine Forge.
Ladd, E. C., Jr. 1966. *Negro Political Leadership in the South*. Ithaca, NY: Cornell University Press.
Lamanna, Marianne, and Agnes Riedmann. 2009. *Marriages and Families: Making Choices in a Diverse Society* (10th ed.). Belmont, CA: Wadsworth.
Lane, Harlan. 1992. *The Mask of Benevolence: Disabling the Deaf Community*. New York: Vintage.
Lapham, Lewis H. 1988. *Money and Class in America: Notes and Observations on Our Civil Religion*. New York: Weidenfeld & Nicolson.
Lapsley, Daniel K., 1990. "Continuity and Discontinuity in Adolescent Social Cognitive Development." In Raymond Montemayor, Gerald R. Adams, and Thomas P. Gullota (Eds.), *From Childhood to Adolescence: A Transitional Period?* (*Advances in Adolescent Development*, vol. 2). Newbury Park, CA: Sage.
Larimer, Tim. 1999. "The Japan Syndrome." *Time* (Oct. 11): 50–51.
Larson, Magali Sarfatti. 1977. *The Rise of Professionalism: A Sociological Analysis*. Berkeley: University of California Press.
Lasch, Christopher. 1977. *Haven in a Heartless World*. New York: Basic.
Lash, Scott, and John Urry. 1994. *Economies of Signs and Space*. London: Sage.
Laumann, Edward O., John H. Gagnon, Robert T. Michael, and Stuart Michaels. 1994. *The Social Organization of Sexuality*. Chicago: University of Chicago Press.
Le Bon, Gustave. 1960. *The Crowd: A Study of the Popular Mind*. New York: Viking (orig. pub. 1895).
Lee, Felicia R. 1993. "Where Guns and Lives Are Cheap." *New York Times* (Mar. 21): 21.
Leenaars, Antoon A. 1988. *Suicide Notes: Predictive Clues and Patterns*. New York: Human Sciences Press.
Leenaars, Antoon A. (Ed.). 1991. *Life Span Perspectives of Suicide: Time-Lines in the Suicide Process*. New York: Plenum.
Lehmann, Jennifer M. 1994. *Durkheim and Women*. Lincoln: University of Nebraska Press.
Leland, John. 2008. "From the Housing Market to the Maternity Ward." *New York Times* (Feb. 1): A12.
Lemert, Charles. 1997. *Postmodernism Is Not What You Think*. Malden, MA: Blackwell.
Lemert, Edwin M. 1951. *Social Pathology*. New York: McGraw-Hill.
Lengermann, Patricia Madoo, and Jill Niebrugge-Brantley. 1998. *The Women Founders: Sociology and Social Theory, 1830–1930*. New York: McGraw-Hill.
Lenzer, Gertrud (Ed.). 1998. *The Essential Writings: Auguste Comte and Positivism*. New Brunswick, NJ: Transaction.
Leonard, Andrew. 1999. "We've Got Mail—Always." *Newsweek* (Sept. 20): 58–61.
Leonard, Wilbert M., and Jonathan E. Reyman. 1988. "The Odds of Attaining Professional Athlete Status: Refining the Computations." *Sociology of Sport Journal*: 162–169.
Lerner, Gerda. 1986. *The Creation of Patriarchy*. New York: Oxford University Press.
Levey, Hilary. 2007. "Here She Is . . . and There She Goes?" *Contexts* (Summer): 70–72.
Levine, Adeline Gordon. 1982. *Love Canal: Science, Politics, and People*. Lexington, MA: Lexington.
Levine, Murray. 1982. "Introduction." In Lois Marie Gibbs, *Love Canal: My Story*. Albany: SUNY Press.
Levine, Peter. 1992. *Ellis Island to Ebbets Field: Sport and the American Jewish Experience*. New York: Oxford University Press.
Levinthal, Charles F. 2002. *Drugs, Behavior, and Modern Society* (3rd ed.). Boston: Allyn & Bacon.
Levy, Janice C., and Eva Y. Deykin. 1989. "Suicidality, Depression, and Substance Abuse in Adolescence." *American Journal of Psychiatry*, 146 (11): 1462–1468.
Lewis, Paul. 1996. "World Bank Moves to Cut Poorest Nations' Debts." *New York Times* (Mar. 16): 17, 18.
Liebow, Elliot. 1993. *Tell Them Who I Am: The Lives of Homeless Women*. New York: Free Press.
Lindberg, Richard, and Vesna Markovic. 2001. "Organized Crime Outlook in the New Russia." Search International. Retrieved Apr. 23, 2005. Online: http://www.search-international.com/Articles/crime/russiacrime.htm
Lindblom, Charles. 1977. *Politics and Markets*. New York: Basic.
Linden, Greg. 2005. "Geeking with Greg: Facebook and Building Social Networks" (Oct. 27). Retrieved Jan. 28, 2006. Online: http://glinden.blogspot.com/2005/10/facebook-and-building-social-networks.html
Linton, Ralph. 1936. *The Study of Man*. New York: Appleton-Century-Crofts.
Lips, Hilary M. 2001. *Sex and Gender: An Introduction* (4th ed.). New York: McGraw-Hill.
Liptak, Adam. 2008. "Same-Sex Marriage and Racial Justice Find Common Ground." *New York Times* (May 17): A10.
Loeb, Paul Rogat. 1994. *Generation at the Crossroads: Apathy and Action on the American Campus*. New Brunswick, NJ: Rutgers University Press.
Lofland, John. 1993. "Collective Behavior: The Elementary Forms." In Russell L. Curtis, Jr., and Benigno E. Aguirre (Eds.), *Collective Behavior and Social Movements*. Boston: Allyn & Bacon, pp. 70–75.
London, Kathryn A. 1991. "Advance Data Number 194: Cohabitation, Marriage, Marital Dissolution, and Remarriage: United States 1988." U.S. Department of Health and Human

Services: Vital and Health Statistics of the National Center, January 4.
Lorber, Judith. 1994. *Paradoxes of Gender.* New Haven, CT: Yale University Press.
Lorber, Judith (Ed.). 2005. *Gender Inequality: Feminist Theories and Politics* (3rd ed.). Los Angeles: Roxbury.
Lott, Bernice. 1994. *Women's Lives: Themes and Variations in Gender Learning* (2nd ed.). Pacific Grove, CA: Brooks/Cole.
Lummis, C. Douglas. 1992. "Equality." In Wolfgang Sachs (Ed.), *The Development Dictionary.* Atlantic Highlands, NJ: Zed, pp. 38–52.
Lundberg, Ferdinand. 1988. *The Rich and the Super-Rich: A Study in the Power of Money Today.* Secaucus, NJ: Lyle Stuart.
Lynch, Jason, and Todd Gold. 2005. "Happy Drew Year." *People* (Apr. 25): 92–96, 98.
Maag, Christopher. 2007. "When the Bullies Turned Faceless." *New York Times* (Dec. 16). Retrieved Dec. 26, 2007. Online: http://www.nytimes.com/2007/12/16/fashion/16meangirls.html?_r=1&oref=slogin
Maccoby, Eleanor E., and Carol Nagy Jacklin. 1987. "Gender Segregation in Childhood." *Advances in Child Development and Behavior,* 20: 239–287.
Mack, Raymond W., and Calvin P. Bradford. 1979. *Transforming America: Patterns of Social Change* (2nd ed.). New York: Random House.
MacSween, Morag. 1993. *Anorexic Bodies: A Feminist and Sociological Perspective on Anorexia Nervosa.* New York: Routledge.
Magid, Larry. 2004. "Talk to Your Kids About Cell Phone Use." Retrieved Mar. 30, 2007. Online: http://www.safekids.com/cellphone.htm
Mahapatra, Rajesh. 2007. "Outsourced Jobs Take Toll on Indians' Health." *Austin American-Statesman* (Dec. 30): H1, H6.
Malinowski, Bronislaw. 1922. *Argonauts of the Western Pacific.* New York: Dutton.
Malthus, Thomas R. 1965. *An Essay on Population.* New York: Augustus Kelley (orig. pub. 1798).
Mangione, Jerre, and Ben Morreale. 1992. *La Storia: Five Centuries of the Italian American Experience.* New York: HarperPerennial.
Mann, Coramae Richey. 1993. *Unequal Justice: A Question of Color.* Bloomington: Indiana University Press.
Mann, Patricia S. 1994. *Micro-Politics: Agency in a Postfeminist Era.* Minneapolis: University of Minnesota Press.
Mantsios, Gregory. 2003. "Media Magic: Making Class Invisible." In Michael S. Kimmel and Abby L. Ferber (Eds.), *Privilege: A Reader.* Boulder, CO: Westview, pp. 99–109.
Marger, Martin N. 1994. *Race and Ethnic Relations: American and Global Perspectives.* Belmont, CA: Wadsworth.
—. 2003. *Race and Ethnic Relations: American and Global Perspectives* (6th ed.). Belmont, CA: Wadsworth.
—. 2009. *Race and Ethnic Relations: American and Global Perspectives* (8th ed.). Belmont, CA: Wadsworth.
Marquart, James W., Sheldon Ekland-Olson, and Jonathan R. Sorensen. 1994. *The Rope, the Chair, and the Needle.* Austin: University of Texas Press.
Marquis, Christopher. 2001. "An American Report Finds the Taliban's Violation of Religious Rights 'Particularly Severe.'" *New York Times* (Oct. 27): B3.
Marshall, Gordon. 1998. *A Dictionary of Sociology* (2nd ed.). New York: Oxford University Press.
Martin, Carol L. 1989. "Children's Use of Gender-Related Information in Making Social Judgments." *Developmental Psychology,* 25: 80–88.
Martin, Susan Ehrlich, and Nancy C. Jurik. 1996. *Doing Justice, Doing Gender.* Thousand Oaks, CA: Sage.
Martineau, Harriet. 1962. *Society in America* (edited, abridged). Garden City, NY: Doubleday (orig. pub. 1837).
Marx, Karl. 1967. *Capital: A Critique of Political Economy.* Ed. Friedrich Engels. New York: International (orig. pub. 1867).
Marx, Karl, and Friedrich Engels. 1967. *The Communist Manifesto.* New York: Pantheon (orig. pub. 1848).
—. 1970. *The German Ideology,* Part 1. Ed. C. J. Arthur. New York: International (orig. pub. 1845–1846).
Massey, Douglas J., G. Hugo Arango, A. Kowasuci, A. Pellegrino, and J. E. Taylor. 1993. "Theories of International Migration: A Review and Appraisal." *Population and Development Review,* 19: 431–466.
Maynard, R. A. 1996. *Kids Having Kids: A Robin Hood Foundation Special Report on the Costs of Adolescent Childbearing.* New York: Robin Hood Foundation.
McAdam, Doug. 1982. *Political Process and the Development of Black Insurgency.* Chicago: University of Chicago Press.
—. 1996. "Conceptual Origins, Current Problems, Future Directions." In Doug McAdam, John D. McCarthy, and Meyer N. Zald (Eds.), *Comparative Perspectives on Social Movements.* New York: Cambridge University Press, pp. 23–40.
McAdam, Doug, John D. McCarthy, and Mayer N. Zald. 1988. "Social Movements." In Neil J. Smelser (Ed.), *Handbook of Sociology.* Newbury Park, CA: Sage, pp. 695–737.
McCann, Lisa M. 1997. "Patrilocal Co-Residential Units (PCUs) in Al-Barba: Dual Household Structure in a Provincial Town in Jordan." *Journal of Comparative Family Studies* (Summer): 113–136.
McCarthy, John D., and Mayer N. Zald. 1977. "Resource Mobilization and Social Movements: A Partial Theory." *American Journal of Sociology,* 82: 1212–1241.
McChesney, Robert W. 1998. "The Political Economy of Global Communication." In Robert W. McChesney, Ellen Meiksins Wood, and John Bellamy Foster (Eds.), *Capitalism and the Information Age: The Political Economy of the Global Communication Revolution.* New York: Monthly Review Press, pp. 1–26.
McDonnell, Janet A. 1991. *The Dispossession of the American Indian, 1887–1934.* Bloomington: Indiana University Press.
McEachern, William A. 2003. *Economics: A Contemporary Introduction.* Mason, OH: Thomson/South-Western.
McGeary, Johanna. 2001. "The Taliban Troubles." *Time* (Oct. 1): 36–43.
McGuire, Meredith B. 2002. *Religion: The Social Context* (5th ed.). Belmont, CA: Wadsworth.
McHale, Susan M., Ann C. Crouter, and C. Jack Tucker. 1999. "Family Context and Gender Role Socialization in Middle Childhood: Comparing Girls to Boys and Sisters to Brothers." *Child Development,* 70: 990–1004.
McKenzie, Roderick D. 1925. "The Ecological Approach to the Study of the Human Community." In Robert Park, Ernest Burgess, and Roderick D. McKenzie, *The City.* Chicago: University of Chicago Press.
McLanahan, Sara, and Karen Booth. 1991. "Mother-Only Families." In Alan Booth (Ed.), *Contemporary Families: Looking Forward, Looking Backward.* Minneapolis: National Council on Family Relations, pp. 405–428.
McPhail, Clark. 1991. *The Myth of the Maddening Crowd.* New York: Aldine de Gruyter.
McPhail, Clark, and Ronald T. Wohlstein. 1983. "Individual and Collective Behavior Within Gatherings, Demonstrations, and Riots." In Ralph H. Turner and James F. Short, Jr. (Eds.), *Annual Review of Sociology,* vol. 9. Palo Alto, CA: Annual Reviews, pp. 579–600.
McPherson, Barry D., James E. Curtis, and John W. Loy. 1989. *The Social Significance of Sport: An Introduction to the Sociology of Sport.* Champaign, IL: Human Kinetics.
McTague, Jim. 2005. "The Underground Economy." *Wall Street Journal* (April). Retrieved Apr. 8, 2007. Online: http://wsjclassroom.com/archive/05apr/econ_underground.htm
Mead, George Herbert. 1934. *Mind, Self, and Society.* Chicago: University of Chicago Press.
Mennell, Stephen. 1996. *All Manners of Food: Eating and Taste in England and France from the Middle Ages to the Present.* Urbana: University of Illinois Press.
Mennell, Stephen, Anne Murcott, and Anneke H. van Otterloo. 1993. *The Sociology of Food: Eating, Diet and Culture.* Thousand Oaks, CA: Sage.
Mental Medicine. 1994. "Wealth, Health, and Status." *Mental Medicine Update,* 3 (2): 7.
Merchant, Carolyn. 1983. *The Death of Nature: Women, Ecology, and the Scientific Revolution.* San Francisco: Harper & Row.
—. 1992. *Radical Ecology: The Search for a Livable World.* New York: Routledge.
Merton, Robert King. 1938. "Social Structure and Anomie." *American Sociological Review,* 3 (6): 672–682.
—. 1949. "Discrimination and the American Creed." In Robert M. MacIver (Ed.), *Discrimination and National Welfare.* New York: Harper & Row, pp. 99–126.
—. 1968. *Social Theory and Social Structure* (enlarged ed.). New York: Free Press.
Messner, Michael A., Margaret Carlisle Duncan, and Kerry Jensen. 1993. "Separating the Men from the Girls: The Gendered Language of Televised Sports." *Gender & Society,* 7 (1): 121–137.
Meyer, David S., and Suzanne Staggenborg. 1996. "Movements, Countermovements, and the Structure of Political Opportunity." *American Journal of Sociology,* 101: 1628–1660.
Miall, Charlene. 1986. "The Stigma of Involuntary Childlessness." *Social Problems,* 33 (4): 268–282.
Michael, Robert T., John H. Gagnon, Edward O. Laumann, and Gina Kolata. 1994. *Sex in America.* Boston: Little, Brown.
Michels, Robert. 1949. *Political Parties.* Glencoe, IL: Free Press (orig. pub. 1911).
Mickelson, Roslyn Arlin, and Stephen Samuel Smith. 1995. "Education and the Struggle Against Race, Class, and Gender Inequality." In Margaret L. Andersen and Patricia Hill Collins (Eds.), *Race, Class, and Gender* (2nd ed.). Belmont, CA: Wadsworth, pp. 289–304.
Mies, Maria, and Vandana Shiva. 1993. *Ecofeminism.* Highlands, NJ: Zed.
Miethe, Terance, and Charles Moore. 1987. "Racial Differences in Criminal Processing: The Consequences of Model Selection on Conclusions About Differential Treatment." *Sociological Quarterly,* 27: 217–237.
Milgram, Stanley. 1963. "Behavioral Study of Obedience." *Journal of Abnormal and Social Psychology,* 67: 371–378.
—. 1974. *Obedience to Authority.* New York: Harper & Row.
Miller, Casey, and Kate Swift. 1991. *Words and Women: New Language in New Times* (updated ed.). New York: HarperCollins.
Miller, Dan E. 1986. "Milgram Redux: Obedience and Disobedience in Authority Relations." In Norman K. Denzin (Ed.), *Studies in Symbolic Interaction.* Greenwich, CT: JAI, pp. 77–106.
Miller, L. Scott. 1995. *An American Imperative: Accelerating Minority Educational Advancement.* New Haven, CT: Yale University Press.
Mills, C. Wright. 1956. *White Collar.* New York: Oxford University Press.
—. 1959a. *The Power Elite.* Fair Lawn, NJ: Oxford University Press.
—. 1959b. *The Sociological Imagination.* London: Oxford University Press.
—. 1976. *The Causes of World War Three.* Westport, CT: Greenwood.
Min, Eungjun (Ed.). 1999. *Reading the Homeless: The Media's Image of Homeless Culture.* Westport, CT: Praeger.
Min, Pyong Gap. 1988. "The Korean American Family." In Charles H. Mindel, Robert W. Habenstein, and Roosevelt Wright, Jr. (Eds.), *Ethnic Families in America: Patterns and Variations* (3rd ed.). New York: Elsevier, pp. 199–229.
Mindel, Charles H., Robert W. Habenstein, and Roosevelt Wright, Jr. (Eds.). 1988. *Ethnic Families in America: Patterns and Variations* (3rd ed.). New York: Elsevier.
mindoh.com. 2007. "I Wish I Knew What to Do?!" Retrieved Feb. 15, 2008. Online: http://www.mindoh.docs/Bullyingbook_excerpt_noCCC.pdf
Mintz, Beth, and Michael Schwartz. 1985. *The Power Structure of American Business.* Chicago: University of Chicago Press.
Mintz, Morton. 2007. "Will Congress Reform Wretched Executive Excess?" Retrieved Mar. 10, 2007. Online: http://www.thenation.com/doc/20070212/mintz
Mishler, Elliot G. 1984. *The Discourse of Medicine: Dialectics of Medical Interviews.* Norwood, NJ: Ablex.
—. 2005. "The Struggle Between the Voice of Medicine and the Voice of the Lifeworld." In Peter Conrad (Ed.), *The Sociology of Health and Illness: Critical Perspectives* (7th ed.). New York: Worth, pp. 319–330.
Miss America. 2008. "Miss America: Key Facts and Figures." Retrieved Jan. 31, 2008. Online: http://www.missamerica.org/organization-info/key-facts-and-figures.asp
Mohai, Paul, and Robin Saha. 2007. "Racial Inequality in the Distribution of Hazardous Waste: A National-Level Reassessment." *Social Problems* (August): 343–370.
Moore, K. A., A. K. Driscoll, and L. D. Lindberg. 1998. *A Statistical Portrait of Adolescent Sex, Contraception, and Childbearing.* Washington, DC: National Campaign to Prevent Teen Pregnancy.
Moore, Patricia, with C. P. Conn. 1985. *Disguised.* Waco, TX: Word.
Morris, Aldon. 1981. "Black Southern Student Sit-In Movement: An Analysis of Internal Organization." *American Sociological Review,* 46: 744–767.
Morselli, Henry. 1975. *Suicide: An Essay on Comparative Moral Statistics.* New York: Arno (orig. pub. 1881).
Murdock, George P. 1945. "The Common Denominator of Cultures." In Ralph Linton (Ed.), *The Science of Man in the World Crisis.* New York: Columbia University Press, pp. 123–142.
Murphy, Robert E., Jessica Scheer, Yolanda Murphy, and Richard Mack. 1988. "Physical Disability and Social Liminality: A Study in the Rituals of Adversity." *Social Science and Medicine,* 26: 235–242.
Mydans, Seth. 1993. "A New Tide of Immigration Brings Hostility to the Surface, Poll Finds." *New York Times* (June 27): A1.
—. 1997. "Its Mood Dark as the Haze, Southeast Asia Aches." *New York Times* (Oct. 26): 3.
Myrdal, Gunnar. 1970. *The Challenge of World Poverty: A World Anti-Poverty Program in Outline.* New York: Pantheon/Random House.
Naffine, Ngaire. 1987. *Female Crime: The Construction of Women in Criminology.* Boston: Allen & Unwin.
NAGIA. 2007. "Graffiti: The Newspaper of the Streets" and "Gang Indicators." National Alliance of Gang Investigators Association. Retrieved Mar. 10, 2007. Online: http://www.nagia.org
National Academy of Sciences. 2003. "Elder Mistreatment, Abuse, Neglect, and Exploitation in an Aging America." Retrieved June 3, 2004. Online: http://books.nap.edu/catalog/10406.html
National Campaign to Prevent Teen Pregnancy. 1997. *Whatever Happened to Childhood? The Problem of Teen Pregnancy in the United States.* Washington, DC: National Campaign to Prevent Teen Pregnancy.
National Center for Education Statistics. 2005. "Highlights from the Trends in International Mathematics and Science Study (TIMSS) 2003." Retrieved Apr. 16, 2005. Online: http://nces.ed.gov/pubs2005/timss03/index.asp
National Center for Health Statistics. 2003. "Self-Inflicted Injury/Suicide." Retrieved June 19, 2004. Online: http://www.cdc.gov/nchs/fastats/suicide.htm
—. 2007. "Births: Preliminary Data for 2006." Retrieved Feb. 13, 2008. Online: http://www.cdc.gov/nchs/data/nvsr/nvsr56/nvsr56_07.pdf

National Center for Injury Prevention and Control. 2003. "Homicide and Suicide Among Native Americans, 1979–1992." Retrieved June 13, 2004. Online: http://www.cdc.gov/ncipc/pub-res/natam.htm
—. 2006. "2004 United States Suicide Injury Deaths and Rates per 100,000." Retrieved Jan. 28, 2007. Online: http://webapp.cdc.gov/sasweb/ncipc/mortrate10_sy.html
National Centers for Disease Control. 2001. "43 Percent of First Marriages Break Up Within 15 Years." Retrieved July 14, 2002. Online: http://www.cdc.gov/nchs/releases/01news/firstmarr.htm
National Council on Crime and Delinquency. 1969. *The Infiltration into Legitimate Business by Organized Crime.* Washington, DC: National Council on Crime and Delinquency.
National Education Association. 2007. "School Safety." Retrieved Apr. 22, 2007. Online: http://www.nea.org/schoolsafety/index.html
National Law Center on Homelessness and Poverty. 2004. "Homelessness and Poverty in America." Retrieved Jan. 21, 2006. Online: http://www.nlchp.org/FA%5FHAPIA
Navarrette, Ruben, Jr. 1997. "A Darker Shade of Crimson." In Diana Kendall (Ed.), *Race, Class, and Gender in a Diverse Society.* Boston: Allyn & Bacon, 1997: 274–279. Reprinted from Ruben Navarrette, Jr., *A Darker Shade of Crimson.* New York: Bantam, 1993.
NCDC. 2007. "About NCDC: Northwestern Community Development Corps." Retrieved Mar. 31, 2007. Online: http://groups.northwestern.edu/ncdc/about.html
Nelson, Margaret K., and Joan Smith. 1999. *Working Hard and Making Do: Surviving in Small Town America.* Berkeley: University of California Press.
Nelson, Mariah Burton. 1994. *The Stronger Women Get, the More Men Love Football: Sexism and the American Culture of Sports.* New York: Harcourt.
Neuborne, Ellen. 1995. "Imagine My Surprise." In Barbara Findlen (Ed.), *Listen Up: Voices from the Next Feminist Generation.* Seattle: Seal, pp. 29–35.
New York Post. 2004. "Death Plunge No. 4: NYU's Grief." (Mar. 10): 1.
New York Times. 1996. "The Megacity Summit." (Apr. 8): A14.
—. 1997. "Shift in Schools' Spending." (Dec. 12): A15.
—. 2006. "Far Down the List of Worries." (Apr. 23): WK14.
—. 2007a. "Editorial: Needed Fixes for No Child Left Behind." (Feb. 15). Retrieved Feb. 11, 2008. Online: http://www.nytimes.com/2007/02/15/opinion/15thur3.html
—. 2007b. "Migrating to Poor Countries, Searching for a Better Life." (Dec. 27): A16.
Newman, Katherine S. 1988. *Falling from Grace: The Experience of Downward Mobility in the American Middle Class.* New York: Free Press.
—. 1993. *Declining Fortunes: The Withering of the American Dream.* New York: Basic.
—. 1999. *No Shame in My Game: The Working Poor in the Inner City.* New York: Knopf and the Russell Sage Foundation.
Nguyen, Bich Minh. 2007. *Stealing Buddha's Dinner: A Memoir.* New York: Viking.
Niebuhr, H. Richard. 1929. *The Social Sources of Denominationalism.* New York: Meridian.
Nielsen, Joyce McCarl. 1990. *Sex and Gender in Society: Perspectives on Stratification* (2nd ed.). Prospects Heights, IL: Waveland.
Nisbet, Robert. 1979. "Conservativism." In Tom Bottomore and Robert Nisbet (Eds.), *A History of Sociological Analysis.* London: Heinemann, pp. 81–117.
Noble, Barbara Presley. 1995. "A Level Playing Field, for Just $121." *New York Times* (Mar. 5): F21.
Noel, Donald L. 1972. *The Origins of American Slavery and Racism.* Columbus, OH: Merrill.
Nuland, Sherwin B. 1997. "Heroes of Medicine." *Time* (Fall special edition): 6–10.
Oakes, Jeannie. 1985. *Keeping Track: How High Schools Structure Inequality.* New Haven, CT: Yale University Press.
Oakes, Jeannie, and Martin Lipton. 2003. *Teaching to Change the World* (2nd ed.) New York: McGraw-Hill.
Oberschall, Anthony. 1973. *Social Conflict and Social Movements.* Englewood Cliffs, NJ: Prentice Hall.
Oboler, Suzanne. 1995. *Ethnic Labels, Latino Lives: Identity and the Politics of (Re)presentation in the United States.* Minneapolis: University of Minnesota Press.
O'Connell, Helen. 1994. *Women and the Family. Prepared for the UN-NGO Group on Women and Development.* Atlantic Highlands, NJ: Zed.
Odendahl, Teresa. 1990. *Charity Begins at Home: Generosity and Self-Interest Among the Philanthropic Elite.* New York: Basic.
Office of Juvenile Justice and Delinquency Prevention. 2002. *2000 National Youth Gang Survey.* Retrieved July 12, 2003. Online: http://www.ncjrs.org/pdffiles1/ojjdp/fs200204.pdf
Ogburn, William F. 1966. *Social Change with Respect to Culture and Original Nature.* New York: Dell (orig. pub. 1922).
Ohio State University. 2007. "Cultural Diversity: Eating in America." Ohio State University Extension Fact Sheet, Family and Consumer Sciences Series. Retrieved Feb. 11, 2007. Online: http://www.ohioline.ag.ohio-state.edu
Omi, Michael, and Howard Winant. 1994. *Racial Formation in the United States: From the 1960s to the 1990s.* New York: Routledge.
Orenstein, Peggy, in association with the American Association of University Women. 1995. *SchoolGirls: Young Women, Self-Esteem, and the Confidence Gap.* New York: Anchor/Doubleday.
Ortner, Sherry B. 1974. "Is Female to Male as Nature Is to Culture?" In Michelle Rosaldo and Louise Lamphere (Eds.), *Women, Culture, and Society.* Stanford, CA: Stanford University Press.
Orum, Anthony M. 1974. "On Participation in Political Protest Movements." *Journal of Applied Behavioral Science,* 10: 181–207.
Orum, Anthony M., and Amy W. Orum. 1968. "The Class and Status Bases of Negro Student Protest." *Social Science Quarterly,* 49 (December): 521–533.
Orzechowski, Shawna, and Peter Sepielli. 2001. "Net Worth and Asset Ownership of Households: 1998 and 2000." U.S. Census Bureau, Current Population Reports, P70–88. Washington, DC: U.S. Government Printing Office.
Ouchi, William. 1981. *Theory Z: How American Business Can Meet the Japanese Challenge.* Reading, MA: Addison-Wesley.
Oxendine, Joseph B. 2003. *American Indian Sports Heritage* (rev. ed.). Lincoln: University of Nebraska Press.
Padilla, Felix M. 1993. *The Gang as an American Enterprise.* New Brunswick, NJ: Rutgers University Press.
Page, Charles H. 1946. "Bureaucracy's Other Face." *Social Forces,* 25 (October): 89–94.
Palen, J. John. 1995. *The Suburbs.* New York: McGraw-Hill.
Parenti, Michael. 1994. *Land of Idols: Political Mythology in America.* New York: St. Martin's.
—. 1998. *America Besieged.* San Francisco: City Lights.
Park, Robert E. 1915. "The City: Suggestions for the Investigation of Human Behavior in the City." *American Journal of Sociology,* 20: 577–612.
—. 1928. "Human Migration and the Marginal Man." *American Journal of Sociology,* 33.
—. 1936. "Human Ecology." *American Journal of Sociology,* 42: 1–15.
Park, Robert E., and Ernest W. Burgess. 1921. *Human Ecology.* Chicago: University of Chicago Press.
Parker, Robert Nash, and Doreen Anderson-Facile. 2000. "Violent Crime Trends." In Joseph F. Sheley (Ed.), *Criminology: A Contemporary Handbook* (3rd ed.). Belmont, CA: Wadsworth, pp. 191–214.
Parrish, Dee Anna. 1990. *Abused: A Guide to Recovery for Adult Survivors of Emotional/Physical Child Abuse.* Barrytown, NY: Station Hill.
Parsons, Talcott. 1951. *The Social System.* Glencoe, IL: Free Press.
—. 1955. "The American Family: Its Relations to Personality and to the Social Structure." In Talcott Parsons and Robert F. Bales (Eds.), *Family, Socialization and Interaction Process.* Glencoe, IL: Free Press, pp. 3–33.
Passel, Jeffrey S. 2006. "The Size and Characteristics of the Unauthorized Migrant Population in the U.S.: Estimates Based on the March 2005 Current Population Survey." Pew Hispanic Center. Retrieved Mar. 8, 2006. Online: http://pewhispanic.org/reports/print.php?ReportID=61
PBS. 2005a. "The Meaning of Food: Food & Culture." Retrieved Feb. 11, 2007. Online: http://www.pbs.org/opb/meaningoffood
—. 2005b. *Online NewsHour:* "The Schiavo Case Receives Strong Media Coverage" (Mar. 24). Retrieved Apr. 10, 2005. Online: http://www.pbs.org/newshour/bb/media/jan-june05/schiavo_3-24.html
—. 2008. "Facts About Global Poverty and Microcredit." Retrieved Jan. 19, 2008. Online: http://www.pbs.org/toourcredit/facts_one.htm
Pear, Robert. 2008. "About Those Health Care Plans by the Democrats . . ." *New York Times* (Mar. 3): A16.
Pearce, Diana. 1978. "The Feminization of Poverty: Women, Work, and Welfare." *Urban and Social Change Review,* 11 (1/2): 28–36.
Pearson, Judy C. 1985. *Gender and Communication.* Dubuque, IA: Brown.
Perry, David C., and Alfred J. Watkins (Eds.). 1977. *The Rise of the Sunbelt Cities.* Beverly Hills, CA: Sage.
Perry, Steven W. 2004. "American Indians and Crime." U.S. Bureau of Justice Statistics. Retrieved Mar. 19, 2005. Online: http://www.ojp.usdoj.gov/bjs/pub/pdf/aic02.pdf
Petersen, John L. 1994. *The Road to 2015: Profiles of the Future.* Corte Madera, CA: Waite Group.
Peterson, Robert. 1992. *Only the Ball Was White: A History of Legendary Black Players and All-Black Professional Teams.* New York: Oxford University Press (orig. pub. 1970).
Pew Charitable Trusts. 2007. "Economic Mobility: Is the American Dream Alive and Well?" Retrieved Dec. 20, 2007. Online: http://www.economicmobility.org
Pew Forum on Religion & Public Life. 2008. "U.S. Religious Landscape Survey." Retrieved Apr. 23, 2008. Online: http://religions.pewforum.org/reports
Phillips, John C. 1993. *Sociology of Sport.* Boston: Allyn & Bacon.
Piaget, Jean. 1954. *The Construction of Reality in the Child.* Trans. Margaret Cook. New York: Basic.
Pierre-Pierre, Garry. 1997. "Traditional Church's New Life." *New York Times* (Nov. 15): A11.
Pietilä, Hilkka, and Jeanne Vickers. 1994. *Making Women Matter: The Role of the United Nations.* Atlantic Highlands, NJ: Zed.
Pillemer, Karl A. 1985. "The Dangers of Dependency: New Findings on Domestic Violence Against the Elderly." *Social Problems,* 33 (December): 146–158.
Pillemer, Karl A., and David Finkelhor. 1988. "The Prevalence of Elder Abuse: A Random Sample Survey." *Gerontologist,* 28 (1): 51–57.
Pinderhughes, Dianne M. 1986. "Political Choices: A Realignment in Partisanship Among Black Voters?" In James D. Williams (Ed.), *The State of Black America 1986.* New York: National Urban League, pp. 85–113.
Pinderhughes, Howard. 1997. *Race in the Hood: Conflict and Violence Among Urban Youth.* Minneapolis: University of Minnesota Press.
Pines, Maya. 1981. "The Civilizing of Genie." *Psychology Today,* 15 (September): 28–29, 31–32, 34.
Pitzer, Ronald. 2003. "Rural Children Under Stress." University of Minnesota Extension Service. Retrieved Aug. 9, 2003. Online: http://www.extension.umn.edu/distribution/familydevelopment/components/7269cm.html
Pokin, Steve. 2007. "Pokin Around: A Real Person, a Real Death." *St. Charles Journal.* Retrieved Dec. 26, 2007. Online: http://stcharlesjournal.stltoday.com/articles/2007/11/10/news/sj2tn20071110-1111stc_pokin_1.ii1.txt
Polakow, Valerie. 1993. *Lives on the Edge: Single Mothers and Their Children in the Other America.* Chicago: University of Chicago Press.
Popenoe, David, and Barbara Dafoe Whitehead. 1999. "The State of Our Unions: The Social Health of Marriage in America" (June). Retrieved Sept. 21, 1999. Online: http://marriage.rutgers.edu/State.html
Population Reference Bureau. 2001. "Human Population: Fundamentals of Growth Patterns of World Urbanization." Retrieved Nov. 22, 2001. Online: http://www.prb.org/Content/NavigationMenu/PRB/E.../Patterns_of_World_Urbanization.html
Postman, Neil, and Steve Powers. 1992. *How to Watch TV News.* New York: Penguin.
Powell, Alvin. 2001. "Partnership Ensures Shelter's Future." *Harvard Gazette* (Jan. 18). Retrieved Feb. 24, 2007. Online: http://www.hno.harvard.edu/gazette/2001/01.18/08-partnership.html
Powell, Brian, and Douglas B. Downey. 1997. "Living in Single-Parent Households: An Investigation of the Same-Sex Hypothesis." *American Sociological Review,* 62 (August): 521–539.
Powell, Michael. 2004. "Evolution Shares a Desk with 'Intelligent Design.'" *Washington Post* (Dec. 26): A1.
Project Censored. 2008. "Top 25 Censored Stories of 2008." Project Censored—Sonoma State University. Retrieved Feb. 16, 2008. Online: http://www.projectcensored.org/censored_2008/index.htm
Puette, William J. 1992. *Through Jaundiced Eyes: How the Media View Organized Labor.* Ithaca, NY: ILR Press.
Puffer, J. Adams. 1912. *The Boy and His Gang.* Boston: Houghton Mifflin.
Queen, Stuart A., and David B. Carpenter. 1953. *The American City.* New York: McGraw-Hill.
Quinney, Richard. 2001. *Critique of the Legal Order.* Piscataway, NJ: Transaction (orig. pub. 1974).
Qvortrup, Jens. 1990. *Childhood as a Social Phenomenon.* Vienna: European Centre for Social Welfare Policy and Research.
Radcliffe-Brown, A. R. 1952. *Structure and Function in Primitive Society.* New York: Free Press.
Raffaelli, Marcela, and Lenna L. Ontai. 2004. "Gender Socialization in Latino/a Families: Results from Two Retrospective Studies." *Sex Roles,* 50: 287–299.
Raffalli, Mary. 1994. "Why So Few Women Physicists?" *New York Times Supplement* (January): Sect. 4A, 26–28.
Ramirez, Marc. 1999. "A Portrait of a Local Muslim Family." *Seattle Times* (Jan. 24). Retrieved Aug. 16, 1999. Online: http://archives.seattletimes.com/cgi-bin/texis.mummy/web/vortex/display?storyID=36d4d218
Ratha, Dilip, and William Shaw. 2007. "South–South Migration and Remittances." New York: World Bank Development Prospects Group. Retrieved Jan. 21, 2008. Online: http://siteresources.worldbank.org/INTPROSPECTS/Resources/SouthSouthMigrationandRemittances.pdf
Reaves, Brian A. 2001. *Felony Defendants in Large Urban Counties, 1998: State Court Processing Statistics.* Washington, DC: U.S. Government Printing Office.
Reckless, Walter C. 1967. *The Crime Problem.* New York: Meredith.

Reich, Robert. 1993. "Why the Rich Are Getting Richer and the Poor Poorer." In Paul J. Baker, Louis E. Anderson, and Dean S. Dorn (Eds.), *Social Problems: A Critical Thinking Approach* (2nd ed.). Belmont, CA: Wadsworth, pp. 145–149. Adapted from *The New Republic*, May 1, 1989.

Reiman, Jeffrey. 1998. *The Rich Get Richer and the Poor Get Prison: Ideology, Class, and Criminal Justice* (5th ed.). Boston: Allyn & Bacon.

Reinharz, Shulamit. 1992. *Feminist Methods in Social Research*. New York: Oxford University Press.

Reinisch, June. 1990. *The Kinsey Institute New Report on Sex: What You Must Know to Be Sexually Literate*. New York: St. Martin's.

religioustolerance.org. 2005. "Science and Religion." Ontario Consultants on Religious Tolerance. Retrieved Apr. 22, 2005. Online: http://www.religioustolerance.org/sci_rel .htm#menu

Relman, Arnold S. 1992. "Self-Referral—What's at Stake?" *New England Journal of Medicine*, 327 (Nov. 19): 1522–1524.

Reskin, Barbara F., and Irene Padavic. 2002. *Women and Men at Work* (2nd ed.). Thousand Oaks, CA: Pine Forge.

Revkin, Andrew C. 2006. "Yelling 'Fire' On a Hot Planet." *New York Times* (Apr. 23): WK1–WK14.

Richardson, Laurel. 1993. "Inequalities of Power, Property, and Prestige." In Virginia Cyrus (Ed.), *Experiencing Race, Class, and Gender in the United States*. Mountain View, CA: Mayfield, pp. 229–236.

Ricketts, Thomas C., III. 1999. "Preface," in Thomas C. Ricketts, III (Ed.), *Rural Health in the United States*. New York: Oxford University Press, pp. vii–viii.

Rigler, David. 1993. "Letters: A Psychologist Portrayed in a Book About an Abused Child Speaks Out for the First Time in 22 Years." *New York Times Book Review* (June 13): 35.

Ritzer, George. 1997. *Postmodern Society Theory*. New York: McGraw-Hill.

—. 2000a. *The McDonaldization of Society*. Thousand Oaks, CA: Pine Forge.

—. 2000b. *Modern Sociological Theory* (5th ed.). New York: McGraw-Hill.

Ritzer, George, and Douglas J. Goodman. 2004. *Modern Sociological Theory* (6th ed.). New York: McGraw-Hill.

Rizzo, Thomas A., and William A. Corsaro. 1995. "Social Support Processes in Early Childhood Friendships: A Comparative Study of Ecological Congruences in Enacted Support." *American Journal of Community Psychology*, 23: 389–418.

Robbins, Alexandra. 2004. *Pledged: The Secret Life of Sororities*. New York: Hyperion.

Roberts, Keith A. 2004. *Religion in Sociological Perspective* (4th ed.). Belmont, CA: Wadsworth.

Robinson, Brian E. 1988. *Teenage Fathers*. Lexington, MA: Lexington.

Robson, Ruthann. 1992. *Lesbian (Out)law: Survival Under the Rule of Law*. New York: Firebrand.

Rocca, Mo. 2007. "TV Drug Ads' Side Effects." CBS News *Sunday Morning* (Oct. 14). Retrieved Feb. 24, 2008. Online: http://www.cbsnews .com/stories/2007/10/14/sunday/main3365346 .shtml

Rodriguez, Clara E. 1989. *Puerto Ricans: Born in the U.S.A.* New York: Unwin Hyman.

Rogers, Harrell R. 1986. *Poor Women, Poor Families: The Economic Plight of America's Female-Headed Households*. Armonk, NY: Sharpe.

Rollins, Judith. 1985. *Between Women: Domestics and Their Employers*. Philadelphia: Temple University Press.

Romero, Mary. 1997. "Introduction." In Mary Romero, Pierrette Hondagneu-Sotelo, and Vilma Ortiz (Eds.), *Challenging Fronteras: Structuring Latina and Latino Lives in the U.S.* New York: Routledge, pp. 3–5.

Roof, Wade Clark. 1993. *A Generation of Seekers: The Spiritual Journeys of the Baby Boom Generation*. San Francisco: HarperSanFrancisco.

Ropers, Richard H. 1991. *Persistent Poverty: The American Dream Turned Nightmare*. New York: Plenum.

Rose, Jerry D. 1982. *Outbreaks*. New York: Free Press.

Rosenbloom, Stephanie. 2008. "Putting Your Best Cyberface Forward." *New York Times* (Jan. 3): E1, E6.

Rosenthal, Naomi, Meryl Fingrutd, Michele Ethier, Roberta Karant, and David McDonald. 1985. "Social Movements and Network Analysis: A Case Study of Nineteenth-Century Women's Reform in New York State." *American Journal of Sociology*, 90: 1022–1054.

Rosenthal, Robert, and Lenore Jacobson. 1968. *Pygmalion in the Classroom: Teacher Expectation and Student's Intellectual Development*. New York: Holt, Rinehart, and Winston.

Rosnow, Ralph L., and Gary Alan Fine. 1976. *Rumor and Gossip: The Social Psychology of Hearsay*. New York: Elsevier.

Ross, Dorothy. 1991. *The Origins of American Social Science*. Cambridge, England: Cambridge University Press.

Rossi, Alice S. 1992. "Transition to Parenthood." In Arlene Skolnick and Jerome Skolnick (Eds.), *Family in Transition*. New York: HarperCollins, pp. 453–463.

Rossides, Daniel W. 1986. *The American Class System: An Introduction to Social Stratification*. Boston: Houghton Mifflin.

Rostow, Walt W. 1971. *The Stages of Economic Growth: A Non-Communist Manifesto* (2nd ed.). Cambridge: Cambridge University Press (orig. pub. 1960).

—. 1978. *The World Economy: History and Prospect*. Austin: University of Texas Press.

Roth, Guenther. 1988. "Marianne Weber and Her Circle." In Marianne Weber, *Max Weber*. New Brunswick, NJ: Transaction, p. xv.

Rothchild, John. 1995. "Wealth: Static Wages, Except for the Rich." *Time* (Jan. 30): 60–61.

Rothman, Robert A. 2001. *Inequality and Stratification: Class, Color, and Gender* (4th ed.). Upper Saddle River, NJ: Prentice Hall.

Rousseau, Ann Marie. 1981. *Shopping Bag Ladies: Homeless Women Speak About Their Lives*. New York: Pilgrim.

Rubin, Lillian B. 1986. "A Feminist Response to Lasch." *Tikkun*, 1 (2): 89–91.

Rural Policy Research Institute. 2004. "Place Matters: Addressing Rural Poverty." Rural Poverty Research Institute. Retrieved Jan. 21, 2008. Online: http://www.rprconline.org/synthesis.pdf

Russo, Nancy Felipe, and Mary A. Jansen. 1988. "Women, Work, and Disability: Opportunities and Challenges." In Michelle Fine and Adrienne Asch (Eds.), *Women with Disabilities: Essays in Psychology, Culture, and Politics*. Philadelphia: Temple University Press.

Rymer, Russ. 1993. *Genie: An Abused Child's Flight from Silence*. New York: HarperCollins.

Sadker, David, and Myra Sadker. 1985. "Is the OK Classroom OK?" *Phi Delta Kappan*, 55: 358–367.

—. 1986. "Sexism in the Classroom: From Grade School to Graduate School." *Phi Delta Kappan*, 68: 512–515.

Sadker, Myra, and David Sadker. 1994. *Failing at Fairness: How America's Schools Cheat Girls*. New York: Scribner.

Safilios-Rothschild, Constantina. 1969. "Family Sociology or Wives' Family Sociology? A Cross-Cultural Examination of Decision-Making." *Journal of Marriage and the Family*, 31 (2): 290–301.

Salt Lake City Sheriff's Department. 2007. "Graffiti: That Writing on the Wall." Retrieved Mar. 10, 2007. Online: http://www.slsheriff .org/html/org/metrogang/graffiti.html

Samovar, Larry A., and Richard E. Porter. 1991. *Communication Between Cultures*. Belmont, CA: Wadsworth.

Sapir, Edward. 1961. *Culture, Language and Personality*. Berkeley: University of California Press.

Sargent, Margaret. 1987. *Sociology for Australians* (2nd ed.). Melbourne, Australia: Longman Cheshire.

Sassen, Saskia. 2001. *The Global City: New York, London, Tokyo* (2nd ed.). Princeton, NJ: Princeton University Press.

Savin-Williams, Ritch C. 2004. "Memories of Same-Sex Attractions." In Michael S. Kimmel and Michael A. Messner (Eds.), *Men's Lives* (6th ed.). Boston: Allyn & Bacon, pp. 116–132.

Schaefer, Jame. 2005. "Reporting Complexity: Science and Religion." In Claire Hoertz Badaracco (Ed.), *Quoting God: How Media Shape Ideas About Religion and Culture*. Waco, TX: Baylor University Press, pp. 211–224.

Schaefer, Richard T., and William W. Zellner. 2007. *Extraordinary Groups: An Examination of Unconventional Lifestyles* (8th ed.). New York: Worth.

Schattschneider, Elmer Eric. 1969. *Two Hundred Americans in Search of a Government*. New York: Holt, Rinehart & Winston.

Schemo, Diana Jean. 1994. "Suburban Taxes Are Higher for Blacks, Analysis Shows." *New York Times* (Aug. 17): A1, A16.

Schneider, Keith. 1993. "The Regulatory Thickets of Environmental Racism." *New York Times* (Dec. 19): E5.

Scholastic Parent & Child. 2007. "How and When to Praise." Retrieved Jan. 3, 2008. Online: http://content.scholastic.com/browse/article .jsp?id=2064

Schor, Juliet B. 1999. *The Overspent American: Upscaling, Downshifting, and the New Consumer*. New York: HarperPerennial.

Schubert, Hans-Joachim (Ed.). 1998. "Introduction." In *On Self and Social Organization—Charles Horton Cooley*. Chicago: University of Chicago Press, pp. 1–31.

Schur, Edwin M. 1983. *Labeling Women Deviant: Gender, Stigma, and Social Control*. Philadelphia: Temple University Press.

Schutske, John. 2002. "Keeping Farm Children Safe." University of Minnesota Extension Service. Retrieved Aug. 9, 2003. Online: http:// www.extension.umn.edu/distribution/youth development/DA6188.html

Schwartz, John. 2003. "Too Much Information, Not Enough Knowledge." *New York Times* (June 9): WK5.

Schwartz, John, and John M. Broder. 2003. "Engineer Warned of Consequences of Liftoff Damage." *New York Times* (Feb. 13): A1–A29.

Schwartz, John (with Matthew L. Wald). 2003. "Costs and Risk Clouding Plans to Fix Shuttles." *New York Times* (June 8): A1–A20.

Schwartz, John, and Matthew L. Wald. 2003. "'Groupthink' Is 30 Years Old, and Still Going Strong." *New York Times* (Mar. 9): WK3.

Schwarz, John E., and Thomas J. Volgy. 1992. *The Forgotten Americans*. New York: Norton.

Scott, Alan. 1990. *Ideology and the New Social Movements*. Boston: Unwin & Hyman.

Seccombe, Karen. 1991. "Assessing the Costs and Benefits of Children: Gender Comparisons Among Childfree Husbands and Wives." *Journal of Marriage and the Family*, 53 (1): 191–202.

Seegmiller, B. R., B. Suter, and N. Duviant. 1980. *Personal, Socioeconomic, and Sibling Influences on Sex-Role Differentiation*. Urbana: ERIC Clearinghouse of Elementary and Early Childhood Education, ED 176 895, College of Education, University of Illinois.

Seid, Roberta P. 1994. "Too 'Close to the Bone': The Historical Context for Women's Obsession with Slenderness." In Patricia Fallon, Melanie A. Katzman, and Susan C. Wooley (Eds.), *Feminist Perspectives on Eating Disorders*. New York: Guilford, pp. 3–16.

Sengoku, Tamotsu. 1985. *Willing Workers: The Work Ethic in Japan, England, and the United States*. Westport, CT: Quorum.

Sengupta, Somini. 1997. "At Holidays, Test of Patience of Muslims." *New York Times* (Dec. 25): A12.

Senna, Joseph J., and Larry J. Siegel. 2002. *Introduction to Criminal Justice* (9th ed.). Belmont, CA: Wadsworth.

Serbin, Lisa A., Phyllis Zelkowitz, Anna-Beth Doyle, Dolores Gold, and Bill Wheaton. 1990.

"The Socialization of Sex-Differentiated Skills and Academic Performance: A Mediational Model." *Sex Roles*, 23: 613–628.

Serrill, Michael S. 1997. "Socialism Dies Again." *Time* (Sept. 22): 44.

Shapin, Steven. 2006. "Paradise Sold: What Are You Buying When You Buy Organic?" *The New Yorker* (May 15). Retrieved Feb. 17, 2007. Online: http://www.newyorker.com/critics/ atlarge/articles/060515crat_atlarge

Shapiro, Joseph P. 1993. *No Pity: People with Disabilities Forging a New Civil Rights Movement*. New York: Times/Random House.

Shaw, Randy. 1999. *Reclaiming America: Nike, Clean Air, and the New National Activism*. Berkeley: University of California Press.

Shawver, Lois. 1998. "Notes on Reading Foucault's *The Birth of the Clinic*." Retrieved Oct. 2, 1999. Online: http://www.california.com/~rathbone/ foucbc.htm

Sheen, Fulton J. 1995. *From the Angel's Blackboard: The Best of Fulton J. Sheen*. Ligouri, MO: Triumph.

Sheff, David. 1995. "If It's Tuesday, It Must Be Dad's House." *New York Times Magazine* (Mar. 26): 64–65.

Sheff, Nick. 1999. "My Long-Distance Life." *Newsweek* (Feb. 15): 16.

Shell, Adam. 2007. "Morgan Stanley Settles Sex-Bias Case." *USA Today* (Apr. 24). Retrieved Feb. 2, 2008. Online: http://www.usatoday.com/ money/companies/2004-07-12-morgan-stanley-suit_x.htm

Sheppard, Ashley Dawn. 2005. "Facebook Includes Finding Friends, Safety Concerns." *Baylor Lariat* (Jan. 25).

Sherman, Suzanne (Ed.). 1992. "Frances Fuchs and Gayle Remick." In *Lesbian and Gay Marriage: Private Commitments, Public Ceremonies*. Philadelphia: Temple University Press, pp. 189–201.

Shevky, Eshref, and Wendell Bell. 1966. *Social Area Analysis: Theory, Illustrative Application and Computational Procedures*. Westport, CT: Greenwood.

Shin, Hyon B. 2005. "School Enrollment—Social and Economic Characteristics of Students: October 2003." U.S. Census Bureau, Current Population Reports, P20–554. Washington, DC: U.S. Government Printing Office.

Shum, Tedd. 1997. "Olympic Gymnast Chow Makes Impact on All Americans." Retrieved Aug. 15, 1999. Online: http://www.dailybruin. ucla.edu/DB/issues/97/05.30/view.shum.html

Siegel, Larry J. 2006. *Criminology* (9th ed.). Belmont, CA: Wadsworth.

—. 2007. *Criminology: Theories, Patterns, and Typologies* (9th ed.). Belmont, CA: Wadsworth.

Simmel, Georg. 1950. *The Sociology of Georg Simmel*. Trans. Kurt Wolff. Glencoe, IL: Free Press (orig. written in 1902–1917).

—. 1957. "Fashion." *American Journal of Sociology*, 62 (May 1957): 541–558 (orig. pub. 1904).

—. 1990. *The Philosophy of Money*. Ed. David Frisby. New York: Routledge (orig. pub. 1907).

Simon, David R. 1996. *Elite Deviance* (5th ed.). Boston: Allyn & Bacon.

Simpson, Sally S. 1989. "Feminist Theory, Crime, and Justice." *Criminology*, 27: 605–632.

Singer, Margaret Thaler, with Janja Lalich. 1995. *Cults in Our Midst*. San Francisco: Jossey-Bass.

Sjoberg, Gideon. 1965. *The Preindustrial City: Past and Present*. New York: Free Press.

Skocpol, Theda, and Edwin Amenta. 1986. "States and Social Policies." In Ralph H. Turner and James F. Short, Jr. (Eds.), *Annual Review of Sociology*, 12: 131–157.

Smelser, Neil J. 1963. *Theory of Collective Behavior*. New York: Free Press.

—. 1988. "Social Structure." In Neil J. Smelser (Ed.), *Handbook of Sociology*. Newbury Park, CA: Sage, pp. 103–129.

Smith, Adam. 1976. *An Inquiry into the Nature and Causes of the Wealth of Nations*. Ed. Roy H. Campbell and Andrew S. Skinner. Oxford, England: Clarendon (orig. pub. 1776).

Smith, Allen C., III, and Sheryl Kleinman. 1989. "Managing Emotions in Medical School:

Students' Contacts with the Living and the Dead." *Social Science Quarterly*, 52 (1): 56–69.

Smith, Dorothy E. 1999. *Writing the Social: Critique, Theory, and Investigations*. Toronto: University of Toronto Press.

Smith, Douglas, Christy Visher, and Laura Davidson. 1984. "Equity and Discretionary Justice: The Influence of Race on Police Arrest Decisions." *Journal of Criminal Law and Criminology*, 75: 234–249.

Smith, Huston. 1991. *The World's Religions*. San Francisco: HarperSanFrancisco.

Smith, Wes. 2001. *Hope Meadows: Real-Life Stories of Healing and Caring from an Inspiring Community*. New York: Berkley.

Smolkin, Rachel. 2007. "What the Mainstream Media Can Learn from Jon Stewart." *American Journalism Review* (June/July 2007). Retrieved Feb. 16, 2008. Online: http://www.ajr.org/Article.asp?id=4329

Snow, David A., and Leon Anderson. 1993. *Down on Their Luck: A Case Study of Homeless Street People*. Berkeley: University of California Press.

Snow, David A., and Robert Benford. 1988. "Ideology, Frame Resonance, and Participant Mobilization." In Bert Klandermans, Hanspeter Kriesi, and Sidney Tarrow (Eds.), *International Social Movement Research*, Vol. 1, From Structure to Action. Greenwich, CT: JAI, pp. 133–155.

Snow, David A., E. Burke Rochford, Jr., Steven K. Worden, and Robert D. Benford. 1986. "Frame Alignment Processes, Micromobilization, and Movement Participation." *American Sociological Review*, 51: 464–481.

Snow, David A., Louis A. Zurcher, and Robert Peters. 1981. "Victory Celebrations as Theater: A Dramaturgical Approach to Crowd Behavior." *Symbolic Interaction*, 4 (1): 21–41.

Snyder, Benson R. 1971. *The Hidden Curriculum*. New York: Knopf.

Sommers, Ira, and Deborah R. Baskin. 1993. "The Situational Context of Violent Female Offending." *Journal of Research in Crime and Delinquency*, 30 (2): 136–162.

Stanley, Alessandra. 2004. "Old-Time Sexism Suffuses New Season." *New York Times* (Oct. 1): B1–B22.

Stannard, David E. 1992. *American Holocaust: Columbus and the Conquest of the New World*. New York: Oxford University Press.

Stapleton-Paff, Katie. 2007. "College Students Prefer 'The Daily Show' to Real News." *The Daily of the University of Washington* (May 21, 2007): 1.

Stark, Rodney, and William Sims Bainbridge. 1981. "American-Born Sects: Initial Findings." *Journal for the Scientific Study of Religion*, 20: 130–149.

Starr, Paul. 1982. *The Social Transformation of Medicine: The Rise of a Sovereign Profession and the Making of a Vast Industry*. New York: Basic.

Steffensmeier, Darrell, and Emilie Allan. 2000. "Looking for Patterns: Gender, Age, and Crime." In Joseph F. Sheley (Ed.), *Criminology: A Contemporary Handbook* (3rd ed.). Belmont, CA: Wadsworth, pp. 85–128.

Stein, Peter J. 1976. *Single*. Englewood Cliffs, NJ: Prentice Hall.

Stein, Peter J. (Ed.). 1981. *Single Life: Unmarried Adults in Social Context*. New York: St. Martin's.

Steinmetz, Erika. 2006. "Americans with Disabilities: 2000." U.S. Census Bureau, Current Population Reports, P70–107. Washington, DC: U.S. Government Printing Office.

Steinmetz, Suzanne K. 1987. "Elderly Victims of Domestic Violence." In Carl D. Chambers, John H. Lindquist, O. Z. White, and Michael T. Harter (Eds.), *The Elderly: Victims and Deviants*. Athens: Ohio University Press, pp. 126–141.

Stevenson, Mary Huff. 1988. "Some Economic Approaches to the Persistence of Wage Differences Between Men and Women." In Ann H. Stromberg and Shirley Harkess (Eds.), *Women Working: Theories and Facts in Perspective* (2nd ed.). Mountain View, CA: Mayfield, pp. 87–100.

Stewart, Abigail J. 1994. "Toward a Feminist Strategy for Studying Women's Lives." In Carol

E. Franz and Abigail J. Stewart (Eds.), *Women Creating Lives: Identities, Resilience, and Resistance*. Boulder, CO: Westview, pp. 11–35.

St. John, Warren. 2007. "A Laboratory for Getting Along." *New York Times* (Dec. 25): A1, A14.

StopGlobalWarming.org. 2006. "Marchers." Retrieved May 6, 2006. Online: http://www.stopglobalwarming.org

Stross, Randall. 2006. "Cellphone as Tracker: X Marks Your Doubts." *New York Times* (Nov. 19). Retrieved Mar. 30, 2007. Online: http://select.nytimes.com/search/restricted/article?res=F30F1FF63F5A0C7A8DDDA80994DE404482

Struck, Doug. 2006. "'Rapid Warming' Spreads Havoc in Canada's Forests." Washingtonpost.com (Mar. 1). Retrieved May 2, 2006. Online: http://www.washingtonpost.com/wp-dyn/content/article/2006/02/28/AR2006022801772

Substance Abuse and Mental Health Services Administration. 2000. *National Household Survey on Drug Abuse, 2000*. Retrieved Nov. 20, 2000. Online: http://www.samhsa.gov/oas/NHSDA/2kNHSDA/chapter2.htm

Sumner, William G. 1959. *Folkways*. New York: Dover (orig. pub. 1906).

Sutherland, Edwin H. 1939. *Principles of Criminology*. Philadelphia: Lippincott.

—. 1949. *White Collar Crime*. New York: Dryden.

Swarns, Rachel L. 2006. "The Immigrant Debate: The Overview; Immigrants Rally in Scores of Cities for Legal Status." *New York Times* (Apr. 11): A1.

Swidler, Ann. 1986. "Culture in Action: Symbols and Strategies." *American Sociological Review*, 51 (April): 273–286.

Tabb, William K., and Larry Sawers. 1984. *Marxism and the Metropolis: New Perspectives in Urban Political Economy* (2nd ed.). New York: Oxford University Press.

Takaki, Ronald. 1993. *A Different Mirror: A History of Multicultural America*. Boston: Little, Brown.

Tannen, Deborah. 1993. "Commencement Address, State University of New York at Binghamton." Reprinted in *Chronicle of Higher Education* (June 9): B5.

Tarbell, Ida M. 1925. *The History of Standard Oil Company*. New York: Macmillan (orig. pub. 1904).

Tavris, Carol. 1993. *The Mismeasure of Woman*. New York: Touchstone.

Tax Policy Center. 2006. "Historical Number of Households, Average Pretax and After-Tax Income and Shares, and Minimum Income." Retrieved Feb. 26, 2006. Online: http://www.taxpolicycenter.org/TaxFacts/TFDB/TFTemplate.cfm?Docid=461

—. 2007. "Historical Number of Households, Average Pretax and After-Tax Income and Shares, and Minimum Income." Retrieved Jan. 19, 2008. Online: http://www.taxpolicycenter.org/taxfacts/displayafact.cfm?Docid=461

Taylor, Steve. 1982. *Durkheim and the Study of Suicide*. New York: St. Martin's.

Teicher, Stacy A. 2006. "Researchers Say That Middle-School Bullying Could Be Curbed by Showing That It's Not Normal." *Christian Science Monitor* (Aug. 17). Retrieved Feb. 12, 2008. Online: http://www.csmonitor.com/2006/0817/p15s02-legn.html

Tergat, Paul. 2005. "Food That Changed My Life." *Guardian* (Apr. 15). Retrieved Feb. 24, 2006. Online: http://www.countercurrents.org/tergat150405.htm

Terkel, Studs. 1990. *Working: People Talk About What They Do All Day and How They Feel About What They Do*. New York: Ballantine (orig. pub. 1972).

Texeira, Erin. 2005. "Multiracial Scenes Now Common in TV Ads But Critics Say Commercials Gloss Over Complicated Racial Realities." MSNBC.com (Feb. 15). Retrieved Mar. 27, 2005. Online: http://www.msnbc.msn.com/id/6975669

That, Sovanny. 2007. "Refugee Women's Alliance." Retrieved Mar. 31, 2007. Online: http://students.washington.edu/sovannyt/communityservice.htm

theadventuresofiman.com 2007. "The Adventures of Iman." Retrieved Mar. 18, 2007. Online: http://www.theadventuresofiman.com/AboutIman.asp

thewritemarket.com 2003. "Building an Internet Community: What Is an 'Internet Community'?" Retrieved July 5, 2003. Online: http://www.thewritemarket.com/promotion/community.htm

Thompson, Becky W. 1994. *A Hunger So Wide and So Deep: American Women Speak Out on Eating Problems*. Minneapolis: University of Minnesota Press.

Thornberry, Terence P., Marvin D. Krohn, Alan J. Lizotte, and Deborah Chard-Wierschem. 1993. "The Role of Juvenile Gangs in Facilitating Delinquent Behavior." *Journal of Research in Crime and Delinquency*, 30 (1): 55–87.

Thorne, Barrie. 1993. *Gender Play: Girls and Boys in School*. New Brunswick, NJ: Rutgers University Press.

Thorne, Barrie, Cheris Kramarae, and Nancy Henley. 1983. *Language, Gender, and Society*. Rowley, MA: Newbury.

Thornton, Russell. 1984. "Cherokee Population Losses During the Trail of Tears: A New Perspective and a New Estimate." *Ethnohistory*, 31: 289–300.

Tierney, Kathleen. 2006. "Foreshadowing Katrina: Recent Sociological Contributions to Vulnerability Science." *Contemporary Sociology*, 35 (3): 207–227.

Tilly, Charles. 1973. "Collective Action and Conflict in Large-Scale Social Change: Research Plans, 1974–78." Center for Research on Social Organization. Ann Arbor: University of Michigan, October.

—. 1975. *The Formation of National States in Western Europe*. Princeton, NJ: Princeton University Press.

—. 1978. *From Mobilization to Revolution*. Reading, MA: Addison-Wesley.

Tilly, Chris, and Charles Tilly. 1998. *Work Under Capitalism*. Boulder, CO: Westview.

Tiryakian, Edward A. 1978. "Emile Durkheim." In Tom Bottomore and Robert Nisbet (Eds.), *A History of Sociological Analysis*. New York: Basic, pp. 187–236.

Tong, Rosemarie. 1989. *Feminist Thought: A Comprehensive Introduction*. Boulder, CO: Westview.

Tönnies, Ferdinand. 1940. *Fundamental Concepts of Sociology* (Gemeinschaft und Gesellschaft). Trans. Charles P. Loomis. New York: American Book Company (orig. pub. 1887).

—. 1963. *Community and Society* (Gemeinschaft and Gesellschaft). New York: Harper & Row (orig. pub. 1887).

Tracy, C. 1980. "Race, Crime and Social Policy: The Chinese in Oregon, 1871–1885." *Crime and Social Justice*, 14: 11–25.

Troeltsch, Ernst. 1960. *The Social Teachings of the Christian Churches*, vols. 1 and 2. Trans. O. Wyon. New York: Harper & Row (orig. pub. 1931).

Tumin, Melvin. 1953. "Some Principles of Stratification: A Critical Analysis." *American Sociological Review*, 18 (August): 387–393.

Turk, Austin. 1969. *Criminality and Legal Order*. Chicago: Rand McNally.

—. 1977. "Class, Conflict and Criminology." *Sociological Focus*, 10: 209–220.

Turner, Jonathan, Leonard Beeghley, and Charles H. Powers. 2002. *The Emergence of Sociological Theory* (5th ed.). Belmont, CA: Wadsworth.

—. 2007. *The Emergence of Sociological Theory* (6th ed.). Belmont, CA: Wadsworth.

Turner, Ralph H., and Lewis M. Killian. 1993. "The Field of Collective Behavior." In Russell L. Curtis, Jr., and Benigno E. Aguirre (Eds.), *Collective Behavior and Social Movements*. Boston: Allyn & Bacon, pp. 5–20.

Turner, Sherry L. 1997. "The Influence of Fashion Magazines on the Body Image Satisfaction of College Women: An Exploratory Analysis." *Adolescence* (Fall). Retrieved Feb. 2, 2008. Online: http://findarticles.com/p/articles/mi_m2248/is_n_127_v32ai_20413253

UNAIDS/WHO. 2000. "Report on the Global HIV/AIDS Epidemic—June 2000." Retrieved

Aug. 5, 2000. Online: http://www.unaids.org/epidemic_update/report/glo_estim.pdf

Union of Concerned Scientists. 2006. "Global Warming: What You Can Do about Global Warming." Retrieved May 6, 2006. Online: http://www.uscusa.org/global_warming/solutions.html

United Nations. 1997. "Global Change and Sustainable Development: Critical Trends." United Nations Department for Policy Coordination and Sustainable Development. Posted online (Jan. 20).

—. 2000. *Long-Range World Population Projections*. New York: United Nations Population Division.

United Nations Conference on Trade and Development. 2002. "The World's 100 Largest Non-Financial TNCs, Ranked by Foreign Assets, 2000." Retrieved Aug. 30, 2003. Online: http://www.unctad.org/sections/dite_dir/docs//Top100WIR2002.pdf

United Nations Development Programme. 1999. *Human Development Report: 1999*. New York: Oxford University Press.

—. 2000. "Higher Taxes May Loosen Tobacco's Hold on Poor Nations." Retrieved Mar. 31, 2004. Online: http://www.undp.org/dpa/frontpagearchive/august00/10aug00/tv081000.pdf

—. 2002. "The Millennium Development Goals and Human Development." Retrieved July 15, 2003. Online: http://hdr.undp.org/docs/mdg/The%20Millennium%20Development%20Goals%20and%20Human%20Development.pdf

—. 2003. *Human Development Report: 2003*. New York: Oxford University Press.

United Nations DPCSD. 1997. "Report of Commission on Sustainable Development, April 1997." New York: United Nations Department for Policy Coordination and Sustainable Development. Posted online.

United Nations Population Division. 1999. Retrieved Oct. 17, 1999. Online: gopher:/gopher.undp.org/00/ungophers/popin/wdtrends

United Press International. 2007. "Indian Women Carry Children for Foreigners." *United Press International.com* (Nov. 11). Retrieved Feb. 9, 2008. Online: http://www.upi.com/NewsTrack/Science/2007/11/11/indian_women_carry_children_for_foreigners/2909

U.S. Bureau of Justice Statistics. 2007. "Criminal Victimization, 2006." Retrieved Jan. 4, 2008. Online: http://www.ojp.usdoj.gov/bjs/pub/pdf/cv06fs.pdf

U.S. Bureau of Labor Statistics. 2007a. "Highlights of Women's Earnings in 2006." Retrieved Jan. 5, 2008. Online: http://www.bls.gov/cps/cpswom2006.pdf

—. 2007b. "Labor Force Statistics from the Current Population Survey." Retrieved Jan. 12, 2008. Online: http://data.bls.gov/cgi-bin/surveymost?ln

U.S. Census Bureau. 2005a. "Disability—Labor Force Status of Civilians 16 to 74 Years Old." Retrieved Apr. 15, 2006. Online: http://www.census.gov/hhes/www/disability/cps/cps205.html

—. 2005b. "Health Insurance Coverage: 2003." Retrieved Apr. 15, 2006. Online: http://www.census.gov/hhes/www/hlthins/hlthin03.html

—. 2005c. "Poverty Tables." Retrieved Mar. 3, 2006. Online: http://pubdb3.census.gov/macro/032005/pov/toc.htm

—. 2006. *Statistical Abstract of the United States: 2006* (125th ed.). Washington, DC: U.S. Government Printing Office.

—. 2007. *Statistical Abstract of the United States: 2007* (126th ed.). Washington, DC: U.S. Government Printing Office.

—. 2008. *Statistical Abstract of the United States* (127th ed.). Washington, DC: U.S. Government Printing Office.

U.S. Conference of Mayors. 2005. "Hunger Homelessness Survey Summary." Retrieved Mar. 12, 2005. Online: http://www.usmayors.org/uscm/us_mayor_newspaper/documents/01_10_05/hunger_survey.asp

U.S. Congress, Joint Economic Committee. 1986. *The Concentration of Wealth in the United States: Trends in the Distribution of Wealth Among American Families.* Washington, DC: U.S. Government Printing Office.
U.S. Department of Education. 2007. "Indicators of School Crime and Safety: 2006." Retrieved Jan. 4, 2008. Online: http://www.ojp.usdoj.gov/bjs/pub/pdf/iscs06.pdf
U.S. Department of Health and Human Services. 2006. "HIV/AIDS: Basic Statistics." Retrieved Apr. 15, 2006. Online: http://www.cdc.gov/hiv/topics/surveillance/basic.htm
—. 2008. "Stop Bullying Now." Retrieved Feb. 12, 2008. Online: http://www.stopbullyingnow.hrsa.gov
U.S. Department of Justice. 2000. *Intimate Partner Violence.* Retrieved Nov. 3, 2001. Online: http://www.ojp.usdoj.gov/bjs/pub/press/ipv.htm
U.S. Department of Labor. 2003. "Business Ownership—Cornerstone of the American Dream." Retrieved July 19, 2003. Online: http://www.dol.gov/odep/pubs/business/business.htm
U.S. Office of Personnel Management. 2004. "2004 Demographic Profile of the Federal Workforce." Retrieved Apr. 27, 2006. Online: http://www.opm.gov/feddata/demograp/demograp.asp#RNOData
Van Ausdale, Debra, and Joe R. Feagin. 2001. *The First R: How Children Learn Race and Racism.* Lanham, MD: Rowman & Littlefield.
Van Biema, David. 1993. "But Will It End the Abortion Debate?" *Time* (June 14): 52–54.
Vaughan, Diane. 1985. "Uncoupling: The Social Construction of Divorce." In James M. Henslin (Ed.), *Marriage and Family in a Changing Society* (2nd ed.). New York: Free Press, pp. 429–439.
Vaughan, George B. 2000. *The Community College Story: A Tale of American Innovation.* Washington, DC: Community College Press.
Veblen, Thorstein. 1967. *The Theory of the Leisure Class.* New York: Viking (orig. pub. 1899).
Venkatesh, Sudhir Alladi. 2006. *Off the Books: The Underground Economy of the Urban Poor.* Cambridge, MA: Harvard University Press.
Vissing, Yvonne. 1996. *Out of Sight, Out of Mind: Homeless Children and Families in Small Town America.* Lexington: University Press of Kentucky.
Vito, Gennaro F., and Ronald M. Holmes. 1994. *Criminology: Theory, Research and Policy.* Belmont, CA: Wadsworth.
Voynick, Steve. 1999. "Living with Ozone." *The World & I* (July): 192–199.
Wadud, Amina. 2002. "A'ishah's Legacy: Amina Wadud Looks at the Struggle for Women's Rights Within Islam." *New Internationalist* (May). Retrieved: Mar. 18, 2007. Online: http://newint.org/features/2002/05/01/aishahs-legacy
Wagner, Elvin, and Allen E. Stearn. 1945. *The Effects of Smallpox on the Destiny of the American Indian.* Boston: Bruce Humphries.
Waldron, Ingrid. 1994. "What Do We Know About Causes of Sex Differences in Mortality? A Review of the Literature." In Peter Conrad and Rachelle Kern (Eds.), *The Sociology of Health and Illness: Critical Perspectives* (4th ed.). New York: St. Martin's.
Wallace, Harvey. 2002. *Family Violence: Legal, Medical, and Social Perspectives* (3rd ed.). Boston: Allyn & Bacon.
Wallace, Walter L. 1971. *The Logic of Science in Sociology.* New York: Aldine de Gruyter.
Wallerstein, Immanuel. 1979. *The Capitalist World-Economy.* Cambridge, England: Cambridge University Press.
—. 1984. *The Politics of the World Economy.* Cambridge, England: Cambridge University Press.
—. 1991. *Unthinking Social Science: The Limits of Nineteenth-Century Paradigms.* Cambridge, England: Polity.
Warner, W. Lloyd, and Paul S. Lunt. 1941. *The Social Life of a Modern Community.* New Haven, CT: Yale University Press.
Warr, Mark. 1993. "Age, Peers, and Delinquency." *Criminology,* 31 (1): 17–40.
—. 2000. "Public Perceptions and Reactions to Crime." In Joseph F. Sheley (Ed.), *Criminology: A Contemporary Handbook* (3rd ed.). Belmont, CA: Wadsworth, pp. 13–55.
Waters, Malcolm. 1995. *Globalization.* London and New York: Routledge.
Watson, Elwood, and Darcy Martin (Eds). 2004. *"There She Is, Miss America": The Politics of Sex, Beauty, and Race in America's Most Famous Pageant.* New York: Palgrave Macmillan.
Watson, Paul. 1997. "Richer, Poorer." *Toronto Star* (Aug. 10): F1, F5.
Watson, Tracey. 1987. "Women Athletes and Athletic Women: The Dilemmas and Contradictions of Managing Incongruent Identities." *Sociological Inquiry,* 57 (Fall): 431–446.
Waxman, Laura, and Sharon Hinderliter. 1996. *A Status Report on Hunger and Homelessness in America's Cities: 1996.* Washington, DC: U.S. Conference of Mayors.
Weber, Max. 1963. *The Sociology of Religion.* Trans. E. Fischoff. Boston: Beacon (orig. pub. 1922).
—. 1968. *Economy and Society: An Outline of Interpretive Sociology.* Trans. G. Roth and G. Wittich. New York: Bedminster (orig. pub. 1922).
—. 1976. *The Protestant Ethic and the Spirit of Capitalism.* Trans. Talcott Parsons. Introduction by Anthony Giddens. New York: Scribner (orig. pub. 1904–1905).
Weeks, John R. 2005. *Population: An Introduction to Concepts and Issues* (9th ed.). Belmont, CA: Wadsworth.
—. 2008. *Population: An Introduction to Concepts and Issues* (10th ed.). Belmont, CA: Wadsworth.
Weigel, Russell H., and P. W. Howes. 1985. "Conceptions of Racial Prejudice: Symbolic Racism Revisited." *Journal of Social Issues,* 41: 124–132.
Weinstein, Michael M. 1997. "'The Bell Curve,' Revisited by Scholars." *New York Times* (Oct. 11): A20.
Weiss, Gregory L., and Lynne E. Lonnquist. 2003. *The Sociology of Health, Healing, and Illness* (4th ed.). Upper Saddle River, NJ: Prentice Hall.
—. 2009. *The Sociology of Health, Healing, and Illness* (6th ed.). Upper Saddle River, NJ: Prentice Hall.
Weiss, Jeffrey. 2004. "Beliefs in UFOs and Ritual Abuse Are Studied as New Religious Movements." *Dallas Morning News* (Aug. 28). Retrieved Jan. 11, 2008. Online: http://www.dallasnews.com
Weiss, Meira. 1994. *Conditional Love: Attitudes Toward Handicapped Children.* Westport, CT: Bergin & Garvey.
Weitz, Rose. 2004. *The Sociology of Health, Illness, and Health Care* (3rd ed.). Belmont, CA: Wadsworth.
Wellhousen, Karyn, and Zenong Yin. 1997. "Peter Pan Isn't a Girls' Part: An Investigation of Gender Bias in a Kindergarten Classroom." *Women and Language,* 20: 35–40.
Wellman, Barry. 2001. "Physical Place and Cyberplace: The Rise of Personalized Networking." *International Journal of Urban and Regional Research* 22 (2): 227–252.
Welner, Kevin Grant, and Jeannie Oakes. 2000. *Navigating the Politics of Detracking.* Arlington Heights, IL: Skylight.
Westrum, Ron. 1991. *Technologies and Society: The Shaping of People and Things.* Belmont, CA: Wadsworth.
Wharton, Amy S. 2004. *The Sociology of Gender: An Introduction to Theory and Research.* London: Blackwell.
White, Jack E. 1997. "I'm Just Who I Am." *Time* (May 5): 32–36.
White, Jonathan R. 2003. *Terrorism: An Introduction* (2002 update). Belmont, CA: Wadsworth.
whitehouse.gov. 2008. "Fact Sheet: Six Years of Student Achievement Under No Child Left Behind." Retrieved Feb. 11, 2008. Online: http://www.whitehouse.gov/news/releases/2008/01/20080107-1.html
Whorf, Benjamin Lee. 1956. *Language, Thought and Reality.* Ed. John B. Carroll. Cambridge, MA: MIT Press.
Wilkerson, Jamie. 1996. "Thoughts of Internet Friendships." Retrieved July 5, 2003. Online: http://web.cetlink.net/~parrothd/Jamie/Poetry1.htm
Williams, Christine. 2004. "The Glass Escalator: Hidden Advantages for Men in the 'Female' Professions." In Michael S. Kimmel and Michael S. Messner (Eds.), *Men's Lives* (6th ed.). Boston: Allyn & Bacon.
Williams, Robin M., Jr. 1970. *American Society: A Sociological Interpretation* (3rd ed.). New York: Knopf.
Willis, Leigh A., David W. Coombs, William C. Cockerham, and Sonja L. Frison. 2002. "Ready to Die: A Postmodern Interpretation of the Increase of African-American Adolescent Male Suicide." *Social Science & Medicine* (September): 907–920.
Wilson, David (Ed.). 1997. "Globalization and the Changing U.S. City." *Annals of the American Academy of Political and Social Sciences,* 551 (May special issue).
Wilson, Edward O. 1975. *Sociobiology: A New Synthesis.* Cambridge, MA: Harvard University Press.
Wilson, Elizabeth. 1991. *The Sphinx in the City: Urban Life, the Control of Disorder, and Women.* Berkeley: University of California Press.
Wilson, William Julius. 1978. *The Declining Significance of Race: Blacks and Changing American Institutions.* Chicago: University of Chicago Press.
—. 1996. *When Work Disappears: The World of the New Urban Poor.* New York: Knopf.
Winik, Lyric Wallwork. 1997. "Oh Nurse, More Beluga Please." *Forbes FYI: The Good Life* (Winter): 157–166.
Winkleby, Marilyn A., and Catherine Cubbin. 2003. "Influence of Individual and Neighbourhood Socioeconomic Status on Mortality Among Black, Mexican-American, and White Women and Men in the United States." *Journal of Epidemiology and Community Health,* 57: 444–452.
Winn, Maria. 1985. *The Plug-in Drug: Television, Children, and the Family.* New York: Viking.
Wirth, Louis. 1938. "Urbanism as a Way of Life." *American Journal of Sociology,* 40: 1–24.
Wischnowsky, Dave. 2005. "Small Town a Bastion of Bigfoot Belief." *Chicago Tribune* (Oct. 10): 2.
Wiseman, Jacqueline. 1970. *Stations of the Lost: The Treatment of Skid Row Alcoholics.* Chicago: University of Chicago Press.
Wonders, Nancy. 1996. "Determinate Sentencing: A Feminist and Postmodern Story." *Justice Quarterly,* 13: 610–648.
Wood, Julia T. 1994. *Gendered Lives: Communication, Gender, and Culture.* Belmont, CA: Wadsworth.
—. 1999. *Gendered Lives: Communication, Gender, and Culture* (3rd ed.). Belmont, CA: Wadsworth.
Wooley, Susan C. 1994. "Sexual Abuse and Eating Disorders: The Concealed Debate." In Patricia Fallon, Melanie A. Katzman, and Susan C. Wooley (Eds.), *Feminist Perspectives on Eating Disorders.* New York: Guilford, pp. 171–211.
World Bank. 2005. *World Development Indicators, 2005.* Retrieved Mar. 4, 2006. Online: http://devdata.worldbank.org/wdi2005/Cover.htm
—. 2007. "Poverty in China: What Do the Numbers Say?" Washington, DC: World Bank. Retrieved Jan. 20, 2008. Online: http://web.worldbank.org
World Health Organization. 2003. *The World Health Report 2003.* Retrieved Jan. 8, 2004. Online: http://www.who.int/whr/2003/en/overview_en.pdf
—. 2004a. "Infant and Under Five Mortality Rates by WHO Region. Year 2000." Retrieved Feb. 28, 2004. Online: http://www.who.int/child-adolescent-health/OVERVIEW/CHILD_HEALTH/Mortality_Rates_00.pdf
—. 2004b. "Suicide Rates and Absolute Numbers of Suicide by Country." Retrieved June 19, 2004. Online: http://www/who.int/mental_health/prevention/suicide/suicideprevent/en
Worster, Donald. 1985. *Natures Economy: A History of Ecological Ideas.* New York: Cambridge University Press.
Wouters, Cas. 1989. "The Sociology of Emotions and Flight Attendants: Hochschild's Managed Heart." *Theory, Culture & Society,* 6: 95–123.
Wright, Erik Olin. 1978. "Race, Class, and Income Inequality." *American Journal of Sociology,* 83 (6): 1397.
—. 1979. *Class Structure and Income Determination.* New York: Academic Press.
—. 1985. *Class.* London: Verso.
—. 1997. *Class Counts: Comparative Studies in Class Analysis.* Cambridge, England: Cambridge University Press.
Wright, Erik Olin, Karen Shire, Shu-Ling Hwang, Maureen Dolan, and Janeen Baxter. 1992. "The Non-Effects of Class on the Gender Division of Labor in the Home: A Comparative Study of Sweden and the U.S." *Gender & Society,* 6 (2): 252–282.
Yablonsky, Lewis. 1997. *Gangsters: Fifty Years of Madness, Drugs, and Death on the Streets of America.* New York: New York University Press.
Yelin, Edward H. 1992. *Disability and the Displaced Worker.* New Brunswick, NJ: Rutgers University Press.
Yinger, J. Milton. 1960. "Contraculture and Subculture." *American Sociological Review,* 25 (October): 625–635.
—. 1982. *Countercultures: The Promise and Peril of a World Turned Upside Down.* New York: Free Press.
Young, John. 1990. *Sustaining the Earth: The Story of the Environmental Movement—Its Past Efforts and Future Challenges.* Cambridge, MA: Harvard University Press.
Young, Michael Dunlap. 1994. *The Rise of the Meritocracy.* New Brunswick, NJ: Transaction (orig. pub. 1958).
Zakaria, Fareed. 2005. "The Wealth of Yet More Nations." *New York Times Book Review* (May 1): 10–11.
Zald, Mayer N., and John D. McCarthy (Eds.). 1987. *Social Movements in an Organizational Society.* New Brunswick, NJ: Transaction.
Zavella, Patricia. 1987. *Women's Work and Chicano Families: Cannery Workers of the Santa Clara Valley.* Ithaca, NY: Cornell University Press.
Zeitlin, Irving M. 1997. *Ideology and Development of Sociological Theory* (6th ed.). Upper Saddle River, NJ: Prentice Hall.
Zellner, William M. 1978. Vehicular Suicide: In Search of Incidence. Unpublished M.A. thesis, Western Illinois University, Macomb. Quoted in Richard T. Schaefer and Robert P. Lamm. 1992. *Sociology* (4th ed.). New York: McGraw-Hill, pp. 54–55.
Zipp, John F. 1985. "Perceived Representativeness and Voting: An Assessment of the Impact of 'Choices' vs. 'Echoes.'" *American Political Science Review,* 60 (3): 738–759.
Zirin, Dave. 2005. "Fade to Black." Edge of Sports (Aug. 31). Retrieved Feb. 26, 2006. Online: http://www.edgeofsports.com/2005-08-31-151/index.html
Zuboff, Shoshana. 1988. *In the Age of the Smart Machine: The Future of Work and Power.* New York: Basic.

Photo Credits

This page constitutes an extension of the copyright page. We have made every effort to trace the ownership of all copyrighted material and to secure permission from copyright holders. In the event of any question arising as to the use of any material, we will be pleased to make the necessary corrections in future printings. Thanks are due to the following authors, publishers, and agents for permission to use the material indicated.

Chapter 1. 2, 3: © AP Images/Tom Gannam **8:** left, © Syracuse Newspapers/John Berry/The Image Works **8:** middle, © Andrew Ward/Life File/Getty Images **8:** right, © DEA/M. BORCHI/Getty Images **10:** top, © Hulton Archive/Getty Images **10:** bottom, Auguste Comte (1798-1857) (oil on canvas), Etex, Louis Jules (1810-1889)/Temple de la Religion de l'Humanite, Paris, France The Bridgeman Art Library International **11:** © Spencer Arnold/Getty Images **12:** top, © Hulton Archive/Getty Images **12:** bottom, © Bettmann/CORBIS **14:** © North Wind Picture Archives. All rights reserved. **15:** top, © Hulton Archive/Getty Images **15:** bottom, © The Granger Collection, New York **16:** © Simon Jarratt/CORBIS **17:** left, © AP Images **17:** right, © Bettmann/CORBIS **18:** top, © The Granger Collection, New York **18:** bottom, © Estate of Robert K. Merton, photo by Sandra Still **19:** © Bob Daemmrich/The Image Works **20:** © Hulton Archive/Getty Images **21:** © AP Images/Denis Farrell **32:** © Michael Newman/PhotoEdit **35:** left, © Addison Geary **35:** right, From CODE OF THE STREET: DECENCY, VIOLENCE, AND THE MORAL LIFE OF THE INNER CITY by Elijah Anderson. Copyright © 1999 by Elijah Anderson. Used by permission of W. W. Norton & Company, Inc. Cover photo © Camilo Jose Vergara. **36:** © Cate Gillon/Getty Images

Chapter 2. 42, 43: © Bob Daemmrich/The Image Works **47:** top, © Gabriela Trojanowska/Shutterstock **47:** middle, © Andresr/Shutterstock **47:** bottom, © Altrendo Images/Getty Images **48:** top left, © Celia Peterson/Getty Images **48:** top right, © Frans Lemmens/Getty Images **48:** bottom, © Eddie Gerald/Alamy **49:** top left, © Spencer Grant/PhotoEdit **49:** top right, © Mark Richards/PhotoEdit **49:** bottom, © Michael Greenlar/The Image Works **50:** © GEORGE FREY/Bloomberg News/Landov **51:** © Ross M. Horowitz/Getty Images **53:** top, © Jim West/The Image Works **53:** bottom, © AP Images/Ann Johansson **57:** © David Falconer/SuperStock **61:** top, © Spencer Grant/PhotoEdit **61:** bottom, © AP Images/Tony Dejak **64:** © Victor Englebert/Time Life Pictures/Getty Images **66:** © Sonda Dawes/The Image Works **67:** © Guilad Kahn/AFP/Getty Images **70:** left, © Ralf-Finn Hestoft/CORBIS **70:** right, © vario images GmbH & Co. KG/Alamy **72:** left, © David McNew/Getty Images **72:** right, © AP Images/John Raoux

Chapter 3. 76, 77: © Frank Trapper/CORBIS **81:** © Martin Rogers/Getty Images **82:** © Bettmann/CORBIS **83:** left, © Bill Aron/PhotoEdit **83:** right, © Jose Luis Pelaez, Inc./Getty Images **84:** top, © Mary Evans/SIGMUND FREUD COPYRIGHTS/The Image Works **84:** bottom, © AFP/Getty Images **86:** both, © Tony Freeman/PhotoEdit **87:** © PhotoAlto/Alamy **89:** top left, © Daniel Bosler/Getty Images **89:** bottom left, © Stella/Getty Images **89:** right, © Digital Vision/Alamy **93:** © Michael Newman/PhotoEdit **94:** top, © SW Productions/Getty Images **94:** bottom, **vii:** © Jeff Greenberg/The Image Works **95:** top left, Courtesy of the National Runaway Switchboard **95:** top right, © Tim Boyle/Getty Images **95:** bottom, © DEAN J. KOEPFLER/Tacoma News Tribune/MCT/Landov **96:** top, © RAFIQUR RAHMAN/Reuters/Landov **97:** © Tony Freeman/PhotoEdit **98:** © Michael Newman/PhotoEdit **101:** © Stu Smucker/OnAsia/Jupiterimages **102:** top, © David Young-Wolff/PhotoEdit **102:** bottom, © AP Images/Mark Humphrey **105:** © Sonda Dawes/The Image Works **107:** © Journal Courier/The Image Works **108:** © Lisette LeBon/SuperStock

Chapter 4. 112, 113: © Bob Collins/The Image Works **117:** © Jeff Brass/Getty Images **118:** left, © Xinhua/Landov **118:** right, © Rachel Epstein/PhotoEdit **121:** © Kim Eriksen/zefa/CORBIS **123:** left, © Gilles Mingasson **123:** right, © MARIO ANZUONI/Reuters/Landov **124:** left, © Bob Daemmrich/PhotoEdit **124:** right, © Ghislain & Marie David de Lossy/Getty Images **125:** © 2007 NBAE, David Dow/NBAE/Getty Images **127:** left, © Popperfoto/Getty Images **127:** right, © Marvin E. Newman/Getty Images **131:** top, © MANDEL NGAN/AFP/Getty Images **131:** bottom, © AP Images/Reed Saxon **133:** top, © AP Images/Carolyn Kaster **133:** bottom, © AP Images/Alex Brandon **136:** left, © Tom Prettyman/PhotoEdit **136:** right, © David Silverman/Getty Images **137:** left, © Dana White/PhotoEdit **137:** top right, © Richard Ross/Getty Images **137:** bottom right, © Krzysztof Mystkowski/AFP/Getty Images **139:** © David Young-Wolff/PhotoEdit **141:** Courtesy of Gretchen Otto

Chapter 5. 144, 145: © AP Images/L. G. Patterson **151:** top, © Duncan Hale-Sutton/Alamy **151:** bottom, © Jeff Greenberg/Alamy **152:** both, © Michael Newman/PhotoEdit **158:** top, © AP Images/Chris O'Meara **158:** middle, NASA Kennedy Space Center (NASA-KSC) **158:** bottom, © AP Images/Dr. Scott Lieberman **161:** left, © John Aikins/CORBIS **161:** right, © Losevsky Pavel/Shutterstock **162:** © Andy Nelson/The Christian Science Monitor/Getty Images **163:** © Sean Justice/Getty Images **164:** left, © Blend Images/Alamy **164:** right, © Mark Richards/PhotoEdit **165:** © Jack Kurtz/The Image Works **167:** © TWPhoto/CORBIS

Chapter 6. 172, 173: © A. Ramey/PhotoEdit **174:** © Paramount/courtesy Everett Collection **178:** © AP Images/Tony Gutierrez **180:** left, © AP Images/John Miller **180:** right, © Paul Hartnett/PYMCA/Jupiterimages **183:** © AP Images/Luke Palmisano **184:** © Nick Koudis/Getty Images **185:** © LEZLIE STERLING/MCT/Landov **187:** © AP Images/Kevork Djansezian **188:** left, © Arthur Tilley/Getty Images **188:** right, © Mark Richards/PhotoEdit **190:** © AP Images/Brett Coomer **195:** FBI **197:** © Sean Cayton/The Image Works **202:** top left, © Morgan Lane Photography/Shutterstock **202:** top middle, © Image99/Jupiterimages **202:** top right, © Image Source/Jupiterimages **202:** bottom, © AP Images/Phil Sandlin

Chapter 7. 212, 213: © Anthony Barboza **217:** © Zubin Shroff/Getty Images **218:** top, © Hulton Archive/Getty Images **218:** bottom left, © Alan Sussman/The Image Works **218:** bottom right, © John Lund/Drew Kelly/Getty Images **219:** © Robyn Beck/AFP/Getty Images **221:** © AP Images/Carlos Osorio **222:** © New Line/courtesy Everett Collection **225:** © Jeff Greenberg/The Image Works **226:** top left, © AP Images/Andrew Shurtleff **226:** top right, Cover of NICKEL AND DIMED : On (Not) Getting By in America. Cover design copyright 2008 by Henry Holt and Company. Text by Barbara Ehrenreich. Reprinted by permission of Henry Holt and Company, LLC. **226:** bottom, © Tony Freeman/PhotoEdit **227:** © Billy Hustace/Getty Images **228:** © ERIC DRAPER/WHITE HOUSE/UPI/Landov **229:** top, © AP Images/Adrian Wyld/CP **229:** bottom, © David Frazier/PhotoEdit **230:** © Michael Newman/PhotoEdit **234,viii:** © AP Images/Michigan Lottery **235:** top, © AP Images/James Branaman/The Kitsap Sun **235:** middle, © Frances M. Roberts/Alamy **235:** bottom, © Liz Hafalia/San Francisco Chronicle/CORBIS **236:** top, © AP Images/Tim Boyd **238:** © Image Source/Getty Images **242:** © Sonda Dawes/The Image Works **243:** © AP Images/Paul Beaty **246:** © Steven Rubin/The Image Works **248:** © Joe Sohm/The Image Works

Chapter 8. 252, 253: © AP Images/Herbert Knosowski **257:** © Viviane Moos/CORBIS **258:** © Reinhard Krause/Reuters/CORBIS **262:** © Dell Inc. for Business Wire via Getty Images **266:** © Jenny Matthews/Alamy **269:** © Chris Hondros/Getty Images **271:** © James Marshall/The Image Works **275:** © Scott Olson/Getty Images **276:** © AP Images/Sandra Boulanger

Chapter 9. 280, 281: © AP Images/Mark J. Terrill **285:** © Jeff Greenberg/PhotoEdit **286:** © AP Images **287:** © Michael Greenlar/The Image Works **292:** left, © Bettmann/CORBIS **292:** right, © TIM JOHNSON/Reuters/Landov **295:** © Bill Aron/PhotoEdit **296:** © Shelly Katz/Getty Images **300:** © AP Images/Mike Yoder/Lawrence Journal-World **301:** © Kevin Fleming/CORBIS **304:** © AP Images/Ron Edmonds **305:** © Purestock/Alamy **306:** © AP Images **308:** © AP Images/Don Ryan **310:** © AP Images/Kathy Willens **311:** © Jim West/The Image Works

Chapter 10. 316, 317: © TIM SHAFFER/Reuters/Landov **323:** left, © Alex Farnsworth/The Image Works **323:** right, © MOHAMMED AMEEN/Reuters/Landov **325:** left, © David Young-Wolff/PhotoEdit **325:** right, © A. Ramey/PhotoEdit **328:** Courtesy of Rima Khorebi **330:** left, © David R. Frazier/The Image Works **330:** right, © Spencer Grant/PhotoEdit **333:** left, © Geri Engberg/The Image Works **333:** right, © Bob Thomas/Getty Images **334:** © Barbara Campbell **335:** © Mary Kate Denny/PhotoEdit **336:** © AP Images **337:** © KAMIL KRZACZYNSKI/Landov **341:** left, © Will Hart/PhotoEdit **341:** right, © Spencer Grant/PhotoEdit **343:** © Peter Dazeley/Getty Images **344:** © AP Images/J. Scott Applewhite **346, x:** © Image Source Pink/Getty Images **347:** top, © Ariel Skelley/Getty Images **347:** middle, © Photodisc/SuperStock **347:** bottom, © Ed Bock/CORBIS **348:** top, © moodboard/Alamy **350:** © Larry Kolvoord/The Image Works **351:** © Mauritius/Photolibrary

Chapter 11. 354, 355: © Myrleen Cate/Index Stock Imagery/Photolibrary **359:** left, © Michael Newman/PhotoEdit **359:** right, © Steven Rubin/The Image Works **363:** © Jeff Randall/Getty Images **364:** left, © Bob Barkany/Getty Images **364:** top right, © Corbis Super RF/Alamy **364:** bottom right, © Michael Newman/PhotoEdit **367:** © Mary Kate Denny/PhotoEdit **368:** © GABRIEL BOUYS/AFP/Getty Images **369:** © Ariel Skelley/Getty Images **373:** © AP Images/Ajit Solanki **374:** © James Devaney/WireImage/Getty Images **375:** © Masterfile **378:** © David Young-Wolff/Getty Images **381:** © Tony Freeman/PhotoEdit **384:** © Michelle D. Bridwell/PhotoEdit

Chapter 12. 388, 389: © AP Images/Jay Laprete **395:** © Radius Images/Jupiterimages **396:** © Enigma/Alamy **397:** © Cleve Bryant/PhotoEdit **398:** © Gabe Palmer/Alamy **401:** left, © Justin Sullivan/Getty Images **401:** right, © Bob Daemmrich/The Image Works **405:** left, Courtesy of Joliet Junior College **405:** right, Courtesy of Miami Dade College, photo by Phil Roche **407:** © BRIAN SNYDER/Reuters/Landov **409:** top left, © James Marshall/Getty Images **409:** top right, © Renaud Visage/Getty Images **409:** bottom left, © DPA/SOA/The Image Works **409:** bottom right, © Richard A. Cooke/CORBIS **413:** © Kimberly White/Getty Images **414:** © David Young-Wolff/PhotoEdit **415:** © Devendra M. Singh/AFP/Getty Images **420:** © Bob Daemmrich/The Image Works

Chapter 13. 426, 427: © AP Images/Jason DeCrow **431:** top, © AP Images/Will Burgess/Pool **431:** middle, © Tim Graham/Getty Images **431:** bottom, © Collection, The Supreme Court Historical Society. Photography by Steve Petteway, The Supreme Court. **434:** © Samir Hussein/WireImage/Getty Images **437:** © AP Images/Stephen J. Boitano **441:** left, © Petrified Collection/Getty Images **441:** right, © HANS DERYK/Reuters/Landov **442:** top, © AP Images/Mary Altaffer **442:** bottom, © AP Images/Lucy Pemoni **446:** © AP Images/Charles Bennett **449:** © Sajjad Hussain/AFP/Getty Images **452:** © Alex Segre/Alamy **456:** © Peter Hvizdak/The Image Works **458:** © AP Images/Mark Lennihan **459:** © PhotoEdit **460:** © Mike Blank/Getty Images

Chapter 14. 464, 465: © ERIK JACOBS/The New York Times/Redux Pictures **469:** left, © Andrew Holbrooke/The Image Works **469:** right, © Lew Lause/SuperStock **471:** © JB REED/Blooomberg News/Landov **472:** © chuck kuhn photography/Getty Images **473:** © David M. Grossman/The Image Works **474:** © Michael Newman/PhotoEdit **478:** left, © Stanley B. Burns, MD & The Burns Archive N.Y./Photo Researchers, Inc. **478:** right, © Peter Ginter/Getty Images **483:** © Wally McNamee/CORBIS **485:** © Royalty Free/CORBIS **488:** © Thinkstock/Jupiterimages **491:** © AP Images/Mike Groll **493:** © ALESSANDRO BIANCHI/Reuters/Landov **496:** © AP Images/Jennifer Graylock/JGRYL, Graylock

Chapter 15. 500, 501: © A. Ramey/PhotoEdit **506:** left, © Chris Hondros/Getty Images **506:** right, © David Young-Wolff/PhotoEdit **508, xii:** © AP Images/Robert F. Bukaty **509:** top, © AP Images/Ann Heisenfelt **509:** middle, © AP Images/Charles Krupa **509:** bottom, © Alexander Tamargo/Getty Images **510:** top, © Tony Freeman/PhotoEdit **518:** © AP Images/George Osodi **520:** top, © JAMIE RECTOR/Bloomberg News/Landov **520:** bottom, © Bob Daemmrich/The Image Works **523:** © Andrew Holbrooke/The Image Works **526:** © Jeff Greenberg/PhotoEdit **532:** left, © THOM BAUR/AFP/Getty Images **532:** right, © AP Images/Mark Duncan

Chapter 16. 538, 539: © Lu Mingxiang/Xinhua/Landov **543:** © William Campbell/Sygma/CORBIS **545:** © AP Images/Jason E. Miczek **546:** © David Young-Wolff/PhotoEdit **548:** © Jonathan Fickies/Getty Images **549:** © AP Images **550:** Copyright 2008, -Kim, Seattle Daily Photo **552:** © AP Images **554:** © Dallas Events Inc./Shutterstock **556:** left, © AP Images/Rich Pedroncelli **556:** right, © Saul Loeb/epa/CORBIS **559:** © AP Images/Wen bao/ICHPL, Imagechina **560:** © Andrew Lichtenstein/The Image Works **563:** left, Photo by Jocelyn Augustino/FEMA Photo Library **563:** right, © AP Images/Bill Haber **564:** © Ted Foxx/Alamy **565:** © AP Images/Steve Yeater **566:** © Jeff Greenberg/PhotoEdi

Name Index

Subject Index

cultural assimilation, 292
cultural capital theory, 65, 396, 397
cultural creation, 67
cultural diversity, 59–63
    in education, 395, 396
    ethnocentrism, 64, 65
    food and, 46, 62
    in the future, 312
    immigration and, 508
    racial/ethnic, 312
    religious tolerance and, 422
cultural imperialism, 66, 67
cultural lag, 58, 59, 565
cultural relativism, 64–65
cultural universals, 47–50
culture, 44, 45–47
    conflict perspective, 67–68, 71
    ethnic subcultures, 61–63
    food and, 42–43, 46, 48
    functionalist perspective, 66–67, 71
    in the future, 72–73
    gender and, 322–327
    global, 65–66
    ideal vs. real, 56
    nonmaterial components, 50–58, 71
    postmodernist perspective, 70–71
    society and, 44–45
    symbolic interactionist perspective, 68, 70, 71
    technology and, 58–59, 72
    transmission of, 392–393
culture shock, 63–64
customs, 49
cyclical unemployment, 457

The Daily Show, 426–428
Darwinism, 12, 13
data presentation, 26
Davis-Moore thesis, 245
de facto segregation, 293
de jure segregation, 293
death, causes of, 505
deductive research method, 27
deference, 137, 138, 246
degree of seriousness, 176
deindustrialization, 243, 262
demedicalization, 488
democracy, 435–436, 442, 460
democratic leaders, 154, 155
democratic socialism, 453
demographic transition theory, 514–515
demography, 502–503
denominations, 418–419
density (population), 506
dependency theory, economic development, 269–271
dependent variable, 28, 29
deprived (urban dwellers), 526
deprofessionalization, 454–455
determinate sentence, 205
deterrence, 204
developing countries, 8, 9
deviance, 174–178. See also crime
    capitalism and, 184–185
    conflict perspective, 183–186, 191
    feminist perspective, 185–186
    functionalist perspective, 176, 179–183, 191
    in the future, 205–207
    gender and, 186
    illness, 468
    medicalization of deviance, 488
    postmodernist perspective, 190–191
    power and, 184
    race and, 186
    social class and, 186
    strain theory, 179–181
    symbolic interactionist perspective, 186–190, 191
diagnostic framing, 557
dictatorship, 436
differential association theory, 186, 187, 191
differential reinforcement theory, 186, 187
diffusion, of culture, 58–59
direct institutionalized discrimination, 289–290
direct participatory democracy, 435
disabled and disabilities, 489, 491
    accommodations for, 459
    acquired, 493
    city life and, 528–529
    conflict perspective, 494
    employment opportunities, 459
    functionalist perspective, 493
    homeless, 125
    learning disabled, 400
    life expectancy and, 491
    population, 491, 492
    rehabilitation, 494
    social inequality, 494–495
    symbolic interactionist perspective, 493–494
    temporary vs. chronic, 492
disasters and tragedies
    Columbia space shuttle, 157, 158
    Hurricane Katrina, 562–563
    natural vs. technological, 542

September 11, 2001, 311, 434, 553
Tokaimura nuclear power plant, 552
Discipline and Punish (Foucault), 190
discovery, 58
discretion, 201, 202
discrimination, 288, 289
    African Americans, 303
    anti-Semitism, 305–306
    disabled and, 494–495, 528
    prejudice, reducing, 291
    sexism, 325–327
    sexual, 256
    in sports, 284–285
    subcultures and, 63
    types of, 289
    typology of prejudice and discrimination, 288
    white ethnic Americans, 305–306
disease agent, 470
distribution (demography), 506
diversity. See cultural diversity
division of labor, 127, 160, 327, 343
divorce, 354–356, 358, 381–383
"doing gender", 346–347
domestic partnership, 367–368
Domhoff, G. William, 438–439
dominant emotion, 544
dominant group, 286, 287
Dominican Republic, 261
double consciousness, 17
dramaturgical analysis, 132–134, 139
drives, 44
drop out rate, in education, 402
drug, 473
drug use and abuse, 94–95, 125, 183, 473–474, 476
Du Bois, W. E. B., 17
dual-earner marriage, 368–369
dumpster diving, 112–113
Durkheim, Emile
    on education, 390–391
    on families, 362
    on government, 436–438
    pluralist model, 436–438, 439
    on religion, 408, 411–414
    The Rules of Sociological Method, 12
    social solidarity, 126–127
    on suicide, 7–8, 12–14
    Suicide, 8, 13, 30
dyad, 152, 153, 160
dysfunction, 19

eating disorders, 324
ecclesia, 416–417
ecofeminism, 558–559
ecological models, urban growth, 521–523, 529
economic development, 256–259, 267–277, 296
Economic Mobility Project, 244
economies and economics
    agrarian societies, 327, 329–333
    capitalism, 446–451
    competition, 447–449
    countries, classification of, 8–9, 254–263
    demographic transition theory, 514–515
    economic development theories, 267–275
    financial crisis (2008), 428
    global perspective, 8–9, 254–256, 445–453
    horticultural and pastoral societies, 327, 328–329
    hunting and gathering societies, 327–328
    industrial society, 330–331
    mixed economy, 452–453
    politics and, 460
    poverty, sources of, 242–243
    preindustrial, 445
    producers and consumers, 10
    social networks of, 124
    socialism, 451–452
    underground, 456–457
edge cities, 533
education, 390, 391. See also schools
    American Dream and, 234–235
    bullying, 394–395
    conflict perspective, 238, 396–399, 417
    for disabled, 492
    dysfunction of, 394–396
    functionalist perspective, 238, 390–396, 417
    in the future, 421, 565–566
    gender bias in, 399
    GI Bill of Rights, 404
    global perspective, 266
    hidden curriculum, 398–399
    immigrants and, 509, 511
    Internet social groups, 168
    IQ tests in, 400–401
    latent functions of, 393
    Latinos, 310
    life chances and, 238
    manifest functions of, 391–393
    marriage and, 362
    multicultural, 72, 73
    poverty and, 238, 244, 266
    presidential election, 2004, 443
    religion and, 388–389, 390, 407–408
    in runaway adolescents, 94–95
    SAT scores, parents and, 454

separation of church and state, 392
    and social class, 223–224
    social inequality and, 238–239
    socialization, 392
    symbolic interactionist perspective, 399–401, 417
    of undocumented immigrants, 504
    of women, 276
egalitarian family, 361
ego, 84, 85
egoistic suicide, 29
Ehrenreich, Barbara, 226
elder abuse, 379–381
elections, 438, 442–443
elite model, of government, 438–439
emancipation theory, 185
emergent norm theory, 547
emigration, 507
emotion management, 139
emotional labor, 134
emotions and emotion management, 134–136, 139
employment
    of disabled, 494–495
    discrimination in, 305–306
    of homeless, 116
    immigrants and, 509
    lifelong, in Japan, 166
    unemployment, 238
endogamy, 362, 363
entrepreneurs, 221–222
environment (epidemiology), 470
environmental activism, 538–539, 542, 551, 558–559
environmental justice, 559, 561
environmental racism, 559–560, 561
epidemiological transition, 516
epidemiology, 468, 470–473
episodic framing, 120
equalitarian pluralism, 293
An Essay on the Principle of Population (Malthus), 513
ethanol, as fuel, 275
ethics, 37, 423
ethnic cleansing, 289
ethnic group, 282
ethnic pluralism, 293
ethnic subcultures, 61–63
ethnic villagers, 526
ethnicity, 9, 282, 283. See also race
    children in foster care, 379
    college enrollment and, 407
    conflict perspective, 68, 293–298
    democracy and, 435
    disabled, 491
    diversity, in U.S., 60
    divorce and, 383
    education level and, 408
    fertility rate, 370
    functionalist perspective, 292–293, 298
    gender socialization, 333
    income distribution, 232, 233
    IQ tests and, 400
    language and, 52–54
    life expectancy and, 472–473
    marital status, 377
    multicultural education, 73
    occupational stratification, 338–341
    pay gap, 339–341
    personal space, 138
    poverty and, 240–241
    presidential election, 2004, 443
    professions and, 454
    school dropout rate, 402
    single persons, 376
    social interaction and, 130
    socialization, 102–103
    sports and, 284, 302
    spouse abuse and, 379
    stereotypes, 68
    suicide rate and, 23
    unemployment and, 243
ethnocentrism, 64, 65, 150, 283, 286–287
ethnography, 35–36
ethnomethodology, 132, 133, 139
evolutionary theory, 12, 388–390
exchange value, 524
exogamy, 362, 363
experiment, 36, 37
experimental group, 36, 37
expressive crowd, 544
expressive leadership, 154, 155
expressive movement, 553
expressive needs, 151
expressive role, 362
expressive task, 18
eye contact, 137, 138

Facebook, 135, 144–145
face-saving behavior, 133
face-time, 146
facial expression, 137
fads, 65–66, 550
families, 356, 357
    adoption into, 372–374
    agent of socialization, 92–93
    characteristics, 357–359

child care, 368–370
childless, 526, 527
conflict perspective, 363–364, 366
descent and inheritance, 360–361
division of labor, 18–19
divorce and, 382
establishing, 365–370
feminist perspective, 364, 366
functionalist perspective, 92–93, 362–363, 366
in the future, 384–385, 565
gender in, 347
gendered division of labor in, 331, 351
homeless, 116
housework, 368–370
income, average after-tax, 232
master status, 118
postmodernist perspective, 365
power and authority in, 361
residence patterns, 362
role conflict in, 121
runaway children, 94–95
second shift, 369
socioeconomic status (SES) and, 363
structure, 357–359
symbolic interactionist perspective, 364–365, 366
transitions in, 376–384
types of, 357–359, 367–368, 384
violence in, 378–379
welfare, 243
working poor, 242–243
Family Educational Right to Privacy Act (FERPA), 148
family of orientation, 357–358, 359
family of procreation, 357–358, 359
fashion, 66, 550
fatalistic suicide, 29
fathers and fatherhood, 375–376
Feagin, Joe R., 289–290
fecundity, 504, 505
fee-for-service, 479–480, 484
feeling rules, 134
felony, 192
femininity, 9, 346–347
feminism, 20–21, 345
    in Christianity, 329
    ecofeminism, 558–559
    Islamic, 328–329
    types of, 345, 349–350
feminist perspective
    deviance, 185–186, 191
    families, 364, 366
    gender inequality, 350–351
    gender stratification, 342
    gendered division of labor, 345, 349–351
    religion, 416
feminity, 333–334
feminization of poverty, 241–242
fertility, 503–505
fertility rate, 370, 371
field research, 35–36
Filipino Americans, 307–308
First World nations, 257
flash mob, 550
Flexner report, 478
folkways, 57
food
    culture and, 42–43, 46, 48
    food recovery, 249
    grain, cost of, 275
    media framing, 69
    religion and, 46
    social stratification and, 328
    socialization and, 19, 62
    supply, perspectives on, 513
    surplus, economic development and, 445
forced assimilation, 300
forced migration, 299–300, 507
formal norms, 56–57
formal operational stage of development, 86
formal organization, 124, 157–166
foster care, 379
Foucault, Michael, 190, 191, 489
fragmentation, cultural, 72, 73
frame alignment, 557
frame analysis, 556–558, 562
Frame Analysis (Goffman), 557
framing, in media, 435
    advertising, 294–295
    culture and food, 69
    frame analysis, 557–558, 562
    immigrants and immigration, 517, 556
    Internet communities, 149
    intuitive, 470
    prescription drug advertising, 470–471
    racial harmony, 294–295
    science and religion, 412–413
    sensationalism, 30
    sympathetic, 470
    thematic and episodic, 120
    types of, 412–413, 557–558
France, 512
free enterprise, 449